ELEMENTARY STATISTICS

EIGHTH EDITION

ELEMENTARY STATISTICS

EIGHTH EDITION

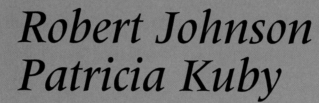

Robert Johnson
Patricia Kuby

Monroe Community College

Duxbury
Thomson Learning™

Pacific Grove • Albany • Belmont • Boston • Cincinnati • Johannesburg • London • Madrid • Melbourne
Mexico City • New York • Scottsdale • Singapore • Tokyo • Toronto

Sponsoring Editor: *Carolyn Crockett*
Marketing Team: *Tom Ziolkowski, Beth Kroenke, Laura Hubrich*
Editorial Assistants: *Kimberly Raburn, Ann Day*
Production Editor: *Tessa McGlasson Avila*
Production Service: *Greg Hubit Bookworks*
Manuscript Editor: *Martha Ghent*
Permissions Editor: *Mary Kay Hancharick*

Interior Design: *Rita Naughton*
Cover Design and Illustration: *Laurie Albrecht*
Art Editor: *Lisa Torri*
Print Buyer: *Vena Dyer*
Typesetting: *Graphic World, Inc.*
Cover Printing: *Phoenix Color Corporation*
Printing and Binding: *R. R. Donnelley/Willard*

COPYRIGHT © 2000 by Brooks/Cole
Duxbury is an imprint of Brooks/Cole, a division of Thomson Learning
The Thomson Learning logo is a trademark used herein under license.

For more information, contact:
DUXBURY
511 Forest Lodge Road
Pacific Grove, CA 93950 USA
www.duxbury.com

Printed in the United States of America

10 9 8 7 6 5 4 3 2 1

Library of Congress Cataloging-in-Publication Data

Johnson, Robert Russell
    Elementary statistics. — 8th ed. / Robert Johnson, Patricia Kuby.
        p.   cm.
    Includes index.
    ISBN 0-534-35676-1
    1. Statistics.   I. Kuby, Patricia.   II. Title.
QA276.12.J64   2000
519.5—dc21   99-23303
                                                    CIP

Preface

Purpose and Prerequisites

This book was written for use as an introductory course for students who need a working knowledge of statistics but do not have a strong mathematical background. Statistics requires the use of many formulas and an occasional solution of a simple algebraic equation. Those students who have not completed intermediate algebra should complete at least one semester of college mathematics as a prerequisite before attempting this course.

Our Objectives

The primary objective of *Elementary Statistics, Eighth Edition,* is to present a truly readable textbook that will promote learning, understanding, and motivation by presenting statistics in a context that relates to personal experiences. Simply, our goal is a clear and interesting introductory statistics textbook.

Statistics is a practical discipline that evolves with the changing needs of our society. Today's student is the product of a particular cultural environment and is motivated differently from students of a few years ago. In this text we present statistics as a useful tool in learning about the world around us. While studying descriptive and inferential concepts, students will become aware of their real-world applications in such fields as the physical and social sciences, business, economics, and engineering.

Important Ongoing Features

This eighth edition continues to feature the following elements:

- A communication style that reflects current student culture;
- A strong computer flavor, with numerous annotated outputs, exercises, and instructions;
- A focus on interpreting computer output;
- Case Studies based on situations of interest and using real data;
- A Chapter 1 that introduces ideas of variability and data collection, as well as basic terms;
- An early descriptive presentation of linear correlation and linear regression in Chapter 3;

- The use of "word algebra" to express formulas in words;
- The use of margins for margin exercises and margin "conversation";
- The natural flow from sampling distribution, to estimation, to hypothesis testing, to p-values;
- Inclusion of both p-value and classical approaches to hypothesis testing;
- Two specially designed tables for determining p-values;
- Large exercise sets;
- Exercises that demonstrate statistical theory through simulation;
- Many exercises designed to be solved using a computer or calculator;
- All data sets of significant number of data are on the data disk, now called *StatSource CD;*
- Brief biographies of four prominent statisticians.

This Revision

In this eighth edition, we hope the users of the previous editions will appreciate the following improvements:

1. The presentation is more approachable, clearer, and more visual throughout. The presentation is *more approachable* because it is often less technical. For example, each chapter begins with an "everyday happening" type case study; it is *clearer* because the side-by-side presentation allows ease of comparison between the technology used in class and the technologies used elsewhere (see page 67); it is *more visual* as demonstrated by the side-by-side presentations of the p-value and classical approaches to hypothesis testing (see page 427).

2. The role of the Chapter Case Study has been expanded to be a more integral part of the chapter; it serves as an introduction and a wrap-up.

3. MINITAB (version 12), EXCEL (97), and TI-83 instructions are displayed side-by-side for ease of comparison (many have access to technology other than just the one used in class, so these three very popular means are shown; see page 423).

4. A rearranged model for the hypothesis test procedure is used to emphasize comparability of the p-value and classical approaches; these changes are reflected in Chapters 8 through 14.

5. Many new exercises with real data sets have been added.

6. Important terms have been highlighted for identification and ease of locating when referenced.

7. An Annotated Instructors Edition is being made available.

8. Chapter 1 was rewritten to include sample designs.

9. The sections making up Chapters 9 and 10 have been rearranged. Inferences about variance is now the last section in these chapters.

10. A CD, *StatSource,* has replaced the disk in the back of the textbook. The CD contains many helpful teaching and learning aids.

To the Instructor: The Text as a Teaching Tool

One primary objective of this book is to offer a truly readable presentation of elementary statistics. The chapters are designed to interest and involve students and to guide them step by step, in a logical manner, through the material.

The *Getting Started* section, page xxi, provides a brief explanation of the components of each chapter and makes suggestions about ways to use this book more effectively. It is suggested reading for both the instructor and the student.

The first three chapters are introductory in nature. Chapter 1 is an introduction to the language of statistics; Chapter 2 covers the descriptive presentation of single-variable data, while Chapter 3 is the descriptive presentation of bivariate data. The bivariate material is presented at this point in the book because students often ask about the relationship between two sets of data (such as heights and weights) while studying Chapter 2.

In the chapters on probability (4 and 5), the concepts of permutations and combinations are deliberately avoided. Instead, this material is contained in Appendix A, "Basic Principles of Counting," so that it may be included as the instructor wishes. The binomial coefficient is introduced in connection with the binomial probability distribution in Chapter 5.

The instructor has several options in the selection of topics to be studied in a given course. We consider Chapters 1 through 9 to be the basic core of a course (some sections of Chapters 2, 4, and 6, and all of Chapter 3 may be omitted without affecting continuity). Following the completion of Chapter 9, any combination of Chapters 3 and 10 through 14 may be studied. However, there are two restrictions: Chapter 3 must be studied prior to chapter 13, and Chapter 10 must precede Chapter 12.

To the Student: The Text as a Learning Tool

Statistics is different from other courses:

1. It has its own extensive technical vocabulary.

2. It is highly cumulative, in that many of the concepts you will be learning at each step become the basis for other concepts learned throughout the rest of the course; therefore, failure to master each concept as presented can cause great difficulty later on.

3. It requires very precise measurements and calculations (a seemingly minor error will often be magnified and lead to wrong answers in some procedures).

4. While it is an academic subject, statistics is also very real and touches each of us frequently in everyday life. Plain talk and emphasis on common sense are the book's main characteristics as a learning tool. This approach should allow you—provided you have the necessary basic mathematics skills—to work your way through the course with relative ease. Examples of this approach are: (a) Illustration 1-1 (p. 13), which is used to reemphasize the meaning of the eight basic definitions presented in Section 1.2, and (b) Chapter Objectives for Chapter 2 (p. 36), which use a familiar situation to motivate the topics of Chapter 2.

Our goal in writing this textbook is to motivate and involve you in the statistics that you are learning. Turn to page xxi and read *Getting Started* for a brief explanation of some of the ways you can use this book more effectively to succeed in learning statistics.

Supplements

The **StatSource CD** contains: Data Sets (263 files), Statistical Concept links, Videos (30- to 60-second film clips showing statistics in action), Video Tutorials (1- to 5-minute film clips demonstrating statistical concepts or calculations), Tutorials (dozens of "smart" questions for each chapter to help study), and PowerPoint slides (for lecture presentation or notes).

The **Statistical Tutor** is a student manual that:
a. Contains the complete solutions to all margin exercises and the odd-numbered exercises (the same exercises whose answers are in the back of the book).
b. Contains many helpful hints and suggestions to serve as a guide through the learning process. It includes many summaries and overviews.
c. Contains several review lessons to help refresh materials studied previously in other courses.

The **Instructor's Manual** is also intended to be uncommonly helpful. It contains:
a. Everything that is in the *Statistical Tutor.*
b. The complete solutions to all exercises.
c. Many helpful teaching suggestions that an instructor might incorporate. The notes specifically intended for the instructor are set in a type different from that for the student material.

The **Test Bank** contains a combination of true-false, multiple-choice, short-answer, matching, and computational test questions for each chapter in the text.

The **MINITAB Lab Manual** is a text-specific guide to the MINITAB statistical analysis system that is keyed to the text discussions and examples. Its purpose is to provide the instructor with a flexible means of integrating technology into their courses. It is also intended to integrate the use of technology to enhance presentation of concepts and to give students a feel for what statistics "is." The laboratory exercises are designed to motivate and involve the student in the statistics they are learning. The overriding goal is to strengthen the student's conceptual view without burdening the student with too many manual calculations.

The **EXCEL Manual** is a text-specific guide to Microsoft Excel, intended to demonstrate the statistical capabilities of Excel using examples and problems in *Elementary Statistics, Eighth Edition.* Microsoft Excel offers a complete repertoire of statistics functions for solving the problems in *Elementary Statistics.* This supplement contains detailed chapter-by-chapter solutions of problems in the textbook, including exercises, tips, and Excel demonstrations of key concepts from the text.

Excel features: 1) a large collection of graph types, 2) a rich assortment of random-number generating functions, and 3) instant connection between worksheet numbers and charts. Therefore, Excel supports interactive problem-solving and student demonstrations, and these are provided in the manual where appropriate. Because students majoring in business or related areas will likely be working with Excel in their careers, the supplement includes numerous business-oriented exercises using data from actual case studies.

The purpose of the **TI-83 Manual** is to explain how to use the Texas Instruments TI-83 Plus in order to solve typical statistics problems. The standard TI-83 Plus built-in statistics functions are discussed and several calculator programs are presented. Examples closely follow the text and are solved using the calculator. Keystrokes and screen illustrations are shown for easy reference.

Thomson World Class Learning Testing Tools™ is a fully integrated suite of test creation, delivery, and course management tools. It consists of three software programs—Test, Test On-line, and Manager—that enable you to quickly and easily assess student performance.

Creating practice exams or final exams on paper, via a local area network, or via the Internet, requires only pointing and clicking. This flexible, powerful software lets you create tests based on testbanks provided by Thomson Learning. Testing Tools is flexible, easy to learn, and provided at no cost to adopters of selected Thomson Learning textbooks.

Acknowledgments

We owe a debt to many other books. Many of the ideas, principles, examples, and developments that appear in this text stem from thoughts provoked by these sources.

It is a pleasure to acknowledge the aid and the encouragement we have received throughout the development of this text from students and colleagues at Monroe Community College. In addition, special thanks to all the reviewers who read and offered suggestions about this and previous editions:

Nancy Adcox
Mt. San Antonio College

Paul Alper
College of St. Thomas

William D. Bandes
San Diego Mesa College

Barbara Jean Blass
Oakland Community College

Austin Bonis
Rochester Institute of Technology

Nancy Bowers
Pennsylvania College of Technology

Louis F. Bush
San Diego City College

Ronnie Catipon
Franklin University

Rodney E. Chase
Oakland Community College

Pinyuen Chen
Syracuse University

David M. Crystal
Rochester Institute of Technology

Joyce Curry and Frank C. Denny
Chabot College

Larry Dorn
Fresno Community College

Shirley Dowdy
West Virginia University

Joan Garfield
University of Minnesota General College

Carol Hall
New Mexico State University

Silas Halperin
Syracuse University

Noal Harbertson
California State University, Fresno

Hank Harmeling
North Shore Community College

Bryan A. Haworth
California State College at Bakersfield

Harold Hayford
Pennsylvania State University, Altoona

Marty Hodges
Colorado Technical University

John C. Holahan
Xerox Corporation

James E. Holstein
University of Missouri

Soon B. Hong
Grand Valley State University

Robert Hoyt
Southwestern Montana State

Peter Intarapanach
Southern Connecticut State University

T. Henry Jablonski, Jr.
East Tennessee State University

Brian Jean
Bakersfield University

Jann-Huei Jinn
Grand Valley State University

Sherry Johnson

Meyer M. Kaplan
The William Patterson College of New Jersey

Michael Karelius
American River College

Anand S. Katiyar
McNeese State University

Jane Keller
Metropolitan Community College

Gayle S. Kent
Southern Florida College

Andrew Kim
Westfield State College

Amy Kimchuk
University of the Sciences in Philadelphia

Raymond Knodel
Bemidji State University

Larry Lesser
University of Northern Colorado

Robert O. Maier
El Camino College

Mark Anthony McComb
Mississippi College

Carolyn Meitler
Concordia University

John Meyer
Muhlenberg College

Jeffrey Mock
Diablo Valley College

David Naccarato
University of New Haven

Harold Nemer
Riverside Community College

Dennis O'Brien
University of Wisconsin, LaCrosse

Daniel Powers
University of Texas, Austin

Janet M. Rich
Miami-Dade Junior College

Larry J. Ringer
Texas A & M University

John T. Ritschdorff
Marist College

John Rogers
California State Polytechnic Institute at
San Luis Obispo

Neil Rogness
Grand Valley State University

Thomas Rotolo
University of Arizona

Barbara F. Ryan and Thomas A. Ryan
Pennsylvania State University

Robert J. Salhany
Rhode Island College

Sherm Sowby
California State University, Fresno

Howard Stratton
State University of New York at
Albany

Larry Stephens
University of Nebraska–Omaha

Paul Stephenson
Grand Valley State University

Thomas Sturm
College of St. Thomas

Edward A. Sylvestre
Eastman Kodak Co.

William Tomhave
University of Minnesota

Bruce Trumbo
California State University, Hayward

Richard Uschold
Canisius College

John C. Van Druff
Fort Steilacoom Community College

Philip A. Van Veidhuizen
University of Alaska

John Vincenzi
Saddleback College

Kenneth D. Wantling
Montgomery College

Mary Wheeler
Monroe Community College

Sharon Whitton
Hofstra University

Don Williams
Austin College

Thanks also to the many authors and publishers who so generously extended reproduction permissions for the news articles and tables used in the text. These acknowledgments are specified individually throughout the text.

And finally, the most significant acknowledgment of all—thank you to our spouses, Barbara and Joe, for your assistance and just "being there."

Robert Johnson
Patricia Kuby

Contents

Part Four MORE INFERENTIAL STATISTICS 536

Sir Ronald A. Fisher 537

GETTING STARTED

Your Guide to Getting the Most out of *Elementary Statistics,* Eighth Edition

Read this section to become familiar with the components of each chapter and their intended purposes.

First realize that studying statistics is more like studying a foreign language than like studying mathematics. Statistics is more than the mathematics of formulas and data. Statistics includes the processes of problem solving, statistical thinking, data collection, obtaining numerical and graphical results, and the follow-up questioning of those results. Sometimes statistics requires the use of mathematics, and sometimes it does not. Everybody (and everything) is an individual and uniquely different from all others. However, when it comes to a single trait, many values of a single variable taken from many individuals will generally form a pattern. Statistical methodologies are used to describe and help explain these patterns. Statistics uses mathematical techniques to quantify the ideas being investigated and to reduce the information to a numeric format, in which it can be treated graphically or algebraically. When concepts become quantified, they become an application of mathematics, not *the* mathematics.

As important as it is to be able to take a set of data and calculate a statistical value (mean, standard deviation, correlation coefficient, etc.) or draw a graphic display (histogram, scatter diagram, etc.), it is far more important that you understand the circumstances being investigated; that you understand the variables involved; that you understand why you are investigating the problem; and that you learn to question the data and the statistical results. Your life experience and understanding of real-life situations form the foundation for understanding statistics. Don't lose sight of the fact that statistics is about describing the world around us. You will see many statistical examples from business, the physical and social sciences, and many other fields and professions as you study from this textbook.

To get the most out of this book, become familiar with its many learning features. On the following pages are examples of several features this book contains and suggestions on how to make the best use of them. Take a moment to look them over and, then, please use them. Active involvement in learning about statistics is the single most important factor in determining success and satisfaction. Now open up your mind and let your imagination and your natural curiosity go to work.

GETTING STARTED

You cannot just get in your car and start driving and expect to arrive at the correct destination. You must know where you are going and what route to take before you start. Studying statistics is much the same. Every chapter in this textbook opens with three important study tools.

Chapter Outlines appear at the beginning of each chapter to give a schematic overview of what is to be presented. These outlines are annotated to give a first impression of key terms and concepts that appear in the chapter.

Chapter

2

Descriptive Analysis and Presentation of Single-Variable Data

CHAPTER OUTLINE

GRAPHIC PRESENTATION OF DATA

2.1 Graphs, Pareto Diagrams, and Stem-and-Leaf Displays
A **picture** is often worth a thousand words.

2.2 Frequency Distributions and Histograms
An **increase** in the amount of data requires us to modify our techniques.

NUMERICAL DESCRIPTIVE STATISTICS

2.3 Measures of Central Tendency
The four measures of central tendency—mean, median, mode, and midrange—are **average** values.

2.4 Measures of Dispersion
Measures of dispersion—range, variance, and standard deviation—assign numerical values to the **amount of spread** in a set of data.

2.5 Mean and Standard Deviation of Frequency Distribution
The **frequency distribution** is an aid in calculating mean and standard deviation.

2.6 Measures of Position
Measures of position allow us to **compare** one piece of data to the set of data.

2.7 Interpreting and Understanding Standard Deviation
A standard deviation is the length of a **standardized yardstick**.

2.8 The Art of Statistical Deception
How the unwitting or the unscrupulous can use **"tricky" graphs** and **insufficient information** to mislead the unwary.

The **Chapter Case Studies** are mostly newspaper articles about everyday kind of phenomenon, with questions that ask the student about the methods to be presented in that chapter; an integral part of the chapter that also serves as an "example introduction" to the Chapter Objectives.

CHAPTER CASE STUDY

Paying Their Own Way

Many students work full-time or part-time, during the academic year or during the summer, or some combination, to earn part of (maybe even all of) their college expenses. Do you work to pay for some part of your college expenses? Do your friends work to pay for part of their college expenses? How much did you or your friends each earn last month? The USA Snapshot®, "Working their way through," that appeared in *USA Today,* March 17, 1998, describes the average monthly earnings of American college students.

Source: Campus Concepts   By Cindy Hall and Jerry Mosemak, USA TODAY

Chapter Objectives prepare the way for new material by describing why the material is important and how it relates to previously studied topics. Use Chapter Objectives to understand the motivation for learning this material.

36   Chapter 2   Descriptive Analysis and Presentation of Single-Variable Data

CHAPTER OBJECTIVES

Imagine that you took an exam at the last meeting of your favorite class. Today your instructor returns your exam paper and it has a grade of 78. If you are like most students in this situation, as soon as you see your grade you want to know how your grade compares to those of the rest of the class and you immediately ask, "What was the average exam grade?" Your instructor replies, "The class average was 68." Since 78 is 10 points above the average, you ask, "How close to the top is my grade?" Your instructor replies that the grades ranged from 42 to 87 points. The accompanying figure summarizes the information we have so far.

A third question that is sometimes asked is "How are the grades distributed?" Your instructor replies that half the class had grades between 65 and 75. With this information you conclude that your grade is fairly good.

MAKING STATISTICS COME ALIVE

As you begin each chapter, view the corresponding short **Videos** on the CD (located inside the back cover) for a glimpse at an interesting illustration of statistics at work in a real-life situation. Answer the questions about the video clip.

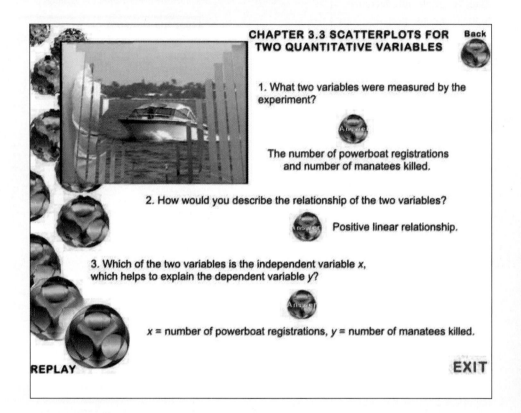

CHAPTER 3.3 SCATTERPLOTS FOR TWO QUANTITATIVE VARIABLES   Back

1. What two variables were measured by the experiment?

The number of powerboat registrations and number of manatees killed.

2. How would you describe the relationship of the two variables?

Positive linear relationship.

3. Which of the two variables is the independent variable *x*, which helps to explain the dependent variable *y*?

x = number of powerboat registrations, *y* = number of manatees killed.

REPLAY   EXIT

Case Studies are designed to teach important concepts and to demonstrate how statistics work in the everyday world. Throughout the text, case studies can be found that incorporate statistical concepts as these concepts are presented. Margin exercises also accompany these case studies and provide an excellent way to try out your new knowledge in a real-world setting.

Section 8.2   Estimation of Mean μ (σ Known)   **359**

C a s e   S t u d y   8.2

Rockies Snow Brings Little Water

When snow melts it becomes water, sometimes more water than at other times. This newspaper article compares the water content of snow from two areas in the United States that typically get about the same amount of snow annually. However, the water content is very different. There are several point estimates for the average included in the USA Today article.

Exercise 8.15
"Rockies snow brings little water" lists "14.36 inches" and "5.07 inches" as point estimates. Describe why these numbers are statistics and why they are also point estimates.

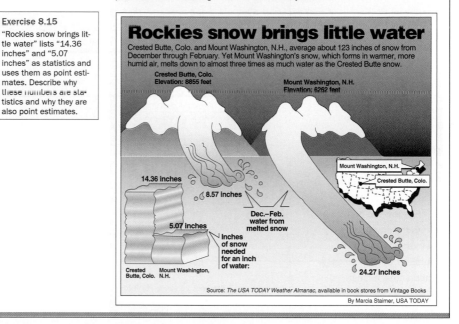

Source: *The USA TODAY Weather Almanac*, available in book stores from Vintage Books

By Marcia Staimer, USA TODAY

UNDERSTANDING STATISTICS

All the important **definitions** and **key terms** appear in boldface; all indexed words are highlighted as they occur. This will assist the reader in identifying important items, and in locating referenced information.

Two Quantitative Variables

When the bivariate data are the result of two quantitative variables, it is customary to express the data mathematically as **ordered pairs** (x, y), where x is the **input variable** (sometimes called the **independent variable**) and y is the **output variable** (sometimes called the **dependent variable**). The data are said to be ordered because one value, x, is always written first. They are called *paired* because for each x value there is a corresponding y value from the same source. For example, if x is height and y is weight, a height and corresponding weight are recorded for each person. The input variable x is

Completely **worked-out examples** present the step-by-step solution process.

---

78   Chapter 2   Descriptive Analysis and Presentation of Single-Variable Data

The variance of our sample 6, 3, 8, 5, 3 is found in Table 2.10 using formula (2.6).

TABLE 2.10   Calculating Variance Using Formula (2.6)

Step 1. Find $\sum x$	Step 2. Find $\bar{x}$	Step 3. Find each $x - \bar{x}$	Step 4. Find $\sum(x - \bar{x})^2$	Step 5. Sample variance
6	$\bar{x} = \dfrac{\sum x}{n}$	$6 - 5 = 1$	$(1)^2 = 1$	$s^2 = \dfrac{\sum(x - \bar{x})^2}{n - 1}$
3		$3 - 5 = -2$	$(-2)^2 = 4$	
8	$\bar{x} = \dfrac{25}{5}$	$8 - 5 = 3$	$(3)^2 = 9$	$s^2 = \dfrac{18}{4}$
5		$5 - 5 = 0$	$(0)^2 = 0$	
3	$\bar{x} = 5$	$3 - 5 = -2$	$(-2)^2 = 4$	$s^2 = \mathbf{4.5}$
$\sum x = 25$		$\sum(x - \bar{x}) = 0$ ⓒⓚ	$\sum(x - \bar{x})^2 = 18$	

Graphical displays can be found in the form of *charts, graphs,* and *tables.* Graphics are very important in statistics. They are the pictures that either demonstrate the theory or condense vast amounts of data in an easy-to-understand format.

QUARTILES

Values of the variable that divide the ranked data into quarters; each set of data has three quartiles. The first quartile, Q_1, is a number such that at most 25% of the data are smaller in value than Q_1 and at most 75% are larger. The second quartile is the median. The third quartile, Q_3, is a number such that at most 75% of the data are smaller in value than Q_3 and at most 25% are larger (see Figure 2.24).

Figure 2.24
Quartiles

Ranked data, increasing order

| 25% | 25% | 25% | 25% |

Q_1 Q_2 Q_3

The procedure for determining the value of the quartiles is the same as that for percentiles and is shown in the following description of *percentiles.*

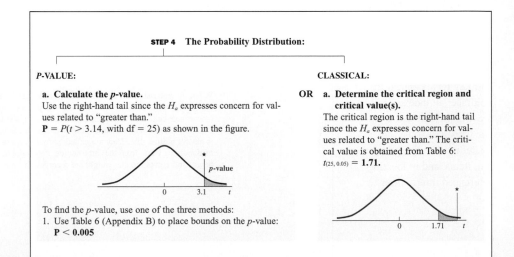

STEP 4   The Probability Distribution:

P-VALUE:

a. Calculate the *p*-value.
Use the right-hand tail since the H_a expresses concern for values related to "greater than."
P = $P(t > 3.14$, with df = 25) as shown in the figure.

p-value

0   3.1   *t*

To find the *p*-value, use one of the three methods:
1. Use Table 6 (Appendix B) to place bounds on the *p*-value:
   P < 0.005

CLASSICAL:

OR   **a. Determine the critical region and critical value(s).**
The critical region is the right-hand tail since the H_a expresses concern for values related to "greater than." The critical value is obtained from Table 6:
$t_{(25,\ 0.05)} = \mathbf{1.71}.$

0   1.71   *t*

Some exercises are located in the book's margin; these **margin exercises** have been placed there for an initial practice on the adjacent concept. Solving these exercises before reading further is an excellent way to begin homework assignments.

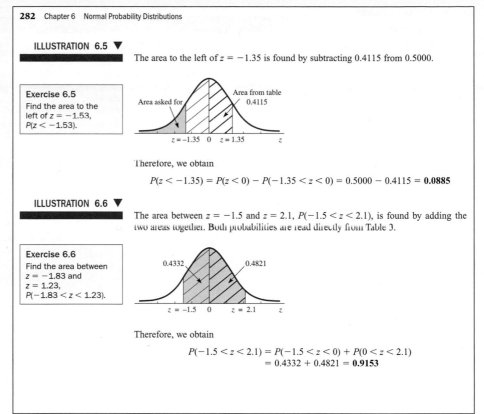

ILLUSTRATION 6.5 ▼

Exercise 6.5
Find the area to the left of $z = -1.53$, $P(z < -1.53)$.

The area to the left of $z = -1.35$ is found by subtracting 0.4115 from 0.5000.

Area asked for

Area from table 0.4115

$z = -1.35$ 0 $z = 1.35$ z

Therefore, we obtain

$$P(z < -1.35) = P(z < 0) - P(-1.35 < z < 0) = 0.5000 - 0.4115 = \mathbf{0.0885}$$

ILLUSTRATION 6.6 ▼

Exercise 6.6
Find the area between $z = -1.83$ and $z = 1.23$, $P(-1.83 < z < 1.23)$.

The area between $z = -1.5$ and $z = 2.1$, $P(-1.5 < z < 2.1)$, is found by adding the two areas together. Both probabilities are read directly from Table 3.

0.4332   0.4821

$z = -1.5$ 0 $z = 2.1$ z

Therefore, we obtain

$$P(-1.5 < z < 2.1) = P(-1.5 < z < 0) + P(0 < z < 2.1)$$
$$= 0.4332 + 0.4821 = \mathbf{0.9153}$$

End-of-Section Exercises give the opportunity to practice concepts that are presented in the section. The best way to learn statistics is to practice, and once a concept is thought to be understood, practice it some more for good measure. This text contains hundreds of exercises with many applications. The *icons* next to some of the exercises indicate the type of application—business, sports, criminal justice, and so on.

EXERCISES • • • • • • • • • • • • •

2.109 The following data are the yields, in pounds, of hops.

| 3.9 | 3.4 | 5.1 | 2.7 | 4.4 | 7.0 | 5.6 | 2.6 | 4.8 | 5.6 |
| 7.0 | 4.8 | 5.0 | 6.8 | 4.8 | 3.7 | 5.8 | 3.6 | 4.0 | 5.6 |

a. Find the first and the third quartiles of the yield.
b. Find the midquartile.
c. Find the following percentiles: (1) P_{15}, (2) P_{33}, (3) P_{90}.

2.110 A research study of manual dexterity involved determining the time required to complete a task. The time required for each of 40 disabled individuals is as follows (data are ranked):

7.1	7.2	7.2	7.6	7.6	7.9	8.1	8.1	8.1	8.3
8.3	8.4	8.4	8.9	9.0	9.0	9.1	9.1	9.1	9.1
9.4	9.6	9.9	10.1	10.1	10.1	10.2	10.3	10.5	10.7
11.0	11.1	11.2	11.2	11.2	12.0	13.6	14.7	14.9	15.5

Find: **a.** Q_1 **b.** Q_2 **c.** Q_3 **d.** P_{95}
e. the 5-number summary   **f.** Draw the box-and-whisker display.

2.111 Consider the following set of ignition times that were recorded for a synthetic fabric.

30.1	30.1	30.2	30.5	31.0	31.1	31.2	31.3	31.3	31.4
31.5	31.6	31.6	32.0	32.4	32.5	33.0	33.0	33.0	33.5
34.0	34.5	34.5	35.0	35.0	35.6	36.0	36.5	36.9	37.0
37.5	37.5	37.6	38.0	39.5					

Find: **a.** the median   **b.** the midrange   **c.** the midquartile
d. the 5-number summary   **e.** Draw the box-and-whisker display.

The **Return to Chapter Case Study** offers an opportunity to revisit the Chapter Case Study, and to answer the questions proposed at the beginning of the chapter, using the knowledge gained from studying the chapter.

Return to CHAPTER CASE STUDY

Many students work full-time or part-time. How much did you or your friends each earn last month? You are prepared to give a full answer to this question now. You learned how to describe data in Chapter 2 and how to make inferences in Chapter 9. (See Chapter Case Study, p. 415.)

Listed below is the amount earned last month by each student in a sample of 35 college students.

0	0	105	0	313	453	769	415	244	0	333	0
0	362	276	158	409	0	0	534	449	281	37	338
240	0	0	0	142	0	519	356	280	161	0	

Exercise 9.123

Use these sample data to describe the amount earned by working college students.
a. Describe the population of interest.
b. How many of the students in the sample above are working?
c. Describe the variable, amount earned by a working college student last month, using one graph, one measure of central tendency, and one measure of dispersion.
d. Find evidence to show that the assumptions for use of Student's t-distribution have been satisfied.
e. Estimate the mean amount earned by a college student per month using a point estimate and a 95% confidence interval.
f. The Statistical Snapshot® in the Chapter Case Study (p. 415) suggests the average amount earned each month by college students is approximately $350. Does the sample show sufficient reason to reject that claim? Use $\alpha = 0.05$.

The **In Retrospect** section summarizes the concepts learned in each chapter, pointing out the relationships and the interrelationships with material covered previously.

IN RETROSPECT

To sum up what we have just learned; there is a distinct difference between the purpose of regression analysis and the purpose of correlation. In regression analysis, we seek a relationship between the variables. The equation that represents this relationship may be the answer desired, or it may be the means to the prediction that is desired. In correlation analysis, we measure the strength of the linear relationship between the two variables.

The case studies show a variety of applications for the techniques of correlation and regression. These articles are worth reading again. When bivariate data appear to fall along a straight line on the scatter diagram, they suggest a linear relationship. But this is not proof of cause and effect. Clearly, if a basketball player commits too many fouls, he will not be scoring more points. He will be in foul trouble and "riding the pine" with no chance to score. It also seems reasonable that the more game time he has, the more points he will score and the more fouls he will commit. Thus, a positive correlation and a positive regression relationship will exist between these two variables.

The bivariate linear methods we have studied thus far have been presented for the purpose of a first, descriptive look. More details must, by necessity, wait until additional developmental work has been completed. After completing this chapter, you should have a basic understanding of bivariate data, how they are different from just two sets of data, how to present them, what correlation and regression analysis are, and how each is used.

Figure skaters practice their skating for many hours in order to make the Olympics. Homework time is the statistics students' practice time. Effort and persistence are the hallmark of every gold medalist and all successful statistics students.

Chapter Exercises are based on all the techniques studied in the chapter. They offer the opportunity to integrate conceptual and computational skills as well as to identify the appropriate procedure needed to produce the desired results.

CHAPTER EXERCISES

10.108 The diastolic blood pressures for 15 patients were determined using two techniques: the standard method used by medical personnel and a method using an electronic device with a digital readout. The results were as follows:

Patient	1	2	3	4	5	6	7	8	9	10	11	12	13	14	15
Standard method	72	80	88	80	80	75	92	77	80	65	69	96	77	75	60
Digital method	70	76	87	77	81	75	90	75	82	64	72	95	80	70	61

Assuming blood pressure is normally distributed, determine the 90% confidence interval for the mean difference in the two readings, where d = standard method − digital readout.

DOING STATISTICS ON THE COMPUTER

The CD provided with this textbook (on the inside back cover), contains a wealth of information and assistance for learning statistics.

1. **Data Sets** contains 263 sets of data from the Case Studies, Exercises, Illustrations, and Working with Your Own Data components of the text.
2. **Concept Simulations** provides several links to Internet Applets that will help you understand the concepts being studied and assist you in finding answers.
3. The **Videos** section has several short film clips showing statistics at work in real-life settings.
4. The **Tutorials** are a collection of questions designed to help you learn.
5. **PowerPoint** contains "slides" outlining the key points of each chapter. Instructors may use these during their class presentations and students may print them before class and use them as a framework for in-class notes.

MINITAB is a statistics computer software package; EXCEL is a spreadsheet computer software package; and the TI-83 is a powerful calculator. All of these contain statistical tools that let the reader focus on learning concepts while it handles the calculations and graphing of data. In today's world, most statistical calculations are accomplished by computer.

MINITAB, EXCEL, and TI-83 commands are introduced in the text, and using them as much as possible is encouraged. You are not expected to use all three technologies. But most likely your situation will use one of these in class and you will have access to one of the others outside of class (in other classes, at work, or even on your own computer).

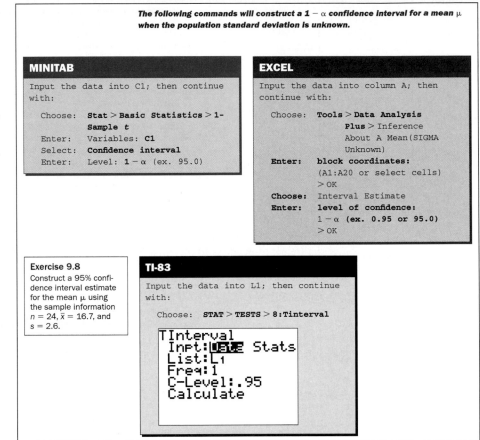

The following commands will construct a 1 − α confidence interval for a mean μ when the population standard deviation is unknown.

MINITAB

Input the data into C1; then continue with:

Choose: Stat > Basic Statistics > 1-
            Sample t
Enter: Variables: C1
Select: Confidence interval
Enter: Level: 1 − α (ex. 95.0)

EXCEL

Input the data into column A; then continue with:

Choose: Tools > Data Analysis
            Plus > Inference
            About A Mean(SIGMA
            Unknown)
Enter: block coordinates:
            (A1:A20 or select cells)
            > OK
Choose: Interval Estimate
Enter: level of confidence:
            1 − α (ex. 0.95 or 95.0)
            > OK

Exercise 9.8
Construct a 95% confidence interval estimate for the mean μ using the sample information $n = 24$, $\bar{x} = 16.7$, and $s = 2.6$.

TI-83

Input the data into L1; then continue with:

Choose: STAT > TESTS > 8:Tinterval

```
TInterval
 Inpt:DATA Stats
 List:L₁
 Freq:1
 C-Level:.95
 Calculate
```

Most of the computer output shown throughout the textbook was generated using MINITAB; however, had Excel or TI-83 been used, the results would be comparable.

Some exercises are also designed to encourage taking advantage of the "friendly" power of the computer.

9.73 Karl Pearson once tossed a coin 24,000 times and recorded 12,012 heads.
      a. Calculate the point estimate for $p = P$(head) based on Pearson's results.
      b. Determine the standard error of proportion.
      c. Determine the 95% confidence interval estimate for $p = P$(head).
      d. It must have taken Mr. Pearson many hours to toss a coin 24,000 times. You can simulate 24,000 coin tosses using the computer and calculator commands listed below. (*Note:* A Bernoulli experiment is like a "single" trial binomial experiment. That is, one toss of a coin is one Bernoulli experiment with $p = 0.5$; and 24,000 tosses of a coin either is a binomial experiment with $n = 24,000$ or is 24,000 Bernoulli experiments. Code: 0 = tail, 1 = head. The sum of the 1's will be the number of heads in the 24,000 tosses.)

```
MINITAB: Choose Calc > Random Data > Bernoulli, then enter
0.5 for the probability. Sum the data and divide by
24,000.

EXCEL: Choose Tools > Data Analysis > Random Number Genera-
tion > Bernoulli, then enter 0.5 for p. Sum the data and
divide by 24,000.

TI-83: Choose MATH > PRB > 5:randInt, then enter 0, 1, num-
ber of trials. The maximum number of elements (trials) in
a list is 999. (slow process for large n's) Sum the data
```

DOING STATISTICS ON YOUR OWN

A **Working with Your Own Data** section is included at the end of each part, encouraging further exploration. These sections provide a personalized learning experience beginning with the collection of data and continuing through the application techniques that have been studied. Experience shows that more concepts are retained by applying methods just learned in regular homework assignments to data that is familiar and understandable.

---

344 Chapter 7 Sample Variability

Working with Your Own Data

The Central Limit Theorem is very important to the development of the rest of this course. Its proof, which requires the use of calculus, is beyond the intended level of this course. However, the truth of the CLT can be demonstrated both theoretically and by experimentation. The following series of questions will help to verify the central limit theorem both ways.

A **T**he Population

Consider the theoretical population that contains the three numbers 0, 3, and 6 in equal proportions.

1. a. Construct the theoretical probability distribution for the drawing of a single number, with replacement, from this population.
   b. Draw a histogram of this probability distribution.
   c. Calculate the mean m and the standard deviation s for this population.

STUDYING AND PREPARING FOR EXAMS

The first step in preparing for an exam should always be: Complete all the reading and exercises assigned.

A **Vocabulary List** near the end of every chapter is an aid in deciding how much of the material is truly understood. Consider this informal self-test: try defining all words on the vocabulary list to a friend. This will determine if more practice is needed.

VOCABULARY LIST

Be able to define each term. Pay special attention to the key terms, which are printed in **red**. In addition, describe in your own words, and give an example of, each term. Your examples should not be the ones given in class or in the textbook.

The bracketed numbers indicate the chapters in which the term first appeared, but you should define the terms again to show increased understanding. Page numbers indicate the first appearance of the term in Chapter 6.

area representation for probability (p. 279)
bell-shaped curve (p. 278)
binomial distribution [5] (p. 304)
binomial probability (p. 304)
continuity correction factor (p. 304)
continuous random variable (p. 278, 304)
discrete random variable [1, 5] (p. 278, 304)

normal distribution (p. 278)
percentage (p. 279)
probability [4] (p. 279)
proportion (p. 279)
random variable [5] (p. 278)
standard normal distribution (p. 279, 286)
standard score [2] (p. 279, 286)

Use the **Tutorial** questions (located on the CD) to prepare for an exam. There are dozens of questions for each chapter, and when you select an incorrect answer, an explanation will appear telling you why that choice is incorrect.

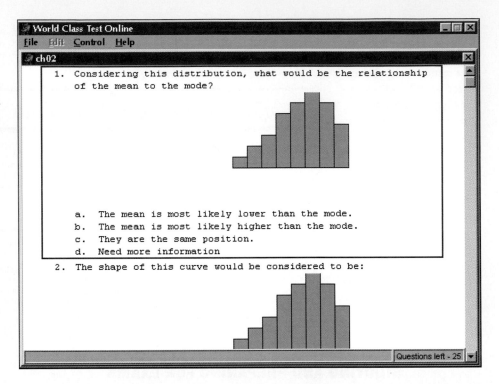

The **Chapter Practice Test** provides a formal self-evaluation of the mastery of the material, to be used before you are tested in class. Correct responses are in the back of the textbook.

CHAPTER PRACTICE TEST

Part I: Knowing the Definitions

Answer "True" if the statement is always true. If the statement is not always true, replace the words shown in bold with words that make the statement always true.

9.1 The Student's *t*-distributions have an approximately normal distribution but are **more** dispersed than the standard normal distribution.

9.2 The **chi-square** distribution is used for inferences about the mean when the σ is unknown.

Part II: Applying the Concepts

Answer all questions, showing all formulas, substitutions, and work.

9.11 Find each of the following:
   a. $z_{(0.02)}$
   b. $t_{(18, 0.95)}$
   c. $\chi^2_{(25, 0.95)}$

9.12 A random sample of 25 data was selected from a normally distributed population for the purpose of estimating the population mean, μ. The sample statistics are $n = 25, \bar{x} = 28.6$, $s = 3.50$.
   a. Find the point estimate for μ.
   b. Find the maximum error of estimate for the 0.95 confidence interval estimate.
   c. Find the lower confidence limit (LCL) and the upper confidence limit (UCL) for the 0.95 confidence interval estimate for μ.

Part III: Understanding the Concepts

9.19 Student B says the range of a set of data may be used to obtain a crude estimate for the standard deviation of a population. Student A is not sure. How will student B correctly explain how and under what circumstances his statement is true?

9.20 Is it (a) the null hypothesis or (b) the alternative hypothesis that the researcher usually believes to be true? Explain.

The most effective way to use a practice test is to: 1) complete the material assigned for the pending test; 2) use the vocabulary list as an informal self-test; 3) take the practice test, on the day before the exam, *under test conditions;* 4) correct it using the answers in the back; and 5) restudy the concepts related to the items missed. Do not use the practice tests as homework; use them to test yourself and they will help eliminate those "silly errors." WARNING: Use of the Practice Test too early in the study process defeats its role in the learning process.

A RECOMMENDED STUDY PROCEDURE

The following suggestions are based on the assumption that efficient and effective use of study time is a high priority. Ask your instructor for a schedule indicating what material (text pages) will be covered in each class and then spend 10 minutes before class reading that day's material. Do not take notes or highlight anything. This is a "warm-up" process—just "skim" read as a preview of the class to come. When you get to class you will be off to a "rolling start" since you have already seen the key words and concepts. When these concepts are discussed in class you will be hearing about them for the second time, and will therefore "hear" more in class. This will, in turn, reduce the study time required to successfully learn the material. (The before-class reading is just a skim read; it is not necessary to study the material at that time.)

As soon as possible after class, read the material again, thoroughly this time, making notes, highlighting the important points, and completing the assigned exercises. Do the margin exercises as you read.

Form a study group with two or three classmates. Friends are not always the best study partners; find study partners that have the same goals for this class and have the same determination to reach those goals as you. Establish a regular once-a-week meeting time and study together. Perhaps add an extra study session the week before a major exam. (True story: At mid-semester a few semesters ago, three students all had grades of D in one of our statistics classes. None of them found this acceptable; they did not know each other and were all determined to do something about their grade. At our suggestion, they formed a study group and met twice a week for the rest of the semester. They also met with us a few times for extra help. Their final grades were B, B, and C+. The study-group approach worked well for them and it can work for you.) In fact, the further you go in education, the more important study partners become.

GETTING ADDITIONAL ASSISTANCE

One of the best forms of assistance is to see your instructor outside of class, preferably during office hours that are set up for that purpose.

The **Statistical Tutor** is a student manual that:

a. Contains the complete solutions to all margin exercises and odd-numbered exercises (the same exercises whose answers are in the back of the book).

b. Contains many overviews, helpful hints, and suggestions to serve as a guide through the learning process.

c. Contains several review lessons to help refresh materials studied previously in other courses.

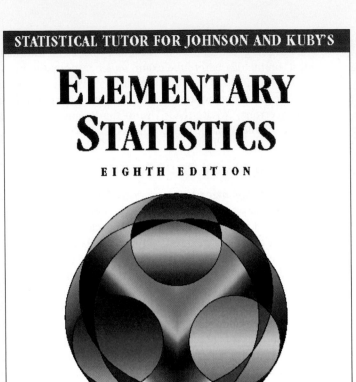

STATISTICAL TUTOR FOR JOHNSON AND KUBY'S

ELEMENTARY STATISTICS

EIGHTH EDITION

Patricia Kuby

Sir Francis Galton

SIR FRANCIS GALTON, English anthropologist and a pioneer of human intelligence studies, was born on February 16, 1822, in a village near Birmingham, England, to Samuel Tertiles and Anne (Violetta) Galton.

Galton's family included men and women of exceptional ability, one of whom was his cousin, Charles Darwin. Although his family life was happy, and he was grateful to his parents for all they had done for him, he felt little use for the conventional religious and classical education he was given.

As a teen, Galton toured a number of medical institutions in Europe and began his medical training in hospitals in Birmingham and London. He continued his medical studies until the death of his father. Having been left a sizable fortune, he decided to discontinue medical training and pursue his love of traveling. Several years of those travels led him to the exploration of primitive parts of southwestern Africa, where he gained valuable information that earned him recognition and a fellowship in the Royal Geographical Society. In 1853, at the age of only 31, Galton was awarded a gold medal from the Society in recognition of his hard work and many achievements. It was also in 1853 that Galton married Louisa Butler and they settled in London; their marriage remained childless.

Although Galton made important contributions to many fields of knowledge, he was best known for his work in eugenics; he spent most of the latter part of his life researching and promoting his belief that inheritance played a major role in the intelligence of man. Galton, a pioneer in the development of some of the refined statistical techniques that we use today, used the laws of probability to support his theory. His application of research techniques (curves of normal distribution, correlation coefficients, and percentile grading) to a large population revealed important facts about the intellectual and physical characteristics that are passed from one generation to the next and the ways in which children differ from their parents. He also discovered, with the use of the graph, that characteristics of two different generations could be plotted against one another, revealing important information.

Sir Francis Galton died in England on January 17, 1911, leaving behind a collection of 9 books, approximately 200 articles and lectures, and a valuable legacy of statistical techniques and knowledge.

The Statistical Process

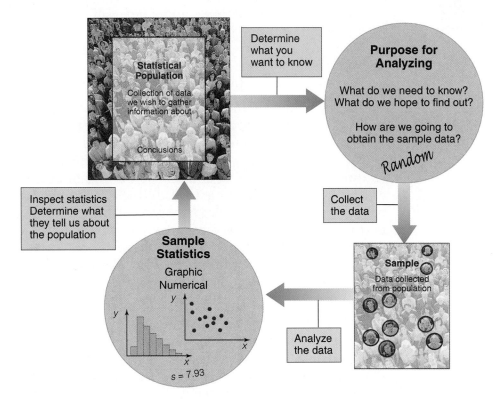

PART

One

Descriptive Statistics

A typical objective in statistics is to describe "the population" based on information obtained by observing relatively few individual elements. We must learn how to sort out the generalizations contained within the clues provided by the sample data and "paint" a picture of the population. We study the sample, but it's the population that is of primary interest to us.

When a statistical solution to a problem is sought, a certain sequence of events must develop:

1. the situation being investigated is carefully and fully defined,
2. a sample is collected from the population following an established and appropriate procedure,
3. the sample data are converted into usable information (this usable information, either numerical or pictorial, is called *sample statistics*), and
4. the theories of statistical inference are applied to the sample information in order to draw conclusions about the sampled population (these conclusions or answers are called inferences).

This sequence of events is illustrated on the Statistical Process diagram on the opposite page.

The first part of this textbook, Chapters 1–3, concentrates on the first three of the four events identified above. The second part, Chapters 4–7, deals with probability theory, the theory on which statistical inferences rely. The third and fourth parts, Chapters 8–10 and Chapters 11–14, survey the various types of inferences that can be made from sample information.

3

Chapter

1

Statistics

CHAPTER CASE STUDY

Americans, Here's Looking at You

The U.S. Census Bureau annually publishes the *Statistical Abstract of the United States,* a book of more than 1000 pages that provides us with a statistical insight into many of the most obscure and unusual facets of our lives. It is only one of thousands of sources for all kinds of things you have always wanted to know about and never thought to ask. Are you interested in how many hours we work and play? How much we spend on snack foods? How the price of Red Delicious apples has gone up? All this and more, much more, can be found in the *Statistical Abstract.*

The collection of statistical tidbits shown below were all reported in *USA Today* but come from separate sources and represent only a tiny sampling of what can be learned about Americans statistically.

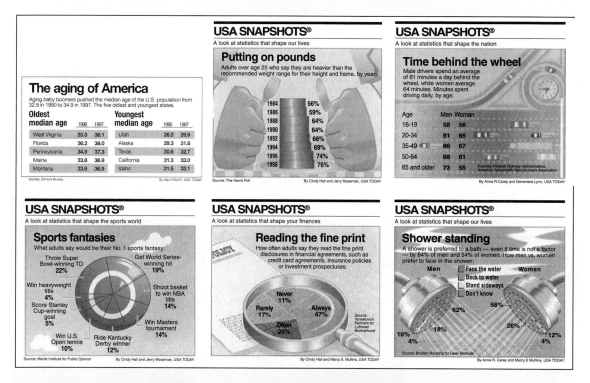

The above and thousands of other measures are used to describe life in the United States. What is a statistical population? Are the above measures data, statistics, or parameters? What is the difference? What variable is related to the median age—38.1 years—of West Virginians? What kind of variable was used to collect information that led to 76% saying they were overweight? We will return to these questions at the end of this chapter, when you will be better prepared to know the answers. (See Exercise 1.64, p. 29.)

CHAPTER OBJECTIVES

The purpose of this introductory chapter is to (1) create an initial image of the field of statistics, an image that will grow and develop; (2) introduce the basic vocabulary used in studying statistics; and (3) present some initial ideas and concerns about the processes used to obtain sample data.

1.1 What Is Statistics?

Statistics is the universal language of the sciences. As potential users of statistics, we need to master both the "science" and the "art" of using statistical methodology correctly. Careful use of statistical methods will enable us to obtain accurate information from data. These methods include (1) carefully defining the situation, (2) gathering data, (3) accurately summarizing the data, and (4) deriving and communicating meaningful conclusions.

Case Study 1.1

Measuring Physical Discomfort

Sitting in an uncomfortable seat for long periods of time is no fun. Slipping into a seat on a jet airliner when you are larger than average can be outright painful.

The world is small for big and tall business travelers. Ask Rosey Grier, a 6-foot-5, 300-pound former NFL defensive lineman, who never met a running back he couldn't tackle. But he can't win against airline seats built for 5-foot-9, 170-pounders. Grier is not alone. Thirteen million other men are at least 6-foot-2 or 225 pounds.

Airline seats have never been roomy. "They were originally designed to fit a very well-known athlete: (jockey) Willie Shoemaker," jokes Ed Perkins, editor of *Consumer Reports Travel Letter,* which measures the distance between seats every 2 years. Over the past 20 years, airlines have gradually squeezed about 10% more seats into their jets. All the while, they have inched the rows on all jets closer together. Rows are 4 inches closer together than they were 20 years ago.

By Del Jones, *USA Today,* 5-27-94

Exercise 1.1

Refer to "Traveling is a problem."

a. Who was surveyed?
b. How many were surveyed?
c. Explain the meaning of "Cramped airline seating: 99%."
d. Why is it possible for all of the reported percentages to be so large?

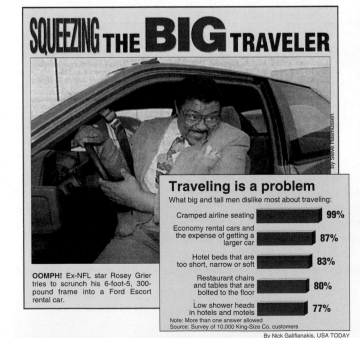

OOMPH! Ex-NFL star Rosey Grier tries to scrunch his 6-foot-5, 300-pound frame into a Ford Escort rental car.

Traveling is a problem
What big and tall men dislike most about traveling:

Cramped airline seating	99%
Economy rental cars and the expense of getting a larger car	87%
Hotel beds that are too short, narrow or soft	83%
Restaurant chairs and tables that are bolted to the floor	80%
Low shower heads in hotels and motels	77%

Note: More than one answer allowed
Source: Survey of 10,000 King-Size Co. customers
By Nick Galifianakis, USA TODAY

Statistics involves information, numbers and visual graphics, to summarize this information, and their interpretation. The word **statistics** has different meanings to people of varied backgrounds and interests. To some people it is a field of "hocus-pocus" in which a person attempts to overwhelm others with incorrect information and conclusions. To others it is a way of collecting and displaying information. And to still another group it is a way of "making decisions in the face of uncertainty." In the proper perspective, each of these points of view is correct.

The field of statistics can be roughly subdivided into two areas: descriptive statistics and inferential statistics. **Descriptive statistics** is what most people think of when they hear the word *statistics*. It includes the collection, presentation, and description of sample data. The term **inferential statistics** refers to the technique of interpreting the values resulting from the descriptive techniques and making decisions and drawing conclusions about the population.

Statistics is more than just numbers: it is data, what is done to data, what is learned from the data, and the resulting conclusions. Let's use the following definition:

STATISTICS

The science of collecting, describing, and interpreting data.

Before we begin our detailed study, let's look at a few illustrations of how and when statistics can be applied.

C a s e   S t u d y   1.2

Explaining One of Life's Mysteries

How much do people earn at their jobs? The U.S. government and its many branches continuously collect information and publish reports about the many aspects of our daily lives. They can explain how we spend our time, how we earn our money, how we spend our money, how they spend our tax money, and so on. The list is endless.

Exercise 1.2

Refer to "1996 Median Wage Samples."

a. Who was surveyed?
b. Who did the surveying?
c. Explain the meaning of Cashiers; 3,262,120; and $5.75.

Log on the Internet site and look at the complete tables. How does your wage rate compare to others of the same occupation?

Sharper Picture of Pay

GANNETT NEWS SERVICE, 1-6-98

The government issues its most detailed report yet of wages for each occupation.

1996 Median Wage Samples

The Labor Department's survey of 1996 pay in 764 occupations showed the majority of 19 managerial and 211 professional occupations paid workers in the upper wage range while service and agricultural jobs paid the lowest.

Occupation	Employment	Hourly median wages
Athletes, coaches, umpires, and related workers	19,710	$8.33
Automotive mechanics	647,560	$12.35
Cashiers	3,262,120	$5.75
Child-care workers	377,980	$6.12
Dentists	85,250	$47.66

Complete tables are available on the Internet at *http://stats.bls.gov/oeshome.htm*

Source: U.S. Bureau of Labor Statistics

Case Study 1.3

Telling Us "What We Think"

Exercise 1.3

Refer to "What we think about the economy."

a. Who was surveyed?
b. When were the surveys taken?
c. How accurate are the reported percentages believed to be?
d. What do you think the 54% combined with the margin of error means?

The newspapers frequently publish articles containing statistics that tell us how we, as a population, think collectively. Did you ever wonder "How much of what we think is directly influenced by the information we read in these articles?"

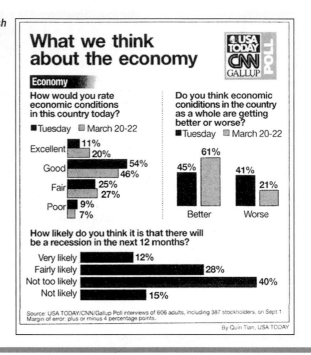

Case Study 1.4

Statistics Is Tricky Business

"One ounce of statistics technique requires one pound of common sense for proper application."

Exercise 1.4

Find a recent newspaper article that illustrates an "apples are bad" type of report.

Harvard Health Letter, Special Supplement, October 1994.

Because we are mortal and . . . live in an imperfect world, risk will always be with us. . . . As we go about our lives, we weigh the relative risks and benefits of our actions all the time. Most often we act on imperfect and incomplete information.

Commercial flight is one of very few areas where the degree of risk has been calculated and reduced about as far as is practical. Once we walk off the airplane, however, our risks vary dramatically and are much more difficult to fathom, so we constantly make decisions based on more or less educated guesses.

Common Sense

So, . . . how can responsible individuals with no special expertise make intelligent decisions about all the information and misinformation that bombards us? The same humble horse sense that keeps us from sticking our hand into the fire is an invaluable tool for sorting out

what we read and hear. It's important to remember that news, by its very definition, is something new and unusual. After all, the hundredth study showing a relationship between cholesterol and heart disease is hardly news, but the one study that fails to make such a connection is likely to become a headline. Clearly it would be silly for people to drastically change their lives on the basis of one newspaper article or a lone study.

That doesn't mean we should throw out everything we read or hear. Once medical experts have reached consensus on a particular health issue, their message is amplified by the popular press and codified in guidelines issued by government agencies and national organizations. Today, for instance, there is widespread agreement that having high blood cholesterol level is a risk factor for heart disease. . . .

Case Study   1.5

Describing Our Romance with the Road

The Department of Transportation takes a snapshot of the country's travelers and finds that America's love affair with the car is still in bloom. Americans are logging 800 billion miles a year on long-distance trips and even on trips up to 2000 miles; the typical traveler would rather drive than fly.

Exercise 1.5

Refer to "Americans' travel habits."

a. Who was surveyed?
b. Based on this information, how would you describe the "typical" long-distance traveler?
c. How do *you* compare to this "typical" long-distance traveler?

Americans' travel habits

When Americans take trips of 100 miles or more, they do it for pleasure and they use their own cars. Findings from the American Travel Survey, a comprehensive look at Americans' long-distance traveling habits:

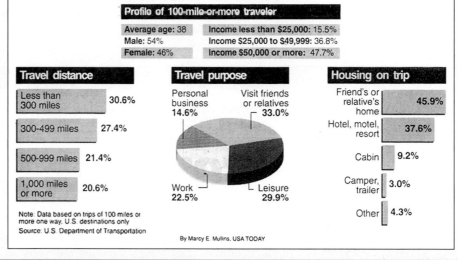

Profile of 100-mile-or-more traveler	
Average age: 38	Income less than $25,000: 15.5%
Male: 54%	Income $25,000 to $49,999: 36.8%
Female: 46%	Income $50,000 or more: 47.7%

Travel distance

Less than 300 miles	30.6%
300-499 miles	27.4%
500-999 miles	21.4%
1,000 miles or more	20.6%

Note: Data based on trips of 100 miles or more one way. U.S. destinations only
Source: U.S. Department of Transportation

Travel purpose

Personal business 14.6%
Visit friends or relatives 33.0%
Work 22.5%
Leisure 29.9%

Housing on trip

Friend's or relative's home	45.9%
Hotel, motel, resort	37.6%
Cabin	9.2%
Camper, trailer	3.0%
Other	4.3%

By Marcy E. Mullins, USA TODAY

And Then There's "Entertainment Statistics"

The 1990s success story for the news media business has been the evolution of "entertainment news." The tabloids have reached new heights of popularity, but the even bigger success story belongs to television's entertainment news stories: newsmagazine shows and talk shows. There are so many of them that they're barely countable. You, of course, realize that the success entertainment news has enjoyed is largely due to "entertainment statistics." (By the way, you will not be tested on these statistics, guaranteed!) On world news:

Statistical Snapshot

© 1995 (Bill Plympton from the Cartoon Bank)

The uses of statistics are unlimited. It is much harder to name a field in which statistics is not used than it is to name one in which statistics plays an integral part. The following are a few examples of how and where statistics are used:

- In education descriptive statistics are frequently used to describe test results.
- In science the data resulting from experiments must be collected and analyzed.
- In government many kinds of statistical data are collected all the time. In fact, the U.S. government is probably the world's greatest collector of statistical data.

A very important part of the statistical process is that of studying the statistical results and formulating appropriate conclusions. These conclusions must then be communicated accurately; nothing is gained from research unless the findings are shared with others. Statistics are being reported everywhere: newspapers, magazines, radio, and television. We read and hear about all kinds of new research results, especially in the health-related fields.

E X E R C I S E S

1.6 Determine which of the following statements is descriptive in nature and which is inferential. Refer to "Traveling is a problem" in Case Study 1.1 (p. 6).
  a. 99% of all big and tall travelers dislike cramped airline seating the most.
  b. 99% of the 10,000 King-Size Co. customers dislike cramped airline seating the most.

1.7 Determine which of the following statements is descriptive in nature and which is inferential. Refer to "Americans' travel habits" in Case Study 1.5 (p. 9).
  a. The average age of the surveyed travelers is 38 years.
  b. 54% of all American long-distance travelers are men.

1.8 Refer to the USA Snapshot® "Cars, trucks older than ever."
  a. Construct a graph of the information; use "years" as the *x*-axis, "age" as the *y*-axis, use red to plot the cars' ages, and blue to plot the trucks' ages.
  b. Does your graph support the claim made by the title, "Cars, trucks older than ever"? Explain.
  c. State an inference about the quality of cars and trucks that is suggested by your graph.

USA SNAPSHOTS®

A look at statistics that shape your finances

Cars, trucks older than ever

The median age of cars on U.S. roads – half are older, half younger – is a record high. How our vehicles have aged:

| | Median age in years | |
	Cars	Trucks[1]
1997	8.1	7.8
1995	7.7	7.6
1990	6.5	6.5
1985	6.9	7.6
1980	6.0	6.3
1975	5.4	5.8
1970	4.9	5.9

1 – Pickups, sport utilities and minivans

Source: Polk Co.   By Anne R. Carey and Jerry Mosemak, USA TODAY

USA SNAPSHOTS®

A look at statistics that shape the nation

Reacting to crime

How people who say they feel less safe due to crime are changing their lifestyles:

Carry less cash	55%
Use charge cards more	29%
Carry a personal protection device	28%
Bought home security device	23%
Bought car security device	22%
Bought a gun	17%

Source: America's Research Group poll of 1,000

By Sam Ward, USA TODAY

1.9 Refer to the USA Snapshot® "Reacting to crime."
   a. What group of people was polled?
   b. How many people were polled?
   c. What information was obtained from each person?
   d. Explain the meaning of "55% Carry less cash."
   e. How many people answered "Carry less cash"?
   f. Why do the reported values (55%, 29%, 28%, . . .) add up to more than 100%?

1.10 Business investors find it quite valuable to compare the performances of competitive firms before deciding which ones to support by buying stock. The table below was extracted from a *Fortune* magazine that reported information gathered from four rental-car companies:

Company	Stock Price	Price/ Earnings Ratio	Earnings Growth Rate
Hertz Corp.	$43\frac{1}{2}$	21.6	17%
Avis Rent A Car	$26\frac{1}{4}$	23.6	17%
Budget Group	$33\frac{3}{4}$	28.4	25%
Dollar Thrifty	$19\frac{1}{4}$	13.0	17%

Note: Stock prices as of April 29, 1998. P/Es are on estimated 1998 earnings.
Source: Fortune, May 25, 1998

   a. Define the nature of the businesses being studied.
   b. What information was collected from each company? Is the *information* descriptive or inferential?
   c. How many representative companies were included in the study?
   d. If you had $5000 to invest in any one of these four companies and no other information available, which stock would you buy? Why?

1.11 During a radio broadcast on August 16, 1998, David Essel reported the following three statistics: (1) The U.S. divorce rate is 55%, and when married adults were asked if they would remarry their spouse, (2) 75% of the women said yes, and (3) 65% of the men said yes.
   a. What is the "stay married" rate?
   b. There seems to be a contradiction in this information. How is it possible for all three of these statements to be correct? Explain.

1.12 If you were an independent researcher, how would you test the accuracy of the statement made in Case Study 1.1 (p. 6) that "airline seats [were] built for 5-foot-9, 170-pounders"? Include whom you would survey and what information you would collect.

1.2   Introduction to Basic Terms

In order to begin our study of statistics we need to first define a few basic terms.

POPULATION

A collection, or set, of individuals or objects or events whose properties are to be analyzed.

The population is the complete collection of individuals or objects that are of interest to the sample collector. The concept of a population is the most fundamental idea in statistics. The population of concern must be carefully defined and is considered fully defined only when its membership list of elements is specified. The set of "all students who have ever attended a U.S. college" is an example of a well-defined population.

Typically, we think of a population as a collection of people. However, in statistics the population could be a collection of animals, or manufactured objects, or whatever. For example, the set of all redwood trees in California could be a population.

There are two kinds of populations, finite and infinite. When the membership of a population can be (or could be) physically listed, the population is said to be **finite.** When the membership is unlimited, the population is **infinite.** The books in your college library form a finite population; the OPAC (Online Public Access Catalog, the computerized card catalog) lists the exact membership. All the registered voters in the United States form a very large finite population; if necessary, a composite of all voter lists from all voting precincts across the United States could be compiled. On the other hand, the population of all people who might use aspirin and the population of all 40-watt light bulbs to be produced by Sylvania are infinite. Large populations are difficult to study; therefore, it is customary to select a *sample* and study the data in the sample.

SAMPLE

A subset of a population.

A sample consists of the individuals, objects, or measurements selected by the sample collector from the population.

VARIABLE (or response variable)

A characteristic of interest about each individual element of a population or sample.

A student's age at entrance into college, the color of the student's hair, the student's height, and the student's weight are four variables.

DATA (singular)

The value of the variable associated with one element of a population or sample. This value may be a number, a word, or a symbol.

For example, Bill Jones entered college at age "23," his hair is "brown," he is "71 inches" tall, and he weighs "183 pounds." These four pieces of data are the values for the four variables as applied to Bill Jones.

DATA (plural)

The set of values collected for the variable from each of the elements belonging to the sample.

The set of 25 heights collected from 25 students is an example of a set of data.

EXPERIMENT

A planned activity whose results yield a set of data.

This includes both the activities for selecting the elements and obtaining the data values.

> Parameters describe the **p**opulation; notice that both begin with the letter **p**.

PARAMETER

A numerical value summarizing all the data of an entire population.

The "average" age at time of admission for all students who have ever attended our college and the "proportion" of students who were over 21 years of age when they entered college are examples of two population parameters. A parameter is a value that describes the entire population. Often a Greek letter is used to symbolize the name of a parameter. These symbols will be assigned as we study specific parameters.

For every parameter there is a *corresponding sample statistic.* The statistic describes the sample the same way the parameter describes the population.

> A **s**tatistic describes the **s**ample; notice, both begin with the letter **s**.

STATISTIC

A numerical value summarizing the sample data.

The "average" height, found by using the set of 25 heights, is an example of a sample statistic. A statistic is a value that describes a sample. Most sample statistics are found with the aid of formulas and are typically assigned symbolic names using letters of the English alphabet (for example, $\bar{x}$, s, and r).

ILLUSTRATION 1.1 ▼

> **Exercise 1.13**
> Thirty-six percent of the adult U.S. population has an allergy. A sample of 1200 randomly selected adults resulted in 33.2% having an allergy. Describe each of the 8 terms.

A statistics student is interested in finding out something about the average dollar value of cars owned by the faculty members of our college. Each of the eight terms just described can be identified in this situation.

1. The *population* is the collection of all cars owned by all faculty members at our college.
2. A *sample* is any subset of that population. For example, the cars owned by members of the mathematics department would be a sample.
3. The *variable* is the "dollar value" of each individual car.
4. One *data* would be the dollar value of a particular car. Mr. Jones's car, for example, is valued at $9400.
5. The *data* would be the set of values that correspond to the sample obtained (9400; 8700; 15,950; . . .).
6. The *experiment* would be the methods used to select the cars forming the sample and determining the value of each car in the sample. It could be carried out by questioning each member of the mathematics department, or in other ways.
7. The *parameter* about which we are seeking information is the "average" value of all cars in the population.

> Parameters are fixed in value, while statistics vary in value.

8. The *statistic* that will be found is the "average" value of the cars in the sample.   ▲

Exercise 1.14

In your own words, explain why the parameter is fixed and the statistic varies.

NOTE If a second sample were to be taken, it would probably result in a different set of people being selected, say the English Department, and therefore a different value would be anticipated for the statistic, average value. The average value for "all faculty owned cars" would not change, however.

There are basically two kinds of variables: (1) variables that result in *qualitative* information, and (2) variables that result in *quantitative* information.

QUALITATIVE, OR ATTRIBUTE, OR CATEGORICAL, VARIABLE

A variable that describes or categorizes an element of a population.

Exercise 1.15

Name two attribute variables about its customers that a newly opened department store might find informative to study.

A sample of four hair-salon customers was surveyed for "hair color," "hometown," and "level of satisfaction" with the results of their salon appointment. All three variables are examples of qualitative (attribute) variables, as they describe some characteristic of the person, and all people with the same attribute belong to the same category. The data collected were {blonde, brown, black, brown}, {Brighton, Columbus, Albany, Jacksonville}, and {very satisfied, satisfied, very satisfied, somewhat satisfied}.

NOTE Arithmetic operations, such as addition and averaging, are not meaningful for data resulting from a qualitative variable.

QUANTITATIVE, OR NUMERICAL, VARIABLE

A variable that quantifies an element of a population.

Exercise 1.16

Name two numerical variables about its customers that a newly opened department store might find informative to study.

"Total cost" of textbooks purchased by each student for this semester's classes is an example of a quantitative (numerical) variable. A sample resulted in the following data: $238.87, $94.57, $139.24. [To find the "average cost," simply add the three numbers and divide by three: (238.87 + 94.57 + 139.24)/3 = $157.56.]

NOTE Arithmetic operations, such as addition and averaging, are meaningful for data resulting from a quantitative variable.

Each of these types of variables (qualitative and quantitative) can be further subdivided.

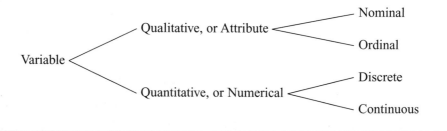

NOMINAL VARIABLE

A qualitative variable that categorizes (or describes, or names) an element of a population. Not only are arithmetic operations not meaningful for data resulting from a nominal variable, an order cannot be assigned to the categories.

In the above survey of four hair-salon customers, two of the variables, "hair color" and "hometown," are examples of nominal variables, as they both name some charac-

Exercise 1.17
Name two nominal variables about its customers that a newly opened department store might find informative to study.

teristic of the person and it would be meaningless to find the sample average by adding and dividing by 4. For example, (blonde + brown + black + brown)/4 is undefined. Further, color of hair and hometown do not have an order to their categories.

ORDINAL VARIABLE

A qualitative variable that incorporates an ordered position, or ranking.

In the above survey of four hair-salon customers, the variable "level of satisfaction" is an example of an ordinal variable, as it does incorporate an ordered ranking. "Very satisfied" ranks ahead of "satisfied" which ranks ahead of "somewhat satisfied." Another illustration of an ordinal variable would be the ranking of five landscape pictures in accordance to someone's preference; first choice, second choice, . . .

Quantitative or numerical variables can also be subdivided into two classifications: *discrete* variables and *continuous* variables.

Exercise 1.18
Name two ordinal variables about its customers that a newly opened department store might find informative to study.

DISCRETE VARIABLE

A quantitative variable that can assume a countable number of values. Intuitively, the discrete variable can assume the values corresponding to isolated points along a line interval. That is, there is a gap between any two values.

CONTINUOUS VARIABLE

A quantitative variable that can assume an uncountable number of values. Intuitively, the continuous variable can assume any value along a line interval, including every possible value between any two values.

In many cases, the two types of variables can be distinguished by deciding whether the variables are related to a count or a measurement. The variable "number of courses for which you are currently registered" is an example of a discrete variable; the values of the variable will be found by counting the courses. [When counting, fractional values cannot occur; thus, there are gaps (fractional numbers) between the values that can occur.] The variable "weight of books and supplies you are carrying as you attend class today" is an example of a continuous random variable; the values of the variable will be found by measuring the weight. (When measuring, any fractional value can occur; thus, every value along the number line is possible.)

When trying to determine whether a variable is discrete or continuous, remember to look at the variable and think about the values that might occur. Do not look at data values that have been recorded; they can be very misleading.

Consider the variable "judge's score" at a figure-skating competition. If we look at some scores that have previously occurred, 9.9, 9.5, 8.8, 10.0, and we see the presence of decimals, we might think that all fractions are possible and conclude the variable is continuous. This is not true, however. A score of 9.134 is not possible; thus, there are gaps between the possible values and the variable is discrete.

Exercise 1.19
a. Explain why the variable "score" for the home team at a basketball game is discrete.
b. Explain why the variable "number of minutes to commute to work" is continuous.

NOTE Don't let the appearance of the data fool you in regard to their type. Qualitative variables are not always easy to recognize; sometimes they appear as numbers. The above sample of hair colors could be coded: 1 = black, 2 = blonde, 3 = brown. The sample data would then appear as {2, 3, 1, 3}, but they are still nominal data.

Calculating the "average hair color" $[(2 + 3 + 1 + 3)/4 = 9/4 = 2.25]$ is still meaningless. The hometowns could have been identified using ZIP codes. The average ZIP code doesn't make sense either; therefore, ZIP code numbers would be nominal, too.

Let's look at another example. Suppose that after surveying a parking lot, I summarized the sample data by reporting 5 red, 8 blue, 6 green, and 2 yellow cars. You must look at each individual source to determine the kind of information being collected. One specific car was red; "red" is the data from that one car, and red is an attribute. Thus, this collection (5 red, 8 blue, and so on) is a summary of the nominal data.

Another example of information that is deceiving is an identification number. Flight #249 and Room #168 both appear to be numerical data. The numeral 249 does not describe any property of the flight: late or on time, quality of snack served, number of passengers, or anything else about the flight. The flight number only identifies a specific flight. Driver's license numbers, Social Security numbers, and bank account numbers are all identification numbers being used in the nominal sense, not in the quantitative sense.

Remember to inspect the individual variable and one individual data, and you should have little trouble distinguishing between the various types of variables.

C a s e   S t u d y   1.6

Eating Together: Still Important

The January 6, 1991, issue of the Democrat and Chronicle *presented the results of a New York* Times/CBS News *poll regarding today's attitudes about family togetherness. The graphic shows the results of four variables.*

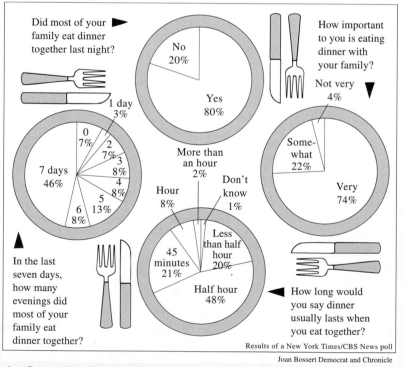

Results of a New York Times/CBS News poll

Joan Bossert Democrat and Chronicle

Exercise 1.20

a. Name the four variables.

b. What kind of variable is each?

Source: Joan Bossert, "Eating Together: Still Important," *Democrat and Chronicle,* January 6, 1991, copyright 1991 by *Democrat and Chronicle.* Reprinted by permission.

EXERCISES   •   •   •   •   •   •   •   •   •   •   •   •   •

1.21 The severity of side effects experienced by patients while being treated with a particular medicine is under study. The severity is measured on a scale of: none, mild, moderate, severe, very severe.
  a. Name the variable of interest.
  b. Identify the type of variable.

1.22 Students are being surveyed about the weight of books and supplies they are carrying as they attend class.
  a. Identify the variable of interest.
  b. Identify the type of variable.
  c. List a few values that might occur in a sample

1.23 A drug manufacturer is interested in the proportion of persons who have hypertension (elevated blood pressure) whose condition can be controlled by a new drug the company has developed. A study involving 5000 individuals with hypertension is conducted, and it is found that 80% of the individuals are able to control their hypertension with the drug. Assuming that the 5000 individuals are representative of the group who have hypertension, answer the following questions:
  a. What is the population?
  b. What is the sample?
  c. Identify the parameter of interest.
  d. Identify the statistic and give its value.
  e. Do we know the value of the parameter?

1.24 The admissions office wants to estimate the cost of textbooks for students at our college. Let the variable x be the total cost of all textbooks purchased by a student this semester. The plan is to randomly identify 100 students and obtain their total textbook costs. The average cost for the 100 students will be used to estimate the average cost for all students. Describe:
  a. the parameter the admissions office wishes to estimate.
  b. the population.
  c. the variable involved.
  d. the sample.
  e. the statistic and how you would use the 100 data collected to calculate the statistic.

1.25 A quality-control technician selects assembled parts from an assembly line and records the following information concerning each part:
  A: defective or nondefective
  B: the employee number of the individual who assembled the part
  C: the weight of the part
  a. What is the population?
  b. Is the population finite or infinite?
  c. What is the sample?
  d. Classify the three variables as either attribute or quantitative.

1.26 Select ten students currently enrolled at your college and collect data for these three variables:
  X: number of courses enrolled in
  Y: total cost of textbooks and supplies for courses
  Z: method of payment used for textbooks and supplies _nominal (certain order)_
  a. What is the population? _students enrolled_
  b. Is the population finite or infinite?
  c. What is the sample? _10 students_
  d. Classify the three variables as nominal, ordinal, discrete, or continuous.

1.27 A study was conducted by Zeneca, Inc. to measure the adverse side effects of Zomig™, a new drug being used for the treatment of migraine headaches. A sample of 2633 migraine headache sufferers was given various dosage amounts of the drug in tablet form for one year. The patients reported whether or not during the period they experienced relief from migraines, and atypical sensations such as hypertension and paresthesia, pain and pressure sensations, digestive disorders, neurological disorders such as dizziness and vertigo, and other adverse side effects (sweating, palpitations, and so on). (*Life,* June 1998, p. 92.)

   a. What is the population being studied?
   b. What is the sample?
   c. What are the characteristics of interest about each element in the population?
   d. Is the data being collected qualitative or quantitative?

1.28 There are many variables that could be used to compare the "size" of Kelly Robbins and John Daly.

   a. Other than weight, what other measures of size might play a role in drive distance?
   b. If Daly is 5′11″ and Robbins is 5′6″, compare them using "yards per inch."

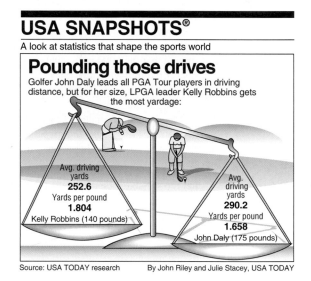

USA SNAPSHOTS®
A look at statistics that shape the sports world

Pounding those drives
Golfer John Daly leads all PGA Tour players in driving distance, but for her size, LPGA leader Kelly Robbins gets the most yardage:

Avg. driving yards
252.6
Yards per pound
1.804
Kelly Robbins (140 pounds)

Avg. driving yards
290.2
Yards per pound
1.658
John Daly (175 pounds)

Source: USA TODAY research   By John Riley and Julie Stacey, USA TODAY

1.29 Identify each of the following as examples of (1) attribute (qualitative) or (2) numerical (quantitative) variables:

   a. the breaking strength of a given type of string
   b. the hair color of children auditioning for the musical *Annie*
   c. the number of stop signs in towns of less than 500 people
   d. whether or not a faucet is defective
   e. the number of questions answered correctly on a standardized test
   f. the length of time required to answer a telephone call at a certain real estate office

1.30 Identify each of the following as examples of (1) nominal, (2) ordinal, (3) discrete, or (4) continuous variables:

   a. a poll of registered voters asking which candidate they support
   b. the length of time required for a wound to heal when using a new medicine
   c. the number of telephone calls arriving at a switchboard per ten-minute period
   d. the distance first-year college women can kick a football
   e. the number of pages per job coming off a computer printer
   f. the kind of tree used as a Christmas tree

1.31 Suppose a 12-year-old asked you to explain the difference between a sample and a population.

   a. What information should your answer include?
   b. What reasons would you give him or her for why one would take a sample instead of surveying every member of the population?

1.32 Suppose a 12-year-old asked you to explain the difference between a statistic and a parameter.
  a. What information should your answer include?
  b. What reasons would you give him or her for why one would report the value of a statistic instead of a parameter?

1.3   Measurability and Variability

Within a set of data, we always expect variation. If little or no variation is found, we would guess that the measuring device is not calibrated with a small enough unit. For example, suppose we take a carton of a favorite candy bar (24) and weigh each bar individually. We observe that each of the 24 candy bars weighs $^7/_8$ ounce, to the nearest $^1/_8$ ounce. Does this mean that the bars are all identical in weight? Not really! Suppose that we were to weigh them on an analytical balance that weighs to the nearest ten-thousandth of an ounce. Now the 24 weights will most likely show **variability.**

It does not matter what the response variable is; there will most likely be variability in the data if the tool of measurement is precise enough. One of the primary objectives in statistical analysis is that of measuring variability. For example, in the study of quality control, measuring variability is absolutely essential. Controlling (or reducing) the variability in a manufacturing process is a field all its own, namely statistical process control.

EXERCISES • • • • • • • • • • • • • • •

1.33 Suppose we measure the weights (in pounds) of the individuals in each of the following groups:

> Group 1: cheerleaders for National Football League teams
> Group 2: players for National Football League teams

For which group would you expect the data to have more variability? Explain why.

1.34 Suppose you were trying to decide which of two machines to purchase. Furthermore, suppose the length to which the machines cut a particular product part was important. If both machines produced parts that had the same length on average, what other consideration regarding the lengths would be important? Why?

1.35 Consumer activist groups for years have encouraged retailers to use unit pricing of products. They argue that food prices, for example, should always be labeled in $/ounce, $/pound, $/gram, $/liter, etc., in addition to $/package, $/can, $/box, $/bottle, and so on. Explain why.

1.36 A coin-operated coffee vending machine dispenses, on the average, 6 oz of coffee per cup. Can this statement be true of a vending machine that occasionally dispenses only enough to barely fill the cup half full (say, 4 oz)? Explain. *yes d/t variability*

1.4   Data Collection

One of the first problems a statistician faces is that of obtaining data. Data doesn't just happen; data must be collected. It is important to obtain "good data" since the inferences ultimately made will be based on the statistics obtained from the data. These inferences can only be as good as the data.

While it is relatively easy to define "good data" as data that accurately represents the population from which it was taken, it is not easy to guarantee that a particular sampling method will produce "good data." We want to use sampling (data collection) methods that are *unbiased*.

BIASED SAMPLING METHOD

A sampling method that produces values which systematically differ from the population being sampled. An **unbiased** sampling method is one that is not biased.

Two commonly used sampling methods that often result in biased samples are the *convenience* and *volunteer samples.*

A **convenience sample** occurs when a sample is selected from elements of a population that are easily accessible, while a **volunteer sample** consists of results collected from those elements of the population that choose to contribute the needed information on their own initiative.

Did you ever buy a basket of fruit at the market based on the "good appearance" of the fruit on top, only to later discover the fruit on the bottom was not as nice? It was not convenient to inspect the bottom fruit, so you trusted a convenience sample.

Have you ever taken part in a volunteer survey? Under what conditions did (would) you take the time to complete such a questionnaire? Most people's immediate attitude is to ignore the survey. Those with strong feelings will make the effort to respond; therefore, representative samples should not be expected when volunteer samples are collected.

The collection of data for statistical analysis is an involved process and includes the following steps:

1. Defining the objectives of the survey or experiment.
   Examples: comparing the effectiveness of a new drug to the effectiveness of the standard drug; estimating the average household income in our county.

2. Defining the variable and the population of interest.
   Examples: length of recovery time for patients suffering from a particular disease; total income for households in our county.

3. Defining the data-collection and data-measuring schemes.
   This includes sampling procedures, sample size, and the data-measuring device (questionnaire, telephone interview, and so on).

4. Determining the appropriate descriptive or inferential data-analysis techniques.

Often an analyst is stuck with data already collected, possibly even data collected for other purposes, which makes it impossible to determine whether or not the data are "good." Using approved techniques to collect your own data is much preferred. Although this text will be concerned chiefly with various data-analysis techniques, you should be aware of the concerns of data collection.

The following illustration describes the population and the variable of interest for a specific investigation:

Exercise 1.37

USA Today regularly asks readers, "Have a complaint about airline baggage, refunds, advertising, customer service? Write: . . . ." What kind of sampling method is this? Are the results likely to be biased? Explain.

ILLUSTRATION 1.2 ▼

The admissions office at our college wishes to estimate the current "average" cost of textbooks per semester, per student. The population of interest is the "currently enrolled student body," and the variable is the "total amount spent for textbooks" by each student this semester. ▲

The two methods used to collect data for a statistical analysis are either an *experiment* or an *observational study.* In an **experiment,** the investigator controls or modifies the environment and observes the effect on the variable under study. We often read about laboratory results obtained by using white rats to test different doses of a new medication and its effect on blood pressure. The experimental treatments were designed specifically to obtain the data needed to study the effect on the variable. In an **observational study,** the investigator does not modify the environment and does not control the process being observed. The data are obtained by sampling some of the population of interest. **Surveys** are often observational studies of people.

If every element in the population can be listed, or enumerated, and observed, then a **census** is compiled. A census is a 100% survey. A census for the population in Illustration 1.2 (p. 20) could be obtained by contacting each student on the registrar's computer printout of all registered students. However, censuses are seldom used because they are often very difficult and time-consuming to compile, and therefore very expensive. Imagine the task of compiling a census of every person who is a potential client at the brokerage firms in Case Study 1.8 (p. 22). Instead of a census, a sample survey is usually conducted.

When selecting a sample for a survey, it is necessary to construct a *sampling frame.*

SAMPLING FRAME

A list of the elements belonging to the population from which the sample will be drawn.

Ideally, the sampling frame should be identical to the population with every element of the population listed once and only once. In Illustration 1.2, the registrar's computer list will serve as a sampling frame for the admissions office. In this case, the list of all registered students becomes the sampling frame. In other situations, a 100% list may not be so easy to obtain. Lists of registered voters or the telephone directory are sometimes used as sampling frames of the general public. Depending on the nature of the information sought, the list of registered voters or the telephone directory may or may not serve as an unbiased sampling frame. Since only the elements in the frame have a chance to be selected as part of the sample, it is important that the sampling frame be **representative** of the population.

Once a representative sampling frame has been established, we proceed with selecting the sample elements from the sampling frame. This selection process is called

C a s e   S t u d y   1.7

Rain Again This Weekend!

It does, in fact, rain more on weekends

The Associated Press, 8-6-98

It's maddening but true: More rain does fall on weekends, a study says. Saturdays receive an average of 22% more precipitation than Mondays, climatologists at Arizona State University report in today's issue of *Nature*. "We were quite surprised to see weekends are substantially wetter than weekdays," said Randall Cerveny, one of the study's principal authors. ...

Exercise 1.38

Was this study an experiment or an observational study? Explain.

the **sample design**. There are many different types of sample designs; however, they all fit into two categories: *judgment samples* and *probability samples*.

JUDGMENT SAMPLES

Samples that are selected on the basis of being "typical."

When a judgment sample is drawn, the person selecting the sample chooses items that he or she thinks are representative of the population. The validity of the results from a judgment sample reflects the soundness of the collector's judgment.

PROBABILITY SAMPLES

Samples in which the elements to be selected are drawn on the basis of probability. Each element in a population has a certain known probability of being selected as part of the sample.

One of the most common probability sampling methods used to collect data is the *simple random sample.*

SIMPLE RANDOM SAMPLE

A sample selected in such a way that every element in the population has an equal probability of being chosen. Equivalently, all samples of size n have an equal chance of being selected. Simple random samples are obtained either by sampling with replacement from a finite population or by sampling without replacement from an infinite population.

Case Study 1.8

Are We Treated Equally?

Exercise 1.39

Refer to "What Every Woman Should Know About Stockbrokers."

a. What is the population of interest?
b. Was this investigation an experiment or an observational study?

What Every Woman Should Know About Stockbrokers

A study by *Money* magazine suggests that "stockbrokers are more courteous to women clients than men," but they "take men more seriously."

Money magazine hired one hundred testers—50 men and 50 women—and sent them into 50 randomly chosen brokerage offices in Los Angeles and Chicago.

Among their findings:
- Sixty percent of the men were asked about their tolerance of risk in investing, against only 48% of the women.
- On the other hand, brokers were less likely to allow interruptions when the prospective client was a woman.

Source: Money, June 1993.

Inherent in the concept of randomness is the idea that the next result (or occurrence) is not predictable. When a simple random sample is drawn, every effort must be made to ensure that each element has an equal probability of being selected and that the next result does not become predictable. The proper procedure for selecting a simple random sample is to use a random-number generator or a table of random numbers. Mistakes are frequently made because the term *random* (equal chance) is confused with **haphazard** (without pattern).

To select a simple random sample, first assign a number to each element in the sampling frame. This is usually done sequentially using the same number of digits for each element. Then go to a table of random numbers and select as many numbers with that number of digits as are needed for the sample size desired. Each numbered element in the sampling frame that corresponds to a selected random number is chosen for the sample.

ILLUSTRATION 1.3 ▼

Let's return to Illustration 1.2 (p. 20). Mr. Clar, who works in the admissions office, has obtained a computer list of this semester's full-time enrollment. There are 4265 student names on the list. He numbered the students 0001, 0002, 0003, and so on, up to 4265; then, using four-digit random numbers, he identified a sample: 1288, 2177, 1952, 2463, 1644, 1004, and so on were selected. (See the *Statistical Tutor* for a discussion of the use of the random-number table.)   ▲

NOTE In this text, all the statistical methods assume that random sampling has been used to collect the data.

There are many ways to approximate random sampling; four of these methods are briefly discussed below. They are presented here to illustrate some of the methodology involved in data collection. The topic of survey sampling is a complete textbook in itself.

One of the easiest-to-use methods for approximating a simple random sample is the *systematic sampling method.*

SYSTEMATIC SAMPLE
A sample in which every kth item of the sampling frame is selected, starting from a randomly selected first element.

To select an $x\%$ systematic sample, first we will randomly select 1 element from the first $\dfrac{100}{x}$ elements, then we proceed to select every $\dfrac{100}{x}$th item thereafter until we have the desired number of data for our sample.

For example, if we desire a 3% systematic sample, we would locate the first element by randomly selecting an integer between 1 and 33 ($\dfrac{100}{x} = \dfrac{100}{3} = 33.33$, which when rounded becomes 33). Suppose 23 was randomly selected. This means that our first data is the value obtained from the subject in the 23rd position in the sampling frame. The second data will come from the subject in the 56th ($23 + 33 = 56$) position. The third from the 89th ($56 + 33$) and so on until our sample is complete.

The systematic technique is easy to describe and execute; however, it has some inherent dangers when the sampling frame is repetitive or cyclical in nature. In these situations the results may not approximate a simple random sample.

When sampling very large populations, sometimes it is possible (and helpful) to divide the population into subpopulations on the basis of some characteristic. These subpopulations are called **strata.** These smaller, easier-to-work-with strata are sampled separately. One of the sample designs that starts by stratifying the sampling frame is the *stratified random sampling method.*

STRATIFIED RANDOM SAMPLE

A sample obtained by stratifying the sampling frame and then selecting a fixed number of items from each of the strata by means of a simple random sampling technique.

> **Exercise 1.40**
> What body of the federal government illustrates a stratified sampling of the people? (A random selection process is not used.)

When a stratified random sample is drawn, the sampling frame is subdivided into various strata, usually some naturally already occurring subdivision, and then a subsample is drawn from each of these strata. These subsamples may be drawn from the various strata by using random or systematic methods. The subsamples are summarized separately first and then combined to draw conclusions about the whole population.

An alternative to selecting the same number of items from each strata is to select from each strata proportionally to the size of the strata; this method is called *proportional sampling.*

PROPORTIONAL SAMPLE

A sample obtained by stratifying the sampling frame and then selecting a number of items in proportion to the size of the strata from each strata by means of a simple random sampling technique.

> **Exercise 1.41**
> What body of the federal government illustrates a proportional sampling of the people? (A random selection process is not used.)

When a proportional random sample is drawn, the sampling frame is subdivided into various strata, then a subsample is drawn from each strata. A convenient way to express the idea of proportional sampling is to establish a proportion. For example, the proportion "one for every 150" directs you to select one (1) element for each 150 elements in the strata. That way, the size of the strata determines the size of the subsample. The subsamples are summarized separately and then combined to draw conclusions about the whole population.

Another sampling method that starts by stratifying the sampling frame is a *cluster sample.*

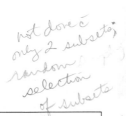
not done c̄ only 2 subsets; random selection of subsets

CLUSTER SAMPLE

A sample obtained by sampling some of, but not all of, the possible subdivisions within a population. These subdivisions, called clusters, often occur naturally within the population.

> **Exercise 1.42**
> Explain why the polls that are so frequently quoted during early returns on election day TV coverage are an example of cluster sampling.

A cluster sample can be obtained by using either random or systematic methods to identify the clusters to be sampled. Either all elements of each cluster, or a simple random sample of some of the elements, can then be selected. The subsamples are summarized separately and the information then combined. A cluster sample becomes a judgment sample when the clusters are selected on the basis of being typical or reliable.

Sample design is not a simple matter. Many colleges and universities offer separate courses in sample surveying and experimental design. It is intended that the above information will provide you with an overview of sampling and put its role in perspective.

EXERCISES

1.43 Consider this question taken from CNN Quick Vote on the Internet on 7/24/98: In this last season of *Seinfeld,* should Jerry Seinfeld have been nominated for an Emmy? The response was 34% yes, 66% no.
   a. What kind of survey was used?
   b. Do you think these results could be biased? Why?

1.44 *USA Today* conducted a survey asking readers, "What is the most hilarious thing that has ever happened to you en route or during a business trip?"
   a. What kind of sampling method is this? *not ..*
   b. Are the results likely to be biased? Explain.

1.45 a. What is a sampling frame?
   b. What did Mr. Clar use for a sampling frame in Illustration 1.3, page 20?
   c. Where did the number 1288 come from, and how was it used?

1.46 Consider a simple population consisting of only the numbers 1, 2, and 3 (an unlimited number of each). There are nine different samples of size two that could be drawn from this population: (1, 1), (1, 2), (1, 3), (2, 1), (2, 2,), (2, 3), (3, 1), (3, 2), (3, 3).
   a. If the population consists of the numbers 1, 2, 3, and 4, list all the samples of size two that could possibly be selected.
   b. If the population consists of the numbers 1, 2, and 3, list all the samples of size three that could possibly be selected.

1.47 A wholesale food distributor in a large metropolitan area would like to test the demand for a new food product. He distributes food through five large supermarket chains. The food distributor selects a sample of stores located in areas where he believes the shoppers are receptive to trying new products. What type of sampling does this represent?

1.48 A random sample could be very difficult to obtain. Why?

1.49 Why is the random sample so important in statistics?

1.50 The Design, Sample, and Method section of an article titled "Making Behavior Changes After a Myocardial Infarction" (*Western Journal of Nursing Research,* Aug. 1993) discusses the selection of 16 informants who constituted 8 family dyads. The article states that "to initiate contact with informants, names of persons who met the criteria were obtained from the medical records of a cardiac rehabilitation center in central Texas. Potential informants were then contacted by telephone to obtain preliminary consent. Confidentiality and anonymity of informants were ensured by coding the data to identify informants and link dyads." *twosome=dyad*
   a. Is this a judgment sample or a probability sample?
   b. Is it appropriate to perform statistical inference using this sample? Justify your answer. *No*

1.51 An article titled "Surface Sampling in Gravel Streams" (*Journal of Hydraulic Engineering,* April 1993) discusses grid sampling and areal sampling. Grid sampling involves the removal by hand of stones found at specific points. These points are established on the gravel surface through the use of a wire mesh or by using predetermined distances on a survey tape. The material collected by grid sampling is usually analyzed as a frequency distribution. An areal sample is collected by removing all the particles found in a predetermined area of a channel bed. The material recovered is most often analyzed as a frequency distribution by weight. Would you categorize these sample designs as judgment samples or probability samples?

1.52 Sheila Jones works for an established marketing research company in Cincinnati, Ohio. Her supervisor just handed her a list of 500 4-digit random numbers extracted from a statistical table of random digits. He told Sheila to conduct a survey by telephoning 500 Cincinnati residents, provided the last 4 digits of their phone number matched one of the numbers on the list. If Sheila follows her supervisor's instructions, is he assured of obtaining a random sample of respondents? Explain.

1.53 One question that people often ask is whether or not the risk of heart disease and other cardiovascular disorders is increased by LDL ("bad") cholesterol levels. LDL is far more prominent in some foods than others. A recent study by the University of Texas Health Sciences Center in San Antonio examined the arteries of 1400 males and females throughout the United States between the ages of 15 and 34 who had died of accidents, homicides, or suicides. The more LDL cholesterol and the less HDL ("good") cholesterol that was found in their blood, the more lesions that were found in their arteries, irrespective of race, sex, or age. What type of sampling does this represent? (*Nutrition Action HealthLetter,* "Kids Can't Wait," April 1997) *Convenience—Why dead people*

1.54 Describe in detail how you would select a 4% systematic sample of the adults in a nearby large city in order to complete a survey about a political issue.

1.55 Suppose that you have been hired by a group of all-sports radio stations to determine the age distribution of their listeners. Describe in detail how you would select a random sample of 2500 from the 35 listening areas involved.

1.56 The telephone book might not be a representative sampling frame. Explain why.

1.57 The election board's voter registration list is not a census of the adult population. Explain why.

1.5 Comparison of Probability and Statistics

Probability and **statistics** are two separate but related fields of mathematics. It has been said that "probability is the vehicle of statistics." That is, if it were not for the laws of probability, the theory of statistics would not be possible.

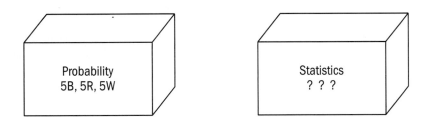

Exercise 1.58

Which of the following illustrates probability? statistics?
A—How likely is heads to occur when a coin is tossed?
B—The weights of 35 babies are studied to estimate weight gain in the first month after birth.

Let's illustrate the relationship and the difference between these two branches of mathematics by looking at two boxes. We know the probability box contains five blue, five red, and five white poker chips. Probability tries to answer questions such as, "If one chip is randomly drawn from this box, what is the chance that it will be blue?" On the other hand, in the statistics box we don't know what the combination of chips is. We draw a sample and, based on the findings in the sample, make conjectures about what we believe to be in the box. Note the difference: Probability asks you about the chance that something specific, like drawing a blue chip, will happen when you know the possibilities (that is, you know the population). Statistics, on the other hand, asks you to draw a sample, describe the sample (descriptive statistics), and then make inferences about the population based on the information found in the sample (inferential statistics).

1.59 Classify each of the following as a probability or a statistics problem.
   a. determining whether a new drug shortens the recovery time from a certain illness
   b. determining the chance that heads will result when a coin is flipped
   c. determining the amount of waiting time required to check out at a certain grocery store
   d. determining the chance that you will be dealt a "blackjack"

1.60 Classify each of the following as a probability or a statistics problem.
   a. determining how long it takes to handle a typical telephone inquiry at a real estate office
   b. determining the length of life for the 100-watt light bulbs a company produces
   c. determining the chance that a blue ball will be drawn from a bowl that contains 15 balls, of which 5 are blue
   d. determining the shearing strength of the rivets that your company just purchased for building airplanes
   e. chance of getting "doubles" when you roll a pair of dice

1.6   Statistics and Technology

In recent years, electronic technology has had a tremendous effect on almost every aspect of life. The field of statistics is no exception. As you will see, the field of statistics uses many techniques that are repetitive in nature: calculations of numerical statistics, procedures for constructing graphic displays of data, and procedures that are followed to formulate statistical inferences. Computers and calculators are very good at performing these sometimes long and tedious operations. If your computer has one of the standard statistical packages on line or you have a statistical calculator, it will make the analysis easy to perform.

Throughout this textbook, as statistical procedures are studied, you will find the information needed to have a computer complete the same procedures using the MINITAB (Release 12) and EXCEL 97 software. Calculator procedures will also be demonstrated using the TI-83 calculator.

An explanation of the most common typographical conventions that will be used in this textbook is given below. As additional explanations or selections are needed, they will be given.

Additional details about the use of MINITAB and EXCEL are available by using the Help system in the MINITAB and EXCEL software. Additional details for the TI-83 are contained in its corresponding *TI-83 Graphing Calculator Guidebook*. Specific details on the use of computers and calculators available to you needs to be obtained from your instructor or from your local computer lab person.

Your local computer center can provide a list of what is available to you. Some of the more readily available packaged programs are: MINITAB, EXCEL, SAS (Statistical Analysis System), JMP, Statgraphics, and SPSS (Statistical Package for the Social Sciences).

There is a great temptation to use the computer or calculator to analyze any and all sets of data and then treat the results as though the statistics are correct. Remember the old adage "Garbage-in, garbage-out!" Responsible use of statistical methodology is very important. The burden is on the user to ensure that the appropriate methods are correctly applied and that accurate conclusions are drawn and communicated to others.

For information about obtaining MINITAB, check the Internet at www.minitab.com

EXCEL is part of Microsoft Office and can be found on many personal computers.

EXCEL 97

Choose: tells you to make a menu or tab selection by a mouse "point and click" entry.

For example: **Choose: Chart Wizard > XY(Scatter) > 1st graph picture > Next** instructs you to, in sequence, "point and click on" the **Chart Wizard** icon, followed by **XY(Scatter)** under Chart type, followed by **1st graph picture** on the Chart subtype, and then followed by **Next** on the dialog window.

Select: indicates that you should click on the small box or circle to the left of a specified item. Often followed by a "point and click on" **Next** or **Finish** on the dialog window

Enter: instructs you to type or select information needed for a specific item.

MINITAB (Release 12)

Choose: tells you to make a menu selection by a mouse "point and click" entry.

For example: **Choose: Stat > SPC > Pareto Chart** instructs you to, in sequence, "point and click on" **Stat** on the menu bar, "followed by" **SPC** on the pulldown, and then "followed by" **Pareto Chart** on the pulldown.

Select: indicates that you should click on the small box or circle to the left of a specified item.

Enter: instructs you to type or select information needed for a specific item.

For information about TI-83, check the Internet at www.ti.com/calc.

TI-83

Choose: tells you which keys to press or menu selections to make.

For example: **Choose: Zoom > 9: Zoom-Stat > Trace > > >** instructs you to press the **Zoom** key, followed by selecting **9:ZoomStat** from the menu, followed by pressing the **Trace** key; **> > >** indicates to press arrow keys repeatedly to move along a graph to obtain important points.

Enter: instructs you to type or select information needed for a specific item.

Screen Capture: pictures of what your calculator screen should look like with chosen specifications highlighted.

EXERCISES • • • • • • • • • • • • • •

1.61 How have computers increased the usefulness of statistics to professionals such as researchers, government workers who analyze data, statistical consultants, and so on?

1.62 How might computers help you in statistics?

1.63 What is meant by the term "Garbage-in, garbage-out!" and how have computers increased the probability that studies may be victimized by the adage?

RETURN TO CHAPTER CASE STUDY

Let's now return to the Chapter Case Study, page 5, as a way to assess what we have learned in this chapter. Study the statistical information presented by the graphs and charts, and ask yourself how the terms (population, sample, variable, statistic, the type of variables) studied in this chapter apply to each and how you compare to the statistical story being told.

EXERCISE 1.64

a. What is the statistical population of interest?

b. Is the statistical information presented data, statistics, or parameters? What is the difference?

c. What variable is related to median age, 38.1 years, of West Virginians?

d. What kind of variable was used to collect information that led to 76% saying they were overweight in 1998?

e. Was an observational study, a survey, or an experiment used to collect the data for "Shower standing"?

f. Explain the meaning of 81 for men in "Behind the wheel." How do you think the values 58, 81, . . . , 55 were obtained? How is it possible for these values to be so different?

g. Do you have a "sports fantasy?" Does your fantasy match one of the categories shown on the pie chart? What population does it appear was surveyed for this information?

h. Were you ever advised to "Read the fine print?" This graph summarizes information gathered using what variable? What type of variable is it?

IN RETROSPECT

You should now have a general feeling of what statistics is about, an image that will grow and change as you work your way through this book. You know what a sample and a population are, the distinction between qualitative (attribute) and quantitative (numerical) variables. You even know the difference between statistics and probability (although we will not study probability in detail until Chapter 4). You should also have an appreciation for and a partial understanding of how important random samples are in statistics.

Throughout the chapter you have seen numerous articles that represent various aspects of statistics. The USA Snapshots® picture a variety of information about ourselves as we describe ourselves and other aspects of the world around us. Statistics can be entertaining—for example, the statistical snapshot "And Then There's 'Entertainment Statistics.'" The examples are endless. Look around and find some examples of statistics in your daily life (see Exercises 1.77 and 1.78, p. 31).

CHAPTER EXERCISES

1.65 We want to describe the so-called typical student at your college. Describe a variable that measures some characteristic of a student and results in
   a. attribute data
   b. numerical data

1.66 A candidate for a political office claims that he will win the election. A poll is conducted and 35 of 150 voters indicate that they will vote for the candidate, 100 voters indicate that they will vote for his opponent, and 15 voters are undecided.
   a. What is the population parameter of interest?
   b. What is the value of the sample statistic that might be used to estimate the population parameter?
   c. What inference can you make about the candidate's chance to win the election based on the results of the poll?

1.67 A researcher studying consumer buying habits asks every 20th person entering Publix Supermarket how many times per week he or she goes grocery shopping. She then records the answer as T.
   a. Is $T = 3$ an example of (1) a sample, (2) a variable, (3) a statistic, (4) a parameter, or (5) a piece of data?

   Suppose the researcher questions 427 shoppers during the survey.
   b. Give an example of a question that can be answered using the tools of descriptive statistics.
   c. Give an example of a question that can be answered using the tools of inferential statistics.

1.68 A researcher studying the attitudes of parents of preschool children interviews a random sample of 50 mothers, each having one preschool child. He asks each mother, "How many times did you compliment your child yesterday?" He records the answer as C.
   a. Is $C = 4$ an example of (1) a piece of data, (2) a statistic, (3) a parameter, (4) a variable, or (5) a sample?
   b. Give an example of a question that can be answered using the tools of descriptive statistics.
   c. Give an example of a question that can be answered using the tools of inferential statistics.

1.69 The August 29/September 5, 1994 issue of *U.S. News & World Report* references a study by health economists at the University of Southern California that indicated that Alzheimer's disease cost the nation $82.7 billion a year in medical expenses and lost productivity. Patients' earning loss was $22 billion, the value of time of unpaid caregivers was $35 billion, and the cost of paid care was $24 billion.
   a. What is the population?
   b. What is the response variable?
   c. What is the parameter?
   d. What is the statistic?

1.70 A USA Snapshot® from *USA Today* (Nov. 1, 1994) described the greatest sources of stress in starting a company. According to the snapshot, the CEOs of *Inc.* magazine's 500 fastest-growing private companies gave the following responses: 50% said company finances, 23% said the need to succeed, 10% said time commitments, 9% said personal relationships, and 8% were classified as "other." Would the data collected and used to determine these percentages be classified as qualitative or quantitative?

1.71 *Ladies' Home Journal,* in a June 1998 article titled "States of Health," presented results of a study that analyzed data collected by the U.S. Census Bureau in 1997. Results reveal that for both men and women in the United States, heart disease remains the number one killer, victimizing 500,000 people annually. Age, obesity, and inactivity all contribute to heart disease, and all three of these factors vary considerably from one location to the next. The highest mortality rates (deaths per 100,000 people) were in New York, Florida, Oklahoma, and Arkansas, whereas the lowest were reported in Alaska, Utah, Colorado, and New Mexico. (*Ladies Home Journal,* "States of Health," June 1998)

a. What is the population?
b. What are the characteristics of interest?
c. What is the parameter?
d. Classify all the variables of the study as either attribute or numerical.

1.72 An article titled "Want a Job in Food?" found in *Parade* magazine (Nov. 13, 1994) references a study at the University of California involving 2000 young men. The study found that in 2000 young men who did not go to college, of those who took restaurant jobs (typically as fast-food counter workers), one in two reached a higher-level blue-collar job and one in four reached a managerial position within four years.

a. What is the population?
b. What is the sample?
c. Is this a judgment sample or a probability sample?

1.73 The June 1994 issue of *Good Housekeeping* reported on a study on rape. The study found that women who screamed, bit, kicked, or ran were more likely to avoid rape than women who tried pleading, crying, or offered no resistance, and they were no more apt to be injured. The authors, however, cautioned that the study could not be interpreted as proof that all women should forcefully resist. The study involved 150 Omaha, Nebraska, police reports of rape or attempted rape.

a. Are the data in this study attribute or numerical?
b. Is this a judgment or a probability sample?

1.74 Teachers use examinations to measure a student's knowledge about their subject. Explain how "a lack of variability in the students' scores might indicate that the exam was not a very effective measuring device."

1.75 Describe, in your own words, and give an example of each of the following terms. Your examples should not be ones given in class or in the textbook.

a. variable   **b.** data   **c.** sample
d. population   **e.** statistic   **f.** parameter

1.76 Describe, in your own words, and give an example of each of the following terms. Your examples should not be ones given in class or in the textbook.

a. random sample   **b.** probability sample   **c.** judgment sample

1.77 Find an article or an advertisement in a newspaper or a magazine that exemplifies the use of statistics. Identify and describe

a. one statistic reported in the article.
b. the variable related to the statistic in (a).
c. the sample related to the statistic in (a).
d. the population from which the sample in (c) was taken.

1.78 a. Find an article in a newspaper or a magazine that exemplifies the use of statistics in a way that might be considered "entertainment" or "recreational." Describe why you think this article fits one of these categories.

b. Find an article in a newspaper or a magazine that exemplifies the use of statistics and is presenting an unusual finding as the result of a study. Describe why these results are (or are not) "newsworthy."

VOCABULARY AND KEY CONCEPTS

Be able to define each term. Pay special attention to the key terms, which are printed in **red.** In addition, describe in your own words, and give an example of, each term. Your examples should not be ones given in class or in the textbook. Page numbers indicate the first appearance of a term.

attribute variable (p. 14)
biased sample (p. 20)
categorical variable (p. 14)
census (p. 21)
cluster sample (p. 24)
continuous variable (p. 15)
convenience sample (p. 20)
data (p. 12)
descriptive statistics (p. 3, 7)
discrete variable (p. 15)
experiment (p. 12, 21)
finite population (p. 12)
haphazard (p. 23)
inferential statistics (p. 7)
infinite population (p. 12)
judgment sample (p. 22)
nominal variable (p. 14)
numerical variable (p. 14)
observational study (p. 21)
ordinal variable (p. 15)
parameter (p. 13)

population (p. 11)
probability (p. 26)
probability sample (p. 22)
proportional sample (p. 24)
qualitative variable (p. 14)
quantitative variable (p. 14)
random sample (p. 22)
representative sampling frame (p. 21)
sample (p. 12)
sample design (p. 22)
sampling frame (p. 21)
statistic (p. 13)
statistics (p. 7, 26)
strata (p. 24)
stratified sample (p. 24)
survey (p. 21)
systematic sample (p. 23)
unbiased (p. 20)
variability (p. 19)
variable (p. 12)
volunteer sample (p. 20)

CHAPTER PRACTICE TEST

Part 1: Knowing the Definitions

Answer "True" if the statement is always true. If the statement is not always true, replace the words shown in bold with words that make the statement always true.

1.1 **Inferential** statistics is the study and description of data that result from an experiment.

1.2 **Descriptive statistics** is the study of a sample that enables us to make projections or estimates about the population from which the sample is drawn.

1.3 A **population** is typically a very large collection of individuals or objects about which we desire information.

1.4 A statistic is the calculated measure of some characteristic of a **population.**

1.5 A parameter is the measure of some characteristic of a **sample.**

1.6 As a result of surveying 50 freshmen, it was found that 16 had participated in interscholastic sports, 23 had served as officers of classes and clubs, and 18 had been in school plays during their high school years. This is an example of **numerical data.**

1.7 The "number of rotten apples per shipping crate" is an example of a **qualitative** variable.

1.8 The "thickness of a sheet of sheet metal" used in a manufacturing process is an example of a **quantitative** variable.

1.9 A **representative** sample is a sample obtained in such a way that all individuals had an equal chance to be selected.

1.10 The basic objective of **statistics** is that of obtaining a sample, inspecting this sample, and then making inferences about the unknown characteristics of the population from which the sample was drawn.

Part II: Applying the Concepts

The owners of Corner Convenience Stores are concerned about the quality of service their customers receive. In order to study the service received, they collected samples for each of several variables.

1.11 Classify each of the following variables as being (A) nominal, (B) ordinal, (C) discrete, or (D) continuous:
   a. method of payment for purchases (cash, credit card, check)
   b. customer satisfaction (very satisfied, satisfied, not satisfied)
   c. amount of sales tax on purchase
   d. number of items purchased
   e. customer's driver's license number

1.12 The mean checkout time for all customers at Corner Convenience Stores is to be estimated by using the mean checkout time of 75 randomly selected customers. Match the items in column 2 with the statistical terms in column 1.

1	2
____ data (one)	(a) the 75 customers
____ data (set)	(b) the mean time for all customers
____ experiment	(c) two minutes, one customer's checkout time
____ parameter	(d) the mean time for the 75 customers
____ population	(e) all customers at Corner Convenience
____ sample	(f) the checkout time for one customer
____ statistic	(g) the 75 times
____ variable	(h) the process used to select 75 customers and measure their times

Part III: Understanding the Concepts

Write a brief paragraph in response to each question.

1.13 The population and the sample are both sets of objects. Describe the relationship between them and give an example.

1.14 The variable and the data for a specific situation are closely related. Explain this relationship and give an example.

1.15 The data, the statistic, and the parameter are all values used to describe a statistical situation. How does one distinguish among these three terms? Give an example.

1.16 What conditions are required in order for a sample to be a random sample? Explain and include an example of a sample that is random and one that is not random.

Chapter

2

Descriptive Analysis and Presentation of Single-Variable Data

CHAPTER OUTLINE

GRAPHIC PRESENTATION OF DATA

2.1 Graphs, Pareto Diagrams, and Stem-and-Leaf Displays
A **picture** is often worth a thousand words.

2.2 Frequency Distributions and Histograms
An **increase** in the amount of data requires us to modify our techniques.

NUMERICAL DESCRIPTIVE STATISTICS

2.3 Measures of Central Tendency
The four measures of central tendency—mean, median, mode, and midrange—are **average** values.

2.4 Measures of Dispersion
Measures of dispersion—range, variance, and standard deviation—assign numerical values to the **amount of spread** in a set of data.

2.5 Mean and Standard Deviation of Frequency Distribution
The **frequency distribution** is an aid in calculating mean and standard deviation.

2.6 Measures of Position
Measures of position allow us to **compare** one piece of data to the set of data.

2.7 Interpreting and Understanding Standard Deviation
A standard deviation is the length of a **standardized yardstick.**

2.8 The Art of Statistical Deception
How the unwitting or the unscrupulous can use **"tricky" graphs** and **insufficient information** to mislead the unwary.

CHAPTER CASE STUDY

Describing the Commute to Work Is Not Simple

Have you ever heard adults describe their commute to work? How much time does it take to get there or back home? How far is it? What means of transportation is used? What conditions exist along the route(s) traveled? Are these conditions stable or do they vary according to day of week? time of day? route traveled? method of travel? The *USA Snapshot*® shown below appeared in *USA Today*, July 13, 1998, and describes the average one-way commuting time for adults who either drive or carpool to work.

USA SNAPSHOTS®

A look at statistics that shape the nation

Commuting time

Average commuting time each way for adults who drive or car pool to work:

Under 10 minutes	27%
10-19	23%
20-29	19%
30-44	13%
45-up	11%
Don't know	6%
No answer	2%

Note: Exceeds 100% due to rounding

Source: Maritz AmeriPoll

By Cindy Hall and Grant Jerding, USA TODAY

Below is a sample of one-way commute times, to the nearest minute, for 50 college students who are gainfully employed.

2	10	3	5	1	19	14	3	15	12
30	10	2	8	19	17	12	29	21	29
15	23	3	24	4	16	49	15	64	5
3	22	19	1	48	4	25	7	22	4
3	10	33	12	14	30	10	13	5	23

How would you summarize this data so that the information contained within the data can best be understood? We will learn various methods of describing data in this chapter and will return to this case study at the end of the chapter. (See Exercise 2.146, p. 112.)

CHAPTER OBJECTIVES

Imagine that you took an exam at the last meeting of your favorite class. Today your instructor returns your exam paper and it has a grade of 78. If you are like most students in this situation, as soon as you see your grade you want to know how your grade compares to those of the rest of the class and you immediately ask, "What was the average exam grade?" Your instructor replies, "The class average was 68." Since 78 is 10 points above the average, you ask, "How close to the top is my grade?" Your instructor replies that the grades ranged from 42 to 87 points. The accompanying figure summarizes the information we have so far.

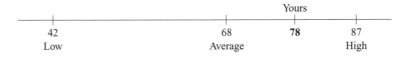

A third question that is sometimes asked is "How are the grades distributed?" Your instructor replies that half the class had grades between 65 and 75. With this information you conclude that your grade is fairly good.

The preceding illustration demonstrates the basic process of statistics and how students make regular use of statistics. The three main components are: (1) a set of data (set of exam scores); (2) the description of the data (a diagram and a few descriptive statistics: average, high, and low grades); and (3) the conclusions drawn by the students about their relative success based on this information.

A large part of this chapter is devoted to learning how to present and describe sets of data. We will learn: (1) several graphical methods for displaying the data and about the types of distribution , and (2) about three types of numerical statistics— measures of central tendency , measures of dispersion (spread), and measures of position .

In this chapter, we deal with **single-variable data,** which is data collected for one numerical variable at a time. In Chapter 3 we will learn to work with two variables at a time and learn about their relationship.

GRAPHIC PRESENTATION OF DATA

2.1   Graphs, Pareto Diagrams, and Stem-and-Leaf Displays

Once the sample data has been collected, we must "get acquainted" with it. One of the most helpful ways to become acquainted with the data is to use an initial exploratory data-analysis technique that will result in a pictorial representation of the data. The resulting display will visually reveal patterns of behavior of the variable being studied. There are several graphic (pictorial) ways to describe data. The type of data and the idea to be presented determines which method is used.

NOTE There is no single correct answer when constructing a graphic display. The analyst's judgment and the circumstances surrounding the problem play a major role in the development of the graphic.

Qualitative Data

CIRCLE GRAPHS AND BAR GRAPHS

Graphs that are used to summarize qualitative, or attribute, or categorical, data. Circle graphs (pie diagrams) show the amount of data that belongs to each category as a proportional part of a circle. Bar graphs show the amount of data that belongs to each category as proportionally sized rectangular areas.

ILLUSTRATION 2.1 ▼

Table 2.1 lists the number of cases of each type of operation performed at General Hospital last year.

Table 2.1   Operations Performed at General Hospital Last Year

Type of Operation	Number of Cases
Thoracic	20
Bones and Joints	45
Eye, ear, nose, and throat	58
General	98
Abdominal	115
Urologic	74
Proctologic	65
Neurosurgey	23
Total	498

Figure 2.1
Circle Graph

Exercise 2.1

Construct a circle graph showing how U.S. consumers paid for goods and services in 1993, as reported in *USA Today,* 6-22-94: cash—70%, check—19%, credit and debit cards—9%, not known—2%.

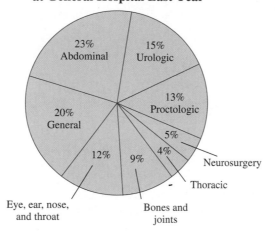

**Operations Performed
at General Hospital Last Year**

NOTE All graphic representations of sets of data need to be completely self-explanatory. That includes a descriptive title, identification of either the population or sample and variable, and scales used.

These data are displayed on a circle graph in Figure 2.1 with each type of operation represented by a relative proportion of the circle, found by dividing the number of cases by the total sample size, namely, 498. They are then reported as a percentage (for example, 25% is $^1/_4$ of the circle). Figure 2.2 displays the same "type of operation" data but in the form of a bar graph. Bar graphs of attribute data should be drawn with a space between bars of equal width.

Figure 2.2
Bar Graph

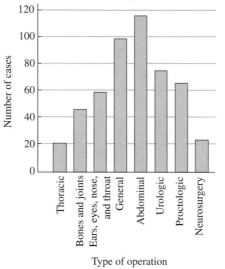

Operations Performed at General Hospital Last Year

Exercise 2.2

Construct a bar graph showing how U.S. consumers paid for goods and services in 1993, as reported in *USA Today,* 6-22-94: cash—70%, check—19%, credit and debit cards—9%, not known—2%.

The following commands will construct a pie chart.

MINITAB

Input the categories into C1 and their corresponding frequencies into C2; then continue with:

Choose: **Graph > Pie Chart . . .**
Select: **Chart table**
Enter: Categories in: **C1**
         Frequencies in: **C2**
         Your title

EXCEL

Input the frequencies for the various categories into column A; then continue with:

Choose: **Chart Wizard > Pie > 1st picture** (usually) **> Next**
Enter: Data range: **(A1:A5 or select cells)**
Check: Series in: **columns > Next**
Choose: **Titles**
Enter: Chart title: **Your title**
Choose: **Data Labels**
Select: **Show label and percent > Next > Finish**

To edit the pie chart:

Click On: Anywhere clear on the chart —use handles to size
          Any category number, click again, click to the right of the category number, backspace and type in category word

Exercise 2.3

In your opinion, does the circle graph (Exercise 2.1) or the bar graph (Exercise 2.2) result in a better representation of the information? Explain.

TI-83

```
Input the frequencies for the various
categories into L1; then continue
with:

    Choose:  PRGM > EXEC > CIRCLE*
    Enter:   LIST: L1 > ENTER
             DATA DISPLAYED?:
             1:PERCENTAGES
                  OR
             2:DATA

*Program 'CIRCLE' is one of the many programs
that are available for downloading from the web-
site //barney.bloom.edu/~skokoska. Copy the file
rjprogs.zip from the site to your computer. Use
WINZIP or PKUNZIP (on your computer) to uncompress
the files. Download the programs to the TI-83 us-
ing the TI-Graph Link software. See the Statisti-
cal Tutor for more details.
```

When the bar graph is presented in the form of a *Pareto diagram,* it presents additional and very helpful information.

PARETO DIAGRAM

A bar graph with the bars arranged from the most numerous category to the least numerous category. It includes a line graph displaying the cumulative percentages and counts for the bars.

ILLUSTRATION 2.2 ▼

The FBI reported (*USA Today,* 6-29-94) the number of hate crimes by category for 1993. The Pareto diagram in Figure 2.3 shows the 6746 categorized hate crimes, their percentages, and cumulative percentages.

Figure 2.3

Pareto Diagram

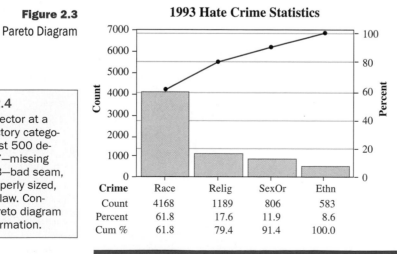

1993 Hate Crime Statistics

Crime	Race	Relig	SexOr	Ethn
Count	4168	1189	806	583
Percent	61.8	17.6	11.9	8.6
Cum %	61.8	79.4	91.4	100.0

Exercise 2.4

A shirt inspector at a clothing factory categorized the last 500 defects as: 67—missing button, 153—bad seam, 258—improperly sized, 22—fabric flaw. Construct a Pareto diagram for this information.

▲

The Pareto diagram is popular in quality-control applications. A Pareto diagram showing types of defects will show the ones that have the greatest effect on the defective rate in order of effect. It will be easy to see which defects should be targeted in order to most effectively lower the defective rate.

The following commands will construct a Pareto diagram.

MINITAB

Input the categories into C1 and the corresponding frequencies into C2; then continue with:

Choose:	**Stat > Quality Tools > Pareto Chart**
Select:	**Chart defects table**
Enter:	Labels in: **C1**
	Frequencies in: **C2**
	Your Title

TI-83

Input the numbered categories into L1 and the corresponding frequencies into L2; then continue with:

Choose:	**PRGM > EXEC > PARETO***
Enter:	LIST: **L2 > ENTER**
	Ymax: **at least the sum of the frequencies > ENTER**
	Yscl: **increment for y-axis > ENTER**

*Program 'PARETO' is one of many programs that are available for downloading from a website. See page 39 for specific instructions.

EXCEL

Input categories into column A and the corresponding frequencies into column B (column headings are optional); then continue with:
First, sorting the table:
  Activate both columns of the distribution

Choose:	**Data > Sort >** Sort by: **Column B** (freq. or rel. freq. col.)
Select:	**Descending >** My list has: **Header row** or **No Header row > OK**
Choose:	**Chart Wizard > Column > 1st picture** (usually) **> Next**
Choose:	**Data Range**
Enter:	Data Range: **(A1:B5 or select cells)**
Select	Series in: **Columns > Next**
Choose:	**Titles**
Enter:	Chart title: **your title**
	Category (x) axis: **title for x-axis**
	Value (y) axis: **title for y-axis > Next > Finish**

To edit the Pareto diagram:
  Click on: Anywhere clear on the Chart
          —use handles to size
          Any title name to change
          Any cell in the category column and type in a different name > Enter

Excel does not include the line graph.

Quantitative Data

One major reason for constructing a graph of quantitative data is to display its *distribution.*

DISTRIBUTION

The pattern of variability displayed by the data of a variable. The distribution displays the frequency of each value of the variable.

One of the simplest graphs used to display a distribution is the *dotplot.*

DOTPLOT DISPLAY

Displays the data of a sample by representing each piece of data with a dot positioned along a scale. This scale can be either horizontal or vertical. The frequency of the values is represented along the other scale.

ILLUSTRATION **2.3** ▼

A sample of 19 exam grades was randomly selected from a large class:

76	74	82	96	66	76	78	72	52	68
86	84	62	76	78	92	82	74	88	

Figure 2.4 is a dotplot of the 19 exam scores.

Figure 2.4
Dotplot

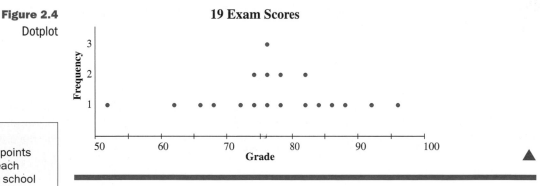

Exercise 2.5
The number of points scored during each game by a high school basketball team last season was:

56	54	61	71
46	61	55	68
60	66	54	61
52	36	64	51

Construct a dotplot of these data.

Notice how the data are "bunched" near the center and more "spread out" near the extremes.

The dotplot display is a convenient technique to use as you first begin to analyze the data. It results in a picture of the data as well as sorts the data into numerical order. (To sort data is to list the data in rank order according to numerical value.)

The following commands will construct a dotplot.

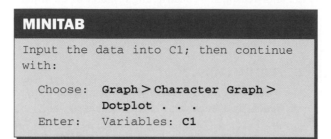

MINITAB

Input the data into C1; then continue with:

Choose: **Graph > Character Graph >**
          Dotplot . . .
Enter: Variables: **C1**

EXCEL

The dotplot display is not available, but the initial step of ranking the data can be done.
Input the data into column A and activate the column of data; then continue with:

Choose: **Data > Sort >** Sort by:
          Column A
Select: **Ascending >** My list has:
          Header row or **No Header row**

Use the sorted data to finish constructing the dotplot display.

TI-83

Input the data into L1; then continue with:

Choose: **PRGM > EXEC > DOTPLOT***
Enter: **LIST: L1 > ENTER**
          Xmin: **at most the lowest x**
          value
          Xmax: **at least the highest**
          x value
          Xscl: **0 or increment**
          Ymax: **at least the highest**
          frequency

*Program 'DOTPLOT' is one of many programs that are available for downloading from a website. See page 39 for specific instructions.

In recent years a technique known as the *stem-and-leaf display* has become very popular for summarizing numerical data. It is a combination of a graphic technique and a sorting technique . These displays are simple to create and use, and are well suited to computer applications.

STEM-AND-LEAF DISPLAY

Displays the data of a sample using the actual digits that make up the data values. Each numerical value is divided into two parts: The leading digit(s) becomes the stem , and the trailing digit(s) becomes the leaf . The stems are located along the main axis, and a leaf for each piece of data is located to display the distribution of the data.

ILLUSTRATION 2.4 ▼

Let's construct a stem-and-leaf display for the 19 exam scores given in Illustration 2.3.

76	74	82	96	66	76	78	72	52	68
86	84	62	76	78	92	82	74	88	

At a quick glance we see that there are scores in the 50s, 60s, 70s, 80s, and 90s. Let's use the first digit of each score as the stem and the second digit as the leaf. Typically, the display is constructed in a vertical position. Draw a vertical line and place the stems, in order, to the left of the line.

```
5 |
6 |
7 |
8 |
9 |
```

Next we place each leaf on its stem. This is done by placing the trailing digit on the right side of the vertical line opposite its corresponding leading digit. Our first data value is 76; 7 is the stem and 6 is the leaf. Thus, we place a 6 opposite the stem 7.

```
7 | 6
```

The next data value is 74, so a leaf of 4 is placed on the 7 stem next to the 6.

```
7 | 6   4
```

The next data value is 82, so a leaf of 2 is placed on the 8 stem.

```
7 | 6   4
8 | 2
```

We continue until each of the other 16 leaves is placed on the display. Figure 2.5 shows the resulting stem-and-leaf display.

Figure 2.5   **19 Exam Scores**
Stem-and-Leaf

```
5 | 2
6 | 6 8 2
7 | 6 4 6 8 2 6 8 4
8 | 2 6 4 2 8
9 | 6 2
```

From Figure 2.5, we see that the scores are centered around the 70s. In this case, all scores with the same tens digit were placed on the same branch, but this may not always be desired. Suppose we reconstruct the display; this time instead of grouping ten possible values on each stem, let's group the values so that only five possible values could fall on each stem. Do you notice a difference in the appearance of Figure 2.6?

Figure 2.6   **19 Exam Scores**
Stem-and-Leaf

```
(50–54)  5 | 2
(55–59)  5 |
(60–64)  6 | 2
(65–69)  6 | 6 8
(70–74)  7 | 4 2 4
(75–79)  7 | 6 6 8 6 8
(80–84)  8 | 2 4 2
(85–89)  8 | 6 8
(90–94)  9 | 2
(95–99)  9 | 6
```

Exercise 2.6
What stem values and leaf values will be used to represent the next pieces of data, 96 and 66, from Illustration 2.4?

Exercise 2.7
Construct a stem-and-leaf display of the number of points scored during each basketball game last season:

56	54	61	71
46	61	55	68
60	66	54	61
52	36	64	51

This data set is on the StatSource CD.

The general shape is approximately symmetrical about the high 70s. Our information is a little more refined, but basically we see the same distribution.   ▲

The following commands will construct a stem-and-leaf diagram.

MINITAB

Input the data into C1; then continue with:

   Choose: **Graph > Char.Graph > Stem-and-Leaf . . .**

   Enter: Variables: **C1**
              Increment value for stem (optional)

EXCEL

The stem-and-leaf diagram is not available, but the initial step of ranking the data can be done.

Use the commands as shown with the dotplot display on page 42, then finish constructing the stem-and-leaf diagram by hand.

TI-83

Input the data into L1; then continue with:

   Choose: **STAT > EDIT > 2:SortA(**
   Enter: **L1**

Use sorted data to finish constructing the stem-and-leaf diagram by hand.

It is fairly typical of many variables to display a distribution that is concentrated (mounded) about a central value and then in some manner dispersed in one or both directions. Often a graphic display reveals something that the analyst may or may not have anticipated. Illustration 2.5 demonstrates what generally occurs when two populations are sampled together.

ILLUSTRATION 2.5 ▼

A random sample of 50 college students was selected. Their weights were obtained from their medical records. The resulting data are listed in Table 2.2.

Notice that the weights range from 98 to 215 pounds. Let's group the weights on stems of ten units using the hundreds and the tens digits as stems and the units digit as the leaf. See Figure 2.7. The leaves have been arranged in numerical order.

Close inspection of Figure 2.7 suggests that two overlapping distributions may be involved. That is exactly what we have: a distribution of female weights and a distribution of male weights. Figure 2.8 shows a "back-to-back" stem-and-leaf display of this set of data and makes it obvious that two distinct distributions are involved.

Figure 2.9, a "side-by-side" dotplot (same scale) of the same 50 weight data, shows the same distinction between the two subsets.

TABLE 2.2   Weights of 50 College Students

Student	1	2	3	4	5	6	7	8	9	10
Male/Female	F	M	F	M	M	F	F	M	M	F
Weight	98	150	108	158	162	112	118	167	170	120
Student	11	12	13	14	15	16	17	18	19	20
Male/Female	M	M	M	F	F	M	F	M	M	F
Weight	177	186	191	128	135	195	137	205	190	120
Student	21	22	23	24	25	26	27	28	29	30
Male/Female	M	M	Γ	M	Γ	Γ	M	M	M	M
Weight	188	176	118	168	115	115	162	157	154	148
Student	31	32	33	34	35	36	37	38	39	40
Male/Female	F	M	M	F	M	F	M	F	M	M
Weight	101	143	145	108	155	110	154	116	161	165
Student	41	42	43	44	45	46	47	48	49	50
Male/Female	F	M	F	M	M	F	F	M	M	M
Weight	142	184	120	170	195	132	129	215	176	183

Exercise 2.8

What do you think "leaf unit = 1.0" means in Figure 2.7?

Figure 2.7
Stem-and-Leaf

**Weights of
50 College Students (lb)
Stem-and-Leaf of WEIGHT
N = 50   Leaf Unit = 1.0**

9	8
10	1 8 8
11	0 2 5 5 6 8 8
12	0 0 0 8 9
13	2 5 7
14	2 3 5 8
15	0 4 4 5 7 8
16	1 2 2 5 7 8
17	0 0 6 6 7
18	3 4 6 8
19	0 1 5 5
20	5
21	5

Figure 2.8
"Back-to-Back" Stem-and-Leaf

Weights of 50 College Students (lb)

Female		Male
8	09	
1 8 8	10	
0 2 5 5 6 8 8	11	
0 0 0 8 9	12	
2 5 7	13	
2	14	3 5 8
	15	0 4 4 5 7 8
	16	1 2 2 5 7 8
	17	0 0 6 6 7
	18	3 4 6 8
	19	0 1 5 5
	20	5
	21	5

Figure 2.9
Dotplots with Common Scale

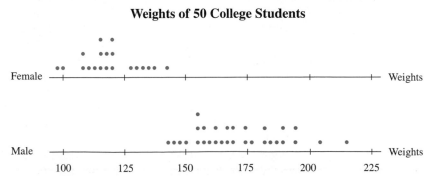

Weights of 50 College Students

Based on the information shown in Figures 2.8 and 2.9, and on what we know about people's weight, it seems reasonable to conclude that female college students weigh less than male college students. Situations involving more than one set of data are discussed further in Chapter 3. ▲

The following commands will construct a *multiple dotplot*, one above the other.

MINITAB

Input the data into C1 and their corresponding numerical categories into C2; then continue with:

Choose: **Graph > Character Graph > Dotplot . . .**
Enter: Variables: **C1**
Select: **By variable:**
Enter: **C2**
Select: **Same scale for all variables**

EXCEL

Multiple dotplots are not available, but the initial step of ranking the data can be done. Use the commands as shown with the dotplot display on page 42, then finish constructing the dotplots by hand.

TI-83

Input the data for the first dotplot into L1 and the data for the second dotplot into L3; then continue with:

Choose: **STAT > EDIT > 2:SortA(**
Enter: **L1 > ENTER**
In L2, enter counting numbers for each category.
Ex.   L1   L2
          15   1
          16   1
          16   2
          17   1
Choose: **STAT > EDIT > 2:SortA(**
Enter: **L3 > ENTER**
In L4, enter counting numbers (a higher set*) for each category; *for example: use 10,10,11,10,10,11,12, . . . (offsets the two dotplots).
Choose: **2nd > FORMAT > AxesOff** (Optional - must return to AxesOn)

Choose: **2nd > STAT PLOT > 1:PLOT1**

Choose: **2nd > STAT PLOT > 2:PLOT2**

Choose: **Window**
Enter: **at most lowest value for both, at least highest value for both, 0 or increment, −2, at least highest counting number,1,1**
Choose: **Graph > Trace > > >** (gives data values)

EXERCISES • • • • • • • • • • • • • •

2.9 An article in *Fortune* magazine entitled "What Really Goes on in Your Doctor's Office?" showed a breakdown of how patients' fees are used to support the operations of various clinics and health care facilities. For each $100 in fees collected by the doctors in the study, the following eight categories of expenses were isolated to show how the $100 was deployed:

Expense Category	Amount
1. Doctor's personal income	$55.60
2. Nonphysician personnel	15.70
3. Office expenses	10.90
4. Medical supplies	4.00
5. Malpractice insurance premiums	3.50
6. Employee physicians	2.30
7. Medical equipment	1.50
8. All other	6.50

Source: Fortune, August 17, 1998, p. 168.

a. Construct a circle graph of this breakdown.
b. Construct a bar graph of this breakdown.
c. Compare the two graphs you constructed in (a) and (b); which one seems to be the most informative? Explain why.

2.10 A sample of student-owned General Motors automobiles was identified and the make of each noted. The resulting sample follows (Ch = Chevrolet, P = Pontiac, O = Oldsmobile, B = Buick, Ca = Cadillac):

Ch	B	Ch	P	Ch	O	B	Ch	Ca	Ch
B	Ca	P	O	P	P	Ch	P	O	O
Ch	B	Ch	B	Ch	P	O	Ca	P	Ch
O	Ch	Ch	B	P	Ch	Ca	O	Ch	B
B	O	Ch	Ch	O	Ch	Ch	B	Ch	B

a. Find the number of cars of each make in the sample.
b. What percentage of these cars were Chevrolets? Pontiacs? Oldsmobiles? Buicks? Cadillacs?
c. Draw a bar graph showing the percentages found in (b).

2.11 The USA Snapshot® "Monster cookies" shows adults' choice of favorite cookie.

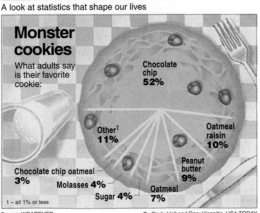

USA SNAPSHOTS®

A look at statistics that shape our lives

Monster cookies

What adults say is their favorite cookie:

Chocolate chip **52%**

Oatmeal raisin **10%**

Peanut butter **9%**

Oatmeal **7%**

Sugar **4%**

Molasses **4%**

Chocolate chip oatmeal **3%**

Other[1] **11%**

1 – all 1% or less

Source: WEAREVER   By Cindy Hall and Gary Visgaitis, USA TODAY

Try the PARETO commands on your computer or calculator.

a. Draw a bar graph picturing the percentages of adults for each kind of cookie.

b. Draw a Pareto diagram picturing adults' favorite cookie. How can the "Other" category be handled so that the resulting graph does not misrepresent the situation?

c. If a store is to stock only four varieties of cookies, which varieties should they carry if they wish to please the greatest number of customers? How does the Pareto diagram show this?

d. If 300 adults are to be surveyed, what frequencies would you expect to occur for each variety based on the "Monster cookies" graph?

2.12 The USA Snapshot® "How to say I love you" reports the results of a David Michaelson & Associates survey for Ethel M Chocolates, on the best way to show affection.

Best way to show affection	Give gift	Hold hands	Hugging/ kissing	Smiling	Other
Percent who said	10%	10%	51%	20%	9%

Draw a Pareto diagram picturing this information.

2.13 The January 10, 1991, USA Snapshot® "What's in U.S. landfills" reports the percentages for each type of waste in our landfills: food—4%, glass—2%, metal—14%, paper—38%, plastic—18%, yard waste—11%, other—13%.

a. Construct a Pareto diagram displaying this information.

b. Because of the size of the "other" category, the Pareto diagram may not be the best graph to use. Explain why, and describe what additional information is needed to make the Pareto diagram more appropriate.

2.14 The final-inspection defect report for assembly line A12 is reported on a Pareto diagram.

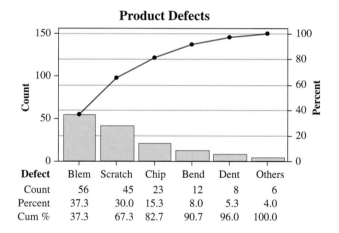

Product Defects

Defect	Blem	Scratch	Chip	Bend	Dent	Others
Count	56	45	23	12	8	6
Percent	37.3	30.0	15.3	8.0	5.3	4.0
Cum %	37.3	67.3	82.7	90.7	96.0	100.0

a. What is the total defect count in the report?

b. Verify the 30.0% listed for "scratch."

c. Explain how the "cum % for bend" value of 90.7% was obtained and what it means.

d. Management has given the production line the goal of reducing their defects by 50%. What two defects would you suggest they give special attention to in working toward this goal? Explain.

2.15 *Sports Illustrated* recently published a mock draft list of 29 future players for the NBA, one week before the 1998 draft actually took place. Shown below are the heights (in inches) of the basketball players who the editorial staff expected to be first round picks by the professional teams:

Many of these data sets are on the StatSource CD.

74	85	79	83	81	77	79	80	83	83
81	82	77	77	73	82	84	78	80	78
83	81	81	75	78	72	82	81	81	

a. Construct a dotplot of the heights of these players.
b. Use the dotplot to uncover the shortest and the tallest players.
c. What is the most common height, and how many players share that height?

2.16 As baseball players, Hank Aaron, Babe Ruth, and Roger Maris were well known for their ability to hit home runs. Mark McGwire, Sammy Sosa, and Ken Griffey, Jr., became well known for their ability to hit home runs during the "great home run chase" of 1998. Listed below is the number of home runs each player hit in each Major League season in which he played in 75 or more games.

If you use your computer or calculator, use the commands on page 46.

Aaron	13	27	26	44	30	39	40	34	45	44	24	32
	44	39	29	44	38	47	34	40	20	12	10	
Ruth	11	29	54	59	35	41	46	25	47	60		
	54	46	49	46	41	34	22					
Maris	14	28	16	39	61	33	23	26	13	9	5	
McGwire	49	32	33	39	22	42	39	52	58	70		
Sosa	15	10	33	25	36	40	36	66				
Griffey	16	22	22	27	45	40	49	56	56			

Data is on the StatSource CD.

a. Construct a dotplot of the data for Aaron, Ruth, Maris, and McGwire, using the same axis.
b. Make a case for each of the following statements with regard to past players: "Aaron is the home run king!" "Ruth is the home run king!" "McGwire is the home run king!" "Maris is not the home run king!"
c. Construct a dotplot of the data for McGwire, Sosa, and Griffey, using the same axis.
d. Make a case for the statement: "Sosa is not currently the home run king."

2.17 A computer was used to construct the following dotplot.

a. How many data are shown?
b. List the values of the five smallest data.
c. What is the value of the largest data?
d. What value occurred the greatest number of times? How many times did it occur?

2.18 Delco Products, a division of General Motors, produces commutators designed to be 18.810 mm in overall length. (A commutator is a device used in the electrical system of an automobile.) The following sample of 35 commutators was taken while monitoring the manufacturing process:

The Overall Length of Commutators

18.802	18.810	18.780	18.757	18.824	18.827	18.825
18.809	18.794	18.787	18.844	18.824	18.829	18.817
18.785	18.747	18.802	18.826	18.810	18.802	18.780
18.830	18.874	18.836	18.758	18.813	18.844	18.861
18.824	18.835	18.794	18.853	18.823	18.863	18.808

Source: With permission of Delco Products Division, GMC.

Use a computer to construct a dotplot of this data.

2.19 The closing prices (nearest dollar) of the first 50 common stocks listed in the New York Stock Exchange with a value between $10 and $99 on July 23, 1998, were as follows:

30	26	41	11	28	47	35	17	19	17
26	72	26	58	16	65	13	22	45	48
13	24	31	17	52	12	17	31	11	52
75	37	36	35	12	75	38	32	14	54
52	90	57	22	21	28	25	52	27	43

Source: USA Today, July 23, 1998. p. 5B.

a. Construct a stem-and-leaf plot of these data.
b. Use the plot to determine the lowest and the highest stock price from those selected.
c. What price interval defined by the plot contains the most values?

2.20 The following amounts are the fees charged by Quik Delivery for the 40 small packages it delivered last Thursday afternoon.

4.03	3.56	3.10	6.04	5.62	3.16	2.93	3.82	4.30	3.86
4.57	3.59	4.57	6.16	2.88	5.03	5.46	3.87	6.81	4.91
3.62	3.62	3.80	3.70	4.15	2.07	3.77	5.77	7.86	4.63
4.81	2.86	5.02	5.24	4.02	5.44	4.65	3.89	4.00	2.99

Construct a stem-and-leaf display.

2.21 Given the following stem-and-leaf display:

```
Stem-and-Leaf of C1 N = 16
Leaf Unit = 0.010
  1    59   7
  4    60   148
 (5)   61   02669
  7    62   0247
  3    63   58
  1    64   3
```

a. What is the meaning of Leaf Unit = 0.010?
b. How many data are shown on this stem-and-leaf?
c. List the first four data values.
d. What is the column of numbers down the left-hand side of the figure?

2.22 A term often used in solar energy research is *heating-degree-days*. This concept is related to the difference between an indoor temperature of 65°F and the average outside temperature for a given day. If the average outside temperature is 5°F, this would give 60 heating-degree-days. The annual heating-degree-day normals for several Nebraska locations are shown on the following stem-and-leaf display constructed by using MINITAB (data are in column C1).
a. What is the meaning of Leaf Unit = 10?
b. List the first four data values.
c. List all the data values that occurred more than once.

```
Stem-and-leaf of C1 N = 25
Leaf Unit = 10
  2    60   78
  7    61   03699
  9    62   69
 11    63   26
 (3)   64   233
 11    65   48
  9    66   8
  8    67   249
  5    68   18
  3    69   145
```

2.2   Frequency Distributions and Histograms

Lists of large sets of data do not present much of a picture. Sometimes we want to condense the data into a more manageable form. This can be accomplished with the aid of a *frequency distribution.*

FREQUENCY DISTRIBUTION

A listing, often expressed in chart form, that pairs each value of a variable with its frequency.

TABLE 2.3 Ungrouped Frequency Distribution

x	f
0	1
1	3
2	8
3	5
4	3

Exercise 2.23

Form an ungrouped frequency distribution of the resulting data: 1, 2, 1, 0, 4, 2, 1, 1, 0, 1, 2, 4

To demonstrate the concept of a frequency distribution, let's use this set of data:

3	2	2	3	2	4	4	1	2	2
4	3	2	0	2	2	1	3	3	1

If we let x represent the variable, we can use a frequency distribution to represent this set of data by listing the x values with their frequencies. For example, the value 1 occurs in the sample three times; therefore, the **frequency** for $x = 1$ is 3. The complete set of data is represented by the frequency distribution shown in Table 2.3 above.

The frequency f is the number of times the value x occurs in the sample. Table 2.3 is an **ungrouped frequency distribution**—"ungrouped" because each value of x in the distribution stands alone. When a large set of data has many different x-values instead of a few repeated values, as in the previous example, we can group the values into a set of classes and construct a **grouped frequency distribution.** The stem-and-leaf display in Figure 2.5 (p. 43) shows, in picture form, a grouped frequency distribution. Each stem represents a class. The number of leaves on each stem is the same as the frequency for that same **class.** The data represented in Figure 2.5 are listed as a frequency distribution in Table 2.4.

TABLE 2.4   Grouped Frequency Distribution

Exercise 2.24

Referring to Table 2.4:

a. Explain what $f = 8$ represents.
b. What is the sum of the frequency column?
c. What does this sum represent?

Class	Frequency
50 or more to less than 60 ➤ $50 \leq x < 60$	1
60 or more to less than 70 ➤ $60 \leq x < 70$	3
70 or more to less than 80 ➤ $70 \leq x < 80$	8
80 or more to less than 90 ➤ $80 \leq x < 90$	5
90 or more to less than 100 ➤ $90 \leq x < 100$	2
	19

The stem-and-leaf process can be used to construct a frequency distribution; however, the stem representation is not compatible with all **class widths.** For example, class widths of 3, 4, or 7 are awkward to use. Thus, sometimes it is advantageous to have a separate procedure for constructing a grouped frequency distribution.

ILLUSTRATION 2.6 ▼

To illustrate this grouping (or classifying) procedure , let's use a sample of 50 final exam scores taken from last semester's elementary statistics class. Table 2.5 lists the 50 scores.

TABLE 2.5   Statistics Exam Scores

60	47	82	95	88	72	67	66	68	98
90	77	86	58	64	95	74	72	88	74
77	39	90	63	68	97	70	64	70	70
58	78	89	44	55	85	82	83	72	77
72	86	50	94	92	80	91	75	76	78

The basic guidelines to follow in constructing a grouped frequency distribution are:

1. Each class should be of the same width.
2. Classes should be set up so that they do not overlap and so that each piece of data belongs to exactly one class.
3. For the exercises given in this textbook, 5 to 12 classes are most desirable since all samples contain fewer than 125 data. (The square root of n is a reasonable guideline for number of classes with samples of fewer than 125 data.)
4. Use a system that takes advantage of a number pattern, to guarantee accuracy. (This is demonstrated below.)
5. When it is convenient, an even class width is often advantageous.

Procedure

1. Identify the high score $(H = 98)$ and the low score $(L = 39)$, and find the range . Range $= H - L = 98 - 39 = 59$.
2. Select a number of classes $(m = 7)$ and a class width $(c = 10)$ so that the product $(mc = 70)$ is a bit larger than the range (range $= 59$).
3. Pick a starting point. This starting point should be a little smaller than the lowest score L. Suppose that we start at 35; counting from there by 10s (the class width), we get 35, 45, 55, 65, . . . , 95, 105. These are called the **class boundaries.**

Our classes for Illustration 2.6 are

35 or more to less than 45	➤	$35 \le x < 45$
45 or more to less than 55	➤	$45 \le x < 55$
55 or more to less than 65	➤	$55 \le x < 65$
65 or more to less than 75	➤	$65 \le x < 75$
⋮		$75 \le x < 85$
		$85 \le x < 95$
95 or more to and including 105		$95 \le x \le 105$

NOTES

1. At a glance you can check the number pattern to determine whether the arithmetic used to form the classes was correct (35, 45, 55, . . . , 105).

2. For the interval, $35 \leq x < 45$, the 35 is the lower class boundary and 45 is the upper class boundary. Observations that fall on the lower class boundary stay in that interval; observations that fall on the upper class boundary go into the next higher interval, except for the last class.

3. The class width is the difference between the upper and lower class boundaries.

4. Many combinations of class widths, number of classes, and starting points are possible when classifying data. There is no one best choice. Try a few different combinations, and use good judgment to decide on the one to use.

Once the classes are set up, we need to sort the data into those classes. The method used to sort will depend on the current format of the data: If the data are ranked, the frequencies can be counted; if the data are not ranked, we will **tally** the data to find the frequency numbers. When classifying data, it helps to use a standard chart (see Table 2.6).

TABLE 2.6   Standard Chart for Frequency Distribution

Class Number	Boundaries	Class Tallies	Frequency
1	$35 \leq x < 45$	||	2
2	$45 \leq x < 55$	||	2
3	$55 \leq x < 65$	⑤ ||	7
4	$65 \leq x < 75$	⑤ ⑤ |||	13
5	$75 \leq x < 85$	⑤ ⑤ |	11
6	$85 \leq x < 95$	⑤ ⑤ |	11
7	$95 \leq x \leq 105$	||||	4
			50

C a s e   S t u d y   2.1

The Age of Talk

Exercise 2.25

a. Express the category "35–44 years" shown in "The age of talk" using the interval notation $a \leq x < b$. (Remember, a person is age 44 for many days, right up to the day of their 45th birthday.)

b. Express the information on the circle graph as a grouped frequency distribution.

A November 1997 issue of USA Today presented this graphic, which shows a relative frequency distribution in the form of a circle graph. Each section of the circle represents a class interval of the values of the variable and the age of listeners. The percentages are the relative frequencies for each interval. This same information could be expressed as a grouped frequency distribution.

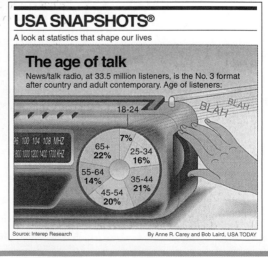

USA SNAPSHOTS®
A look at statistics that shape our lives

The age of talk
News/talk radio, at 33.5 million listeners, is the No. 3 format after country and adult contemporary. Age of listeners:

18-24 — 7%
25-34 — 16%
35-44 — 21%
45-54 — 20%
55-64 — 14%
65+ — 22%

Source: Interep Research   By Anne R. Carey and Bob Laird, USA TODAY

NOTES

1. If the data have been ranked (list form, dotplot, or stem-and-leaf), tallying is unnecessary; just count the data belonging to each class.
2. If the data are not ranked, be careful as you tally.
3. The frequency f for each class is the number of pieces of data that belong in that class.
4. The sum of the frequencies should be exactly equal to the number of pieces of data n ($n = \Sigma f$). This summation serves as a good check.

NOTE See the *Statistical Tutor* for information about the Σ-**notation** (read "**summation notation**").

Each class needs a single numerical value to represent all the data values that fall into that class. The **class midpoint** (sometimes called *class mark*) is the numerical value that is exactly in the middle of each class and is found by adding the class boundaries and dividing by 2. Table 2.7 shows an additional column for the class midpoint, x.

TABLE 2.7   Frequency Distribution with Class Midpoints

Class Number	Class Boundaries	Frequency f	Class Midpoints x
1	$35 \leq x < 45$	2	40
2	$45 \leq x < 55$	2	50
3	$55 \leq x < 65$	7	60
4	$65 \leq x < 75$	13	70
5	$75 \leq x < 85$	11	80
6	$85 \leq x < 95$	11	90
7	$95 \leq x \leq 105$	4	100
		50	

Exercise 2.26

a. 65 belongs to which class?

b. Explain the meaning of "$65 \leq x < 75$."

c. Explain what "class width" is, and describe four ways that it shows up.

As a check of your arithmetic, successive class midpoints should be a class width apart, which is 10 in this illustration (40, 50, 60, . . . , 100 is a recognizable pattern).   ▲

NOTE Now you can see why it is helpful to have an even class width. An odd class width would have resulted in a class midpoint with an extra digit. (For example, the class 45–54 is 9 wide and the class midpoint is 49.5.)

Note that when we classify data into classes, we lose some information. Only when we have all the raw data do we know the exact values that were actually observed for each class. For example, we put a 47 and a 50 into class number 2, with class boundaries of 45 and 55. Once they are placed in the class, their values are lost to us and we use the class midpoint, 50, as their representative value.

HISTOGRAM

A bar graph representing a frequency distribution of a quantitative variable. A histogram is made up of the following components:

1. A title, which identifies the population or sample of concern.
2. A vertical scale, which identifies the frequencies in the various classes.

(continued)

3. A horizontal scale, which identifies the variable *x*. Values for the class boundaries or class midpoint may be labeled along the *x*-axis. Use whichever method of labeling the axis that best presents the variable.

The frequency distribution from Table 2.7 appears in histogram form in Figure 2.10.

Figure 2.10

Frequency Histogram

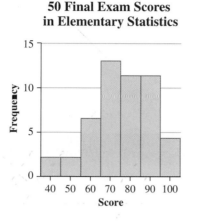

50 Final Exam Scores in Elementary Statistics

> Be sure to identify both scales so that the histogram tells the complete story

---

Exercise 2.27

Draw a frequency histogram of the annual salaries for resort-club managers. Label class boundaries.

Ann. Sal. ($1000)	No. Mgrs.
15 to 25	12
25 to 35	37
35 to 45	26
45 to 55	19
55 to 65	6

Sometimes the **relative frequency** of a value is important. The relative frequency is a proportional measure of the frequency for an occurrence. It is found by dividing the class frequency by the total number of observations. Relative frequency can be expressed as a common fraction, in decimal form, or as a percentage. For example, in Illustration 2.6 the frequency associated with the third class (55–65) is 7. The relative frequency for the third class is $\frac{7}{50}$, or 0.14, or 14%. Relative frequencies are often useful in a presentation because nearly everybody understands fractional parts when expressed as percents. Relative frequencies are particularly useful when comparing the frequency distributions of two different size sets of data. Figure 2.11 is a **relative frequency histogram** of the sample of the 50 final exam scores from Table 2.7.

Figure 2.11

Relative Frequency Histogram

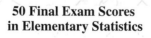

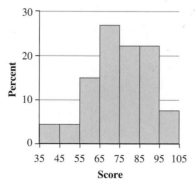

50 Final Exam Scores in Elementary Statistics

Exercise 2.28

Explain the similarities and the differences between Figures 2.10 and 2.11.

A stem-and-leaf display contains all the information in a histogram. Figure 2.5 (p. 43) shows the stem-and-leaf display constructed in Illustration 2.4. Figure 2.5 has been rotated 90° and labels have been added to form the histogram shown in Figure 2.12.

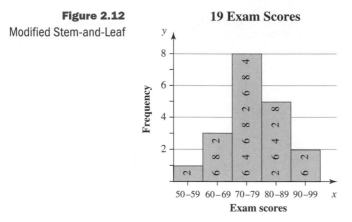

Figure 2.12
Modified Stem-and-Leaf

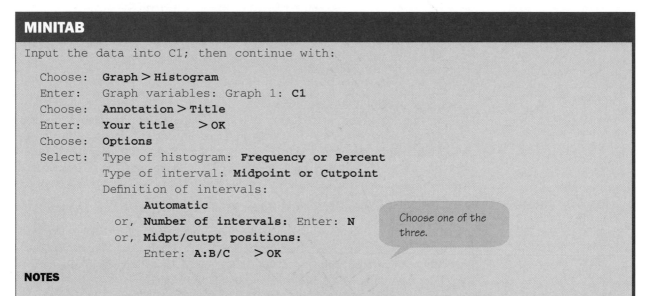

The following commands will construct a histogram.

MINITAB

Input the data into C1; then continue with:

Choose: **Graph > Histogram**
Enter: Graph variables: Graph 1: **C1**
Choose: **Annotation > Title**
Enter: **Your title   > OK**
Choose: **Options**
Select: Type of histogram: **Frequency or Percent**
          Type of interval: **Midpoint or Cutpoint**
          Definition of intervals:
               Automatic
           or, **Number of intervals:** Enter: **N**
           or, **Midpt/cutpt positions:**
               Enter: **A:B/C   > OK**

Choose one of the three.

NOTES

1. Midpoints are the class midpoints, and cutpoints are the class boundaries.
2. Percent is relative frequency.
3. Automatic means MINITAB will make all the choices; N = number of intervals, that is, the number of classes you want used.
4. A = smallest class midpoint or boundary, B = largest class midpoint or boundary, C = class width you want to specify.

The following commands will draw the histogram of a frequency distribution. The end classes can be made full width by adding an extra class with frequency zero to each end of the frequency distribution. Input the class midpoints into C1 and the corresponding frequencies into C2.

Choose: **Graph > Plot**
Enter: Graph variables: Graph 1: Y: **C2**
                                    X: **C1**
          Data display: Item 1: Display: **AREA**
Choose: **Edit Attributes**
Enter: Fill Type: **None**
Select: Connection Function: **Step   > OK**

EXCEL

Input the data into column A and the upper class limits* in column B (optional) and (column headings are optional); then continue with:

* If boundary = 50, then limit = 49.9 (depending on the number of decimal places in the data).

Choose: **Tools > Data Analysis** ** > **Histogram** > **OK**
Enter: Input Range: **Data (A1:A6 or select cells)**
          Bin Range: **upper class limits (B1:B6 or select cells)**
                     [leave blank if EXCEL determines the intervals]
Select: **Labels** (if column headings are used)
          Output Range
Enter: **area for freq. distr. & graph (C1 or select cell)**
Select: **Chart Output**

**If Data Analysis does not show on the Tools menu:

Choose: **Tools > Add-Ins . . .**
Select: **Analysis ToolPak**
          Analysis Toolpak-VBA

To remove gaps between bars

Click on: **Any bar on graph**
Click on: **Right mouse button**
Choose: **Format Data Series > Options**
Enter: Gap Width: **0**

To edit histogram

Click on: Anywhere clear on the chart
           —use handles to size
           Any title or axis name to change
           Any upper class limit* or frequency in the frequency distribution to change value > Enter

*Note that the upper class limits appear in the center of the bars. Replace with class midpoints. The "More" cell in the frequency distribution may also be deleted.

For tabled data, input the classes in column A (ex. 30–40) and the frequencies in column B; then continue with:

Choose: **Chart Wizard > Column > 1st picture** (usually) > **Next**
Enter: Data Range: **(A1:B4 or select cells)**
Select: Series in: **Columns > Next**
Choose: **Titles**
Enter: Chart title: **your title**
          Category (x) axis: **title for x-axis**
          Value (y) axis: **title for yaxis** > **Next > Finish**

Do as above to remove gaps and adjust.

TI-83

Input the data into L1; then continue with:

    Choose: **2nd > STAT PLOT > 1:Plot1**

```
Plot1  Plot2  Plot3
On Off
Type: 📊 📈 📉
      📊 📊 📈
Xlist:L1
Freq:1
```

Calculator selects classes:

    Choose: **Zoom > 9:ZoomStat > Trace > > >**

Individual selects classes:

    Choose: **Window**
    Enter: **at most lowest value, at least highest value, class width, −1, at least highest frequency, 1 (depends on frequency numbers), 1**
    Choose: **Graph > Trace** (use values to construct frequency distribution)

For tabled data, input the class marks into L1 and the frequencies into L2; then continue with:

    Choose: **2nd > STAT PLOT > 1:Plot1**

```
Plot1  Plot2  Plot3
On Off
Type: 📊 📈 📉
      📊 📊 📈
Xlist:L1
Freq:L2
```

    Choose: **Window**
    Enter: **smallest lower class boundary, largest upper class boundary, class width, −ymax/4, highest frequency, 0 (for no tick marks), 1**
    Choose: **Graph > Trace > > >**

To obtain a relative frequency histogram of tabled data instead:

    Choose: **STAT > EDIT > 1:EDIT . . .**
    Highlight: **L3**
    Enter: **L3 =: L2/SUM(L2)**
             [SUM − 2nd LIST > MATH > 5:sum]
    Choose: **2nd > STAT PLOT > 1PLOT1**

```
Plot1  Plot2  Plot3
On Off
Type: 📊 📈 📉
      📊 📊 📈
Xlist:L1
Freq:L3
```

    Choose: **Window**
    Enter: **smallest lower class boundary, largest upper class boundary, class width, −ymax/4, highest rel. frequency, 0 (for no tick marks), 1**
    Choose: **Graph > Trace > > >**

Histograms are valuable tools. For example, the histogram of a sample should have a distribution shape that is very similar to that of the population from which the sample was drawn. If the reader of a histogram is at all familiar with the variable involved, he or she will usually be able to interpret several important facts. Figure 2.13 presents histograms with descriptive labels resulting from their geometric shape.

Briefly, the terms used to describe histograms are as follows:

Symmetrical: Both sides of this distribution are identical (halves are mirror images).

Uniform (rectangular): Every value appears with equal frequency.

Figure 2.13
Shapes of Histograms

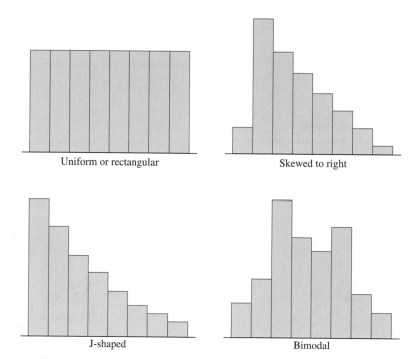

Symmetrical, normal, or triangular   Uniform or rectangular   Skewed to right

Skewed to left   J-shaped   Bimodal

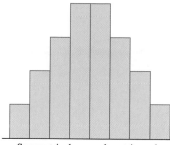

Exercise 2.29

Can you think of variables whose distribution might yield histograms like each of these?

Skewed: One tail is stretched out longer than the other. The direction of skewness is on the side of the longer tail.

J-shaped: There is no tail on the side of the class with the highest frequency.

Bimodal: The two most populous classes are separated by one or more classes. This situation often implies that two populations are being sampled.(See Figure 2.7, p. 45.)

Normal: A symmetrical distribution is mounded up about the mean and becomes sparse at the extremes. (Additional properties are discussed later.)

NOTES

1. The **mode** is the value of the piece of data that occurs with the greatest frequency. (Mode will be discussed in Section 2.3, p. 69)

2. The **modal class** is the class with the greatest frequency.

3. A **bimodal distribution** has two high-frequency classes separated by classes with lower frequencies. It is not necessary for the two high frequencies to be the same.

Another way to express a frequency distribution is to use a *cumulative frequency distribution.*

CUMULATIVE FREQUENCY DISTRIBUTION

A frequency distribution that pairs cumulative frequencies with values of the variable.

The **cumulative frequency** for any given class is the sum of the frequency for that class and the frequencies of all classes of smaller values. Table 2.8 shows the cumulative frequency distribution from Table 2.7 (p. 54).

TABLE 2.8   Using Frequency Distribution to Form a Cumulative Frequency Distribution

Class Number	Class Boundaries	Frequency f	Cumulative Frequency
1	$35 \leq x < 45$	2	2 (2)
2	$45 \leq x < 55$	2	4 (2 + 2)
3	$55 \leq x < 65$	7	11 (7 + 4)
4	$65 \leq x < 75$	13	24 (13 + 11)
5	$75 \leq x < 85$	11	35 (11 + 24)
6	$85 \leq x < 95$	11	46 (11 + 35)
7	$95 \leq x \leq 105$	4	50 (4 + 46)
		50	

Exercise 2.30

Express this frequency distribution as a cumulative frequency distribution:

Ann. Sal. ($1000)	No. Mgrs.
15 to 25	12
25 to 35	37
35 to 45	26
45 to 55	19
55 to 65	6

The same information can be presented by using a *cumulative relative frequency distribution* (see Table 2.9). This combines the cumulative frequency and the relative frequency ideas.

TABLE 2.9   Cumulative Relative Frequency Distribution

Class Number	Class Boundaries	Cumulative Relative Frequency	Cumulative frequencies are for interval 35 up to the upper boundary of that class
1	$35 \leq x < 45$	2/50, or 0.04	◄— *from 35 up to less than 45*
2	$45 \leq x < 55$	4/50, or 0.08	◄— *from 35 up to less than 55*
3	$55 \leq x < 65$	11/50, or 0.22	◄— *from 35 up to less than 65*
4	$65 \leq x < 75$	24/50, or 0.48	·
5	$75 \leq x < 85$	35/50, or 0.70	·
6	$85 \leq x < 95$	46/50, or 0.92	·
7	$95 \leq x \leq 105$	50/50, or 1.00	◄— *from 35 up to and including 105*

Exercise 2.31

Express this frequency distribution as a cumulative relative frequency distribution:

Ann. Sal. ($1000)	No. Mgrs.
15 to 25	12
25 to 35	37
35 to 45	26
45 to 55	19
55 to 65	6

OGIVE (pronounced o′jīv)

A line graph of a cumulative frequency or cumulative relative frequency distribution. An ogive has the following components:

1. A title, which identifies the population or sample.
2. A vertical scale, which identifies either the cumulative frequencies or the cumulative relative frequencies. (Figure 2.14 shows an ogive with cumulative relative frequencies.)
3. A horizontal scale, which identifies the upper class boundaries . Until the upper boundary of a class has been reached, you cannot be sure you have accumulated all the data in that class. Therefore, the horizontal scale for an ogive is always based on the upper class boundaries.

Exercise 2.32

Construct an ogive for the cumulative relative frequency distribution found in answering Exercise 2.31.

NOTE Every ogive starts on the left with a relative frequency of zero at the lower class boundary of the first class and ends on the right with a relative frequency of 100% at the upper class boundary of the last class.

Figure 2.14

Ogive

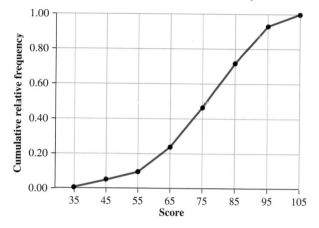

REMEMBER All graphic representations of sets of data need to be completely self-explanatory. That includes a descriptive meaningful title and proper identification of the vertical and horizontal scales.

The following commands will construct an ogive.

MINITAB

Input the class boundaries into C1 and the cumulative percentages into C2 [enter 0 (zero) for the percentage paired with the lower boundary of the first class and pair each cumulative percentage with the class upper boundary]. Use percentages; that is, use 25% in place of .25.

Choose: **Graph > Plot**
Enter: Graph variables: Graph 1:
          Y, **C2** and X, **C1**
          Data display: Item 1: Display: **Connect**
Choose: **Annotation > Title**
Enter: **your title**   **> OK**

TI-83

Input the class boundaries into L1 and the frequencies into L2 (include an extra class boundary at the beginning with a frequency of 0); then continue with:

Choose: **STAT > EDIT > 1:EDIT . . .**
Highlight: **L3**
Enter: L3 = **2nd > LIST > OPS > 6:cum sum(L2)**
Highlight: **L4**
Enter: L4 = **L3/2nd > LIST > Math > 5:sum (L2)**
Choose: **2nd STAT > PLOT > 1:Plot**

Choose: **Zoom > 9:ZoomStat > Trace > > >**

Adjust window if needed for better readability.

EXCEL

Input the data into column A and the upper class limits* into column B (include an additional class at the beginning).

* If the boundary = 50, then the limit = 49.9 (depending on the number of decimal places in the data).

Choose: **Tools > Data Analysis > Histogram   > OK**
Enter: Input Range: **data (A1:A6 or select cells)**
         Bin Range: **upper class limits (B1:B6 or select cells)**
Select: **Labels** (if column headings were used)
         Output Range
         Enter: **area for freq. distr. & graph (C1 or select cell)**
         Cumulative Percentage
         Chart Output

To close gaps and for editing, see histogram commands on page 57.

For tabled data input the upper class boundaries in column A and the cumulative relative frequencies in column B (include an additional class boundary at the beginning with a cumulative relative frequency equal to 0(zero)); then continue with:

Choose: **Chart Wizard > Line > 4th picture (usually) > Next**
Choose: **Series** (have more control on input) **> Remove** (remove all columns except column B
Enter: Name: **(B1 or select name cell—cum. rel.freq.)**
         Values: **(B2:B6 or select cells)**
         Category (*x*) axis labels: **(A2:A8 or select cells) > Next**
Choose: **Titles**
Enter: Chart title: **your title**
         Category (*x*) axis: **title for *x*-axis**
         Value (*y*) axis: **title for *y*-axis > Next > Finish**

For editing, see histogram commands on page 57.

EXERCISES • • • • • • • • • • • • •

2.33 In the May 1990 issue of the journal *Social Work,* the following ungrouped frequency distribution was used to represent the "size of families":

Number of Children Living at Home	Number of Mexican-American Women
0	23
1	22
2	17
3	7
4	1

Source: Copyright 1990, National Association of Social Workers, Inc. *Social Work.*

(continued)

a. Which column represents the "size" of families? The "frequency" of those sizes?
b. Construct a frequency histogram of this distribution.

(Retain these answers for use in Exercises 2.43, p. 65 and 2.162, p. 117.)

2.34 The success of a batter against a pitcher in baseball varies with the count (balls and strikes) and, in particular, the number of strikes already called against a batter. *Sports Illustrated* (7-6-98) in their article, "The Book on Maddux," tabulated the existing strike count against the number of hits batters obtained on the next pitch and the number of times the at bat ended on the next pitch. This information is for batters who faced Greg Maddux during the first half of the 1998 season. Results are summarized in the table:

Strikes	Hits	At Bats
0	38	137
1	43	138
2	25	223

The writer concluded, ". . . as the numbers show, let Maddux get two strikes on you, and you don't stand a chance."
a. Construct a frequency histogram for the strike count when the player made a hit.
b. Construct a frequency histogram of the strike count when the player's turn at bat ended.
c. Do you agree or disagree with the writer's conclusion? Explain why or why not.

2.35 The ages of 50 dancers who responded to a call to audition for a musical comedy are:

21	19	22	19	18	20	23	19	19	20
19	20	21	22	21	20	22	20	21	20
21	19	21	21	19	19	20	19	19	19
20	20	19	21	21	22	19	19	21	19
18	21	19	18	22	21	24	20	24	17

> Many of these data sets are on the StatSource CD.

a. Prepare an ungrouped frequency distribution of these ages.
b. Prepare an ungrouped relative frequency distribution of the same data.
c. Prepare a relative frequency histogram of these data.
d. Prepare a cumulative relative frequency distribution of the same data.
e. Prepare an ogive of these data.

2.36 The opening-round scores for the Ladies' Professional Golf Association tournament at Locust Hill Country Club were posted as follows:

69	73	72	74	77	80	75	74	72	83	68	73
75	78	76	74	73	68	71	72	75	79	74	75
74	74	68	79	75	76	75	77	74	74	75	75
72	73	73	72	72	71	71	70	82	77	76	73
72	72	72	75	75	74	74	74	76	76	74	73
74	73	72	72	74	71	72	73	72	72	74	74
67	69	71	70	72	74	76	75	75	74	73	74
74	78	77	81	73	73	74	68	71	74	78	70
68	71	72	72	75	74	76	77	74	74	73	73
70	68	69	71	77	78	68	72	73	78	77	79
79	77	75	75	74	73	73	72	71	68	70	71
78	78	76	74	75	72	72	72	75	74	76	77
78	78										

a. Form an ungrouped frequency distribution of these scores.
b. Draw a histogram of the first-round golf scores. Use the frequency distribution from (a).

2.37 The KSW computer-science aptitude test was given to 50 students. The frequency distribution that is given in the table resulted from their scores.

KSW Test Score	Frequency
0–4	4
4–8	8
8–12	8
12–16	20
16–20	6
20–24	3
24–28	1

a. What are the class boundaries for the class of largest frequency?
b. Give all class midpoints associated with this frequency distribution.
c. What is the class width?
d. Give the relative frequencies for the classes.
e. Draw a relative frequency histogram of the test scores.

2.38 The USA Snapshot® "Nuns an aging order" reports that the median age of the USA's 94,022 Roman Catholic nuns is 65 years and that the percentage of nuns in the country by age group is:

Under 50	51–70	Over 70	Refused to give age
16%	42%	37%	5%

This information is based on a survey of 1049 Roman Catholic nuns.

Suppose the survey had resulted in the frequency distribution shown below (52 ages unknown).

Age	20–30	30–40	40–50	50–60	60–70	70–80	80–90
Freq.	34	58	76	187	254	241	147

> Use the computer or calculator commands on page 56 to construct a histogram of a frequency distribution.

a. Draw and completely label a frequency histogram.
b. Draw and completely label a relative frequency histogram of the same distribution.
c. Carefully examine the two histograms, (a) and (b), and explain why one of them might be easier to understand. (Retain these solutions for use in answering Exercise 2.101 p. 91.)

2.39 The speeds of 55 cars were measured by a radar device on a city street:

27	23	22	38	43	24	35	26	28	18	20
25	23	22	52	31	30	41	45	29	27	43
29	28	27	25	29	28	24	37	28	29	18
26	33	25	27	25	34	32	36	22	32	33
21	23	24	18	48	23	16	38	26	21	23

> Use the computer or calculator commands on page 56 to construct a histogram for a given set of data.

a. Classify these data into a grouped frequency distribution by using class boundaries 12, 18, 24, . . . , 54.
b. Find the class width.
c. For the class 24–30, find (1) the class midpoint, (2) the lower class boundary, and (3) the upper class boundary.
d. Construct a frequency histogram of these data.

2.40 The hemoglobin A_{1c} test, a blood test given to diabetics during their periodic checkups, indicates the level of control of blood sugar during the past two to three months. The following data were obtained for 40 different diabetics at a university clinic that treats diabetic patients.

6.5	5.0	5.6	7.6	4.8	8.0	7.5	7.9	8.0	9.2
6.4	6.0	5.6	6.0	5.7	9.2	8.1	8.0	6.5	6.6
5.0	8.0	6.5	6.1	6.4	6.6	7.2	5.9	4.0	5.7
7.9	6.0	5.6	6.0	6.2	7.7	6.7	7.7	8.2	9.0

 a. Classify these A_{1c} values into a grouped frequency distribution using the classes 3.7–4.7, 4.7–5.7, and so on.
 b. What are the class midpoints for these classes?
 c. Construct a frequency histogram of these data.

2.41 All of the third-graders at Roth Elementary School were given a physical-fitness strength test. These data resulted:

12	22	6	9	2	9	5	9	3	5	16	1	22
18	6	12	21	23	9	10	24	21	17	11	18	19
17	5	14	16	19	19	18	3	4	21	16	20	15
14	17	4	5	22	12	15	18	20	8	10	13	20
6	9	2	17	15	9	4	15	14	19	3	24	

 a. Construct a dotplot.
 b. Prepare a grouped frequency distribution using classes 1–4, 4–7, and so on, and draw a histogram of the distribution. (Retain solution for use in answering Exercise 2.59, p. 73.)
 c. Prepare a grouped frequency distribution using classes 0–3, 3–6, 6–9, and so on, and draw a histogram of the distribution.
 d. Prepare a grouped frequency distribution using class boundaries −2.5, 2.5, 7.5, 12.5 and so on, and draw a histogram of the distribution.
 e. Prepare a grouped frequency distribution using classes of your choice, and draw a histogram of the distribution.
 f. Describe the shape of the histograms found in (b), (c), (d), and (e) separately. Relate the distribution seen in the histogram to the distribution seen in the dotplot.
 g. Discuss how the number of classes used and the choice of class boundaries used affect the appearance of the resulting histogram.

2.42 The following 40 amounts are the fees that Fast Delivery charged for delivering small freight items last Thursday afternoon.

4.03	3.56	3.10	6.04	5.62	3.16	2.93	3.82	4.30	3.86
4.57	3.59	4.57	5.16	2.88	5.02	5.46	3.87	6.81	4.91
3.62	3.62	3.80	3.70	4.15	4.07	3.77	5.77	7.86	4.63
4.81	2.86	5.02	5.24	4.02	5.44	4.65	3.89	4.00	2.99

> Adjust computer and calculator commands for a histogram (pages 56–58) to construct a relative frequency histogram for the set of data.

 a. Classify these data into a grouped frequency distribution.
 b. Construct a relative frequency histogram of these data.

2.43 **a.** Prepare a cumulative relative frequency distribution for the variable "number of children living at home" in Exercise 2.33.
 b. Construct an ogive of the distribution.

2.44 **a.** Prepare a cumulative relative frequency distribution for the variable "KSW test score" in Exercise 2.37.
 b. Construct an ogive of the distribution.

NUMERICAL DESCRIPTIVE STATISTICS

2.3   Measures of Central Tendency

Measures of central tendency are numerical values that locate, in some sense, the center of a set of data. The term average is often associated with all measures of central tendency.

MEAN (Arithmetic mean)

The average with which you are probably most familiar. The sample mean is represented by $\bar{x}$ (read "**x bar**" or "**sample mean**"). The mean is found by adding all the values of the variable x (this sum of x values is symbolized $\sum x$) and dividing by the number of these values, n ("sample size") . We express this in formula form as

$$\text{sample mean:} \quad x\text{-}bar = \frac{\textit{sum of all x}}{\textit{number of x}}$$

$$\bar{x} = \frac{\sum x}{n} \tag{2.1}$$

NOTES

1. See the *Statistical Tutor* for information about the $\sum$ notation (**"summation notation"**).
2. The population mean , μ (lowercase mu , Greek alphabet), is the mean of all x values for the entire population.

ILLUSTRATION 2.7 ▼

A set of data consists of the five values 6, 3, 8, 6, and 4. Find the mean.

Solution

Using formula (2.1), we find

$$\bar{x} = \frac{\sum x}{n} = \frac{6 + 3 + 8 + 6 + 4}{5} = \frac{27}{5} = 5.4$$

Therefore, the mean of this sample is **5.4**.   ▲

Figure 2.15
Physical Representation of Mean

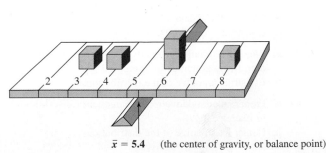

$\bar{x} = \textbf{5.4}$ (the center of gravity, or balance point)

Exercise 2.45

The number of children, x, belonging to each of eight families registering for swimming was 1, 2, 1, 3, 2, 1, 5, 3. Find the mean, $\bar{x}$.

A physical representation of the mean can be constructed by thinking of a number line balanced on a fulcrum. A weight is placed on a number on the line corresponding to each number in the sample of Illustration 2.7. In Figure 2.15 there is one weight each on the 3, 8, and 4 and two weights on the 6, since there were two 6s in the sample. The mean is the value that balances the weights on the number line—in this case, 5.4.

The following commands will calculate the mean.

MINITAB

Input the data into C1; then continue with:

    Choose: Calc > Column Statistics
    Select: Mean
    Enter: Input variable: C1

EXCEL

Input the data into column A and activate a cell for the answer; then continue with:

    Choose: Paste Function, f_x* > Statis-
             tical > AVERAGE > OK
    Enter: Number 1: (A2:A6 or select
             cells)
             [Start at A1 if no header
             row (column title) is used]
    *Alternative Method-Choose: Insert > Function . . .

TI-83

Input the data into L1; then continue with:

    Choose: 2nd > LIST > Math > 3:mean(
    Enter: L1

MEDIAN

The value of the data that occupies the middle position when the data are ranked in order according to size. The sample median is represented by $\tilde{x}$ (read "x tilde" or "sample median").

NOTE The population median, M (uppercase mu, Greek alphabet), is the data value in the middle position of the entire ranked population.

Procedure for Finding the Median

STEP 1:   Rank the data.

STEP 2:   Determine the depth, or position, of the median.
The **depth** (number of positions from either end) of the median is determined by the formula

depth of median: *depth of median* $= \dfrac{\textit{number} + 1}{2}$

$$d(\tilde{x}) = \frac{n+1}{2}$$

(2.2)

(continued)

The median's depth (or position) is found by adding the position numbers of the smallest data (1) and largest data (n) and dividing by 2. (n is the number of pieces of data.)

STEP 3: Determine the value of the median.

Count over the ranked data, locating the data in the $d(\tilde{x})$th position. The median will be the same regardless of which end of the ranked data (high or low) you count from. In fact counting from both ends will serve as an excellent check.

The following two illustrations will demonstrate this procedure as it applies to both odd-numbered and even-numbered sets of data.

ILLUSTRATION 2.8 ▼

Find the median for the set of data {6, 3, 8, 5, 3}.

Solution

STEP 1 The data, ranked in order of size, are 3, 3, 5, 6, and 8.

STEP 2 Depth of the median: $d(\tilde{x}) = \dfrac{n+1}{2} = \dfrac{5+1}{2} = 3$ (the "3rd" position)

STEP 3 That is, the median is the third number from either end in the ranked data, or $\tilde{x} = 5$. Notice that the median essentially separates the ranked set of data into two subsets of equal size (see Figure 2.16).

> The value of $d(\tilde{x})$ is the depth of the median, NOT the value of the median, $\tilde{x}$.

Figure 2.16
Median of {3, 3, 5, 6, 8}

$\tilde{x} = 5$ (the middle value; **2 data are smaller, 2 are larger**)

Exercise 2.46
Find the median height of a basketball team: 73, 76, 72, 70, and 74 inches.

As in Illustration 2.8, when n is odd, the depth of the median $d(\tilde{x})$, will always be an integer. However, when n is even, the depth of the median, $d(\tilde{x})$, will always be a half-number, as shown in Illustration 2.9.

ILLUSTRATION 2.9 ▼

Find the median for the sample 9, 6, 7, 9, 10, 8.

Solution

STEP 1 The data, ranked in order of size, are 6, 7, 8, 9, 9, and 10.

STEP 2 Depth of the median: $d(\tilde{x}) = \dfrac{n+1}{2} = \dfrac{6+1}{2} = 3.5$ (the "3.5th" position)

Figure 2.17
Median of {6, 7, 8, 9, 9, 10}

$\tilde{x} = 8.5$ (value in middle; **3 data smaller, 3 larger**)

Exercise 2.47
Find the median rate paid at Jim's Burgers if the workers' hourly rates are 4.25, 2.15, 4.90, 4.25, 4.60, 4.50, 4.60, 4.75.

STEP 3 That is, the median is halfway between the third and fourth pieces of data. To find the number halfway between any two values, add the two values together and divide by 2. In this case, add the third value (8) and the fourth value (9), then divide by 2. The median is $\tilde{x} = \dfrac{8+9}{2} = 8.5$, a number halfway between the "middle" two numbers (see Figure 2.17). Notice that the median again separates the ranked set of data into two subsets of equal size.   ▲

The following commands will calculate the median.

MINITAB

Input the data into C1; then continue with:

Choose: **Calc > Column Statistics**
Select: **Median**
Enter: Input variable: **C1**

EXCEL

Input the data into column A and activate a cell for the answer; then continue with:

Choose: **Paste Function,** f_x **> Statistical > MEDIAN > OK**
Enter: Number 1: **(A2:A6 or select cells)**

TI-83

Input the data into L1; then continue with:

Choose: **2nd > LIST > Math > 4:median(**
Enter: **L1**

MODE

The mode is the value of x that occurs most frequently.

In the set of data from Illustration 2.8, {3, 3, 5, 6, 8}, the mode is 3 (see Figure 2.18).

Figure 2.18
Mode of {3, 3, 5, 6, 8}

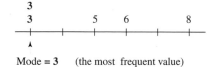

Mode = **3**   (the most frequent value)

Exercise 2.48

The number of cars per apartment, owned by a sample of tenants in a large complex is 1, 2, 1, 2, 2, 2, 1, 2, 3, 2. What is the mode?

In the sample 6, 7, 8, 9, 9, 10, the mode is 9. In this sample, only the 9 occurs more than once; in the data from Illustration 2.8, only the 3 occurs more than once. If two or more values in a sample are tied for the highest frequency (number of occurrences), we say there is no mode . For example, in the sample 3, 3, 4, 5, 5, 7, both the 3 and the 5 appear an equal number of times. There is no one value that appears most often; thus, this sample has no mode.

MIDRANGE

The number exactly midway between a lowest valued data *L* and a highest valued data *H*. It is found by averaging the low and the high values.

$$\text{midrange} = \frac{\textit{low value + high value}}{2}$$

$$\text{midrange} = \frac{L + H}{2} \tag{2.3}$$

For the set of data from Illustration 2.8, {3, 3, 5, 6, 8}, $L = 3$ and $H = 8$ (see Figure 2.19). Therefore, the midrange is

$$\text{midrange} = \frac{L + H}{2} = \frac{3 + 8}{2} = 5.5$$

Figure 2.19

Midrange of {3, 3, 5, 6, 8}

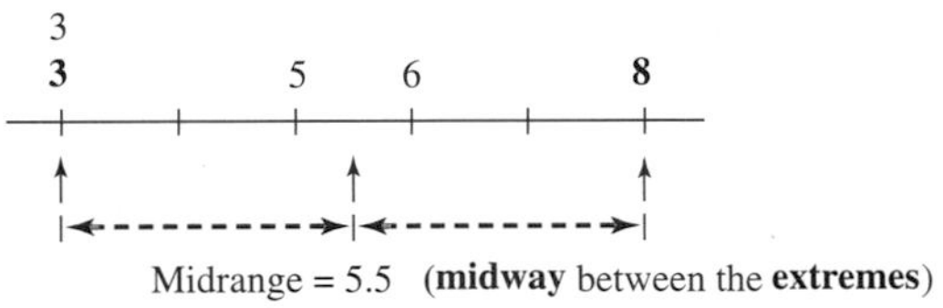

3
3   5   6   8

Midrange = 5.5   (**midway** between the **extremes**)

Exercise 2.49

USA Today, July 1998, reported on "What airlines spent on food" per passenger for the 1st quarter '98: least, $0.20 and most, $10.76. Find the midrange.

The four measures of central tendency represent four different methods of describing the middle. These four values may be the same, but more likely they will result in different values. For the sample data from Illustration 2.9, the mean $\bar{x}$ is 8.2, the median $\tilde{x}$ is 8.5, the mode is 9, and the midrange is 8. Their relationship to each other and to the data is shown in Figure 2.20.

Figure 2.20

Measures of Central Tendency for {6, 7, 8, 9, 9, 10}

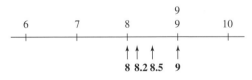

6   7   8   9   10
9
8 8.2 8.5   9

Exercise 2.50

Find the mean, median, mode, and midrange for the sample data 9, 6, 7, 9, 10, 8.

ROUND-OFF RULE

When rounding off an answer, let's agree to keep one more decimal place in our answer than was present in our original information. To avoid round-off buildup, round off only the final answer, not the intermediate steps. That is, avoid use of a rounded value to do further calculations. In our previous examples, the data were composed of whole numbers; therefore, those answers that have decimal values should be rounded to the nearest tenth. See the *Statistical Tutor* for specific instructions on how to perform the rounding off.

Case Study 2.2

"Average" Means Different Things

*When it comes to convenience, few things can match that wonderful mathematical device called averaging. With an **average** you can take a fistful of figures on any subject (temperatures, incomes, velocities, light-years, anything that can be measured) and compute one figure that will represent the whole fistful.*

But there is one thing to remember. There are several kinds of measures ordinarily known as averages. And each gives a different picture of the figures it is called on to represent.

Take an example. Here are the annual incomes of ten families:

$54,000	$39,000	$37,500	$36,750	$35,250
$31,500	$31,500	$31,500	$31,500	$25,500

What would this group's "typical" income be? Averaging would provide the answer, so let's compute the typical income by the simpler and most frequently used kinds of averaging.

The **arithmetic mean**. When anyone cites an average without specifying which kind, you can probably assume that he or she has the arithmetic mean in mind. It is the most common form of average, obtained by adding items in the series, then dividing by the number of items. In our example, the sum of the ten incomes divided by 10 is $35,400. The mean is representative of the series in the sense that the sum of the amounts by which the higher figures exceed the mean is exactly the same as the sum of the amounts by which the lower figures fall short of the mean.

The **median**. As you may have observed, six families earn less than the mean, four earn more. You might very well wish to represent this varied group by the income of the family that is right smack dab in the middle of the whole bunch. To do this, you need to find the median. It would be easy if there were 11 families in the group. The family sixth from highest (or sixth from lowest) would be in the middle and have the median income. But with ten families there is no middle family. So you add the two central incomes ($31,500 and $35,250 in this case) and divide by 2. The median works out to $33,375.

The **midrange**. Another number that might be used to represent the group is the midrange, computed by calculating the figure that lies halfway between the highest and lowest incomes. To find this figure, add the highest and lowest incomes ($54,000 and $25,500), divide by 2, and you have the amount that lies halfway between the extremes, $39,750.

The **mode**. So, three kinds of averages, and not one family actually has an income matching any of them. Say you want to represent the group by stating the income that occurs most frequently. That kind of representativeness is called a mode. In this example $31,500 would be the modal income. More families earn that income than any other.

Four different averages, each valid, correct, and informative in its way. But how they differ!

arithmetic mean	$35,400
median	$33,375
midrange	$39,750
mode	$31,500

And they would differ still more if just one family in the group were a millionaire—or one were jobless!

So there are three lessons to take away from today's class in averages. First, when you see or hear an average, find out which average it is. Then you'll know what kind of picture you are being given. Second, think about the figures being averaged so you can judge whether the average used is appropriate. And third, don't assume that a literal mathematical quantification is intended every time somebody says "average." It isn't. All of us often say "the average person" with no thought of implying a mean, median, or mode. All we intend to convey is the idea of other people who are in many ways a great deal like the rest of us.

Source: Reprinted by permission from CHANGING TIMES magazine (March 1980 issue). Copyright by The Kiplinger Washington Editors.

Exercise 2.51

Case Study 2.2 uses a sample of ten annual incomes to discuss the four averages.

a. Calculate the mean, median, mode, and midrange for the ten incomes. Compare your results with those found in the article.

b. What is there about the distribution of these ten data that causes the values of these four averages to be so different?

EXERCISES   •   •   •   •   •   •   •   •   •   •   •   •   •   •

2.52 Consider the sample 2, 4, 7, 8, 9. Find the following:
   a. the mean $\bar{x}$ **b.** the median $\tilde{x}$
   c. the mode   **d.** the midrange

2.53 Consider the sample 6, 8, 7, 5, 3, 7. Find the following:
   a. the mean $\bar{x}$ **b.** the median $\tilde{x}$
   c. the mode   **d.** the midrange

2.54 Fifteen randomly selected college students were asked to state the number of hours they slept last night. The resulting data are 5, 6, 6, 8, 7, 7, 9, 5, 4, 8, 11, 6, 7, 8, 7. Find the following:
   a. the mean $\bar{x}$ **b.** the median $\tilde{x}$
   c. the mode   **d.** the midrange

2.55 An article titled "Financing Your Kids' College Education" (*Farming,* Sept./Oct. 1994) listed the following in-state tuition and fees per school year for 14 land-grant universities: 1554, 2291, 2084, 4443, 2884, 2478, 3087, 3708, 2510, 2055, 3000, 2052, 2550, 2013.
   a. Find the mean in-state tuition and fees per school year.
   b. Find the median in-state tuition and fees per school year.
   c. Find the midrange in-state tuition and fees per school year.
   d. Find the mode, if one exists, per school year.

2.56 *Atlantic Monthly* (Nov. 1990) contains an article titled "The Case for More School Days." The number of days in the standard school year is given for several different countries and provinces as follows:

Country	n(days)/yr	Country	n(days)/year
Japan	243	New Zealand	190
West Germany	226–240	Nigeria	190
South Korea	220	British Columbia	185
Israel	216	France	185
Luxembourg	216	Ontario	185
Soviet Union	211	Ireland	184
Netherlands	200	New Brunswick	182
Scotland	200	Quebec	180
Thailand	200	Spain	180
Hong Kong	195	Sweden	180
England/Wales	192	United States	180
Hungary	192	French Belgium	175
Swaziland	191	Flemish Belgium	160
Finland	190		

   a. Find the mean and median number of days per year of school for the countries and provinces listed. (Use the midpoint of the 226–240 interval for West Germany when computing your answers.)
   b. Construct a stem-and-leaf display of these data.
   c. Describe the relationship between the mean and the median and what properties of the data cause the mean to be larger than the median.

   (Retain these solutions for use in answering Exercise 2.77, p. 83.)

2.57 *USA Today* (July 22, 1998) reported the following statistics about the average in-season daily greens fees by state at US golf courses.

   Highest: Hawaii, $85.70   Lowest: South Dakota, $23.80

   a. Based on this information, find the "average" of the 50 state average fees.
   b. Explain why your answer in part (a) is the only average value you can determine from the given information.   *(continued)*

c. If you were told the mean value of the 50 state averages is $37.30, what can you tell about their distribution?

2.58 *USA Today,* 10-28-94, reported on the average annual pay received by all workers covered by state and federal unemployment insurance for the 50 states. Connecticut had the highest with $33,169; South Dakota had the lowest with $18,613.
   a. Estimate the national average with the midrange for the states.
   b. The national average was reported to be $26,362. What can you conclude about the distribution of the state averages based on the relationship between the midrange and the national average?

2.59 All of the third-graders at Roth Elementary School were given a physical-fitness strength test. The following data resulted:

Many of these data sets are on the StatSource CD.

12	22	6	9	2	9	5	9	3	5	16	1	22
18	6	12	21	23	9	10	24	21	17	11	18	19
17	5	14	16	19	19	18	3	4	21	16	20	15
14	17	4	5	22	12	15	18	20	8	10	13	20
6	9	2	17	15	9	4	15	14	19	3	24	

   a. Construct a dotplot.
   b. Find the mode.
   c. Prepare a grouped frequency distribution using classes 1–4, 4–7, and so on, and draw a histogram of the distribution. (See Exercise 2.41, p. 65.)
   d. Describe the distribution; specifically, is the distribution bimodal (about what values)?
   e. Compare your answers in (a) and (c), and comment on the relationship between the mode and the modal classes that occurred in these data.
   f. Could a discrepancy like this occur when using an ungrouped frequency distribution? Explain.
   g. Explain why, in general, the mode of a set of data does not necessarily tell us the same information as the modal classes do.

2.60 Consumers are frequently cautioned against eating too much food that is high in calories, fat, and sodium for numerous health and fitness reasons. *Nutrition in Action* published a list of popular low-fat brands of hot dogs commonly labeled "fat-free," "reduced fat," "low-fat," "light," and so on, together with their calories, fat content, and sodium. All measured quantities are for one hot dog:

Hot Dog Brand	Calories	Fat (g)	Sodium (mg)
Ball Park Fat Free Beef Franks	50	0	460
Butterball Fat Free Franks	40	0	490
Oscar Meyer Free Hot Dogs	40	0	490
Jennie-O Fat Free Turkey Franks	40	0	520
Ball Park Fat Free Smoked White Turkey Franks	40	0	530
Eckrich Fat Free Beef Franks	50	0	570
Hormel Fat Free Beef Hot Dogs	50	0	590
Healthy Choice	60	2	430
Empire Kosher Turkey Franks	50	2	320
Hebrew National 97% Fat Free	50	2	440
Hillshire Farm Lean & Hearty Roast Turkey Hot Dogs	80	5	650
Eckrich Lite Bunsize Franks	110	7	480
Ball Park Smart Creations Lite Beef Franks	100	7	630
Louis Rich Bun-Length Cheese Franks	90	7	540
Gwaltney Great Dogs	140	10	730

(continued)

Hot Dog Brand	Calories	Fat (g)	Sodium (mg)
Hillshire Farm Lean & Hearty Hot Dogs	80	6	610
Oscar Meyer Light Beef Franks	110	8	610
Butterball Bun Size Franks	130	10	600
Shelton's Smoked Uncured Turkey Franks	200	16	800
Mr. Turkey Cheese Franks	110	9	540
Hebrew National Reduced Fat	120	10	350
Mr. Turkey Bun Size Franks	130	11	670
Jennie-O Jumbo Turkey Franks	130	11	600
Shelton's Uncured Franks	80	7	350
Jennie-O Turkey Franks	80	7	360
Wampler Foods Turkey Franks	90	8	430
Frankfurter, beef, typical	180	16	580
Wampler Foods Chicken Franks	120	11	480

Source: Nutrition Action HealthLetter, "On the Links," July/August, 1998, pp. 12–13.

a. Find the mean, median, mode, and midrange of the calories, fat, and sodium content of all the frankfurters listed. Use a table to summarize your results.

b. Construct a dotplot of the fat content. Locate the mean, median, mode, and midrange on the plot.

c. In the summer of 1997, the winner of Nathan's Famous Fourth of July Hot Dog Eating Contest consumed 24.5 hot dogs in 12 minutes. If he had been served the median hot dog, how many calories, grams of fat, and milligrams of sodium did he consume in the single sitting? If the recommended daily allowance for sodium intake is 2400 mg, did he likely exceed it? Explain.

2.61 The number of runs that baseball teams score is likely influenced by the ballparks in which they play. In an attempt to measure differences between stadiums, *Sports Illustrated* collected data for nearly every major league team over a 3-year span that included the number of runs scored per game by teams and their opponents while playing on their home field, and compared them to the number of runs scored per game while playing away (at the opponents' fields). The following table summarizes the data:

Runs per Game

Stadium / (Team)	Home	Away
Coors Field (Rockies)	13.65	8.76
The Ballpark (Rangers)	10.90	9.78
Three Rivers Stadium (Pirates)	9.65	8.95
County Stadium (Brewers)	10.17	9.63
Wrigley Field (Cubs)	9.59	9.15
Metrodome (Twins)	10.65	10.20
Fenway Park (Red Sox)	10.82	10.50
Veterans Stadium (Phillies)	9.26	9.03
Kauffman Stadium (Royals)	9.60	9.43
Olympic Stadium (Expos)	8.78	8.67
Tiger Stadium (Tigers)	10.55	10.44
Kingdome (Mariners)	10.97	10.93
Jacobs Field (Indians)	10.33	10.33
Cinergy Field (Reds)	9.18	9.24
Skydome (Blue Jays)	9.29	9.35

(continued)

Edison International Field (Angels)	10.20	10.29
Busch Stadium (Cardinals)	8.83	9.09
Camden Yards (Orioles)	9.82	10.18
Yankee Stadium (Yankees)	9.77	10.17
Oakland Coliseum (A's)	10.36	10.83
3Com Park (Giants)	9.43	10.03
Pro Player Stadium (Marlins)	8.70	9.32
Shea Stadium (Mets)	8.47	9.44
Comiskey Park (White Sox)	9.85	11.06
Astrodome (Astros)	8.63	10.19
Qualcomm Stadium (Padres)	8.66	10.23
Dodger Stadium (Dodgers)	7.64	9.38

Source: *Sports Illustrated*, "The Coors Curse," July 20, 1998.

a. Find the mean, median, maximum, minimum, and midrange of the combined runs scored by the teams while playing at home and away.
b. Subtract the combined runs scored away from the combined runs scored at home to obtain the difference for each stadium in the list. Then find the mean, median, maximum, minimum, and midrange of the difference between the combined runs scored.
c. Compare each of the measures you found in (a) and (b) against the data collected for just Coors Field, home of the Rockies. What can you conclude?

2.62 You are responsible for planning the parking needed for a new 256-unit apartment complex and you're told to base the needs on the statistic, "average number of vehicles per household is 1.9."
a. Which average (mean, median, mode, midrange) will be helpful to you? Explain.
b. Explain why "1.9" cannot be the median, the mode, or the midrange for the variable "number of vehicles."
c. If the owner wants parking that will accommodate 90% of all the tenants who own vehicles, how many spaces must you plan for?

2.63 Starting with the data values 70 and 100, add three data values to your sample so that the sample has (justify your answer in each case):
a. a mean of 100
b. a median of 70
c. a mode of 87
d. a midrange of 70
e. a mean of 100 and a median of 70
f. a mean of 100 and a mode of 87
g. a mean of 100 and a midrange of 70
h. a mean of 100, a median of 70, and a mode of 87

2.4   Measures of Dispersion

Having located the "middle" with the measures of central tendency, our search for information from data sets now turns to the measures of dispersion (spread). The **measures of dispersion** include the *range, variance,* and *standard deviation*. These numerical values describe the amount of spread, or variability, that is found among the data: closely grouped data have relatively small values, and more widely spread-out data have larger values. The closest possible grouping occurs when the data have no dispersion (all data are the same value); in this situation the measure of dispersion will be zero. There is no limit to how widely spread out the data can be; therefore, measures of dispersion can be very large.

Exercise 2.64
USA Today, 9-23-98, reported on the airline rates for reports of lost luggage per 1000 passengers in July: fewest on Continental, 3.50 and most on Alaska, 8.64. Find the range.

RANGE

The difference in value between the highest-valued (*H*) and the lowest-valued (*L*) pieces of data:

$$range = high\ value - low\ value$$

$$range = H - L \tag{2.4}$$

The sample 3, 3, 5, 6, 8 has a range of

$$H - L = 8 - 3 = 5$$

The range of 5 tells us that these data all fall within a 5-unit interval (see Figure 2.21).

Figure 2.21

Range of {3, 3, 5, 6, 8}

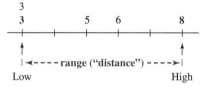

The other measures of dispersion to be studied in this chapter are measures of dispersion about the mean. To develop a measure of dispersion about the mean, let's first answer the question "How far is each *x* from the mean?"

DEVIATION FROM THE MEAN

A deviation from the mean, $x - \bar{x}$, is the difference between the value of *x* and the mean $\bar{x}$.

Each individual value *x* deviates from the mean by an amount equal to $(x - \bar{x})$. This deviation $(x - \bar{x})$ is zero when *x* is equal to the mean $\bar{x}$. The deviation $(x - \bar{x})$ is positive if *x* is larger than $\bar{x}$ and negative if *x* is smaller than $\bar{x}$.

Consider the sample 6, 3, 8, 5, 3. Using formula (2.1), $\bar{x} = \dfrac{\sum x}{n}$, we find that the mean is 5. Each deviation, $(x - \bar{x})$, is then found by subtracting 5 from each *x*-value.

Data	x	6	3	8	5	3
Deviation	$x - \bar{x}$	1	−2	3	0	−2

Figure 2.22 shows the four nonzero deviations from the mean.

Figure 2.22

Deviations from the Mean

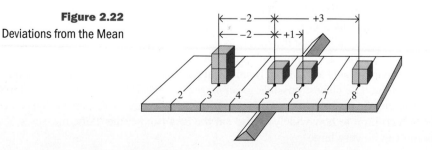

To describe the "average" value of these deviations we might use the mean deviation, the sum of the deviations divided by n, $\dfrac{\sum(x-\bar{x})}{n}$. However, since the sum of the deviations, $\sum(x-\bar{x})$, is exactly zero, the mean deviation will also be zero. As a matter of fact, it will always be zero. Since the mean deviation will always be zero, it will not be a useful statistic.

The sum of the deviations $\sum(x-\bar{x})$ is always zero, because of the neutralizing effect between the deviations of x-values smaller than the mean (which are negative) and of x-values larger than the mean (which are positive). This neutralizing effect can be removed if we do something to make all the deviations positive. This can be accomplished in two ways. First, by using the absolute value of the deviation, $|x-\bar{x}|$, we can treat each deviation for its "size" or distance only. For our illustration we obtain the following *absolute deviations*.

Exercise 2.65

a. The data value $x = 45$ has a deviation value of 12. Explain the meaning of this.
b. The data value $x = 84$ has a deviation value of -20. Explain the meaning of this.

Data	x	6	3	8	5	3
Absolute Value of Deviation:	$\|x-\bar{x}\|$	1	2	3	0	2

Exercise 2.66

The summation $\sum(x-\bar{x})$ is always zero. Why? Think back to the definition of the mean (p. 66) and see if you can justify this statement.

MEAN ABSOLUTE DEVIATION

The mean of the absolute values of the deviations from the mean.

$$\text{mean absolute deviation} = \frac{\textit{sum of (absolute values of deviations)}}{\textit{number}}$$

$$\text{mean absolute deviation} = \frac{\sum|x-\bar{x}|}{n} \qquad (2.5)$$

For our example, the sum of the absolute deviations is 8 $(1 + 2 + 3 + 0 + 2)$ and

$$\text{mean absolute deviation} = \frac{\sum|x-\bar{x}|}{n} = \frac{8}{5} = 1.6$$

Although this particular measure of spread is not used too frequently, it is a measure of dispersion. It tells us the mean "distance" the data is from the mean.

A second way to eliminate the positive-negative neutralizing effect is to square each of the deviations; squared deviations will all be nonnegative (positive or zero) values. The squared deviations are used to find the *variance*.

SAMPLE VARIANCE

The sample variance, s^2, is the mean of the squared deviations, calculated using $n-1$ as the divisor.

$$\text{sample variance:} \quad s\text{-squared} = \frac{\textit{sum of (deviation)}^2}{\textit{number} - 1}$$

$$s^2 = \frac{\sum(x-\bar{x})^2}{n-1} \qquad (2.6)$$

where n is the sample size, that is, the number of data in the sample.

The variance of our sample 6, 3, 8, 5, 3 is found in Table 2.10 using formula (2.6).

TABLE 2.10   Calculating Variance Using Formula (2.6)

Step 1. Find Σx	Step 2. Find $\bar{x}$	Step 3. Find each $x - \bar{x}$	Step 4. Find $\Sigma(x - \bar{x})^2$	Step 5. Sample variance
6	$\bar{x} = \dfrac{\Sigma x}{n}$	$6 - 5 = 1$	$(1)^2 = 1$	$s^2 = \dfrac{\Sigma(x - \bar{x})^2}{n - 1}$
3		$3 - 5 = -2$	$(-2)^2 = 4$	
8	$\bar{x} = \dfrac{25}{5}$	$8 - 5 = 3$	$(3)^2 = 9$	$s^2 = \dfrac{18}{4}$
5		$5 - 5 = 0$	$(0)^2 = 0$	
3	$\bar{x} = 5$	$3 - 5 = -2$	$(-2)^2 = 4$	
$\Sigma x = 25$		$\Sigma(x - \bar{x}) = 0$ ⓒⓚ	$\Sigma(x - \bar{x})^2 = 18$	$s^2 = \textbf{4.5}$

Exercise 2.67
Use formula (2.6) to find the variance for the sample {1, 3, 5, 6, 10}.

NOTES

1. The sum of all the x's is used to find $\bar{x}$.

2. The sum of the deviations, $\Sigma(x - \bar{x})$, is always zero, provided the exact value of $\bar{x}$ is used. Use this as a check in your calculations, as was done in Table 2.10 (denoted by ⓒⓚ).

3. If a rounded value of $\bar{x}$ is used, then the $\Sigma(x - \bar{x})$ will not always be exactly zero. It will, however, be reasonably close to zero.

4. The sum of the squared deviations is found by squaring each deviation, then adding the squared values.

The set of data in Exercise 2.67 is more dispersed than the data in Table 2.10, and therefore its variance is larger. A comparison of these two samples is shown in Figure 2.23.

Figure 2.23
Comparison of Data

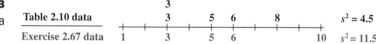

The set of data in Exercise 2.67 is more dispersed than the data in Table 2.10, and therefore its variance is larger.

SAMPLE STANDARD DEVIATION

The standard deviation of a sample, s, is the positive square root of the variance:

sample standard deviation: $s = $ *square root of sample variance*

$$s = \sqrt{s^2} \tag{2.7}$$

For the samples shown in Figure 2.23, the standard deviations are $\sqrt{4.5}$, or **2.1,** and $\sqrt{11.5}$, or **3.4.**

NOTE The numerator for sample variance, $\Sigma(x - \bar{x})^2$, is often called the "sum of squares for x" and symbolized by $SS(x)$. Thus, formula (2.6) can be expressed

sample variance: $s^2 = \dfrac{SS(x)}{n - 1},$ where $SS(x) = \Sigma(x - \bar{x})^2$ (2.8)

The formulas for variance can be modified into other forms for easier use in various situations. For example, suppose that we have the sample 6, 3, 8, 5, 2. The variance for this sample is computed in Table 2.11.

TABLE 2.11   Calculating Variance Using Formula (2.6)

Step 1. Find $\sum x$	Step 2. Find $\bar{x}$	Step 3. Find each $x - \bar{x}$	Step 4. Find $\sum(x - \bar{x})^2$	Step 5. Sample variance
6	$\bar{x} = \dfrac{\sum x}{n}$	$6 - 4.8 = \ \ 1.2$	$(1.2)^2 = \ \ 1.44$	$s^2 = \dfrac{\sum(x - \bar{x})^2}{n - 1}$
3		$3 - 4.8 = -1.8$	$(-1.8)^2 = \ \ 3.24$	
8	$\bar{x} = \dfrac{24}{5}$	$8 - 4.8 = \ \ 3.2$	$(3.2)^2 = 10.24$	$s^2 = \dfrac{22.80}{4}$
5		$5 - 4.8 = \ \ 0.2$	$(0.2)^2 = \ \ 0.04$	
2	$\bar{x} = 4.8$	$2 - 4.8 = -2.8$	$(-2.8)^2 = \ \ 7.84$	$s^2 = \mathbf{5.7}$
$\sum x = 24$		$\sum(x - \bar{x}) = \ \ 0$ ⓒⓚ	$\sum(x - \bar{x})^2 = 22.80$	

The arithmetic for this example has become more complicated because the mean contains nonzero digits to the right of the decimal point. However, the "sum of squares for x," the numerator of formula (2.6), can be rewritten so that $\bar{x}$ is not included:

$$\text{sum of squares:}\quad SS(x) = \sum x^2 - \frac{(\sum x)^2}{n} \qquad\qquad \textbf{(2.9)}$$

Combining formulas (2.8) and (2.9) yields the "shortcut formula" for sample variance:

See page 81 for explanation of icons.

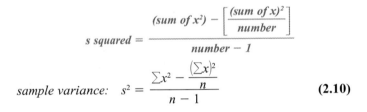

$$s \text{ squared} = \frac{(\text{sum of } x^2) - \left[\dfrac{(\text{sum of } x)^2}{\text{number}}\right]}{\text{number} - 1}$$

$$\text{sample variance:}\quad s^2 = \frac{\sum x^2 - \dfrac{(\sum x)^2}{n}}{n - 1} \qquad\qquad \textbf{(2.10)}$$

Formulas (2.9) and (2.10) are called "shortcut" because they bypass the calculation of $\bar{x}$. The computations for $SS(x)$, s^2, and s using formulas (2.9), (2.10), and (2.7) are performed as shown in Table 2.12.

TABLE 2.12   Calculating Standard Deviation Using the Shortcut Method

Step 1. Find $\sum x$	Step 2. Find $\sum x^2$	Step 3. Find $SS(x)$	Step 4. Find variance	Step 5. Find standard deviation
6	$6^2 = 36$	$SS(x) = \sum x^2 - \dfrac{(\sum x)^2}{n}$	$s^2 = \dfrac{\sum x^2 - \dfrac{(\sum x)^2}{n}}{n - 1}$	$s = \sqrt{s^2}$
3	$3^2 = \ \ 9$			
8	$8^2 = 64$	$SS(x) = 138 - \dfrac{(24)^2}{5}$		
5	$5^2 = 25$		$s^2 = \dfrac{22.80}{4}$	$s = \sqrt{5.7}$
2	$2^2 = \ \ 4$	$SS(x) = 138 - 115.2$		
$\sum x = 24$	$\sum x^2 = 138$	$SS(x) = \mathbf{22.8}$	$s^2 = \mathbf{5.7}$	$s = \mathbf{2.4}$

Exercise 2.68

Use formula (2.10) to find the variance of the sample {1, 3, 5, 6, 10}. Compare the results to Exercise 2.67.

The unit of measure for the standard deviation is the same as the unit of measure for the data. For example, if our data are in pounds, then the standard deviation s will also be in pounds. The unit of measure for variance might then be thought of as *units squared*. In our example of pounds, this would be *pounds squared*. As you can see, the unit has very little meaning.

The following commands will calculate the standard deviation.

MINITAB

Input the data into C1; then continue with:

    Choose: **Calc > Column Statistics**
    Select: **Standard deviation**
    Enter: Input variable: **C1**

EXCEL

Input the data into column A and activate a cell for the answer; then continue with:

    Choose: **Paste Function,** f_x **> Statistical > STDEV > OK**
    Enter: Number 1: **(A2:A6 or select cells)**

TI-83

Input the data into L1; then continue with:

    Choose: **2nd > LIST > Math > 7:StdDev(**
    Enter: **L1**

The following commands will calculate additional statistics.

MINITAB

Input the data into C1, then continue with:

    Choose: **Calc > Column Statistics**
              Then one at a time select the desired statistic
    Select: **N total** Number of data in column
              Sum   Sum of the data in column
              Minimum Smallest value in column
              Maximum Largest value in column
              Range   Range of values in column
              SSQ   Sum of squared x-values, $\sum x^2$
    Enter: Input variable: **C1**

EXCEL

Input the data into column A and activate a cell for the answer; then continue with:

    Choose: **Paste Function,** f_x
              > Statistical > COUNT
                               > MIN
                               > MAX
              OR   **> All > SUM**
                          > SUMSQ
    Enter: Number 1: **(A2:A6 or select cells)**

For range, write a formula:
Max( ) — Min( )

TI-83

Input the data into L1; then continue with:

    Choose: **2nd > LIST > Math > 5:sum(**
                               > 1:min(
                               > 2:max(
    Enter: **L1**

MULTIPLE FORMULAS

The reason we have multiple formulas is for convenience; that is, convenience relative to the situation. The following statements will help you decide which formula to use.

1. When you are working on a computer and using statistical software, you will generally store all the data values first. The computer is very capable with repeated operations and can "revisit" the stored data as often as necessary to complete a procedure. The computations for sample variance will be done using formula (2.6) following the process shown in Table 2.10.

2. When you are working on a calculator with built-in statistical functions, the calculator must perform all necessary operations on each piece of data as the data are entered (most handheld nongraphing calculators do not have the ability to store data). Then after all data have been entered, the computations will be completed using the appropriate summations. The computations for sample variance will be completed using formula (2.10), following the procedure shown in Table 2.12.

3. If you are doing the computations either by hand or with the aid of a calculator, but not using statistical functions, the most convenient formula to use will depend on how many data there are and how convenient the numerical values are to work with.

When a formula has multiple forms, look for one of these icons:

will be used to identify the formula most likely to be used by a computer.

will be used to identify the formula most likely to be used by a calculator.

will be used to identify the formula most likely to be convenient for hand calculations.

will be used to identify the "definition" formula.

EXERCISES

2.69 Consider the sample 2, 4, 7, 8, 9. Find the following:
a. range   **b.** variance s^2, using formula (2.6)   **c.** standard deviation, s

2.70 Consider the sample 6, 8, 7, 5, 3, 7. Find the following:
a. range   **b.** variance s^2, using formula (2.6)   **c.** standard deviation, s

2.71 Given the sample 7, 6, 10, 7, 5, 9, 3, 7, 5, 13, find the following:
a. variance s^2, using formula (2.6)   **b.** variance s^2, using formula (2.10)
c. standard deviation s

2.72 Fifteen randomly selected college students were asked to state the number of hours they slept last night. The resulting data are 5, 6, 6, 8, 7, 7, 9, 5, 4, 8, 11, 6, 7, 8, 7. Find the following:
a. variance s^2, using formula (2.6)   **b.** variance s^2, using formula (2.10)
c. standard deviation s

2.73 An article titled "Financing Your Kids' College Education" (*Farming,* Sept./Oct. 1994) listed the following in-state tuition and fees per school year for 14 land-grant universities: 1554, 2291, 2084, 4443, 2884, 2478, 3087, 3708, 2510, 2055, 3000, 2052, 2550, 2013. Find the following:
a. variance s^2 **b.** standard deviation s

2.74 Adding (or subtracting) the same number from each value in a set of data does not affect the measures of variability for that set of data.

    a. Find the variance of the following set of annual heating-degree-day data:

        6017, 6173, 6275, 6350, 6001, 6300

    b. Find the variance of the following set of data [obtained by subtracting 6000 from each value in (a)]:

        17, 173, 275, 350, 1, 300

2.75 Recruits for a police academy were required to undergo a test that measures their exercise capacity. The exercise capacity (measured in minutes) was obtained for each of 20 recruits:

25	27	30	33	30	32	30	34	30	27
26	25	29	31	31	32	34	32	33	30

    a. Draw a dotplot of the data.
    b. Find the mean.
    c. Find the range.
    d. Find the variance.
    e. Find the standard deviation.
    f. Using the dotplot from part (a): (1) draw a line representing the range, and (2) draw a line starting at the mean whose length represents the value of the standard deviation.
    g. Describe how the distribution of data, the range, and the standard deviation are related.

2.76 Since 1981, *Fortune* magazine has been tracking what they judge to be the "best 100 companies to work for." The companies must be at least ten years old and employ no less than 500 people. Below are the top 25 from the list compiled in 1998, together with each company's percentage of females, percentage of job growth over a 2-year span, and number of hours of professional training required each year by the employer.

Many of these data sets are on the StatSource CD.

Company Name	Women (%)	Job Growth (%)	Training (hr/yr.)
Southwest Airlines	55	26	15
Kingston Technology	48	54	100
SAS Institute	53	34	32
FEL-Pro	36	10	60
TDIndustries	10	31	40
MBNA	58	48	48
W. L. Gore	43	26	27
Microsoft	29	22	8
Merck	52	24	40
Hewlett-Packard	37	10	—
Synovus Financial	65	23	13
Goldman Sachs	40	13	20
MOOG	19	17	25
DeLoitte & Touche	45	23	70
Corning	38	9	80
Wegmans Food Products	54	3	30
Harley-Davidson	22	15	50
Federal Express	32	11	40
Procter & Gamble	40	1	25
Peoplesoft	44	122	—
First Tennessee Bank	70	1	60
J. M. Smucker	48	1	24
Granite Rock	17	29	43
Patagonia	52	−5	62
Cisco Systems	25	189	80

Source: Fortune, "The 100 Best Companies to Work For in America," January 12, 1998.

    a. Find the mean, range, variance, and standard deviation for each of the three variables shown in the list. Present your results in a table.

    b. Using your results from (a), compare the distributions for job growth percentage and percentage of women employed. What can you conclude?

2.77 **a.** Find the range and the standard deviation for the number of days per year of school, using the data in Exercise 2.56 (p. 72).

    b. Draw lines on the stem-and-leaf diagram drawn in answering Exercise 2.56 that represent the range and the standard deviation. Remember, the standard deviation is a measure of the spread about the mean.

    c. Describe the relationship among the distribution of the data, the range, and the standard deviation.

2.78 *Financial Times* publishes the midday temperatures at well-known cities and nations located throughout the world. The table below shows the midday temperature in °F on August 4, 1998, in 100 world locations:

Location	°F	Location	°F	Location	°F	Location	°F
Abu Dhabi	115	Cologne	73	Lisbon	99	Paris	75
Accra	82	Dakar	88	London	73	Perth	63
Algiers	90	Dallas	98	Los Angeles	81	Prague	75
Amsterdam	68	Delhi	86	Luxembourg	72	Rangoon	88
Athens	90	Dubai	111	Lyon	75	Reykjavik	55
Atlanta	90	Dublin	70	Madeira	81	Rio	79
Bangkok	91	Dubrovnik	95	Madrid	93	Rome	90
Barcelona	86	Edinburgh	78	Majorca	86	San Francisco	80
Beijing	86	Faro	91	Malta	93	Seoul	90
Belfast	70	Frankfurt	77	Manchester	68	Singapore	88
Belgrade	99	Geneva	77	Manila	91	Stockholm	68
Berlin	77	Gibraltar	81	Melbourne	61	Strasbourg	77
Bermuda	88	Glasgow	63	Mexico City	86	Sidney	66
Birmingham	70	Hamburg	68	Miami	92	Tangiers	86
Bogota	68	Helsinki	68	Milan	88	Tel Aviv	91
Bombay	90	Hong Kong	91	Montreal	82	Tokyo	86
Brussels	70	Honolulu	90	Moscow	79	Toronto	83
Budapest	97	Istanbul	95	Munich	75	Vancouver	76
Buenos Aires	59	Jakarta	90	Nairobi	73	Venice	91
Copenhagen	68	Jersey	66	Naples	93	Vienna	91
Cairo	99	Johannesburg	59	Nassau	99	Warsaw	77
Caracas	90	Karachi	99	New York	84	Washington	86
Cardiff	68	Kuwait	120	Nice	82	Wellington	54
Casablanca	84	Las Palmas	82	Nicosia	93	Winnipeg	79
Chicago	79	Lima	66	Oslo	70	Zurich	73

Source: Financial Times, August 4, 1998.

    Find the mean, range, variance, and standard deviation for the temperatures of the cities in the published list. (Retain solutions for use in answering Exercise 2.100, p. 91.)

2.79 Consider the following two sets of data:

    Set 1:   46   55   50   47   52
    Set 2:   30   55   65   47   53

    Both sets have the same mean, which is 50. Compare these measures for both sets: $\sum(x - \bar{x})$, $\sum|x - \bar{x}|$, SS(x), and range. Comment on the meaning of these comparisons.

2.80 Comment on the following statement: "The mean loss for customers at First State Bank (which was not insured) was $150. The standard deviation of the losses was −$125."

2.81 Start with $x = 100$ and add 4 x-values, making a sample of five data such that
    a. $s = 0$.   **b.** $0 < s < 1$.   **c.** $5 < s < 10$.   **d.** $20 < s < 30$.

2.82 Each of two samples has a standard deviation of 5. If the two sets of data are made into one set of ten data, will the new sample have a standard deviation that is: (a) less than, (b) about the same as , or (c) larger than the original standard deviation of 5? Make up two sets of five data each whose standard deviation is 5 and justify your answer. Include the calculations.

2.5 $\mathbf{M}$ean and Standard Deviation of Frequency Distribution

When the sample data are in the form of a frequency distribution , we will need to make a slight adaptation to formulas (2.1) and (2.10) in order to find the mean, the variance, and the standard deviation.

ILLUSTRATION 2.10 ▼

Find the mean, the variance, and the standard deviation for the sample data represented by the frequency distribution in Table 2.13.

TABLE 2.13 Ungrouped Frequency Distribution

x	f
1	5
2	9
3	8
4	6
	$\Sigma f = 28$

REMEMBER This frequency distribution represents a sample of 28 values: five 1's, nine 2's, eight 3's, and six 4's.

In order to calculate the sample mean $\bar{x}$ and sample variance s^2 using formulas (2.1) and (2.10), we need the sum of the 28 x-values, Σx, and the sum of the 28 x-squared values, Σx^2. The summations, Σx and Σx^2, could be found as follows:

$$\Sigma x = \underbrace{1 + 1 + \cdots + 1}_{5 \text{ of them}} + \underbrace{2 + 2 + \cdots + 2}_{9 \text{ of them}} + \underbrace{3 + 3 + \cdots + 3}_{8 \text{ of them}} + \underbrace{4 + 4 + \cdots + 4}_{6 \text{ of them}}$$

$$\Sigma x = (5)(1) + (9)(2) + (8)(3) + (6)(4)$$

$$\Sigma x = 5 + 18 + 24 + 24 = \mathbf{71}$$

$$\Sigma x^2 = \underbrace{1^2 + \cdots + 1^2}_{5 \text{ of them}} + \underbrace{2^2 + \cdots + 2^2}_{9 \text{ of them}} + \underbrace{3^2 + \cdots + 3^2}_{8 \text{ of them}} + \underbrace{4^2 + \cdots + 4^2}_{6 \text{ of them}}$$

$$\Sigma x^2 = (5)(1) + (9)(4) + (8)(9) + (6)(16)$$

$$\Sigma x^2 = 5 + 36 + 72 + 96 = \mathbf{209}$$

However, we will use the frequency distribution to determine these summations by expanding it to become an extensions table . The extensions xf and $x^2 f$ are formed by multiplying across the columns row by row; then adding to find three column totals. The objective of the extensions table is to obtain these three column totals. (See Table 2.14.)

TABLE 2.14 Ungrouped Frequency Distribution: Extensions xf and $x^2 f$

x	f	xf	$x^2 f$
1	5	5	5
2	9	18	36
3	8	24	72
4	6	24	96
	$\Sigma f = 28$	$\Sigma xf = 71$	$\Sigma x^2 f = 209$ ◄── *sum of x^2, using frequencies*

number of data   *sum of x, using frequencies*

Exercise 2.83

A survey asking the "number of telephones" per household, x, was conducted; results are shown here as a frequency distribution.

x	f
0	1
1	3
2	8
3	5
4	3

a. Complete the extensions table.
b. Find the three summations, Σf, Σxf, $\Sigma x^2 f$, for the frequency distribution.
c. Describe what each of the following represents: $x = 4$, $f = 8$, Σf, Σxf.

NOTES

1. The extensions in the xf column are the subtotals of the like values.
2. The extensions in the x^2f column are the subtotals of the like squared values.
3. The three column totals, Σf, Σxf, and $\Sigma x^2 f$, are the same values as were previously known as n, Σx, and Σx^2, respectively. That is, $\Sigma f = n$, the number of pieces of data; $\Sigma xf = \Sigma x$, the sum of the data; and $\Sigma x^2 f = \Sigma x^2$, the sum of the squared data.
4. Think of the f in the summation expressions Σxf and $\Sigma x^2 f$ as an indication that the sums were obtained with the use of a frequency distribution.
5. The sum of the x-column is NOT a meaningful number. The x-column lists each possible value of x once, not accounting for the repeated values.

> **Exercise 2.84**
>
> Explain why (a) the "sum of the x column" is not the same as the "sum of the data" and (b) the "Σxf" represents the "sum of the data" represented by the frequency distribution in Exercise 2.83.

To find the **mean** of a frequency distribution, formula (2.1) on page 66 is modified to indicate the use of the frequency distribution.

$$x\ bar = \frac{sum\ of\ x,\ using\ frequencies}{number,\ using\ frequencies}$$

mean of frequency distribution: $\quad \bar{x} = \dfrac{\Sigma xf}{\Sigma f}$ **(2.11)**

The mean value of x for the frequency distribution in Table 2.14 is found by using formula (2.11).

$$\text{mean:} \quad \bar{x} = \frac{\Sigma xf}{\Sigma f} = \frac{71}{28} = 2.536 = \mathbf{2.5}$$

> **Exercise 2.85**
>
> Find the mean of the data shown in the frequency distribution in Exercise 2.83.

To find the **variance** of the frequency distribution, formula (2.10) is modified to indicate the use of the frequency distribution.

$$s\ squared = \frac{(sum\ of\ x^2,\ using\ frequencies) - \left[\dfrac{(sum\ of\ x,\ using\ frequencies)^2}{number,\ using\ frequencies}\right]}{number,\ using\ frequencies - 1}$$

$$s^2 = \frac{\Sigma x^2 f - \dfrac{(\Sigma xf)^2}{\Sigma f}}{\Sigma f - 1} \tag{2.12}$$

> **Exercise 2.86**
>
> Find the variance for the data shown in the frequency distribution in Exercise 2.83.

The variance of x for the frequency distribution in Table 2.14 is found by using formula (2.12).

$$\text{variance: } s^2 = \frac{\Sigma x^2 f - \dfrac{(\Sigma xf)^2}{\Sigma f}}{\Sigma f - 1} = \frac{209 - \dfrac{(71)^2}{28}}{28 - 1} = \frac{28.964}{27} = 1.073 = \mathbf{1.1}$$

> **Exercise 2.87**
>
> Find the standard deviation for the data shown in the frequency distribution in Exercise 2.83.

The **standard deviation** of x for the frequency distribution in Table 2.14 is found by using formula (2.7), the positive square root of variance.

$$\text{standard deviation: } s = \sqrt{s^2} = \sqrt{1.073} = 1.036 = \mathbf{1.0}$$

ILLUSTRATION 2.11 ▼

Find the mean, variance, and standard deviation of the sample of 50 exam scores using the grouped frequency distribution found in Table 2.7 (p. 54).

Solution

We will use an extensions table to find the three summations in the same manner we did in Illustration 2.10. The class midpoints will be used as the representative values for the classes.

TABLE 2.15   Frequency Distribution of 50 Exam Scores

Class Number	Class Midpoints, x	f	xf	x^2f
1	40	2	80	3,200
2	50	2	100	5,000
3	60	7	420	25,200
4	70	13	910	63,700
5	80	11	880	70,400
6	90	11	990	89,100
7	100	4	400	40,000
		$\sum f = 50$	$\sum xf = 3780$	$\sum x^2f = 296,600$

The mean value of x for the frequency distribution in Table 2.15 is found by using formula (2.11).

$$\text{mean: } \bar{x} = \frac{\sum xf}{\sum f} = \frac{3780}{50} = \textbf{75.6}$$

The variance of x for the frequency distribution in Table 2.15 is found by using formula (2.12).

$$\text{variance: } s^2 = \frac{\sum x^2f - \dfrac{(\sum xf)^2}{\sum f}}{\sum f - 1} = \frac{296,600 - \dfrac{3780^2}{50}}{50 - 1} = \frac{10,832}{49} = 221.0612 = \textbf{221.1}$$

The standard deviation of x for the frequency distribution in Table 2.15 is found by using formula (2.7).

$$\text{standard deviation: } s = \sqrt{s^2} = \sqrt{221.0612} = 14.868 = \textbf{14.9}$$

▲

Exercise 2.88

Find the mean, variance, and standard deviation of the data shown in the following frequency distribution.

Class	f
2–6	2
6–10	10
10–14	12
14–18	9
18–22	7

The following commands will find the mean, variance, and standard deviation of a frequency distribution when class midpoints or specific data values are used for x.

MINITAB

Input the class midpoints or data values into C1 and the corresponding frequencies into C2; then continue with the following commands to obtain the extensions table:

Choose: **Calc > Calculator . . .**
Enter: Store result in
          variable: **C3**
          Expression: **C1*C2 > OK**

Repeat above commands replacing the variable with **C4** and the expression with **C1*C3.**

Choose: **Calc > Column Statistics**
Select: **Sum**
Enter: Input variable: **C2**
          Store result in: **K1 > OK**

Repeat above 'sum' commands replacing variable with **C3** and result with **K2.**
Repeat above 'sum' commands replacing variable with **C4** and result with **K3.**

Choose: **Manip > Display data**
Enter: Columns to display: **C1-C4**
          K1-K3

To find the mean, variance, and standard deviation respectively; continue with:

Choose: **Calc > Calculator**
Enter: Store result in variable: **K4**
          Expression: **K2/K1**

Repeat above 'mean' commands replacing variable with **K5** and expression with **(K3-(K2**2/K1))/(K1-1).**
Repeat above 'mean' commands replacing variable with **K6** and expression with **SQRT(K5).**
(select square root from functions)

Choose: **Manip > Display data**
Enter: Columns to display: **K4-K6**

EXCEL

Input the class midpoints or data values into column A and the corresponding frequencies into column B; activate C1 or C2 (depending on whether column headings are used or not), then continue with the following commands to obtain the extensions table:

Choose: **Edit Formula (=)**
Enter: **A2*B2** (if column headings are used)
Drag: **Bottom right corner of C2 down to give other products**

Activate **D2** and repeat above commands replacing the formula with **A2*C2.**
Activate the data in columns B, C, and D.

Choose: **AutoSum** (sums will appear at the bottom of the columns)

To find the mean, activate **E2;** then continue with:

Choose: **Edit Formula (=)**
Enter: **(column C total/column B total) (ex. C9/B9)**

To find the variance, activate **E3** and repeat above 'mean' commands replacing the formula with **(D9-(C9^2/B9))/(B9-1).**

To find the standard deviation, activate **E4** and repeat above 'mean' commands replacing the formula with **SQRT(E3).**

TI-83

Input the class midpoints or data values into L1 and the frequencies into L2; then continue with:

Highlight: **L3**
Enter: L3=: **L1*L2**
Highlight: **L4**
Enter: L4=: **L1*L3**
Highlight: **L5(1)** (first position in L5 column)
Enter: L5(1)=: **sum(L2)** $[\sum f]$ [sum = 2nd LIST > MATH >
                                  5:sum(]
            L5(2)=: **sum(L3)** $[\sum xf]$
            L5(3)=: **sum(L4)** $[\sum x^2 f]$
            L5(4)=: **L5(2)/L5(1)**   [to find mean]
            L5(5)=: **(L5(3)-((L5(2))²/L5(1)))/(L5(1)-1)**
                    [to find variance]
            L5(6)=: **2nd $\sqrt{\ }$ (L5(5))** [to find standard
                    deviation]

If the extensions table is not needed, just use:

Choose: **STAT > CALC > 1:1-VAR STATS**
Enter: **L1, L2**

EXERCISES • • • • • • • • • • • • •

2.89 A survey of medical doctors asked the number of children each had fathered. The results are summarized by this ungrouped frequency distribution:

Number of Children	0	1	2	3	4	6
Number of Doctors	15	12	26	14	4	2

Calculate the sample mean, variance, and standard deviation for the number of children the doctors had fathered.

2.90 The weight gains (in grams) for chicks fed on a high-protein diet were as follows:

Weight Gain	Frequency
12.5	2
12.7	6
13.0	22
13.1	29
13.2	12
13.8	4

a. Find the mean.
b. Find the variance.
c. Find the standard deviation.

2.91 A 1993 issue of *Library Journal* (Vol. 118, No. 17) gives the following table for the salaries of minority placements by type of library.   *(continued)*

Library Type	Number	Average Salary
Academic	46	$27,825
Public	34	24,657
School	23	30,336
Special	16	29,406
Other	4	25,200

  a. Find the total of all salaries for the above 123 individuals.
  b. Find the mean salary for the above 123 individuals.
  c. What is the modal library type? Explain.
  d. Find the standard deviation for the above 123 salaries.

2.92 Pediatric dentists say a child's first dental exam should occur between age 6 months and 1 year. The ages at first dental exam for a sample of children is shown in the distribution.

Age at first dental exam, x	1	2	3	4	5
Number of children, f	9	11	23	16	21

  a. Find the mean age of first dental exam for these children.
  b. Find the median age.
  c. Find the standard deviation.

2.93 Find the mean, variance, and standard deviation for the following grouped frequency distribution.

Class Boundaries	f
3–6	2
6–9	10
9–12	12
12–15	9
15–18	7

2.94 Find the mean and the variance for the following grouped frequency distribution.

Class Boundaries	f
2–6	7
6–10	15
10–14	22
14–18	14
18–22	2

2.95 The following distribution of commuting distances was obtained for a sample of Mutual of Nebraska employees.

Distance (miles)	Frequency
1.0–3.0	2
3.0–5.0	6
5.0–7.0	12
7.0–9.0	50
9.0–11.0	35
11.0–13.0	15
13.0–15.0	5

Find the mean and standard deviation for the commuting distances.

2.96 A quality-control technician selected 25 one-pound boxes from a production process and found the following distribution of weights (in ounces).

Weight	Frequency
15.95–15.98	2
15.98–16.01	4
16.01–16.04	15
16.04–16.07	3
16.07–16.10	1

Find the mean and standard deviation weight for this distribution.

2.97 There are 35.2 million Americans 16 years and older who fish our waters. A sample of fresh water fisherman produced the following distribution for their ages.

Age of fisherman, x	15–25	25–35	35–45	45–55	55–65	65–75
Number of fisherman, f	13	20	28	20	10	9

Find the mean and standard deviation for this distribution.

2.98 Private industry reported that over 31,000 workers were absent from work for carpal tunnel syndrome (a nerve disorder causing arm, wrist, and hand pain) in 1995. The lengths of time (in days) workers are absent due to this problem vary greatly.

Length of absence, x	0–10	10–20	20–30	30–40	40–50
Number of workers	37	24	38	32	27

Find the mean and standard deviation for this distribution.

2.99 A Senior PGA professional golfer is not expected to play in all the available tournaments in the course of a season. Midway through the 1998 tour, *USA Today* published a list of the top 50 money leaders that showed the number of tournaments each had played:

Senior PGA Player	Tournaments	Senior PGA Player	Tournaments
Hale Irwin	13	Tom Wargo	19
Gil Morgan	15	Jim Dent	18
Larry Nelson	16	Walter Morgan	21
Jay Sige	19	Bob Murphy	18
Dave Stockton	18	Terry Dill	20
J. M. Canizares	18	Dave Eichelberger	19
B. Summerhays	21	George Archer	18
Jim Colbert	19	Bob Eastwood	18
V. Fernandez	15	Brian Barnes	16
David Graham	18	Walter Hall	13
John Jacobs	19	John Bland	16
Isao Aoki	11	Frank Conner	20
M. McCullough	21	Simon Hobday	17
Dana Quigley	22	Gibby Gilbert	17
Bob Duval	19	Bob Charles	17
Ray Floyd	14	Tom Jenkins	13
Hugh Baiocchi	20	Gary Player	11
Hubert Green	15	Larry Ziegler	17
Jim Albus	20	Tom Shaw	21
L. Thompson	21	Kermit Zarley	16
Graham Marsh	17	D. Lundstrom	19

(continued)

J. C. Snead	19	Bud Allin	19
Bob Dickson	20	John Morgan	14
Dale Douglass	17	Al Geiberger	14
Lee Trevino	16	Ed Dougherty	9

Source: USA Today, July 23, 1998.

a. Construct a grouped frequency distribution showing the number of tournaments played using group intervals 9–11, 11–13, . . . , 21–23, the class midpoints, and the associated frequency counts.

b. Find the mean, variance, and standard deviation of the number of tournaments played both with and without using the grouped distribution.

c. Compare the two sets of answers you obtained in (b). What percent is the error in each case?

 2.100 a. Using the worldwide midday temperature readings data shown in Exercise 2.78, page 83, construct a grouped frequency distribution using 10 °F class boundaries of 51, 61, 71, . . . , 121. Show the class midpoints and the associated frequency counts in your table.

b. Find the mean, standard deviation, and variance of the grouped frequency distribution.

c. Compare these values against the corresponding ungrouped statistics using the percent error in each case, and present all results in a table.

2.101 The USA Snapshot® "Nuns an aging order" reports that the median age of the USA's 94,022 Roman Catholic nuns is 65 years and that the percentage of nuns in the country by age group is:

Under 50	51–70	Over 70	Refused to give age
16%	42%	37%	5%

This information is based on a survey of 1049 Roman Catholic nuns.

Suppose the survey had resulted in the frequency distribution shown below (52 ages unknown).

Age	20–30	30–40	40–50	50–60	60–70	70–80	80–90
Freq.	34	58	76	187	254	241	147

(See histogram drawn as answer to Exercise 2.38, p. 64.)

a. Find the mean, median, mode, and midrange for this distribution of ages.

b. Find the variance and standard deviation.

2.102 The amount of money adults say they will spend on gifts during this holiday was described in "What 'Santa' Will Spend," *USA Today,* 11-23-94.

Amount	Nothing	$1–$300	$301–$600	$601–$1000	Over $1000	Didn't Know
Percent who said	1%	24%	30%	20%	14%	11%
			Average: $734			

don't know # of people

Let the following distribution represent that part of the sample who did know:

Amount, x	0	150	450	800	1500 ← from midpt)
Frequency	1	24	30	20	14

 # people

a. Find the mean of the frequency distribution.

b. Do you believe the average reported could have been the mean? Explain.

c. Find the median of the frequency distribution.

(continued)

Many of these data sets are on the StatSource CD.

d. Do you believe the average reported could have been the median? Explain.
e. Find the mode of the frequency distribution.
f. Do you believe the average reported could have been the mode? Explain.
g. Could the average reported have been the midrange? If so, what was the largest amount of money reported?

2.6   Measures of Position

Measures of position are used to describe the position a specific data value possesses in relation to the rest of the data. *Quartiles* and *percentiles* are two of the most popular measures of position.

QUARTILES

Values of the variable that divide the ranked data into quarters; each set of data has three quartiles. The first quartile, Q_1, is a number such that at most 25% of the data are smaller in value than Q_1 and at most 75% are larger. The second quartile is the median. The third quartile, Q_3, is a number such that at most 75% of the data are smaller in value than Q_3 and at most 25% are larger (see Figure 2.24).

Figure 2.24
Quartiles

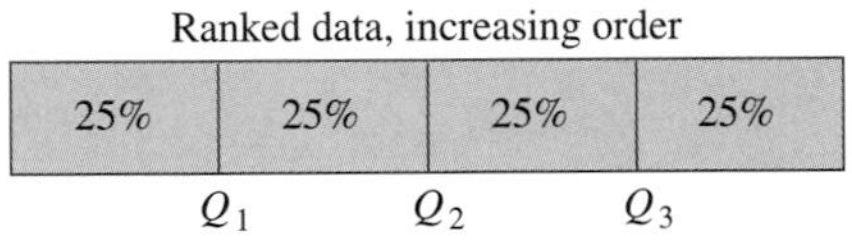

The procedure for determining the value of the quartiles is the same as that for percentiles and is shown in the following description of *percentiles*.

PERCENTILES

Values of the variable that divide a set of ranked data into 100 equal subsets; each set of data has 99 percentiles (see Figure 2.25). The kth percentile, P_k, is a value such that at most k% of the data are smaller in value than P_k and at most $(100 - k)$% of the data are larger (see Figure 2.26).

Figure 2.25
Percentiles

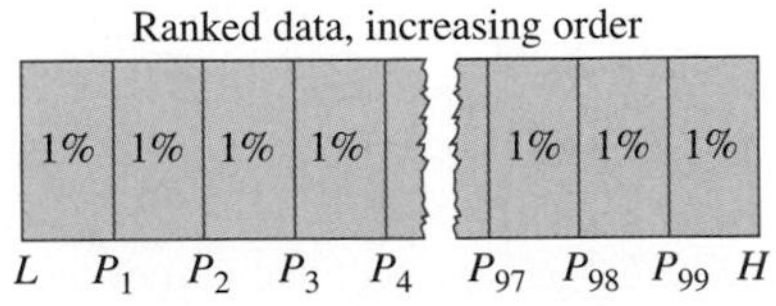

Figure 2.26
kth Percentile

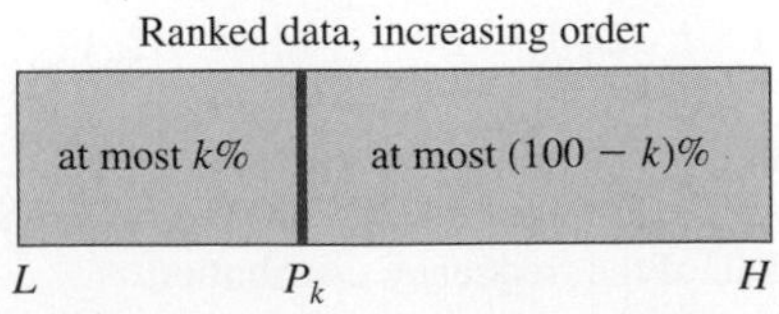

NOTES

1. The 1st quartile and the 25th percentile are the same; that is, $Q_1 = P_{25}$. Also, $Q_3 = P_{75}$.
2. The median, the 2nd quartile, and the 50th percentile are all the same: $\tilde{x} = Q_2 = P_{50}$. Therefore, when asked to find P_{50} or Q_2, use the procedure for finding the median.

The procedure for determining the value of any kth percentile (or quartile) involves four basic steps as outlined on the diagram in Figure 2.27. Illustration 2.12 demonstrates the procedure.

Figure 2.27
Finding P_k Procedure

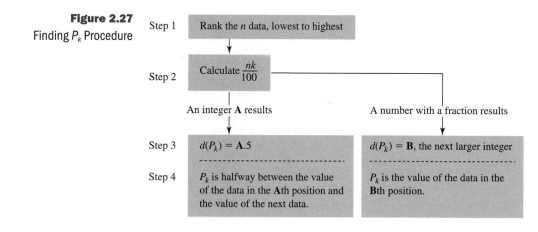

ILLUSTRATION 2.12 ▼

Using the sample of 50 elementary statistics final exam scores listed in Table 2.16, find the first quartile Q_1, the 58th percentile P_{58}, and the third quartile Q_3.

TABLE 2.16   Raw Scores for Elementary Statistics Exam

60	47	82	95	88	72	67	66	68	98
90	77	86	58	64	95	74	72	88	74
77	39	90	63	68	97	70	64	70	70
58	78	89	44	55	85	82	83	72	77
72	86	50	94	92	80	91	75	76	78

Solution

Exercise 2.103

Using the concept of depth, describe the position of 91 in the set of 50 exam scores in two different ways.

STEP 1   Rank the data: A ranked list may be formulated (see Table 2.17), or a graphic display showing the ranked data may be used. The dotplot or the stem-and-leaf are handy for this purpose. The stem-and-leaf is especially helpful since it gives depth numbers counted from both extremes when it is computer generated (see Figure 2.28).

Find Q_1:

STEP 2   Find $\dfrac{nk}{100}$: $\dfrac{nk}{100} = \dfrac{(50)(25)}{100} = \mathbf{12.5}$ ($n = 50$ and $k = 25$, since $Q_1 = P_{25}$.)

STEP 3   Find the depth of Q_1: $d(Q_1) = \mathbf{13}$ (Since 12.5 contains a fraction, **B** is the next larger integer, 13.)

(continued)

TABLE 2.17   Ranked Data: Exam Scores

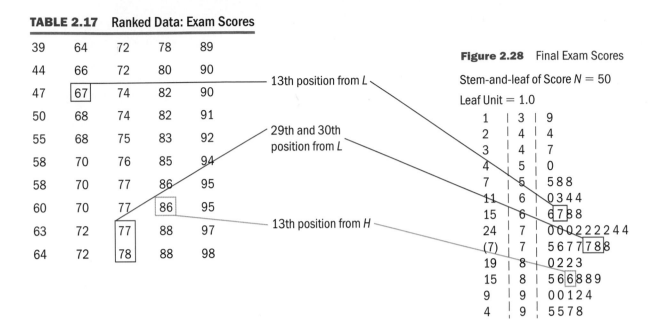

Figure 2.28   Final Exam Scores

Exercise 2.104
Find P_{20} and P_{35} for the exam scores in Table 2.17.

STEP 4   Find Q_1: Q_1 **is the 13th value, counting from L** (see Table 2.17 or Figure 2.28).
$Q_1 = 67$

Find P_{58}:

STEP 2   Find $\dfrac{nk}{100}$: $\dfrac{nk}{100} = \dfrac{(50)(58)}{100} = 29$ ($n = 50$ and $k = 58$, since P_{58}.)

STEP 3   Find the depth of P_{58}: $d(P_{58}) = 29.5$ [Since **A** = 29 (an integer), add .5, use 29.5.]

STEP 4   Find P_{58}: P_{58} is the value halfway between the values of the 29th and the 30th pieces of data counting from L (see Table 2.17 or Figure 2.28).

$$P_{58} = \frac{77 + 78}{2} = 77.5$$

Optional technique: When k is greater than 50, subtract k from 100 and use $(100 - k)$ in place of k in step 2. The depth is then counted from the largest valued data H.

Find Q_3 using the optional technique:

STEP 2   Find $\dfrac{nk}{100}$: $\dfrac{nk}{100} = \dfrac{(50)(25)}{100} = 12.5$ ($n = 50$ and $k = 75$, since $Q_3 = P_{75}$, and $k > 50$; use $100 - k = 100 - 75 = 25$.)

Exercise 2.105
Find P_{80} and P_{95} for the exam scores in Table 2.17.

STEP 3   Find the depth of Q_3 from H: $d(Q_3) = 13$

STEP 4   Find Q_3: Q_3 is the 13th value, counting from H (see Table 2.17 or Figure 2.28).
$Q_3 = 86$ ▲

An additional measure of central tendency, the *midquartile,* can now be defined.

MIDQUARTILE

The numerical value midway between the first quartile and the third quartile.

$$\text{midquartile} = \frac{Q_1 + Q_3}{2} \qquad\qquad\qquad \textbf{(2.13)}$$

ILLUSTRATION 2.13 ▼

Find the midquartile for the set of 50 exam scores given in Illustration 2.12.

Solution

$Q_1 = 67$ and $Q_3 = 86$, as found in Illustration 2.12. Thus,

$$\text{midquartile} = \frac{Q_1 + Q_3}{2} = \frac{67 + 86}{2} = \textbf{76.5}$$ ▲

The median, the midrange, and the midquartile are not necessarily the same value. They are each middle values, but by different definitions of the middle. Figure 2.29 summarizes the relationship of these three statistics as applied to the 50 exam scores from Illustration 2.12.

Figure 2.29
Final Exam Scores

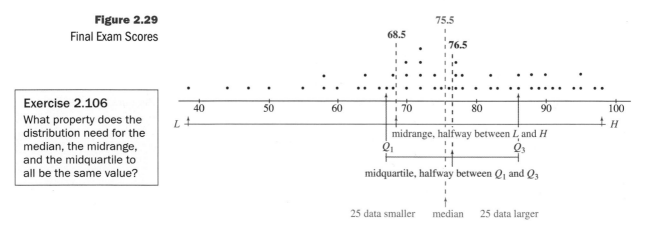

Exercise 2.106
What property does the distribution need for the median, the midrange, and the midquartile to all be the same value?

A 5-*number summary* is very effective in describing a set of data. It is easy information to obtain and is very informative to the reader.

5-NUMBER SUMMARY

The 5-number summary is composed of:
1. *L*, the smallest value in the data set,
2. Q_1, the first quartile (also called P_{25}, 25th percentile),
3. $\tilde{x}$, the median,
4. Q_3, the third quartile (also called P_{75}, 75th percentile),
5. *H*, the largest value in the data set.

The 5-number summary for the set of 50 exam scores in Illustration 2.12 is

39	67	75.5	86	98
L	Q_1	$\tilde{x}$	Q_3	H

Notice that these five numerical values divide the set of data into four subsets, with one-quarter of the data in each subset. From the 5-number summary we can observe how much the data are spread out in each of the quarters. An additional measure of dispersion can now be defined.

INTERQUARTILE RANGE

The difference between the first and third quartiles. It is the range of the middle 50% of the data.

The 5-number summary is even more informative when it is displayed on a diagram drawn to scale. One of the computer-generated graphic displays that accomplishes this is known as the *box-and-whiskers display.*

BOX-AND-WHISKERS DISPLAY

A graphic representation of the 5-number summary. The five numerical values (smallest, first quartile, median, third quartile, and largest) are located on a scale, either vertical or horizontal. The box is used to depict the middle half of the data that lies between the two quartiles. The whiskers are line segments used to depict the other half of the data: One line segment represents the quarter of the data that is smaller in value than the first quartile, and a second line segment represents the quarter of the data that is larger in value than the third quartile.

Figure 2.30 pictures a box-and-whiskers display of the 50 exam scores.

Figure 2.30
Box-and-Whiskers

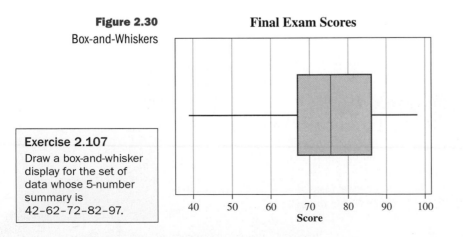

Final Exam Scores

Exercise 2.107

Draw a box-and-whisker display for the set of data whose 5-number summary is 42–62–72–82–97.

The following commands will find various percentiles.

MINITAB

Input the data into C1; then continue
with:

Choose: **Manip > Sort . . .**
Enter: Sort column(s): **C1**
         Store sorted column(s) in: **C2**
         Sort by column: **C1**
A ranked list of data will be obtained in C2. Determine the depth
position and locate desired percentile.

EXCEL

Input the data into column A and activate a cell for the answer; then continue with:

Choose: **Paste Function,** f_x **> Statis-
         tical > PERCENTILE > OK**
Enter: **(A2:A6 or select cells)**
         K (desired percentile;
         ex. .95, .47)

TI-83

Input the data into L1; then continue
with:

Choose: **STAT > EDIT > 2:SortA(**
Enter: **L1**
Enter: **percentile × sample size**
         (ex. .25 × 100)
Based on product, determine the
depth position; then continue
with:
Enter: **L1(depth position) > Enter**

The following commands will find the 5-number summary values.

MINITAB

Input the data into C1; then continue
with:

Choose: **Stat > Basic Statistics >
         Display Desc. Statistics ...**
Enter: Variables: **C1**

TI-83

Input the data into L1; then continue
with:

Choose: **STAT > CALC > 1:1-VAR STATS**
Enter: **L1**

EXCEL

Input the data into column A; then
continue with:

Choose: **Tools > Data Analysis* > De-
         scriptive Statistics > OK**
Enter: Input Range: **(A2:A6 or se-
         lect cells)**
Select: **Labels in First Row** (if
         necessary)
         Output Range
Enter: **(B1:C16 or select cells)**
         (need 2 columns)
Select: **Summary Statistics > OK**
To make output readable:
Choose: **Format > Column > Autofit
         Selection**

*If Data Analysis does not show on the Tool Menu:
see page 57

The following commands will construct a box-and-whisker diagram.

MINITAB

Input the data into C1; then continue with:

Choose: **Graph > Boxplot**
Enter: Graph 1, Y: **C1**
Optional:
Choose: **Annotations > Title**
Enter: **your title > OK**
Choose: **Options**
Select: **Transpose > OK**

For multiple boxplots, enter additional set of data into C2; then do as above plus:

Enter: Graph 2, Y: **C2**
Choose: **Frame > Multiple Graphs**
Select: **Overlay graphs on the same page > OK**

EXCEL

Input the data into column A; then continue with:

Choose: **Tools > Data Analysis Plus* > BoxPlot > OK**
Enter: **(A2:A6 or select cells)**

To edit the boxplot, review options shown with editing histograms on page 57.

*Data Analysis Plus is a collection of statistical macros for EXCEL. They can be downloaded onto your computer from your StatSource CD. Any new updates to Data Analysis Plus are available for downloading from the website: www.globalserve.net/~gkeller.

TI-83

Input the data into L1; then continue with:

Choose: **2nd > STAT PLOT > 1:Plot1**

Choose: **ZOOM > 9:ZoomStat > TRACE > > >**

If class midpoints are in L1 and the frequencies in L2, do as above except for:

Enter: Freq: **L2**

For multiple boxplots, enter additional set of data into L2 or L3; do as above plus:

Choose: **2nd > STAT PLOT > 2:Plot2**

The position of a specific value can be measured in terms of the mean and standard deviation using the *standard score,* commonly called the *z-score.*

STANDARD SCORE OR z-SCORE

The position a particular value of x has relative to the mean, measured in standard deviations. The z-score is found by the formula

$$z = \frac{\text{value} - \text{mean}}{\text{st. dev.}} = \frac{x - \bar{x}}{s} \qquad (2.14)$$

ILLUSTRATION 2.14 ▼

Find the standard scores for (a) 92 and (b) 72 with respect to a sample of exam grades that have a mean score of 75.9 and a standard deviation of 11.1.

Solution

a. $x = 92, \bar{x} = 75.9; s = 11.1$. Thus, $z = \dfrac{x - \bar{x}}{s} = \dfrac{92 - 75.9}{11.1} = \dfrac{16.1}{11.1} = \mathbf{1.45}$.

b. $x = 72, \bar{x} = 75.9; s = 11.1$. Thus, $z = \dfrac{x - \bar{x}}{s} = \dfrac{72 - 75.9}{11.1} = \dfrac{-3.9}{11.1} = \mathbf{-0.35}$

Exercise 2.108

Find the z-score for test scores of 92 and 63 on a test with a mean of 72 and a standard deviation of 12.

This means that the score 92 is approximately one-and-one-half standard deviations above the mean, while the score 72 is approximately one-third of a standard deviation below the mean.   ▲

NOTES

1. Typically, the calculated value of z is rounded to the nearest hundredth.
2. z-scores typically range in value from approximately -3.00 to $+3.00$.

Since the z-score is a measure of relative position with respect to the mean, it can be used to help make a comparison of two raw scores that come from separate populations. For example, suppose that you want to compare a grade you received on a test with a friend's grade on a comparable exam in her course. You received a raw score of 45 points; she obtained 72 points. Is her grade better? We need more information before we can draw such a conclusion. Suppose that the mean on the exam you took was 38 and the mean on her exam was 65. Your grades are both 7 points above the mean, so we still can't draw a definite conclusion. However, the standard deviation on the exam you took was 7 points, and it was 14 points on your friend's exam. This means that your score is one (1) standard deviation above the mean ($z = 1.0$), whereas your friend's grade is only one-half of a standard deviation above the mean ($z = 0.5$). Since your score has the "better" relative position, we would conclude that your score is slightly better than your friend's score. (Again, this is speaking from a relative point of view.)

The following additional commands may be helpful in analyzing data.

EXCEL

Input the data into column A; then continue with the following to sort the data:

Choose: **Data > Sort**
Enter: Sort by: **(A2:A6 or select cells)**
Select: **Ascending** or **Descending**
         Header row or **No header row**

MINITAB

Input the data into C1; then:
To sort the data into ascending order
and store it in C2; continue with

Choose: **Manip > Sort**
Enter: Sort column(s): **C1**
         Store sorted column(s)
         in: **C2**
         Sort by column: **C1**

To form an ungrouped frequency distribution of integer data; continue with

Choose: **Stat > Tables > Tally**
Enter: Variables: **C1**
Select: **Counts**

To print data on the session window;
continue with

Choose: **Manip > Display Data**
Enter: Columns to display: **C1** or
         C1 C2 or **C1-C4**

TI-83

Input the data into L1; then continue
with the following to sort the data:

Choose: **2nd > STAT > OPS > 1:SortA(**
Enter: **L1**

To form a frequency distribution of
the data in L1; continue with:

Choose: **PRGM > EXEC > FREQDIST***
Enter: **L1 > ENTER**
         LW BOUND = : **first lower
         class boundary**
         UP BOUND = : **last upper
         class boundary**
         WIDTH = : **class width** (use 1
         for ungrouped distribution)

*Program 'FREQDIST' is one of many programs that
are available for downloading from a website. See
page 39 for specific instructions.

The following commands will generate random samples of K data from a specified theoretical population.

MINITAB

The data will be put into C1:

Choose: **Calc > Random Data > {Normal,
         Uniform, etc.}**
Enter: Generate: **K** rows of data
         Store in column(s): **C1**
         Populations parameters
         needed: (μ, σ, **L, H, A,
         or B**)
            (Required parameters will
            vary depending on the dis-
            tribution)

EXCEL

Choose: **Tools > Data Analysis > Ran-
         dom Number Generation > OK**
Enter: Number of Variables: **1**
         Number of Random Numbers:
         (desired quantity)
Select: Distribution: **Normal or
         others**
Enter: Parameters: (μ, σ, **L, H, A,
         or B**)
            (Required parameters will
            vary depending on the dis-
            tribution)
Select: **Output Range**
Enter: **(A1 or select cell)**

TI-83

Choose: **STAT > 1:EDIT**
Highlight: **L1**
Choose: **MATH > PRB > 6:randNorm(**
           or **5:randInt(**
Enter: μ, σ, **# of trials** or **L,
           H, # of trials**

The following commands will select random samples of K data from a specified column of existing data.

MINITAB

The existing data to be selected from should be in C1; then continue with:

Choose: **Calc > Random Data > Sample from Columns**

Enter: Sample: **K** rows from column(s): **C1**
Store samples in: **C2**

Select: **Sample with replacement** (optional)

EXCEL

The existing data to be selected from should be in column A; then continue with:

Choose: **Tools > Data Analysis > Sampling > OK**

Enter: Input range: **(A2:A10 or select cells)**

Select: **Labels** (optional)
Random
Enter: Number of Samples: **K**
Output range:
Enter: **(B1 or select cell)**

EXERCISES • • • • • • • • • • • • •

2.109 The following data are the yields, in pounds, of hops.

| 3.9 | 3.4 | 5.1 | 2.7 | 4.4 | 7.0 | 5.6 | 2.6 | 4.8 | 5.6 |
| 7.0 | 4.8 | 5.0 | 6.8 | 4.8 | 3.7 | 5.8 | 3.6 | 4.0 | 5.6 |

a. Find the first and the third quartiles of the yield.
b. Find the midquartile.
c. Find the following percentiles: (1) P_{15}, (2) P_{33}, (3) P_{90}.

2.110 A research study of manual dexterity involved determining the time required to complete a task. The time required for each of 40 disabled individuals is as follows (data are ranked):

7.1	7.2	7.2	7.6	7.6	7.9	8.1	8.1	8.1	8.3
8.3	8.4	8.4	8.9	9.0	9.0	9.1	9.1	9.1	9.1
9.4	9.6	9.9	10.1	10.1	10.1	10.2	10.3	10.5	10.7
11.0	11.1	11.2	11.2	11.2	12.0	13.6	14.7	14.9	15.5

Find: **a.** Q_1 **b.** Q_2 **c.** Q_3 **d.** P_{95}
e. the 5-number summary   **f.** Draw the box-and-whisker display. *— don't do*

2.111 Consider the following set of ignition times that were recorded for a synthetic fabric.

30.1	30.1	30.2	30.5	31.0	31.1	31.2	31.3	31.3	31.4
31.5	31.6	31.6	32.0	32.4	32.5	33.0	33.0	33.0	33.5
34.0	34.5	34.5	35.0	35.0	35.6	36.0	36.5	36.9	37.0
37.5	37.5	37.6	38.0	39.5					

Find: **a.** the median   **b.** the midrange   **c.** the midquartile
d. the 5-number summary   **e.** Draw the box-and-whisker display.

2.112 Consider the worldwide temperature data shown in Exercise 2.78, p. 83.
a. Omit names and build a 10 × 10 table of ranked temperatures in ascending order, reading vertically in each column.
b. Construct a 5-number summary table.
c. Find the midquartile temperature reading and the interquartile range.
d. What are the z-scores for Kuwait, Berlin, Washington, London, and Bangkok?

2.113 A sample has a mean of 50 and a standard deviation of 4.0. Find the z-score for each value of x.
   a. $x = 54$ **b.** $x = 50$ **c.** $x = 59$ **d.** $x = 45$

2.114 An exam produced grades with a mean score of 74.2 and a standard deviation of 11.5. Find the z-score for each of the following test scores, x:
   a. $x = 54$ **b.** $x = 68$ **c.** $x = 79$ **d.** $x = 93$

2.115 A nationally administered test has a mean of 500 and a standard deviation of 100. If your standard score on this test was 1.8, what was your test score?

2.116 A sample has a mean of 120 and a standard deviation of 20.0. Find the value of x that corresponds to each of these standard scores:
   a. $z = 0.0$ **b.** $z = 1.2$ **c.** $z = -1.4$ **d.** $z = 2.05$

2.117 **a.** What does it mean to say that $x = 152$ has a standard score of $+1.5$?
   b. What does it mean to say that a particular value of x has a z-score of -2.1?
   c. In general, the standard score is a measure of what?

 2.118 In a study involving mastery learning (*Research in Higher Education,* Vol. 20, No. 4, 1984), 34 students took a pretest. The mean score was 11.04, and the standard deviation was 2.36. Find the z-score for scores of 9 and 15 on the 20-question pretest.

2.119 Which x-value has the higher value relative to the set of data from which it comes?

   A: $x = 85$, where mean $= 72$ and standard deviation $= 8$

   B: $x = 93$, where mean $= 87$ and standard deviation $= 5$

2.120 Which x-value has the lower relative position with respect to the set of data from which it comes?

   A: $x = 28.1$, where $\bar{x} = 25.7$ and $s = 1.8$

   B: $x = 39.2$, where $\bar{x} = 34.1$ and $s = 4.3$

2.7 Interpreting and Understanding Standard Deviation

Standard deviation is a measure of fluctuation (dispersion) in the data. It has been defined as a value calculated with the use of formulas. But you may wonder what it really is. It is a kind of yardstick by which we can compare the variability of one set of data with another. This particular "measure" can be understood further by examining two statements: *Chebyshev's theorem* and the *empirical rule.*

Exercise 2.121

Instructions for an essay assignment include the statement "The length is to be within 25 words of 200." What values of x, number of words, satisfy these instructions?

CHEBYSHEV'S THEOREM

The proportion of any distribution that lies within k standard deviations of the mean is at least $1 - \dfrac{1}{k^2}$, where k is any positive number larger than 1. This theorem applies to all distributions of data.

This theorem says that within two standard deviations of the mean ($k = 2$) you will always find at least 75% (that is, 75% or more) of the data.

$$1 - \frac{1}{k^2} = 1 - \frac{1}{2^2} = 1 - \frac{1}{4} = \frac{3}{4} = 0.75, \textbf{ at least 75\%}$$

Figure 2.31 shows a mounded distribution that illustrates at least 75%.

Figure 2.31

Chebyshev's Theorem with $k = 2$

Figure 2.32

Chebyshev's Theorem with $k = 3$

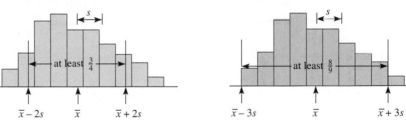

Exercise 2.122

According to Chebyshev's theorem, what proportion of a distribution will be within $k = 4$ standard deviations of the mean?

If we consider the interval enclosed by three standard deviations on either side of the mean ($k = 3$), the theorem says that we will always find at least 89% (that is, 89% or more) of the data.

$$1 - \frac{1}{k^2} = 1 - \frac{1}{3^2} = 1 - \frac{1}{9} = \frac{8}{9} = 0.89, \textbf{ at least 89\%}$$

Figure 2.32 shows a mounded distribution that illustrates at least 89%.

EMPIRICAL RULE

If a variable is normally distributed, then: within one standard deviation of the mean there will be approximately 68% of the data; within two standard deviations of the mean there will be approximately 95% of the data; and within three standard deviations of the mean there will be approximately 99.7% of the data. [This rule applies specifically to a **normal distribution (bell-shaped distribution)**, but it is frequently applied as an interpretive guide to any mounded distribution.]

Figure 2.33 shows the intervals of one, two, and three standard deviations about the mean of an approximately normal distribution. Usually these proportions do not occur exactly in a sample, but your observed values will be close when a large sample is drawn from a normally distributed population.

Figure 2.33

Empirical Rule

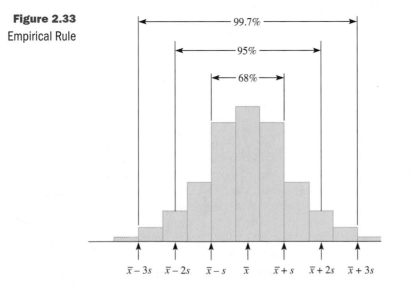

Test for Normality

Exercise 2.123

a. What proportion of a normal distribution is greater than the mean?

b. What proportion is within one standard deviation of the mean?

c. What proportion is greater than a value that is one standard deviation below the mean?

The empirical rule can be used to determine whether or not a set of data is approximately normally distributed. Let's demonstrate this application by working with the distribution of final exam scores that we have been using throughout this chapter. The mean, $\bar{x}$, was found to be 75.6, and the standard deviation, s, was 14.9. The interval from one standard deviation below the mean, $\bar{x} - s$, to one standard deviation above the mean, $\bar{x} + s$, is $75.6 - 14.9 = 60.7$ to $75.6 + 14.9 = 90.5$. This interval (60.7 to 90.5) includes 61, 62, 63, . . . , 89, 90. Upon inspection of the ranked data (Table 2.17, p. 94), we see that 35 of the 50 data pieces, or 70%, lie within one standard deviation of the mean. Further, $\bar{x} - 2s = 75.6 - (2)(14.9) = 75.6 - 29.8 = 45.8$, to $\bar{x} + 2s = 75.6 + 29.8 = 105.4$, gives the interval from 45.8 to 105.4. Forty-eight of the 50 data pieces, or 96%, lie within two standard deviations of the mean. All 50 data, or 100%, are included within three standard deviations of the mean (from 30.9 to 120.3). This information can be placed in a table for comparison with the values given by the empirical rule (see Table 2.18).

TABLE 2.18 Observed Percentages Interval Versus the Empirical

Interval	Empirical Rule Percentage	Percentage Found
$\bar{x} - s$ to $\bar{x} + s$	≈ 68	70
$\bar{x} - 2s$ to $\bar{x} + 2s$	≈ 95	96
$\bar{x} - 3s$ to $\bar{x} + 3s$	≈ 99.7	100

The percentages found are reasonably close to those of the empirical rule. By combining this evidence with the shape of the histogram, we can safely say that the final exam data are approximately normally distributed.

If a distribution is approximately normal, it will be nearly symmetrical and the mean will divide the distribution in half (the mean and the median are the same in a symmetrical distribution). This allows us to refine the empirical rule. Figure 2.34 shows this refinement.

Figure 2.34

Refinement of Empirical Rule

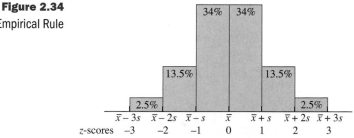

There is also a graphic way to test for normality. This is accomplished by drawing a relative frequency ogive of the grouped data on probability paper (which can be purchased at some college bookstores). On this paper the vertical scale is measured in percentages and is placed on the right side of the graph paper. All the directions and guidelines given on pages 60–61 for drawing an ogive must be followed. This graph, also known as a *probability plot,* can be constructed using a computer or graphing calculator. An ogive of the statistics final exam scores is drawn and labeled on a piece of probability paper in Figure 2.35. (Dashed lines are for later references to this drawing.)

To test for normality, first draw a straight line from the lower-left corner to the upper-right corner of the graph connecting the "next-to-end" points of the ogive. Then, if the ogive lies close to this straight line, the distribution is said to be approximately normal. The ogive for an exactly normal distribution will trace the straight line. The red dashed line (┊) in Figure 2.35 is the straight-line test for normality. The ogive suggests that the distribution of exam scores is approximately normal. (*Warning:* This graphic technique is very sensitive to the scale used along the horizontal axis.)

Figure 2.35

Ogive on Probability Paper

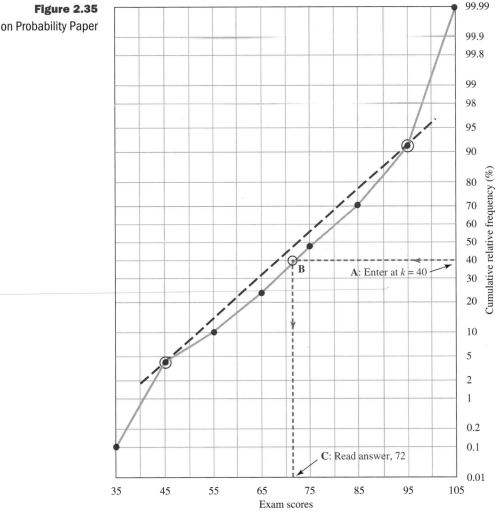

The ogive is a handy tool to use for finding percentiles of a frequency distribution. If the cumulative frequency distribution is drawn on probability paper, as in Figure 2.35, it is a simple matter to determine any kth percentile. Locate the value k on the vertical scale along the right-hand side, and follow the horizontal line for k until it intersects the line of the ogive. Then follow the vertical line that passes through this point of intersection to the bottom of the graph. Read the value of x from the horizontal scale. This value of x is the value of the kth percentile. For example, let's find the value of P_{40} for the data shown in Figure 2.35. First locate the value k (40) along the right-hand side of the graph (point A). Follow the horizontal line at k (40) (green dashed line) until it intersects the graph of the ogive (point B). Then follow the vertical green line to determine the value of x that corresponds to this point of intersection by reading it from the

Exercise 2.124

Using the ogive in Figure 2.35, estimate the value of P_{90}.

scale on the *x*-axis (point *C*): $P_{40} = 72$. This method can be used to find percentiles, quartiles, or the median. (*Note:* Because of the grouping into classes, the results may differ slightly from the answers obtained when using ranked data.)

2.125   Chebyshev's theorem guarantees what proportion of a distribution will be included between the following?
  a. $\bar{x} - 2s$ and $\bar{x} + 2s$ **b.** $\bar{x} - 3s$ and $\bar{x} + 3s$

2.126   The empirical rule indicates that we can expect to find what proportion of the sample to be included between the following?
  a. $\bar{x} - s$ and $\bar{x} + s$ **b.** $\bar{x} - 2s$ and $\bar{x} + 2s$ **c.** $\bar{x} - 3s$ and $\bar{x} + 3s$

2.127   Why is it that the *z*-score for a value belonging to a normal distribution usually lies between -3 and $+3$?

2.128   The mean mileage per tire is 30,000 miles and the standard deviation is 2500 miles for a certain tire.
  a. If we assume that the mileage is normally distributed, approximately what percentage of all such tires will give between 22,500 and 37,500 miles?
  b. If we assume nothing about the shape of the distribution, approximately what percentage of all such tires will give between 22,500 and 37,500 miles?

2.129   According to the empirical rule, practically all the data should lie between $(\bar{x} - 3s)$ and $(\bar{x} + 3s)$. The range also accounts for all the data.
  a. What relationship should hold (approximately) between the standard deviation and the range?
  b. How can you use the results of (a) to estimate the standard deviation in situations when the range is known?

2.130   The average clean-up time for a crew of a medium-size firm is 84.0 hours and the standard deviation is 6.8 hours. Assuming that the empirical rule is appropriate,
  a. What proportion of the time will it take the clean-up crew 97.6 or more hours to clean the plant?
  b. The total clean-up time will fall within what interval 95% of the time?

2.131   Chebyshev's theorem can be stated in an equivalent form to that given on page 102. For example, to say "at least 75% of the data fall within two standard deviations of the mean" is equivalent to stating that "at most, 25% will be more than two standard deviations away from the mean."
  a. At most, what percentage of a distribution will be three or more standard deviations from the mean?
  b. At most, what percentage of a distribution will be four or more standard deviations from the mean?

2.132   Using the empirical rule, determine the approximate percentage of a normal distribution that is expected to fall within the interval described.
  a. Greater than the mean
  b. Greater than one standard deviation above the mean.
  c. Less than one standard deviation above the mean.
  d. Between one standard deviation below the mean and two standard deviations above the mean.

2.133   In an article titled "Development of the Breast-Feeding Attrition Prediction Tool" (*Nursing Research*, March/April 1994, Vol. 43, No. 2), the results of administering the Breast-Feeding Attrition Prediction Tool (BAPT) to 72 women with prior successful experience breast-feeding are reported. The mean score for the Positive Breast-Feeding Sentiment (PBS) score for this sample is reported to equal 356.3 with a standard deviation of 65.9.

a. According to Chebyshev's theorem, at least what percent of the PBS scores are between 224.5 and 488.1?

b. If it is known that the PBS scores are normally distributed, what percent of the PBS scores are between 224.5 and 488.1?

2.134 An article titled "Computer-Enhanced Algebra Resources: The Effects on Achievement and Attitudes" (*International Journal of Math Education in Science and Technology,* 1980, Vol. 11, No. 4) compared algebra courses that use computer-assisted instruction with courses that do not. The scores that the computer-assisted instruction group made on an achievement test consisting of 50 problems had these summary statistics: $n = 57$, $\bar{x} = 23.14$, $s = 7.02$.

a. Find the limits within which at least 75% of the scores fell.

b. If the scores are normally distributed, what percentage of the scores will be below 30.16?

2.135 The top 50 Nike Tour money leaders through the Dakota Dunes Open, which ended August 2, 1998, together with their total earnings for the year playing golf are reproduced below:

Name of Player	Money ($)	Name of Player	Money ($)
Joe Ogilvie	130,381	Brian Bateman	60,349
Emlyn Aubrey	112,272	Craig Bowden	59,228
Doug Dunakey	108,618	Chris Riley	55,619
Eric Booker	106,267	Mike Sullivan	54,683
John Maginnes	104,630	Don Walsworth	54,662
John Wilson	102,865	Craig Kanada	53,843
Dennis Paulson	100,921	Joey Snyder	52,736
Bob Burns	98,790	Steve Lamontagne	50,743
Robin Freeman	97,365	Carl Paulson	50,513
Ryan Howison	91,084	Michael Allen	49,469
Chris Zambri	88,935	Jay Williamson	44,851
Sean Murphy	81,100	Charles Raulerson	44,497
Woody Austin	77,338	Jeff Brehaut	43,946
Gene Sauers	75,999	Mike Sposa	43,131
Tom Scherrer	73,963	Geoffrey Sisk	42,930
Eric Johnson	72,754	Chris Starkjohann	42,581
Michael Clark	71,112	Vance Veazey	41,681
Greg Lesher	68,747	David Berganio, Jr.	39,968
Notah Begay, III	68,421	Charlie Rymer	38,544
Deane Pappas	68,272	Jeff Gove	35,702
Jimmy Green	65,928	John Kernohan	35,057
Jeff Julian	61,576	John Elliott	34,984
Pat Bates	60,943	Steve Flesch	34,050
Perry Moss	60,887	Sam Randolph	33,016
Casey Martin	60,401	J. L. Lewis	31,819

Source: Omaha World Herald, "Nike Tour Leaders," August 5, 1998.

a. Calculate the mean and standard deviation of the earnings of the Nike Tour golf players.

b. Find the values of $\bar{x} - s$ and $\bar{x} + s$.

c. How many of the 50 pieces of data have values between $\bar{x} - s$ and $\bar{x} + s$? What percentage of the sample is this?

d. Find the values of $\bar{x} - 2s$ and $\bar{x} + 2s$.

e. How many of the 50 pieces of data have values between $\bar{x} - 2s$ and $\bar{x} + 2s$? What percentage of the sample is this?

Many of these data sets are on the StatSource CD.

f. Find the values of $\bar{x} - 3s$ and $\bar{x} + 3s$.

g. What percentage of the sample has values between $\bar{x} - 3s$ and $\bar{x} + 3s$?

h. Compare the answers found in (f) and (g) to the results predicted by Chebyshev's theorem.

i. Compare the answers found in (c), (e), and (g) to the results predicted by the empirical rule. Does the result suggest an approximately normal distribution?

 2.136 On the first day of class last semester, 50 students were asked for the one-way distance from home to college (to the nearest mile). The resulting data follow:

6	5	3	24	15	15	6	2	1	3
5	10	9	21	8	10	9	14	16	16
10	21	20	15	9	4	12	27	10	10
3	9	17	6	11	10	12	5	7	11
5	8	22	20	13	1	8	13	4	18

a. Construct a grouped frequency distribution of the data by using 1–4 as the first class.

b. Calculate the mean and standard deviation.

c. Determine the values of $\bar{x} \pm 2s$, and determine the percentage of data within two standard deviations of the mean.

2.137 The empirical rule states that the one, two, and three standard deviation intervals about the mean will contain approximately 68%, 95%, and 99.7% of the data, respectively.

a. Use the computer and/or calculator commands on page 100 to randomly generate a sample of 100 data from a normal distribution with mean 50 and standard deviation 10. Construct a histogram using class boundaries that are multiples of the standard deviation 10; that is, use boundaries from 10 to 90 in intervals of 10 (see commands on page 56). Calculate the mean and standard deviation using the commands found on pages 67 and 80, then inspect the histogram to determine the percentage of the data that fell within each of the one, two, and three standard deviation intervals. How closely do the three percentages compare to the percentages claimed in the empirical rule?

b. Repeat part (a). Did you get results similar to those in part (a)? Explain.

c. Consider repeating (a) several more times. Are the results similar each time? If so, in what way?

d. What do you conclude about the truth of the empirical rule?

2.138 Chebyshev's theorem states that "at least $1 - \dfrac{1}{k^2}$" of the data of a distribution will lie within k standard deviations of the mean.

a. Use the computer commands on page 100 to randomly generate a sample of 100 data from a uniform (nonnormal) distribution that has a low value of 1 and a high value of 10. Construct a histogram using class boundaries of 0 to 10 in increments of 1 (see commands on page 56). Calculate the mean and standard deviation using the commands found on pages 67 and 80, then inspect the histogram to determine the percentage of the data that fell within each of the one, two, three, and four standard deviation intervals. How closely do these percentages compare to the percentages claimed in Chebyshev's theorem and the empirical rule?

b. Repeat part (a). Did you get results similar to those in part (a)? Explain.

c. Consider repeating (a) several more times. Are the results similar each time? If so, in what way are they similar?

d. What do you conclude about the truth of Chebyshev's theorem and the empirical rule?

2.8   The Art of Statistical Deception

"There are three kinds of lies—lies, damned lies, and statistics." These remarkable words spoken by Benjamin Disraeli (19th-century British prime minister) represent the cynical view of statistics held by many people. Most people are on the consumer end of statistics and therefore have to "swallow" them.

Good Arithmetic, Bad Statistics

Let's explore an outright statistical lie. Suppose that a small business firm employs eight people who earn between $300 and $350 per week. The owner of the business pays himself $1250 per week. He reports to the general public that the average wage paid to the employees of his firm is $430 per week. That may be an example of good arithmetic, but it is also an example of bad statistics. It is a misrepresentation of the situation, since only one employee, the owner, receives more than the mean salary. The public will think that most of the employees earn about $430 per week.

Graphic representations can be tricky and misleading. The frequency scale (which is usually the vertical axis) should start at zero in order to present a total picture. Usually, graphs that do not start at zero are used to save space. Nevertheless, this can be deceptive. Graphs in which the frequency scale starts at zero tend to emphasize the size of the numbers involved, whereas graphs that are chopped off may tend to emphasize the variation in the number without regard to the actual size of the number. The labeling of the horizontal scale can be misleading also. You need to inspect graphic presentations very carefully before you draw any conclusions from the "story being told." The following four case studies will demonstrate some of these misrepresentations.

> **Exercise 2.139**
>
> Is it possible for eight employees to earn between $300 and $350, and one earn $1250 per week, and for the mean to be $430? Verify your answer.

Insufficient Information

Case Study 2.3

The Information That Was Almost Presented

An attractive colorful graph about the age of dog bite victims, I guess. But what does the height of the bars on the graph represent?

> The bite taken out of the graph looks more like a human bite mark than a dog's. Just what is this graph depicting?

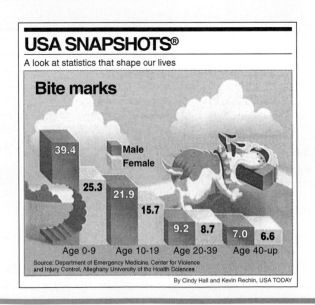

Truncated Vertical Scale

Case Study 2.4

Claiming Rapid Growth

Exercise 2.140

a. What characteristic(s) of the graph in Case Study 2.4 make the increase seem so pronounced?

b. Draw this graph on graph paper using a vertical scale that starts at "zero."

c. Does your graph present the same image as the one in Case Study 2.4?

An increase from $12.20 to $12.73 is a 4.3% increase over the one-year interval from May 1997 to May 1998. The graph seems to show the increase to be out of control.

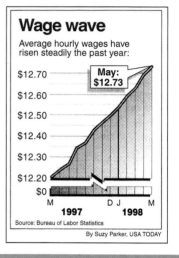

Wage wave

Average hourly wages have risen steadily the past year:

May: $12.73

$12.70
$12.60
$12.50
$12.40
$12.30
$12.20
$0

M   D J   M
1997   **1998**

Source: Bureau of Labor Statistics

By Suzy Parker, USA TODAY

Nonhorizontal Horizontal Axis

Case Study 2.5

Increasing the Increase

The size of the flock seems to be increasing dramatically, but upon inspection of the horizontal axis, you will see that it has an amazing upward trend also. How much of the trend line's visual increase is due to the horizontal axis? The four-year interval along the horizontal axis also changed.

Exercise 2.141

a. Redraw the graph in Case Study 2.5 using a horizontal axis that is spaced uniformly by years.

b. Comment on the effect that your graph has on the impression presented.

USA SNAPSHOTS®

A look at statistics that shape the nation

Bigger flock per priest

The number of Catholics per priest continues to rise from its historic low in 1942. U.S. Catholics per priest:

1,228

1,400
1,200
1,000
800
600
400
200
0

'42 '46 '50 '54 '58 '62 '66 '70 '74 '78 '82 '86 '90 '94 '96[1]

1 – Latest year compiled
Source: Dr. Robert G. Kennedy, University of St. Thomas/*Official Catholic Directory*

By Anne R. Carey and Sam Ward, USA TODAY

Variable Scales

C a s e   S t u d y   2.6

S-t-r-e-t-c-h-ing the Graph

Exercise 2.142

a. Plot the points (62,23), (74,23.5), (80,24) and (94,33) on a coordinate axis system using graph paper. Connect the points with line segments.

b. Compare graph (a) to the Overweight adult male graph in Case Study 2.6. Comment on the difference in appearance.

Sometimes the deception is unintentional and sometimes it is difficult to tell intent from mistake. The "Overweight adults" graph is such a graph. Study it carefully and find its errors, then do Exercise 2.142 to be sure you understand the deception that has occurred and how it was accomplished.

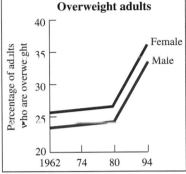

What it all comes down to is that statistics, like all languages, can be and is abused. In the hands of the careless, the unknowledgeable, or the unscrupulous, statistical information can be as false as "damned lies."

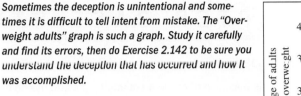

Darrell Huff's *How to Lie with Statistics* is an easy, fun-to-read, clever book that every student of statistics should read.

EXERCISES • • • • • • • • • • • •

2.143 Exercise 2.102 on page 91 displays a relative frequency distribution and a frequency distribution. A histogram of either distribution would qualify as a "tricky graph." How does this graph violate the guidelines for drawing histograms?

$ **2.144** *PC World* magazine periodically publishes their performance test results and overall ratings of the top 20 power desktop computers. No machine in the top 20 received an overall rating below 75, and the maximum rating any machine could have received was 100. The top four manufacturers, together with the performance ratings of their machines, are shown below:

Manufacturer	Overall Rating
1. Gateway	88
2. Dell	87
3. Quantex	84
4. NEC	82

Prepare two bar graphs to depict the performance rating data. Scale the vertical axis on the first graph from 80 to 89. Scale the second graph from 0 to 100. What is your conclusion concerning how the performance ratings of the four computers stack up based on the two bar graphs, and what would you recommend, if anything, to improve the presentations?

2.145 Find an article or an advertisement containing a graph that in some way misrepresents the information of statistics. Describe how this graph misrepresents the facts.

RETURN TO CHAPTER CASE STUDY

Let's return to the Chapter Case Study, page 35. Have you ever tried to statistically describe the commute to work for your employed fellow students? Now you have an opportunity to try out many of the descriptive statistics techniques that we have learned in this chapter.

Return to CHAPTER CASE STUDY

EXERCISE 2.146

Below is a sample of one-way commute times, to the nearest minute, for 50 college students who are gainfully employed.

2	10	3	5	1	19	14	3	15	12
30	10	2	8	19	17	12	29	21	29
15	23	3	24	4	16	49	15	64	5
3	22	19	1	48	4	25	7	22	4
3	10	33	12	14	30	10	13	5	23

a. Construct three different kinds of graphs for these data.
b. Find the five measures of central tendency.
c. Find the three measures of dispersion.
d. Using one graph from part (a), one measure of central tendency from part (b), and one measure of dispersion from part (c), write a short presentation of the 50 data that you believe best describes the typical employed college student's commute to work.
e. Explain how the graph and the two statistics used in answering part (d) best describe the situation.

IN RETROSPECT

You have been introduced to some of the more common techniques of descriptive statistics. There are far too many specific types of statistics used in nearly every specialized field of study for us to review here. We have outlined the uses of only the most universal statistics. Specifically, you have seen several basic graphic techniques (circle and bar graphs, Pareto diagrams, dotplots, stem-and-leaf displays, histograms, and box-and-whiskers) that are used to present sample data in picture form. You have also been introduced to some of the more common measures of central tendency (mean, median, mode, midrange, and midquartile), measures of dispersion (range, variance, and standard deviation), and measures of position (quartiles, percentiles, and z-scores).

You should now be aware that an average can be any one of five different statistics, and you should understand the distinctions among the different types of averages. The article "'Average' Means Different Things" in Case Study 2.2 (p. 71) discusses four of the averages studied in this chapter. You might reread it now and find that it has more meaning and is of more interest. It will be time well spent!

You should also have a feeling for, and an understanding of, the concept of a standard deviation. You were introduced to Chebyshev's theorem and the empirical rule for this purpose.

The exercises in this chapter (as in others) are extremely important; they will help you to reinforce the concepts studied before you go on to learn how to use these ideas in later chapters. A good understanding of the descriptive techniques presented in this chapter is fundamental to your success in the later chapters.

CHAPTER EXERCISES

2.147 Near the end of the 20th century the corporate world predicted that thousands of desktop computer programs would fail to handle the year 2000 because of installed instruction codes that depended upon a dating scheme designed for the previous century. For example, it was estimated by Tangram Enterprise Solutions, Inc. that 4593 desktops running CalcuPro software prior to version 3.0 under MS-Windows were at risk to fail in the year 2000. The breakdown within corporate divisions was as follows:

Corporate Area	At-Risk Frequency
1. Sales	700
2. Research & Development	1550
3. Finance	843
4. Administration	1100
5. Other	400

Source: Fortune, "Teaching Chipmunks to Dance," May 25, 1998. Extracted from *Teaching Chipmunks to Dance,* Kendall/Hunt Publishing.

   a. Construct a relative frequency distribution of these data.
   b. Construct a bar graph using these data.
   c. Explain why the graph drawn in (b) is not a histogram.

2.148 Identify each of the following as examples of (1) attribute (qualitative) or (2) numerical (quantitative) variables:
   a. scores registered by people taking their written state automobile driver's license examination
   b. whether or not a motorcycle operator possesses a valid motorcycle operator's license
   c. the number of television sets installed in a house
   d. the brand of bar soap being used in a bathroom
   e. the value of a cents-off coupon used with the purchase of a box of cereal
   f. the amount of weight lost in the past month by a person following a strict diet
   g. batting averages of major league baseball players
   h. decisions by the jury in felony trials
   i. sunscreen usage before going in the sun {always, often, sometimes, seldom, never}
   j. reason manager failed to act against an employee's poor performance

2.149 Samples A and B are shown in the following table. Notice that the two samples are the same except that the 8 in A has been replaced by a 9 in B.

A:	2	4	5	5	7	8
B:	2	4	5	5	7	9

What effect does changing the 8 to a 9 have on each of the following statistics?
   a. mean   **b.** median   **c.** mode   **d.** midrange
   e. range   **f.** variance   **g.** standard deviation

2.150 Samples C and D are shown in the following table. Notice that the two samples are alike except for two values.

C:	20	60	60	70	90
D:	20	30	70	90	90

What effect does changing the two 60's to 30 and 90 have on each of the following statistics?
   a. mean   **b.** median   **c.** mode   **d.** midrange
   e. range   **f.** variance   **g.** standard deviation

2.151 The addition of a new accelerator is claimed to decrease the drying time of latex paint by more than 4%. Several test samples were conducted with the following percentage decreases in drying time:

5.2   6.4   3.8   6.3   4.1   2.8   3.2   4.7

   a. Find the sample mean.
   b. Find the sample standard deviation.
   c. Do you think these percentages average 4 or more? Explain.
      (Retain these solutions for use in answering Exercise 9.32.)

2.152 Gasoline pumped from a supplier's pipeline is supposed to have an octane rating of 87.5. On 13 consecutive days, a sample was taken and analyzed with the following results:

88.6   86.4   87.2   88.4   87.2   87.6   86.8
86.1   87.4   87.3   86.4   86.6   87.1

> Many of these data sets are on the StatSource CD.

   a. Find the sample mean.
   b. Find the sample standard deviation.
   c. Do you think these readings seem to average 87.5? Explain.
      (Retain these solutions for use in answering Exercise 9.44.)

2.153 The following set of data gives the ages of 118 known offenders who committed an auto theft last year in Garden City, Michigan.

11	14	15	15	16	16	17	18	19	21	25	36
12	14	15	15	16	16	17	18	19	21	25	39
13	14	15	15	16	17	17	18	20	22	26	43
13	14	15	15	16	17	17	18	20	22	26	46
13	14	15	16	16	17	17	18	20	22	27	50
13	14	15	16	16	17	17	19	20	23	27	54
13	14	15	16	16	17	18	19	20	23	29	59
13	15	15	16	16	17	18	19	20	23	30	67
14	15	15	16	16	17	18	19	21	24	31	
14	15	15	16	16	17	18	19	21	24	34	

   a. Find the mean.   **b.** Find the median.   **c.** Find the mode.
   d. Find Q_1 and Q_3.   **e.** Find P_{10} and P_{95}.

2.154 A survey of 32 workers at building 815 of Eastman Kodak Company was taken last May. Each worker was asked: "How many hours of television did you watch yesterday?" The results were as follows:

0	0	$1/2$	1	2	0	3	$2^1/_2$
0	0	1	$1^1/_2$	5	$2^1/_2$	0	2
$2^1/_2$	1	0	2	0	$2^1/_2$	4	0
6	$2^1/_2$	0	$1/2$	1	$1^1/_2$	0	2

   a. Construct a stem-and-leaf display.
   b. Find the mean.
   c. Find the median.
   d. Find the mode.
   e. Find the midrange.
   f. Which one of the measures of central tendency would best represent the average viewer if you were trying to portray the typical television viewer? Explain.
   g. Which measure of central tendency would best describe the amount of television watched? Explain.
   h. Find the range.
   i. Find the variance.
   j. Find the standard deviation.

2.155 The stopping distance on a wet surface was determined for 25 cars each traveling at 30 miles per hour. The data (in feet) are shown on the following stem-and-leaf display:

```
 6 |  3 7 6 3 9
 7 |  4 2 0 1 1 2 0 5
 8 |  5 4 5 5 6
 9 |  4 1 0 0 5
10 |  5 4
```

Find the mean and the standard deviation of these stopping distances.

2.156 Compute the mean and the standard deviation for the following set of data. Then find the percentage of the data that is within two standard deviations of the mean (Leaf unit = 0.1).

```
1 |  4 7 1
2 |  4 5
3 |  5 0 4 1
4 |  4
5 |  5 8 7
6 |  8 8 2 8 6
7 |  5
8 |
9 |  4
```

 2.157 The number of lost luggage reports filed per 1000 airline passengers during July 1998 was reported by *USA Today*. On September 23, 1998, they reported the airlines with the fewest: Continental 3.50, US Airways 3.95, Delta 4.07, American 4.22. On September 24, 1998, they reported the airlines with the most: Alaska 8.64, United 7.63, Northwest 6.76, TWA 5.12. The industry average was 5.09.
 a. Define the terms *population* and *variable* with regard to this information.
 b. Are the numbers reported (3.50, 3.95, . . . , 8.64) data or statistics? Explain.
 c. Is the average, 5.09, a data, a statistic, or a parameter value? Explain why.
 d. Is the average value the midrange?

 2.158 Ask one of your instructors for a list of exam grades (15 to 25 grades) from a class.
 a. Find five measures of central tendency.
 b. Find the three measures of dispersion.
 c. Construct a stem-and-leaf display. Does this diagram suggest that the grades are normally distributed?
 d. Find the following measures of location: (1) Q_1 and Q_3, (2) P_{15} and P_{60}, (3) the standard score z for the highest grade.

2.159 The distribution of credit hours, per student, taken this semester at a certain college was as follows:

Credit Hours	Frequency
3	75
6	150
8	30
9	50
12	70
14	300
15	400
16	1050
17	750
18	515
19	120
20	60

(continued)

 a. Draw a histogram of the data.
 b. Find the five measures of central tendency.
 c. Find Q_1 and Q_3.
 d. Find P_{15} and P_{12}.
 e. Find the three measures of dispersion (range, s^2, and s).

 2.160 An article in *Therapeutic Recreation Journal* reports a distribution for the variable, number of persistent disagreements. Sixty-six patients and their therapeutic recreation specialist each answered a checklist of problems with yes or no. Disagreement occurs when the specialist and the patient did not respond identically to an item on the checklist. It becomes a persistent disagreement if the item remains in disagreement after a second interview.
 a. Draw a dotplot of this sample data.
 b. Find the median number of persistent disagreements.
 c. Find the mean number of persistent disagreements.
 d. Find the standard deviation for the number of persistent disagreements.
 e. Draw a vertical line on the dotplot at the mean.
 f. Draw a horizontal line segment on the dotplot whose length represents the standard deviation (start at the mean).

Number of Items	Frequency
0	2
1	2
2	4
3	10
4	7
5	9
6	8
7	11
8	7
9	3
10	1
11	2

Source: Data reprinted with permission of the National Recreation and Park Association, Alexandria, VA, from Pauline Petryshen and Diane Essex-Sorlie, "Persistent Disagreement Between Therapeutic Recreation Specialists and Patients in Psychiatric Hospitals," *Therapeutic Recreation Journal,* Vol. XXIV, Third Quarter, 1990.

 2.161 *USA Today,* 10-25-94, reported in the Snapshot® "Mystery of the remote" that 44% of the families surveyed never misplaced the family television remote control, while 38% misplaced it one to five times weekly and 17% misplaced it more than five times weekly. One percent of the families surveyed didn't know. Suppose you took a survey that resulted in the following data. Let x be the number of times per week that the family's television remote control gets misplaced. The survey resulted in the following data.

x	0	1	2	3	4	5	6	7	8	9
f	220	92	38	21	24	30	34	20	16	5

 a. Construct a histogram.
 b. Find the mean, median, mode, and midrange.
 c. Find the variance and standard deviation.
 d. Find Q_1, Q_3, and P_{90}.
 e. Find the midquartile.
 f. Find the 5-number summary and draw a box-and-whiskers display.

2.162 In the May 1990 issue of the journal *Social Work,* Marlow reports the following results:

Number of Children Living at Home	Mexican- American Women	Anglo- American Women
0	23	38
1	22	9
2	17	15
3	7	9
4	1	1

Copyright 1990, National Association of Social Workers, Inc. *Social Work.*

a. Construct a frequency histogram for each of the preceding distributions. Draw them on the same axis, using two different colors, so that you can compare their distributions. (See Exercise 2.33, p. 62.)
b. Calculate the mean and standard deviation for the Mexican-American data.
c. Calculate the mean and standard deviation for the Anglo-American data.
d. Do these two distributions seem to be different? Cite specific reasons for your answer.

2.163 The length of life of 220 incandescent 60-watt lamps was obtained and yielded the frequency distribution shown in the following table.

Class Limit	Frequency
500–600	3
600–700	7
700–800	14
800–900	28
900–1000	64
1000–1100	57
1100–1200	23
1200–1300	13
1300–1400	7
1400–1500	4

a. Construct a histogram of these data using a vertical scale for relative frequencies.
b. Find the mean length of life.
c. Find the standard deviation.

 2.164 The following table shows the age distribution of heads of families.

Age of Head of Family (years)	Number
20–25	23
25–30	38
30–35	51
35–40	55
40–45	53
45–50	50
50–55	48
55–60	39
60–65	31
65–70	26
70–75	20
75–80	16
	450

(continued)

   a. Find the mean age of the heads of families.
   b. Find the standard deviation.

 2.165 Mutual funds attract millions of private investors every year. *USA Today* assembles a
   Mutual Fund Scoreboard that lists 15 of the largest stock funds available and a Mutual
   Fund Spotlight that shows the 14 top-yielding growth and income funds during the past
   four weeks. Three more funds were selected randomly from the general listing of all mutual funds. Midway through 1998, the combined lists were shown below, together with
   each fund's total return for the year:

Fund Name	Total Return (%)	Fund Name	Total Return (%)
Fidelity Magellan	22.0	Berkshire Cap Gro & Value	41.1
Vanguard Index 500 Port	20.9	Marsico Growth & Income	39.3
Washington Mutual Inv	14.6	Nations Mars Gro & Inc	39.2
Investment Co of America	15.7	Strong Blue Chip 100 Fund	30.9
Fidelity Growth & Income	19.5	Janus Growth & Income	29.8
Fidelity Contrafund	23.0	Excelsior Inst Value Equity	22.3
Vanguard Windsor II	16.2	Schroder Large Cap Equity	26.7
Amer Cent/20th C Ultra	30.4	Strong Total Return Fund	24.6
Vanguard Wellington	9.7	Reynolds Blue Chip Growth	35.1
Fidelity Puritan	13.0	Diversified Investors	30.1
Fidelity Equity-Inc	12.9	Strong Growth & Income	26.9
Fidelity Adv Growth Oppty	14.5	Newpoint Equity Fund	23.6
Vanguard Windsor	10.1	New Providence Cap Gro	10.9
Income Fund of America	6.9	Weitz Partners Value Fund	30.2
Janus Fund	26.2	Putnam High Yield	4.6
Pioneer Capital Growth	8.0	Standish Internat'l Equity	30.1

Source: USA Today, July 23, 1998.

   a. Omit names and build a 4 × 8 table of ranked total return percentages in ascending
   order, reading vertically in each column.
   b. Construct a 5-number summary table.
   c. Find the midquartile total return percentage and the interquartile range.
   d. What are the *z*-scores for Berkshire Capital Growth & Value, Income Fund of
   America, and Janus Fund?
   e. Based on the data and your calculations, can you pinpoint the three funds selected at
   random?

 2.166 The lengths (in millimeters) of 100 brown trout in pond 2-B at Happy Acres Fish Hatchery on June 15 of last year were as follows:

15.0	15.3	14.4	10.4	10.2	11.5	15.4	11.7	15.0	10.9
13.6	10.5	13.8	15.0	13.8	14.5	13.7	13.9	12.5	15.2
10.7	13.1	10.6	12.1	14.9	14.1	12.7	14.0	10.1	14.1
10.3	15.2	15.0	12.9	10.7	10.3	10.8	15.3	14.9	14.8
14.9	11.8	10.4	11.0	11.4	14.3	15.1	11.5	10.2	10.1
14.7	15.1	12.8	14.8	15.0	10.4	13.5	14.5	14.9	13.9
10.1	14.8	13.7	10.9	10.6	12.4	14.5	10.5	15.1	15.8
12.0	15.5	10.8	14.4	15.4	14.8	11.4	15.1	10.3	15.4
15.0	14.0	15.0	15.1	13.7	14.7	10.7	14.5	13.9	11.7
15.1	10.9	11.3	10.5	15.3	14.0	14.6	12.6	15.3	10.4

   a. Find the mean.   **b.** Find the median.   **c.** Find the mode.
   d. Find the midrange.   **e.** Find the range.   **f.** Find Q_1 and Q_3.
   g. Find the midquartile.   **h.** Find P_{35} and P_{64}.
   i. Construct a grouped frequency distribution that uses 10.0–10.5 as the first class.
   j. Construct a histogram of the frequency distribution.

k. Construct a cumulative relative frequency distribution.
l. Construct an ogive of the cumulative relative frequency distribution.
m. Find the mean of the frequency distribution.
n. Find the standard deviation of the frequency distribution.

2.167 "Grandparents are grand" appeared as a USA Snapshot® in *USA Today,* 12-9-94, and reported on "how much grandparents say they spend annually on gifts and entertainment for each of their grandchildren."

Dollar Interval	$0–$100	$101–$200	$201–$500	$501 or more
Percent	54.4%	16.5%	15.7%	4.7%

Suppose that this information was obtained from a sample of 1000 grandparents and that the 8.7% who did not answer spent nothing. Use values of $0, $50, $150, $350, and $750 as class midpoints, and estimate the sample mean and the standard deviation for the variable x, amount spent.

2.168 The earnings per share for 40 firms in the radio and transmitting equipment industry follow:

4.62	0.10	1.29	7.25	6.04	3.20	9.56	4.90	4.22	3.71
0.25	1.34	2.11	5.39	0.84	-0.19	3.72	2.27	2.08	1.12
1.07	2.50	2.14	3.46	1.91	7.05	5.10	1.80	0.91	0.50
5.56	1.62	1.36	1.93	2.05	2.75	3.58	0.44	3.15	1.93

a. Prepare a frequency distribution and a frequency histogram for these data.
b. Which class of your frequency distribution contains the median?

2.169 For a normal (or bell-shaped) distribution, find the percentile rank that corresponds to
a. $z = 2$ **b.** $z = -1$

2.170 For a normal (or bell-shaped) distribution, find the z-score that corresponds to the kth percentile:
a. $k = 20$ **b.** $k = 95$

2.171 Bill and Rob are good friends, although they attend different high schools in their city. The city school system uses a battery of fitness tests to test all high school students. After completing the fitness tests, Bill and Rob are comparing their scores to see who did better in each event. They need help.

	Sit-ups	Pull-ups	Shuttle Run	50-Yard Dash	Softball Throw
Bill	$z = -1$	$z = -1.3$	$z = 0.0$	$z = 1.0$	$z = 0.5$
Rob	61	17	9.6	6.0	179 ft
Mean	70	8	9.8	6.6	173 ft
Standard Deviation	12	6	0.6	0.3	16 ft

Bill received his test result in z-scores, whereas Rob was given raw scores. Since both boys understand raw scores, convert Bill's z-scores to raw scores in order to make an accurate comparison.

2.172 Twins Jean and Joan Wong are in fifth grade (different sections), and the class has been given a series of ability tests.

	Results	
Skill	**Jean: z-Score**	**Joan: Percentile**
Fitness	2.0	99
Posture	1.0	69
Agility	1.0	88
Flexibility	−1.0	35
Strength	0.0	50

(continued)

If the scores for these ability tests are approximately normally distributed, which girl has the higher relative score on each of the skills listed? Explain your answers.

2.173 Manufacturing specifications are often based on the results of samples taken from satisfactory pilot runs. The following data resulted from just such a situation in which eight pilot batches were completed and sampled. The resulting particle sizes, in angstroms (where $1\text{Å} = 10^{-8}$ cm) were

| 3923 | 3807 | 3786 | 3710 | 4010 | 4230 | 4226 | 4133 |

a. Find the sample mean.
b. Find the sample standard deviation.
c. Assuming that particle size has an approximately normal distribution, determine the manufacturing specs that bound 95% of the particle sizes (that is, find the 95% interval, $\bar{x} \pm 2s$).

2.174 Delco Products, a division of General Motors, produces a bracket that is used as part of a power doorlock assembly. The length of this bracket is constantly being monitored. A sample of 30 power door brackets resulted in the following lengths (in millimeters).

11.86	11.88	11.88	11.91	11.88	11.88	11.88	11.88	11.88	11.86
11.88	11.88	11.88	11.88	11.86	11.83	11.86	11.86	11.88	11.88
11.88	11.83	11.86	11.86	11.86	11.88	11.88	11.86	11.88	11.83

Source: With permission of Delco Products Division, GMC.

a. Without doing any calculations, what would you estimate for a sample mean?
b. Construct an ungrouped frequency distribution.
c. Draw a histogram of this frequency distribution.
d. Use the frequency distribution and calculate the sample mean and standard deviation.
e. Determine the limits of the $\bar{x}$ and $3s$ interval and mark this interval on the histogram.
f. The product specification limits are 11.7–12.3. Does the sample indicate that production is within these requirements? Justify your answer.

2.175 Americans love soups, and they remain one of the most popular foods for lunch and as an appetizer before dinner. The manufacturers provided the calorie and sodium content of 50 popular brands of soups that were published in the December 1997 issue of *Nutrition in Action*. The data for multi-serving (8 oz.) cans and mixes, most of which were low-fat varieties, appear in the table below:

Soup Brand	Calories	Sodium (mg)	Soup Brand	Calories	Sodium (mg)
Health Valley Organic	90	150	Healthy Choice Clam Chowder	120	480
Taste Adventure Minestrone	140	210	Campbell's HR Chicken with Rice	60	480
Health Valley Fat-Free	90	230	Campbell's HR Clam Chowder	120	480
Pritkin Lentil or Split Pea	160	290	Campbell's HR Cream of Chicken	70	480
Health Valley Healthy Pasta	110	290	Taste Adventure Golden Pea	210	490
Pritkin Hearty Vegetable	90	290	Westbrae Natural	90	580
Arrowhead Mills Red Lentil	100	230	Lipton Recipe Secrets Onion	20	610
Taste Adventure Navy Bean	160	390	Progresso Chicken Noodle	90	620
Healthy Choice Minestrone	110	400	Campbell's FF Ramen, Chicken	140	720
Healthy Choice Chicken Noodle	120	410	Lipton Noodle with Chicken Broth	60	720
Baxters Mediterranian Tomato	70	420	Campbell's Tomato	100	730
Shari's Organic	130	420	Knorr Vegetable	30	730
Healthy Choice Vegetable Beef	130	420	Progresso Lentil	140	750
Healthy Choice Bean and Ham	160	430	Maruchan Ramen, Chicken	190	780
Baxters Italian Bean & Pasta	80	430	Campbell's Cream of Mushroom	110	870
Mayacamas Black Bean	60	450	Campbell's Cream of Chicken	130	890
Healthy Choice Cream of Mushroom	80	450	Nissin Top Ramen, Chicken	190	910

(continued)

Healthy Choice Beef & Potato	120	450	Campbell's Vegetable	90	920
Progresso Beef Barley or Lentil	140	460	Progresso Minestrone	120	960
Campbell's HR Bean w/Ham & Bac	150	480	Campbell's 98% FF Clam Chowder	150	970
Campbell's HR Southwestern Veg	160	480	Campbell's Chicken Noodle	70	980
Hain Healthy Naturals	120	480	Campbell's French Onion	70	980
Campbell's HR Hearty Minestrone	120	480	Knorr French Onion	45	980
Hain Home Style Chunky Tomato	120	480	Progresso Clam Chowder	200	1050
Campbell's HR Chicken Noodle	120	480	Campbell's Cream of Mushroom (jar)	260	1130

Source: Nutrition in Action, "Soups: The Middle Ground," December, 1997.

    a. Compute the mean and standard deviation of both calorie and sodium content of the soups listed in the table.

    b. Use your answers in (a) to test Chebyshev's theorem that at least 75% of the soups' calorie and sodium content will fall within ± 2 standard deviations from the mean. Is this the case?

    c. Find the limits for ± 1 standard deviation from the mean for the soups' sodium content. Does the sodium content of soups appear to be normally distributed? Explain.

2.176 The manager of Jerry's Barber Shop recently asked his last 50 customers to punch a time card when they first arrived at the shop and to punch out right after they paid for their hair cut. He then used the data on the cards to measure how long it took Jerry and his barbers to cut hair in order to schedule their appointment intervals. The following data were tabulated:

Length of Time Required to Service Customer (min.)

50	21	36	35	35	27	38	51	28	35
32	32	27	25	24	38	43	46	29	45
40	27	36	38	35	31	28	38	33	46
35	31	38	48	23	35	43	31	32	38
43	32	18	43	52	52	49	53	46	19

    a. Construct a stem-and-leaf plot of these data.

    b. Compute the mean, median, mode, range, midrange, variance, and standard deviation of the hair cut service times.

    c. Construct a 5-number summary table.

    d. According to Chebyshev's theorem, at least 75% of the hair cut service times will fall between what two amounts? Is this true? Explain why or why not.

    e. How far apart would you recommend that Jerry schedule his appointments to keep his shop operating at a comfortable pace?

2.177 Each year stock car drivers compete for the NASCAR Winston Cup. Points are earned on the basis of finishes in sanctioned races scheduled on the circuit. Midway through the 1998 season, the standings were published in *USA Today*. The top 40 drivers are shown in the table below:

NASCAR Winston Cup Standings

Driver	Points	Driver	Points
Jeff Gordon	2527	Ted Musgrave	1698
Mark Martin	2475	Chad Little	1684
Dale Jarrett	2429	Dick Trickle	1598
Jeremy Mayfield	2390	Darrell Waltrip	1565
Rusty Wallace	2307	Ricky Rudd	1490
Bobby Labonte	2205	Steve Grissom	1461
Terry Labonte	2166	Kenny Irwin	1459
Jeff Burton	2134	Joe Nemechek	1434

(continued)

Driver	Points	Driver	Points
Jimmy Spencer	1968	Robert Pressley	1431
Dale Earnhardt	1961	Rick Mast	1404
Ken Schrader	1924	Kenny Wallace	1380
Bill Elliott	1896	Geoff Bodine	1332
Bobby Hamilton	1845	Kyle Petty	1310
John Andretti	1808	Lake Speed	1297
Michael Waltrip	1794	Mike Skinner	1192
Ward Burton	1759	Derrike Cope	997
Ernie Irvan	1740	Kevin Lepage	977
Johnny Benson	1728	Jerry Nadeau	953
Sterling Marlin	1704	David Green	803
Brett Bodine	1700	Randy LaJoie	768

Source: USA Today, July 23, 1998.

  a. Calculate the mean and standard deviation of the points accumulated by the Winston Cup stock car drivers.

  b. Construct a 5-number summary table.

  c. According to Chebyshev's theorem, at least 75% of the points will fall between what two amounts? Is this the case?

  d. According to the empirical rule, approximately 68% of the points will fall between what two amounts? Is this the case?

  e. Compare your answers in (c) and (d) to the results predicted by the empirical rule. Does your comparison suggest that the distribution of Winston Cup points approximates the normal distribution? Explain.

2.178 The dotplot below shows the number of attempted passes thrown by the quarterbacks for 22 of the NFL teams that played on one particular Sunday afternoon

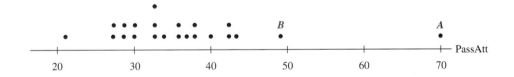

  a. Describe the distribution, including how points *A* and *B* seem to relate to the others.

  b. If you remove point *A*, and maybe *B*, would you say the remaining data have an approximately normal distribution? Explain.

  c. Based on the information about distributions that Chebyshev's theorem and the empirical rule give us, how typical an event do you feel point *A* represents? Explain.

2.179 Starting with the data values of 70 and 85, add three data values to form a sample so that the sample has (justify your answer in each case):

  a. a standard deviation of 5.

  b. a standard deviation of 10.

  c. a standard deviation of 15.

  d. compare your three samples and the variety of values needed to obtain each of the required standard deviations.

2.180 Make up a set of 18 data (think of them as exam scores) so that the sample meets each of these sets of criteria.

  a. Mean is 75, standard deviation is 10.

  b. Mean is 75, maximum is 98, minimum is 40, standard deviation is 10.

  c. Mean is 75, maximum is 98, minimum is 40, standard deviation is 15.

  d. How are the data in the sample for (b) different from those in (c)?

2.181 Construct two different graphs of the points (62,2), (74,14), (80,20), (94,34).
   a. On the first graph, along the horizontal axis, lay off equal intervals and label them 62, 74, 80, and 94; lay off equal intervals along the vertical axis and label them 0, 10, 20, 30, and 40. Plot the points and connect with line segments.
   b. On the second graph, along the horizontal axis, lay off equally spaced intervals and label them 60, 65, 70, 75, 80, 85, 90, and 95; mark off the vertical axis in equal intervals and label them 0, 10, 20, 30, and 40. Plot the points and connect with line segments.
   c. Compare the effect that scale has on the appearance of graphs (a) and (b). Explain the impression presented by each graph.
   d. Explain how your graphs demonstrate the inappropriateness of the graph in Case Study 2.6, page 111.

2.182 Use a computer to generate a random sample of 500 values of a normally distributed variable x with a mean of 100 and a standard deviation of 20. Construct a histogram of the 500 values.
   a. Use the computer commands on page 100 to randomly generate a sample of 500 data from a normal distribution with mean 100 and standard deviation 20. Construct a histogram using class boundaries that are multiples of the standard deviation 20; i.e., use boundaries from 20 to 180 in intervals of 20 (see commands on page 56).

Let's consider the 500 x-values found in (a) as a population.
   b. Use the computer commands on page 101 to randomly select a sample of 30 values from the population found in (a). Construct a histogram of the sample using the same class intervals as used in (a).
   c. Repeat part (b) three times.
   d. Calculate several values (mean, median, maximum, minimum, standard deviation, etc.) describing the population and each of the four samples. (See page 97 for commands.)
   e. Do you think a sample of 30 data adequately represent a population? [Compare each of the four samples found in (b) and (c) to the population.]

2.183 Repeat Exercise 2.182 using a different sample size. You might try a few different sample sizes; $n = 10$, $n = 15$, $n = 20$, $n = 40$, $n = 50$, $n = 75$. What effect does increasing the sample size have on the effectiveness of the sample to depict the population? Explain.

2.184 Repeat Exercise 2.182 using populations with different shaped distributions.
   a. Use a uniform or rectangular distribution. (Replace the subcommands used in Exercise 2.182; in place of NORMAL use: UNIFORM with a low of 50 and a high of 150, and use class boundaries of 50 to 150 in increments of 10.)
   b. Use a skewed distribution. (Replace the subcommands used in Exercise 2.182; in place of NORMAL use: POISSON 50 and use class boundaries of 20 to 90 in increments of 5.)
   c. Use a J-shaped distribution. (Replace the subcommands used in Exercise 2.182; in place of NORMAL use: EXPONENTIAL 50 and use class boundaries of 0 to 250 in increments of 10.)
   d. Does the shape of the distribution of the population have an effect on how well a sample of size 30 represents the population? Explain.
   e. What effect do you think changing the sample size would have on the effectiveness of the sample to depict the population? Try a few different sample sizes. Do the results agree with your expectations? Explain.

VOCABULARY LIST

Be able to define each term. Pay special attention to the key terms, which are printed in **red.** In addition, describe in your own words, and give an example of, each term. Your examples should not be ones given in class or in the textbook. Page numbers indicate the first appearance of the term.

bar graph (p. 37)
bell-shaped distribution (p. 103)
bimodal frequency distribution (p. 59)
box-and-whiskers display (p. 96)
Chebyshev's theorem (p. 102)
circle graph (p. 37)
class (p. 51)
class boundary (p. 52)
class midpoint (p. 54)
class width (p. 52)
cumulative frequency (p. 59)
depth (p. 67)
deviation from the mean (p. 76)
distribution (p. 41)
dotplot (p. 41)
empirical rule (p. 103)
5-number summary (p. 95)
frequency (p. 51)
frequency distribution (p. 51)
frequency histogram (p. 55)
grouped frequency distribution (p. 51)
histogram (p. 54)
interquartile range (p. 96)
J-shaped (p. 59)
mean (p. 66, 85)
measure of central tendency (p. 66)
measure of dispersion (p. 75)
measure of position (p. 92)
median (p. 67)

midquartile (p. 95)
midrange (p. 70)
modal class (p. 59)
mode (p. 59, 69)
normal distribution (p. 59, 103)
ogive (p. 60)
Pareto diagrams (p. 39)
percentile (p. 92)
qualitative data (p. 37)
quantitative data (p. 41)
quartile (p. 92)
range (p. 76)
rectangular distribution (p. 58)
relative frequency (p. 55)
relative frequency histogram (p. 55)
single-variable data (p. 36)
skewed distribution (p. 59)
standard deviation (p. 78, 85)
standard score (p. 98)
stem-and-leaf display (p. 42)
summation (p. 54, 66)
symmetrical (p. 58)
tally (p. 53)
ungrouped frequency distribution (p. 51)
uniform (p. 58)
variance (p. 77, 85)
x bar ($\overline{x}$) (p. 66)
z-score (p. 98)

CHAPTER PRACTICE TEST

Part I: Knowing the Definitions

Answer "True" if the statement is always true. If the statement is not always true, replace the words in bold with the words that make the statement always true.

2.1 The **mean** of a sample always divides the data into two equal halves (half larger and half smaller in value than itself).

2.2 A measure of **central tendency** is a quantitative value that describes how widely the data are dispersed about a central value.

2.3 The sum of the squares of the deviations from the mean, $\sum(x - \overline{x})^2$, will **sometimes** be negative.

2.4 For any distribution, the sum of the deviations from the mean equals **zero.**

2.5 The standard deviation for the set of values 2, 2, 2, 2, and 2 is **2.**

2.6 On a test John scored at the 50th percentile and Jorge scored at the 25th percentile; therefore, John's test score was **twice** Jorge's test score.

2.7 The frequency of a class is the number of pieces of data whose values fall within the **boundaries** of that class.

2.8 **Frequency distributions** are used in statistics to present large quantities of repeating values in a concise form.

2.9 The unit of measure for the standard score is always in **standard deviations**.

2.10 For a bell-shaped distribution, the range will be approximately equal to **six standard deviations.**

Part II: Applying the Concepts

2.11 The results of a consumer study completed at Corner Convenience Store were reported in the accompanying histogram. Find the answer to each of the following.

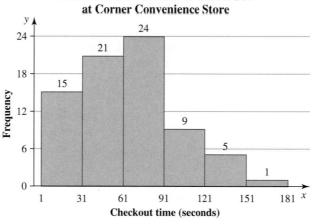

Amount of Time Needed to Check Out at Corner Convenience Store

a. What is the class width?
b. What is the class midpoint for the class 31–61?
c. What is the upper class boundary for the class 61–91?
d. What is the frequency of the class 1–31?
e. What is the frequency of the class containing the largest observed value of x?
f. What is the lower class boundary of the class with the largest frequency?
g. How many pieces of data are shown in this histogram?
h. What is the value of the mode?
i. What is the value of the midrange?
j. Estimate the value of the 90th percentile, P_{90}.

2.12 A sample of the purchases for several Corner Convenience Store customers resulted in the following sample data:

x = number of items purchased per customer

x	f
1	6
2	10
3	9
4	8
5	7

a. What does the "2" represent?
b. What does the "9" represent?
c. How many customers were used to form this sample?
d. How many items were purchased by the customers in this sample?   *(continued)*

e. What is the largest number of items purchased by one customer?

Find each of the following (show formulas and work):
f. mode   **g.** median   **h.** midrange
i. mean   **j.** variance   **k.** standard deviation

2.13 Given the set of data 4, 8, 9, 8, 6, 5, 7, 5, 8, find each of the following sample statistics.
  a. mean   **b.** median   **c.** mode
  d. midrange   **e.** first quartile   **f.** P_{40}
  g. variance   **h.** standard deviation   **i.** range

2.14 a. Find the standard score for the value $x = 452$ relative to its sample, where the sample mean is 500 and the standard deviation is 32.
  b. Find the value of x that corresponds to the standard score of 1.2, where the mean is 135 and the standard deviation is 15.

Part III: Understanding the Concepts

Answer all questions.

2.15 The Corner Convenience Store kept track of the number of paying customers it had during the noon hour each day for 100 days. The following are the resulting statistics rounded to the nearest integer:

mean = 95	median = 97	mode = 98
first quartile = 85	third quartile = 107	midrange = 93
range = 56	standard deviation = 12	

  a. The Corner Convenience Store served what number of paying customers during the noon hour more often than any other number? Explain how you determined your answer.
  b. On how many days were there between 85 and 107 paying customers during the noon hour? Explain how you determined your answer.
  c. What was the greatest number of paying customers during any one noon hour? Explain how you determined your answer.
  d. For how many of the 100 days was the number of paying customers within three standard deviations of the mean $(\bar{x} \pm 3s)$? Explain how you determined your answer.

2.16 Mr. VanCott started his own machine shop several years ago. His business has grown and become very successful in recent years. Currently he employs 14 people, including himself, and pays the following annual salaries:

Owner, President	$80,000	Worker	$25,000
Business Manager	50,000	Worker	25,000
Production Manager	40,000	Worker	25,000
Shop Foreman	35,000	Worker	20,000
Worker	30,000	Worker	20,000
Worker	30,000	Worker	20,000
Worker	28,000	Worker	20,000

  a. Calculate the four "averages": mean, median, mode, midrange.
  b. Draw a dotplot of these salaries and locate each of the four averages on it.
  c. Suppose you were the feature writer assigned to write this week's feature story on Mr. VanCott's machine shop, one of a series on local small businesses that are prospering. You plan to interview Mr. VanCott, his business manager, the shop foreman, and one of his newer workers. Which statistical average do you think each will give as their answer when asked, "What is the average annual salary paid to the employees here at VanCott's?" Explain why each person interviewed has a different perspective and why this viewpoint will cause each to cite a different statistical average.
  d. What is there about the distribution of these salaries that causes the four "average values" to be so different from each other?

2.17 Create a set of data containing three or more values:
   a. where the mean is 12 and the standard deviation is zero.
   b. where the mean is 20 and the range is 10.
   c. where the mean, median, and mode are all equal.
   d. where the mean, median, and mode are all different.
   e. where the mean, median, and mode are all different and the median is the largest and the mode is the smallest.
   f. where the mean, median, and mode are all different and the mean is the largest and the median is the smallest.

2.18 A set of test papers was machine scored. Later it was discovered that two points should be added to each score. Student A said, "The mean score should also be increased by two points." Student B added, "The standard deviation should also be increased by 2 points." Who is right? Justify your answer.

2.19 Student A stated that "Both the standard deviation and the variance preserved the same unit of measurement as the data." Student B disagreed, arguing that "The unit of measurement for variance was a meaningless unit of measurement." Who is right? Justify your answer.

Chapter 3

Descriptive Analysis and Presentation of Bivariate Data

CHAPTER OUTLINE

3.1 Bivariate Data
Two variables are paired together for analysis.

3.2 Linear Correlation
Does an increase in the value of one variable **indicate a change** in the value of the other?

3.3 Linear Regression
The **line of best fit** is a mathematical expression for the relationship between two variables.

CHAPTER CASE STUDY

How Do You Spend Your Time?

Many of us worry about spending too much time working (or studying) and not enough time at the pleasures of life. How many hours did you, personally, spend working last week? How many hours did you spend at leisure pursuits? Is there a relationship between these two quantities? If you spend more time at one, do you spend less time at the other? The USA Snapshot®, "Balancing work and play," that appeared in *USA Today* December 26, 1996, lists the average hours per week for each activity for several different years.

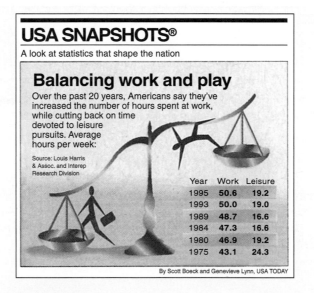

USA SNAPSHOTS®

A look at statistics that shape the nation

Balancing work and play

Over the past 20 years, Americans say they've increased the number of hours spent at work, while cutting back on time devoted to leisure pursuits. Average hours per week:

Source: Louis Harris & Assoc. and Interep Research Division

Year	Work	Leisure
1995	50.6	19.2
1993	50.0	19.0
1989	48.7	16.6
1984	47.3	16.6
1980	46.9	19.2
1975	43.1	24.3

By Scott Boeck and Genevieve Lynn, USA TODAY

How would you picture this information on a graph so that the viewer will be able to visualize any relationship that might be contained in these numbers? How else could you summarize this information? We will learn various methods of describing bivariate data in this chapter and will return to this case study at the end of the chapter. (See Exercise 3.59, p. 166.)

CHAPTER OBJECTIVES

In the field of statistics there are many problems that require a combined analysis of two variables. In business, in education, and in many other fields, we often want to answer such questions as "Are these two variables related?" "If so, how are they related?" "Are these variables correlated?" The relationships being discussed are not cause-and-effect relationships, but rather mathematical relationships that allow us to predict the behavior of one variable based on knowledge about another variable.

Let's look at a few specific illustrations.

ILLUSTRATION 3.1 ▼

Are female voter opinions on the U.S. president's current position regarding tax increases related to political party affiliation?

ILLUSTRATION 3.2 ▼

As a person grows taller, he or she usually gains weight. Someone might ask, "Is there a relationship between height and weight?"

ILLUSTRATION 3.3 ▼

Research doctors test new drugs (old ones, too) by prescribing different dosages and observing the responses of their patients. One question we could ask here is "Does the drug dosage prescribed determine the amount of recovery time the patient needs?"

ILLUSTRATION 3.4 ▼

A high school guidance counselor would like to predict the academic success that students graduating from her school will have in college. In cases like this, the predicted value (grade-point average at college) depends on many traits of the students: (1) how well they did in high school, (2) their intelligence, (3) their desire to succeed at college, and so on.   ▲

These questions all require the analysis of bivariate data to obtain the answers. In this chapter we take a first look at the techniques of tabling and graphing bivariate data and the descriptive aspects of correlation and regression analysis. The objectives of this chapter are (1) to be able to represent bivariate data in tabular and graphic form, (2) to gain an understanding of the distinction between the basic purposes of correlation analysis and regression analysis, and (3) to become familiar with the ideas of descriptive presentation. With these objectives in mind, we will restrict our discussion to the simplest and most basic form of correlation and regression analysis—the bivariate linear case.

3.1   Bivariate Data

BIVARIATE DATA

Bivariate data consist of the values of two different variables that are obtained from the same population element.

Each of the two variables may be either *qualitative* or *quantitative* in nature. As a result, three combinations of variable types can form bivariate data:

1. Both variables are qualitative (attribute).
2. One variable is qualitative (attribute) and the other is quantitative (numerical).
3. Both variables are quantitative (both numerical).

In this section we study tabular and graphic methods for displaying each of these combinations of bivariate data.

Two Qualitative Variables

When bivariate data result from two qualitative (attribute or categorical) variables, the data are often arranged on a **cross-tabulation** or **contingency table.** Let's look at an illustration.

ILLUSTRATION 3.5 ▼

Thirty students from our college were randomly identified and classified according to two variables: (1) gender (M/F) and (2) major (liberal arts, business administration, technology), as shown in Table 3.1. These 30 bivariate data can be summarized on a 2 × 3 cross-tabulation table where the two rows represent the two gender categories of male and female, and the three columns represent the three majors liberal arts,

TABLE 3.1   Gender and Major of 30 College Students

Name	Gender	Major	Name	Gender	Major
Adams	M	LA	Kee	M	BA
Argento	F	BA	Kleeberg	M	LA
Baker	M	LA	Light	M	BA
Bennett	F	LA	Linton	F	LA
Brock	M	BA	Lopez	M	T
Brand	M	T	McGowan	M	BA
Chun	F	LA	Mowers	F	BA
Crain	M	T	Ornt	M	T
Cross	F	BA	Palmer	F	LA
Ellis	F	BA	Pullen	M	T
Feeney	M	T	Rattan	M	BA
Flanigan	M	LA	Sherman	F	LA
Hodge	F	LA	Small	F	T
Holmes	M	T	Tate	M	BA
Jopson	F	T	Yamamoto	M	LA

business administration, and technology. The entry in each cell is found by determining how many students fit categorically into each cell. Adams is male (M) and liberal arts (LA) and is classified in the cell in the first row, first column. See the red tally mark in Table 3.2. The other 29 students are classified (tallied, shown in black) in a similar fashion.

TABLE 3.2   Cross-Tabulation of Gender and Major (tallied)

		Major					
		Liberal Arts		Business Ad.		Technology	
Gender	Male	⦀	(5)	⦀⎮	(6)	⦀⎮⎮	(7)
	Female	⦀⎮	(6)	⎮⎮⎮⎮	(4)	⎮⎮	(2)

The resulting 2 × 3 cross-tabulation (contingency table), Table 3.3, shows the frequency for each cross category of the two variables along with the row and column totals, called *marginal totals* (or marginals). The total of the marginal totals is the *grand total* and is equal to *n*, the *sample size*.

TABLE 3.3   Cross-Tabulation of Gender and Major (frequencies)

		Major			
		Liberal Arts	Business Ad.	Technology	Row Totals
Gender	Male	5	6	7	18
	Female	6	4	2	12
	Column Totals	11	10	9	30

Contingency tables often show percentages (relative frequencies). These percentages can be based on the entire sample or on the subsample (row or column) classifications.

Percentages Based on the Grand Total (Entire Sample)   The contingency table shown in Table 3.3 can easily be converted to percentages of the grand total by dividing each frequency by the grand total and multiplying by 100. For example, 6 becomes 20% $\left[\left(\dfrac{6}{30}\right) \times 100 = 20\right]$. See Table 3.4.

TABLE 3.4   Cross-Tabulation of Gender and Major (relative frequencies; % of grand total)

		Major			
		Liberal Arts	Business Ad.	Technology	Row Totals
Gender	Male	17%	20%	23%	60%
	Female	20%	13%	7%	40%
	Column Totals	37%	33%	30%	100%

With the table expressed in percentages of the grand total, we can easily see that 60% of the sample was male, 40% female, 30% technology majors, and so on. These same statistics (numerical values describing sample results) can be shown in a bar graph (see Figure 3.1).

Figure 3.1

Bar Graph

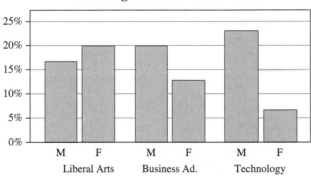

Percentage Based on Grand Total

Table 3.4 and Figure 3.1 show the distribution of male liberal arts students, female liberal arts students, male business administration students, and so on relative to the entire sample.

Percentages Based on Row Totals The entries in the same contingency table, Table 3.3, can be expressed as percentages of the row totals (or gender) by dividing each row entry by that row's total and multiplying by 100. Table 3.5 is based on row totals.

TABLE 3.5 Cross-Tabulation of Gender and Major (% of row totals)

		Major			
		Liberal Arts	**Business Ad.**	**Technology**	**Row Totals**
Gender	**Male**	28%	33%	39%	100%
	Female	50%	33%	17%	100%
	Column Totals	37%	33%	30%	100%

From Table 3.5 we see that 28% of the male students were majoring in liberal arts while 50% of the female students were majoring in liberal arts. These same statistics are shown in the bar graph in Figure 3.2.

Figure 3.2

Bar Graph

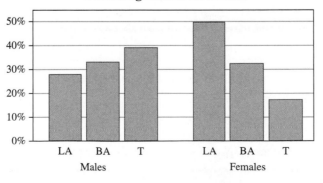

Percentage Based on Gender

Exercise 3.1

In a national survey of 500 business and 500 leisure travelers, each was asked where they would most like "more space."

	On Airplane	Hotel Room	All Other
Business	355	95	50
Leisure	250	165	85

Express the table as percentages of the total.

Exercise 3.2

Express the table in Exercise 3.1 as percentages of the row totals. Why might one prefer the table to be expressed this way?

Table 3.5 and Figure 3.2 show the distribution of the three majors for male and female students separately.

Percentages Based on Column Totals   The entries in the contingency table in Table 3.3 can also be expressed as percentages of the column totals (or major) by dividing each column entry by that column's total and multiplying by 100. Table 3.6 is based on column totals.

TABLE 3.6   Cross-Tabulation of Gender and Major (% of column totals)

		Major			
		Liberal Arts	Business Ad.	Technology	Row Totals
Gender	Male	45%	60%	78%	60%
	Female	55%	40%	22%	40%
	Column Totals	100%	100%	100%	100%

From Table 3.6 we see that 45% of the liberal arts students were male while 55% of the liberal arts students were female. These same statistics are shown in the bar graph in Figure 3.3.

Figure 3.3
Bar Graph

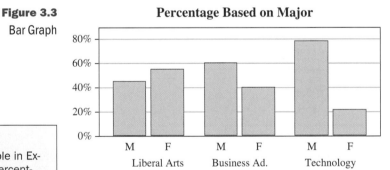

Percentage Based on Major

Exercise 3.3

Express the table in Exercise 3.1 as percentages of the column totals. Why might one prefer the table to be expressed this way?

Table 3.6 and Figure 3.3 show the distribution of male and female students for each major separately.   ▲

One Qualitative and One Quantitative Variable

When bivariate data result from one qualitative and one quantitative variable, the quantitative values are viewed as separate samples, each set identified by levels of the qualitative variable. Each sample is described using the techniques from Chapter 2, and the results are displayed side-by-side for easy comparison.

(continued on p. 136)

The following commands will construct a cross-tabulation table.

MINITAB

Input the row-variable categorical values into C1 and the corresponding column-variable categorical values into C2; then continue with:

Choose: **Stat > Tables**
              > CrossTabulation
Enter: Classification variables:
             C1 C2
Select: **Counts**
             Row Percents
             Column Percents
             Total Percents

Suggestion: The four subcommands that are available for selection can be used together; however, the resulting table will be much easier to read if one subcommand at a time is used.

EXCEL

Using column headings or titles, input the row-variable categorical values into column A and corresponding column-variable categorical values in column B; then continue with:

Choose: **Data > Pivot Table Report ...**
Select: **Microsoft Excel list or**
            database > Next
Enter: Range: **(A1:B5 or select**
            cells) > Next
Drag: **Headings to row or column**
            (depends on preference)
            One heading into data area*
            > Next
Select: **Existing Worksheet**
Enter: **(C1 or select cell) > Finish**

*For other summations, double click "Count of " in data area box; then continue with:
Choose: Summarize by: **Count**
            > Options
            Show data as: **% of row** or **% of column** or **% of total > OK**

TI-83

The categorical data must be numerically coded first; use 1, 2, 3,...for the various row variables and 1, 2, 3,...for the various column variables. Input the numeric row-variable values into L1 and the corresponding numeric column-variable values into L2; then continue with:

Choose: **PRGM > EXEC > CROSSTAB***
Enter: **ROWS: L1 > ENTER**
           COLS: L2 > ENTER

The cross-tabulation table showing frequencies is stored in matrix [A]. The cross-tabulation table showing row percentages is in matrix [B], column percentages in matrix [C], and percentages based on the grand total in matrix [D]. All matrices contain marginal totals. To view the matrices, continue with:

Choose: **MATRX > NAMES**
Enter: **1:[A]** or **2:[B]** or **3:[C]** or **4:[D] > ENTER**

*Program 'CROSSTAB' is one of many programs that are available for downloading from a website. See page 39 for specific instructions.

ILLUSTRATION 3.6 ▼

The distance required to stop a 3000-pound automobile on wet pavement was measured to compare the stopping capability of three different tread designs. Tires of each design were tested repeatedly on the same automobile on a controlled wet pavement.

TABLE 3.7 Stopping Distances for Three Tread Designs

Design A (n = 6)		Design B (n = 6)		Design C (n = 6)	
37	36	33	35	40	39
34	40	34	42	41	41
38	32	38	34	40	43

The design of the tread is a qualitative variable with three levels of response, and the stopping distance is a quantitative variable. The distribution of the stopping distances for tread design A is to be compared with the distribution of stopping distances for each of the other tread designs. This comparison may be accomplished with both numerical and graphical techniques. Some of the available options are shown in Figure 3.4, Table 3.8, and Table 3.9.

Figure 3.4

Dotplot and Box-and-Whisker Using a Common Scale

Stopping Distances

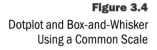

TABLE 3.8 5-Number Summary for Each Design

	Design A	Design B	Design C
High	40	42	43
Q_3	38	38	41
Median	36.5	34.5	40.5
Q_1	34	34	40
Low	32	33	39

TABLE 3.9 Mean and Standard Deviation for Each Design

	Design A	Design B	Design C
Mean	36.2	36.0	40.7
St. Dev.	2.9	3.4	1.4

Exercise 3.4

The January unemployment rate for U.S. cities:

Eastern: 5.5, 6.6, 5.3, 5.4, 5.8

Western: 5.4, 6.0, 6.4, 5.6, 6.9.

Display these rates as two dotplots using the same scale; compare means and medians.

Much of the information described above can be demonstrated using many other statistical techniques such as stem-and-leaf displays, histograms, and other numerical statistics.   ▲

The following commands will construct side-by-side boxplots or dotplots for comparison purposes.

MINITAB

Input the numerical values into C1 and their corresponding categories into C2; then continue with:

Choose: **Graph > Boxplot**
Enter: Y: **C1**   X: **C2**

MINITAB commands to construct side-by side dotplots for data in this form are located on page 46.

If the data for the various categories are in separate columns, use the MINITAB commands for multiple boxplots on page 98. If side-by-side dotplots are needed for data in this form, continue with:

Choose: **Graph > Character Graphs
          > Dotplot**
Enter: Variables: **C1 C2 C3...**
Select: **Same scale for all
          variables**

EXCEL

EXCEL commands to construct a single boxplot are located on page 42.

TI-83

TI-83 commands to construct multiple boxplots are located on page 98

TI-83 commands to construct multiple dotplots are located on page 46.

Two Quantitative Variables

When the bivariate data are the result of two quantitative variables, it is customary to express the data mathematically as **ordered pairs** (x, y), where x is the **input variable** (sometimes called the **independent variable**) and y is the **output variable** (sometimes called the **dependent variable**). The data are said to be ordered because one value, x, is always written first. They are called *paired* because for each x value there is a corresponding y value from the same source. For example, if x is height and y is weight, a height and corresponding weight are recorded for each person. The input variable x is

Exercise 3.5

Which variable, height or weight, would you use as the input variable? Explain why.

measured or controlled in order to predict the output variable y. For example, in Illustration 3.3 (p. 130) the researcher can control the amount of drug prescribed. Therefore, the amount of drug would be referred to as x. In the case of height and weight, either variable could be treated as input, the other as output, depending on the question being asked. However, different results will be obtained from the regression analysis, depending on the choice made.

In problems that deal with two quantitative variables, we will present our sample data pictorially on a *scatter diagram*.

SCATTER DIAGRAM

A plot of all the ordered pairs of bivariate data on a coordinate axis system. The input variable x is plotted on the horizontal axis, and the output variable y is plotted on the vertical axis.

Exercise 3.6

Draw a coordinate axis and plot the points (0, 6), (3, 5), (3, 2), (5, 0).

NOTE When constructing a scatter diagram, it is convenient to construct scales so that the range of the y-values along the vertical axis is equal to or slightly shorter than the range of the x-values along the horizontal axis. This creates a "window of data" that is approximately square.

C a s e   S t u d y   3.1

Paid Vacation Days

The number of paid vacation days salaried workers receive each year generally varies according to the company worked for and the length of employment. The USA Snapshot® "Trading years for days" displays the percentages based on column totals.

Exercise 3.7

Referring to "Trading years for days":

a. Name the two variables used.
b. Does this information suggest a relationship between the two variables? Explain.

USA SNAPSHOTS®

A look at statistics that shape your finances

Trading years for days

Paid vacation days large companies give salaried workers:

Days	Years service 1	5	10	15	20	30
5-9	3%					
10-14	83%	20%	1%			
15-19	13%	75%	62%	9%	2%	2%
20-24	1%	5%	37%	88%	62%	32%
25-29				3%	35%	48%
30-plus					1%	18%

Source: Hewitt Associates' 1996 survey   By Anne R. Carey and Marcia Staimer, USA TODAY

ILLUSTRATION 3.7 ▼

In Mr. Chamberlain's physical fitness course, several fitness scores were taken. The following sample is the number of push-ups and sit-ups done by ten randomly selected students:

(27, 30), (22, 26), (15, 25), (35, 42), (30, 38),
(52, 40), (35, 32), (55, 54), (40, 50), (40, 43)

Table 3.10 shows these sample data, and Figure 3.5 shows a scatter diagram of these data.

TABLE 3.10   Data for Push-ups and Sit-ups

Student	1	2	3	4	5	6	7	8	9	10
Push-ups (x)	27	22	15	35	30	52	35	55	40	40
Sit-ups (y)	30	26	25	42	38	40	32	54	50	43

Figure 3.5

Scatter Diagram

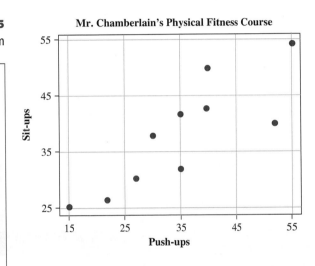

Exercise 3.8

Does studying for an exam pay off?

a. Draw a scatter diagram of the number of hours studied, x, compared to the exam grade, y, received.

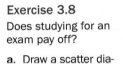

x	2	5	1	4	2
y	80	80	70	90	60

b. Explain what you can conclude based on the pattern of data shown on the scatter diagram drawn in (a).

The following commands will construct a scatter diagram.

MINITAB

```
Input the x-variable values into C1
and the corresponding y-variable values
into C2; then continue with:

  Choose:  Graph > Plot
  Enter:   Graph 1: Y: C2 X: C1
  Choose:  Annotation > Title
  Enter:   your title > OK
```

EXCEL

Input the *x*-variable values into column A and the corresponding *y*-variable values into column B; then continue with:

Choose: **Chart Wizard > XY(Scatter) > 1st picture** (usually) **> Next**
Enter: Data Range: **(A1:B12 or select cells(if necessary)) > Next**
Choose: Titles
Enter: Chart title: **your title**;
Value(*x*) axis: **title for x axis**;
Value(*y*) axis: **title for y axis > Finish**

To remove gridlines:

Choose: **Gridlines**
Unselect: **Value(y) axis: Major Gridlines > Finish**

To edit the scatter diagram, follow the basic editing commands as shown for a histogram on page 57.

To change the scale, double click on the axis; then continue with:

Choose: **Scale**
Unselect: **any automatic values**
Enter: **new values > OK**

TI-83

Input the *x*-variable values into L1 and the corresponding *y*-variable values into L2; then continue with:

Choose: **2nd > STAT PLOT > 1:Plot1**

Choose: **ZOOM > 9:ZoomStat > TRACE > > >** or **WINDOW**
Enter: **at most lowest x value, at least highest x value, x-scale, - y-scale, at least highest y-value, y-scale,1 TRACE > > >**

E X E R C I S E S • • • • • • • • • • • • • •

3.9 The USA Snapshot® "Can't get enough of school" shows the results from a 2 × 3 contingency table of two qualitative variables.
   a. Identify the population and name the two variables.
   b. Construct the contingency table using entries of percentages based on row totals.

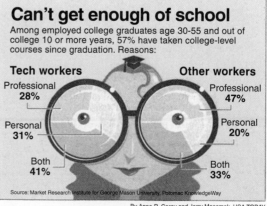

USA SNAPSHOTS®

A look at statistics that shape your finances

Can't get enough of school

Among employed college graduates age 30-55 and out of college 10 or more years, 57% have taken college-level courses since graduation. Reasons:

Tech workers
Professional 28%
Personal 31%
Both 41%

Other workers
Professional 47%
Personal 20%
Both 33%

Source: Market Research Institute for George Mason University, Potomac KnowledgeWay

By Anne R. Carey and Jerry Mosemak, USA TODAY

3.10 The USA Snapshot® "I don't want to grow up" shows the results from a 9 × 2 contingency table for one qualitative and one quantitative variable.
  a. Identify the population, name the qualitative and the quantitative variable.
  b. Construct a histogram showing the two distributions side by side.
  c. Does there seem to be a big difference between the genders on this subject?

USA SNAPSHOTS®
A look at statistics that shape our lives

I don't want to grow up
The age adults say they'd like to remain for the rest of their lives if they could:

Age	Men	Women
1-4	0%	2%
5-10	8%	8%
11-14	4%	6%
15-20	34%	20%
21-25	29%	28%
26-30	8%	10%
31-35	7%	10%
36 40	3%	7%
41-up	7%	9%

Source: IRC Research for Walt Disney   By Cindy Hall and Genevieve Lynn, USA TODAY

3.11 Under the National Highway System Designation Act passed in 1995, states were allowed to set their own highway speed limits. Most of the states raised the limits. As of early 1998, the maximum speed limits on interstate highways for cars and trucks by each state are given in the table below (in miles per hour):

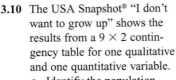

If you are using a computer or a calculator, try the cross-tabulation table commands on page 135.

State	Cars	Trucks	State	Cars	Trucks
Alabama	70	70	Montana	65	65
Alaska	65	65	Nebraska	75	75
Arizona	75	75	Nevada	75	75
Arkansas	70	65	New Hampshire	65	65
California	70	55	New Jersey	65	65
Colorado	75	75	New Mexico	75	75
Connecticut	55	55	New York	65	65
Delaware	65	65	North Carolina	70	70
Florida	70	70	North Dakota	70	70
Georgia	70	70	Ohio	65	55
Hawaii	55	55	Oklahoma	75	75
Idaho	75	75	Oregon	65	55
Illinois	65	55	Pennsylvania	65	65
Indiana	65	60	Rhode Island	65	65
Iowa	65	65	South Carolina	65	65
Kansas	70	70	South Dakota	75	65
Kentucky	65	65	Tennessee	65	65
Louisiana	70	70	Texas	70	60
Maine	65	65	Utah	75	75
Maryland	65	65	Vermont	65	65
Massachusetts	65	65	Virginia	65	65
Michigan	70	55	Washington	70	60
Minnesota	70	70	West Virginia	70	70
Mississippi	70	70	Wisconsin	65	65
Missouri	70	70	Wyoming	75	75

Source: The World Almanac and Book of Facts 1998, p. 214. Data provided by National Motorists Assoc.

Many of these sets are on the StatSource CD.

  a. Build a cross-tabulation of the two variables, vehicle type and maximum speed limit on interstate highway. Express the results in frequencies, showing marginal totals.

b. Express the contingency table you derived in (a) in percentages based on the grand total.

c. Draw a bar graph showing the results from (b).

3.12 a. Express the contingency table you derived in Exercise 3.11 (a) in percentages based on the marginal total for speed limit.

b. Draw a bar graph showing the results from (b).

3.13 A statewide survey was conducted to investigate the relationship between viewers' preferences for ABC, CBS, NBC, or PBS for news information and their political party affiliation. The results are shown in tabular form:

	ABC	CBS	NBC	PBS
Democrat	200	200	250	150
Republican	450	350	500	200
Other	150	400	100	50

a. How many viewers were surveyed?

b. Why is this bivariate data? What type of variable is each one?

c. How many preferred to watch CBS?

d. What percentage of the survey was Republican?

e. What percentage of the Democrats preferred ABC?

3.14 Consider the accompanying contingency table, which presents the results of an advertising survey about the use of credit by Martan Oil Company customers:

Preferred Method of Payment	Number of Purchases at Gasoline Station Last Year					
	0–4	5–9	10–14	15–19	20 and Over	Sum
Cash	150	100	25	0	0	275
Oil company card	50	35	115	80	70	350
National or bank credit card	50	60	65	45	5	225
Sum	250	195	205	125	75	850

a. How many customers were surveyed?

b. Why is this bivariate data? What type of variable is each one?

c. How many customers preferred to use an oil-company credit card?

d. How many customers made 20 or more purchases last year?

e. How many customers preferred to use an oil-company credit card and made only between five and nine purchases last year?

f. What does the 80 in the fourth cell in the second row mean?

3.15 What effect does the minimum amount have on the interest rate being offered on six-month CDs? The following are advertised rate of return, y, for a minimum deposit of $500, $1000, or $2500, x in $100.

x	25	25	10	25	25	10	25	25	10	25	25	10	25	25
y	5.13	5.00	5.45	5.10	5.50	5.11	5.04	4.91	5.00	5.30	5.10	5.17	5.06	5.39

x	25	10	10	10	10	10	10	5	5	5	5	5	5
y	5.00	5.32	5.10	5.32	5.28	5.06	5.13	5.06	5.12	4.00	5.26	5.16	5.06

x	5	5	5	10	25	25	10	5	10	5	5	10	25
y	4.55	5.00	5.12	5.10	5.11	5.22	5.06	5.20	5.32	4.94	5.25	5.28	4.75

If you are using a computer or calculator for Ex. 3.15, try the commands on pages 97 and 137.

a. Prepare a dotplot of the three sets of data using a common scale.

b. Prepare a 5-number summary and a boxplot of the three sets of data. Use the same scale for the boxplots.

c. Describe any differences you see between the three sets of data.

3.16 Can a woman's height be predicted using her mother's height? The heights of some mother-daughter pairs are listed; x is the mother's height and y is the daughter's height.

all same info

x	63	63	67	65	61	63	61	64	62	63	
y	63	65	65	65	64	64	63	62	63	64	
x	64	63	64	64	63	67	61	65	64	65	66
y	64	64	65	65	62	66	62	63	66	66	65

$(x + y)$

a. Draw two dotplots using the same scale showing the two sets of data side by side.
b. What can you conclude from seeing the two sets of heights shown as separate sets this way? Explain.
c. Draw a scatter diagram of these data as ordered pairs.
d. What can you conclude from seeing the data presented as ordered pairs? Explain.

3.17 Listed below are the height, weight, and age of the players on the two teams that played in the 1994 World Cup finals, from Italy and Brazil:

Player	Ht-Ital	Wt-Ital	Age-Ital	Ht-Braz	Wt-Braz	Age-Braz
1	73	192	27	71	176	28
2	72	168	27	69	152	29
3	67	144	25	67	163	31
4	70	163	28	74	183	29
5	73	170	26	71	176	26
6	69	155	34	71	167	30
7	73	165	27	70	145	30
8	71	161	30	70	164	30
9	70	163	34	67	158	27
10	69	159	27	74	191	29
11	69	159	22	66	154	28
12	74	170	28	74	194	29
13	73	159	22	71	164	28
14	73	174	27	69	163	24
15	69	157	24	73	169	24
16	68	150	30	66	156	24
17	69	161	31	67	147	27
18	72	173	25	67	158	25
19	70	163	33	70	158	28
20	67	150	26	67	165	17
21	66	143	28	70	174	25
22	71	176	25	72	169	35

a. Compare each of the three variables—height, weight, and age—using either a dotplot or a histogram (use the same scale).
b. Based on what you see in the graphs in (a), can you detect a substantial difference between the two teams in regard to these three variables? Explain.
c. Explain why the data, as used in (a), was not bivariate data.

> If you are using a computer or calculator, try the commands on page 139.

3.18 a. Draw a scatter diagram showing height, x, and weight, y, for the Italian World Cup soccer team using the data in Exercise 3.17.
b. Draw a scatter diagram showing height, x, and weight, y, for the Brazilian World Cup soccer team using the data in Exercise 3.17.
c. Explain why the data, as used in (a) and (b), are bivariate data.

3.19 The accompanying data show the number of hours x studied for an exam and the grade y received in the exam. (y is measured in 10s; that is, $y = 8$ means that the grade, rounded to the nearest 10 points, is 80.) Draw the scatter diagram. (Retain this solution for use in answering Exercise 3.33, p. 152.)

x	2	3	3	4	4	5	5	6	6	6	7	7	7	8	8
y	5	5	7	5	7	7	8	6	9	8	7	9	10	8	9

3.20 An experimental psychologist asserts that the older a child is, the fewer irrelevant answers he or she will give during a controlled experiment. To investigate this claim, the following data were collected. Draw a scatter diagram. (Retain this solution for use in answering Exercise 3.34, p 152.)

Age (x)	2	4	5	6	6	7	9	9	10	12
Number of Irrelevant Answers (y)	12	13	9	7	12	8	6	9	7	5

3.21 In a study involving children's fear related to being hospitalized, the age and the score each child made on the Child Medical Fear Scale (CMFS) were:

Age (x)	8	9	9	10	11	9	9	9	11	11
CMFS score (y)	31	25	40	27	35	29	25	34	27	36

Construct a scatter diagram of these data. (Retain for use in answering Exercise 3.31, p. 152.)

3.22 A sample of 15 upper-class students who commute to classes was selected at registration. They were asked to estimate the distance (x) and the time (y) required to commute each day to class (see the following table). Construct a scatter diagram depicting these data.

Distance, x (nearest mile)	Time, y (nearest 5 minutes)	Distance, x (nearest mile)	Time, y (nearest 5 minutes)
18	20	2	5
8	15	15	25
20	25	16	30
5	20	9	20
5	15	21	30
11	25	5	10
9	20	15	20
10	25		

3.23 Walter Payton was one of the NFL's greatest running backs. Below are listed the number of carries and the total yards gained in each of his 13 years with the Chicago Bears.

Number Carries	196	311	339	333	369	317	339	148	314	381	324	321	146
Total Yards	679	1390	1852	1359	1610	1460	1222	596	1421	1684	1551	1333	586

a. Construct a scatter diagram depicting these data.

b. How would you describe this scatter diagram? What do you see that is unusual about the scatter diagram?

c. What circumstances might explain this "two group" appearance of the data points? Explain.

3.24 Total solar eclipses actually take place nearly as often as total lunar eclipses, but they are visible over a much narrower path. Both the path width and the duration vary substantially from one eclipse to the next. The table below shows the duration (seconds) and path width (miles) of 44 total solar eclipses measured in the past and those projected to the year 2010:

Date	Duration (s)	Width (mi)	Date	Duration (s)	Width (mi)
1950	73	83	1983	310	123
1952	189	85	1984	119	53
1954	155	95	1985	118	430
1955	427	157	1986	1	1
1956	284	266	1987	7	3
1958	310	129	1988	216	104
1959	181	75	1990	152	125
1961	165	160	1991	413	160
1962	248	91	1992	320	182
1963	99	63	1994	263	117
1965	315	123	1995	129	48
1966	117	52	1997	170	221
1968	39	64	1998	248	94
1970	207	95	1999	142	69
1972	155	109	2001	296	125
1973	423	159	2002	124	54
1974	308	214	2003	117	338
1976	286	123	2005	42	17
1977	157	61	2006	247	114
1979	169	185	2008	147	144
1980	248	92	2009	399	160
1981	122	67	2010	320	160

Source: The World Almanac and Book of Facts 1998.

a. Draw a scatter diagram showing duration, y, and path width, x, for the total solar eclipses.
b. How would you describe this diagram?

3.2   Linear Correlation

The primary purpose of **linear correlation analysis** is to measure the strength of a linear relationship between two variables. Let's examine some scatter diagrams demonstrating different relationships between input, or independent variables, x, and output, or dependent variables, y. If as x increases there is no definite shift in the values of y, we say there is **no correlation,** or no relationship between x and y. If as x increases there is a shift in the values of y, there is a correlation. The correlation is **positive** when y tends to increase and **negative** when y tends to decrease. If the ordered pairs (x, y) tend to follow a straight-line path, there is a linear correlation. The preciseness of the shift in y as x increases determines the strength of the **linear correlation.** The scatter diagrams in Figure 3.6 demonstrate these ideas.

Perfect linear correlation occurs when all the points fall exactly along a straight line, as shown in Figure 3.7. This can be either positive or negative, depending on whether y increases or decreases as x increases. If the data form a straight horizontal or vertical line, there is no correlation, since one variable has no effect on the other, as shown in Figure 3.8.

Figure 3.6
Scatter Diagram and Correlation

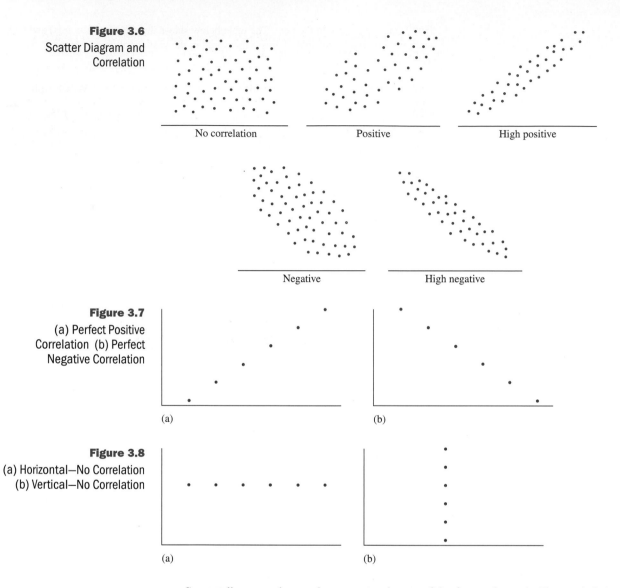

No correlation

Positive

High positive

Negative

High negative

Figure 3.7
(a) Perfect Positive Correlation (b) Perfect Negative Correlation

(a)

(b)

Figure 3.8
(a) Horizontal—No Correlation (b) Vertical—No Correlation

(a)

(b)

Scatter diagrams do not always appear in one of the forms shown in Figures 3.6, 3.7, and 3.8. Sometimes they suggest relationships other than linear, as in Figure 3.9. There appears to be a definite pattern; however, the two variables are not related linearly, and therefore there is no linear correlation.

Figure 3.9
No Linear Correlation

The **coefficient of linear correlation** r is the numerical measure of the strength of the linear relationship between two variables. The coefficient reflects the consistency of the effect that a change in one variable has on the other. The value of the linear correlation coefficient helps us to answer the question, "Is there a linear correlation between the two variables under consideration?" The linear correlation coefficient r always has a value between -1 and $+1$. A value of $+1$ signifies a perfect positive correlation, and a value of -1 shows a perfect negative correlation. If as x increases there is a general increase in the value of y, then r will be positive in value. For example, a positive value of r would be expected for age and height of children, because as children grow older, they grow taller. Also, consider the age x and resale value y of an automobile. As the car ages, its resale value decreases. Since as x increases, y decreases, the relationship results in a negative value for r.

The value of r is defined by **Pearson's product moment formula:**

$$r = \frac{\sum(x - \bar{x})(y - \bar{y})}{(n - 1)s_x s_y} \qquad (3.1)$$

NOTES

1. s_x and s_y are the standard deviations of the x and y variables.

2. The development of this formula is discussed in Chapter 13.

To calculate r, we will use an alternative formula, formula (3.2), that is equivalent to formula (3.1). As preliminary calculations, we will separately calculate three sums of squares and then substitute them into formula (3.2) to obtain r.

$$r = \frac{\textit{sum of squares for xy}}{\sqrt{(\textit{sum of squares for x})(\textit{sum of squares for y})}}$$

$$r = \frac{\text{SS}(xy)}{\sqrt{\text{SS}(x)\text{SS}(y)}} \qquad (3.2)$$

Recall the SS(x) calculation from formula 2.9 on page 79.

> SS(x) is the numerator of the variance, page 78.

$$\textit{sum of squares for x} = \textit{sum of } x^2 - \frac{(\textit{sum of x})^2}{n}$$

$$\text{where SS}(x) = \sum x^2 - \frac{(\sum x)^2}{n} \qquad (2.9)$$

$$\textit{sum of squares for y} = \textit{sum of } y^2 - \frac{(\textit{sum of y})^2}{n}$$

$$\text{SS}(y) = \sum y^2 - \frac{(\sum y)^2}{n} \qquad (3.3)$$

$$\textit{sum of squares for xy} = \textit{sum of xy} - \frac{(\textit{sum of x})(\textit{sum of y})}{n}$$

$$\text{SS}(xy) = \sum xy - \frac{\sum x \sum y}{n} \qquad (3.4)$$

ILLUSTRATION 3.8 ▼

Find the linear correlation coefficient for the push-up/sit-up data in Illustration 3.7, page 138.

Solution

First, we need to construct an extension table (Table 3.11) listing all the pairs of values (x, y) to aid in finding the extensions x^2, xy, and y^2 and the five column totals.

TABLE 3.11 Extensions Table for Finding Five Summations

Student	Push-ups (x)	x^2	Sit-ups (y)	y^2	xy
1	27	729	30	900	810
2	22	484	26	676	572
3	15	225	25	625	375
4	35	1,225	42	1,764	1,470
5	30	900	38	1,444	1,140
6	52	2,704	40	1,600	2,080
7	35	1,225	32	1,024	1,120
8	55	3,025	54	2,916	2,970
9	40	1,600	50	2,500	2,000
10	40	1,600	43	1,849	1,720
	$\sum x = 351$	$\sum x^2 = 13{,}717$	$\sum y = 380$	$\sum y^2 = 15{,}298$	$\sum xy = 14{,}257$
	↑	↑	↑	↑	↑
	sum of x	*sum of x²*	*sum of y*	*sum of y²*	*sum of xy*

Second, to complete the preliminary calculations, substitute the five summations (the five column totals) from the extensions table into formulas (2.9), (3.3), and (3.4), and calculate the three sums of squares.

> The $\sum$'s and SS's will be needed later for regression in Section 3.3. Be sure to save them!

$$SS(x) = \sum x^2 - \frac{(\sum x)^2}{n} = 13{,}717 - \frac{(351)^2}{10} = 1396.9$$

$$SS(y) = \sum y^2 - \frac{(\sum y)^2}{n} = 15{,}298 - \frac{(380)^2}{10} = 858.0$$

$$SS(xy) = \sum xy - \frac{\sum x \sum y}{n} = 14{,}257 - \frac{(351)(380)}{10} = 919.0$$

Third, substitute the three sums of squares into formula (3.2) and obtain the value of the correlation coefficient.

$$r = \frac{SS(xy)}{\sqrt{SS(x)SS(y)}} = \frac{919.0}{\sqrt{(1396.9)(858.0)}} = 0.8394 = \mathbf{0.84}$$

NOTE Typically, r is rounded to the nearest hundredth. ▲

Exercise 3.25

Does studying for an exam pay off? The number of hours studied, x, is compared to the exam grade, y:

x	2	5	1	4	2
y	80	80	70	90	60

a. Complete the preliminary calculations: extensions, five sums, and SS(x), SS(y), SS(xy).
b. Find r.

The value of the calculated linear correlation coefficient helps us answer the question "Is there a linear correlation between the two variables under consideration?" When the calculated value of r is close to zero, we conclude that there is little or no linear correlation. As the calculated value of r changes from 0.0 toward either $+1.0$ or -1.0, it indicates an increasingly stronger linear correlation between the two variables. From a graphical viewpoint, when we calculate r, we are measuring how well a straight line describes the scatter diagram of ordered pairs. As the value of r changes from 0.0 toward $+1.0$ or -1.0, the data points creating a pattern move closer to a straight line.

The following commands will calculate the correlation coefficient.

MINITAB

Input the *x*-variable data into C1 and the corresponding *y*-variable data into C2; then continue with:

Choose: **Stat > Basic Statistics > Correlation ...**

Enter: Variables: **C1 C2**

EXCEL

Input the *x*-variable data into column A and the corresponding *y*-variable data into column B, activate a cell for the answer; then continue with:

Choose: **Paste function, f$_x$ > Statistical > CORREL > OK**

Enter: Array 1: **x data range**
          Array 2: **y data range**

TI-83

Input the *x*-variable data into L1 and the corresponding *y*-variable data into L2; then continue with:

*Choose: **2nd > CATALOG > DiagnosticOn***
          > ENTER > ENTER

Choose: **STAT > CALC > 8:LinReg(a + bx)**

Enter: **L1, L2**

*DiagnosticOn must be selected for *r* and *r^2* to show. Once set, omit this step.

Understanding the Linear Correlation Coefficient

The following method will create: (1) a visual meaning for correlation, (2) a visual meaning for what the linear coefficient is measuring, and (3) an estimate for *r*. It is quick and generally yields a reasonable estimate when the "window of data" is approximately square.

NOTE This estimation technique *does not* replace the calculation of *r*. It is very sensitive to the "spread" of the diagram. However, if the "window of data" is approximately square, this approximation will be helpful when used as a mental estimate or check.

Procedure

1. Lay two pencils on your scatter diagram. Keeping them parallel, move them to a position so that they are as close together as possible yet have all the points on the scatter diagram between them. (See Figure 3.10).

Figure 3.10

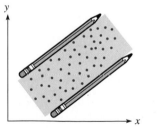

2. Visualize a rectangular region that is bounded by the two pencils and that ends just beyond the points on the scatter diagram. (See the shaded portion of Figure 3.10.)

3. Estimate how many times longer the rectangle is than it is wide. An easy way to do this is to mentally mark off squares in the rectangle. (See Figure 3.11.) Call this number of multiples k.

4. The value of r may be estimated as $\pm(1 - \frac{1}{k})$.

Case Study 3.2

Points and Fouls in Basketball

Albert Shulte discusses the relationship between the number of personal fouls committed and total points scored during a season by the members of a junior varsity basketball team. It appears, from Figure 1, the players score about three points for every personal foul committed. The data for the second year, Table 2, seem to indicate approximately the same relationship.

Sometimes you may think that two different sets of numbers are related in some way, although not perfectly. How can you decide if they are related? If they are related, is there any reason why they should be, or is it purely accidental? Does a change in one of the variables cause a change in the other?

The data come from the records of a junior varsity basketball team . . . for two separate years. . . . A quick look at Table 2 makes one feel that a strong relationship exists between the number of points . . . and the number of personal fouls. . . . In Figure 1, the data for the first year have been plotted. . . .

Table 2   Second Year

Player	Total Points	Personal Fouls
Brummett	2	1
Cooper	75	24
Felice	0	1
Hook	59	18
Hurd	9	9
Kampsen	7	3
McPartlin	35	5
Pointer	46	20
Schuback	0	1
Wilson	2	3
Zuelch	57	22

Figure 1
First Year

Exercise 3.26

a. Draw a scatter diagram of the Case Study 3.2 data in Table 2.

b. Calculate the correlation coefficient for the data in Table 2.

c. Does this mean that an increase in number of fouls will cause the player to score more points? What does a strong correlation coefficient mean? Explain in your own words.

Now let's think a bit more about the relationship that seems to exist. . . . The table and the figure both indicate that the more fouls [a player] commits, the more points he scores. . . . Surely no coach would coach a player to make lots of fouls in the hope that he would therefore score more points. The crucial word in the previous sentence is "therefore." Is the fact that a player commits more fouls the reason that the he scores more points? Of course not! But maybe both of these . . . are the results of some other act. Perhaps . . . more game time.

A graph such as that in Figure 1 can show that a high degree of correlation exists, but it does not tell us why.

Source: Albert P. Shulte, "Points and Fouls in Basketball," in *Exploring Data*, from *Statistics by Example*. Edited and prepared by the Joint Committee on the Curriculum in Statistics and Probability of the American Statistical Association and the National Council of Teachers of Mathematics. Copyright 1973 by Addison-Wesley Publishing Co., Inc. Reprinted by permission.

Figure 3.11

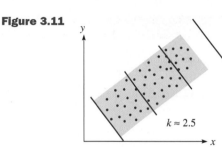

$k \approx 2.5$

5. The sign assigned to r is determined by the general position of the length of the rectangular region. If it lies in an increasing position (see Figure 3.12), r will be positive; if it lies in a decreasing position, r will be negative. If the rectangle is in either a horizontal or a vertical position, then r will be zero, regardless of the length–width ratio.

Exercise 3.27

Estimate the correlation coefficient for each of the following:

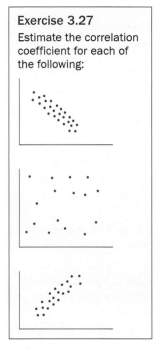

Figure 3.12

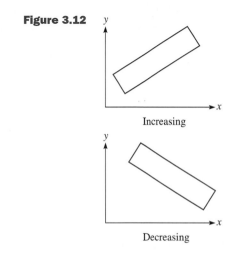

Increasing

Decreasing

Let's use this method to estimate the value of the linear correlation coefficient for the relationship between the number of push-ups and sit-ups. As shown in Figure 3.13, we find that the rectangle is approximately 3.5 times longer than it is wide; that is, $k \approx 3.5$, and the rectangle lies in an increasing position. Therefore, our estimate for r is

$$r \approx +(1 - \frac{1}{3.5}) \approx +0.7$$

Figure 3.13

Push-ups Versus Sit-ups for Ten Students

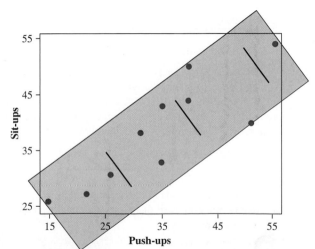

3.28 How would you interpret the findings of a correlation study that reported a linear correlation coefficient of -1.34?

3.29 How would you interpret the findings of a correlation study that reported a linear correlation coefficient of $+0.3$?

3.30 Explain why it makes sense for a set of data to have a correlation coefficient of zero when the data show a very definite pattern as in Figure 3.9 on page 146.

3.31 In an article titled "Self-Reported Fears of Hospitalized School-Age Children" (*Journal of Pediatric Nursing,* Vol. 9, No. 2, 1994), the authors report a correlation of 0.10 between the ages of children and the score they made on the Child Medical Fear Scale (CMFS). Suppose the ages and CMFS scores for ten children were as follows (same data as Exercise 3.21):

CMFS Score	31	25	40	27	35	29	25	34	27	36
Age	8	9	9	10	11	9	9	9	11	11

Find: **a.** SS(x) **b.** SS(y) **c.** SS(xy) **d.** the value of r for these data

3.32 Consider the following data, which give the weight (in thousands of pounds) x and gasoline mileage (miles per gallon) y for ten different automobiles.

x	2.5	3.0	4.0	3.5	2.7	4.5	3.8	2.9	5.0	2.2
y	40	43	30	35	42	19	32	39	15	44

Find: **a.** SS(x) **b.** SS(y) **c.** SS(xy) **d.** Pearson's product moment r

> Have you tried the correlation commands on your computer or calculator?

3.33 a. Use the scatter diagram you drew in answering Exercise 3.19 (p. 143) to estimate r for the sample data relating the number of hours studied and the exam grade.
  b. Calculate r.

3.34 a. Use the scatter diagram you drew in answering Exercise 3.20 (p. 144) to estimate r for the sample data relating the number of irrelevant answers and the child's age.
  b. Calculate r.

3.35 A marketing firm wished to determine whether or not the number of television commercials broadcast were linearly correlated to the sales of its product. The data, obtained from each of several cities, are shown in the following table.

City	A	B	C	D	E	F	G	H	I	J
No. TV Commercials (x)	12	6	9	15	11	15	8	16	12	6
Sales Units (y)	7	5	10	14	12	9	6	11	11	8

a. Draw a scatter diagram. **b.** Estimate r. **c.** Calculate r.

3.36 An article titled "Leader Power, Commitment Satisfaction, and Propensity to Leave a Job Among U.S. Accountants" (*Journal of Social Psychology,* Vol. 133, No. 5, Oct. 1993) reported a linear correlation coefficient of -0.61 between satisfaction with work scores and propensity to leave a job scores. Suppose similar assessments of work satisfaction, x, and propensity to leave a job, y, gave the following scores.

x	12	24	17	28	24	36	20
y	44	36	25	23	32	17	24

a. Find the linear correlation between x and y.
b. What does the value of this correlation coefficient seem to be telling us? Explain.

3.37 Cable television video networks can be measured on the basis of both their number of noncable affiliates and number of subscribers. Below is a table of 18 video networks, their affiliates, and subscribers (in millions):

Network	Affiliates	Subscribers
ESPN-1	27,600	71.1
CNN	11,528	71.0
TNT	10,538	70.5
TBS	11,668	69.9
C-SPAN	6,003	69.7
USA Network	12,500	69.7
TNN	17,636	68.9
LIFETIME Television	8,300	67.0
The Family Channel	13,352	66.9
Arts & Entertainment	12,000	66.9
MTV: Music Television	9,176	66.7
Nickelodeon	11,788	66.0
Nick at Nite	11,711	66.0
The Weather Channel	6,500	64.2
Headline News	6,470	64.0
CNBC	11,711	60.0
QVC Network	5,895	58.2
VH-1	6,088	56.3

Source: Cable Television Developments, National Cable Television Assn., Jan.–Mar. 1997.

a. Draw a scatter diagram of the two variables, affiliates, and subscribers.
b. Find the linear correlation coefficient and interpret the results.

3.38 An article titled "College Recreation Facility Survey" (*Athletic Business,* April 1994) reported the following results from 358 four-year colleges and universities in the United States and Canada.

Enrollment	Number of Schools	Total Square Feet Devoted to Recreation per School
0–1,249	58	47,864
1,250–2,499	53	71,828
2,500–4,999	53	89,716
5,000–9,999	62	101,016
10,000–17,999	68	127,952
18,000 or over	64	200,896

a. Using the midpoint of the first five enrollment classes and 25,000 in place of the class 18,000 or over for the *x*-values and the total square feet devoted to recreation per school for the *y*-values, find the linear correlation coefficient between *x* and *y*.
b. What does the value of this correlation seem to be telling us? Explain.

3.3   Linear Regression

Although the correlation coefficient measures the strength of a linear relationship, it does not tell us about the mathematical relationship between the two variables. In Section 3.2, the correlation coefficient for the push-up/sit-up data was found to be 0.84 (see p. 148). This implies that there is a linear relationship between the number of push-ups and the number of sit-ups a student does. The correlation coefficient does not help us predict the number of sit-ups a person can do based on knowing he or she can do 28 push-ups. **Regression analysis** finds the equation of the line that best describes the relationship between the two variables. One use of this equation is to make predictions. There are many situations in which we make use of these predictions regularly: for example, predicting the success a student will have in college based on high school results and predicting the distance required to stop a car based on its speed. Generally, the exact value of y is not predictable and we are usually satisfied if the predictions are reasonably close.

The relationship between these two variables will be an algebraic expression describing the mathematical relationship between x and y. Here are some examples of various possible relationships, called *models* or **prediction equations:**

Linear (straight-line): $\hat{y} = b_0 + b_1 x$

Quadratic: $\hat{y} = a + bx + cx^2$

Exponential: $\hat{y} = a(b^x)$

Logarithmic: $\hat{y} = a \log_b x$

Figures 3.14, 3.15, and 3.16 show patterns of bivariate data that appear to have a relationship, whereas in Figure 3.17, the variables do not seem to be related.

Figure 3.14

Linear Regression with Positive Slope

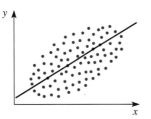

Figure 3.15

Linear Regression with Negative Slope

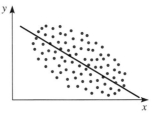

Figure 3.16

Curvilinear Regression (Quadratic)

Figure 3.17

No Relationship

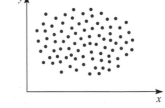

If a straight-line model seems appropriate, the best-fitting straight line is found by using the **method of least squares**. Suppose that $\hat{y} = b_0 + b_1 x$ is the equation of a straight line, where $\hat{y}$ (read "y hat") represents the **predicted value of y** that corresponds to a particular value of x. The **least squares criterion** requires that we find the constants b_0 and b_1 such that the sum $\sum(y - \hat{y})^2$ is as small as possible. (See Figure 3.18.)

Figure 3.18 shows the distance of an observed value of y from a **predicted value of $\hat{y}$**. The length of this distance represents the value $(y - \hat{y})$ (shown as the red line segment in Figure 3.18). Note that $(y - \hat{y})$ is positive when the point (x, y) is above the line and negative when (x, y) is below the line.

Figure 3.18

Observed and
Predicted Values of y

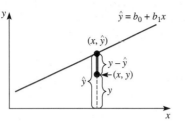

Figure 3.19 shows a scatter diagram with what might appear to be the **line of best fit,** along with the ten individual $(y - \hat{y})$'s. (Positive ones are shown in red, negative ones in green.) The sum of the squares of these differences is minimized (made as small as possible) if the line is indeed the line of best fit.

Figure 3.19

The Line of Best Fit

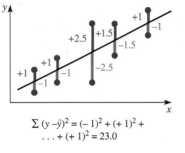

Figure 3.20 shows the same data points as Figure 3.19 with the ten individual $(y - \hat{y})$'s associated with a line that is definitely not the line of best fit. (The value of $\sum (y - \hat{y})^2$ is 149, much larger than 23 from Figure 3.19.) Every different line drawn through this set of ten points will cause a different value for $\sum(y - \hat{y})^2$. Our job is to find the one line that will result in $\sum (y - \hat{y})^2$ being the smallest possible value.

Figure 3.20

Not the Line of Best Fit

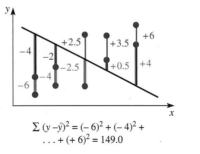

The equation of the line of best fit is determined by its **slope (b_1)** and its ***y*-intercept (b_0).** (See the *Statistical Tutor* for a review of the concepts of slope and intercept of a straight line.) The values of the constants—slope and y-intercept—that satisfy the least squares criterion are found by using these formulas:

$$\text{slope: } b_1 = \frac{\sum(x - \bar{x})(y - \bar{y})}{\sum(x - \bar{x})^2} \tag{3.5}$$

We will use a mathematical equivalent of formula (3.5) to find slope b_1 that uses the sums of squares found in the preliminary calculations for correlation:

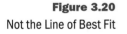

$$\text{slope: } b_1 = \frac{SS(xy)}{SS(x)} \tag{3.6}$$

Notice that the numerator of formula (3.6) is the SS(xy) formula (3.4) and the denominator is formula (2.9) from the correlation coefficient calculations. Thus, if you have previously calculated the linear correlation coefficient using the procedure outlined on pages 147 and 148, you can easily find the slope of the line of best fit. If you did not previously calculate r, set up a table similar to Table 3.11 (p. 148) and complete the necessary preliminary calculations.

$$y\text{-intercept} = \frac{(sum\ of\ y) - [(slope)(sum\ of\ x)]}{number}$$

$$y\text{-intercept:} \quad b_0 = \frac{\sum y - (b_1 \cdot \sum x)}{n} \qquad \textbf{(3.7)}$$

alternative for b_0: $b_0 = \bar{y} - (b_1 \cdot \bar{x})$ $\qquad$ **(3.7a)**

(The derivation of these formulas is beyond the scope of this text.)

Now let's consider the data in Illustration 3.7 (p. 138) and the question of predicting a student's sit-ups based on the number of push-ups. We want to find the line of best fit, $\hat{y} = b_0 + b_1 x$. The preliminary calculations have already been completed in Table 3.11 and on page 148. To calculate the slope, b_1, using formula (3.6), recall SS(xy) = 919.0 and SS(x) = 1396.9.

$$\text{slope:} \quad b_1 = \frac{\text{SS}(xy)}{\text{SS}(x)} = \frac{919.0}{1396.9} = 0.6579 = \textbf{0.66}$$

To calculate the y-intercept, b_0, using formula (3.7), recall $\sum x = 351$ and $\sum y = 380$ from the extensions table.

$$y\text{-intercept:} \quad b_0 = \frac{\sum y - (b_1 \cdot \sum x)}{n} = \frac{380 - (0.6579)(351)}{10}$$

$$= \frac{380 - 230.9229}{10} = 14.907 = \textbf{14.9}$$

By placing the two values just found into the model, $\hat{y} = b_0 + b_1 x$, the equation of the line of best fit is

$$\hat{y} = \textbf{14.9} + \textbf{0.66}x$$

NOTES

1. Remember to keep at least three extra decimal places while doing the calculations to ensure an accurate answer.

2. When rounding off the calculated values of b_0 and b_1, always keep at least two significant digits in the final answer.

Now that we know the equation for the line of best fit, let's draw the line on the scatter diagram so that we can visualize the relationship between the line and the data. Two points will be needed in order to draw the line on the diagram. Select two convenient x-values, one near each extreme of the domain ($x = 10$ and $x = 60$ are good choices for this illustration), and find their corresponding y-values.

For $x = 10$: $\hat{y} = 14.9 + 0.66x = 14.9 + 0.66(10) = 21.5$; **(10, 21.5)**

For $x = 60$: $\hat{y} = 14.9 + 0.66x = 14.9 + 0.66(60) = 54.5$; **(60, 54.5)**

These two points (10, 21.5) and (60, 54.5) are then located on the scatter diagram (use a blue + to distinguish them from data points) and the line of best fit is drawn (shown in red in Figure 3.21 on page 157).

Exercise 3.39

Show that formula 3.7a is equivalent to formula 3.7.

Exercise 3.40

The formulas for finding the slope and the y-intercept of the line of best fit use both summations, $\sum$'s, and sums of squares, SS( )'s. It is important to know the difference. In reference to Illustration 3.8 (p. 147):

a. Find three pairs of values: $\sum x^2$, SS(x); $\sum y^2$, SS(y) and $\sum xy$, SS(xy).
b. Explain the difference between the numbers for each pair of numbers.

Exercise 3.41

The values of x used to find points for graphing the line $\hat{y} = 14.9 + 0.66x$ are arbitrary. Suppose you choose to use $x = 20$ and then $x = 50$.

a. What are the corresponding $\hat{y}$-values?
b. Locate these two points on Figure 3.21. Are these points on the line of best fit? Explain why or why not.

There are some additional facts about the least squares method that we need to discuss.

1. The slope b_1 represents the predicted change in y per unit increase in x. In our example, where $b_1 = 0.66$, if a student can do an additional ten push-ups (x), we would predict that he or she would be able to do approximately an additional seven (0.66×10) sit-ups (y).

2. The y-intercept is the value of y where the line of best fit intersects the y-axis. (The y-intercept is easily seen on the scatter diagram, shown in a green + in Figure 3.21, when the vertical scale is located above $x = 0$.) However, in interpreting b_0 you first must consider whether $x - 0$ is a realistic x-value before you conclude that you would predict $\hat{y} = b_0$ if $x = 0$. To predict that if a student did no push-ups he or she would still do approximately 15 sit-ups ($b_0 = 14.9$) is probably incorrect. Second, the x-value of zero is outside the domain of the data on which the regression line is based. In predicting y based on an x-value, check to be sure that the x-value is within the domain of the x-values observed.

3. The line of best fit will always pass through the centroid, the point $(\bar{x}, \bar{y})$.

   When drawing the line of best fit on your scatter diagram, use this point as a check. For our illustration,

$$\bar{x} = \frac{\Sigma x}{n} = \frac{351}{10} = 35.1, \qquad \bar{y} = \frac{\Sigma y}{n} = \frac{380}{10} = 38.0;$$

therefore, $(\bar{x}, \bar{y}) = (35.1, 38.0)$, as shown in green $\oplus$ in Figure 3.21.

Exercise 3.42

Does it pay to study for an exam? The number of hours studied, x, is compared to the exam grade, y:

x	2	5	1	4	2
y	80	80	70	90	60

a. Find the equation for the line of best fit.
b. Draw the line of best fit on the scatter diagram of the data drawn in Exercise 3.8 (p. 139).
c. Based on what you see in answers (a) and (b), does it pay to study for an exam? Explain.

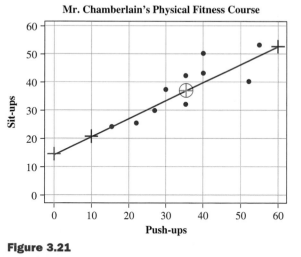

Figure 3.21

Line of Best Fit for Push-ups Versus Sit-ups

ILLUSTRATION 3.9 ▼

In a random sample of eight college women, each was asked for her height (to the nearest inch) and her weight (to the nearest five pounds). The data obtained are shown in Table 3.12. Find an equation to predict the weight of a college woman based on her height (the equation of the line of best fit), and draw it on the scatter diagram in Figure 3.22.

TABLE 3.12 Data for College Women's Heights and Weights

	1	2	3	4	5	6	7	8
Height (x)	65	65	62	67	69	65	61	67
Weight (y)	105	125	110	120	140	135	95	130

Solution

Before we start the process of finding the equation for the line of best fit, it is often helpful to draw the scatter diagram, which will give you visual insight about the relationship between the two variables. The scatter diagram for the data on the height and weight of college women, shown in Figure 3.22, indicates that the linear model is appropriate.

Figure 3.22

Scatter Diagram

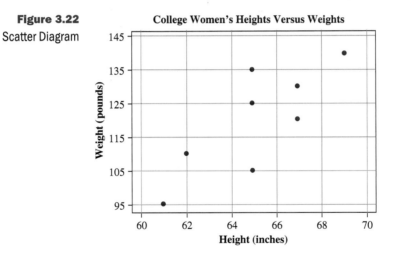

To find the equation for the line of best fit, we first need to complete the preliminary calculations, as shown in Table 3.13.

TABLE 3.13 Preliminary Calculations Needed to Find b_1 and b_0

Student	Height (x)	x^2	Weight (y)	xy
1	65	4,225	105	6,825
2	65	4,225	125	8,125
3	62	3,844	110	6,820
4	67	4,489	120	8,040
5	69	4,761	140	9,660
6	65	4,225	135	8,775
7	61	3,721	95	5,795
8	67	4,489	130	8,710
	$\sum x = 521$	$\sum x^2 = 33,979$	$\sum y = 960$	$\sum xy = 62,750$

The other preliminary calculations include finding SS(x), formula (2.9), and SS(xy), formula (3.4).

$$SS(x) = \sum x^2 - \frac{(\sum x)^2}{n} = 33{,}979 - \frac{(521)^2}{8} = 48.875$$

$$SS(xy) = \sum xy - \frac{\sum x \sum y}{n} = 62{,}750 - \frac{(521)(960)}{8} = 230.0$$

Second, we need to find the slope and the y-intercept using formulas (3.6) and (3.7).

slope: $b_1 = \dfrac{SS(xy)}{SS(x)} = \dfrac{230.0}{48.875} = 4.706 = \mathbf{4.71}$

y-intercept: $b_0 = \dfrac{\sum y - (b_1 \cdot \sum x)}{n} = \dfrac{960 - (4.706)(521)}{8} = -186.478 = \mathbf{-186.5}$

Thus, the equation of the line of best fit is

$$\hat{y} = -186.5 + 4.71x$$

To draw the line of best fit on the scatter diagram, we need to locate two points. Substitute two values for x, for example, 60 and 70, into the equation for the line of best fit and obtain two corresponding values for $\hat{y}$:

$\hat{y} = -186.5 + 4.71x = -186.5 + (4.71)(60) = -186.5 + 282.6 = 96.1 = 96$ and

$\hat{y} = -186.5 + 4.71x = -186.5 + (4.71)(70) = -186.5 + 329.7 = 143.2 = 143$

The values (60, 96) and (70, 143) represent two points (designated by a + and shown in red in Figure 3.23) that enable us to draw the line of best fit.

Figure 3.23

Scatter Diagram with Line of Best Fit

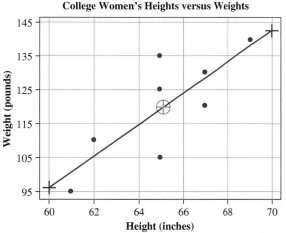

College Women's Heights versus Weights

Exercise 3.43

For Illustration 3.9 and the scatter diagram in Figure 3.23:

a. Explain how the slope of 4.71 can be seen.
b. Explain why the y-intercept of -186.5 cannot be seen.

NOTE In Figure 3.23, $(\overline{x}, \overline{y}) = (65.1, 120)$ is also on the line of best fit. It is the green cross in the circle. Use $(\overline{x}, \overline{y})$ as a check on your work.

Making Predictions

One of the main purposes for obtaining a regression equation is for making predictions. Once a linear relationship has been established and the value of the input variable x is known, we can predict a value of y, $\hat{y}$. For example, in the physical fitness illustration,

Exercise 3.44

If all students who can
do 40 push-ups are
asked to do as many sit-
ups as possible:

a. How many sit-ups
   do you expect each
   can do?
b. Will they all be able
   to do the same
   number?
c. Explain the meaning
   of answer (a).

the equation was found to be $\hat{y} = 14.9 + 0.66x$. If student A can do 25 push-ups, how many sit-ups do you therefore predict that A will be able to do? The predicted value is

$$\hat{y} = 14.9 + 0.66x = 14.9 + (0.66)(25) = 14.9 + 16.5 = 31.4 = \mathbf{31}$$

You should not expect this predicted value to occur exactly; rather, it is the average number of sit-ups that you would expect from all students who could do 25 push-ups.

When making predictions based on the line of best fit, observe the following restrictions:

1. The equation should be used to make predictions only about the population from which the sample was drawn. For example, using the relationship between the height and the weight of college women to predict the weight of professional athletes given their height would be questionable.

2. The equation should be used only within the sample domain of the input variable. We know the data demonstrate a linear trend within the domain of the x data, but we do not know what the trend is outside this interval. Hence predictions can be very dangerous outside of the domain of the x data. For example, in Illustration 3.9 to predict that a college woman of height zero will weigh -186.5 pounds is nonsense. Do not use a height outside the sample domain of 61 to 69 inches to predict weight. On occasion you might wish to use the line of best fit to estimate values outside the domain interval of the sample. This can be done, but you should do it with caution and only for values close to the domain interval.

3. If the sample was taken in 1998, do not expect the results to have been valid in 1929 or to hold in 2010. The women of today may be different from the women of 1929 and the women of 2010.

The following commands will find the equation of the line of best fit.

MINITAB

```
Input the x-values into C1 and the corresponding y-values
into C2; then continue with:

    Choose:  Stat > Regression > Regression…
    Enter:   Response (y): C2
             Predictors (x): C1
    Choose:  Storage
    Select:  FITS (calculates predicted y values and stores
             them in C3 or the first available column)

To draw the scatter diagram with the line of best fit super-
imposed on the data points, FITS must have been selected
above; then continue with:

    Choose:  Graph > Plot
    Enter:   Graph 1: Y: C2 X2: C1
    Choose:  Annotation > Title
    Enter:   your title > OK
    Choose:  Annotation > Line
    Enter:   Points: C1 C3
             Type: Solid > OK
```

EXCEL

Input the x-variable data into column A and the corresponding y-variable data into column B; then continue with:

Choose: **Tools > Data Analysis > Regression > OK**
Enter: Input Y Range: **(B1:B10 or select cells)**
           Input X Range: **(A1:A10 or select cells)**
Select: **Labels** (if necessary)
           Output Range
           Enter: **(C1 or select cell)**
           Line Fits Plots

To make the output readable, continue with:

Choose: **Format > Column > Autofit Selection**

To form the regression equation, the y-intercept is located
at the intersection of intercept and coefficients columns,
whereas the slope is located at the intersection of the x
variable and the coefficients columns.

To draw the line of best fit on the scatter diagram, activate
the chart; then continue with:

Choose: **Chart > Add Trendline > Linear**

(This command also works with the scatter diagram EXCEL commands on page 140.)

TI-83

Input the x-variable data into L1 and the corresponding y-
variable data into L2; then continue with:

If just the equation is desired:

Choose: **STAT > CALC > 8:LinReg(a + bx)**
Enter: **L1, L2***

*If the equation and graph on the scatter diagram are de-
sired, use:
Enter: **L1, L2, Y1****

then continue with the same commands for a scatter diagram
as shown on page 139.
**To enter Y1; use:
Choose: **VARS > Y-VARS > 1:Function > 1:Y1 > ENTER**

Understanding the Line of Best Fit

The following method will create: (1) a visual meaning for the line of best fit, (2) a visual meaning for what the line of best fit is describing, and (3) an estimate for the slope
and y-intercept of the line of best fit. As with the approximation of r, estimations of the
slope and y-intercept of the line of best fit should be used only as a mental estimate or
check.

NOTE This estimation techniques *does not* replace the calculations for b_1 and b_0.

Case Study 3.3

Thirty-five Years of Car Prices and Incomes

The following table lists the average price of a new car and the median annual family income for each of several years. The average price of a new car seems to be approximately one-half the annual median family income for the same year. Do you think there is a linear relationship?

Exercise 3.45

a. Sketch a scatter diagram for family income vs. price of car, year vs. price of car, year vs. family income (*x* vs. *y*).

b. Are the relationships linear? Explain.

c. Does the average price of a new car seem to be approximately one-half the annual median family income? Explain.

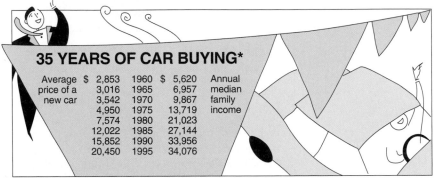

35 YEARS OF CAR BUYING*

Average price of a new car				Annual median family income
$	2,853	1960	$ 5,620	
	3,016	1965	6,957	
	3,542	1970	9,867	
	4,950	1975	13,719	
	7,574	1980	21,023	
	12,022	1985	27,144	
	15,852	1990	33,956	
	20,450	1995	34,076	

*Statistics from the U.S. Department of Commerce, supplied by the Motor Vehicle Manufacturers Association.

Source: New Woman, November 1990. Statistics from the U.S. Department of Commerce, supplied by the Motor Vehicles Manufacturers Association of the United States, and *Statistical Abstract of the United States 1995.* Graphics by Laura Wallace.

Procedure

1. On the scatter diagram of the data, draw the straight line that appears to be the line of best fit. [*Hint:* If you draw a line parallel to and halfway between the two pencils, whose location was described in Section 3.2, p. 149, you will have a reasonable estimate for the line of best fit.] The two pencils border the "path" demonstrated by the ordered pairs, and the line down the center of this path approximates the line of best fit. Figure 3.24 shows the pencils and the resulting estimated line for illustration 3.9.

Figure 3.24

Estimate the Line of Best Fit for the College Women Data

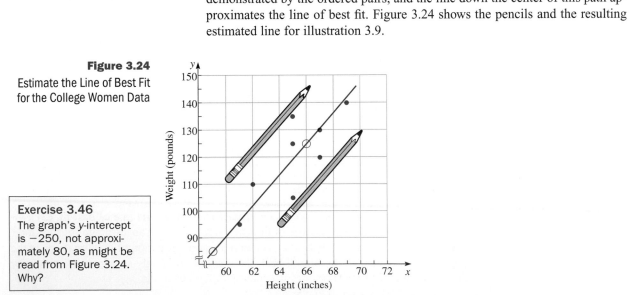

Exercise 3.46

The graph's *y*-intercept is −250, not approximately 80, as might be read from Figure 3.24. Why?

Exercise 3.47

The choice of the two points, (x_1, y_1) and (x_2, y_2), is somewhat arbitrary. When different points are selected, slightly different values for b_0 and b_1 will result, but they should be approximately the same. Use points (61, 95) and (67, 130) and find the slope and the y-intercept.

2. This line can now be used to approximate the equation. First, locate two points, (x_1, y_1) and (x_2, y_2), along the line and determine their coordinates. Two such points, circled in Figure 3.24, have the coordinates (59, 85) and (66, 125). These two pairs of coordinates can now be used in the following formula to obtain an estimate for the slope b_1.

$$\text{Estimation for the slope, } b_1 \quad b_1 \approx \frac{y_2 - y_1}{x_2 - x_1} = \frac{125 - 85}{66 - 59} = \frac{40}{7} = 5.7$$

3. Using this result, the coordinates of one of the points, and the following formula, we can determine an estimation for the y-intercept b_0:

$$\text{Estimate for the y-intercept, } b_0: b_0 \approx y - b_1 \cdot x = 85 - (5.7)(59) =$$
$$85 - 336.3 = -251.3; \text{ thus } b_0 \text{ is approximately } -250$$

4. We now can write the estimated equation for the line of best fit to be $\hat{y} = -250 + 5.7x$. This should serve as a crude estimate.

EXERCISES • • • • • • • • • • • • • •

3.48 Draw a scatter diagram for these data:

x	2	12	4	6	9	4	11	3	10	11	3	1	13	12	14	7	2	8
y	4	8	10	9	10	8	8	5	10	9	8	3	9	8	8	11	6	9

Would you be justified in using the techniques of linear regression on these data to find the line of best fit? Explain.

3.49 Ajay used linear regression to help him understand his monthly telephone bill. The line of best fit was $\hat{y} = 23.65 + 1.28x$; x is the number of long-distance calls made during a month and y is the total telephone cost for a month. In terms of number of long distance calls and cost:
   a. Explain the meaning of the y-intercept, 23.65.
   b. Explain the meaning of the slope, 1.28.

3.50 A study investigating the relationship between the cost y (in tens of thousands of dollars) per unit of equipment manufactured and the number of units produced per run x was conducted. The resulting equation for the line of best fit was $\hat{y} = 7.31 - 0.01x$, with x being observed for values between 10 and 200. If a production run were scheduled to produce 50 units, what would you predict the cost per unit to be?

3.51 A study was conducted to investigate the relationship between resale price y (in hundreds of dollars) and the age x (in years) of midsize luxury American automobiles. The equation of the line of best fit was $\hat{y} = 185.7 - 21.52x$.
   a. Find the resale value of such a car when it is three years old.
   b. Find the resale value of such a car when it is six years old.
   c. What is the average annual decrease in the resale price of these cars?

3.52 An article titled "Microbioluminometry Assay (MBA): Determination of Erythromycin Activity in Plasma or Serum" (*Journal of Pharmaceutical Sciences,* Dec. 1989) compares the MBA assay with another method, agar diffusion plate assay, for determining the erythromycin activity in plasma or serum. The MBA technique is economical because it requires less sample and reagents.

$$Y = 0.054128 + 0.92012X \qquad r = 0.9525 \qquad n = 206$$

If the agar diffusion plate assay determined the level of erythromycin to be 1200, predict what the MBA assay would be.

3.53 In the article "Beyond Prediction: The Challenge of Minority Achievement in Higher Education" (Lunneborg and Lunneborg, *Journal of Multicultural Counseling and Development*), the relationship between high school GPA and first-year university GPA was investigated for the following groups: Asian Americans, blacks, chicanos, Native Americans, and whites. For the 43 Native Americans, the correlation was found to be 0.26 and the equation of the line of best fit was found to be university GPA = 1.85 + 0.30 × high school GPA. Both GPAs were 4.0 scales.

Use the equation of the line of best fit to predict the mean first-year university GPA for all Native Americans who had a high school GPA equal to 3.0.

3.54 People not only live longer today but also live longer independently. The May/June 1989 issue of *Public Health Reports* published an article titled "A Multistate Analysis of Active Life Expectancy." Two of the variables studied were a person's current age and the expected number of years remaining.

Age x	65	67	69	71	73	75	77	79	81	83
Years Remaining y	16.5	15.1	13.7	12.4	11.2	10.1	9.0	8.4	7.1	6.4

a. Draw a scatter diagram.
b. Calculate the equation of best fit.
c. Draw the line of best fit on the scatter diagram.
d. What are the expected years remaining for a person who is 70 years old? Find the answer in two different ways: Use the equation from (b) and use the line on the scatter diagram from (c).
e. Are you surprised that the data all lie so close to the line of best fit? Explain why the ordered pairs follow the line of best fit so closely.

> Have you tried the computer or calculator commands yet?

3.55 A record of maintenance costs is kept for each of several cash registers throughout a department store chain. A sample of 14 registers gave the following data:

Age x (years)	Maintenance Cost y (dollars)	Age x (years)	Maintenance Cost y (dollars)
6	142	2	99
7	231	1	114
1	73	9	191
3	90	3	160
6	176	8	155
4	132	9	231
5	167	8	202

a. Draw a scatter diagram that shows these data.
b. Calculate the equation of the line of best fit.
c. A particular cash register is eight years old. How much maintenance (cost) do you predict it will require this year?
d. Interpret your answer to (c).

3.56 The following data are the ages and the asking prices for 19 used foreign compact cars:

Age x (years)	Price y (× $100)	Age x (years)	Price y (× $100)
3	68	6	42
5	52	8	22
3	63	5	50
6	24	6	36
4	60	5	46
4	60	7	36
6	28	4	48
7	36	7	20
2	68	5	36
2	64		

a. Draw a scatter diagram.
b. Calculate the equation of the line of best fit.
c. Graph the line of best fit on the scatter diagram.
d. Predict the average asking price for all such foreign cars that are five years old. Obtain this answer in two ways: Use the equation from (b) and use the line drawn in (c).

3.57 The success of a professional golfer can be measured along a number of dimensions. The bottom line is probably how much money a golfer earns in a given year, but golfers are also given a world ranking by points for each event that they enter. Below is a combined table extracted from *Golf* magazine that shows both earnings and world ranking of 20 players on the PGA tour, midway through the 1998 season:

Player	Earnings ($)	World Ranking
David Duval	1,272,305	8.77
Fred Couples	1,056,533	6.47
Tiger Woods	1,056,086	11.91
Justin Leonard	1,052,346	8.76
Mark O'Meara	894,724	7.49
Phil Mickelson	788,800	7.72
Mark Calcavecchia	766,224	5.72
Davis Love III	621,987	10.67
Ernie Els	601,363	12.35
Jesper Parnevik	573,855	5.49
Jeff Maggert	563,376	5.04
Tom Lehman	531,562	7.66
Scott Hoch	502,689	5.73
Lee Westwood	501,340	6.36
Jim Furyk	486,615	5.67
Tom Watson	418,385	5.25
Vijay Singh	383,979	6.59
Nick Price	296,668	7.79
Steve Jones	285,416	4.79
Colin Montgomery	272,000	9.03

Source: Golf, "Stats+," July 1998.

a. Draw a scatter diagram with money earnings as the dependent variable, y, and world ranking as the predictor variable, x.
b. Calculate the equation of best fit.
c. Draw the line of best fit on the scatter diagram you obtained in (a).
d. Suppose a player not shown on this list had obtained a world ranking of 7.0 from the experts. What would you predict his mid-year money earnings to be?

3.58 An article titled "Women, Work and Well-Being: The Importance of Work Conditions (*Health and Social Behavior,* Vol. 35, No. 3, Sept. 1994) studied 202 full-time homemakers and 197 employed wives. The linear correlation coefficient between family income and education was reported to equal 0.43 for the participants in the study. A similar study involving eight individuals gave the following results (x represents the years of education, and y represents the family income in thousands of dollars):

x	12	13	10	14	11	14	16	16
y	34	45	36	47	43	35	50	42

a. Find the linear correlation between x and y.
b. Find the equation of the line of best fit.

Return to CHAPTER CASE STUDY

As a way of assessing the statistical techniques for bivariate data that we have learned in this chapter, let's return to the Chapter Case Study. You have probably thought about the number of hours you spend at work or at leisure, but have you ever wondered about the relationship between the number of hours that all Americans spend per week at work and at leisure?

EXERCISE 3.59

a. How many hours did you, personally, spend working last week? How many hours did you spend at leisure pursuits?

Using the information in the USA Snapshot,® "Balancing work and play," on page 129:

b. Construct one graph using years as a horizontal axis, hours as the vertical axis, and plot hours of work and hours of leisure using two different colored dots. Connect the dots.
c. Do you see any pattern(s) to the graph? Explain.
d. Is there a relationship between the two variables, hours spent at work and hours devoted to leisure? When people spend more time at one, do they spend less time at the other? Use the techniques learned in this chapter to present graphic and numerical statistics that aid in the explanation of your answers to these questions.

IN RETROSPECT

To sum up what we have just learned; there is a distinct difference between the purpose of regression analysis and the purpose of correlation. In regression analysis, we seek a relationship between the variables. The equation that represents this relationship may be the answer desired, or it may be the means to the prediction that is desired. In correlation analysis, we measure the strength of the linear relationship between the two variables.

The case studies show a variety of applications for the techniques of correlation and regression. These articles are worth reading again. When bivariate data appear to fall along a straight line on the scatter diagram, they suggest a linear relationship. But this is not proof of cause and effect. Clearly, if a basketball player commits too many fouls, he will not be scoring more points. He will be in foul trouble and "riding the pine" with no chance to score. It also seems reasonable that the more game time he has, the more points he will score and the more fouls he will commit. Thus, a positive correlation and a positive regression relationship will exist between these two variables.

The bivariate linear methods we have studied thus far have been presented for the purpose of a first, descriptive look. More details must, by necessity, wait until additional developmental work has been completed. After completing this chapter, you should have a basic understanding of bivariate data, how they are different from just two sets of data, how to present them, what correlation and regression analysis are, and how each is used.

CHAPTER EXERCISES

3.60 "Fear of the dentist" (or the dentist's chair) is an emotion felt by many people of all age groups. A survey of 100 individuals in each of five age groups was conducted about this fear, and the results were as follows:

	Elementary	Jr. High	Sr. High	College	Adult
No. Who Fear	37	28	25	27	21
No. Who Do Not Fear	63	72	75	73	79

   a. Find the marginal totals.
   b. Express the table as percentages of the grand total.
   c. Express the table as percentages of each age group's marginal totals.
   d. Express the table as percentages of those who fear and those who do not fear.
   e. Draw a bar graph based on age groups.

3.61 The USA Snapshot® "Rainy day savings" lists in percentages the distributions for the amount both genders have saved for emergencies.

   a. Identify the population, the variables, and the type of variables.
   b. Construct a bar graph showing the two distributions side by side.
   c. Do the distributions seem to be different for the genders? Explain.

USA SNAPSHOTS®

A look at statistics that shape the nation

Rainy day savings

Among workers ages 25-64, 62% of men and 53% of women have savings set aside for emergencies. What they have:

	Men	Women
Less than a month's income	12%	18%
1 to less than 3 months	31%	24%
3 to less than 6 months	21%	29%
6 or more months income	36%	26%
Don't know	0%	3%

Source: Merrill Lynch   By Anne R. Carey and Grant Jerding, USA TODAY

3.62 Six breeds of dogs have been rather popular in the United States over the past few years. The table below lists the breed coupled with the number of registrations filed with the American Kennel Club in 1995 and 1996:

Breed	1995	1996
Labrador Retriever	132,051	149,505
Rottweiler	93,656	89,867
German Shepherd	78,088	79,076
Golden Retriever	64,107	68,993
Beagle	57,063	58,946
Poodle	54,784	56,803

Source: American Kennel Club, New York, NY.

   a. Build a cross-tabulation of the two variables, year (rows) and dog breed (columns). Express the results in frequencies, showing marginal totals.
   b. Express the contingency table you derived in (a) in percentages based on the grand total.
   c. Draw a bar graph showing the results from (b).
   d. Express the contingency table you derived in (a) in percentages based on the marginal total for the year.
   e. Draw a bar graph showing the results from (d).

3.63 When was the last time you saw your doctor? That was the question asked for the survey summarized below:

		Time Since Last Consultation with Your Physician		
		Less Than 6 Months	6 Months to Less Than 1 Year	1 Year or More
Age	Under 28 years	413	192	295
	28–40	574	208	218
	Over 40	653	288	259

a. Find the marginal totals.
b. Express the table as percentages of the grand total.
c. Express the table as percentages of each age group's marginal totals.
d. Express the table as percentages of each time period.
e. Draw a bar graph using percentages based on the grand total.

3.64 Part of quality control is keeping track of what is occurring. The contingency table below shows the number of rejected castings that occurred last month.

	Causes for Rejection of Casting		
	1st Shift	2nd Shift	3rd Shift
Sand	87	110	72
Shift	16	17	4
Drop	12	17	16
Corebreak	18	16	33
Broken	17	12	20
Other	8	18	22

a. Find the marginal totals.
b. Express the table as percentages of the grand total.
c. Express the table as percentages of each shift's marginal totals.
d. Express the table as percentages of each type of rejection.
e. Draw a bar graph based on the shifts.

3.65 Determine whether each of the following questions requires correlation analysis or regression analysis to obtain an answer.
a. Is there a correlation between the grades a student attained in high school and the grades he or she attained in college?
b. What is the relationship between the weight of a package and the cost of mailing it first class?
c. Is there a linear relationship between a person's height and shoe size?
d. What is the relationship between the number of worker-hours and the number of units of production completed?
e. Is the score obtained on a certain aptitude test linearly related to a person's ability to perform a certain job?

3.66 An automobile owner records the number of gallons of gasoline, x, required to fill the gasoline tank and the number of miles traveled, y, between fill-ups.
a. If she does a correlation analysis on the data, what would be her purpose and what would be the nature of her results?
b. If she does a regression analysis on the data, what would be her purpose and what would be the nature of her results?

3.67 The following data were generated using the equation $y = 2x + 1$.

x	0	1	2	3	4
y	1	3	5	7	9

(continued)

A scatter diagram of these data results in five points that fall perfectly on a straight line. Find the correlation coefficient and the equation of the line of best fit.

3.68 Consider this set of bivariate data:

x	1	1	3	3
y	1	3	1	3

a. Draw a scatter diagram.
b. Calculate the correlation coefficient.
c. Calculate the line of best fit.

3.69 Start with the point (5, 5) and add at least four ordered pairs, (x, y), to make a set of ordered pairs that display the following properties. Show that your sample satisfies the requirements.
a. The correlation of x and y is 0.0.
b. The correlation of x and y is +1.0,
c. The correlation of x and y is −1.0.
d. The correlation of x and y is between −0.2 and 0.0.
e. The correlation of x and y is between +0.5 and +0.7.

3.70 Start with the point (5, 5) and add at least four ordered pairs, (x, y), to make a set of ordered pairs that display the following properties. Show that your sample satisfies the requirements.
a. The correlation of x and y is between +0.9 and +1.0, and the slope of the line of best fit is 0.5.
b. The correlation of x and y is between +0.5 and +0.7, and the slope of the line of best fit is 0.5.
c. The correlation of x and y is between −0.7 and −0.9, and the slope of the line of best fit is −0.5.
d. The correlation of x and y is between +0.5 and +0.7, and the slope of the line of best fit is −1.0.

3.71 "Fast-Food Fat Counts Full of Surprises," in *USA Today,* 10-20-94, compared some of the popular fast-food items in calories and fat.

Calories (x)	270	420	210	450	130	310	290	450	446	640	233
Fat (y)	9	20	10	22	6	25	7	20	20	38	11

x	552	360	838	199	360	345	552
y	55	6	20	12	36	28	22

a. Draw a scatter diagram of these data.
b. Calculate the linear coefficient, r.
c. Find the equation of the line of best fit.
d. Explain the meaning of the above answers.

3.72 A biological study of a minnow called the blacknose dace was conducted. The length, y, in millimeters and the age, x, to the nearest year were recorded.

x	0	3	2	2	1	3	2	4	1	1
y	25	80	45	40	36	75	50	95	30	15

a. Draw a scatter diagram of these data.
b. Calculate the correlation coefficient.
c. Find the equation of the line of best fit.
d. Explain the meaning of the above answers.

3.73 Lakes are bodies of water surrounded by land and may include seas. The following table lists the area (sq. mi.) and maximum depth (ft) of 32 lakes throughout the world:

Lake	Area (sq mi)	Max. Depth (ft)	Lake	Area (sq mi)	Max. Depth (ft)
Caspian Sea	143,244	3,363	Chad	6,300	24
Superior	31,700	1,330	Maracaibo	5,217	115
Victoria	26,828	270	Onega	3,710	328
Aral Sea	24,904	220	Titicaca	3,200	922
Huron	23,000	750	Nicaragua	3,100	230
Michigan	22,300	923	Athabasca	3,064	407
Tanganyika	12,700	4,823	Reindeer	2,568	720
Baykal	12,162	5,315	Turkana	2,473	240
Great Bear	12,096	1,463	Issyk Kul	2,355	2,303
Nyasa	11,150	2,280	Vanern	2,156	328
Great Slave	11,031	2,015	Winnipegosis	2,075	38
Erie	9,910	210	Albert	2,075	168
Winnipeg	9,417	60	Kariba	2,050	390
Ontario	7,340	802	Nipigon	1,872	540
Balkhash	7,115	85	Urmia	1,815	49
Ladoga	6,835	738	Manitoba	1,799	12

Source: Geological Survey, U.S. Dept. of the Interior.

a. Draw a scatter diagram showing area, x, and maximum depth, y, for the lakes.
b. Find the linear correlation coefficient between area and maximum depth. What does the value of this linear correlation imply?

$ **3.74** Investors in mutual funds keep a sharp eye on the total return on their money. They also are aware of the risk involved in their investment, commonly measured by a fund's volatility (the greater the volatility, the higher the risk). Below is a list of 30 mutual funds randomly selected in 1998 from *Fortune*'s list of stock and bond funds, together with their five-year total return (%) and risk assessment:

Fund Name	Total Return	Risk	Fund Name	Total Return	Risk
MFS Emerging Growth	21.5	20.6	AIM Balanced A	15.9	10.8
Kaufmann	19.7	18.4	Greenspring	14.0	7.2
AIM Constellation A	17.6	18.4	Delaware A	13.6	8.6
Weitz Hickory	29.9	19.7	Calamos Convertible A	14.3	9.9
Oak Value	25.6	13.0	Managers Bond	10.3	5.4
Gabelli Westwood Equity	23.0	12.3	Harbor Bond	7.3	4.4
Nationwide	24.3	12.0	Northeast Investors	13.6	5.5
Fidelity Growth/Income	22.6	13.0	Strong Gov't. Securities	7.0	4.4
Stratton Growth	21.3	11.8	Lexington GNMA Income	6.9	3.5
GAM International A	22.6	19.9	Marshall Gov't. Income	5.8	3.7
Scudder International	14.3	13.7	Wright U.S. Treasury	6.3	7.5
Janus Worldwide	23.6	13.7	Excelsior Tax-Exempt	7.6	6.7
Oppenheimer Global A	19.0	14.4	Vanguard Municipal	6.5	5.5
New Perspective	18.8	12.1	Goldman Sachs Global	7.2	4.1
Putnam Europe Growth A	22.7	14.6	Capital World Bond	5.9	4.9

Source: Fortune, "The Best Mutual Funds," August 17, 1998.

a. Draw a scatter diagram with five-year total return as the y-axis and risk as the x-axis.
b. Calculate the correlation coefficient.
c. Would you conclude that in order to obtain higher total returns, mutual fund investors must take greater risks? Explain.

(continued)

3.75 a. Verify, algebraically, that formula (3.2) for calculating r is equivalent to the definition formula (3.1).

  b. Verify, algebraically, that formula (3.6) is equivalent to formula (3.5).

3.76 The following equation gives a relationship that exists between b_1 and r:

$$r = b_1 \sqrt{\frac{SS(x)}{SS(y)}}$$

  a. Verify this equation for these data:

x	4	3	2	3	0
y	11	8	6	7	4

  b. Verify this equation using formulas (3.2) and (3.6).

VOCABULARY LIST

Be able to define each term. Pay special attention to key terms, which are printed in **red.** In addition, describe in your own words, and give an example of, each term. Your examples should not be the ones given in class or in the textbook. Page numbers indicate the first appearance of the term.

bivariate data (p. 131)
coefficient of linear correlation (p. 146)
contingency table (p. 131)
correlation (p. 145)
correlation analysis (p. 145)
cross-tabulation (p. 131)
dependent variable (p. 137)
equation for line of best fit (p. 155)
independent variable (p. 137)
input variable (p. 137)
least squares criterion (p. 154)
line of best fit (p. 155)
linear correlation (p. 145)
linear regression (p. 153)

method of least squares (p. 154)
negative correlation (p. 145)
no correlation (p. 145)
ordered pair (p. 137)
output variable (p. 137)
Pearson's product moment r (p. 147)
positive correlation (p. 145)
predicted value (p. 154)
prediction equation (p. 154)
regression (p. 154)
regression analysis (p. 154)
scatter diagram (p. 138)
slope, b_1 (p. 155)
y-intercept, b_0 (p. 155)

CHAPTER PRACTICE TEST

Part I: Knowing the Definitions

Answer "True" if the statement is always true. If the statement is not always true, replace the words shown in bold with words that make the statement always true.

~~F~~ *Regression*

3.1 **Correlation** analysis is a method of obtaining the equation that represents the relationship between two variables.

F **3.2** The linear correlation coefficient is used to determine the **equation that represents** the relationship between two variables.   *strength or direction*

F **3.3** A correlation coefficient of **zero** means that the two variables are perfectly correlated.   *+1 or −1*

T **3.4** Whenever the slope of the regression line is zero, the **correlation coefficient** will also be zero.

F **3.5** When r is positive, b_1 will always be **negative.**   *positive*

T **3.6** The **slope** of the regression line represents the amount of change expected to take place in y when x increases by one unit.

3.7 When the calculated value of r is positive, the calculated value of b_1 will be **negative**.

3.8 Correlation coefficients range between **0 and +1**.

3.9 The value being predicted is called the **input variable**.

3.10 The line of best fit is used to predict the **average value of y** that can be expected to occur at a given value of x.

Part II: Applying the Concepts

3.11

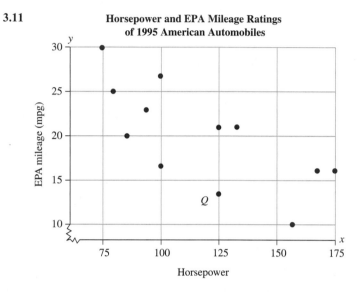

Horsepower and EPA Mileage Ratings of 1995 American Automobiles

a. Match the items described in column 2 with the terms in column 1.

Column 1	Column 2
___ Population	A. The horsepower rating for an automobile
___ Sample	B. All 1995 American-made automobiles
___ Input Variable	C. The EPA mileage rating for an automobile
___ Output Variable	D. The 1995 automobiles whose ratings are shown on the scatter diagram

b. Find the sample size.
c. What is the smallest value reported for the output variable?
d. What is the largest value reported for the input variable?
e. Does the scatter diagram suggest a positive (P), negative (N), or zero (Z) linear correlation coefficient?
f. What are the coordinates of point Q?
g. Will the slope for the line of best fit be positive (P), negative (N), or zero (Z)?
h. Will the intercept for the line of best fit be positive (P), negative (N), or zero (Z)?

3.12 A research group reports the correlation coefficient for two variables to be 2.3. What can you conclude from this information?

3.13 For the bivariate data, the extensions, and the totals shown on the table, find the following:

x	y	x^2	xy	y^2
2	6	4	12	36
3	5	9	15	25
3	7	9	21	49
4	7	16	28	49
5	7	25	35	49
5	9	25	45	81
6	8	36	48	64
28	49	124	204	353

  a. SS(x)
  b. SS(y)
  c. SS(xy)
  d. The linear correlation coefficient, r
  e. The slope, b_1
  f. The y-intercept, b_0
  g. The equation of the line of best fit $\hat{y} = b_0 + b_1 x$

Part III: Understanding the Concepts

3.14 A test was administered to measure the mathematics ability of the people in a certain town. Some of the townspeople were totally surprised to find out that their test results and their shoe sizes correlated strongly. Explain why a strong positive correlation should not have been such a surprise.

3.15 Student A collected a set of bivariate data and calculated r, the linear correlation coefficient. The resulting value was -1.78. Student A proclaimed that this indicated that there was no correlation between the two variables since the value of r was not between -1.0 and $+1.0$. Student B argues that -1.78 was impossible and that only values of r near zero implied no correlation. Who is correct? Justify your answer. *Student B   cannot   get # > -1*

3.16 The linear correlation coefficient, r, is a numerical value that ranges from -1.0 to $+1.0$. Write a sentence or two describing the meaning of r for each of these values:

  a. -0.93 **b.** $+0.89$ **c.** -0.03 **d.** $+0.08$ **e.** -2.3
  strong   *strong*   *veryweak*   *veryweak*   *cannot have*

3.17 Make up a set of three or more ordered pairs such that *won't ask on test*

  a. $r = 0.0$ **b.** $r = +1.0$ **c.** $r = -1.0$ **d.** $b_1 = 0.0$

Working with Your Own Data

Each semester, new students enter your college environment. You may have wondered, "What will the student body be like this semester?" As a beginning statistics student, you have just finished studying three chapters of basic descriptive statistical techniques. Let's use some of these techniques to describe some characteristics of your college's student body.

A Single Variable Data

1. Define the population to be studied.
2. Choose a variable to define. (You may define your own variable, or you may use one of the variables in the accompanying table if you are not able to collect your own data. Ask your instructor for guidance.)
3. Collect 35 pieces of data for your variable.
4. Construct a stem-and-leaf display of your data. Be sure to label it.
5. Calculate the value of the measure of central tendency that you believe best answers the question "What is the average value of your variable?" Explain why you chose this measure.
6. Calculate the sample mean for your data (unless you used the mean in Question 5).
7. Calculate the sample standard deviation for your data.
8. Find the value of the 85th percentile, P_{85}.
9. Construct a graphic display (other than a stem-and-leaf) that you believe "best" displays your data. Explain why the graph best presents your data.

B Bivariate Data

1. Define the population to be studied.
2. Choose and define two quantitative variables that will produce bivariate data. (You may define your own variables, or you may use two of the variables in the accompanying table if you are not able to collect your own data. Ask your instructor for guidance.)
3. Collect 15 ordered pairs of data.
4. Construct a scatter diagram of your data. (Be sure to label it completely.)
5. Using a table to assist with the organization, calculate the extensions x^2, xy, and y^2, and the summations of x, y, x^2, xy, and y^2.
6. Calculate the linear correlation coefficient r.
7. Calculate the equation of the line of best fit.
8. Draw the line of best fit on your scatter diagram.

The following table of data was collected on the first day of class last semester. You may use it as a source for your data if you are not able to collect your own.

Variable A = student's gender (male/female)
Variable B = student's age at last birthday
Variable C = number of completed credit hours toward degree
Variable D = "Do you have a job (full/part time)?" (yes/no)
Variable E = number of hours worked last week, if D = yes
Variable F = wages (before taxes) earned last week, if D = yes

The computer will select your random sample, see p. 101.

Student	A	B	C	D	E	F	Student	A	B	C	D	E	F
1	M	21	16	No			51	F	42	34	Yes	40	244
2	M	18	0	Yes	10	34	52	M	25	60	Yes	60	503
3	F	23	18	Yes	46	206	53	M	39	32	Yes	40	500
4	M	17	0	No			54	M	29	13	Yes	39	375
5	M	17	0	Yes	40	157	55	M	19	18	Yes	51	201
6	M	40	17	No			56	M	25	0	Yes	48	500
7	M	20	16	Yes	40	300	57	F	18	0	No		
8	M	18	0	No			58	M	32	68	Yes	44	473
9	F	18	0	Yes	20	70	59	F	21	0	No		
10	M	29	9	Yes	8	32	60	F	26	0	Yes	40	320
11	M	20	22	Yes	38	146	61	M	24	11	Yes	45	330
12	M	34	0	Yes	40	340	62	F	19	0	Yes	40	220
13	M	19	31	Yes	29	105	63	M	19	0	Yes	10	33
14	M	18	0	No			64	F	35	59	Yes	25	88
15	M	20	0	Yes	48	350	65	F	24	6	Yes	40	300
16	F	27	3	Yes	40	130	66	F	20	33	Yes	40	170
17	M	19	10	Yes	40	202	67	F	26	0	Yes	52	300
18	F	18	16	Yes	40	140	68	F	17	0	Yes	27	100
19	M	19	4	Yes	6	22	69	M	25	18	Yes	41	355
20	F	29	9	No			70	M	24	0	No		
21	F	21	0	Yes	20	80	71	M	21	0	Yes	30	150
22	F	39	6	No			72	M	30	12	Yes	48	555
23	M	23	34	Yes	42	415	73	F	19	0	Yes	38	169
24	F	31	0	Yes	48	325	74	M	32	45	Yes	40	385
25	F	22	7	Yes	40	195	75	M	26	90	Yes	40	340
26	F	27	75	Yes	20	130	76	M	20	64	Yes	10	45
27	F	19	0	No			77	M	24	0	Yes	30	150
28	M	22	20	Yes	40	470	78	M	20	14	No		
29	F	60	0	Yes	40	390	79	M	21	70	Yes	40	340
30	M	25	14	No			80	F	20	13	Yes	40	206
31	F	24	45	No			81	F	33	3	Yes	32	246
32	M	34	4	No			82	F	25	68	Yes	40	330
33	M	29	48	No			83	F	29	48	Yes	40	525
34	M	22	80	Yes	40	336	84	F	40	0	Yes	40	400
35	M	21	12	Yes	26	143	85	F	36	3	Yes	40	300
36	F	18	0	No			86	F	35	0	Yes	40	280
37	M	18	0	Yes	13	65	87	F	28	0	Yes	40	350
38	M	40	64	Yes	40	390	88	F	27	9	Yes	40	260
39	F	31	0	Yes	40	200	89	F	26	3	Yes	40	240
40	F	32	0	Yes	40	270	90	F	23	9	Yes	40	330
41	F	37	0	Yes	24	150	91	M	41	3	Yes	23	253
42	F	35	0	Yes	40	350	92	M	39	0	Yes	40	110
43	M	21	72	Yes	45	470	93	M	21	0	Yes	40	246
44	F	27	0	Yes	40	550	94	F	32	0	Yes	40	350
45	F	42	47	Yes	37	300	95	F	48	58	Yes	40	714
46	F	41	21	Yes	40	250	96	F	26	0	Yes	32	200
47	M	36	0	Yes	40	400	97	F	27	0	Yes	40	350
48	M	25	16	Yes	40	480	98	F	52	56	Yes	40	390
49	F	18	0	Yes	45	189	99	F	34	27	Yes	8	77
50	M	22	0	Yes	40	385	100	F	49	3	Yes	24	260

Two

Probability

Before continuing our study of statistics, we take a slight detour and study some basic probability. Probability is often called the "vehicle" of statistics; that is, the probability associated with chance occurrences is the underlying theory for statistics. Recall that in Chapter 1 we described probability as the science of making statements about what will occur when samples are drawn from known populations. Statistics was described as the science of selecting a sample and making inferences about the unknown population from which the sample is drawn. To make these inferences, we need to study sample results in situations in which the population is known so that we will be able to understand the behavior of chance occurrences.

In Part Two we study the basic theory of probability (Chapter 4), probability distributions of discrete variables (Chapter 5), and probability distributions for continuous random variables (Chapter 6). Following this brief study of probability, we will study the techniques of inferential statistics in Part Three.

Karl Pearson

KARL PEARSON, known as one of the fathers of modern statistics, was born March 27, 1857, in London, the second son of prominent attorney William Pearson and his wife, Fanny Smith. Karl was tutored at home until, at age nine, he entered University College School in London. In 1875, following a year of illness that required him to be privately tutored, he was awarded a scholarship to King's College, Cambridge. There, in May 1879, he earned his B.A. (with honors) in mathematics; he then went on to earn an M.A. in law in 1882.

After receiving his law degree, he moved to Heidelberg, Germany, where he became proficient in literature, philosophy, physics, and metaphysics, as well as in German history and folklore.

Pearson returned to University College, where, in 1884, he was appointed Goldsmid Professor of Applied Mathematics and Mechanics. In addition, Pearson lectured in geometry at Gresham College, London, from 1891 to 1894. Later, in 1911, he relinquished the Goldsmid chair to become the first Galton Professor of Eugenics.

In 1896 Pearson was elected to the Royal Society and was awarded the Society's Darwin Medal in 1898. In 1900 Pearson developed the chi-square (denoted by χ^2); it is the oldest inference procedure still used in its original form and is often used in today's economics and business applications. Around that time, he also verified the concept of random phenomena (or probability), by tossing a coin 24,000 times to determine the frequency of its landing "heads up" as opposed to "tails up." Result: 12,012 heads, a relative frequency of 0.5005.

It was during Pearson's association with Sir Francis Galton that he developed the linear correlation coefficient (sometimes referred to as the Pearson product moment correlation coefficient, in his honor). He was editor of, and a major contributor to, the statistical journal *Biometrica*, which he co-founded with fellow statisticians Galton and Weldon. Pearson returned to London in 1933, where he died on April 27, 1936.

Pearson's only son, Egon S. (second oldest of three children), born in 1895 to Karl and his wife, Maria Sharpe, also became a well-known statistician.

Probability

CHAPTER CASE STUDY

Lotteries Are Big Business

State-run lotteries have become BIG, in fact, VERY BIG business. There are over 200 different games run by over 35 different states and at least 6 interstate games. Some states run only one game each, while some run over 10 different games. Did you ever wonder if the numbers that get drawn are in fact randomly selected? The listing below appeared in the *Pittsburgh Post-Gazette, December 24, 1994*, and reported an analysis of Pennsylvania's "Big 4" lottery. This game had been played 3345 times from November 22, 1980, to December 24, 1994.

Lottery Results for Friday, Dec. 23, 1994

PENNSYLVANIA LOTTERY

Big 4

9-6-3-3

Number of winners: 559
Money paid out: $147,900

Last time numbers hit straight:
Never

Big 4 Analysis

(Times that each number has been picked in the first, second, third, or fourth positions, and total times drawn since the game began Nov. 22, 1980.)

No.	First	Second	Third	Fourth	Total
0	343	312	352	328	1335
1	326	330	351	357	1364
2	347	323	315	344	1329
3	320	327	350	351	1348
4	304	345	331	318	1298
5	321	348	322	343	1334
6	339	306	329	316	1290
7	348	346	351	311	1356
8	337	367	329	350	1383
9	360	341	315	327	1343

What is the likelihood of picking the correct four single-digit numbers? If the numbers selected at each drawing are equally likely, how many times would you expect each number to have been drawn in 3345 games? Does it appear that the numbers have appeared as winners with equal frequency? (See Exercise 4.110, p. 225.)

You may already be familiar with some ideas of probability, because probability is part of our everyday culture. We constantly hear people making probability-oriented statements such as:

- Our team will probably win the game tonight."
- "There is a 40% chance of rain this afternoon."
- "I will most likely have a date for the winter weekend."
- "If I park in the faculty parking area, I will probably get a ticket."
- "I have a 50-50 chance of passing today's chemistry exam."

Everyone has made or heard these kinds of statements. What exactly do they mean? Do they, in fact, mean what they say? Some statements may be based on scientific information and others on subjective prejudice. Whatever the case may be, they are probabilistic inferences—not fact but conjectures.

In this chapter we study the basic concept of probability and the rules that apply to the probability of both simple and compound events.

CONCEPTS OF PROBABILITY

4.1 The Nature of Probability

Let's consider an experiment in which we toss two coins simultaneously and record the number of heads that occur. The only possible outcomes are 0H (zero heads), 1H (one head), and 2H (two heads). Let's toss the two coins 10 times and record our findings.

2H,   1H,   1H,   2H,   1H,   0H,   1H,   1H,   1H,   2H

Summary:

Outcome	Frequency
2H	3
1H	6
0H	1

Suppose that we repeat this experiment 19 times. Table 4.1 shows the totals for 20 sets of 10 tosses. (Trial 1 shows the totals from our first experiment.)

TABLE 4.1 Experimental Results of Tossing Two Coins

Outcome	Trial																				Total

Outcome	1	2	3	4	5	6	7	8	9	10	11	12	13	14	15	16	17	18	19	20	Total
2H	3	3	5	1	4	2	4	3	1	1	2	5	6	3	1	4	1	0	3	1	53
H	6	5	5	5	5	7	5	5	5	5	8	4	3	7	5	1	5	4	5	9	104
0H	1	2	0	4	1	1	1	2	4	4	0	1	1	0	4	5	4	6	2	0	43

The total of 200 tosses of the pair of coins resulted in 2H on 53 occasions, 1H on 104 occasions, and 0H on 43 occasions. We can express these results in terms of relative frequencies and show the results using a histogram, as in Figure 4.1.

Figure 4.1

Relative Frequency Histogram for Coin-Tossing Experiment

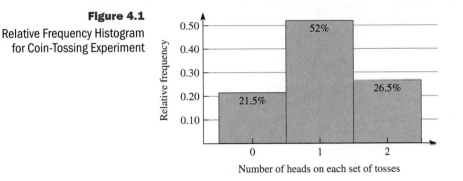

What conclusions can be reached? If we look at the individual sets of 10 tosses, we notice a large variation in the number of times each of the events (2H, 1H, and 0H) occurred. In both the 0H and the 2H categories, there were as many as 6 occurrences and as few as 0 occurrences in a set of 10 tosses. In the 1H category there were as few as 1 occurrence and as many as 9 occurrences.

If we were to continue this experiment for several hundred more tosses, what would you expect to happen in terms of the relative frequencies of these three events? It looks as if we have approximately a $1:2:1$ ratio in the totals of Table 4.1. We might therefore expect to find the **relative frequency** for 0H to be approximately $\frac{1}{4}$, or 25%; the relative frequency for 1H to be approximately $\frac{1}{2}$, or 50%; and the relative frequency for 2H to be approximately $\frac{1}{4}$, or 25%. These relative frequencies accurately reflect the concept of probability.

Many probability experiments can be simulated by using random numbers tables or by having a computer/calculator randomly generate number values representing the various experimental outcomes. For example, the preceding tossing of two coins experiment could be simulated by letting odd integers represent heads (H) and even integers, tails (T). Since we are tossing two coins, we will need a two-digit random integer, the first digit for the first coin and the second digit for the second coin, for each toss of the two coins. The key to either method is to maintain the probabilities. $[P(\text{H}) = \frac{1}{2}$ and $P(\text{odd}) = \frac{1}{2}$, $P(\text{T}) = \frac{1}{2}$ and $P(\text{even}) = \frac{1}{2}$; therefore, we have assigned random digits to the events so as to maintain the probabilities of heads and tails.] Thus, the random integer 45 would represent T, H, or 1H. This two-digit method would work with both the random numbers table and the computer. However, if the computer is used, by letting $1 = \text{H}$ and $0 = \text{T}$ on each toss of a coin, and by randomly generating two columns, one for each coin, the computer will do all the tally work, too. By adding the two columns together, we will find the total to be the number of heads seen in each toss of the two coins. See the commands that follow.

The following commands will generate random integers 0 and 1, to obtain the number of heads on the two coins. The findings will then be tallied to give the results of the experiment.

MINITAB

```
Choose:   Calc > Random Data > Integer
Enter:    Generate: 200
          Store in: C1 C2
          Minimum value: 0
          Maximum value: 1 > OK
Choose:   Calc > Calculator
Enter:    Store result in variable:
          C3
          Expression: C1 + C2 > OK
Choose:   Stat > Tables > Tally
Enter:    Variable: C3
Check:    Counts and Percents
```

TI-83

```
Choose:   MATH > PRB > 5:randInt(
Enter:    0,1,200)
Choose:   STO→ > 2nd L1
```

Repeat above commands but store data in L2.

```
Choose:   STAT > EDIT > 1:Edit
Highlight: L3 (column heading)
Enter:    L1 + L2
Choose:   2nd > STAT PLOT > 1:Plot1
```

```
Choose:   WINDOW
Enter:    −.5, 2.5, 1, −1, 120, 20,1
Choose:   TRACE > > > (use to find
          counts; percents can be
          calculated)
```

EXCEL

Enter 0,1 into column A, and 0.5, 0.5 into column B. Label column C as Coin1, column D as Coin2, and column E as Sum.

```
Choose:   Tools > Data Analysis
          > Random Number Generation
Enter:    Number of Variables: 2
          Number of Random Numbers:
          200
          Distribution: Discrete
          Value and Probability Input
          Range: (A1:B2 or select
          cells)
Select:   Output Range
Enter:    (C2:D201 or select cells)
          > OK
```

Activate the cell E2.

```
Choose:   Edit formula (=)
Enter:    C2 + D2 > OK
Drag:     Bottom right corner of E2
          down to give other sums
Choose:   Data > Pivot Table
          Report. . .
Select:   Microsoft Excel list or
          database > Next
Enter:    Range: (E1:E201 or select
          cells) > Next
Drag:     "Sum" heading into row &
          data areas
```

Double click the "sum of" in data area box; then continue with:

```
Choose:   Summarize by: Count
          > Options
          Show data as: % of totals
          or Normal > OK > Next
Select:   Existing Worksheet
Enter:    (F1 or select cell)
          > Finish
```

EXERCISES · · · · · · · · · · · ·

4.1   Toss a single coin 10 times and record H (head) or T (tail) after each toss. Using your results, find the relative frequency of
a. heads   **b.** tails

4.2   Roll a single die 20 times and record a 1, 2, 3, 4, 5, or 6 after each roll. Using your results, find the relative frequency of
a. 1   **b.** 2   **c.** 3   **d.** 4   **e.** 5   **f.** 6

4.3   Place three coins in a cup, shake and dump them out, and observe the number of heads showing. Record 0H, 1H, 2H, or 3H after each trial. Repeat the process 25 times. Using your results, find the relative frequency of
a. 0H   **b.** 1H   **c.** 2H   **d.** 3H

4.4   Place a pair of dice in a cup, shake and dump them out. Observe the sum of dots. Record 2, 3, 4, . . . , 12. Repeat the process 25 times. Using your results, find the relative frequency for each of the values: 2, 3, 4, 5, . . . , 12.

4.5   Use either the random numbers table (Appendix B) or a computer/calculator to simulate:
a. The rolling of a die 50 times; express your results as relative frequencies.
b. The tossing of a coin 100 times; express your results as relative frequencies.

4.6   Use either the random numbers table (Appendix B) or a computer/calculator to simulate the random selection of 100 single-digit numbers, 0 through 9.
a. List your 100 digits.
b. Prepare a relative frequency distribution of the 100 digits.
c. Prepare a relative frequency histogram of the distribution in (b).

4.2   Probability of Events

We are now ready to define what is meant by probability. Specifically, we talk about "the probability that a certain event will occur."

> ### PROBABILITY THAT AN EVENT WILL OCCUR
> The relative frequency with which that event can be expected to occur.

The probability of an event may be obtained in three different ways: (1) *empirically*, (2) *theoretically*, and (3) *subjectively*. The first method was illustrated in the experiment in Section 4.1 and might be called **experimental,** or **empirical probability.** This probability is the *observed relative frequency with which an event occurs*. In our coin-tossing illustration, we observed exactly one head (1H) on 104 of the 200 tosses of the pair of coins. The observed empirical probability for the occurrence of 1H was 104/200, or 0.52.

When the value assigned to the probability of an event results from experimental data, we will identify the probability of the event with the symbol $P'(\ )$.

NOTE The *prime notation* is used to denote empirical probabilities.

The value assigned to the probability of event A as a result of experimentation can be found by means of the formula

$$P'(A) = \frac{n(A)}{n} \tag{4.1}$$

where $n(A)$ is the number of times that event A is observed and n is the number of times the experiment is attempted.

Exercise 4.7

If you roll a die 40 times and 9 of the rolls result in a "5," what empirical probability was observed for the event "5"?

Exercise 4.8

Explain why an empirical probability, an observed proportion, and a relative frequency are actually three different names for the same thing.

Consider the rolling of a die. Define event A as the occurrence of a 1. In a single roll of a die, there are six possible outcomes. Assuming that the die is symmetrical, each number should have an equal likelihood of occurring. Intuitively, the probability of A, or the expected relative frequency of a 1, is $\frac{1}{6}$. (Later we will formalize this calculation.)

What does this mean? Does it mean that once in every six rolls a 1 will occur? No, it does not. Saying that the probability of a 1, $P(1)$, is $\frac{1}{6}$ means that in the long run the proportion of times that a 1 occurs is approximately $\frac{1}{6}$. How close to $\frac{1}{6}$ can we expect the observed relative frequency to be?

TABLE 4.2 Experimental Results of Rolling a Die Six Times in Each Trial

Trial	Column 1: Number of 1's Observed	Column 2: Relative Frequency	Column 3: Cumulative Relative Frequency
1	1	1/6	1/6 = 0.17
2	2	2/6	3/12 = 0.25
3	0	0/6	3/18 = 0.17
4	1	1/6	4/24 = 0.17
5	0	0/6	4/30 = 0.13
6	1	1/6	5/36 = 0.14
7	2	2/6	7/42 = 0.17
8	2	2/6	9/48 = 0.19
9	0	0/6	9/54 = 0.17
10	0	0/6	9/60 = 0.15
11	1	1/6	10/66 = 0.15
12	0	0/6	10/72 = 0.14
13	2	2/6	12/78 = 0.15
14	1	1/6	13/84 = 0.15
15	1	1/6	14/90 = 0.16
16	3	3/6	17/96 = 0.18
17	0	0/6	17/102 = 0.17
18	1	1/6	18/108 = 0.17
19	0	0/6	18/114 = 0.16
20	1	1/6	19/120 = 0.16

Table 4.2 (above) shows the number of 1's observed in each set of six rolls of a die (column 1), an observed relative frequency for each set of six rolls (column 2), and a cumulative relative frequency (column 3). Each trial is a set of six rolls. Figure 4.2a shows the fluctuation of the observed probability for event A on each of the 20 trials (column 2, Table 4.2). Figure 4.2b shows the fluctuation of the cumulative relative frequency (column 3, Table 4.2). Notice that the observed relative frequency on each trial of six rolls of a die tends to fluctuate about $\frac{1}{6}$. Notice also that the observed values on the cumulative graph seem to become more stable; in fact, they become relatively close to the expected $\frac{1}{6}$, or $0.166\overline{6} = 0.167$.

Figure 4.2

Fluctuations Found in the
Die-Tossing Experiment
(a) Relative Frequency

(b) Cumulative relative
frequency

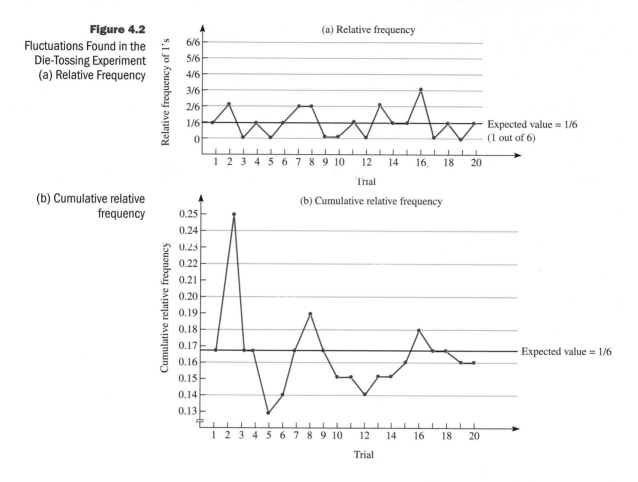

A cumulative graph such as Figure 4.2b demonstrates the idea of long-term average. When only a few rolls were observed (as on each trial, Figure 4.2a), the probability $P'(A)$ fluctuated between 0 and $\frac{1}{2}$. As the experiment was repeated, however, the cumulative graph suggests a stabilizing effect on the observed cumulative probability. This stabilizing effect, or **long-term average** value, is often referred to as the *law of large numbers*.

LAW OF LARGE NUMBERS

If the number of times an experiment is repeated is increased, the ratio of the number of successful occurrences to the number of trials will tend to approach the theoretical probability of the outcome for an individual trial.

The law of large numbers is telling us that the larger the number of experimental trials n, the closer the empirical probability $P'(A)$ is expected to be to the true or theoretical probability $P(A)$. There are many applications of this concept. The die-tossing experiment is an example in which we can easily compare the results to what we expect to happen; this gives us a chance to verify the claim of the law of large numbers. Illustration 4.1 is an illustration in which we live with the results obtained from large sets of data when the theoretical expectation is unknown.

ILLUSTRATION 4.1 ▼

The key to establishing proper life insurance rates is using the probability that the insureds will live one, two, or three years, and so forth, from the time they purchase policies. These probabilities are derived from actual life and death statistics and hence are empirical probabilities. They are published by the government and are extremely important to the life insurance industry. ▲

EXERCISES • • • • • • • • • • • • •

4.9 Explain what is meant by the statement "When a single die is rolled, the probability of a 1 is $\frac{1}{6}$."

4.10 Explain what is meant by the statement "When one coin is tossed one time, there is a 50-50 chance of getting a tails."

 4.11 According to the National Climatic Data Center (NESDIS, NOAA, U.S. Dept. of Commerce), San Juan, Puerto Rico, experienced 197 days of measurable precipitation (rain) in 1995. This was more days of rain than any of the other weather stations reported that year but was not considered unusual for the city. By comparison, the station at Phoenix, Arizona reported 31 days of precipitation during the same year (again, not unusual).
  a. Suppose you were taking a one-day business trip to San Juan. What is the probability that it will rain while you are there?
  b. Suppose you decide instead to take your trip to Phoenix. What is the probability that measurable precipitation will occur while you are there?

4.12 The September 1998 issue of *Visual Basic Programmer's Journal* reported the results of a survey of 500 subscribers who indicated the number of hours worked per week at their jobs. Results are summarized in the following table:

Hours Worked per Week	Number of Respondents	Percentage of Respondents
30 or less	20	4.0
31–40	105	21.0
41–50	300	60.0
51–60	65	13.0
61 or more	20	4.0
Total	500	100.0

Source: *Visual Basic Programmer's Journal*, "Fall 1998 Salary Survey Results," September, 1998.

Suppose one of the respondents from the sample of returns is selected at random. Find the probability of the following events:
  a. The respondent works less than 41 hours per week.
  b. The respondent works over 50 hours per week.
  c. The respondent works between 31 and 60 hours per week.

4.13 Take two dice (one white and one colored) and roll them 50 times, recording the results as ordered pairs [(white, color), for example; (3, 5) represents 3 on the white die and 5 on the colored die]. (You could simulate these 50 rolls using a random numbers table or a computer/calculator.) Then calculate the following observed probabilities:
  a. P'(white die is an odd number)
  b. P'(sum is 6)
  c. P'(both dice show odd number)
  d. P'(number on colored die is larger than number on white die)

4.14 Use a random numbers table or a computer/calculator to simulate the rolling of a pair of dice 100 times.

    a. List the results of each roll as an ordered pair, and record the sum of the ordered pair.

    b. Prepare an ungrouped frequency distribution for the sums and a frequency histogram of the sums.

    c. Describe how these results compare to what you expect to occur when two dice are rolled.

The following commands will simulate the rolling of a pair of dice 100 times.

EXCEL

Enter 1,2,3,4,5,6 into column A, label C1: **Die1**; D1: **Die2**; E1: **Dice**, and activate B1.

Choose:	**Format > Cells > Number > Number**
Enter:	Decimal places: **8** **> OK**
Enter:	**1/6** in B1
Drag:	**Bottom right corner of B1 down for 6 entries**
Choose:	**Tools > Data Analysis > Random Number Generation > OK**
Enter:	Number of Variables: **2**
	Number of Random Numbers: **100**
	Distribution: **Discrete**
	Value and Probability Input Range: **(A1:B6 or select cells)**
Select:	Output Range
Enter:	**(C2:D101 or select cells) > OK**

Activate the **E2** cell.

Choose:	**Edit formula (=)**
Enter:	**C2 + D2 > OK**
Drag:	**Bottom right corner of E2 down for 100 entries**
Choose:	**Data > Pivot Table Report . . .**
Select:	**Microsoft Excel list or database > Next**
Enter:	Range: **(E1:E101 or select cells) > Next**
Drag:	**"Dice" heading into row & data areas**

Double click the "sum of" in data area box; then continue with:

Choose:	Summarize by: **Count > OK > Next**
Select:	**Existing Worksheet**
Enter:	**(F1 or select cell) > Finish**

Label column J "sums" and input the numbers 2, 3, 4, . . ., 12 into it. Use the EXCEL histogram commands on page 57 with column E as the input range and column J as the bin range.

MINITAB

Choose:	**Calc > Random Data > Integer**
Enter:	Generate: **100**
	Store in: **C1 C2**
	Minimum value: **1**
	Maximum value: **6 > OK**
Choose:	**Calc > Calculator**
Enter:	Store result in variable:
	C3
	Expression: **C1 + C2 > OK**
Choose:	**Stat > Tables > Tally**
Enter:	Variable: **C3**
Select:	**Counts**

Use the MINITAB commands on page 56 to construct a frequency histogram of the data in C3 with the midpoints 2:12/1.

TI-83

Choose:	**MATH > PRB > 5:randInt(**
Enter:	**1,6,100)**
Choose:	**STO→ > 2ⁿᵈ L1**
Repeat above for L2	
Choose:	**STAT > EDIT > 1:Edit**
Highlight:	**L3**
Enter:	L3 =: **L1 + L2**
Choose:	**2ⁿᵈ STAT PLOT > 1:Plot1**

```
Plot1  Plot2  Plot3
On Off
Type:
Xlist:L3
Freq:1
```

Choose:	**WINDOW**
Enter:	**−5, 12.5, 1, −10, 40, 10,1**
Choose:	**TRACE > > >**

4.15 Using a coin, perform the die experiment discussed on pages 184–185. Toss a coin 10 times, observing the number of heads (or put 10 coins in a cup, shake and dump them into a box, and use each toss for a block of 10), and record the results. Repeat until you have 200 tosses. Chart and graph the data as individual sets of 10 and as cumulative relative frequencies. Do your data tend to support the claim that $P(\text{head}) = \frac{1}{2}$? Explain.

4.16 Let's estimate the probability that a thumbtack lands "point up," ⌣ (as opposed to "point down" ◓) when tossed and lands on a hard surface. Using a thumbtack, perform the die experiment discussed on pages 184–185. Toss the thumbtack 10 times, observing the number of "point up" (or put 10 identical thumbtacks in a cup, shake, and dump them into a box, and use each toss for a block of 10), and record the results. Repeat until you have 200 tosses. Chart and graph the data as individual sets of 10 and as cumulative relative frequencies. What do you believe the $P'(⌣)$ to be? Explain.

4.3   Simple Sample Spaces

Let's return to an earlier question: What values might we expect to be assigned to the three events (0H, 1H, 2H) associated with our coin-tossing experiment? As we inspect these three events, we see that they do not tend to happen with the same relative frequency. Why? Suppose that the experiment of tossing two pennies and observing the number of heads had actually been carried out using a penny and a nickel—two distinct coins. Would this have changed our results? No, it would have had no effect on the experiment. However, it does show that there are more than three possible outcomes.

When a penny is tossed, it may land as heads or tails. When a nickel is tossed, it may also land as heads or tails. If we toss them simultaneously, we see that there are actually four different possible outcomes. These four outcomes match up with the previous events as follows:

1. heads on penny and heads on nickel—2H

2. heads on penny and tails on nickel—1H

3. tails on penny and heads on nickel—1H

4. tails on penny and tails on nickel—0H

In this experiment with the penny and the nickel, let's use an **ordered pair** notation. The first listing will correspond to the penny and the second will correspond to the nickel. Thus, (H, T) represents the event that a head occurs on the penny and a tail occurs on the nickel. Our listing of events for the tossing of a penny and a nickel looks like this:

$$(H, H), (H, T), (T, H), (T, T)$$

What we have accomplished here is a listing of what is known as the sample space for this experiment.

EXPERIMENT

Any process that yields a result or an observation.

OUTCOME

A particular result of an experiment.

SAMPLE SPACE

The set of all possible outcomes of an experiment. The sample space is typically denoted by S and may take any number of forms: a list, a tree diagram, a lattice grid system, and so on. The individual outcomes in a sample space are called **sample points.** $n(S)$ is the number of sample points in sample space S.

EVENT

Any subset of the sample space. If A is an event, then $n(A)$ is the number of sample points that belong to event A.

Regardless of the form in which they are presented, the outcomes in a sample space can never overlap. Also, all possible outcomes must be represented. These characteristics are called **mutually exclusive** and **all inclusive,** respectively. A more detailed explanation of these characteristics will be presented later; for the moment, however, an intuitive grasp of their meaning is sufficient.

Now let's look at some illustrations of probability experiments and their associated sample spaces.

EXPERIMENT 4.1 ▼

A single coin is tossed once and the outcome—a head (H) or a tail (T)—is recorded.

Sample space: $S = \{H, T\}$ and $n(S) = 2$

EXPERIMENT 4.2 ▼

Two coins, one penny and one nickel, are tossed simultaneously and the outcome for each coin is recorded using ordered pair notation: (penny, nickel). The sample space is shown here in two different ways:

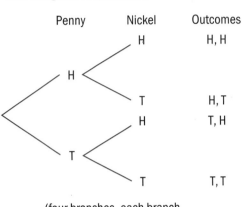

Tree Diagram Representation *

Penny	Nickel	Outcomes
H	H	H, H
	T	H, T
T	H	T, H
	T	T, T

(four branches, each branch shows a possible outcome)

Listing

$S = \{(H, H), (H, T), (T, H), (T, T)\}$

and $n(S) = 4$

* See the *Statistical Tutor* for information about tree diagrams.

Exercise 4.17

You are to select one single-digit number randomly. List the sample space.

Notice that both representations show the same four possible outcomes. For example, the top branch on the tree diagram shows heads on both coins, as does the first ordered pair in the listing.

EXPERIMENT 4.3 ▼

A die is rolled one time and the number of spots on the top face observed. The sample space is

$$S = \{1, 2, 3, 4, 5, 6\} \text{ and } n(S) = 6$$

EXPERIMENT 4.4 ▼

A box contains three poker chips (one red, one blue, one white), and two are drawn *with replacement*. (This means that one chip is selected, its color is observed, and then the chip is replaced in the box.) The chips are scrambled before a second chip is selected and its color observed. The sample space is shown in two different ways:

Tree Diagram Representation

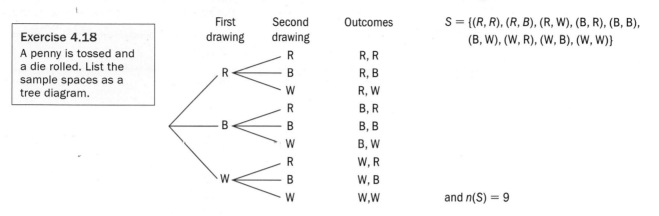

First drawing	Second drawing	Outcomes
R	R	R, R
	B	R, B
	W	R, W
B	R	B, R
	B	B, B
	W	B, W
W	R	W, R
	B	W, B
	W	W,W

Listing

$S = \{(R, R), (R, B), (R, W), (B, R), (B, B),$
$(B, W), (W, R), (W, B), (W, W)\}$

and $n(S) = 9$

Exercise 4.18

A penny is tossed and a die rolled. List the sample spaces as a tree diagram.

EXPERIMENT 4.5 ▼

A box contains one each of a red, a blue, and a white poker chip. Two chips are drawn simultaneously or one at a time *without replacement* (meaning one chip is selected, and then a second is selected without replacing the first). The sample space is shown in two ways:

Exercise 4.19

Draw a tree diagram representing the possible arrangements of boys and girls from oldest to youngest for a family of:

a. two children
b. three children

Tree Diagram Representation

First drawing	Second drawing	Outcomes
R	B	R, B
	W	R, W
B	R	B, R
	W	B, W
W	R	W, R
	B	W, B

Listing

$S = \{(R, B), (R, W), (B, R),$
$(B, W), (W, R), (W, B)\}$

and $n(S) = 6$

This experiment is the same as Experiment 4.4 except that the first chip is not replaced before the second selection is made.

EXPERIMENT 4.6 ▼

Three coins are tossed, or one coin is tossed three times, with head (H) or tail (T) observed on each coin. The sample space is shown as follows:

Tree Diagram Representation

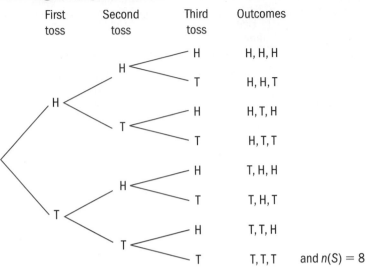

<table>
<tr><td>First
toss</td><td>Second
toss</td><td>Third
toss</td><td>Outcomes</td></tr>
</table>

H, H, H

H, H, T

H, T, H

H, T, T

T, H, H

T, H, T

T, T, H

T, T, T   and $n(S) = 8$

EXPERIMENT 4.7 ▼

Two dice (one white, one black) are each rolled one time and the number of dots showing on each die is observed. The sample space is shown by a chart representation:

Exercise 4.20

A box contains one each of $1, $5, $10, and $20 bills.

a. One is selected at random; list the sample space.
b. Two bills are drawn at random (without replacement); list the sample space as a tree diagram.
c. Two bills are drawn at random (without replacement); list the sample space as a chart.

Chart Representation

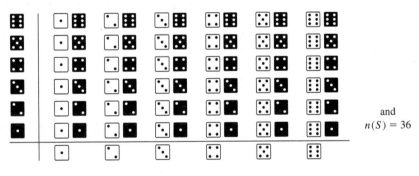

and
$n(S) = 36$

EXPERIMENT 4.8 ▼

Two dice are rolled and the sum of their dots is observed. The sample space is

$$S = \{2, 3, 4, 5, 6, 7, 8, 9, 10, 11, 12\} \text{ and } n(S) = 11$$

(or the 36-point sample space listed in Experiment 4.7).

You will notice that two different sample spaces are suggested for Experiment 4.8. Both of these sets satisfy the definition of a sample space and thus either could be used. We will learn later why the 36-point sample space is more useful than the other.

EXPERIMENT 4.9 ▼

A weather forecaster predicts that there will be a measurable amount of precipitation or no precipitation on a given day. The sample space is

$$S = \{\text{precipitation, no precipitation}\} \text{ and } n(S) = 2$$

EXPERIMENT 4.10 ▼

The 6024 students at a nearby college have been cross-tabulated according to gender and their college status:

	Full-time Students	Part-time Students	
Female	2136	548	
Male	2458	882	and $n(S) = 6024$

One student is to be picked at random from the student body.

EXPERIMENT 4.11 ▼

A lucky customer will get to randomly select one key from a barrel containing a key to each car on Used Car Charlie's lot. The accompanying **Venn diagram*** summarizes Charlie's inventory.

used to show events not mutually exclusive

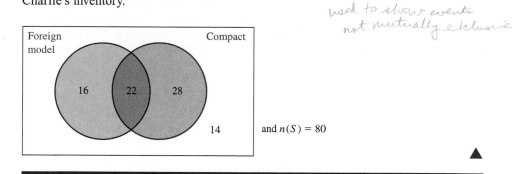

and $n(S) = 80$

▲

Special attention should always be given to the sample space. Like the statistical population, the sample space must be well defined. Once the sample space is defined, you will find the remaining work much easier.

*See the *Statistical Tutor* for information about Venn diagrams.

EXERCISES • • • • • • • • • • • • • • •

4.21 The face cards are removed from a regular deck and then 1 card is selected from this set of 12 face cards. List the sample space for this experiment.

💲 **4.22** An experiment consists of drawing one marble from a box that contains a mixture of red, yellow, and green marbles. There are at least two marbles of each color.
  a. List the sample space.
  b. Can we be sure that each outcome in the sample space is equally likely?
  c. If two marbles are drawn from the box, list the sample space.

4.23 The September 1998 issue of *Visual Basic Programmer's Journal* reported the results of a survey of 500 subscribers who were asked whether the training they had undergone at work affected their current salary level. Results: 54.9% said "yes," and 45.1% said "no." Suppose three respondents are selected randomly from the returns. Give the sample space and the possible combinations of selected responses to the question.

4.24 Marty Wilson just cut out four equilateral triangles the same size from a sheet of cardboard. He labeled each triangle A, B, C, and D. Then he taped them together at their edges to form a perfect tetrahedron. If Marty throws the tetrahedron into the air, the outcome is that any one of the four sides will be face down when it lands.
  a. Draw a tree diagram depicting two straight throws. Show all the outcomes and the size of the sample space.
  b. What is the size of the sample space after six successive throws?

4.25 a. A balanced coin is tossed twice. List a sample space showing the possible outcomes.
  b. A biased coin (it favors heads in a ratio of 3 to 1) is tossed twice. List a sample space showing the possible outcomes.

4.26 A computer generates (in random fashion) pairs of integers. The first integer is between 1 and 5, inclusive, and the second is between 1 and 4, inclusive. Represent the sample space on a coordinate axis system where x is the first number and y is the second number. (Retain your answer for use in answering Exercise 4.30.)

4.27 An experiment consists of two trials. The first is tossing a penny and observing heads or tails; the second is rolling a die and observing a 1, 2, 3, 4, 5, or 6. Construct the sample space. (Retain your answer for use in answering Exercise 4.29.)

4.28 A box stored in a warehouse contains 100 identical parts, of which 10 are defective and 90 are nondefective. Three parts are selected without replacement. Construct a tree diagram representing the sample space.

4.29 Use a computer/calculator (or random numbers table) to simulate 200 trials of the experiment described in Exercise 4.27, the tossing of a penny and the rolling of a die. Let 1 = H, 2 = T for the penny, and 1, 2, 3, 4, 5, 6 for the die. Report your results using a cross-tabulated table showing the frequency of each outcome.
  a. Find the relative frequency for heads.
  b. Find the relative frequency for 3.
  c. Find the relative frequency for (H, 3).

> Adjust the computer and calculator commands on pages 187 and 188.

4.30 Use a computer/calculator (or random numbers table) to simulate the experiment described in Exercise 4.26; x is an integer 1 to 5, and y is an integer 1 to 4. Generate a list of 100 random x-values and 100 y-values. Make a list of the resulting 100 ordered pairs of integers.
  a. Find the relative frequency for $x = 2$.
  b. Find the relative frequency for $y = 3$.
  c. Find the relative frequency for the ordered pair (2, 3).

4.4 Rules of Probability

Let's return now to the concept of probability and relate it to the sample space. Recall that the probability of an event was defined as the relative frequency with which the event could be expected to occur.

In the sample space associated with Experiment 4.1, the tossing of one coin, we find two possible outcomes: heads (H) and tails (T). We have an intuitive feeling that these two events will occur with approximately the same frequency. The coin is a symmetrical object and, therefore, would not be expected to favor either of the two outcomes. We would expect heads to occur $\frac{1}{2}$ of the time. Thus, the probability that a head will occur on a single toss of a coin is thought to be $\frac{1}{2}$.

This description is the basis for the second technique for assigning the probability of an event. In a sample space containing *sample points that are* **equally likely** *to occur,* the probability $P(A)$ of an event A is the ratio of the number $n(A)$ of points that satisfy the definition of event A to the number $n(S)$ of sample points in the entire sample space. That is,

$$P(A) = \frac{n(A)}{n(S)} \qquad (4.2)$$

This formula gives a **theoretical probability** value of event A's occurrence.

If we apply formula (4.2) to the equally likely sample space for the tossing of two coins (Experiment 4.2), we find the theoretical probabilities $P(0H)$, $P(1H)$, and $P(2H)$ discussed previously.

$$P(0H) = \frac{n(0H)}{n(S)} = \frac{1}{4}, \quad P(1H) = \frac{n(1H)}{n(S)} = \frac{2}{4} = \frac{1}{2}, \quad P(2H) = \frac{n(2H)}{n(S)} = \frac{1}{4}$$

The use of formula (4.2) requires the existence of a sample space in which each outcome is equally likely. Thus, when dealing with experiments that have more than one possible sample space, it is helpful to construct a sample space in which the sample points are equally likely.

Consider Experiment 4.8, where two dice were rolled. If you list the sample space as the 11 sums, the sample points are not equally likely. If you use the 36-point sample space, all the sample points are equally likely as in Experiment 4.7.

For example, the sum of 2 represents $\{(1, 1)\}$; the sum of 3 represents $\{(2, 1), (1, 2)\}$; and the sum of 4 represents $\{(1, 3), (3, 1), (2, 2)\}$. Thus, we can use formula (4.2) and the 36-point sample space to obtain the probabilities for the 11 sums.

> **Exercise 4.31**
>
> Find the probabilities $P(5)$, $P(6)$, $P(7)$, $P(8)$, $P(9)$, $P(10)$, $P(11)$, $P(12)$ for the sum of two dice.

$$P(2) = \frac{n(2)}{n(S)} = \frac{1}{36}, \quad P(3) = \frac{n(3)}{n(S)} = \frac{2}{36}, \quad P(4) = \frac{n(4)}{n(S)} = \frac{3}{36}$$

and so forth.

In many cases the assumption of equally likely events does not make sense. The sample points in Experiment 4.8 are not equally likely, and there is no reason to believe that the sample points in Experiment 4.9 are equally likely.

What do we do when the sample space elements are not equally likely or not a combination of equally likely events? We could use empirical probabilities. But what do we do when no experiment has been done or can be performed?

Let's look again at Experiment 4.9. The weather forecaster often assigns a probability to the event "precipitation." For example, "There is a 20% chance of rain today," or "There is a 70% chance of snow tomorrow." In some cases the method for assigning probabilities is personal judgment. These probability assignments are called **subjective probabilities.** The accuracy of subjective probabilities depends on the individual's ability to correctly assess the situation.

Often, personal judgment of the probability of the possible outcomes of an experiment is expressed by comparing the likelihood among the various outcomes. For example, the weather forecaster's personal assessment might be that "it is five times more likely to rain (R) tomorrow than not rain (NR)"; $P(R) = 5 \cdot P(NR)$. If this is the case, what values should be assigned to $P(R)$ and $P(NR)$? To answer this question, we need to review some of the ideas about probability that we've already discussed.

1. Probability represents a relative frequency.
2. $P(A)$ is the ratio of the number of times an event can be expected to occur divided by the number of trials. (same as sample space)
3. The numerator of the probability ratio must be a positive number or zero.
4. The denominator of the probability ratio must be a positive number (greater than zero).
5. The number of times an event can be expected to occur in n trials is always less than or equal to the total number of trials, n.

Thus, it is reasonable to conclude that a probability is always a numerical value between zero and one.

Property 1 $0 \le$ Each $P(A) \le 1$

NOTES

1. The probability is zero if the event cannot occur.
2. The probability is one if the event occurs every time.

Property 2 $$\sum_{\text{all outcomes}} P(A) = 1$$

Property 2 states that if we add the probabilities of each sample point in the sample space, the total probability must equal one. This property makes sense because when we sum all the probabilities, we are asking, "What is the probability the experiment will yield an outcome?" which will happen every time.

Now we are ready to assign probabilities to $P(R)$ and $P(NR)$. The events R and NR cover the sample space, and the weather forecaster's personal judgment was

$$P(R) = 5 \cdot P(NR)$$

From Property 2, we know that

$$P(R) + P(NR) = 1$$

By substituting $5 \cdot P(NR)$ for $P(R)$, we get

$$5 \cdot P(NR) + P(NR) = 1$$

$$6 \cdot P(NR) = 1$$

$$P(NR) = \frac{1}{6}$$

$$P(R) = 5 \cdot P(NR) = 5\left(\frac{1}{6}\right) = \frac{5}{6}$$

> **Exercise 4.32**
>
> If four times as many students pass a statistics course as fail, and one statistics student is selected at random, what is the probability that the student will pass statistics?

Odds

The statement "It is five times more likely to rain tomorrow (R) than not rain (NR)" is often expressed as "The odds are 5 to 1 in favor of rain tomorrow" (also written 5:1). Odds is simply another way of expressing probabilities. The relationships among odds for an event, odds against an event, and the probability of an event are expressed in the following rules.

> If the odds in favor of an event A are **a to b**, then
>
> 1. The odds against event A are **b to a** (or **b:a**).
> 2. The probability of event A is $P(A) = \dfrac{a}{a+b}$
> 3. The probability that event A will not occur is $P(A \text{ does not occur}) = \dfrac{b}{a+b}$

> **Exercise 4.33**
>
> The odds for the Broncos winning next year's Super Bowl are 2 to 7.
>
> a. What is the probability the Broncos will win next year's Super Bowl?
> b. What are the odds against the Broncos winning next year's Super Bowl?

To illustrate these rules, consider the statement "The odds favoring rain tomorrow are 5 to 1." Using the preceding notation, $a = 5$ and $b = 1$. Therefore, the probability of rain tomorrow is $\dfrac{5}{5+1}$, or $\dfrac{5}{6}$. The odds against rain tomorrow are 1 to 5 (or 1:5), and the probability that there is no rain tomorrow is $\dfrac{1}{5+1}$, or $\dfrac{1}{6}$.

Case Study   4.1

Trying to Beat the Odds

not onled

Exercise 4.34

Referring to "Trying to beat the odds," find:

a. The probability that a high school senior basketball player makes a pro team.
b. The odds that a player who makes a college basketball team plays as a senior.
c. The odds against a college senior basketball player making a pro team.

Many young men aspire to become professional athletes. Only a few make it to the big time as indicated in the following graph. For every 2400 college senior basketball players, only 64 make a professional team; that translates to a probability of only 0.027 (64/2400).

Source: Copyright 1990, USA Today. Reprinted with permission.

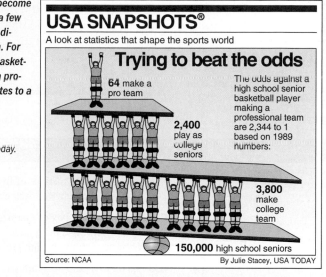

USA SNAPSHOTS®

A look at statistics that shape the sports world

Trying to beat the odds

64 make a pro team

The odds against a high school senior basketball player making a professional team are 2,344 to 1 based on 1989 numbers:

2,400 play as college seniors

3,800 make college team

150,000 high school seniors

Source: NCAA   By Julie Stacey, USA TODAY

COMPLEMENT OF AN EVENT

The set of all sample points in the sample space that do not belong to event A. The complement of event A is denoted by $\overline{A}$ (read "A complement").

For example, the complement of the event "success" is "failure"; the complement of "heads" is "tails" for the tossing of one coin; the complement of "at least one head" on 10 tosses of a coin is "no heads."

By combining the information in the definition of complement with Property 2 (p. 196), we can say that

$$P(A) + P(\overline{A}) = 1.0 \text{ for any event A}$$

It then follows that

$$P(\overline{A}) = 1 - P(A) \tag{4.3}$$

Exercise 4.35

Find the probability of drawing a non-face card from a well-shuffled deck of 52 playing cards.

NOTE Every event A has a complementary event $\overline{A}$. Complementary probabilities are very useful when the question asks for the probability of "at least one." Generally this represents a combination of several events, but the complementary event "none" is a single outcome. It is easier to solve for the complementary event and get the answer by using formula (4.3).

ILLUSTRATION 4.2 ▼

Two dice are rolled. What is the probability that the sum is at least 3 (that is, 3 or larger)?

Solution

Rather than finding the probability for each of the sums 3 and larger, it will be much simpler to find the probability that the sum is 2 (less than 3) and then use formula (4.3), letting "at least 3" be $\overline{A}$.

$$P(A) = \frac{1}{36} \text{ (as found on page 195)}$$

$$P(\overline{A}) = 1 - P(A) = 1 - \frac{1}{36} = \frac{35}{36} \text{ [using formula (4.3)]} \qquad \blacktriangle$$

EXERCISES • • • • • • • • • • • • •

4.36 A box contains marbles of five different colors: red, green, blue, yellow, and purple. There is an equal number of each color. Assign probabilities to each color in the sample space.

4.37 Suppose that a box of marbles contains an equal number of red marbles and yellow marbles but twice as many green marbles as red marbles. Draw one marble from the box and observe its color. Assign probabilities to the elements in the sample space.

4.38 As of 1998, the two professional football coaches who won the most games during their careers were Don Shula and George Halas. Shula's teams (Colts and Dolphins) won 347 and tied 6 of the 526 games that he coached, whereas Halas's team (Bears) won 324 and tied 31 of the 506 games that he coached. *(Source: World Almanac and Book of Facts 1998)*

Suppose one filmstrip from every game each man coached is thrown into a bin and mixed. You select one filmstrip from the bin and load it into a projector. What is the probability that the film you select shows:
a. a tie game.
b. a losing game.
c. one of Shula's teams winning a game.
d. Halas's team winning a game.
e. one of Shula's teams losing a game.
f. Halas's team losing a game.
g. one of Shula's teams playing to a tie.
h. Halas's team playing to a tie.
i. a game coached by Halas.
j. a game coached by Shula.

4.39 A single die is rolled. Find the probability the number on top is:
a. a 3.
b. an odd number.
c. a number less than 5.
d. a number no more than 3.

4.40 A transportation engineer in charge of a new traffic-control system expresses the subjective probability that the system functions correctly 99 times as often as it malfunctions.
a. Based on this belief, what is the probability that the system functions properly?
b. Based on this belief, what is the probability that the system malfunctions?

4.41 Events A, B, and C are defined on sample space S. Their corresponding sets of sample points do not intersect and their union is S. Further, event B is twice as likely to occur as event A, and event C is twice as likely to occur as event B. Determine the probability of each of these three events.

4.42 Three coins are tossed and the number of heads observed is recorded. Find the probability for each of the possible results, 0H, 1H, 2H, and 3H.

4.43 Let x be the success rating of a new television show. The accompanying table lists the subjective probabilities assigned to each x for a particular new show as assigned by three different media people. Which of these sets of probabilities are inappropriate because they violate a basic rule of probability? Explain. *(continued)*

Success Rating (x)	Judge		
	A	**B**	**C**
Highly successful	0.5	0.6	0.3
Successful	0.4	0.5	0.3
Not successful	0.3	-0.1	0.3

4.44 Two dice are rolled (Experiment 4.7). Find the probabilities in parts (b) through (e). Use the sample space given on page 199.
  a. Why is the set {2, 3, 4, . . ., 12} not a useful sample space?
  b. P(white die is an odd number)
  c. P(sum is 6)
  d. P(both dice show odd numbers)
  e. P(number on black die is larger than number on white die)
  f. Explain why these answers and the answers found in Exercise 4.13 (p. 186) are not exactly the same.

4.45 A group of files in a medical clinic classifies the patients by gender and by type of diabetes (I or II). The groupings may be shown as follows. The table gives the number in each classification.

Gender	Type of Diabetes	
	I	**II**
Male	25	20
Female	35	20

If one file is selected at random, find the probability that
  a. the selected individual is female.
  b. the selected individual is a Type II.

4.46 The odds against being dealt a contract bridge hand containing 13 cards of the same suit are 158,753,389,899 to 1. The odds against being dealt a royal flush while playing poker are 649,739 to 1.
  a. What is the probability of being dealt a contact bridge hand containing 13 cards all of the same suit?
  b. What is the probability of being dealt a royal flush poker hand?
  c. Express the answers to parts (a) and (b) in scientific notation (powers of 10). (Save answers for use in answering Exercise 4.80.)

4.47 Worldwide the rate for maternal deaths (a woman's risk of dying from pregnancy and childbirth) is 1 in 233. By regions around the world this rate is: North America—1 in 3700, Northern Europe—1 in 4000, Africa—1 in 16, Asia—1 in 65, and Latin America/Caribbean—1 in 130. Express the risk of maternal death as (i) odds in favor of dying, (ii) odds against dying, and (iii) probability of dying for each of the following:
  a. Worldwide   **b.** North America   **c.** Northern Europe
  d. Africa   **e.** Asia   **f.** Latin America/Caribbean

4.48 In the USA Shapshot® "Who is killed by firearms?" *USA Today* (1-17-95) reported: per 100,000 population: total—14.8, male—26.0, female—4.1.
  a. Express the three rates as probabilities.
  b. Explain why "male rate" + "female rate" ≠ "total rate."

4.49 According to an article in *Glamour* (April 1991), one out of every nine people diagnosed with AIDS during 1991 will be female. Based on this information, find the probability that an individual diagnosed with AIDS in 1991 will be male.

4.50 According to *Science News* (November 1990), sleep apnea affects 2 million individuals in the United States. The sleep disorder interrupts breathing and can awaken its sufferers as

(continued)

often as five times an hour. Many people do not recognize the condition even though it causes loud snoring. Assuming there are 200 million people in the United States, what is the probability that an individual chosen at random will not be affected by sleep apnea?

CALCULATING PROBABILITIES OF COMPOUND EVENTS

Compound events are formed by combining several simple events. We will study the following three compound events in the remainder of this chapter:

1. the probability that either event A or event B will occur, $P(A \text{ or } B)$
2. the probability that both events A and B will occur, $P(A \text{ and } B)$
3. the probability that event A will occur given that event B has occurred, $P(A \mid B)$

NOTE In determining which compound probability we are seeking, it is not enough to look for the words *either/or, and,* or *given.* We must carefully examine the question asked to determine what combination of events is called for.

4.5 Mutually Exclusive Events and the Addition Rule

Mutually Exclusive Events

MUTUALLY EXCLUSIVE EVENTS

Events defined in such a way that the occurrence of one event precludes the occurrence of any of the other events. (In short, if one of them happens, the others cannot happen.)

The following illustrations give examples of events to help you understand the concept of mutually exclusive.

ILLUSTRATION 4.3 ▼

A group of 200 college students is known to consist of 140 full-time (80 female and 60 male) students and 60 part-time (40 female and 20 male) students. From this group one student is to be selected at random.

Exercise 4.51

One student is selected from the student body of your college. The student is: M—male, F—female, S—registered for statistics.

a. Are events M and F mutually exclusive? Explain.
b. Are events M and S mutually exclusive? Explain.
c. Are events F and S mutually exclusive? Explain.

(continued)

200 College Students

	Full-time	Part-time	Total
Female	80	40	120
Male	60	20	80
Total	140	60	200

Two events related to this selection are defined. Event A is "the student selected is full-time" and event B is "the student selected is a part-time male." Since no student is both "full-time" and "part-time male," the two events A and B are mutually exclusive events.

A third event, event C, is defined to be "the student selected is female." Now let's consider the two events A and C. Since there are 80 students that are "full-time" and "female," the two events A and C are not mutually exclusive events. This **"intersection"** of A and C can be seen on the accompanying Venn diagram or in the preceding table.

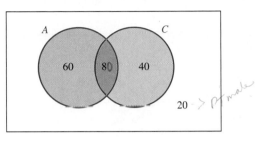

Exercise 4.51—cont'd

d. Are events M and F complementary? Explain.

e. Are events M and S complementary? Explain.

f. Are complementary events also mutually exclusive events? Explain.

g. Are mutually exclusive events also complementary events? Explain.

ILLUSTRATION 4.4 ▼

Consider an experiment in which two dice are rolled. Three events are defined:

A: The sum of the numbers on the two dice is 7.
B: The sum of the numbers on the two dice is 10.
C: Each of the two dice shows the same number.

Let's determine whether these three events are mutually exclusive.

We can show three events are mutually exclusive by showing that each pair of events is mutually exclusive. Are events A and B mutually exclusive? Yes, they are, since the sum on the two dice cannot be both 7 and 10 at the same time. If a sum of 7 occurs, it is impossible for the sum to be 10.

Figure 4.3 presents the sample space for this experiment. This is the same sample space shown in Experiment 4.7, except that ordered pairs are used in place of the pictures. The ovals, diamonds, and rectangles show the ordered pairs that are in events A, B, and C, respectively. We can see that events A and B do not intersect. Therefore, they are mutually exclusive. Point (5, 5) below satisfies both events B and C. Therefore, B and C are not mutually exclusive. Two dice can each show a 5, which satisfies C, and the total satisfies B. Since we found one pair of events that are not mutually exclusive, events A, B, and C are not mutually exclusive.

Figure 4.3
Sample Space for the Roll of Two Dice

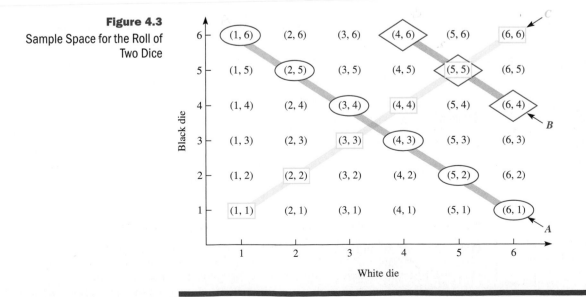

Addition Rule

Let us now consider the compound probability $P(A \text{ or } B)$, where A and B are mutually exclusive events.

ILLUSTRATION 4.5 ▼

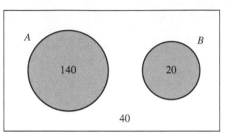

In Illustration 4.3 we considered an experiment in which one student was to be selected at random from a group of 200. Event A was "student selected is full-time," and event B was "student selected is part-time and male." The probability of event A, $P(A)$, is $\frac{140}{200}$, or 0.7, and the probability of event B, $P(B)$, is $\frac{20}{200}$, or 0.1. Let's find the probability of A or B, $P(A \text{ or } B)$. It seems reasonable to add the two probabilities 0.7 and 0.1 to obtain an answer of $P(A \text{ or } B) = 0.8$. This probability is further justified by looking at the sample space. We see there is a total of 160 that either are full-time or are male part-time students ($\frac{160}{200} = 0.8$). Recall that these two events, A and B, are mutually exclusive.

When events are *not mutually exclusive,* we cannot find the probability that one or the other occurs by simply adding the individual probabilities. Why not? Let's look at an illustration and see what happens when events are not mutually exclusive.

ILLUSTRATION 4.6 ▼

Using the sample space and the events defined in Illustration 4.3, find the probability that the student selected is "full-time" or "female," $P(A \text{ or } C)$.

Solution

If we look at the sample space, we see that $P(A) = \frac{140}{200}$, or 0.7, and that $P(C) = \frac{120}{200}$, or 0.6. If we add the two numbers 0.7 and 0.6, we will get 1.3, a number larger than 1. We also know, from basic properties of probability, that probability of an event can never be larger than 1. So what happened? If we take another look at the sample space, we will see that 80 of the 200 students have been counted twice if we add $\frac{140}{200}$ and $\frac{120}{200}$. Only 180 students are "full-time" or "female." Thus, the probability of A or C is

$$P(A \text{ or } C) = \frac{180}{200} = 0.9$$

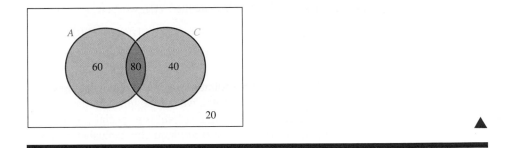

We can add probabilities to find the probability of an "or" compound event, but we must make an adjustment in situations such as the previous example.

GENERAL ADDITION RULE

Let A and B be two events defined in a sample space S.

$$P(\text{A or B}) = P(\text{A}) + P(\text{B}) - P(\text{A and B}) \qquad \textbf{(4.4a)}$$

SPECIAL ADDITION RULE

Let A and B be two events defined in a sample space. If A and B are *mutually exclusive events*, then

$$P(\text{A or B}) = P(\text{A}) + P(\text{B}) \qquad \textbf{(4.4b)}$$

This can be expanded to consider more than two mutually exclusive events:

$$P(\text{A or B or C or \ldots or E}) = P(\text{A}) + P(\text{B}) + P(\text{C}) + \cdots + P(\text{E}) \qquad \textbf{(4.4c)}$$

The key to this formula is the property "mutually exclusive." If two events are mutually exclusive, there is no double counting of sample points. If events are not mutually exclusive, then when probabilities are added, the double counting will occur. Let's look at some examples.

In Figure 4.4, events A and B are not mutually exclusive. The probability of the event "A or B," $P(\text{A or B})$, is represented by the union (total) of the shaded regions. The probability of the event "A and B," $P(\text{A and B})$, is represented by the area contained in region II. The probability $P(\text{A})$ is represented by the area of circle A. That is, $P(\text{A}) = P(\text{region I}) + P(\text{region II})$. Furthermore, $P(\text{B}) = P(\text{region II}) + P(\text{region III})$. And $P(\text{A or B})$ is the sum of the probabilities associated with the three regions:

Figure 4.4
Nonmutually Exclusive Events

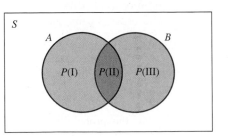

$$P(\text{A or B}) = P(\text{I}) + P(\text{II}) + P(\text{III})$$

However, if $P(A)$ is added to $P(B)$, we have

$$P(A) + P(B) = [P(\text{I}) + P(\text{II})] + [P(\text{II}) + P(\text{III})]$$

$$= P(\text{I}) + 2P(\text{II}) + P(\text{III})$$

This is the double count previously mentioned. However, if we subtract one measure of region II from this total, we will be left with the correct value.

The addition formula (4.4b) is a special case of the more general rule stated in formula (4.4a). If A and B are mutually exclusive events, $P(A \text{ and } B) = 0$. (They cannot both happen at the same time.) Thus, the last term in formula (4.4a) is zero when events are mutually exclusive.

ILLUSTRATION 4.7 ▼

One white die and one black die are rolled. Find the probability that the white die shows a number smaller than 3 or the sum of the dice is greater than 9.

Solution 1

A = white die shows a 1 or a 2; B = sum of both dice is 10, 11, or 12.

$$P(A) = \frac{12}{36} = \frac{1}{3} \text{ and } P(B) = \frac{6}{36} = \frac{1}{6}$$

see pg 192

$(5,5)\,(6,4)\,(6,5)$
$(4,6)\,(5,6)\,(6,6)$

$$P(A \text{ or } B) = P(A) + P(B) - P(A \text{ and } B)$$

$$= \frac{1}{3} + \frac{1}{6} - 0 = \frac{1}{2}$$

$[P(A \text{ and } B) = 0$, since the events do not intersect.]

Solution 2

$$P(A \text{ or } B) = \frac{n(A \text{ or } B)}{n(S)} = \frac{18}{36} = \frac{1}{2}$$

(Look at the sample space, Figure 4.3, p. 201, and count.)

ILLUSTRATION 4.8 ▼

A pair of dice is rolled. Event T is defined as the occurrence of a "total of 10 or 11," and event D is the occurrence of "doubles." Find the probability $P(T \text{ or } D)$.

Solution

Look at the sample space of 36 ordered pairs for the rolling of two dice in Figure 4.3. Event T occurs if any one of 5 ordered pairs occurs: (4, 6), (5, 5), (6, 4), (5, 6), (6, 5). Therefore, $P(T) = \frac{5}{36}$. Event D occurs if any one of 6 ordered pairs occurs: (1, 1), (2, 2), (3, 3), (4, 4), (5, 5), (6, 6). Therefore, $P(D) = \frac{6}{36}$. Notice, however, that these two events are not mutually exclusive. The two events "share" the point (5, 5). Thus, the probability $P(T \text{ and } D) = \frac{1}{36}$. As a result, the probability $P(T \text{ or } D)$ will be found using formula (4.4a).

Exercise 4.52

A pair of dice is rolled. Define events as:
A—sum of 7, C—doubles, E—sum of 8.

a. Which pairs of events, A and C, A and E, C and E, are mutually exclusive? Explain.
b. Find the probabilities $P(A \text{ or } C)$, $P(A \text{ or } E)$, and $P(C \text{ or } E)$.

$$P(\text{T or D}) = P(\text{T}) + P(\text{D}) - P(\text{T and D})$$

$$= \frac{5}{36} + \frac{6}{36} - \frac{1}{36} = \frac{10}{36} = \frac{5}{18}$$

(Look at the sample space in Figure 4.3 and verify $P(\text{T or D}) = \frac{5}{18}$.)   ▲

EXERCISES

4.53 Determine whether or not each of the following pairs of events is mutually exclusive.
  a. Five coins are tossed: "one head is observed," "at least one head is observed."
  b. A salesperson calls on a client and makes a sale: "the sale exceeds $100," "the sale exceeds $1000."
  c. One student is selected at random from a student body: the person selected is "male," the person selected is "over 21 years of age."
  d. Two dice are rolled: the total showing is "less than 7," the total showing is "more than 9."

4.54 Determine whether each of the following sets of events is mutually exclusive.
  a. Five coins are tossed: "no more than one head is observed," "two heads are observed," "three or more heads are observed."
  b. A salesperson calls on a client and makes a sale: the amount of the sale is "less than $100," is "between $100 and $1000," is "more than $500."
  c. One student is selected at random from the student body: the person selected is "female," is "male," or is "over 21."
  d. Two dice are rolled: the number of dots showing on each die are "both odd," are "both even," are "total seven," or are "total eleven."

4.55 Explain why $P(\text{A and B}) = 0$ when events A and B are mutually exclusive.

4.56 Explain why $P(\text{A occurring, when B has occurred}) = 0$ when events A and B are mutually exclusive.

4.57 If $P(\text{A}) = 0.3$ and $P(\text{B}) = 0.4$, and if A and B are mutually exclusive events, find the following:
  a. $P(\overline{\text{A}})$ b. $P(\overline{\text{B}})$ c. $P(\text{A or B})$ d. $P(\text{A and B})$

4.58 If $P(\text{A}) = 0.4$, $P(\text{B}) = 0.5$, and $P(\text{A and B}) = 0.1$, find $P(\text{A or B})$.

4.59 One student is selected at random from a student body. Suppose the probability that this student is female is 0.5 and the probability that this student is working a part-time job is 0.6. Are the two events "female" and "working" mutually exclusive events? Explain.

4.60 The college board's SAT scores are divided between verbal and math, both of which average near 500 year after year. The following table shows the frequency counts from the 50 states and the District of Columbia for verbal and math scores that averaged above and below 500 in 1997:

	500 or less	Over 500	Total
Verbal	13	38	51
Math	10	41	51
Total	23	79	102

Source: College Board, 1998.

Let A be the event that a state (or DC), selected at random, "has an average score above 500" and B be the event "math score."   *(continued)*

a. Find $P(A)$.

b. Find $P(B)$.

c. Find $P(A \text{ and } B)$.

d. Find $P(A \text{ or } B)$ and $P(\text{not } A \text{ or } B)$.

e. Draw the Venn diagram and show the probabilities in the diagram in (a) through (c) and $P(\text{not } A \text{ or } B)$. Are the events A and B mutually exclusive? Explain.

4.61 The population of the United States doubled between 1930 and 1990. The following table shows the frequencies of the 50 states and the District of Columbia for 1930 and 1990 that were either above or below 2,000,000:

	Under 2,000,000	Over 2,000,000	Total
1930 Census	28	23	51
1990 Census	18	33	51
Total	46	56	102

Source: Bureau of Census, U.S. Dept. of Commerce.

Let A be the event that a state (or DC) selected at random, "has a population under 2,000,000"and B be the event "1990 census."

a. Find $P(A)$.   4.5

b. Find $P(B)$. $\frac{51}{102} = .5$

c. Find $P(A \text{ and } B)$.

d. Find $P(A \text{ or } B)$ and $P(\text{not } A \text{ or } B)$.

e. Draw the Venn diagram and show the probabilities in the diagram in (a) through (c) and $P(\text{not } A \text{ or } B)$. Are the events A and B mutually exclusive? Explain.

4.62 An article titled "Pain and Pain-Related Side Effects in an ICU and on a Surgical Unit: Nurses' Management" (*American Journal of Critical Care,* January 1994, Vol. 3, No. 1) gave the following table summarizing the study participants. (Note: ICU is an acronym for intensive care unit.)

Gender	ICU	Surgical Unit
Female	9	6
Male	11	18

One of these participants is randomly selected. Answer the following questions.

a. Are the events "being a female" and "being in the ICU" mutually exclusive?

b. Are the events "being in the ICU" and "being in the surgical unit" mutually exclusive?

c. Find $P(\text{ICU or female})$.

d. Find $P(\text{ICU or male})$.

4.63 A parts store sells both new and used parts. Sixty percent of the parts in stock are used. Sixty-one percent are used or defective. If 5% of the store's parts are defective, what percentage are both used and defective?

4.64 Union officials report that 60% of the workers at a large factory belong to the union, 90% make over $12 per hour, and 40% belong to the union and make over $12 per hour. Do you believe these percentages? Explain.

4.6   Independence, the Multiplication Rule, and Conditional Probability

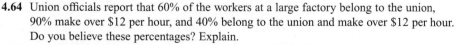

Consider this example. The event that a 2 shows on a white die is A, and the event that a 2 shows on a black die is B. If both dice are rolled once, what is the probability that two 2's occur?

$$P(A) = \frac{1}{6} \text{ and } P(B) = \frac{1}{6}$$

$$P(A \text{ and } B) = \frac{n(A \text{ and } B)}{n(S)} = \frac{1}{36}$$

Notice that by multiplying the probabilities of the simple events the correct answer is found for $P(A \text{ and } B)$. Multiplication does not always work, however. For example, $P(\text{sum of 7 and double})$ when two dice are rolled is zero (as seen in Figure 4.3). However, if $P(7)$ is multiplied by $P(\text{double})$, we obtain $\left(\frac{6}{36}\right)\left(\frac{6}{36}\right) = \left(\frac{1}{6}\right)\left(\frac{1}{6}\right) = \frac{1}{36}$.

Multiplication does not work for $P(\text{sum of 10 and double})$, either. By definition and by inspection of the sample space, we know that $P(10 \text{ and double}) = \frac{1}{36}$ (the point $(5, 5)$ is the only element). However, if we multiply $P(10)$ by $P(\text{double})$, we obtain $\left(\frac{3}{36}\right)\left(\frac{6}{36}\right) = \frac{1}{72}$. The probability of this event cannot be both values.

The property that is required for multiplying probabilities is **independence.** Multiplication worked in the one foregoing example because the events were independent. In the other two examples, the events were not independent and multiplication gave us incorrect answers.

NOTE There are several situations that result in the compound event "and." Some of the more common ones are: (1) A followed by B, (2) A and B occurred simultaneously, (3) the intersection of A and B, (4) both A and B, and (5) A but not B (equivalent to A and not B).

Independence and Conditional Probabilities

INDEPENDENT EVENTS

Two events A and B are independent events if and only if the occurrence (or nonoccurrence) of one does not affect the probability assigned to the occurrence of the other.

Sometimes independence is easy to determine, for example, if the two events being considered have to do with unrelated trials, such as the tossing of a penny and a nickel. The results on the penny in no way affect the probability of heads or tails on the nickel. Similarly, the results on the nickel have no effect on the probability of heads or tails on the penny. Therefore, the results on the penny and the results on the nickel are independent. However, if events are defined as combinations of outcomes from separate trials, the independence of the events may or may not be so easy to determine. The separate results of each trial (dice in the next illustration) may be independent, but the compound events defined using both trials (both dice) may or may not be independent.

Lack of independence, called **dependence,** is demonstrated by the following illustration. Reconsider the experiment of rolling two dice and observing the two events "sum of 10" and "double." As stated previously, $P(10) = \frac{3}{36} = \frac{1}{12}$ and $P(\text{double}) = \frac{6}{36} = \frac{1}{6}$. Does the occurrence of 10 affect the probability of a double? Think of it this way. A sum of 10 has occurred; it must be one of the following: $\{(4, 6), (5, 5), (6, 4)\}$. One of these three possibilities is a double. Therefore, we must conclude that the P(double, knowing 10 has occurred), written $P(\text{double} \mid 10)$, is $\frac{1}{3}$. Since $\frac{1}{3}$ does not equal the original probability of a double, $\frac{1}{6}$, we can conclude that the event "10" has an effect on the probability of a double. Therefore, "double" and "10" are dependent events.

Whether or not events are independent often becomes clear by examining the events in question. Rolling one die does not affect the outcome of a second roll. However, in many cases, independence is not self-evident, and the question of independence itself may be of special interest. Consider the events "having a checking account at a bank" and "having a loan account at the same bank." Having a checking account at a bank may increase the probability that the same person has a loan account. This situation has practical implications. For example, it would make sense to advertise loan programs to checking-account clients if they are more likely to apply for loans than are people who are not customers of the bank.

One approach to the problem is to *assume* independence or dependence. The correctness of the probability analysis depends on the truth of the assumption. In practice, we often assume independence and compare calculated probabilities with actual frequencies of outcomes in order to infer whether the assumption of independence is warranted.

CONDITIONAL PROBABILITY

The symbol $P(A \mid B)$ represents the probability that A will occur given that B has occurred. This is called a conditional probability.

The previous definition of independent events can now be written in a more formal manner.

INDEPENDENT EVENTS

Two events A and B are independent events if and only if

$$P(A \mid B) = P(A) \quad \text{or} \quad P(B \mid A) = P(B). \tag{4.5}$$

Let's consider conditional probability. Take, for example, the experiment in which a single die is rolled: $S = \{1, 2, 3, 4, 5, 6\}$. Two events that can be defined for this experiment are A = "a 4 occurs," and B = "an even number occurs." Then $P(A) = \frac{1}{6}$. Event A, A = $\{4\}$, is satisfied by exactly one of the six equally likely sample points in S. The conditional probability of A given B, $P(A \mid B)$, is found in a similar manner, but the list of possible events is no longer the sample space. Think of it this way. A die is rolled out of your sight, and you are told the condition, the number showing is even, that is, event B, B = $\{2, 4, 6\}$, has occurred. Knowing this condition, you are asked to assign a probability to the event that the even number is a 4. There are only three possibilities in the *current or reduced sample space*, $\{2, 4, 6\}$. Each of the three outcomes is equally likely; thus, $P(A \mid B) = \frac{1}{3}$.

We can write this as

$$P(A \mid B) = \frac{n(A \cap B)}{n(B)} \tag{4.6a}$$

or equivalently,

$$P(A \mid B) = \frac{P(A \text{ and } B)}{P(B)} \tag{4.6b}$$

Thus, using the formula to verify the answer of $\frac{1}{3}$,

$$P(A \mid B) = \frac{\frac{1}{6}}{\frac{1}{2}} = \frac{1}{3}$$

ILLUSTRATION 4.9 ▼

In a sample of 150 residents, each person was asked if he or she favored the concept of having a single countywide police agency. The county is composed of one large city and many suburban townships. The residence (city or outside the city) and the responses of the residents are summarized in Table 4.3. If one of these residents was to be selected at random, what is the probability that the person will (a) favor the concept? (b) favor the concept if the person selected is a city resident? (c) favor the concept if the person selected is a resident from outside the city? (d) Are the events F (favor the concept) and C (reside in city) independent?

TABLE 4.3 Sample Results for Illustration 4.9

	Opinion		
Residence	Favor (F)	Oppose ($\overline{F}$)	Total
In city (C)	80	40	120
Outside of city ($\overline{C}$)	20	10	30
Total	100	50	150

Exercise 4.65

300 viewers were asked if they were satisfied with coverage of a recent disaster.

Gender	Female	Male
Satisfied	80	55
Not satisfied	120	45

One viewer is to be randomly selected from those surveyed.

a. Find 208P(Satisfied).
b. Find $P(S \mid \text{female})$.
c. Find $P(S \mid \text{male})$.
d. Is event S independent of gender? Explain.

Solution

(a) $P(F)$ is the proportion of the total sample that favor the concept. Therefore,

$$P(F) = \frac{n(F)}{n(S)} = \frac{100}{150} = \frac{2}{3}$$

(b) $P(F \mid C)$ is the probability that the person selected favors the concept given that he or she lives in the city. The sample space is reduced to the 120 city residents in the sample. Of these, 80 favored the concept; therefore,

$$P(F \mid C) = \frac{n(F \text{ and } C)}{n(C)} = \frac{80}{120} = \frac{2}{3}$$

(c) $P(F \mid \overline{C})$ is the probability that the person selected favors the concept, knowing that the person lives outside the city. The sample space is reduced to the 30 non-city residents; therefore,

$$P(F \mid \overline{C}) = \frac{n(F \text{ and } \overline{C})}{n(\overline{C})} = \frac{20}{30} = \frac{2}{3}$$

(d) All three probabilities have the same value, $\frac{2}{3}$. Therefore, we can say that the events F (favor) and C (reside in city) are independent. The location of residence did not affect $P(F)$.   ▲

Multiplication Rule

GENERAL MULTIPLICATION RULE

Let A and B be two events defined in sample space S. Then

$$P(A \text{ and } B) = P(A) \cdot P(B \mid A) \tag{4.7a}$$

or

$$P(A \text{ and } B) = P(B) \cdot P(A \mid B) \tag{4.7b}$$

Exercise 4.66

R and H are events with $P(R) = 0.6$ and $P(H \mid R) = 0.25$. Find $P(R \text{ and } H)$.

If events A and B are independent, then the general multiplication rule [formula (4.7)] reduces to the *special multiplication rule,* formula (4.8).

SPECIAL MULTIPLICATION RULE

Let A and B be two events defined in sample space S. A and B are independent events if and only if

$$P(A \text{ and } B) = P(A) \cdot P(B) \tag{4.8a}$$

This formula can be expanded. A, B, C, . . ., G are independent events if and only if

$$P(A \text{ and } B \text{ and } C \text{ and } \ldots \text{ and } G) = P(A) \cdot P(B) \cdot P(C) \cdot \ldots \cdot P(G) \tag{4.8b}$$

Exercise 4.67

A and B are independent events, and $P(A) = 0.7$ and $P(B) = 0.4$. Find $P(A \text{ and } B)$.

ILLUSTRATION 4.10 ▼

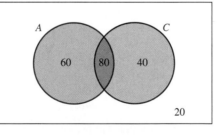

One student is selected at random from a group of 200 known to consist of 140 full-time (80 female and 60 male) students and 60 part-time (40 female and 20 male) students. (See Illustration 4.3.) Event A is "the student selected is full-time," and event C is "the student selected is female."

(a) Are events A and C independent?

(b) Find the probability $P(A \text{ and } C)$ using the multiplication rule.

This is a test for independence

Solution 1

(a) First find the probabilities $P(A)$, $P(C)$, and $P(A \mid C)$.

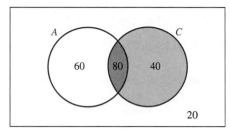

$$P(A) = \frac{n(A)}{n(S)} = \frac{140}{200} = 0.7$$

$$P(C) = \frac{n(C)}{n(S)} = \frac{120}{200} = 0.6$$

$$P(A\,|\,C) = \frac{n(A \text{ and } C)}{n(C)} = \frac{80}{120} = 0.67$$

A and C are dependent events since $P(A) \neq P(A\,|\,C)$.

(b) $P(A \text{ and } C) = P(C) \cdot P(A\,|\,C) = \dfrac{120}{200} \cdot \dfrac{80}{120} = \dfrac{80}{200} = \mathbf{0.4}$

Solution 2

(a) First find the probabilities $P(A)$, $P(C)$, and $P(C\,|\,A)$.

Exercise 4.68

a. Find $P(A \text{ and } C)$ in Illustration 4.10 using the sample space and formula (4.2).

b. Does your answer in (a) agree with the solution in Illustration 4.10? Explain.

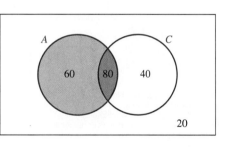

$$P(A) = \frac{n(A)}{n(S)} = \frac{140}{200} = 0.7$$

$$P(C) = \frac{n(C)}{n(S)} = \frac{120}{200} = 0.6$$

$$P(C\,|\,A) = \frac{n(C \text{ and } A)}{n(A)} = \frac{80}{140} = 0.57$$

A and C are dependent events since $P(C) \neq P(C\,|\,A)$.

(b) $P(A \text{ and } C) = P(A) \cdot P(C\,|\,A) = \dfrac{140}{200} \cdot \dfrac{80}{140} = \dfrac{80}{200} = \mathbf{0.4}$

ILLUSTRATION 4.11 ▼

One white and one black die are rolled. Find the probability that the sum of their numbers is 7 and that the number on the black die is larger than the number on the white die.

Solution

A = "sum is 7"; B = "black number larger than white number." The "and" requires the use of the multiplication rule. However, we do not yet know whether events A and B are independent. (Refer to Figure 4.3 for the sample space of this experiment.) We see that

(continued)

C a s e   S t u d y   4.2

New York PICK 10

Exercise 4.69

Case Study 4.2 describes rules for playing New York State's PICK 10 lottery. Many states have similar lottery games.

a. If someone plays PICK 10, that person has a 1 in 8,911,711 chance of winning a half-million-dollar prize. Verify this probability.
b. Verify that the chance of winning the $6000 prize is 1 in 163,381.
c. Verify that the overall chance of winning something is 1 in 17.

New York State, like many other states and cities, runs several lottery games for the purpose of raising money to finance education. The PICK 10 game is currently one of several New York State games that a person can play. The game is explained and the chances of winning each of the various prizes is stated on the ticket. Did you know that you could win a lottery by being totally wrong? The rules say that if you pick ten numbers and zero of them match the numbers drawn, you win $4 on a $1 ticket with a 1 in 22 chance. Can you verify the probability of $\frac{1}{22}$? To have zero matches means that all ten numbers picked did not match the ten numbers picked by the lottery.

$P(\text{zero matches}) = P(A \text{ and } B \text{ and } C \text{ and } D \text{ and } \ldots \text{ and } J), \text{ (all nonmatches)}$

$$= \left(\frac{60}{80}\right)\left(\frac{59}{79}\right)\left(\frac{58}{78}\right)\left(\frac{57}{77}\right)\left(\frac{56}{76}\right)\cdots\left(\frac{51}{71}\right) = 0.04579 = \frac{1}{21.839},$$

or approximately $\frac{1}{22}$

Source: New York State Pick 10 Play Card courtesy of NYS Division of the Lottery.

HOW TO PLAY PICK 10
(a) Each play card has 5 games, you may play 1, 2, 3, 4 or all 5 games. Each game costs $1.00 to play.
(b) Select 10 numbers in each game you want to play.
(c) Select the number of days you want to play your numbers.
(d) Use only blue or black ballpoint pen or pencil for marking. Red ink will not be accepted.
(e) If you can't think of 10 numbers, just mark the Quick Pick option to have the computer randomly select all or some of your numbers for you.
(f) If you make a mistake, mark the void box for that game.
(g) Present your completed play card to any licensed On-Line Lottery Agent for processing.

HOW TO WIN
(a) Each night, the Lottery randomly selects 20 numbers from a field of 80; these 20 are the winning numbers.
(b) You may win if all, some or none of your selections match any of the winning numbers.

PRIZE LEVELS AND CHANCES OF WINNING

WINNING NUMBERS MATCHED PER PANEL	PRIZE	CHANCES OF WINNING ON ONE GAME PANEL
10	$500,000	1:8,911,711
9	$ 6,000	1:163,381
8	$ 300	1:7,384
7	$ 40	1:621
6	$ 10	1:87
0	$ 4	1:22

Overall Chances of Winning on a $1 play: 1 in 17

$P(A) = \dfrac{6}{36} = \dfrac{1}{6}$. Also, $P(A \mid B)$ is obtained from the reduced sample space, which includes 15 points above the gray diagonal line. Of the 15 equally likely points, 3 of them, namely (1, 6), (2, 5), and (3, 4), satisfy event A. Therefore,

$$P(A \mid B) = \frac{n(A \text{ and } B)}{n(B)} = \frac{3}{15} = \frac{1}{5}$$

Since this is a different value than $P(A)$, the events are dependent. So we must use formula (4.7b) to obtain $P(A \text{ and } B)$.

$$P(A \text{ and } B) = P(B) \cdot P(A \mid B) = \frac{15}{36} \cdot \frac{3}{15} = \frac{3}{36} = \frac{1}{12} \qquad \blacktriangle$$

NOTES

1. Independence and mutually exclusive are two very different concepts.
   a. Mutually exclusive says the two events cannot occur together; that is, they have no intersection.
   b. Independence says each event does not affect the other event's probability.
2. $P(A \text{ and } B) = P(A) \cdot P(B)$ when A and B are independent.
   a. Since $P(A)$ and $P(B)$ are not zero, $P(A \text{ and } B)$ is nonzero.
   b. Thus, independent events have an intersection.
3. Events cannot be both mutually exclusive and independent. Therefore,
   a. if two events are independent, then they are not mutually exclusive.
   b. if two events are mutually exclusive, then they are not independent.

EXERCISES   •   •   •   •   •   •   •   •   •   •   •   •   •   •

4.70 **a.** Describe in your own words what it means for two events to be mutually exclusive.
    b. Describe in your own words what it means for two events to be independent.
    c. Explain how mutually exclusive and independence are two very different properties.

4.71 Determine whether or not each of the following pairs of events is independent:
    a. rolling a pair of dice and observing a "1" on the first die and a "1" on the second die
    b. drawing a "spade" from a regular deck of playing cards and then drawing another "spade" from the same deck without replacing the first card
    c. same as (b) except the first card is returned to the deck before the second drawing
    d. owning a red automobile and having blonde hair
    e. owning a red automobile and having a flat tire today
    f. studying for an exam and passing the exam

4.72 Determine whether or not the following pairs of events are independent:
    a. rolling a pair of dice and observing a "2" on one of the dice and having a "total of 10"
    b. drawing one card from a regular deck of playing cards and having a "red" card and having an "ace"
    c. raining today and passing today's exam
    d. raining today and playing golf today
    e. completing today's homework assignment and being on time for class

4.73 If $P(A) = 0.3$ and $P(B) = 0.4$ and A and B are independent events, what is the probability of each of the following?
    a. $P(A \text{ and } B)$ **b.** $P(B \mid A)$ **c.** $P(A \mid B)$

4.74 Suppose that $P(A) = 0.3$, $P(B) = 0.4$, and $P(A \text{ and } B) = 0.12$.
    a. What is $P(A \mid B)$?   **b.** What is $P(B \mid A)$?   **c.** Are A and B independent?

4.75 Suppose that $P(A) = 0.3$, $P(B) = 0.4$, and $P(A \text{ and } B) = 0.20$.
  a. What is $P(A \mid B)$?   **b.** What is $P(B \mid A)$?   **c.** Are A and B independent?

4.76 Suppose that A and B are events and that the following probabilities are known:
  $P(A) = 0.3$, $P(B) = 0.4$, and $P(A \mid B) = 0.2$. Find $P(A \text{ or } B)$.

4.77 A single card is drawn from a standard deck. Let A be the event that "the card is a face card" (jack, queen, or king), B be the occurrence of a "red card," and C represent "the card is a heart." Check to determine whether the following pairs of events are independent or dependent:
  a. A and B   **b.** A and C   **c.** B and C

4.78 A box contains four red and three blue poker chips. What is the probability when three are selected randomly that all three will be red if we select each chip
  a. with replacement?   **b.** without replacement?

4.79 Excluding job benefit coverage, approximately 49% of adults have purchased life insurance. The likelihood for those ages 18 to 24 without life insurance purchasing life insurance in the next year is 15% and for those ages 25 to 34 it is 26%. (*Source:* Opinion Research)
  a. Find the probability that a randomly selected adult has not purchased life insurance.
  b. What is the probability that an adult of age 18 to 24 will purchase life insurance within the next year?
  c. Find the probability that a randomly selected adult will be from 25 to 34 years old, does not currently have life insurance, and will purchase it within the next year.

4.80 Tracy Shark walked into the card room at her country club and played one hand of bridge at one table and one hand of poker at another table. She was dealt a bridge hand with all 13 cards in the same suit, and then she was dealt a poker hand with a royal flush. As she left the room shaking her head, she was overheard mumbling, "I doubt that will happen again in a million years." Assuming the decks were shuffled and the dealers were straight, compute the probability of the joint occurrence of the two events. Do you agree with Tracy? (See Exercise 4.46 on page 199.)

4.81 The December 1994 issue of *The American Spectator* quotes a poll by the Times-Mirror Center for the People and the Press as finding that 71% of Americans believe that the press "gets in the way of society solving its problems."
  a. If two Americans are randomly selected, find the probability that both will believe that the press "gets in the way of society solving its problems."
  b. If two Americans are selected, find the probability that neither of the two will believe that the press "gets in the way of society solving its problems."
  c. If three are selected, what is the probability that all three believe the press gets in the way?

4.82 The August 1, 1994, issue of *The New Republic* gives the results of a U.S. Justice Department study which states that among white spousal murder victims, 62% are female. If the records of three victims are randomly selected from a large data base of such murder victims, what is the probability that all three victims are male?

4.83 An article involving smoking cessation intervention in *Heart & Lung* (March/April 1994) divided 80 subjects into a two-way classification:

Group	Diagnosis		
	Cardiovascular	**Oncology**	**General Surgery**
Experimental	10	14	13
Usual Care	12	16	15

Suppose one of these 80 subjects is selected at random. Find the probabilities of the following events:

(continued)

a. The subject is not in the experimental group.

b. The subject is in the experimental group and has an oncology diagnosis.

c. The subject is in the experimental group or has a cardiovascular diagnosis.

4.84 A study concerning coping strategies of abstainers from alcohol appeared in *Image, the Journal of Nursing Scholarship* (Vol. 25, No. 1, Spring 1993). The study involved 23 subjects who were classified according to sex as well as marital status as shown in the table.

Marital Status	Men	Women
Currently married	10	3
Divorced/separated	3	6
Never married	1	0

One of the subjects is selected at random. Find:

a. The probability the subject is currently married given that the individual is a man.

b. The probability the subject is a woman given that the individual is divorced/separated.

c. The probability the subject is a man given that the individual has never married.

4.85 The owners of a two-person business make their decisions independently of each other, then compare their decisions. If they agree, the decision is made; if they do not agree, then further consideration is necessary before a decision is reached. If they each have a history of making the right decision 60% of the time, what is the probability that together they

a. make the right decision on the first try?

b. make the wrong decision on the first try?

c. delay the decision for further study?

4.86 The odds against throwing a pair of dice and getting a total of 5 are 8 to 1. The odds against throwing a pair of dice and getting a total of 10 are 11 to 1. What is the probability of throwing the dice twice and getting a total of 5 on the first throw and 10 on the second throw?

4.87 Consider the set of integers 1, 2, 3, 4, and 5.

a. One integer is selected at random. What is the probability that it is odd?

b. Two integers are selected at random (one at a time with replacement so that each of the five integers is available for a second selection). Find the probability that (1) neither is odd; (2) exactly one of them is odd; (3) both are odd.

4.88 A box contains 25 parts, of which 3 are defective and 22 are nondefective. If 2 parts are selected without replacement, find the following probabilities:

a. *P*(both are defective)   **b.** *P*(exactly one is defective)   **c.** *P*(neither is defective)

4.89 Graduation rates reached a record low in 1997. The percentage of students graduating within five years was 44.2% for public and 56.6% for private colleges. One of the reasons for this might be that 42% of the students attend part-time. (*Source:* ACT)

a. What additional information do you need in order to determine the probability that a student selected at random is part-time and will graduate within five years?

b. Is it likely that these two events have the needed property? Explain.

c. If appropriate, find the probability that a student selected at random is part-time and will graduate within five years.

4.90 Forty-eight percent of adults plan to buy candy this year at Easter. The types of candy they will buy:

Chocolate	Non-chocolate	Jellybeans	Cream-filled	Marshmallow	Malted	Don't know
30%	25%	13%	11%	8%	7%	6%

Source: International Mass Retail Association

(continued)

a. What additional information do you need in order to determine the probability that a customer selected at random will buy candy and it will be chocolate?
b. Is it likely that these two events have the needed property? Explain.
c. If appropriate, find the probability that a customer selected at random will buy candy and it will be chocolate.

4.7 Combining the Rules of Probability

Tree diagrams can be used to represent many probability problems. In these instances, the addition and multiplication rules can be applied quite readily. To illustrate the use of tree diagrams in solving probability problems, let's use Experiment 4.5. Two poker chips are drawn from a box containing one each of red, blue, and white chips. The tree diagram representing this experiment (Figure 4.5) shows a first drawing and then a second drawing. One chip was drawn on each drawing and not replaced.

Figure 4.5

All Possible Combinations That Can be Drawn

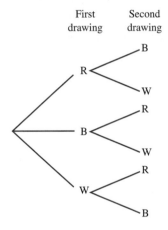

After the tree has been drawn and labeled, we need to assign probabilities to each branch of the tree. If we assume that it is equally likely that any chip would be drawn at each stage, we can assign a probability to each branch segment of the tree, as shown in Figure 4.6. Notice that a set of branches that initiate from a single point has a total probability of 1. In this diagram there are four such sets of branch segments. The tree diagram shows six different outcomes. Reading down: branch (1) shows (R, B), branch (2) shows (R, W), and so on. (*Note:* Each outcome for the experiment is represented by a branch that begins at the common starting point and ends at the terminal points at the right.)

Figure 4.6

Probabilities of All Possible Combinations

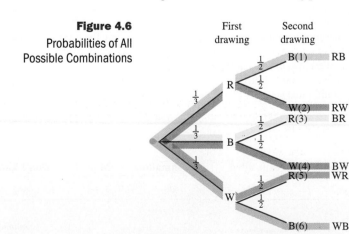

The probability associated with outcome (R, B), P(R on first drawing and B on second drawing), is found by multiplying P(R on first drawing) by P(B on second drawing | R on first drawing). These are the two probabilities $\frac{1}{3}$ and $\frac{1}{2}$ shown on the two branch segments of branch (1) in Figure 4.6. The $\frac{1}{2}$ is the conditional probability asked for by the multiplication rule. Thus, we will multiply along the branches.

Some events will be made up of more than one outcome from our experiment. For example, suppose that we had asked for the probability that one red chip and one blue chip are drawn. You will find two outcomes that satisfy this event, branch (1) or branch (3). With "or" we will use the addition rule (4.4b). Since the branches of a tree diagram represent mutually exclusive events, we have

$$P(\text{one R and one B}) = P[(R_1 \text{ and } B_2) \text{ or } (B_1 \text{ and } R_2)]$$

$$= \left(\frac{1}{3}\right)\left(\frac{1}{2}\right) + \left(\frac{1}{3}\right)\left(\frac{1}{2}\right) = \frac{1}{6} + \frac{1}{6} = \frac{1}{3}$$

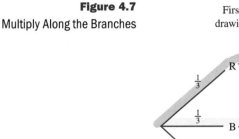

Figure 4.7

Multiply Along the Branches

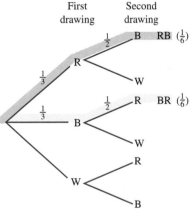

NOTES

1. Multiply along the branches (on Figure 4.7; $\frac{1}{3} \cdot \frac{1}{2}$).

2. Add across the branches (on Figure 4.7; $\frac{1}{6} + \frac{1}{6}$).

Now let's consider an example that places all the rules in perspective.

ILLUSTRATION 4.12 ▼

A firm plans to test a new product in one randomly selected market area. The market areas can be categorized on the basis of location and population density. The number of markets in each category is presented in Table 4.4.

TABLE 4.4   Number of Markets by Location and by Population Density

	Population Density		
Location	**Urban (U)**	**Rural (R)**	**Total**
East (E)	25	50	75
West (W)	20	30	50
Total	45	80	125

What is the probability that the test market selected is in the East, $P(E)$? In the West, $P(W)$? What is the probability that the test market is in an urban area, $P(U)$? In a rural area, $P(R)$? What is the probability that the market is a western rural area, $P(W \text{ and } R)$? What is the probability it is an eastern or urban area, $P(E \text{ or } U)$? What is the probability that if it is in the East, it is an urban area, $P(U \mid E)$? Are "location" and "population density" independent? (What do we mean by independence or dependence in this situation?)

Solution

The first four probabilities, $P(E)$, $P(W)$, $P(U)$, and $P(R)$, represent "or" questions. For example, $P(E)$ means that the area is an eastern urban area or an eastern rural area. Since in this and the other three cases, the two components are mutually exclusive (an area can't be both urban and rural), the desired probabilities can be found by simply adding. In each case the probabilities are added across all the rows or columns of the table. Thus, the totals are found in the total column or row.

$$P(E) = \frac{75}{125} \text{ (total for East divided by total number of markets)}$$

$$P(W) = \frac{50}{125} \text{ (total for West divided by total number of markets)}$$

$$P(U) = \frac{45}{125} \text{ (total for urban divided by total number of markets)}$$

$$P(R) = \frac{80}{125} \text{ (total for rural divided by total number of markets)}$$

Now we solve for $P(W \text{ and } R)$. There are 30 western rural markets and a total of 125 markets. Thus,

$$P(W \text{ and } R) = \frac{n(W \text{ and } R)}{n(S)} = \frac{30}{125}$$

Note that $P(W) \cdot P(R)$ does not give the right answer $[(\frac{50}{125})(\frac{80}{125}) = \frac{32}{125}]$. Therefore, "location" and "population density" are dependent events.

$P(E \text{ or } U)$ can be solved in several different ways. The most direct way is to simply examine the table and count the number of markets that satisfy the condition that they are in the East or they are urban. We find 95 [25 + 50 + 20]. Thus,

$$P(E \text{ or } U) = \frac{n(E \text{ or } U)}{n(S)} = \frac{95}{125}$$

Note that the first 25 markets were both in the East and urban; thus, E and U are not mutually exclusive events.

Another way to solve for $P(E \text{ or } U)$ is to use the addition formula:

$$P(E \text{ or } U) = P(E) + P(U) - P(E \text{ and } U)$$

which yields

$$\frac{75}{125} + \frac{45}{125} - \frac{25}{125} = \frac{95}{125}$$

A third way to solve the problem is to recognize that the complement of (E or U) is (W and R). Thus, $P(E \text{ or } U) = 1 - P(W \text{ and } R)$. Using the previous calculation, we get $1 - \frac{30}{125} = \frac{95}{125}$.

Finally, we solve for $P(U \mid E)$. Looking at Table 4.4, we see that there are 75 markets in the East. Of the 75 eastern markets, 25 are urban. Thus,

$$P(U \mid E) = \frac{n(U \text{ and } E)}{n(E)} = \frac{25}{75}$$

The conditional probability formula could also be used:

$$P(U \mid E) = \frac{P(U \text{ and } E)}{P(E)} = \frac{\dfrac{25}{125}}{\dfrac{75}{125}} = \frac{25}{75}$$

"Location" and "population density" are not independent events. They are dependent. This means that the probability of these events is affected by the occurrence of each other.

Although each rule for computing compound probabilities has been discussed separately, you should not think they are only used separately. In many cases they are combined to solve problems. Consider the following two illustrations.

ILLUSTRATION 4.13 ▼

A production process produces an item. On the average, 20% of all items produced are defective. Each item is inspected before being shipped. The inspector misclassifies an item 10% of the time; that is,

$$P(\text{classified good} \mid \text{defective item}) = P(\text{classified defective} \mid \text{good item})$$
$$= 0.10$$

> Misclassification can happen two ways!

What proportion of the items will be "classified good"?

Solution

What do we mean by the event "classified good"?

$$G = \text{item good}$$

$$D = \text{item defective}$$

$$CG = \text{item classified good by inspector}$$

$$CD = \text{item classified defective by inspector}$$

Figure 4.8

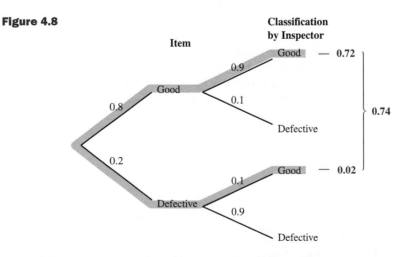

(continued)

CG consists of two possibilities: "the item is good and is correctly classified good" or "the item is defective and misclassified good." Thus,

$$P(CG) = P[(CG \text{ and } G) \text{ or } (CG \text{ and } D)]$$

Since the two possibilities are mutually exclusive, we can start by using the addition rule, formula (4.4b).

$$P(CG) = P(CG \text{ and } G) + P(CG \text{ and } D)$$

The condition of an item and its classification by the inspector are not independent. The multiplication rule for dependent events must be used. Therefore,

$$P(CG) = [P(G) \cdot P(CG \mid G)] + [P(D) \cdot P(CG \mid D)]$$

Substituting the known probabilities, we get

$$P(CG) = [(0.8)(0.9)] + [(0.2)(0.1)]$$
$$= 0.72 + 0.02$$
$$= \mathbf{0.74}$$

That is, 74% of the items are classified good.

ILLUSTRATION 4.14 ▼

Reconsider Illustration 4.13. Suppose that only items that pass inspection are shipped. Items not classified good are scrapped. What is the quality of the shipped items? That is, what percentage of the items shipped are good, $P(G \mid CG)$?

Solution

Using the conditional probability formula (4.6b),

$$P(G \mid CG) = \frac{P(G \text{ and } CG)}{P(CG)}$$

$$P(G \mid CG) = \frac{P(G) \cdot P(CG \mid G)}{P(CG)} \quad \text{(See Illustration 4.13.)}$$

$$= \frac{(0.8)(0.9)}{0.74} = 0.97297 = \mathbf{0.973}$$

In other words, 97.3% of all items shipped will be good. Inspection increases the quality of items shipped from 80% good to 97.3% good.   ▲

The Reverend Thomas Bayes (1702–1761), an English Presbyterian minister and mathematician, developed an expanded form for conditional probabilities. This expanded rule, called **Bayes's rule,** allows us to revise (or adjust) the probabilities assigned to events in accordance with new information.

BAYES'S RULE

$$P(A_i \mid B) = \frac{P(A_i) \cdot P(B \mid A_i)}{\sum [P(A_i) \cdot P(B \mid A_i)]} \tag{4.9}$$

where $A_1, \dots, A_n$ is an all-inclusive set of possible outcomes given B.

Let's take another look at Illustration 4.14 to explain Bayes's rule.

ILLUSTRATION 4.15 ▼

Consider the situation described in Illustration 4.13 and suppose that only items classified as good after inspection are shipped. What percentage of the items shipped are good, $P(G \mid CG)$? What percentage of those shipped are defective, $P(D \mid CG)$?

Solution

First, let's match up the events of the problem to the Bayes's rule notation. The events Bayes identifies as A_i are $A_1 = G$ (good) and $A_2 = D$ (defective), an all-inclusive set of events. The given or conditional event B is CG, classified good.

A tabular approach will be used to help organize the solution. To construct the table, start by listing all possible outcomes A_i that can occur given event B (that is, "shipped") in the first column. See Table 4.5. In the second column, list the initial probabilities of the A_i outcomes. In the third column we list the conditional probability that B happened for each A_i, $P(B \mid A_i)$. For our illustration, $P(B \mid A_1) = P(CG \mid G)$ and $P(B \mid A_2) = P(CG \mid D)$. These first three columns represent the information obtained from the problem.

TABLE 4.5   Tabular Presentation of Given Information

(1) A$_i$, Possible Outcomes	(2) $P(A_i)$	(3) $P(B \mid A_i)$
A$_1$, item good	0.8	0.9
A$_2$, item defective	0.2	0.1
Total	1.0 ⓒⓚ	

To solve for the conditional probabilities $P(A_i \mid B)$, the first calculation is to multiply each number of column (2) by the number in the same row of column (3). This product is placed in column (4) of the table (see Table 4.6). The column is labeled $P(A_i$ and B). The values calculated represent the probability that both A_i and B will occur. Thus, 72% of the items produced will be good and classified good; 2% of the items will be defective and classified good.

TABLE 4.6   Tabular Solution of Bayes's Rule

(1) A$_i$, Possible Outcomes	(2) $P(A_i)$	(3) $P(B \mid A_i)$	(4) $P(A_i$ and B) $=$ $P(A_i)(P(B \mid A_i))$	(5) $P(A_i \mid B)$
A$_1$, item good	0.8	0.9	0.72	$\dfrac{0.72}{0.74} = \mathbf{0.973} = P(G \mid shipped)$
A$_2$, item defective	0.2	0.1	0.02	$\dfrac{0.02}{0.74} = \mathbf{0.027} = P(D \mid shipped)$
Total	1.0 ⓒⓚ		$0.74 = P(B)$	1.000 ⓒⓚ

The second step is to add column (4). The sum represents $P(B)$. Thus, 74% of the items produced will be classified good.

Finally, the answers we are looking for, the conditional probabilities $P(A_i \mid B)$, are obtained by dividing each number in column (4) by the total of column (4). The results

Exercise 4.91

Describe the similarities between the methods used in Illustrations 4.13 and 4.14, and the methods used in Illustration 4.15.

are placed in column (5) and are the answers. Thus 0.973 is the proportion of items classified good that are good, and 0.027 is the proportion of items classified good that are actually defective.

NOTE The totals of columns (2) and (5) must equal 1. The total of column (3) need not equal 1.

Bayes's rule is of special interest because it gives us a mechanism to review initial probability estimates when new information is learned, as we see in the next illustration.

ILLUSTRATION 4.16 ▼

Consider the situation in which we feel that the probability that a stock is a good buy is 0.4. That is, our **prior** (before new information) probabilities are $P(\text{good buy}) = 0.4$ and $P(\text{bad buy}) = 0.6$. Now we find out that an investment service that has a record of being right 80% of the time recommends the stock. What should be our **revised,** or **posterior** (after new information), probability that the stock is a good buy, that is, $P(\text{good buy} \mid \text{investment service recommends it})$, or $P(A_i \mid B)$?

Solution

Using the Bayesian analysis, we find that the revised, or posterior, probability is 0.727 that the stock is a good buy and 0.273 that it is a bad buy (see Table 4.7).

TABLE 4.7 Tabular Analysis for Illustration 4.16

(1) A_i	(2) $P(A_i)$	(3) $P(B \mid A_i)$	(4) $P(A_i)(P(B \mid A_i)$	(5) $P(A_i \mid B)$
A_1, good buy	0.4	0.8	0.32	$\dfrac{0.32}{0.44} = \mathbf{0.727} = P(\text{good buy}\mid\text{recommended})$
A_2, bad buy	0.6	0.2	0.12	$\dfrac{0.12}{0.44} = \mathbf{0.273} = P(\text{bad buy}\mid\text{recommended})$
Total	1.0 ⓒⓚ		$0.44 = P(B)$	1.000 ⓒⓚ

(continued)

Case Study 4.3

Is What You Read, What Was Printed?

USA Today published a profile of affluence that showed 17 million Americans live in households with annual incomes of at least $100,000. Of these, 75% owned a house, 70% were married, and 40% of the houses occupied were valued at over $200,000.

Source: USA Today, July 23, 1998.

Suppose one person from this group is selected at random. What is the probability that the person selected:

a. owns a house valued at over $200,000?

b. is a married homeowner?

c. is a married homeowner living in a house valued at over $200,000?

Exercise 4.92

What conditional probabilities would you need in order to find the answers to the Case Study 4.3 questions?

The answer to all these questions cannot be determined from the data supplied because there is no way of ascertaining the conditional probabilities. The individual events, owning a house and being married, are not independent. Houses can be rented or owned, and married couples can rent or own a house, irrespective of value. This problem illustrates the potential pitfall of misinterpreting survey results.

In reviewing the use of Bayes's rule to revise prior probability estimates in light of new information, we note the following relationship: The stronger the prior probability, the less effect the new information has on changing the probabilities. Also, the more conclusive the new information, the greater the impact on the revised probabilities.

EXERCISES • • • • • • • • • • • • •

4.93 If $P(A) = 0.4$ and $P(B) = 0.5$, and if A and B are independent events, find $P(A$ or $B)$.

4.94 $P(G) = 0.5$, $P(H) = 0.4$, and $P(G$ and $H) = 0.1$ (see diagram)

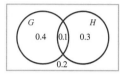

a. Find $P(G \mid H)$.   **b.** Find $P(H \mid G)$.   **c.** Find $P(\overline{H})$.
d. Find $P(G$ or $H)$.   **e.** Find $P(G$ or $\overline{H})$.
f. Are events G and H mutually exclusive? Explain.
g. Are events G and H independent? Explain.

4.95 $P(R) = 0.5$, $P(S) = 0.3$, and events R and S are independent.
a. Find $P(R$ and $S)$.   **b.** Find $P(R$ or $S)$.   **c.** Find $P(\overline{S})$.
d. Find $P(R \mid S)$.   **e.** Find $P(\overline{S} \mid R)$.
f. Are events R and S mutually exclusive? Explain.

4.96 $P(M) = 0.3$, $P(N) = 0.4$, and events M and N are mutually exclusive.
a. Find $P(M$ and $N)$.   **b.** Find $P(M$ or $N)$.   **c.** Find $P(M$ or $\overline{N})$.
d. Find $P(M \mid N)$.   **e.** Find $P(M \mid \overline{N})$.
f. Are events M and N independent? Explain.

4.97 Two flower seeds are randomly selected from a package that contains five seeds for red flowers and three seeds for white flowers.
a. What is the probability that both seeds will result in red flowers?
b. What is the probability that one of each color is selected?
c. What is the probability that both seeds are for white flowers?

Draw a tree diagram.

4.98 The probability that a certain door is locked is 0.6. The key to the door is one of five unidentified keys hanging on a key rack. Two keys are randomly selected before approaching the door. What is the probability that the door may be opened without returning for another key?

4.99 Alex, Bill, and Chen each in turn toss a balanced coin. The first one to throw a head wins.
a. What are their respective chances of winning if each tosses only one time?
b. What are their respective chances of winning if they continue, given a maximum of two tosses each?

Draw a tree diagram.

4.100 A coin is flipped three times.
a. Draw a tree diagram that represents all possible outcomes.
b. Identify all branches that represent the event "exactly one head occurred."
c. Find the probability of "exactly one head occurred."

4.101 Box 1 contains two red balls and three green balls, and Box 2 contains four red balls and one green ball. One ball is randomly selected from Box 1 and placed in Box 2. Then one ball is randomly selected from Box 2. What is the probability that the ball selected from Box 2 is green?

4.102 A company that manufactures shoes has three factories. Factory 1 produces 25% of the company's shoes, Factory 2 produces 60%, and Factory 3 produces 15%. One percent of the shoes produced by Factory 1 are mislabeled, 0.5% of those produced by Factory 2 are mislabeled, and 2% of those produced by Factory 3 are mislabeled. If you purchase one pair of shoes manufactured by this company, what is the probability that the shoes are mislabeled?

4.103 An article titled "A Puzzling Plague" found in the January 14, 1991, issue of *Time,* stated that one out of every ten American women will get breast cancer. It also states that of those who do, one out of four will die of it. Use these probabilities to find the probability that a randomly selected American woman will
   a. never get breast cancer
   b. get breast cancer and not die of it.
   c. get breast cancer and die from it.

4.104 The *Regional Economic Digest* listed the total deposits in commercial banks and thrifts within the states of Missouri and Nebraska during the fourth quarter of 1997. The deposits were further subdivided into two major categories, checkable deposits and time/savings deposits. Results are shown in the following table in millions of dollars:

| | Missouri | | Nebraska | | |
	Commercial Banks	Thrifts	Commercial Banks	Thrifts	Total
Checkable	5,111	1,407	5,017	406	11,941
Time/Savings	12,428	128	15,447	5,312	33,315
Total	17,539	1,535	20,464	5,718	45,256

Source: Regional Economic Digest, First Quarter, 1998, Vol. 9, No. 1.

Let the event A = "checkable deposit," the event B = "Nebraska deposit," and the event C = "deposit in a commercial bank." Suppose $1 is withdrawn randomly from the grand total of all deposits. Find each of the following probabilities and describe in your own words what they mean.
   a. $P(A)$, $P(B)$, and $P(C)$
   b. $P(A \text{ and } B)$, $P(A \text{ and } C)$, and $P(B \text{ and } C)$, and $P(A \text{ and } B \text{ and } C)$
   c. $P(A \text{ or } B)$, $P(A \text{ or } C)$, $P(B \text{ or } C)$, and $P(\text{not } A \text{ or } B \text{ or } C)$.
   d. Sketch the Venn diagram and show all related probabilities. Are A, B, and C all independent events? Explain.

4.105 One thousand employees at the Russell Microprocessor Company were polled about worker satisfaction.

| | Male | | Female | | |
	Skilled	Unskilled	Skilled	Unskilled	Total
Satisfied	350	150	25	100	625
Unsatisfied	150	100	75	50	375
Total	500	250	100	150	1000

One employee is selected at random.
   a. Find the probability that an unskilled worker is satisfied with work.
   b. Find the probability that a skilled woman employee is satisfied with work.
   c. Is satisfaction for women employees independent of their being skilled or unskilled?

4.106 Given the following:

$P(A_1) = 0.2$ $P(A_2) = 0.4$ $P(A_3) = 0.3$ $P(A_4) = 0.1$

$P(B|A_1) = 0.5$ $P(B|A_2) = 0.4$ $P(B|A_3) = 0.2$ $P(B|A_4) = 0.1$

Find: **a.** $P(A_1|B)$ **b.** $P(A_2|B)$ **c.** $P(A_3|B)$ **d.** $P(A_4|B)$.

4.107 In an article titled "Why Quitting Means Gaining" (*Time,* March 25, 1991), it was reported that giving up cigarette smoking often results in gaining weight. In examining a group of quitters, the following data were found.

| | **Weight Gain** | | | |
	Major	**Significant**	**Moderate**	**Slight**
Men	9%	14%	22%	55%
Women*	12%	11%	26%	50%

*Due to rounding, numbers for women do not total 100%.

Suppose the group were 60% men and 40% women. If a participant were randomly selected and found to have experienced
a. a major weight gain, find the probability that it was a man.
b. a slight weight gain, find the probability that it was a woman.

4.108 Given the information in the accompanying table, compute $P(A_1|UF)$ and $P(A_2|UF)$ by filling in the rest of the table. (UF = unfavorable survey results)

	$P(A_i)$	$P(UF|A_i)$	$P(A_i \text{ and } UF)$	$P(A_i|UF)$
$P(A_1|UF)$, profitable	0.6	0.4		
$P(A_1|UF)$, not profitable	0.4	0.7		

4.109 Immediately following the scandals that scarred the Clinton administration, many experts predicted that the Republicans would regain control of the executive branch of the federal government in the year 2000 election. At one time, the general consensus among political scientists was that if the Republicans fielded their best presidential candidate, they would have a 65% chance of winning the 2000 election. In addition, a poll of several thousand voters from across the nation indicated that 54% intended to vote Republican.

Revise the experts' opinion in light of the information from the poll using a Bayesian tabular analysis.

Return to CHAPTER CASE STUDY

The Pennsylvania Lottery game "Big 4" has been played for more than 18 years. The chapter case study on page 179 lists the number of times each single-digit number was the winning number for each of the four positions. The frequencies for each number in each position range from 304 to 367.

EXERCISE 4.110

a. Do the frequencies of each number as a winner in the first position appear to indicate that the numbers occur equally likely as first position winners? What statistical evidence can you find to justify your answer? Present a convincing case.
b. Do the frequencies of each number as a winner in the second, third and fourth positions appear to indicate that the numbers occur equally likely as position winners? What statistical evidence can you find to justify your answer? Present a convincing case.
c. Each single-digit number has appeared as a winning number a different number of times ranging from 1290 to 1383. Do you think these numbers vary sufficiently to make a case that the digits do not occur with equal probability? Present evidence to support your answer.

IN RETROSPECT

You have been studying the basic concepts of probability. These fundamentals need to be mastered before we continue with our study of statistics. Probability is the vehicle of statistics, and we have begun to see how probabilistic events occur. We have explored theoretical and experimental probabilities for the same event. Does the experimental probability turn out to have the same value as the theoretical? Not exactly, but we have seen that over the long run it does have approximately the same value.

Upon completion of this chapter you should understand the properties of mutually exclusive and independence, and be able to apply the multiplication and addition rules to "and" and "or" compound events. You should also be able to calculate conditional probabilities.

In the next three chapters we will look at distributions associated with probabilistic events. This will prepare us for the statistics that will follow. We must be able to predict the variability that the sample will show with respect to the population before we will be successful at "inferential statistics," in which we describe the population based on the sample statistics available.

CHAPTER EXERCISES

4.111 The Federal Highway Administration periodically tracks the number of licensed vehicle drivers by sex and by age. The following table shows the results of the administration's findings in 1995:

Age Group (yr)	Male	Female
19 or less	4,761,567	4,362,558
20–24	8,016,601	7,508,844
25–29	9,234,547	8,822,290
30–34	10,255,668	10,028,055
35–39	10,381,712	10,227,348
40–44	9,512,860	9,465,126
45–49	8,469,713	8,401,960
50–54	6,493,069	6,397,959
55–59	5,167,725	5,057,785
60–64	4,530,005	4,428,256
65–69	4,248,092	4,234,797
70–74	3,582,678	3,702,020
75–79	2,465,550	2,577,527
80 and over	2,094,581	2,149,589
Total	89,214,367	87,414,115

Source: Federal Highway Administration, U.S. Department of Transportation.

Suppose you encountered a driver of a vehicle at random. Find the probabilities of the following events:
a. The driver is a male, over the age of 59.
b. The driver is a female, under the age of 30.
c. The driver is under the age of 25.
d. The driver is a female.
e. The driver is a male between the ages of 35 and 49.
f. The driver is a female between the ages of 25 and 44.
g. The driver is over the age of 69.

4.112 Probabilities for events A, B, and C are distributed as shown on the following figure.

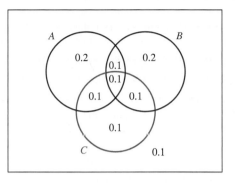

Find
a. P(A and B)   **b.** P(A or C)   **c.** P(A | C)

  4.113 Suppose a certain ophthalmic trait is associated with eye color. Three hundred randomly selected individuals are studied with results as follows:

Trait	Eye Color			Totals
	Blue	**Brown**	**Other**	
Yes	70	30	20	120
No	20	110	50	180
Totals	90	140	70	300

a. What is the probability that a person selected at random has blue eyes?
b. What is the probability that a person selected at random has the trait?
c. Are events A (has blue eyes) and B (has the trait) independent? Justify your answer.
d. How are the two events A (has blue eyes) and C (has brown eyes) related (independent, mutually exclusive, complementary, or all-inclusive)? Explain why or why not each term applies.

4.114 A study was conducted in 1997 to measure the total fat content, calories, and sodium content of vegetable burgers available at supermarkets and commonly used as a meat substitute. Measurements were taken on 54 different brands of "veggie burgers," and the results were used to develop the contingency table shown below.

Total Fat Content	Under 130 Calories		130 Calories or More	
	Under 320 g Sodium	**320 g Sodium or More**	**Under 320 g Sodium**	**320 g Sodium or More**
Under 3 g	6	12	1	1
3–4 g	7	4	3	5
Over 4 g	1	1	7	6

Source: Nutrition Action Health Letter, "Where's the Beef?" Vol. 24, No. 1, 1997.

Let the event A = "Under 130 calories," B = "320 g Sodium or more," and C = "Over 4 g total fat." A vegetable burger is selected randomly from the group.
a. Find P(A), P(B), and P(C).
b. Find P(A and B), P(A and C), and P(B and C) and P(A and B and C).
c. Find P(A or B), P(A or C), and P(B or C).
d. Sketch the Venn diagram and show all related probabilities. Are A, B, and C mutually independent events? Explain.

4.115 Events R and S are defined on the sample space. If $P(R) = 0.2$ and $P(S) = 0.5$, explain why each of the following statements is either true or false.
 a. If R and S are mutually exclusive, then $P(R \text{ or } S) = 0.10$.
 b. If R and S are independent, then $P(R \text{ or } S) = 0.6$.
 c. If R and S are mutually exclusive, then $P(R \text{ and } S) = 0.7$.
 d. If R and S are mutually exclusive, then $P(R \text{ or } S) = 0.6$.

4.116 Show that if event A is a subset of event B, then $P(A \text{ or } B) = P(B)$.

4.117 Let's assume that there are three traffic lights between your house and a friend's house. As you arrive at each light, it may be red (R) or green (G).
 a. List the sample space showing all possible sequences of red and green lights that could occur on a trip from your house to your friend's. (RGG represents red at the first light and green at the other two.)

 Assuming that each element of the sample space is equally likely to occur,

 b. What is the probability that on your next trip to your friend's house you will have to stop for exactly one red light?
 c. What is the probability that you will have to stop for at least one red light?

4.118 Assuming that a woman is equally likely to bear a boy as a girl, use a tree diagram to compute the probability that a four-child family consists of one boy and three girls.

4.119 Suppose that when a job candidate comes to interview for a job at RJB Enterprises, the probability that he or she will want the job (A) after the interview is 0.68. Also, the probability that RJB wants the candidate (B) is 0.36. The probability $P(A \mid B)$ is 0.88.
 a. Find $P(A \text{ and } B)$.   **b.** Find $P(B \mid A)$.
 c. Are events A and B independent? Explain.
 d. Are events A and B mutually exclusive? Explain.
 e. What would it mean to say A and B are mutually exclusive events in this exercise?

4.120 A traffic analysis at a busy traffic circle in Washington, DC, showed that 0.8 of the autos using the circle entered from Connecticut Avenue. Of those entering the traffic circle from Connecticut Avenue, 0.7 continued on Connecticut Avenue at the opposite side of the circle. What is the probability that a randomly selected auto observed in the traffic circle entered from Connecticut and will continue on Connecticut?

4.121 According to the National Cancer Data Base report for Hodgkin's disease (*CA—A Cancer Journal for Clinicians,* Jan./Feb. 1991), the highest percentage of patients (31%) were 20 to 29 years of age, and they had a three-year observed survival rate of 91%. What is the probability that an individual who has been diagnosed with Hodgkin's disease is between 20 and 29 years of age and will survive for three years?

4.122 The probability that thunderstorms are in the vicinity of a particular midwestern airport on an August day is 0.70. When thunderstorms are in the vicinity, the probability that an airplane lands on time is 0.80. Find the probability that thunderstorms are in the vicinity and the plane lands on time.

4.123 Tires salvaged from a train wreck are on sale at the Getrich Tire Company. Of the 15 tires offered in the sale, 5 tires have suffered internal damage and the remaining 10 are damage free. If you were to randomly select and purchase 2 of these tires,
 a. what is the probability that the tires you purchase are both damage free?
 b. what is the probability that exactly 1 of the tires you purchase is damage free?
 c. what is the probability that at least 1 of the tires you purchase is damage free?

4.124 According to automobile accident statistics, one out of every six accidents results in an insurance claim of $100 or less in property damage. Three cars insured by an insurance company are involved in different accidents. Consider these two events:

 A: The majority of claims exceed $100.
 B: Exactly two claims are $100 or less.

 a. List the sample points for this experiment.
 b. Are the sample points equally likely?   *(continued)*

    c. Find $P(A)$ and $P(B)$.

    d. Are A and B independent? Justify your answer.

4.125 One thousand persons screened for a certain disease are given a clinical exam. As a result of the exam, the sample of 1000 persons is distributed according to height and disease status.

	Disease Status				
Height	**None**	**Mild**	**Moderate**	**Severe**	**Totals**
Tall	122	78	139	61	400
Medium	74	51	90	35	250
Short	104	71	121	54	350
Totals	300	200	350	150	1000

Use this information to estimate the probability of being medium or short in height and of having moderate or severe disease status.

4.126 The following table shows the sentiments of 2500 wage-earning employees at the Spruce Company on a proposal to emphasize fringe benefits rather than wage increases during their impending contract discussions.

	Opinion			
Employee	**Favor**	**Neutral**	**Opposed**	**Totals**
Male	800	200	500	1500
Female	400	100	500	1000
Totals	1200	300	1000	2500

    a. Calculate the probability that an employee selected at random from this group will be opposed.

    b. Calculate the probability that an employee selected at random from this group will be female.

    c. Calculate the probability that an employee selected at random from this group will be opposed, given that the person is male.

    d. Are the events "opposed" and "female" independent? Explain.

4.127 A shipment of grapefruit arrived containing the following proportions of types: 10% pink seedless, 20% white seedless, 30% pink with seeds, 40% white with seeds. A grapefruit is selected at random from the shipment. Find the probability that

    a. it is seedless.   b. it is white.

    c. it is pink and seedless.   d. it is pink or seedless.

    e. it is pink, given that it is seedless.   f. it is seedless, given that it is pink.

4.128 Salespersons Adams and Jones call on three and four customers, respectively, on a given day. Adams could make 0, 1, 2, or 3 sales, whereas Jones could make 0, 1, 2, 3, or 4 sales. The sample space listing the number of possible sales for each person on a given day is given in the following table:

	Jones				
Adams	**0**	**1**	**2**	**3**	**4**
0	0, 0	1, 0	2, 0	3, 0	4, 0
1	0, 1	1, 1	2, 1	3, 1	4, 1
2	0, 2	1, 2	2, 2	3, 2	4, 2
3	0, 3	1, 3	2, 3	3, 3	4, 3

(continued)

(3, 1 stands for: 3 sales by Jones and 1 sale by Adams.) Assume that each sample point is equally likely. Let's define the events:

$$A = \text{at least one of the salespersons made no sales}$$
$$B = \text{together they made exactly three sales}$$
$$C = \text{each made the same number of sales}$$
$$D = \text{Adams made exactly one sale}$$

Find the following probabilities by counting sample points:

a. $P(A)$ **b.** $P(B)$ **c.** $P(C)$

d. $P(D)$ **e.** $P(A \text{ and } B)$ **f.** $P(B \text{ and } C)$

g. $P(A \text{ or } B)$ **h.** $P(B \text{ or } C)$ **i.** $P(A \mid B)$

j. $P(B \mid D)$ **k.** $P(C \mid B)$ **l.** $P(B \mid \overline{A})$

m. $P(C \mid \overline{A})$ **n.** $P(A \text{ or } B \text{ or } C)$

Are the following pairs of events mutually exclusive? Explain.

o. A and B   **p.** B and C   **q.** B and D

Are the following pairs of events independent? Explain.

r. A and B   **s.** B and C   **t.** B and D

4.129 A testing organization wishes to rate a particular brand of television. Six TVs are selected at random from stock. If nothing is found wrong with any of the six, the brand is judged satisfactory.
  a. What is the probability that the brand will be rated satisfactory if 10% of the TVs actually are defective?
  b. What is the probability that the brand will be rated satisfactory if 20% of the TVs actually are defective?
  c. What is the probability that the brand will be rated satisfactory if 40% of the TVs actually are defective?

4.130 Coin A is loaded in such a way that $P(\text{heads})$ is 0.6. Coin B is a balanced coin. Both coins are tossed. Find the following:
  a. the sample space that represents this experiment; assign a probability measure to each outcome
  b. $P(\text{both show heads})$
  c. $P(\text{exactly one head shows})$
  d. $P(\text{neither coin shows a head})$
  e. $P(\text{both show heads} \mid \text{coin A shows a head})$
  f. $P(\text{both show heads} \mid \text{coin B shows a head})$
  g. $P(\text{heads on coin A} \mid \text{exactly one head shows})$

4.131 Professor French forgets to set his alarm with a probability of 0.3. If he sets the alarm it rings with a probability of 0.8. If the alarm rings, it will wake him on time to make his first class with a probability of 0.9. If the alarm does not ring, he wakes in time for his first class with a probability of 0.2. What is the probability that Professor French wakes in time to make his first class tomorrow?

4.132 A two-page typed report contains an error on one of the pages. Two proofreaders review the copy. Each has an 80% chance of catching the error. What is the probability that the error will be identified if
  a. each reads a different page?
  b. they each read both pages?
  c. the first proofreader randomly selects a page to read, then the second proofreader randomly selects a page unaware of which page the first selected?

4.133 Solve this exercise using Bayes's rule in tabular form. The treasurer's initial opinion is that there is a 30% chance that an investment will exceed expectations, a 50% chance that it will equal expectations, and a 20% chance that it will return less than expected. A private investment consulting service reviews the investment and reports that it should equal expectations. In the past the consultants were correct 60% of the time, underestimated the return 10% of the time, and overestimated the return 30% of the time. What should be the treasurer's revised probabilities?

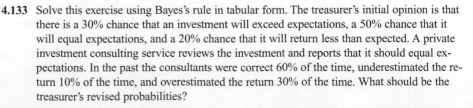

4.134 All her life, Cindy George has enjoyed drinking Fizzle, her favorite brand of soda. For years she has dreamed that on the day her savings account obtained a $10,000 balance, she would buy stock in the company that bottles Fizzle. Cindy is convinced that there is only a 20% chance that the stock will drop during the five-year planning horizon that she intends to keep the stock before selling it at what she feels will be a substantial profit. When she called her broker to move forward on her investment, she was horrified to hear that the company was not recommended by his investment advisory committee. The committee, he explained, has a 65% track record of selecting stock investments, and they predict firmly that the corporation that manufactures Fizzle will likely fail sometime during the next three years due to the highly competitive nature of the carbonated soft drink industry.

Revise Cindy's probability estimates in light of the broker's information using a Bayesian tabular analysis.

4.135 Ninety percent of the insulators produced by Superior Insulator Company are satisfactory. The firm hires an inspector. The inspector inspects all the insulators and correctly classifies an item 90% of the time; $P(\text{classify good} \mid \text{good}) = P(\text{classify defective} \mid \text{defective}) = 0.9$. Items classified good are shipped and those classified defective are scrapped.
a. What percentage of items shipped can be expected to be good?
b. What percentage of items scrapped can be expected to be good?

4.136 The firm in Exercise 4.135 hires a second inspector, who has the same accuracy record. The second inspector inspects all insulators independently of the first inspector. What percentage of items shipped and what percentage of items scrapped can be expected to be good if items are shipped only if
a. both inspectors independently say they are good?
b. at least one inspector says they are good?

4.137 In sports, championships are often decided by two teams playing each other in a championship series. Often the fans of the losing team claim they were unlucky and their team is actually the better team. Suppose team A is the better team, and the probability it will beat team B in any one game is 0.6. What is the probability that the better team A will lose the series if it is
a. a one-game series?
b. a best out of three series?
c. a best out of seven series?
d. Suppose the probability that A would beat B in any given game were actually 0.7. Recompute (a) through (c).
e. Suppose the probability that A would beat B in any given game were actually 0.9. Recompute (a) through (c).
f. What is the relationship between the "best" team winning and the number of games played? The best team winning and the probabilities that each will win?

4.138 A woman and a man (unrelated) each have two children. At least one of the woman's children is a boy, and the man's older child is a boy. Determine whether the probability that the woman has two boys is (greater than, equal to, or less than) the probability that the man has two boys.
a. Demonstrate the truth of your answer using a simple sample to represent each family.
b. Demonstrate the truth of your answer by taking two samples, one from men with two-children families and one from women with two-children families.
c. Demonstrate the truth of your answer using computer simulation. (one simulation) Use the Bernoulli probability function with $p = 0.5$ (let 0 = girl and 1 = boy), generate 500 "families of two children" for the man and the woman. Determine which of the 500 satisfy the condition for each and determine the observed proportion with two boys.
d. Demonstrate the truth of your answer by repeating the computer simulation several times. Repeat the simulation in (c) several times.
e. Do the above procedures seem to yield the same results? Explain.

VOCABULARY LIST

Be able to define each term. Pay special attention to the key terms, which are printed in **red.** In addition, describe in your own words, and give an example of, each term. Your examples should not be ones given in class or in the textbook.

The bracketed numbers indicate the chapter in which the term first appeared, but you should define the terms again to show increased understanding of their meaning. Page numbers indicate the first appearance of the term in Chapter 4.

addition rule (p. 203)
all-inclusive events (p. 189)
Bayes's rule (p. 220)
complementary event (p. 197)
compound event (p. 200)
conditional probability (p. 208)
dependent events (p. 207)
empirical probability (p. 183)
equally likely events (p. 194)
event (p. 189)
experiment [1] (p. 189)
experimental probability (p. 183)
general addition rule (p. 203)
general multiplication rule (p. 210)
independence (p. 207)
independent events (p. 207, 208)
intersection (p. 201)
law of large numbers (p. 185)
long-term average (p. 185)

multiplication rule (p. 210)
mutually exclusive events (p. 189, 200)
odds (p. 196)
ordered pair (p. 189)
outcome (p. 189)
posterior probability (p. 222)
prior probability (p. 222)
probability of an event (p. 183)
relative frequency [2] (p. 181)
revised probability (p. 222)
sample point (p. 189)
sample space (p. 189)
special addition rule (p. 203)
special multiplication rule (p. 210)
subjective probability (p. 195)
theoretical probability (p. 194)
tree diagram (p. 190)
Venn diagram (p. 193)

CHAPTER PRACTICE TEST

Part I: Knowing the Definitions

Answer "True" if the statement is always true. If the statement is not always true, replace the words shown in bold with the words that make the statement always true.

4.1 The probability of an event is a **whole number.**

4.2 The concepts of probability and relative frequency as related to an event are **very similar.**

4.3 The **sample space** is the theoretical population for probability problems.

4.4 The sample points of a sample space are **equally likely** events.

4.5 The value found for experimental probability will **always be** exactly equal to the theoretical probability assigned to the same event.

4.6 The probabilities of complementary events always **are equal.**

4.7 If two events are mutually exclusive, they are also **independent.**

4.8 If events A and B are **mutually exclusive,** the sum of their probabilities must be exactly one.

4.9 If the sets of sample points belonging to two different events do not intersect, the events are **independent.**

4.10 A compound event formed by use of the word *and* requires the use of the **addition rule.**

Part II: Applying the Concepts

4.11 A computer is programmed to generate the eight single-digit integers 1, 2, 3, 4, 5, 6, 7, and 8 with equal frequency. Consider the experiment—"the next integer generated." Define:

Event A = "Odd number" = {1, 3, 5, 7}

Event B = "Number more than 4" = {5, 6, 7, 8}

Event C = "1 or 2" = {1, 2}

Find:

a. $P(A)$ **b.** $P(B)$ **c.** $P(C)$ **d.** $P(\overline{C})$

e. $P(A \text{ and } B)$ **f.** $P(A \text{ or } B)$ **g.** $P(B \text{ and } C)$ **h.** $P(B \text{ or } C)$

i. $P(A \text{ and } C)$ **j.** $P(A \text{ or } C)$ **k.** $P(A \mid B)$ **l.** $P(B \mid C)$

m. $P(A \mid C)$

n. Are events A and B mutually exclusive? Explain.

o. Are events B and C mutually exclusive? Explain.

p. Are events A and C mutually exclusive? Explain.

q. Are events A and B independent? Explain.

r. Are events B and C independent? Explain.

s. Are events A and C independent? Explain.

4.12 Given that events A and B are mutually exclusive and $P(A) = 0.4$ and $P(B) = 0.3$, find

    a. $P(A \text{ and } B)$ **b.** $P(A \text{ or } B)$

    c. $P(A \mid B)$ **d.** Are A and B independent? Explain.

4.13 Given that events C and D are independent and $P(C) = 0.2$ and $P(D) = 0.7$:

    a. Find $P(C \text{ and } D)$.   **b.** Find $P(C \text{ or } D)$.

    c. Find $P(C \mid D)$.   **d.** Are C and D mutually exclusive? Explain.

4.14 Given events E and F with probabilities $P(E) = 0.5$, $P(F) = 0.4$, and $P(E \text{ and } F) = 0.2$:

    a. Find $P(E \text{ or } F)$.   **b.** Find $P(E \mid F)$.

    c. Are E and F mutually exclusive? Explain.

    d. Are E and F independent? Explain.

4.15 Given events G and H with probabilities $P(G) = 0.3$, $P(H) = 0.2$, and $P(G \text{ and } H) = 0.1$:

    a. Find $P(G \text{ or } H)$.   **b.** Find $P(G \mid H)$.

    c. Are G and H mutually exclusive? Explain.

    d. Are G and H independent? Explain.

4.16 Janice wants to become a police officer. She must pass a physical exam and then a written exam. Records show the probability of passing the physical exam is 0.85 and that once the physical is passed the probability of passing the written exam is 0.60. What is the probability that Janice passes both exams?

Part III: Understanding the Concepts

4.17 Explain briefly how you decide which of the following two events is the more unusual:

    A: a 90-degree day in Vermont   or   B: a 100-degree day in Florida.

4.18 Student A says that "independence" and "mutually exclusive" are basically the same thing; namely, both mean "neither event has anything to do with the other one." Student B argues that although Student A's statement has some truth in it, Student A has missed the point of these two properties. Student B is correct. Carefully explain why.

4.19 Using complete sentences, describe in your own words.

    a. mutually exclusive events.   **b.** independent events.

    c. the probability of an event.   **d.** a conditional probability.

4.20 The probability that there are no winners on any one Powerball Lottery game is approximately 0.15.

    a. Interpret the meaning of that probability in terms of how often one might expect a drawing to have no winner.

    b. Explain why you would or would not consider "no winner on one Powerball game" to be a rare occurrence.

    c. What is the approximate likelihood of no winners for two consecutive Powerball games? Interpret the meaning of your answer.

    d. Explain why you would or would not consider no winners on two consecutive Powerball games to be a rare occurrence.

5

Probability Distributions (Discrete Variables)

CHAPTER CASE STUDY

Family Values and Family Togetherness

Family values and family togetherness are very important aspects of our everyday lives and have often been the topics of newspaper headlines in recent years. The number of activities that a family participates in together is a measure of a family's strength and well-being. One daily activity that many families pay particular attention to is the evening meal and for many families the food served at dinner is the focal point. The USA Snapshot®, "What's for dinner?" that appeared in *USA Today*, October 7, 1996, shows the frequency of home-cooked meals each week in many American homes.

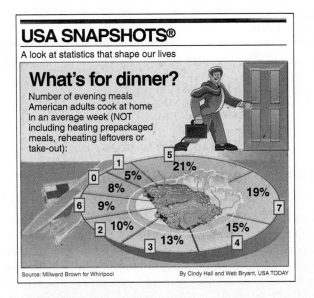

USA SNAPSHOTS®
A look at statistics that shape our lives

What's for dinner?
Number of evening meals American adults cook at home in an average week (NOT including heating prepackaged meals, reheating leftovers or take-out):

5 — 21%
1 — 5%
0 — 8%
6 — 9%
2 — 10%
3 — 13%
4 — 15%
7 — 19%

Source: Millward Brown for Whirlpool

By Cindy Hall and Web Bryant, USA TODAY

What percentage of families eat home-cooked meals on all seven evenings? What is the percentage for no home-cooked meals? What number of nights has the highest likelihood of occurrence? What variable could be used to describe all eight of the events shown on the graph? What other statistical graph could be used to picture this information? What other statistical methods can be used to describe the information shown on the pie chart? We will be learning about various statistical methods for describing information of the above type in this chapter and will return to these questions at the end of the chapter. (See Exercise 5.97, p. 269.)

CHAPTER OBJECTIVES

Chapter 2 dealt with frequency distributions of data sets, and Chapter 4 dealt with the fundamentals of probability. Now we are ready to combine these ideas to form probability distributions, which are much like relative frequency distributions. The basic difference between probability and relative frequency distributions is that probability distributions are theoretical probabilities (populations) whereas relative frequency distributions are empirical probabilities (samples).

In this chapter we investigate discrete probability distributions and study measures of central tendency and dispersion for such distributions. Special emphasis is given to the binomial random variable and its probability distribution, since it is an extremely important discrete random variable encountered in many fields of application.

5.1 Random Variables

If each outcome of a probability **experiment** is assigned a numerical value, then as we observe the results of the experiment we are observing values of a random variable. This numerical value is the *random variable value.*

RANDOM VARIABLE

A variable that assumes a unique numerical value for each of the outcomes in the sample space of a probability experiment.

> **Exercise 5.1**
> You are to survey your classmates regarding the number of siblings and the length of the last conversation with their mother. Identify the two random variables of interest and list their possible values.

In other words, a random variable is used to denote the outcomes of a probability experiment. The random variable can take on any numerical value that belongs to the set of all possible outcomes of the experiment. (It is called "random" because the value it assumes is the result of a chance, or random event.) Each event in a probability experiment must also be defined in such a way that only one value of the random variable is assigned to it (**mutually exclusive events**), and every event must have a value assigned to it (all-inclusive events).

The following illustrations demonstrate several random variables.

ILLUSTRATION 5.1 ▼

We toss five coins and observe the "number of heads" visible. The random variable x is the number of heads observed and may take on integer values from 0 to 5.

ILLUSTRATION 5.2 ▼

Let the "number of phone calls received" per day by a company be the random variable. Integer values ranging from zero to some very large number are possible values.

ILLUSTRATION 5.3 ▼

Let the "length of the cord" on an electric appliance be a random variable. The random variable will be a numerical value between 12 and 72 inches for most appliances.

ILLUSTRATION 5.4 ▼

Let the "qualifying speed" for race cars trying to qualify for the Indianapolis 500 be a random variable. Depending on how fast the driver can go, the speeds will be approximately 220 and faster and be measured in miles per hour (to the nearest thousandth of a mile).   ▲

Numerical random variables can be subdivided into two classifications: *discrete random variables* and *continuous random variables*.

> Discrete and continuous variables were defined on pages 15 and 16.

Exercise 5.2
a. Explain why the variables in Illustrations 5.1 and 5.2 are discrete.
b. Explain why the variables in Illustrations 5.3 and 5.4 are continuous.

DISCRETE RANDOM VARIABLE

A quantitative random variable that can assume a countable number of values.

CONTINUOUS RANDOM VARIABLE

A quantitative random variable that can assume an uncountable number of values.

EXERCISES • • • • • • • • • • • • •

5.3 The variables in Exercise 5.1 are either discrete or continuous. Which are they and why?

5.4 a. Explain why the variable "number of dinner guests for Thanksgiving dinner" is discrete.
b. Explain why the variable "number of miles to your grandmother's house" is continuous.

5.5 A social worker is involved in a study about family structure. She obtains information regarding the number of children per family for a certain community from the census data. Identify the random variable of interest, determine whether it is discrete or continuous, and list its possible values.

5.6 The staff at *Fortune* recently isolated what they considered to be the 100 best companies in America to work for. Of these, the top four had the most new jobs in the past two years:

Company	New Jobs
Lowe's	13,000
Intel	11,196
FedEx	6,000
Marriott	5,936

Source: Fortune, "The 100 Best Companies to Work for in America," Jan. 12, 1998.

a. What is the random variable involved in this study?   *new jobs*
b. Is the random variable discrete or continuous? Explain.   *discrete*

5.7 An archer shoots arrows at a bull's-eye of a target and measures the distance from the center of the target to the arrow. Identify the random variable of interest, determine whether it is discrete or continuous, and list its possible values.   *sn #'s*

5.8 "How women define holiday shopping," a USA Snapshot® (12-9-94) reported that 50% said "a pleasure," 22% said "a chore," 19% said "no big deal," and 8% said "a nightmare." The percentages do not sum to 100% due to round-off error.
a. What is the variable involved, and what are the possible values?
b. Why is this variable not a random variable?

5.9 The year 1998 may go down in baseball history as the year of the home run. When Albert Belle ripped 16 home runs in July, it marked the third straight month that a record for most homers in that month was equaled or broken. In June, Sammy Sosa broke the record for home runs hit in any month of the season. A summary table is on the next page:

(continued)

Month	Player, *Team*	Home Runs	Year
March	Vinnie Castilla, *Rockies*	2	1998
April	Ken Griffey Jr., *Mariners*	13	1997
May	Mickey Mantle, *Yankees*	16	1956
	Mark McGwire, *Cardinals*	16	1998
June	Sammy Sosa, *Cubs*	20	1998
July	Albert Belle, *White Sox*	16	1998
August	Rudy York, *Tigers*	18	1937
September	Babe Ruth, *Yankees*	17	1927
	Albert Belle, *Indians*	17	1995
October	Six players tied	4	

Source: Sports Illustrated, "Inside Baseball," Aug. 10, 1998.

    a. What is the random variable involved in this study?
    b. Is the random variable discrete or continuous? Explain.

5.10 Midway through 1998 there were three players chasing the single season home run records set by Babe Ruth in 1927 with 60 and Roger Maris in 1961 with 61. Did any one of the five sluggers have an unfair advantage by facing weaker pitchers? The percentage of wins and the earned run average (ERA) of opposing pitchers who gave up home runs to each player are shown in the table below:

	Opposing Pitcher Statistics	
Player, *Year*	**Win/Loss %**	**ERA**
Babe Ruth, *1927*	48.3	4.10
Roger Maris, *1961*	48.2	4.06
Sammy Sosa, *1998*	53.7	4.59
Ken Griffey Jr., *1998*	48.1	4.81
Mark McGwire, *1998*	55.9	4.10

Source: Sports Illustrated, "Servin' Up Taters," July 6, 1998.

    a. What are the two random variables involved in this study?
    b. Are these random variables discrete or continuous? Explain.

5.2   Probability Distributions of a Discrete Random Variable

Recall the coin-tossing experiment we used at the beginning of Section 4.1. Two coins were tossed and no heads, one head, or two heads were observed. If we define the random variable x to be the number of heads observed when two coins are tossed, x can take on the values 0, 1, or 2. The probability of each of these three events is the same as we calculated in Chapter 4 (p. 181):

$$P(x = 0) = P(0\text{H}) = \frac{1}{4}$$

$$P(x = 1) = P(1\text{H}) = \frac{1}{2}$$

$$P(x = 2) = P(2\text{H}) = \frac{1}{4}$$

These probabilities can be listed in any number of ways. One of the most convenient is a table format known as a *probability distribution* (see Table 5.1).

TABLE 5.1   Probability Distribution: Tossing Two Coins

x	P(x)
0	0.25
1	0.50
2	0.25

PROBABILITY DISTRIBUTION

A distribution of the probabilities associated with each of the values of a random variable. The probability distribution is a theoretical distribution; it is used to represent populations.

In an experiment in which a single die is rolled and the number of dots on the top surface is observed, the random variable is the number observed. The probability distribution for this random variable is shown in Table 5.2.

TABLE 5.2   Probability Distribution: Rolling a Die

x	P(x)
1	$\frac{1}{6}$
2	$\frac{1}{6}$
3	$\frac{1}{6}$
4	$\frac{1}{6}$
5	$\frac{1}{6}$
6	$\frac{1}{6}$

Sometimes it is convenient to write a rule that algebraically expresses the probability of an event in terms of the value of the random variable. This expression is typically written in formula form and is called a *probability function*.

> **Exercise 5.11**
>
> Express the tossing of one coin as a probability distribution of x, the number of heads occurring (that is, H = 1 and T = 0).

PROBABILITY FUNCTION

A rule that assigns probabilities to the values of the random variables.

A probability function can be as simple as a list, pairing the values of a random variable with their probabilities. Tables 5.1 and 5.2 show two such listings. However, a probability function is most often expressed in formula form.

Consider a die that has been modified so that it has one face with one dot, two faces with two dots, and three faces with three dots. Let x be the number of dots observed when this die is rolled. The probability distribution for this experiment is presented in Table 5.3.

TABLE 5.3   Probability Distribution: Rolling the Modified Die

x	$P(x)$
1	$\frac{1}{6}$
2	$\frac{2}{6}$
3	$\frac{3}{6}$

Exercise 5.12

Express $P(x) = \frac{1}{6}$; for $x = 1, 2, 3, 4, 5,$ or 6, in distribution form.

Exercise 5.13

Explain how the various values of x in a probability distribution form a set of mutually exclusive events.

Each of the probabilities can be represented by the value of x divided by 6. That is, each $P(x)$ is equal to the value of x divided by 6, where $x = 1, 2,$ or 3. Thus,

$$P(x) = \frac{x}{6}; \text{ for } x = 1, 2, \text{ or } 3$$

is the formula expression for the probability function of this experiment.

The probability function for the experiment of rolling one ordinary die is

$$P(x) = \frac{1}{6}; \text{ for } x = 1, 2, 3, 4, 5, \text{ or } 6$$

This particular function is called a **constant function** because the value of $P(x)$ does not change as x changes.

Every probability function must display the two basic properties of probability (see p. 195 and 196). These two properties are (1) the probability assigned to each value of the random variable must be between 0 and 1, inclusive, that is,

Property 1 $0 \leq$ each $P(x) \leq 1$

These properties were previously presented in Chapter 4.

and (2) the sum of the probabilities assigned to all the values of the random variable must equal 1, that is,

Property 2 $\sum_{\text{all } x} P(x) = 1$

ILLUSTRATION 5.5 ▼

Is $P(x) = \frac{x}{10}$; for $x = 1, 2, 3,$ or 4, a probability function?

Solution

To answer this question we need only test the function in terms of the two basic properties. The probability distribution is shown in Table 5.4.

The values of the random variable are all inclusive.

TABLE 5.4   Probability Distribution for $P(x) = \frac{x}{10}$; for $x = 1, 2, 3,$ or 4

x	$P(x)$
1	$\frac{1}{10} = 0.1$ ✓
2	$\frac{2}{10} = 0.2$ ✓
3	$\frac{3}{10} = 0.3$ ✓
4	$\frac{4}{10} = 0.4$ ✓
	$\frac{10}{10} = 1.0$ (ck)

Exercise 5.14

How does the property "all inclusive" relate to the probability distribution shown in Table 5.4?

Property 1 is satisfied, since 0.1, 0.2, 0.3, and 0.4 are all numerical values between 0 and 1. (See the ✓ showing each value was checked.) Property 2 is also satisfied, since the sum of all four probabilities is exactly 1. (See the (ck) showing the sum was checked). Since both properties are satisfied, we can conclude that $P(x) = \dfrac{x}{10}$; for $x = 1, 2, 3,$ or 4 is a probability function.

What about $P(x = 5)$—or any value other than $x = 1, 2, 3,$ or 4—for the function $P(x) = \dfrac{x}{10}$; for $x = 1, 2, 3,$ or 4? $P(x = 5)$ is considered to be zero. That is, the probability function provides a probability of zero for all values of x other than the values specified as part of a domain.   ▲

Probability distributions can be presented graphically. Regardless of the specific graphic representation used, the values of the random variable are plotted on the horizontal scale, and the probability associated with each value of the random variable is plotted on the vertical scale. The probability distribution of a discrete random variable could be presented by a set of line segments drawn at the values of x and whose lengths represent the probability of each x. Figure 5.1 shows the probability distribution of

$$P(x) = \frac{x}{10}; \text{ for } x = 1, 2, 3, \text{ or } 4.$$

Figure 5.1

Line Representation: Probability Distribution for $P(x) = \dfrac{x}{10}$; for $x = 1, 2, 3,$ or 4

The graph in Figure 5.1 is sometimes called a "needle graph."

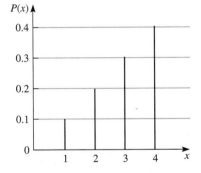

However, a regular histogram is more frequently used to present probability distributions. Figure 5.2 shows the probability distribution of Figure 5.1 as a **probability histogram.**

Figure 5.2

Histogram: Probability Distribution for $P(x) = \dfrac{x}{10}$; for $x = 1, 2, 3,$ or 4

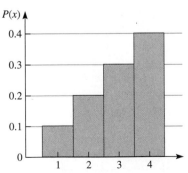

Exercise 5.15

a. Construct a histogram of the probability distribution
$$P(x) = \frac{1}{6}; \text{ for } x = 1,$$
2, 3, 4, 5, or 6.

b. Describe the shape of the histogram in (a).

The histogram of a probability distribution uses the physical area of each bar to represent its assigned probability. The bar for $x = 2$ is 1 unit wide (from 1.5 to 2.5) and is 0.2 unit high. Therefore, its area (length × width) is (1)(0.2) = 0.2, the probability assigned to $x = 2$. The areas of the other bars can be determined in similar fashion. This area representation will be an important concept in Chapter 6 when we begin to work with continuous random variables.

The following commands will generate random data from a discrete probability distribution.

MINITAB

Input the possible values of the random variable in C1 and their corresponding probabilities in C2; then continue with:

```
Choose:    Calc > Random Data > Discrete
Enter:     Generate: 25 (number wanted)
           Store in: C3
           Values (of x) in: C1
           Probabilities in: C2
```

Case Study 5.1

Who Needs the Ambulance?

Robert Giordana used a relative frequency histogram to help him explain how his ambulance services are used. The histogram is of a discrete variable (number of trips per day) and shows, in percentages, the relative frequency of days with various numbers of service trips.

The Austin City Ambulance Company last week appealed to the City Council for additional municipal funding. Mr. Robert Giordana, the company's business manager, stated that while people see ambulances on the streets occasionally, they rarely have any real concept of the frequency with which an ambulance is called upon for assistance.

In surveying the company's records, the Council found that one ambulance responds to between one and six calls for help on a typical day. The records for a recent six-month period showed that an ambulance made three trips on 25% of the days and four trips on 21% of the days. It was further revealed that there was only one day in every three weeks when no trips were made. But on one day in the same three weeks, they made seven or more trips.

Mr. Giordana reminded the Council that the company was working hard for the community all year long.

Exercise 5.16

a. Express the histogram in Case Study 5.1 as a probability distribution.

b. Explain how the distribution implies the conclusion "that there was only one day every three weeks when no trips were made."

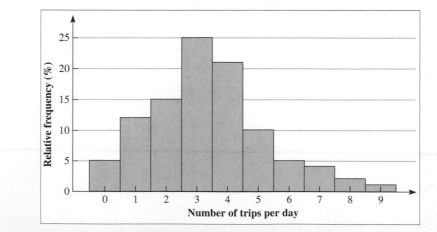

EXCEL

Input the possible values of the random variable in column A and their corresponding probabilities in column B; then continue with:

Choose: **Tools > Data Analysis > Random Number Generation > OK**

Enter: Number of Variables: **1**
          Number of Random Numbers:
          25 (# wanted)
          Distribution: **Discrete**
          Value & Prob. Input Range:
          **(A2:B5 select data cells,
          not labels)**

Select: **Output Range**

Enter   **(C1 or select cell)**

EXERCISES

5.17 Census data are often used to obtain probability distributions for various random variables. Census data for families with a combined income of $50,000 or more in a particular state show that 20% have no children, 30% have one child, 40% have two children, and 10% have three children. From this information, construct the probability distribution for x, where x represents the number of children per family for this income group.

5.18 Test the following function to determine whether it is a probability function. If it is not, try to make it into a probability function. $R(x) = 0.2$ for $x = 0, 1, 2, 3$, or 4.
a. List the distribution of probabilities.
b. Sketch a histogram.

5.19 Test the following function to determine if it is a probability function.

$$P(x) = \frac{x^2 + 5}{50}; \text{ for } x = 1, 2, 3, \text{ or } 4$$

a. List the probability distribution.
b. Sketch a histogram.

5.20 Test the following function to determine whether it is a probability function. If it is not, try to make it into a probability function.

$$S(x) = \frac{6 - |x - 7|}{36}; \text{ for } x = 2, 3, 4, 5, 6, 7, \ldots, 11, \text{ or } 12$$

a. List the distribution of probabilities and sketch a histogram.
b. Do you recognize $S(x)$? If so, identify it.

5.21 Commissions have often emphasized the importance of strong working relationships between audit committees and internal auditing in preventing financial reporting problems. A study published in 1998 was conducted to analyze the number of audit committee meetings per year that companies held with their chief internal auditor. From a survey of 71 responding companies, the results are shown in the table on the next page:

(continued)

Meetings per Year	Percentage (%)
0	8.5
1	11.3
2	21.1
3	5.6
4	35.2
5 or more	18.3

Source: Accounting Horizons, Vol. 12, No. 1, March 1998.

a. Is this a probability distribution? Explain.

b. Draw a relative frequency histogram to depict the results shown in the table.

5.22 "Kids who smoke," a USA Snapshot® (4-25-94), reports the percentage of children in each age group who smoke.

Age, x	Percent Who Smoke
12	1.7
13	4.9
14	8.9
15	16.3
16	25.2
17	37.2

Is this a probability distribution? Explain why or why not.

5.23 Accounting auditing results tend to vary from one occasion to the next because of inherent differences between companies, industries, and methods used when the audit is conducted. A 1998 study provided a classification table of audit differences based upon the industry in which 2221 audits were conducted:

Industry	Number of audits	Percentage (%)
Agriculture	104	4.7
High technology	205	9.2
Manufacturing	714	32.1
Merchandising	476	21.4
Real estate	83	3.7
Other	639	28.8

Source: Auditing: A Journal of Practice and Theory, Spring 1998, Vol. 17, No. 1.

a. Is this a probability distribution? Explain.

b. Draw a graph picturing the results shown in the table.

5.24 A USA Snapshot® (*USA Today,* 7-23-98) presented a table depicting a profile of affluence in today's society. Statistics were derived for 17 million adults living in households with annual incomes of at least $100,000:

Characteristic	Percentage (%)
Own house	75
Married	70
Age 35–54	58
Children under age 18	45
Value of home over $200,000	40

Is this a probability distribution? Explain.

5.25 **a.** Use a computer (or random numbers table) to generate a random sample of 25 observations drawn from the discrete probability distribution.

x	1	2	3	4	5
$P(x)$	0.2	0.3	0.3	0.1	0.1

Compare the resulting data to your expectations.

b. Form a relative frequency distribution of the random data.

c. Construct a probability histogram of the given distribution and a relative frequency histogram of the observed data using class marks of 1, 2, 3, 4, and 5.

d. Compare the observed data with the theoretical distribution. Describe your conclusions.

e. Repeat parts (a) through (d) several times with $n = 25$. Describe the variability you observe between samples.

f. Repeat parts (a) through (d) several times with $n = 250$. Describe the variability you see between samples of this much larger size.

MINITAB

a. Input the x values of the random variable in C1 and their corresponding probabilities $P(x)$ in C2; then continue with the generating random data MINITAB commands on page 242.

b. To obtain the frequency distribution, continue with:

Choose: **Stat > Tables > Cross Tabulation**
Enter: Classification variables: **C3**
Select: Display: **Total percents**

c. To construct the histogram of the generated data in C3, continue with the histogram MINITAB commands on page 56, selecting the options: percent and midpoint with intervals 1:5/1.

To construct a bar graph of the given distribution, continue with:

Choose: **Graph > Chart**
Enter: Y: **C2**   X: **C1**

EXCEL

a. Input the x values of the random variable in column A and their corresponding probabilities $P(x)$ in column B; then continue with the generating random data EXCEL commands on page 243 for an $n = 25$.

b. & c. The frequency distribution is given with the histogram of the generated data. Use the histogram EXCEL commands on page 57 using the data in column C and the bin range in column A.

To construct a histogram of the given distribution, continue with:

Choose: **Chart Wizard > Column > 1ˢᵗ picture(usually) > Next**
Enter: Data range: **(A1:B6 or select cells)**
Choose: **Series > Remove** (Series 1: x column) **> Next > Titles**
Enter: **Chart and axes titles > Finish**   (Edit as needed)

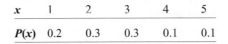
Use the computer commands in Exercise 5.25; change the arguments.

5.26 **a.** Use a computer (or random numbers table) and generate a random sample of 100 observations drawn from the discrete probability population $P(x) = \dfrac{5 - x}{10}$, for $x = 1, 2, 3,$ or 4. List the resulting sample.

b. Form a relative frequency distribution of the random data.

c. Form a probability distribution of the expected probability distribution. Compare the resulting data to your expectations.

(continued)

d. Construct a probability histogram of the given distribution and a relative frequency histogram of the observed data using class marks of 1, 2, 3, and 4.

e. Compare the observed data with the theoretical distribution. Describe your conclusions.

f. Repeat parts (a) through (d) several times with $n = 100$. Describe the variability you observe between samples.

5.3   Mean and Variance of a Discrete Probability Distribution

Recall that in Chapter 2 we calculated several numerical sample statistics (mean, variance, standard deviation, and others) to describe empirical sets of data. Probability distributions represent theoretical populations, the counterpart to samples. We use the **population parameters** (mean, variance, and standard deviation) to describe these probability distributions just as we use **sample statistics** to describe samples.

NOTES

1. $\bar{x}$ is the mean of the sample .

2. s^2 and s are the variance and standard deviation of the sample , respectively.

3. $\bar{x}$, s^2, and s are called *sample statistics.*

4. μ (lowercase Greek letter " mu ") is the mean of the population .

5. σ^2 (" sigma squared ") is the variance of the population .

6. σ (lowercase Greek letter " sigma ") is the standard deviation of the population .

7. μ, σ^2, and σ are called *population parameters.* (A parameter is a constant; μ, σ^2, and σ are typically unknown values.)

The *mean of the probability* distribution of a discrete random variable , or the *mean of a discrete random variable,* is found in a manner somewhat similar to that used to find the mean of a frequency distribution.

MEAN OF A DISCRETE RANDOM VARIABLE

The mean, μ, of a discrete random variable x is found by multiplying each possible value of x by its own probability and then adding all the products together.

mean of x: mu = sum of (each x multiplied by its own probability)

$$\text{mean of } x: \quad \mu = \sum[xP(x)] \tag{5.1}$$

The variance of a discrete random variable is defined in much the same way as the variance of sample data, the "mean of the squared deviations from the mean."

VARIANCE OF A DISCRETE RANDOM VARIABLE

Variance, σ^2, of a discrete random variable x is found by multiplying each possible value of the squared deviation from the mean, $(x - \mu)^2$, by its own probability and then adding all the products together.

Variance: sigma squared = sum of (squared deviation times probability)

$$\text{variance:} \quad \sigma^2 = \sum[(x - \mu)^2 \, P(x)] \tag{5.2}$$

Formula 5.2 is often not convenient to use; it can be reworked into the following form(s):

Variance: *sigma squared* $-$ *sum of* (x^2 *times probability*) $-$ *[sum of* (x *times probability*)$]^2$

variance: $\sigma^2 = \sum[x^2\, P(x)] - \{\sum[xP(x)]\}^2$ **(5.3a)**

or $\sigma^2 = \sum[x^2 P(x)] - \mu^2$ **(5.3b)**

Exercise 5.27

Verify that formulas 5.3a and 5.3b are equivalent to formula 5.2.

STANDARD DEVIATION OF A DISCRETE RANDOM VARIABLE

The positive square root of variance.

standard deviation: $\sigma = \sqrt{\sigma^2}$ **(5.4)**

ILLUSTRATION 5.6 ▼

Find the mean, variance, and standard deviation of the probability function

$$P(x) = \frac{x}{10}; \text{ for } x = 1, 2, 3, \text{ or } 4$$

Solution

We will find the mean using formula (5.1), the variance using formula (5.3a), and the standard deviation using formula (5.4). The most convenient way to organize the products and find the totals needed is to expand the probability distribution into an exten-sions table   (see Table 5.5).

TABLE 5.5   Extensions Table: Probability Distribution,
$P(x) = \dfrac{x}{10}$, for $x = 1, 2, 3,$ or 4

x	$P(x)$	$xP(x)$	x^2	$x^2P(x)$
1	$\dfrac{1}{10} = 0.1$ ✓	0.1	1	0.1
2	$\dfrac{2}{10} = 0.2$ ✓	0.4	4	0.8
3	$\dfrac{3}{10} = 0.3$ ✓	0.9	9	2.7
4	$\dfrac{4}{10} = 0.4$ ✓	1.6	16	6.4
	$\dfrac{10}{10} = 1.0$ ⓒⓚ	$\sum[xP(x)] = 3.0$		$\sum\left[x^2P(x)\right] = 10.0$

Exercise 5.28

a. Form the probability distribution table for $P(x) = \dfrac{x}{6}$, for $x = 1, 2, 3.$

b. Find the extensions $xP(x)$ and $x^2P(x)$ for each x.

c. Find $\sum[xP(x)]$ and $\sum[x^2P(x)]$.

Exercise 5.29

Find the mean for $P(x) = \dfrac{x}{6}$, for $x = 1, 2, 3$ (use the results of Exercise 5.28).

Find the mean of x: The $xP(x)$ column contains the value of each x multiplied by its corresponding probability, and the sum at the bottom is the value needed by for-mula (5.1).

$$\mu = \sum[xP(x)] = \textbf{3.0}$$

Exercise 5.30
Find the variance for
$P(x) = \dfrac{x}{6}$, for x = 1, 2,
or 3 (use the results of
Exercise 5.28).

Exercise 5.31
Find the standard
deviation for $P(x) = \dfrac{x}{6}$,
for x = 1, 2, 3 (use the
results of Exercise 5.30).

Find the variance of x: The totals at the bottom of the $xP(x)$ and $x^2\,P(x)$ columns are substituted into formula (5.3a).

$$\sigma^2 = \sum[x^2P(x)] - \{\sum[xP(x)]\}^2$$

$$\sigma^2 = 10.0 - \{3.0\}^2 = \mathbf{1.0}$$

Find the standard deviation of x: Use formula (5.4).

$$\sigma = \sqrt{\sigma^2} = \sqrt{1.0} = \mathbf{1.0}$$ ▲

NOTES

1. The purpose of the extensions table is to organize the process of finding the three column totals: $\sum[P(x)]$, $\sum[xP(x)]$, and $\sum[x^2P(x)]$.
2. The other columns, x and x^2, *should not* be totaled; they are not used.
3. $\sum[P(x)]$ will always be 1.0; use this only as a check.
4. $\sum[xP(x)]$ and $\sum[x^2P(x)]$ are used to find the mean and variance of x.

ILLUSTRATION 5.7 ▼

A coin is tossed three times. Let the "number of heads" occurring in those three tosses be the random variable x. Find the mean, variance, and standard deviation of x.

Solution

There are eight possible outcomes to this experiment: {HHH, HHT, HTH, HTT, THH, THT, TTH, TTT}. One results in $x = 0$, three in $x = 1$, three in $x = 2$, and one in $x = 3$. Therefore, the probabilities for this random variable are $\dfrac{1}{8}, \dfrac{3}{8}, \dfrac{3}{8}$, or $\dfrac{1}{8}$. The probability distribution associated with this experiment is shown in Figure 5.3 and in Table 5.6. The necessary extensions and summations for the calculation of its mean, variance, and standard deviation are also shown in Table 5.6.

Figure 5.3
Probability Distribution:
Three Coins

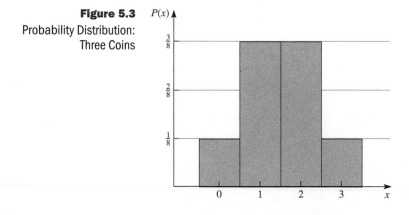

The mean is found using formula (5.1):

$$\mu = \sum[xP(x)] = \mathbf{1.5}$$

This result, 1.5, is the mean number of heads expected per experiment of three coins.

Exercise 5.32

If you find the sum of the x and the x^2 columns on the extensions table, exactly what have you found?

TABLE 5.6   Extensions Table of Probability Distribution of Three Coins

x	$P(x)$	$xP(x)$	x^2	$x^2P(x)$
0	$\dfrac{1}{8}$ ✓	$\dfrac{0}{8}$	0	$\dfrac{0}{8}$
1	$\dfrac{3}{8}$ ✓	$\dfrac{3}{8}$	1	$\dfrac{3}{8}$
2	$\dfrac{3}{8}$ ✓	$\dfrac{6}{8}$	4	$\dfrac{12}{8}$
3	$\dfrac{1}{8}$ ✓	$\dfrac{3}{8}$	9	$\dfrac{9}{8}$
Totals	$\sum[P(x)] = \dfrac{8}{8} = 1.0$ (ck)	$\sum[xP(x)] = \dfrac{12}{8} = 1.5$		$\sum\left[x^2P(x)\right] = \dfrac{24}{8} = 3.0$

The variance is found using formula (5.3a):

$$\sigma^2 = \sum[x^2P(x)] - \{\sum[xP(x)]\}^2$$

$$\sigma^2 = 3.0 - \{1.5\}^2 = 3.0 - 2.25 = \mathbf{0.75}$$

The standard deviation is found using formula (5.4):

$$\sigma = \sqrt{\sigma^2} = \sqrt{0.75} = 0.866 = \mathbf{0.87}$$

That is, 0.87 is the standard deviation expected among the number of heads observed per experiment of three coins.   ▲

EXERCISES

5.33 Given the probability function $P(x) = \dfrac{5-x}{10}$ for $x = 1, 2, 3,$ or 4, find the mean and standard deviation.

5.34 Given the probability function $R(x) = 0.2$ for $x = 0, 1, 2, 3,$ or 4, find the mean and standard deviation.

5.35 **a.** Draw a histogram of the probability distribution for the single-digit random numbers $(0, 1, 2, \ldots, 9)$.
b. Calculate the mean and standard deviation associated with the population of single-digit random numbers.
c. Represent (1) the location of the mean on the histogram with a vertical line and (2) the magnitude of the standard deviation with a line segment.
d. How much of this probability distribution is within two standard deviations of the mean?

5.36 The Air Transport Association of America tracks the number of fatal accidents experienced by all commercial airlines each year. The following table shows the frequency distribution from 1981–96:

Fatal Accidents	Frequency		Relative Frequency
1	2	$\frac{2}{16}$ =	0.1250
2	2		0.1250
3	2		0.1250
4	8	$\frac{8}{16}$ =	0.5000
6	1	$\frac{1}{16}$ =	0.0625
8	1		0.0625

Source: The World Almanac and Book of Facts 1998.   *(continued)*

a. Build an extensions table of the probability distribution and use it to find the mean and standard deviation of the number of fatal accidents experienced annually by commercial airlines.

b. Draw the histogram of the relative frequencies.

5.37 The random variable A has the following probability distribution.

A	1	2	3	4	5
$P(A)$	0.6	0.1	0.1	0.1	0.1

a. Find the mean and standard deviation of A.

b. How much of the probability distribution is within two standard deviations of the mean?

c. What is the probability that A is between $\mu - 2\sigma$ and $\mu + 2\sigma$?

5.38 The random variable $\bar{x}$ has the following probability distribution.

$\bar{x}$	1	2	3	4	5
$P(\bar{x})$	0.6	0.1	0.1	0.1	0.1

a. Find the mean and standard deviation of $\bar{x}$.

b. What is the probability that $\bar{x}$ is between $\mu - \sigma$ and $\mu + \sigma$?

5.39 A USA Snapshot® (11-1-94) titled "How many telephones we have" reported that 1% have none, 11% have one, 31% have two, and 57% have three or more. Let x equal the number of phones per home, and replace the category "three or more" with exactly "three."

a. Find the mean and standard deviation for the random variable x.

b. Explain the effect that replacing the category "three or more" with "three" had on the distribution of x, the mean, and the standard deviation.

5.40 a. Use the probability distribution shown below and describe in your own words how the mean of the variable x is found.

x	1	2	3	4
$P(x)$	0.1	0.2	0.3	0.4

b. Find the mean of x.

c. Find the deviation from the mean for each x-value.

d. Find the value of each "squared deviation from the mean."

e. Recalling your answer to (a), find the mean of the variable "squared deviation."

f. Your answer should be the same as found in Illustration 5.6, page 247. "Variance" was the name given to the "mean of the squared deviations." Explain how formula (5.2) expresses the variance as a mean.

5.4   The Binomial Probability Distribution

Consider the following probability experiment. Your instructor gives the class a surprise four-question multiple-choice quiz. You have not studied the material being quizzed and therefore decide to answer the four questions by randomly guessing the answers without reading the questions or the answers.

ANSWER PAGE TO QUIZ

Directions: Circle the best answer to each question.

1.   A   B   C
2.   A   B   C
3.   A   B   C
4.   A   B   C

That's right, guess!

Circle your answers before continuing.

Before we look at the correct answers to the quiz and find out how you did, let's think about some of the things that might happen if a quiz is answered this way.

1. How many of the four questions are you likely to have answered correctly?
2. How likely are you to have more than half of the answers correct?
3. What is the probability that you selected the correct answers to all four questions?
4. What is the probability that you selected wrong answers for all four questions?
5. If an entire class answers the quiz by guessing, what do you think the class "average" number of correct answers will be?

To find the answers to all of these questions, let's start with a tree diagram picturing the sample space showing all 16 possible ways to answer the four-question quiz. Each of the four questions is answered with the correct answer (C) or is answered with a wrong answer (W). See Figure 5.4 on p. 252.

Let's convert the information on the tree diagram into a probability distribution. Let x be the "number of correct answers" on one person's quiz when the quiz was taken by randomly guessing. The random variable x may take on any one of the values 0, 1, 2, 3, 4 for each quiz. Figure 5.4 shows 16 branches representing five different values of x. Notice that the event $x = 4$, "four correct answers," is represented by the top branch of the tree diagram, and that event $x = 0$, "zero correct answers," is shown on the bottom branch. The other events, "one correct answer," "two correct answers," and "three correct," are each represented by several branches of the tree. We find that the event $x = 1$ occurs on four different branches, event $x = 2$ occurs on six branches, and event $x = 3$ occurs on four branches.

Since each individual question has only one correct answer among the three possible answers, the probability of selecting the correct answer to an individual question is $\frac{1}{3}$. The probability that a wrong answer is selected on each question is $\frac{2}{3}$. The probabilities of each value of x can be found by calculating the probabilities of all branches and then combining the probabilities for branches of like x-values. The calculations follow, and the resulting probability distribution appears in Table 5.7.

$P(x = 0)$ is the probability that zero questions are answered correctly and four are answered wrong (one branch, WWWW):

> **Exercise 5.41**
> Explain why the four questions represent four independent trials.

$$P(x = 0) \times \frac{2}{3} \times \frac{2}{3} \times \frac{2}{3} \times \frac{2}{3} = \left(\frac{2}{3}\right)^4 = \frac{16}{81} = \mathbf{0.198}$$

+ look @ tree diagram
p 252
see
handout

NOTE The answering of each individual question is a separate and independent event, thereby allowing us to use formula (4.8b) and multiply the probabilities.

$P(x = 1)$ is the probability that exactly one question is answered correctly and the other three are answered wrong (there are four branches on Figure 5.4 where this occurs, namely: CWWW, WCWW, WWCW, WWWC, and each has the same probability):

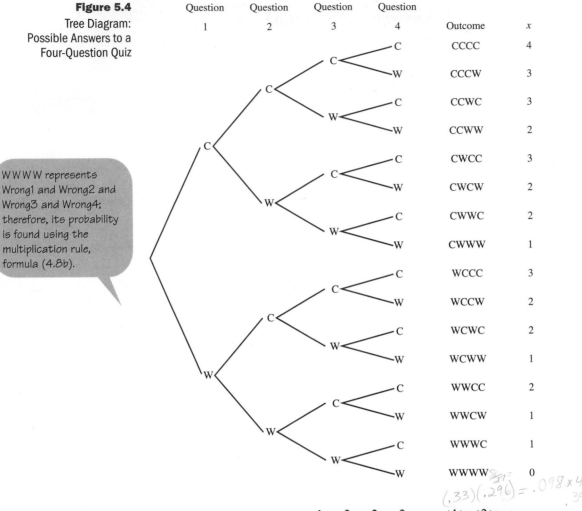

Figure 5.4

Tree Diagram: Possible Answers to a Four-Question Quiz

WWWW represents Wrong1 and Wrong2 and Wrong3 and Wrong4; therefore, its probability is found using the multiplication rule, formula (4.8b).

$(.33)(.296) = .098 \times 4 = .392$

$$P(x = 1) = (4) \times \frac{1}{3} \times \frac{2}{3} \times \frac{2}{3} \times \frac{2}{3} = (4)\left(\frac{1}{3}\right)^1\left(\frac{2}{3}\right)^3 = \textbf{0.395}$$

Exercise 5.42

Explain why the number 4 is part of the calculation for $P(x = 1)$.

$P(x = 2)$ is the probability that exactly two questions are answered correctly and the other two are answered wrong (there are six branches on Figure 5.4 where this occurs: CCWW, CWCW, CWWC, WCCW, WCWC, WWCC, and each has the same probability):

$$P(x = 2) = (6) \times \frac{1}{3} \times \frac{1}{3} \times \frac{2}{3} \times \frac{2}{3} = (6)\left(\frac{1}{3}\right)^2\left(\frac{2}{3}\right)^2 = \textbf{0.296}$$

$P(x = 3)$ is the probability that exactly three questions are answered correctly and the other one is answered wrong (there are four branches on Figure 5.4 where this occurs: CCCW, CCWC, CWCC, WCCC, and each has the same probability):

$$P(x = 3) = (4) \times \frac{1}{3} \times \frac{1}{3} \times \frac{1}{3} \times \frac{2}{3} = (4)\left(\frac{1}{3}\right)^3\left(\frac{2}{3}\right)^1 = \textbf{0.099}$$

$P(x = 4)$ is the probability that all four questions are answered correctly (there is only one branch where all four are correct, CCCC):

$$P(x = 4) = \frac{1}{3} \times \frac{1}{3} \times \frac{1}{3} \times \frac{1}{3} = \left(\frac{1}{3}\right)^4 = \frac{1}{81} = \textbf{0.012}$$

TABLE 5.7   Probability Distribution for the Four-Question Quiz

x	$P(x)$
0	0.198
1	0.395
2	0.296
3	0.099
4	0.012
	1.000 (ck)

I got .995

Now we can answer the five questions, on page 251, that were asked about the four-question quiz.

Answer 1: The most likely occurrence would be to get one answer; it has a probability of 0.395. Zero, one, or two correct answers are expected to result approximately 89% of the time ($0.198 + 0.395 + 0.296 = 0.889$).

Answer 2: To have more than half correct is represented by $x = 3$, or 4; their total probability is 0.111. (You will pass this quiz only 11% of the time by random guessing.)

Answer 3: $P(\text{all 4 correct}) = P(x = 4) = 0.012$ (All correct only 1% of the time.)

Answer 4: $P(\text{all 4 wrong}) = P(x = 0) = 0.198$. (That's almost 20% of the time.)

Answer 5: The class average would be expected to be $\frac{1}{3}$ of 4, or 1.33, correct answers.

The correct answers to the quiz are B, C, B, A. How many correct answers did you have? Which branch of the tree in Figure 5.4 represents your quiz results? You might ask several people to answer the quiz for you by guessing the answers. Then construct an observed relative frequency distribution and compare it to the distribution shown in Table 5.7.

Many experiments are composed of repeated trials whose outcomes can be classified in one of two categories, **success** or **failure.** Examples of such experiments include the previous experiments of tossing coins, right/wrong quiz answers, and other more practical experiments, such as determining whether a product did its prescribed job or did not do it, and whether a candidate gets elected or not. There are experiments in which the trials have many outcomes that, under the right conditions, may fit this general description of being classified in one of two categories. For example, when we roll a single die, we usually consider six possible outcomes. However, if we are interested only in knowing whether a "one" shows or not, there are really only two outcomes: the "one" shows or "something else" shows. The experiments just described are called *binomial probability experiments*.

Exercise 5.43

Where did $\frac{1}{3}$ and 4 come from? Why multiply them to find an expected average?

Exercise 5.44

Ask 20 or more people to take the quiz independently; grade their quizzes; form a frequency distribution of the variable "number of correct answers on each quiz."

BINOMIAL PROBABILITY EXPERIMENT

An experiment that is made up of repeated trials that possess the following properties:

1. There are n repeated (identical) **independent trials.**
2. Each **trial** has two possible outcomes (success , failure).
3. $P(\text{success}) = p$, $P(\text{failure}) = q$, and $p + q = 1$.
4. The **binomial random variable** x is the count of the number of successful trials that occur; x may take on any integer value from zero to n.

NOTES

1. Properties 1 and 2 are the two basic properties of any binomial experiment.
2. Property 3 concerns the algebraic notation for each trial.
3. Property 4 concerns the algebraic notation for the complete experiment.
4. It is of utmost importance that both x and p be associated with "success."

The four-question quiz qualifies as a binomial experiment made up of four trials when all four of the answers are obtained by random guessing.

Property 1: A <u>trial</u> is the <u>answering of one question</u>, and is repeated <u>$n = 4$</u> times. The trials are <u>independent</u> since the probability of a correct answer on any one question is not affected by the answers on other questions.

Property 2: Two outcomes on each trial: <u>success = C</u>, correct answer,
<u>failure = W</u>, wrong answer.

Property 3: For each trial (each question): <u>$p = P(\text{correct})$</u> $= \dfrac{1}{3}$

and <u>$q = P(\text{wrong})$</u> $= \dfrac{2}{3}$. $[p + q = 1;$ ⓒⓚ$]$

Property 4: For the total experiment (the quiz): <u>$x =$ number of correct answers</u> and can be any integer value <u>from 0 to $n = 4$.</u>

NOTE Independent trials mean that the result of one trial does not affect the probability of success of any other trial in the experiment. In other words, the probability of "success" remains constant throughout the entire experiment.

ILLUSTRATION 5.8 ▼

Consider the experiment of rolling a die 12 times and observing a "one" or "something else." When all 12 rolls have been completed, the number of "ones" is reported. The random variable x would be the number of times that a "one" is observed in the $n = 12$ trials. Since "one" is the outcome of concern, it is considered "success"; therefore, $p = P(\text{one}) = \dfrac{1}{6}$ and $q = P(\text{not one}) = \dfrac{5}{6}$. This experiment is binomial.

ILLUSTRATION 5.9 ▼

Exercise 5.45

Identify the properties that make flipping a coin 50 times and keeping track of heads a binomial experiment.

If you were an inspector on a production line in a plant where television sets were manufactured, you would be concerned with identifying the number of defective sets. You probably would define "success" as the occurrence of a defective television. This is not what we normally think of as success, but if we count "defective" sets in a binomial experiment, we must define "success" as a "defective." The random variable x indicates the number of defective sets found per lot of n sets, while $p = P(\text{television is defective})$ and $q = P(\text{television is good})$. ▲

The key to working with any probability experiment is its probability distribution. All binomial probability experiments have the same properties, and therefore the same organization scheme could be used to represent all of them. The *binomial probability function* will allow us to find the probabilities for each possible value of x.

BINOMIAL PROBABILITY FUNCTION

For a binomial experiment, let p represent the probability of a "success" and q represent the probability of a "failure" on a single trial; then $P(x)$, the probability that there will be exactly x successes in n trials, is

$$P(x) = \binom{n}{x} (p^x) (q^{n-x}), \text{ for } x = 0, 1, 2, \ldots, \text{ or } n \qquad \text{(5.5)}$$

When you look at the probability function, you notice that it is the product of three basic factors:

1. the number of ways that exactly x successes can occur in n trials, $\binom{n}{x}$
2. the probability of exactly x successes, p^x
3. the probability that failure will occur on the remaining $(n-x)$ trials, q^{n-x}

The number of ways that exactly x successes can occur in a set of n trials is represented by the symbol $\binom{n}{x}$ which must always be a positive integer. This term is called the **binomial coefficient** and is found by using the formula

$$\binom{n}{x} = \frac{n!}{x!(n-x)!} \qquad \text{(5.6)}$$

NOTES

1. $n!$ ("n factorial") is an abbreviation for the product of the sequence of integers starting with n and ending with 1. For example, $3! = 3 \cdot 2 \cdot 1 = 6$, and $5! = 5 \cdot 4 \cdot 3 \cdot 2 \cdot 1 = 120$. There is one special case, $0!$, which is defined to be 1. For further information about **factorial notation,** see the *Statistical Tutor.*

2. The values for $n!$ and $\binom{n}{x}$ can readily be found using most scientific calculators.

3. The binomial coefficient $\binom{n}{x}$ is equivalent to the number of combinations $_nC_x$, the symbol most likely on your calculator.

4. Also, see the *Statistical Tutor* for general information on the binomial coefficient.

Let's reconsider Illustration 5.7 (p. 248); a coin is tossed three times and we observe the number of heads that occur in the three tosses. This is a binomial experiment because it displays all the properties of a binomial experiment:

1. There are $\underline{n = 3}$ repeated independent trials (each coin toss is a separate trial, and the outcome of any one trial has no affect on the probability of another).

2. Each trial (each toss of the coin) has two outcomes: success = heads (what we are counting) and failure = tails.

3. The probability of success is $p = P(H) = 0.5$, and the probability of failure is $q = P(T) = 0.5$. $[p + q = 0.5 + 0.5 = 1; \text{ck}]$

4. The random variable x is the number of heads that occur in the three trials. x will assume exactly one of the values 0, 1, 2, or 3 when the experiment is complete.

Exercise 5.46

Find the value of

a. $4!$

b. $\binom{4}{3}$

Exercise 5.47

Use the probability function for three coins and verify the probabilities for $x = 0, 2,$ and 3.

The binomial probability function for the tossing of three coins is

$$P(x) = \binom{n}{x} (p)^x (q)^{n-x} = \binom{3}{x} (0.5)^x (0.5)^{n-x}, \text{ for } x = 0, 1, 2, \text{ or } 3$$

Let's find the probability of $x = 1$ using the preceding binomial probability function.

In Table 5.6 (p. 249),

$P(1) = \dfrac{3}{8}$. Here,

$P(1) = 0.375$.

$\dfrac{3}{8} = 0.375$.

$$P(x = 1) = \binom{3}{1} (0.5)^1 (0.5)^2 = 3(0.5)(0.25) = \mathbf{0.375}$$

Compare this to the value found in Illustration 5.7 (p. 248).

ILLUSTRATION 5.10 ▼

Consider an experiment that calls for drawing five cards, one at a time with replacement, from a well-shuffled deck of playing cards. The drawn card is identified as a spade or not a spade, returned to the deck, the deck reshuffled, and so on. The random variable x is the number of spades observed in the set of five drawings. Is this a binomial experiment? Let's identify the various properties.

1. There are <u>five repeated drawings</u>; <u>$n = 5$</u>. These individual trials are <u>independent</u>, since the drawn card is returned to the deck and the deck reshuffled before the next drawing.
2. Each drawing is a trial, and each drawing has two possible outcomes, "<u>spade</u>" or "<u>not spade</u>."
3. $p = P(\underline{\text{spade}}) = \dfrac{13}{52}$ and $q = P(\underline{\text{not spade}}) = \dfrac{39}{52}$. $[p + q = 1;$ (ck)]
4. x is the <u>number of spades</u> recorded upon completion of the five trials; possible values <u>0, 1, 2, . . ., 5</u>.

The binomial probability function is

$$P(x) = \binom{5}{x} \left(\frac{13}{52}\right)^x \left(\frac{39}{52}\right)^{5-x} = \binom{5}{x} \left(\frac{1}{4}\right)^x \left(\frac{3}{4}\right)^{5-x}$$

$$= \binom{5}{x} (0.25)^x (0.75)^{5-x}, \text{ for } x = 0, 1, \dots, \text{ or } 5$$

Exercise 5.48

a. Calculate $P(4)$ and $P(5)$ for Illustration 5.10.
b. Verify that the six probabilities $P(0)$, $P(1)$, $P(2)$, . . . , $P(5)$ form a probability distribution.

$$P(0) = \binom{5}{0} (0.25)^0 (0.75)^5 = (1)(1)(0.2373) = \mathbf{0.2373}$$

$$P(1) = \binom{5}{1} (0.25)^1 (0.75)^4 = (5)(0.25)(0.3164) = \mathbf{0.3955}$$

$$P(2) = \binom{5}{2} (0.25)^2 (0.75)^3 = (10)(0.0625)(0.421875) = \mathbf{0.2637}$$

$$P(3) = \binom{5}{3} (0.25)^3 (0.75)^2 = (10)(0.015625)(0.5625) = \mathbf{0.0879}$$

The two remaining probabilities are left for you (Exercise 5.48).   ▲

The preceding distribution of probabilities indicates that the single most likely value of x is 1, the event of observing exactly one spade in a hand of five cards. What is the least likely number of spades that would be observed?

ILLUSTRATION 5.11 ▼

The manager of Steve's Food Market guarantees that none of his cartons of eggs containing a dozen eggs will contain more than one bad egg. If a carton contains more than one bad egg, he will replace the whole dozen and allow the customer to keep the original eggs. If the probability that an individual egg is bad is 0.05, what is the probability that the manager will have to replace a given carton of eggs?

Solution

Assuming that this is a binomial experiment, let x be the number of bad eggs found in a carton of a dozen eggs, $p = P(\text{bad}) = 0.05$, and let the inspection of each egg be a trial resulting in finding a "bad" or "not bad" egg. There will be $n = 12$ trials. To find the probability that the manager will have to make good on his guarantee, we need the probability function associated with this experiment:

$$P(x) = \binom{12}{x}(0.05)^x(0.95)^{12-x}, \text{ for } x = 0, 1, 2, \ldots, \text{ or } 12$$

The probability that the manager will replace a dozen eggs is the probability that $x = 2, 3, 4, \ldots,$ or 12. Recall that $\sum P(x) = 1$, that is,

$$\mathbf{P(0) + P(1) + P(2) + \cdots + P(12) = 1}$$

$$P(\text{replacement}) = P(2) + P(3) + \cdots + P(12) = 1 - \mathbf{[P(0) + P(1)]}$$

Finding $P(x = 0)$ and $P(x = 1)$ and subtracting their total from 1 is easier than finding each of the other probabilities:

$$P(x) = \binom{12}{x}(0.05)^x(0.95)^{12-x}$$

$$P(0) = \binom{12}{0}(0.05)^0(0.95)^{12} = \mathbf{0.540}$$

$$P(1) = \binom{12}{1}(0.05)^1(0.95)^{11} = \mathbf{0.341}$$

see notes 11/29 (2)

NOTE The value of many binomial probabilities, for values of $n \leq 15$ and common values of p, are found in Table 2 of Appendix B. In this example, we have $n = 12$ and $p = 0.05$, and we want the probabilities for $x = 0$ and 1. We need to locate the section of Table 2 where $n = 12$, find the column marked $p = 0.05$, and read the numbers opposite $x = 0$ and 1. We find .540 and .341, as shown in Table 5.8 below. (Look up these values in Table 2 in Appendix B.)

TABLE 5.8   Abbreviated Portion of Table 2 in Appendix B, Binomial Probabilities

								p								
n	x	0.01	0.05	0.10	0.20	0.30	0.40	0.50	0.60	0.70	0.80	0.90	0.95	0.99	x
	⋮														
	⋮														
12	0	.886	.540	.282	.069	.014	.002	0+	0+	0+	0+	0+	0+	0+	0
	1	.107	.341	.377	.206	.071	.017	.003	0+	0+	0+	0+	0+	0+	1
	2	.006	.099	.230	.283	.168	.064	.016	.002	0+	0+	0+	0+	0+	2
	3	0+	.017	.085	.236	.240	.142	.054	.012	.001	0+	0+	0+	0+	3
	4	0+	.002	.021	.133	.231	.213	.121	.042	.008	.001	0+	0+	0+	4
	⋮														
	⋮														

Now let's return to our illustration:

$$P(\text{replacement}) = 1 - (0.540 + 0.341) = \mathbf{0.119}$$

If $p = 0.05$ is correct, the manager of Steve's Food Market will be busy replacing cartons of eggs. If he replaces 11.9% of all the cartons of eggs he sells, he certainly will be giving away a substantial proportion of his eggs. This suggests that he should adjust his guarantee (or market better eggs). For example, if he were to replace a carton of eggs only when four or more were found to be bad, he would expect to replace only 3 out of 1000 cartons $[1.0 - (0.540 + 0.341 + 0.099 + 0.017)]$, or 0.3% of the cartons sold. Notice that he will be able to control his "risk" (probability of replacement) if he adjusts the value of the random variable stated in his guarantee.

NOTE A convenient notation to identify the binomial probability distribution for a binomial experiment with $n = 12$ and $p = 0.30$ is $B(12, 0.30)$. ▲

Exercise 5.49

What would be the manager's "risk" if he bought "better" eggs, say with $P(\text{bad}) = 0.01$ using the "more than one" guarantee?

Exercise 5.50

a. Use a calculator or computer to find the probability that $x = 3$ in a binomial experiment where $n = 12$ and $p = 0.30$:
$P(x = 3 \mid B(12, 0.30))$.

b. Use Table 5.8 to verify the answer in (a).

The following commands will determine binomial probabilities or cumulative binomial probabilities for a particular number of trials, n, and particular probability, p; B(n, p).

MINITAB

For binomial probabilities, input x values into C1; then continue with:

Choose: **Calc > Probability Distributions > Binomial**

Select: **Probability***

Enter: Number of trials: **n**
Probability of success: **p**

Select: **Input column**

Enter: **C1**
Optional Storage: **C2** (not necessary)

Or

Select: **Input constant**

Enter: **One single x value**

*For cumulative binomial probabilities, repeat the commands above replacing the probability selection with:

Select: **Cumulative Probability**

EXCEL

For binomial probabilities, input x values into column A and activate the column B cell across from the first x value; then continue with:

Choose: **Paste function, f$_x$ > Statistical > BINOMDIST > OK**

Enter: Number_s: **(A1:A4 or select 'x value' cells)**
Trials: **n**
Probability_s: **p**
Cumulative: **false*** (gives individual probabilities) **> OK**

Drag: **Bottom right corner of probability value cell in column B down to give other probabilities**

*For cumulative binomial probabilities, repeat the commands above replacing the false cumulative with:

Cumulative: **true** (gives cumulative probabilities) **> OK**

```
TI-83

To obtain a complete list of probabilities for a particular
n and p:

    Choose:   2nd > DISTR > 0:binompdf(
    Enter:    n , p)

Use the right arrow key to scroll through the probabilities.

To scroll through a vertical list in L1:

    Choose:   STO→ > L1 > ENTER
              STAT > EDIT > 1:Edit

To obtain individual probabilities for a particular n, p,
and x:

    Choose:   2nd > DISTR > 0:binompdf(
    Enter:    n, p, x)

To obtain cumulative probabilities for x = 0 to x = n for a
particular n and p:

    Choose:   2nd > DISTR > A:binomcdf(
    Enter:    n, p)*    (see above for scrolling through
                        probabilities)

*To obtain individual cumulative probabilities for a particular n, p, and x,
repeat the commands above replacing the enter with:
Enter:        n, p, x)
```

EXERCISES   •   •   •   •   •   •   •   •   •   •   •   •   •   •

5.51 Evaluate each of the following.

  a. $4!$ **b.** $7!$ **c.** $0!$ **d.** $\dfrac{6!}{2!}$

  e. $\dfrac{5!}{2!3!}$ **f.** $\dfrac{6!}{4!(6-4)!}$ **g.** $(0.3)^4$ **h.** $\dbinom{7}{3}$

  i. $\dbinom{5}{2}$ **j.** $\dbinom{3}{0}$ **k.** $\dbinom{4}{1}(0.2)^1(0.8)^3$ **l.** $\dbinom{5}{0}(0.3)^0(0.7)^5$

5.52 Show that each of the following is true for any values of n and k. Use two specific sets of values for n and k to show that each is true.

  a. $\dbinom{n}{0} = 1$ and $\dbinom{n}{n} = 1$ **b.** $\dbinom{n}{1} = n$ and $\dbinom{n}{n-1} = n$ **c.** $\dbinom{n}{k} = \dbinom{n}{n-k}$

5.53 A carton containing 100 T-shirts is inspected. Each T-shirt is rated "first quality" or "irregular." After all 100 T-shirts have been inspected, the number of irregulars is reported as a random variable. Explain why x is a binomial random variable.

5.54 A die is rolled 20 times and the number of "fives" that occurred is reported as being the random variable. Explain why x is a binomial random variable.

5.55 Four cards are selected, one at a time, from a standard deck of 52 cards. Let x represent the number of aces drawn in the set of 4 cards.   *(continued)*

a. If this experiment is completed without replacement, explain why x is not a binomial random variable.

b. If this experiment is completed with replacement, explain why x is a binomial random variable.

5.56 The employees at a General Motors assembly plant are polled as they leave work. Each is asked, "What brand of automobile are you riding home in?" The random variable to be reported is the number of each brand mentioned. Is x a binomial random variable? Justify your answer.

5.57 Consider a binomial experiment made up of three trials with outcomes of success S and failure F, where $P(S) = p$ and $P(F) = q$.

a. Complete the accompanying tree diagram. Label all branches completely.

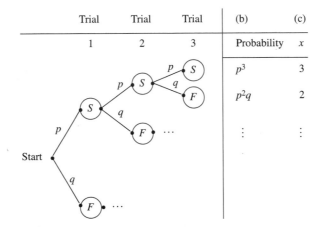

b. In column (b) of the tree diagram, express the probability of each outcome represented by the branches as a product of powers of p and q.

c. Let x be the random variable, the number of successes observed. In column (c), identify the value of x for each branch of the tree diagram.

d. Notice that all the products in column (b) are made up of three factors and that the value of the random variable is the same as the exponent for the number p.

e. Write the equation for the binomial probability function for this situation.

5.58 Draw a tree diagram picturing a binomial experiment of four trials.

5.59 If x is a binomial random variable, calculate the probability of x for each case.

a. $n = 4, x = 1, p = 0.3$ **b.** $n = 3, x = 2, p = 0.8$

c. $n = 2, x = 0, p = \dfrac{1}{4}$ **d.** $n = 5, x = 2, p = \dfrac{1}{3}$

e. $n = 4, x = 2, p = 0.5$ **f.** $n = 3, x = 3, p = \dfrac{1}{6}$

5.60 If x is a binomial random variable, use Table 2 in Appendix B to determine the probability of x for each case.

a. $n = 10, x = 8, p = 0.3$ **b.** $n = 8, x = 7, p = 0.95$
c. $n = 15, x = 3, p = 0.05$ **d.** $n = 12, x = 12, p = 0.99$
e. $n = 9, x = 0, p = 0.5$ **f.** $n = 6, x = 1, p = 0.01$
g. Explain the meaning of the symbol $0+$ that appears in Table 2.

5.61 Test the following function to determine whether or not it is a binomial probability function. List the distribution of probabilities and sketch a histogram.

$$T(x) = \binom{5}{x}\left(\frac{1}{2}\right)^x\left(\frac{1}{2}\right)^{5-x} \quad \text{for } x = 0, 1, 2, 3, 4, \text{ or } 5$$

5.62 Let x be a random variable with the following probability distribution.

x	0	1	2	3
$P(x)$	0.4	0.3	0.2	0.1

Does x have a binomial distribution? Justify your answer.

5.63 Ninety percent of the trees planted by a landscaping firm survive. What is the probability that eight or more of the ten trees they just planted will survive? (Find the answer by using a table.)

5.64 In California, 30% of the people have a certain blood type. What is the probability that exactly 5 out of a randomly selected group of 14 Californians will have that blood type? (Find the answer by using a table.)

5.65 On the average, 1 out of every 10 boards purchased by a cabinet manufacturer is unusable for building cabinets. What is the probability that 8, 9, or 10 of a set of 11 such boards are usable? (Find the answer by using a table.)   *see homework* $n=11, \ x=8,9,10 \ p=.1 \ q=.9$

5.66 A local polling organization maintains that 90% of the eligible voters have never heard of John Anderson, who was a presidential candidate in 1980. If this is so, what is the probability that in a randomly selected sample of 12 eligible voters, 2 or fewer have heard of John Anderson?

5.67 In the biathlon event of the Olympic Games, a participant skis cross-country and on four intermittent occasions stops at a rifle range and shoots a set of five shots. If the center of the target is hit, no penalty points are assessed. If a particular man has a history of hitting the center of the target with 90% of his shots, what is the probability that he will hit the center of the target with   *$n=5$*   *$p=.9$*   *$q=.1$*
a. all five of his next set of five shots?   *$x=5$*
b. at least four of his next set of five shots? (Assume independence.)   *$x=4$*

5.68 A basketball player has a history of making 80% of the foul shots taken during games. What is the probability that he will miss three of the next five foul shots he takes?

5.69 A machine produces parts of which 0.5% are defective. If a random sample of ten parts produced by this machine contains two or more defectives, the machine is shut down for repairs. Find the probability that the machine will be shut down for repairs based on this sampling plan.

5.70 The survival rate during a risky operation for patients with no other hope of survival is 80%. What is the probability that exactly four of the next five patients survive this operation?

5.71 According to an article in the February 1991 issue of *Reader's Digest*, Americans face a 1 in 20 chance of acquiring an infection while hospitalized. If the records of 15 randomly selected hospitalized patients are examined, find the probability that
a. none of the 15 acquired an infection while hospitalized.
b. 1 or more of the 15 acquired an infection while hospitalized.

5.72 An article in the *Omaha World-Herald* (12-1-94) stated that only about 60% of the individuals needing a bone marrow transplant find a suitable donor when they turn to registries of unrelated donors. In a group of ten individuals needing a bone marrow transplant,
a. what is the probability that all ten will find a suitable donor among the registries of unrelated donors?
b. what is the probability that exactly eight will find a suitable donor among the registries of unrelated donors?
c. what is the probability that at least eight will find a suitable donor among the registries of unrelated donors?
d. what is the probability that no more than five will find a suitable donor among the registries of unrelated donors?

5.73 If boys and girls are equally likely to be born, what is the probability that in a randomly selected family of six children, there will be boys? (Find the answer using a formula.)

5.74 One-fourth of a certain breed of rabbits are born with long hair. What is the probability that in a litter of six rabbits, exactly three will have long hair? (Find the answer by using a formula.)

5.75 Colorado Rockies baseball player Larry Walker's league-leading batting average reached .344 after 415 times at bat during the 1998 season (ratio of hits to at bats). Suppose Walker has five official times at bat during his next game. Assuming no extenuating circumstances and that the binomial model will produce reasonable approximations, what is the probability that Walker:
a. gets less than two hits?
b. gets more than three hits?
c. goes five-for-five (all hits)?

5.76 As a quality-control inspector for toy trucks, you have observed that wooden wheels bored off-center occur about 3% of the time. If six wooden wheels are used on each toy truck produced, what is the probability that a randomly selected set of wheels has no off-center wheels?

5.77 According to the USA Snapshot® "Knowing drug addicts," 45% of Americans know somebody who became addicted to a drug other than alcohol. Assuming this to be true, what is the probability that
a. exactly 3 of a random sample of 5 know someone who became addicted? Calculate the value.
b. exactly 7 of a random sample of 15 know someone who became addicted? Estimate using Table 2 in Appendix B.
c. at least 7 of a random sample of 15 know someone who became addicted? Estimate using Table 2.
d. no more than 7 of a random sample of 15 know someone who became addicted? Estimate using Table 2.

5.78 According to *Financial Executive* (July/August 1993) disability causes 48% of all mortgage foreclosures. Given that 20 mortgage foreclosures are audited by a large lending institution,
a. find the probability that five or fewer of the foreclosures are due to a disability.
b. find the probability that at least three are due to a disability.

5.79 Use a computer/calculator to find the probabilities for all possible *x*-values for a binomial experiment where $n = 30$ and $p = 0.35$.

MINITAB

```
Choose:   Calc > Make Patterned Data > Simple Set of
          Numbers
Enter:    Store patterned data in: C1
          From first value: 0
          To last value: 30
          In steps of: 1      > OK
```

Continue with the binomial probability MINITAB commands on page 258, using $n = 30$, $p = 0.35$ and C2 for optional storage.

EXCEL

```
Enter:   0,1,2, . . ., 30 into column A
```

Continue with the binomial probability EXCEL commands on page 259, using $n = 30$ and $p = 0.35$.

(continued)

TI-83

Use the binomial probability TI-83 commands on page 259,
using $n = 30$ and $p = 0.35$.

5.80 Use a computer/calculator to find the cumulative probabilities for all possible *x*-values for
a binomial experiment where $n = 45$ and $p = 0.125$.
  a. Explain why there are so many 1.000's listed.
  b. Explain what is represented by each number listed.

MINITAB

> Choose: **Calc > Make Patterned Data > Simple Set of**
>   **Numbers**
> Enter: Store patterned data in: **C1**
>   From first value: **0**
>   To last value. **45**
>   In steps of: **1**   **> OK**

Continue with the <u>cumulative</u> binomial probability MINITAB
commands on page 258, using $n = 45$, $p = 0.125$ and C2 as optional storage.

EXCEL

> Enter: **0,1,2, . . ., 45** into column A

Continue with the <u>cumulative</u> binomial probability EXCEL
commands on page 258, using $n = 45$ and $p = 0.125$.

TI-83

Use the <u>cumulative</u> binomial probability TI-83 commands on
page 259, using $n = 45$ and $p = 0.125$.

5.81 Results of a 1997 study conducted by Scarborough Research and published by *Fortune*
("Where Lotto Is King," 1-12-98) showed that San Antonio leads all major cities in lottery participation. When adult respondents in San Antonio were asked whether they had
purchased a lottery ticket within the last seven days, 50% said "yes." On the other hand,
only 35% said "yes" to the same question in Salt Lake City. *Note*: The state of Utah has
no lottery, so more travel is required.

Suppose you randomly select 25 adults in San Antonio and 25 adults in Salt Lake City.
  a. What is the probability that the San Antonio sample contains 10 or fewer lottery
  players?
  b. What is the probability that the San Antonio sample and the Salt Lake City sample
  both contain 13 or fewer lottery players?

5.82 An article titled "Mom, I Want to Live with My Boyfriend" appeared in the February
1994 issue of *Reader's Digest*. The article quoted a Columbia University study that found
that only 19% of the men who lived with their girlfriends eventually walked down the
aisle with them. Suppose 25 men are interviewed who have lived with a girlfriend in the
past. What is the probability that 5 or fewer of them married the girlfriend?

5.83 State a very practical reason why the defective item in an industrial situation would be
defined to be the "success" in a binomial experiment.

5.84 If the binomial $(q + p)$ is squared, the result is $(q + p)^2 = q^2 + 2qp + p^2$. For the binomial
experiment with $n = 2$, the probability of no successes in two trials is q^2 (the first term in
the expansion), the probability of one success in two trials is $2qp$ (the second term in the
expansion), and the probability of two successes in two trials is p^2 (the third term). Find
$(q + p)^3$ and compare its terms to the binomial probability for $n = 3$ trials.

5.5   Mean and Standard Deviation of the Binomial Distribution

The mean and standard deviation of a theoretical binomial probability distribution can be found by using these two formulas:

$$\mu = np \tag{5.7}$$

$$\sigma = \sqrt{npq} \tag{5.8}$$

The formula for the mean seems appropriate, the number of trials multiplied by the probability of "success." [Recall that the mean number of correct answers on the binomial quiz (Answer 5, p. 253) was expected to be $\frac{1}{3}$ of 4, $4(\frac{1}{3})$, or np.] The formula for the standard deviation is not as easily understood. Thus, at this point it is appropriate to look at an example that demonstrates that formulas (5.7) and (5.8) yield the same results as formulas (5.1) and (5.3a) and (5.4).

Returning to Illustration 5.7 (p. 248), x is the number of heads seen when tossing three coins, $n = 3$ and $p = \frac{1}{2} = 0.5$. Using formula (5.7), we find the mean of x to be

$$\mu = np = (3)(0.5) = \mathbf{1.5}$$

Using formula (5.8), we find the standard deviation of x to be

$$\sigma = \sqrt{npq} = \sqrt{(3)(0.5)(0.5)} = \sqrt{0.75} = 0.866 = \mathbf{0.87}$$

Look back at the solution for illustration 5.7 (p. 248–249). Note that the results are the same, regardless of the formula you use. However, formulas (5.7) and (5.8) are much easier to use when x is a binomial random variable.

ILLUSTRATION 5.12 ▼

Find the mean and standard deviation of the binomial distribution when $n = 20$ and $p = \frac{1}{5}$ (or 0.2, in decimal form). Recall that the "binomial distribution where $n = 20$ and $p = 0.2$" has a probability function

$$P(x) = \binom{20}{x}(0.2)^x(0.8)^{20-x} \text{ for } x = 0, 1, 2, \ldots, \text{ or } 20$$

and a corresponding distribution with 21 x-values and 21 probabilities as shown in the following distribution chart, Table 5.9, and on the histogram in Figure 5.5.

Exercise 5.85

Find the mean and standard deviation for the binomial random variable x with $n = 30$ and $p = 0.6$.

Figure 5.5

Histogram of Binomial Distribution $B(20, 0.2)$

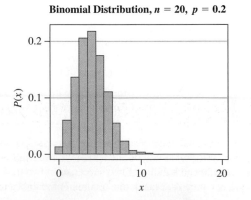

Binomial Distribution, $n = 20$, $p = 0.2$

TABLE 5.9   Binomial $n = 20, p = 0.2$

x	$P(x)$
0	0.012
1	0.058
2	0.137
3	0.205
4	0.218
5	0.175
6	0.109
7	0.055
8	0.022
9	0.007
10	0.002
11	0+
12	0+
13	0+
.	.
:	:
20	0+

Exercise 5.86

Consider the binomial distribution where $n = 11$ and $p = 0.05$.

a. Find the mean and standard deviation using formulas (5.7) and (5.8).

b. Using Table 2 in Appendix B, list the probability distribution and draw a histogram.

c. Locate μ and σ on the histogram.

Let's find the mean and the standard deviation of this distribution of x using formulas (5.7) and (5.8).

$$\mu = np = (20)(0.2) = \mathbf{4.0}$$

$$\sigma = \sqrt{npq} = \sqrt{(20)(0.2)(0.8)} = \sqrt{3.2} = \mathbf{1.79}$$ ▲

Figure 5.6 shows the mean (vertical blue line) and the size of the standard deviation (horizontal red line segment) relative to the probability distribution of the variable x.

Figure 5.6

Histogram of Binomial Distribution $B(20, 0.2)$

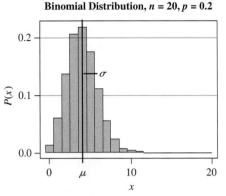

Binomial Distribution, $n = 20, p = 0.2$

EXERCISES • • • • • • • • • • • • •

5.87 Consider the binomial distribution where $n = 11$ and $p = 0.05$ (see Exercise 5.86).

a. Use the distribution [Exercise 5.86(b) or Table 2] and find the mean and standard deviation using formulas (5.1), (5.3a), and (5.4).

b. Compare the results of (a) to the answers found in Exercise 5.86(a).

5.88 Given the binomial probability function

$$T(x) = \binom{5}{x} \cdot \left(\frac{1}{2}\right)^x \cdot \left(\frac{1}{2}\right)^{5-x} \text{ for } x = 0, 1, 2, 3, 4, 5$$

   a. calculate the mean and standard deviation of the random variable by using formulas (5.1) and (5.3a and 5.4).
   b. calculate the mean and standard deviation using formulas (5.7) and (5.8).
   c. compare the results of (a) and (b).

5.89 Find the mean and standard deviation of x for each of the following binomial random variables:
   a. the number of tails seen in 50 tosses of a quarter.
   b. the number of aces seen in 100 draws from a well-shuffled bridge deck (with replacement).
   c. the number of cars found to have unsafe tires among the 400 cars stopped at a road-block for inspection. Assume that 6% of all cars have one or more unsafe tires.
   d. the number of melon seeds that germinate when a package of 50 seeds is planted. The package states that the probability of germination is 0.88.

5.90 Find the mean and standard deviation for each of the following binomial random variables:
   a. the number of sixes seen in 50 rolls of a die.
   b. the number of defective televisions in a shipment of 125. The manufacturer claimed that 98% of the sets were operative.
   c. the number of operative televisions in a shipment of 125. The manufacturer claimed that 98% of the sets were operative.
   d. How are questions (b) and (c) related? Explain.

5.91 A binomial random variable has a mean equal to 200 and a standard deviation of 10. Find the values of n and p.

5.92 The probability of success on a single trial of a binomial experiment is known to be $\frac{1}{4}$.

The random variable x, number of successes, has a mean value of 80. Find the number of trials involved in this experiment and the standard deviation of x.

 5.93 Seventy-five percent of the foreign-made autos sold in the United States in 1984 are now falling apart.
   a. Determine the probability distribution of x, the number of these autos that are falling apart in a random sample of five cars.
   b. Draw a histogram of the distribution.
   c. Calculate the mean and standard deviation of this distribution.

5.94 A binomial random variable x is based on 15 trials with the probability of success equal to 0.3. Find the probability that this variable will take on a value more than two standard deviations from the mean.

5.95 A 1998 survey conducted by *Fortune* ("The Diversity Elite," 8-3-98) revealed that the Marriott International workforce was composed of 50.3% minorities. A further subdivision revealed 6.0% Asian, 24.2% black, and 19.6% Hispanic. Find the mean and standard deviation of all samples of 25 randomly selected employees of the Marriott International workforce for each of the three minority groups. Present your statistics in a table.

5.96 Imprints Galore buys T-shirts (to be imprinted with item of customer's choice) from a manufacturer with the guarantee that the shirts have been inspected and that no more than 1% are imperfect in any way. The shirts arrive in boxes of 12. Let x be the number of imperfect shirts found in any one box.
   a. List the probability distribution and draw the histogram of x.
   b. What is the probability that any one box has no imperfect shirts?
   c. What is the probability that any one box has no more than one imperfect shirt?
   d. Find the mean and standard deviation of x.   *(continued)*

e. What proportion of the distribution is between $\mu - \sigma$ and $\mu + \sigma$?

f. What proportion of the distribution is between $\mu - 2\sigma$ and $\mu + 2\sigma$?

g. How does this information relate to the empirical rule and Chebyshev's theorem? Explain.

h. Use a computer/calculator to simulate Imprints Galore buying 200 boxes of shirts and observing x, the number of imperfect shirts per box of 12. Describe how the information from the simulation compares to what was expected [answers (a) through (g) describe the expected results].

i. Repeat (h) several times. Describe how these results compare to those of (a)–(g) and of (h).

MINITAB

a. Choose: `Calc > Make Patterned Data > Simple Set of`
                 `Numbers`

   Enter: `Store patterned data in: C1`
            `From first value: -1 (see note)`
            `To last value: 12`
            `In steps of: 1    > OK`

`Continue with the binomial probability MINITAB commands on page 258, using` $n = 12$`,` $p = 0.01$ `and C2 for optional storage.`

   Choose: `Graph > Plot`
   Enter: `Graph variable: Y: C2 X: C1`
             `Data display: Display: Area`
   Choose: `Edit Attributes`
   Enter: `Graph: Fill Type: None`
   Select: `Connection function: Step    > OK`

c. `Continue with the` <u>`cumulative`</u> `binomial probability MINITAB commands on page 258, using` $n = 12$`,` $p = 0.01$ `and C3 for optional storage.`

h. Choose: `Calc > Random Data > Binomial`
   Enter: `Generate: 200 rows of data`
             `Store in column C4`
             `Number of trials: 12`
             `Probability: .01    > OK`
   Choose: `Stat > Tables > Cross Tabulation`
   Enter: `Classification variables: C4`
   Select: `Display: Total percents    > OK`
   Choose: `Calc > Column Statistics`
   Select: `Statistic: Mean`
   Enter: `Input variable: C4    > OK`
   Choose: `Calc > Column Statistics`
   Select: `Statistic: Standard deviation`
   Enter: `Input variable: C4    > OK`

`Continue with the histogram MINITAB commands on page 56, using the data in C4 and selecting the options: percent and midpoint with intervals 0:12/1.`

NOTE `The binomial variable` x `cannot take on the value -1. The use of -1 (the next would-be class midpoint to the left of 0) allows MINITAB to draw the histogram of a probability distribution. Without -1, PLOT will draw only half of the bar representing` $x = 0$`.`

EXCEL

a. Enter: `0,1,2, . . ., 12` into column A

Continue with the binomial probability EXCEL commands on page 258, using $n = 12$ and $p = 0.01$.

Activate columns A and B, then continue with:

Choose: **Chart Wizard > Column > 1ˢᵗ picture**(usually) >
          Next > Series
Choose: **Series 1 > Remove**
Enter: Category (x)axis labels: **(A1:A13 or select**
          'x value' cells)
Choose: **Next > Finish**

c. Continue with the <u>cumulative</u> binomial probability EXCEL commands on page 258, using $n = 12$, $p = 0.01$ and column C for the activated cell.

h. Choose: **Tools > Data Analysis > Random Number Generation**
          > OK
Enter: Number of Variables: **1**
          Number of Random Numbers: **200**
          Distribution: **Binomial**
          p Value = **0.01**
          Number of Trials = **12**
Select: Output Options: **Output Range**
Enter   **(D1 or select cell) > OK**
Activate the E1 cell, then:
Choose: **Paste function, fₓ > Statistical > AVERAGE > OK**
Enter: Number 1: **D1:D200 > OK**
Activate the E2 cell, then:
Choose: **Paste function, fₓ > Statistical > STDEV > OK**
Enter: Number 1: **D1:D200 > OK**

Continue with the histogram EXCEL commands on page 57, using the data in column D and the bin range in column A.

TI-83

a. Choose: **STAT > EDIT > 1:Edit**
Enter: L1: `0,1,2,3,4,5,6,7,8,9,10,11,12`
Choose: **2ⁿᵈ QUIT > 2ⁿᵈ DISTR > 0:binompdf(**
Enter: **12, 0.01) > ENTER**
Choose: **STO→ > L2 > ENTER**
Choose: **2ⁿᵈ > STAT PLOT > 1:Plot1**

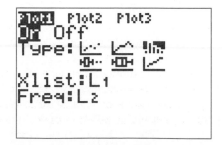

```
          Choose:   WINDOW
          Enter:    0, 13, 1, −.1, .9, .1, 1
          Choose:   TRACE > > >
    c.    Choose:   2ⁿᵈ > DISTR > A:binomcdf(
          Enter:    12, 0.01)
          Choose:   STO→ > L3 > ENTER
                    STAT > EDIT > 1:Edit
    h.    Choose:   MATH > PRB > 7:randBin(
          Enter:    12, .01, 200) (takes a while to process)
          Choose:   STO→ > L4 > ENTER
          Choose:   2ⁿᵈ LIST > Math > 3:mean(
          Enter:    L4
          Choose:   2ⁿᵈ LIST > Math > 7:stdDev(
          Enter:    L4
```

Continue with the histogram TI-83 commands on page 58, using the data in column L4 and adjusting the window after the initial look using ZoomStat.

Return to C H A P T E R   C A S E   S T U D Y

Let's take a second look at the questions asked in the Chapter Case Study, the USA Snapshot®, "What's for dinner?" that appeared in *USA Today,* October 7, 1996, on page 235 and test our knowledge of the material presented in this chapter.

EXERCISE 5.97

  a. What variable could be used to describe all eight of the events shown on the graph?
  b. Express the information on the circle graph as a probability distribution.
  c. What percentage of families eat home-cooked meals on all seven evenings?
  d. What is the percentage for no home-cooked meals?
  e. What number of nights has the highest likelihood of occurrence?
  f. What other statistical graph could be used to display the information shown on the pie chart? Construct it.
  g. Describe the histogram drawn in (f). Is it a normal distribution? Explain.
  h. Find the mean and the standard deviation of the variable described in (a).
  i. Locate the mean and standard deviation found in (g) on the histogram drawn in (f).
  j. Do the empirical and Chebyshev rules apply? Justify your answer.

IN RETROSPECT

In this chapter we combined concepts of probability with some of the ideas presented in Chapter 2. We now are able to deal with distributions of probability values and to find means and standard deviations.

  In Chapter 4 we explored the concepts of mutually exclusive events and independent events. The addition and multiplication rules were used on several occasions in this chapter, but very little was said about mutual exclusiveness or independence. Recall that every time we add probabilities, as we did in each of the probability distributions, we need to know that the associated events are mutually exclusive. If you look back over the chapter, you will notice that the random variable actually requires events to be mutually exclusive; therefore, no real emphasis was placed on this concept. The same basic comment can be made in reference to the multiplication of probabilities and the concept of independent events. Throughout this chapter, probabilities were multiplied and occasionally independence was mentioned. Independence, of course, is necessary in order to be able to multiply probabilities.

Now, after completing Chapter 5, if we were to take a closer look at some of the sets of data in Chapter 2, we would see that several problems could be reorganized to form probability distributions. Some examples might be: (1) Let x be the number of credit hours for which a student is registered in this semester, with the percentage of the student body being reported for each value of x; (2) let x be the number of correct passageways through which an experimental laboratory animal passes before taking a wrong one, paired with the probability of each x-value; (3) let x be the number of trips per day for the ambulance service (Case Study 5.1), paired with the probability of each x-value. The list of examples is endless.

We are now ready to extend these concepts to continuous random variables, which we will do in Chapter 6.

CHAPTER EXERCISES

5.98 What are the two basic properties of every probability distribution?

5.99 **a.** Explain the difference and the relationship between a probability distribution and a probability function.
b. Explain the difference and the relationship between a probability distribution and a frequency distribution, and explain how they relate to a population and a sample.

5.100 Verify whether or not each of the following is a probability function. State your conclusion and explain.

a. $f(x) = \dfrac{\dfrac{3}{4}}{x!(3-x)!}$ for $x = 0, 1, 2, 3$

b. $f(x) = 0.25$ for $x = 9, 10, 11, 12$

c. $f(x) = \dfrac{(3-x)}{2}$ for $x = 1, 2, 3, 4$

d. $f(x) = \dfrac{(x^2 + x + 1)}{25}$ for $x = 0, 1, 2, 3$

5.101 The number of ships to arrive at a harbor on any given day is a random variable represented by x. The probability distribution for x is

x	10	11	12	13	14
$P(x)$	0.4	0.2	0.2	0.1	0.1

Find the probability that on a given day
a. exactly 14 ships arrive.   **b.** at least 12 ships arrive.   **c.** at most 11 ships arrive.

5.102 Did you ever wonder "How many times buyers see an infomercial before they purchase its product or service?" The USA Snapshot® "Television's hard sell" (10-21-94) answers that question. According to the National Infomercial Marketing Association:

Times Watched Before Buy	1	2	3	4	5 or more
Percentage of Buyers	0.27	0.31	0.18	0.09	0.15

(continued)

a. What is the probability that a buyer watched only once before buying?
b. What is the probability that a viewer watching for the first time will buy?
c. What percentage of the buyers watched the infomercial three or more times before purchasing?
d. Is this a binomial probability experiment?
e. Let x be the number of times a buyer watched before making a purchase. Is this a probability distribution?
f. Assign $x = 5$ for "5 or more," and find the mean and standard deviation of x.

5.103 A doctor knows from experience that 10% of the patients to whom he gives a certain drug will have undesirable side effects. Find the probabilities that among the ten patients to whom he gives the drug,
a. at most two will have undesirable side effects.
b. at least two will have undesirable side effects.

5.104 In a recent survey of women, 90% admitted that they had never looked at a copy of *Vogue* magazine. Assuming that this is accurate information, what is the probability that a random sample of three women will show that fewer than two have looked at the magazine?

5.105 Seventy percent of those seeking a driver's license admitted that they would not report someone if he or she copied some answers during the written exam. You have just entered the room and see ten people waiting to take the written exam. What is the probability that, if the incident happened, five of the ten would not report what they saw?

5.106 The engines on an airliner operate independently. The probability that an individual engine operates for a given trip is 0.95. A plane will be able to complete a trip successfully if at least one-half of its engines operate for the entire trip. Determine whether a four-engine or a two-engine plane has the higher probability of a successful trip.

5.107 A USA Snapshot® titled "Stress does not love company" (11-3-94) answered the question "How people say they prefer to spend stressful times." Forty-eight percent responded "alone," 29% responded "with family," 18% responded "with friends," and 5% responded "other/don't know." Ten individuals are randomly selected and asked the question "How do you prefer to spend stressful times?"
a. What is the probability that two or fewer will respond by saying "alone"?
b. Explain why this question can be answered using binomial probabilities.

5.108 Americans go on vacation for a number of reasons and do all sorts of things while vacationing. Results of a study published in *Newsweek* during the summer of 1998 reveal what people do when they go away on vacations:

What People Do on Vacations	Percentage (%)
Shopping	32
Outdoor activity	17
Historical/Museum	14
Beach	11
Visit national/state parks	10
Cultural events / festivals	9
Theme/Amusement parks	8
Night life/Dancing	8
Gambling	7
Sporting events	6
Golfing/Tennis/Skiing	4

The study also revealed that 60% of U.S. adults under 30 years of age feel that going away on vacation is very important.
a. Is this a probability distribution? Explain why or why not.
b. If you sampled 30 adults under the age of 30 who completed the questionnaire, what is the probability that fewer than 16 would agree with the survey's results that going away on vacation is very important?

5.109 The town council has nine members. A proposal must have at least two-thirds of the votes to be accepted. A proposal to establish a new industry in this town has been tabled. If we know that two members of the town council are opposed and that the others randomly vote "in favor" and "against," what is the probability that the proposal will be accepted?

5.110 The "Health Update" section of *Better Homes and Gardens* (July 1990) reported that patients who take long half-life tranquilizers are 70% more likely to suffer falls resulting in hip fractures than those taking similar drugs with a short half-life. It was also reported that in a Massachusetts study, 30% of nursing-home patients who used tranquilizers used the long half-life ones. Suppose that in a survey of 15 nursing-home patients in New York who used tranquilizers, it was found that 10 of the 15 used long half-life tranquilizers.
  a. If the 30% figure for Massachusetts also holds in New York, find the probability of finding 10 or more in a random sample of 15 who use long half-life tranquilizers.
  b. What might you infer from your answer in (a)?

5.111 A box contains ten items of which three are defective and seven are nondefective. Two items are selected without replacement, and x is the number of defectives in the sample of two. Explain why x is not a binomial random variable.

5.112 A large shipment of radios is accepted upon delivery if an inspection of ten randomly selected radios yields no more than one defective radio.
  a. Find the probability that this shipment is accepted if 5% of the total shipment is defective.
  b. Find the probability that this shipment is not accepted if 20% of this shipment is defective.
  c. The binomial probability distribution is often used in situations similar to this one, namely, large populations sampled without replacement. Explain why the binomial yields a good estimate.

5.113 A discrete random variable has a standard deviation equal to 10 and a mean equal to 50. Find $\sum x^2 P(x)$.

5.114 A binomial random variable is based on $n = 20$ and $p = 0.4$. Find $\sum x^2 P(x)$.

5.115 For years, the manager of a certain company had sole responsibility for making decisions with regard to company policy. This manager has a history of making the correct decision with a probability of p. Recently company policy has changed, and now all decisions are to be made by majority rule of a three-person committee.
  a. Each member makes a decision independently, and each has a probability of p of making the correct decision. What is the probability that the committee's majority decision will be correct?
  b. If $p = 0.1$, what is the probability that the committee makes the correct decision?
  c. If $p = 0.8$, what is the probability that the committee makes the correct decision?
  d. For what values of p is the committee more likely to make the correct decision by majority rule than the former manager?
  e. For what values (there are three) of p is the probability of a correct decision the same for the manager and for the committee? Justify your answer.

5.116 Suppose one member of the committee in Exercise 5.115 always makes the decision by rolling a die. If the die roll results in an even number, he/she votes for the proposal, and if an odd number occurs, he/she votes against it. The other two members still decide independently and have a probability of p of making the correct decision.
  a. What is the probability that the committee's majority decision will be correct?
  b. If $p = 0.1$, what is the probability that the committee makes the correct decision?
  c. If $p = 0.8$, what is the probability that the committee makes the correct decision?
  d. For what value of p is the committee more likely to make the correct decision by majority rule than the former manager?

(continued)

e. For what values of p is the probability of a correct decision the same for the manager and for the committee? Justify your answer.

f. Why is the answer to (e) different than the answer to Exercise 5.115 (e)?

5.117 A business firm is considering two investments. It will choose the one that promises the greater payoff. Which of the investments should it accept? (Let the mean profit measure the payoff.)

Invest in Tool Shop		Invest in Book Store	
Profit	**Probability**	**Profit**	**Probability**
$100,000	0.10	$400,000	0.20
50,000	0.30	90,000	0.10
20,000	0.30	−20,000	0.40
−80,000	0.30	−250,000	0.30
Total	1.00	*Total*	1.00

5.118 Bill has completed a ten-question multiple-choice test on which he answered seven questions correctly. Each question had one correct answer to be chosen from five alternatives. Bill says that he answered the test by randomly guessing the answers without reading the questions or answers.

a. Define the random variable x to be the number of correct answers on this test, and construct the probability distribution if the answers were obtained by random guessing.

b. If Bill really did guess, what is the probability he has exactly seven of the ten answers correct?

c. What is the probability that anybody can guess six or more answers correctly?

d. Do you believe that Bill actually randomly guessed as he claims? Explain.

5.119 A random variable that can assume any one of the integer values 1, 2, . . ., n with equal probabilities of $\frac{1}{n}$ is said to have a uniform distribution. The probability function is written $P(x) = \frac{1}{n}$, for $x = 1, 2, 3, . . ., n$. Show that: $\mu = \frac{(n + 1)}{2}$.

(*Hint:* $1 + 2 + 3 + \cdots + n = [n(n + 1)]/2$.)

VOCABULARY LIST AND KEY CONCEPTS

Be able to define each term. Pay special attention to the key terms, which are printed in **red.** In addition, describe in your own words, and give an example of, each term. Your examples should not be the ones given in class or in the textbook.

The bracketed numbers indicate the chapter in which the term first appeared, but you should define the terms again to show increased understanding of their meaning. Page numbers indicate the first appearance of the term in Chapter 5.

binomial coefficient (p. 255)
binomial experiment (p. 253)
binomial probability function (p. 255)
binomial random variable (p. 253)
constant function (p. 240)
continuous random variable (p. 237)
discrete random variable (p. 237)
experiment [1, 4] (p. 236)
factorial notation (p. 255)
failure (p. 253)
independent trials (p. 253)
mean of discrete random variable (p. 246)
mutually exclusive events [4] (p. 236)

population parameter [1] (p. 246)
probability distribution (p. 239)
probability function (p. 239)
probability histogram (p. 241)
random variable (p. 236)
sample statistic [1] (p. 246)
standard deviation of discrete random variable (p. 247)
success (p. 253)
trial (p. 253)
variance of discrete random variable (p. 246)

CHAPTER PRACTICE TEST

Part I: Knowing the Definitions

Answer "True" if the statement is always true. If the statement is not always true, replace the words shown in bold with words that make the statement always true.

5.1 The number of hours you waited in line to register this semester is an example of a **discrete** random variable.

5.2 The number of automobile accidents you were involved in as a driver last year is an example of a **discrete** random variable.

5.3 The sum of all the probabilities in any probability distribution is always exactly **two.**

5.4 The various values of a random variable form a list of **mutually exclusive events.**

5.5 A binomial experiment always has **three or more** possible outcomes to each trial.

5.6 The formula $\mu = np$ may be used to compute the mean of a **discrete** population.

5.7 The binomial parameter p is the probability of **one success occurring in n trials** when a binomial experiment is performed.

5.8 A parameter is a statistical measure of some aspect of a **sample.**

5.9 **Sample statistics** are represented by letters from the Greek alphabet.

5.10 The probability of event A or B is equal to the sum of the probability of event A and the probability of event B when A and B are **mutually exclusive events.**

Part II: Applying the Concepts

5.11 **a.** Show that the following is a probability distribution:

x	$P(x)$
1	0.2
3	0.3
4	0.4
5	0.1

b. Find $P(x = 1)$. **c.** Find $P(x = 2)$. **d.** Find $P(x > 2)$.
e. Find the mean of x. **f.** Find the standard deviation of x.

5.12 A T-shirt manufacturing company advertises that the probability of an individual T-shirt being irregular is 0.1. A box of 12 such T-shirts is randomly selected and inspected.
a. What is the probability that exactly 2 of these 12 T-shirts are irregular?
b. What is the probability that exactly 9 of these 12 T-shirts are not irregular?

Let x be the number of T-shirts that are irregular in all such boxes of 12 T-shirts.
c. Find the mean of x.
d. Find the standard deviation of x.

Part III: Understanding the Concepts

5.13 What properties must an experiment possess in order for it to be a binomial probability experiment?

5.14 Student A uses a relative frequency distribution for a set of sample data and calculates the mean and standard deviation using formulas learned in Chapter 5. Student A justifies her choice of formulas by saying that since relative frequencies are empirical probabilities, her sample is represented by a probability distribution and therefore her choice of formulas was correct. Student B argues that since the distribution represented a sample, then the mean and standard deviation involved are known as $\bar{x}$ and s and must be calculated using the corresponding frequency distribution and formulas learned in Chapter 2. Who is correct, A or B? Justify your choice.

5.15 Student A and Student B were discussing one entry in a probability distribution chart.

x	$P(x)$
-2	0.1

Student B felt that this entry was okay since the $P(x)$ was a value between 0.0 and 1.0. Student A argued that this entry was impossible for a probability distribution since x was a -2, and negatives are not possible. Who is correct, A or B? Justify your choice.

Chapter 6

Normal Probability Distributions

CHAPTER CASE STUDY

Measures of Intelligence

Did you ever wonder just how "smart" you are? Did you ever wonder how intelligent your friends or classmates are? Chances are, you were tested while in elementary school. Measuring intelligence is no simple task; in fact, the **I.Q. (Intelligence Quotient) score** discussed here is only one of many ways to measure a person's intelligence. This excerpt was taken from the *Encyclopedia of Psychology* and gives only some of the very most basic information about an I.Q. score.

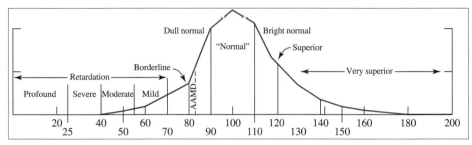

Figure 1.

The distribution of Wechsler Adult Intelligence Scale I.Q. categories. These categories of measured intelligence serve as only one of several indices of intelligence. (From J. D. Matarazzo, *Wechsler's measurement and appraisal of adult intelligence.* 5th ed. Baltimore: Williams and Wilkins, 1976). p. 124. © Oxford University Press, New York. Reprinted by permission.

Is the I.Q. score a continuous variable? Does the I.Q. score discussed here seem to have a normal distribution? What is the mean and the standard deviation for this distribution of I.Q. scores? What percentage of the adult population has "superior" intelligence? What I.Q. score is two standard deviations above the mean? In this chapter we will learn how to use the standard normal probability distribution to help us answer these and similar questions. At the end of the chapter we will return to the above questions. (See Exercise 6.88, p. 310.)

CHAPTER OBJECTIVES

Until now we have considered distributions of discrete variables only. In this chapter we examine one particular family of continuous probability distributions of major importance whose domain is the set of all real numbers. These distributions are called the normal, the Gaussian distributions, or the bell-shaped. "Normal" is simply the traditional title of this particular type of distribution and is not a descriptive name meaning "typical." Although there are many other types of continuous distributions (uniform, triangular, skewed, and so on), many variables have an approximately normal distribution. For example, several of the histograms drawn in Chapter 2 suggested a normal distribution. A mounded histogram that is approximately symmetric can be an indication of such a distribution.

In addition to learning what a normal distribution is, we consider (1) how probabilities are found, (2) how they are represented, and (3) how normal distributions are used.

6.1 Normal Probability Distributions

The **normal probability distribution** is considered to be the single most important probability distribution. There are an unlimited number of **continuous random variables** that have either a normal or an approximately normal distribution. There are also other probability distributions of both discrete and continuous random variables that are approximately normal under certain conditions.

Recall that in Chapter 5 we learned how to use a probability function to calculate probabilities associated with **discrete random variables.** The normal probability distribution has a continuous random variable and it uses two functions: one function to determine the ordinates (y-values) of the graph picturing the distribution, and a second to determine probabilities. Formula (6.1) expresses the ordinate (y-value) that corresponds to each abscissa (x-value).

Normal probability distribution function :

$$y = f(x) = \frac{e^{-\frac{1}{2}\left(\frac{x-\mu}{\sigma}\right)^2}}{\sigma\sqrt{2\pi}}, \text{ for all real } x \tag{6.1}$$

NOTE Each different pair of values for mean (μ) and standard deviation (σ) will result in a different normal probability distribution function.

When a graph of all such points is drawn, the **normal (bell-shaped) curve** will appear as shown in Figure 6.1.

Figure 6.1
The Normal Probability
Distribution

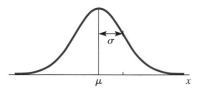

Formula (6.2) yields the probability associated with the interval from $x = a$ to $x = b$.

$$P(a \le x \le b) = \int_a^b f(x)dx \tag{6.2}$$

The probability that x is within the interval from $x = a$ to $x = b$ is shown as the shaded area in Figure 6.2.

Figure 6.2
Shaded Area: $P(a \leq x \leq b)$

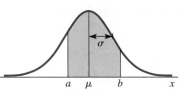

We will not be using the preceding formulas to calculate probabilities for normal distributions. The definite integral of formula (6.2) is a calculus topic and is mathematically beyond what is expected in elementary statistics. (These formulas often appear at the top of normal probability tables as identification.) Instead of using formulas (6.1) and (6.2), we will use a table to find probabilities for normal distributions. Before we learn to use the table, it must be pointed out that the table is expressed in "standardized" form. It is standardized so that this one table can be used to find probabilities for all combinations of mean μ and standard deviation σ values. That is, the normal probability distribution with mean 38 and standard deviation 7 is very similar to the normal probability distribution with mean 123 and standard deviation 32. Recall the empirical rule and the percentage of the distribution that falls within certain intervals of the mean (see page 103). The three percentages held true for all normal distributions.

NOTE **Percentage, proportion,** and **probability** are basically the same concepts. Percentage (25%) is usually used when talking about a proportion ($\frac{1}{4}$) of a population.

Probability is usually used when talking about the chance that the next individual item will possess a certain property. Area is the graphic representation of all three when we draw a picture to illustrate the situation. The empirical rule is a fairly crude measuring device; with it we are able to find probabilities associated only with whole-number multiples of the standard deviation (within one, two, or three standard deviations of the mean). We will often be interested in the probabilities associated with fractional parts of the standard deviation. For example, we might want to know the probability that x is within 1.37 standard deviations of the mean. Therefore, we must refine the empirical rule so that we can deal with more precise measurements. This refinement is discussed in the next section.

6.2   The Standard Normal Distribution

There are an unlimited number of normal probability distributions, but fortunately they are all related to one distribution, the **standard normal distribution.** The standard normal distribution is the normal distribution of the standard variable z (called **"standard score"** or **"z-score"**).

PROPERTIES OF THE STANDARD NORMAL DISTRIBUTION

1. The total area under the normal curve is equal to 1.
2. The distribution is mounded and symmetric; it extends indefinitely in both directions, approaching but never touching the horizontal axis.
3. The distribution has a mean of 0 and a standard deviation of 1.
4. The mean divides the area in half, 0.50 on each side.
5. Nearly all the area is between $z = -3.00$ and $z = 3.00$.

Table 3 in Appendix B lists the probabilities associated with the intervals from the mean (located at $z = 0.00$) to a specific value of z. Probabilities of other intervals will be found by using the table entries and the operations of addition and subtraction, in accordance with the preceding properties. Let's look at several illustrations demonstrating how to use Table 3 to find probabilities of the standard normal score, z.

ILLUSTRATION 6.1 ▼
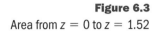

Find the area under the standard normal curve between $z = 0$ and $z = 1.52$.

Figure 6.3
Area from $z = 0$ to $z = 1.52$

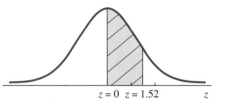

$z = 0$ $z = 1.52$ z

Solution

Table 3 is designed to read the area between $z = 0$ and $z = 1.52$ directly. The z-score is located on the margin, with the units and tenths digit located along the left side and the hundredths digit located along the top margin. For $z = 1.52$, locate the row labeled 1.5 and the column labeled 0.02; at their intersection you will find 0.4357, the measure of the area or probability for the interval $z = 0.00$ to $z = 1.52$. Expressed as a probability: $P(0.00 < z < 1.52) = \mathbf{0.4357}$. (See Table 6.1, below.)

TABLE 6.1   A Portion of Table 3

z	0.00	0.01	0.02	...
⋮				
1.5			0.4357	...
⋮				

Exercise 6.1
Find the area under the standard normal curve between $z = 0$ and $z = 1.37$.

▲

Recall that one of the basic properties of probability is that the sum of all probabilities is exactly 1.0. Since the area under the normal curve represents the measure of probability, the total area under the bell-shaped curve is exactly 1 unit. This distribution is also symmetric with respect to the vertical line drawn through $z = 0$, which cuts the area exactly in half at the mean. Can you verify this fact by inspecting formula (6.1)? That is, the area under the curve to the right of the mean is exactly one-half unit, 0.5, and the area to the left is also one-half unit, 0.5. Areas (probabilities) not given directly in the table can be found by relying on these facts.

Now let's look at some illustrations.

ILLUSTRATION 6.2 ▼

Find the area under the normal curve to the right of $z = 1.52$; $P(z > 1.52)$.

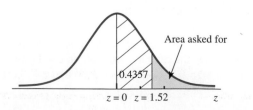

Area asked for

0.4357

$z = 0$ $z = 1.52$ z

Solution

The area to the right of the mean (all shading in the figure) is exactly 0.5000. The question asks for the shaded area that is not included in the 0.4357. Therefore, subtract 0.4357 from 0.5000.

$$P(z > 1.52) = 0.5000 - 0.4357 = \textbf{0.0643}$$ ▲

> **Exercise 6.2**
> Find the area to the right of $z = 2.03$, $P(z > 2.03)$.

SUGGESTION As we have done here, always draw and label a sketch. It is most helpful.

ILLUSTRATION 6.3 ▼

Find the area to the left of $z = 1.52$; $P(z < 1.52)$.

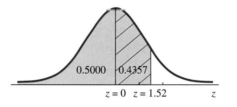

Solution

The total shaded area is made up of 0.4357 found in the table and the 0.5000 that is to the left of the mean. Therefore, add 0.4357 to 0.5000.

$$P(z < 1.52) = P(z < 0) + P(0 < z < 1.52) = 0.5000 + 0.4357 = \textbf{0.9357}$$ ▲

> **Exercise 6.3**
> Find the area to the left of $z = 1.73$, $P(z < 1.73)$.

NOTE The addition and subtraction done in Illustrations 6.2 and 6.3 are correct because the "areas" represent mutually exclusive events (discussed in Section 4.5).

The symmetry of the normal distribution is a key factor in determining probabilities associated with values below (to the left of) the mean. The area between the mean and $z = -1.52$ is exactly the same as the area between the mean and $z = +1.52$. This fact allows us to find values related to the left side of the distribution.

ILLUSTRATION 6.4 ▼

The area between the mean ($z = 0$) and $z = -2.1$ is the same as the area between $z = 0$ and $z = +2.1$; that is,

$$P(-2.1 < z < 0) = P(0 < z < 2.1).$$

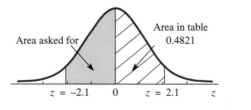

> **Exercise 6.4**
> Find the area between -1.39 and the mean, $P(-1.39 < z < 0.00)$.

Thus, we have

$$P(-2.1 < z < 0) = P(0 < z < 2.1) = \textbf{0.4821}$$ ▲

ILLUSTRATION 6.5 ▼

The area to the left of $z = -1.35$ is found by subtracting 0.4115 from 0.5000.

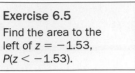

Exercise 6.5

Find the area to the left of $z = -1.53$, $P(z < -1.53)$.

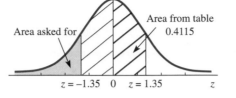

Therefore, we obtain

$$P(z < -1.35) = P(z < 0) - P(-1.35 < z < 0) = 0.5000 - 0.4115 = \mathbf{0.0885}$$

ILLUSTRATION 6.6 ▼

The area between $z = -1.5$ and $z = 2.1$, $P(-1.5 < z < 2.1)$, is found by adding the two areas together. Both probabilities are read directly from Table 3.

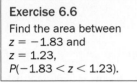

Exercise 6.6

Find the area between $z = -1.83$ and $z = 1.23$, $P(-1.83 < z < 1.23)$.

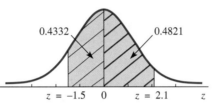

Therefore, we obtain

$$P(-1.5 < z < 2.1) = P(-1.5 < z < 0) + P(0 < z < 2.1)$$
$$= 0.4332 + 0.4821 = \mathbf{0.9153}$$

ILLUSTRATION 6.7 ▼

The area between $z = 0.7$ and $z = 2.1$, $P(0.7 < z < 2.1)$, is found by subtracting. The area between $z = 0$ and $z = 2.1$ includes all the area between $z = 0$ and $z = 0.7$. Therefore, the area between $z = 0$ and $z = 0.7$ is subtracted from the area between $z = 0$ and $z = 2.1$.

Exercise 6.7

Find the area between $z = 0.75$ and $z = 2.25$, $P(0.75 < z < 2.25)$.

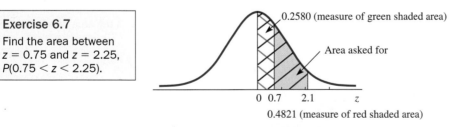

Thus, we have

$$P(0.7 < z < 2.1) = P(0 < z < 2.1) - P(0 < z < 0.7)$$
$$= 0.4821 - 0.2580 = \mathbf{0.2241} \qquad \blacktriangle$$

The normal distribution table can also be used to determine a z-score if we are given an area. The next illustration considers this idea.

ILLUSTRATION 6.8 ▼

What is the z-score associated with the 75th percentile? (Assume the distribution is normal.) See Figure 6.4.

Figure 6.4
P_{75} and Its
Associated z-Score

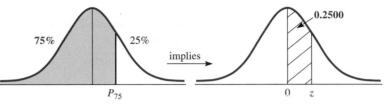

Exercise 6.8

Find the z-score for the 80th percentile of a normal distribution.

Solution

To find this z-score, look in Table 3, Appendix B, and find the "area" entry that is closest to 0.2500; this area entry is 0.2486. Now read the z-score that corresponds to this area.

z	...	0.07		0.08	...
⋮					
0.6		**0.2486**	**0.2500**	**0.2517**	...
⋮					

From the table the z-score is found to be $z = 0.67$. This says that the 75th percentile in a normal distribution is 0.67 (approximately $\frac{2}{3}$) standard deviation above the mean.

ILLUSTRATION 6.9 ▼

What z-scores bound the middle 95% of a normal distribution? See Figure 6.5.

Figure 6.5
Middle 95% of Distribution
and Its Associated z-scores

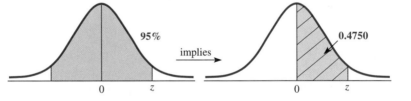

Exercise 6.9

Find the z-scores that bound the middle 75% of a normal distribution.

Solution

The 95% is split into two equal parts by the mean; 0.4750 is the area (percentage) between $z = 0$, the mean, and the z-score at the right boundary. Since we have the area, we look for the entry in Table 3 closest to 0.4750 (it happens to be exactly 0.4750) and read the z-score in the margin. We obtain $z = 1.96$.

z	...	**0.06**	...
⋮			
1.9		**0.4750**	...
⋮			

Therefore, $z = -1.96$ and $z = 1.96$ bound the middle 95% of a normal distribution.

EXERCISES • • • • • • • • • • • • •

6.10 a. Describe the distribution of the standard normal score z.
   b. Why is this distribution called standard normal?

6.11 Find the area under the normal curve that lies between the following pairs of z-values.
  a. $z = 0$ to $z = 1.30$ **b.** $z = 0$ to $z = 1.28$
  c. $z = 0$ to $z = -3.20$ **d.** $z = 0$ to $z = -1.98$

6.12 Find the probability that a piece of data picked at random from a normal population will have a standard score (z) that lies between the following pairs of z-values.
  a. $z = 0$ to $z = 2.10$ **b.** $z = 0$ to $z = 2.57$
  c. $z = 0$ to $z = -1.20$ **d.** $z = 0$ to $z = -1.57$

6.13 Find the area under the standard normal curve that corresponds to the following z-values.
  a. between 0 and 1.55   **b.** to the right of 1.55
  c. to the left of 1.55   **d.** between -1.55 and 1.55

6.14 Find the probability that a piece of data picked at random from a normal population will have a standard score (z) that lies
  a. between 0 and 0.84.   **b.** to the right of 0.84.
  c. to the left of 0.84.   **d.** between -0.84 and 0.84.

6.15 Find the area under the normal curve that lies between the following pairs of z-values.
  a. $z = -1.20$ to $z = 1.22$ **b.** $z = -1.75$ to $z = 1.54$
  c. $z = -1.30$ to $z = 2.58$ **d.** $z = -3.5$ to $z = -0.35$

6.16 Find the probability that a piece of data picked at random from a normal population will have a standard score (z) that lies between the following pairs of z-values:
  a. $z = -2.75$ to $z = 1.38$ **b.** $z = 0.67$ to $z = 2.95$ **c.** $z = -2.95$ to $z = -1.18$

6.17 Find the following areas under the normal curve.
  a. to the right of $z = 0.00$ **b.** to the right of $z = 1.05$ **c.** to the right of $z = -2.30$
  d. to the left of $z = 1.60$ **e.** to the left of $z = -1.60$

6.18 Find the probability that a piece of data picked at random from a normally distributed population will have a standard score that is
  a. less than 3.00   **b.** greater than -1.55 **c.** less than -0.75
  d. less than 1.25   **e.** greater than -1.25

6.19 Find the following:
  a. $P(0.00 < z < 2.35)$ **b.** $P(-2.10 < z < 2.34)$
  c. $P(z > 0.13)$ **d.** $P(z < 1.48)$

6.20 Find the following:
  a. $P(-2.05 < z < 0.00)$ **b.** $P(-1.83 < z < 2.07)$
  c. $P(z < -1.52)$ **d.** $P(z < -0.43)$

6.21 Find the z-score for the standard normal distribution shown on each of the following diagrams.

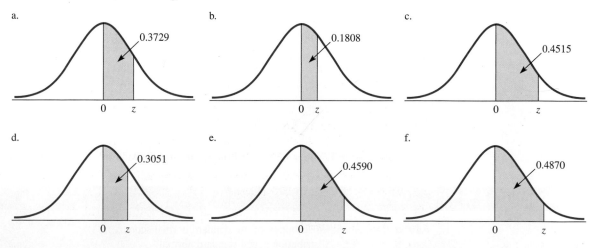

6.22 Find the z-score for the standard normal distribution shown in each of the following diagrams:

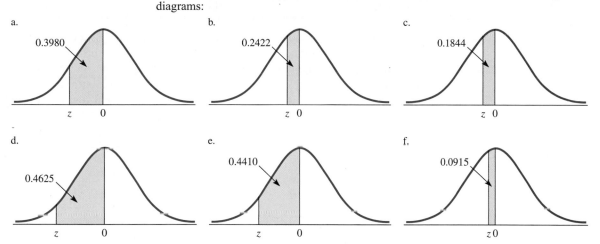

a.   0.3980

b.   0.2422

c.   0.1844

d.   0.4625

e.   0.4410

f.   0.0915

6.23 Find the standard score z shown on each of the following diagrams.

a.   0.05

b.   0.025

c.   0.01

6.24 Find the standard score z shown on each of the following diagrams.

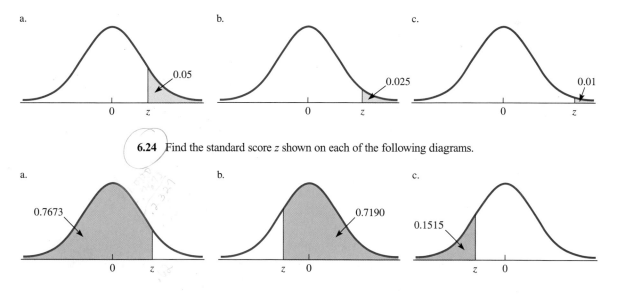

a.   0.7673

b.   0.7190

c.   0.1515

6.25 Find a value of z such that 40% of the distribution lies between it and the mean. (There are two possible answers.)

6.26 Find the standard z-score such that
   a. 80% of the distribution is below (to the left of) this value.
   b. the area to the right of this value is 0.15.

6.27 Find the two z-scores that bound the middle 50% of a normal distribution.

6.28 Find the two standard scores z such that
   a. the middle 90% of a normal distribution is bounded by them.
   b. the middle 98% of a normal distribution is bounded by them.

6.29 Assuming a normal distribution, what is the z-score associated with the 90th percentile? the 95th percentile? the 99th percentile?

6.30 Assuming a normal distribution, what is the z-score associated with the 1st quartile? 2nd quartile? 3rd quartile?

6.3 Applications of Normal Distributions

In Section 6.2 we learned how to use Table 3 in Appendix B to convert information about the standard normal variable z into probability, or the opposite, to convert probability information about the standard normal distribution into z-scores. Now we are ready to apply this methodology to all normal distributions. The key is the standard score, z. The information associated with a normal distribution will be in terms of x-values or probabilities. We will use the z-score and Table 3 as the tools to "go between" the given information and the desired answer.

Recall that the standard score, z, was defined in Chapter 2.

$$\text{Standard score, } z: \quad z = \frac{x - (\text{mean of } x)}{(\text{standard deviation of } x)}$$

$$z = \frac{x - \mu}{\sigma} \tag{6.3}$$

(Note, when $x = \mu$, the standard score $z = 0$.)

ILLUSTRATION 6.10 ▼

Consider the intelligence quotient (I.Q.) scores for people. I.Q. scores are normally distributed with a mean of 100 and a standard deviation of 16. If a person is picked at random, what is the probability that his or her I.Q. is between 100 and 115; that is, what is $P(100 < x < 115)$?

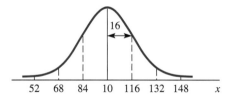

Solution

$P(100 < x < 115)$ is represented by the shaded area in the following figure.

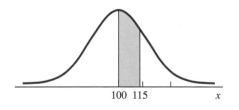

The variable x must be standardized using formula (6.3). The z-values are shown on the figure at the top of page 287.

$$z = \frac{x - \mu}{\sigma}$$

	Exercise 6.31

Exercise 6.31
Given $x = 58$, $\mu = 43$, and $\sigma = 5.2$, find z.

When $x = 100$:
$$z = \frac{100 - 100}{16} = \mathbf{0.00}$$

When $x = 115$:
$$z = \frac{115 - 100}{16} = \mathbf{0.94}$$

Exercise 6.32
Find the probability that a randomly selected person will have an I.Q. score between 100 and 120 (Illus. 6.10).

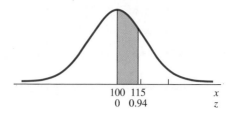

100 115
 0   0.94

Therefore,

$$P(100 < x < 115) - P(0.00 < z < 0.94) - \mathbf{0.3264}$$

The value 0.3264 is found by using Table 3 in Appendix B.

Thus, the probability is 0.3264 that a person picked at random has an I.Q. between 100 and 115.

ILLUSTRATION 6.11 ▼

Find the probability that a person selected at random will have an I.Q. greater than 90.

Solution

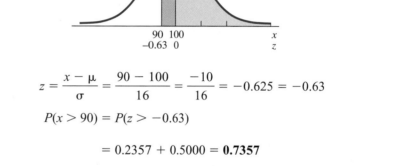

90 100
−0.63 0

$$z = \frac{x - \mu}{\sigma} = \frac{90 - 100}{16} = \frac{-10}{16} = -0.625 = -0.63$$

$$P(x > 90) = P(z > -0.63)$$

$$= 0.2357 + 0.5000 = \mathbf{0.7357}$$

Exercise 6.33
Find the probability that a randomly selected person will have an I.Q. score above 80. (Illus. 6.10)

Thus, the probability is 0.7357 that a person selected at random will have an I.Q. greater than 90.   ▲

The normal table can be used to answer many kinds of questions that involve a normal distribution. Many times a problem will call for the location of a "cutoff point," that is, a particular value of x such that there is exactly a certain percentage in a specified area. The following illustrations concern some of these problems.

ILLUSTRATION 6.12 ▼

In a large class, suppose your instructor tells you that you need to obtain a grade in the top 10% of your class to get an A on a particular exam. From past experience she is able to estimate that the mean and standard deviation on this exam will be 72 and 13, respectively. What will be the minimum grade needed to obtain an A? (Assume that the grades will be approximately normally distributed.)

Solution

Start by converting the 10% to information that is compatible with Table 3 by subtracting: (continued on p. 290)

Case Study 6.1

A Predictive Failure Cost Tool

The standard normal probability distribution can be and is used in connection with many variables. This article shows its application in a predictive role.

The normal curve is commonly used for analysis of process capability and percentage out of specification. Additionally, some engineers use it to calculate probability of occurrence of an event. But there is another use for the normal curve that is seldom taken advantage of—process cost analysis. Using a normal curve to estimate scrap and rework is an accurate method of estimating those costs, and the calculations are simple.

A factory, for example, makes a process change that is expected to reduce scrap and save the company $60,000 a year. Interestingly enough, neither the new or old process is capable of producing 100 percent of product to specifications, and the company has accepted sorting as a way of doing business.

The new process was started at the beginning of a week. Within three days, there was enough data for normal curve cost analysis to show that the process would not save $60,000. In fact, it would cost an additional $30,000 over the older process because, even though scrap had been reduced as predicted, the new process generated more rework at higher rework cost. But cost accounting didn't recognize the problem for a full month until it had enough data to identify the problem with the process' performance.

To use normal curve analysis for estimating cost, measurement error must be disregarded and the process must:

- Not be capable of meeting specifications.
- Be in statistical control.
- Produce a normal distribution.

A widget manufacturer, for example, shows how this statistical technique works. Table 1 shows the parameters for the widget process and cost information.

Table 1. Process parameters

Spec	0.5530 ± 0.0025
Process average	0.5535
Standard deviation	0.0015
Production rate	100,000 per year

Rework all undersize parts at a cost of $5.25 each.
Scrap all oversize parts at a cost of $15.34 each.

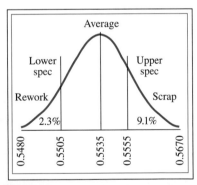

Figure 1

Figure 1 shows that the process is not capable of meeting specification. The process parameters indicate that all undersize parts are reworked and oversize parts are scrapped.

To determine rework costs, calculate the area under the normal curve (z-score) that represents the percent of undersized parts.

$$z\text{-score} = \frac{\text{lower spec–average}}{\text{std.dev.}}$$

where:

lower spec	= 0.5505	average	= 0.5535
std. dev.	= 0.0015	z-score	= −2.0.

A z-score of -2.0 represents 2.3 percent of parts undersized.
The total annual production of undersized parts is calculated by:

$$\text{Parts reworked annually} = (\text{fraction defective})(\text{annual production})$$

$$= (0.023)(100{,}000)$$

$$= 2{,}300 \text{ parts.}$$

The annual cost of reworking parts is:

$$\text{Annual rework cost} = (\text{parts reworked annually})(\text{cost/part})$$

$$= (2{,}300)(\$5.25)$$

$$= \$12{,}075.00.$$

The preceding calculations represent only the part production that can be reworked. Additionally, the cost of making scrap must also be considered. To determine the cost of making scrap, calculate the z-score for scrapped parts.

$$z\text{-score} = \frac{\text{upper spec--average}}{\text{std.dev.}}$$

where:

$$\text{upper spec} = 0.5555$$

$$\text{average} = 0.5535$$

$$\text{std. dev.} = 0.0015$$

$$z\text{-score} = 1.3.$$

A z-score of 1.3 represents 9.1 percent scrap.
The annual scrap production is calculated by:

$$\text{Annual scrap production} = (\text{fraction defective})(\text{annual production})$$

$$= (0.091)(100{,}000)$$

$$= 9{,}100.$$

The annual cost of scrap is:

$$\text{Annual cost of scrap} = (\text{annual scrap production})(\text{cost/scrapped part})$$

$$= (9{,}100)(\$15.34)$$

$$= \$139{,}594.00.$$

The widget prediction shows that the process will generate $12,075.00 in rework and $139,594.00 in scrap or a total of $151,669.00 of failure cost. If the process is centered in the tolerance zone and the cost recalculated, the process will yield $25,090.00 in rework and $73,310.00 in scrap—a total failure cost of $98,400.00. This one step yields a $53,269.00 cost reduction over the previous process setting.

Obviously, some time needs to be spent to determine the process setting that will minimize scrap and rework cost. This requires a balance between scrap and rework to minimize cost impact.

Source: Gregory Roth, "A Predictive Failure Cost Tool," *Quality,* October 1993.

Exercise 6.34

Refer to Case Study 6.1:

a. Verify the z-score for the lower spec and fraction of reworkable parts.

b. Verify the z-score for the upper spec and fraction of scrap parts.

c. Verify the recalculated amounts of $25,090 in rework and $73,310 in scrap.

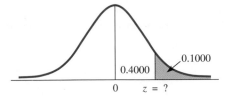

Why is 0.5000 used?

$$10\% = 0.1000; \quad 0.5000 - 0.1000 = 0.4000$$

Look in Table 3 to find the value of z associated with the area entry closest to 0.4000; it is $z = 1.28$. Thus,

$$P(z > 1.28) = 0.10$$

Now find the x-value that corresponds to $z = 1.28$ by using formula (6.3):

Exercise 6.35

Suppose the instructor in Illustration 6.12 said the top 15% were to get A's. Find the minimum score to receive an A.

$$z = \frac{x - \mu}{\sigma}: \quad 1.28 = \frac{x - 72}{13}$$

$$x - 72 = (13)(1.28)$$

$$x = 72 + (13)(1.28) = 72 + 16.64 = 88.64, \text{ or } \mathbf{89}$$

Thus, if you receive an 89 or higher, you can expect to be in the top 10% (which means an A).

ILLUSTRATION 6.13 ▼

Find the 33rd percentile for I.Q. scores ($\mu = 100$ and $\sigma = 16$ from illustration 6.10, p. 286).

Solution

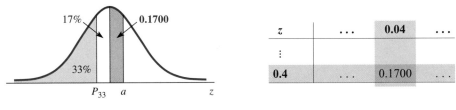

$$P(0 < z < a) = 0.17$$

$$a = 0.44 \text{ (cutoff value of } z \text{ from Table 3)}$$

$$33\text{rd percentile of } z = -0.44 \text{ (below mean)}$$

Now we convert the 33rd percentile of the z-scores, -0.44, to an x-score:

Exercise 6.36

Find the 25th percentile for I.Q. scores in illustration 6.10.

$$\text{Formula (6.3); } z = \frac{x - \mu}{\sigma}: \quad -0.44 = \frac{x - 100}{16}$$

$$x - 100 = 16(-0.44)$$

$$x = 100 - 7.04 = \mathbf{92.96}$$

Thus, 92.96 is the 33rd percentile for I.Q. scores. ▲

Illustration 6.14 concerns a situation in which you are asked to find the mean μ when given the related information.

ILLUSTRATION 6.14 ▼

The incomes of junior executives in a large corporation are normally distributed with a standard deviation of $1200. A cutback is pending, at which time those who earn less than $28,000 will be discharged. If such a cut represents 10% of the junior executives, what is the current mean salary of the group of junior executives?

Solution

If 10% of the salaries are below $28,000, then 40% (or 0.4000) are between $28,000 and the mean μ. Table 3 indicates that $z = -1.28$ is the standard score that occurs at $x - $28,000$. Using formula (6.3) we can find the value of μ:

Exercise 6.37

If 20% of the salaries in Illustration 6.14 are below $28,000, find the current mean salary.

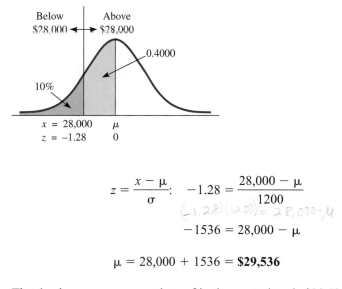

$$z = \frac{x - \mu}{\sigma}: \quad -1.28 = \frac{28,000 - \mu}{1200}$$

$(-1.28)(1200) = 28,000 - \mu$

$$-1536 = 28,000 - \mu$$

$$\mu = 28,000 + 1536 = \textbf{\$29,536}$$

That is, the current mean salary of junior executives is $29,536.   ▲

Referring again to the I.Q. scores, what is the probability that a person picked at random has an I.Q. of 125, $P(x = 125)$? (I.Q. scores are normally distributed with a mean of 100 and a standard deviation of 16.) This situation has two interpretations: theoretical and practical. Let's look at the theoretical interpretation first. Recall that the probability associated with an interval for a continuous random variable is represented by the area under the curve. That is, $P(a < x < b)$ is equal to the area between a and b under the curve. $P(x = 125)$ (that is, x is exactly 125) is then $P(125 \leq x \leq 125)$, or the area of the vertical line segment at $x = 125$. This area is zero. However, this is not the practical meaning of $x = 125$. It generally means 125 to the nearest integer value. Thus, $P(x = 125)$ would most likely be interpreted as

$$P(124.5 < x < 125.5)$$

The interval from 124.5 to 125.5 under the curve has a measurable area and is then nonzero. In situations of this nature, you must be sure of the meaning being used.

NOTE A standard notation used to abbreviate "normal distribution with mean μ and standard deviation σ" is $N(\mu, \sigma)$. That is, $N(58, 7)$ represents "normal distribution, mean = 58 and standard deviation = 7."

The following commands will generate n random data from a normal distribution with a given mean μ and standard deviation σ.

MINITAB

```
Choose:    Calc > Random Data > Normal
Enter:     Generate:       n  rows of data
           Store in column(s): C1
           Mean:           μ
           Stand. dev.: σ
```

If multiple samples (say, 12), all of the same size are wanted, modify the above commands: Store in column(s): C1-C12

Note: To find descriptive statistics for each of these samples, use the commands: Stat > Basic Statistics > Display Descriptive Statistics for C1-C12.

Exercise 6.38

Generate a random sample of 100 data from a normal distribution with mean 50 and standard deviation 12.

EXCEL

```
Choose:    Tools > Data Analysis > Random Number Generation >
           OK
Enter:     Number of Variables: 1
           Number of Random Numbers: n
           Distribution: Normal
           Mean = :  μ
           Standard Deviation = : σ
Select:    Output Options: Output Range
Enter:     (A1 or select cell) > OK
```

If multiple samples (say, 12), all of the same size are wanted, modify the above commands: Number of variables: 12.

Note: To find descriptive statistics for each of these samples, use the commands: Tools > Data Analysis > Descriptive Statistics for columns A through L.

Exercise 6.39

Generate 10 random samples, each of size 25, from a normal distribution with mean 75 and standard deviation 14.

TI-83

```
Choose:    MATH > PRB > 6:randNorm(
Enter:     μ, σ, # of trials)
Choose:    STO→ > L1 > ENTER
```

If multiple samples (say, 6), all of the same size are wanted, repeat the above commands six times and store in L1 through L6.

Note: To find descriptive statistics for each of these samples, use the commands: STAT > CALC > 1:1-Var Stats for L1-L6.

The following commands will find the ordinate values for given abscissas from a normal distribution with a given mean μ and standard deviation σ.

MINITAB

Input the desired abscissas into C1; then continue with:

```
Choose:  Calc > Probability Distribu-
         tions > Normal
Select:  Probability Density
Enter:   Mean:               μ
         Stand. dev.:        σ
         Input column:       C1
         Optional Storage: C2
```

To draw the graph of a normal probability curve with the x-values in C1 and the y-values in C2; continue with:

```
Choose:  Graph > Plot
Enter:   Graph variables: Y: C2 X: C1
         Data display: Display:
         Connect
```

EXCEL

Input the desired abscissas into column A and activate B1; then continue with:

```
Choose:  Paste function fₓ > Statisti-
         cal > NORMDIST > OK
Enter:   X: (A1:A100 or select
         'x value' cells)
         Mean: μ
         Standard dev.: σ
         Cumulative: False   > OK
Drag:    Bottom right corner of the
         ordinate value box down to
         give other ordinates
```

To draw the graph of a normal probability curve with the x-values in column A and the y-values in column B; continue with:

```
Choose:  Chart Wizard > XY(Scatter) >
         1ˢᵗ picture > Next > Data Range
Enter:   Data range: (A1:B100 or se-
         lect x & y cells)
Choose:  Next > Finish
```

Exercise 6.40

Use the random sample of 100 data found in Exercise 6.38 and find the 100 corresponding y-values for the normal distribution curve with mean 50 and standard deviation 12.

TI-83

The ordinate values can be calculated for individual abscissa values, 'x'.

```
Choose:  2ⁿᵈ > DISTR > 1:normalpdf(
Enter:   x, μ, σ)
```

To draw the graph of the normal probability curve for a particular μ and σ; continue with:

```
Choose:  WINDOW
Enter:   μ − 3σ, μ + 3σ, σ, −.05,
         .1, 1, 1
Choose:  Y = > 2ⁿᵈ > DISTR >
         1:normalpdf(
Enter:   x, μ, σ) (x is the variable
         'x', not an individual
         value as above)
```

After an initial graph, adjust with 0:ZoomFit from the ZOOM menu.

Exercise 6.41

Use the 100 ordered pairs found in Exercise 6.39 and draw the curve for the normal distribution with mean 50 and standard deviation 12.

The following commands will find the cumulative probabilities paired with the listed abscissas, from a normal distribution with a given mean μ and standard deviation σ.

MINITAB

Input the desired abscissas into C1; then continue with:

Choose: **Calc > Probability Distributions > Normal**

Select: **Cumulative probability**

Enter: Mean: **μ**
Stand. dev.: **σ**
Input column: **C1**
Optional Storage: **C3**

Note 1. To find the probability between two x-values, enter the two values in C1, use the above commands, and subtract using the numbers in C3.

2. To draw a graph of the cumulative probability distribution (ogive), use the PLOT commands on p. 293 with C3 as the y-variable.

TI-83

The cumulative probabilities can be calculated for individual abscissa values, 'x'.

Choose: **2ⁿᵈ > DISTR > 2:normalcdf(**

Enter: **−1EE99, x, μ, σ)** [Use key with EE, but only E will appear on screen.]

Note 1. To find the probability between two x-values, enter the two values in place of −1EE99 and the x.

2. To draw a graph of the cumulative probability distribution (ogive), use either the Scatter command under STATPLOTS, with the x values and their cumulative probabilities in a pair of lists or normalcdf(−1EE99, x, μ, σ) in the Y=editor.

EXCEL

Input the desired abscissas into column A and activate C1; then continue with:

Choose: **Paste function fₓ > Statistical > NORMDIST > OK**

Enter: X: **(A1:A100 or select 'x value' cells)**
Mean: μ
Standard dev.: σ
Cumulative: **True > OK**

Drag: **Bottom right corner of the cumulative probability box down to give other cumulative probabilities**

Note 1. To find the probability between two x-values, enter the two values in column A, use the above commands, and subtract using the numbers in column C.

2. To draw a graph of the cumulative probability distribution (ogive), use the Chart Wizard commands on p. 293, choosing the subcommand Series with column C as the y-values and column A as the x-values.

Exercise 6.42

Find the probability that a randomly selected value from a normal distribution with mean 50 and standard deviation 12 will be between 55 and 65. Verify your results by using Table 3.

EXERCISES

6.43 Given that x is a normally distributed random variable with a mean of 60 and a standard deviation of 10, find the following probabilities.

   a. $P(x > 60)$ **b.** $P(60 < x < 72)$ **c.** $P(57 < x < 83)$

   d. $P(65 < x < 82)$ **e.** $P(38 < x < 78)$ **f.** $P(x < 38)$

6.44 According to the November 1993 issue of *Harper's* magazine, our kids spend from 1200 to 1800 hours a year in front of the television set. Suppose the time spent by kids in front of the television set per year is normally distributed with a mean equal to 1500 hours and a standard deviation equal to 100 hours.

   a. What percentage spend between 1400 and 1600 hours?

   b. What percentage spend between 1300 and 1700 hours?

   c. What percentage spend between 1200 and 1800 hours?

   d. Compare the results (a) through (c) with the empirical rule. Explain the relationship.

6.45 For a particular age group of adult males, the distribution of cholesterol readings, in mg/dl, is normally distributed with a mean of 210 and a standard deviation of 15.

   a. What percentage of this population would have readings exceeding 250?

   b. What percentage would have readings less than 150?

6.46 Electronic books are becoming a reality thanks to the computer age and especially flat panel, touch-sensitive screens. The books can be underlined, notes can be written on them, searches can be made on key words, and definitions of words can be obtained instantly while reading the book. Compared to bound paper books, E-books are not cheap, and the average price in 1998 was $400, although prices are expected to fall in the future as production levels increase. (*Fortune,* "Electronic Books Are Coming at Last!," 7-6-98.) Suppose the price of E-books is normally distributed with a standard deviation of $150. What percentage of the E-books being sold in 1998 were priced:

   a. between $300 and $600?   **b.** between $200 and $800?

   c. less than $250?   **d.** over $900?

6.47 Finding a source of supply for factory workers is especially difficult when a nation's unemployment rate is relatively low. Under these conditions, available labor is scarce, and wages also tend to increase. According to *Fortune* ("The Hunt for Good Factory Workers," 6-22-98), in April 1998 the average production worker in the United States earned $13.47 an hour, up 3.1% from a year earlier. That percentage increase was more than all of 1997 and 1996 combined. Suppose production worker's wages are normally distributed with a standard deviation of $4.75. What percentage of the nation's production workers earn:

   a. between $11.00 and $15.00 an hour?   **b.** between $8.00 and $19.00 an hour?

   c. over $20.00 an hour?   **d.** less than $6.00 an hour?

6.48 At Pacific Freight Lines, bonuses are given to billing clerks when they complete 300 or more freight bills during an eight-hour day. The number of bills completed per clerk per eight-hour day is approximately normally distributed with a mean of 270 and a standard deviation of 16. What proportion of the time should a randomly selected billing clerk expect to receive a bonus?

6.49 The waiting time x at a certain bank is approximately normally distributed with a mean of 3.7 min and a standard deviation of 1.4 min.

   a. Find the probability that a randomly selected customer has to wait less than 2.0 min.

   b. Find the probability that a randomly selected customer has to wait more than 6 min.

   c. Find the value of the 75th percentile for x.

6.50 A brewery filling machine is adjusted to fill quart bottles with a mean of 32.0 oz of ale and a variance of 0.003. Periodically, a bottle is checked and the amount of ale is noted.

   a. Assuming the amount of fill is normally distributed, what is the probability that the next randomly checked bottle contains more than 32.02 oz?

   b. Let's say you buy 100 quart bottles of this ale for a party; how many bottles would you expect to find containing more than 32.02 oz of ale?

6.51 Final averages are typically approximately normally distributed with a mean of 72 and a standard deviation of 12.5. Your professor says that the top 8% of the class will receive an A; the next 20%, a B; the next 42%, a C; the next 18%, a D; and the bottom 12%, an F.
  a. What average must you exceed to obtain an A?
  b. What average must you exceed to receive a grade better than a C?
  c. What average must you obtain to pass the course? (You'll need a D or better.)

6.52 A radar unit is used to measure the speed of automobiles on an expressway during rush-hour traffic. The speeds of individual automobiles are normally distributed with a mean of 62 mph.
  a. Find the standard deviation of all speeds if 3% of the automobiles travel faster than 72 mph.
  b. Using the standard deviation found in (a), find the percentage of these cars that are traveling less than 55 mph.
  c. Using the standard deviation found in (a), find the 95th percentile for the variable "speed."

6.53 The weights of ripe watermelons grown at Mr. Smith's farm are normally distributed with a standard deviation of 2.8 lb. Find the mean weight of Mr. Smith's ripe watermelons if only 3% weigh less than 15 lb.

6.54 A machine fills containers with a mean weight per container of 16.0 oz. If no more than 5% of the containers are to weigh less than 15.8 oz, what must the standard deviation of the weights equal? (Assume normality.)

6.55 According to a USA Snapshot® (10-26-94), the average annual salary for a worker in the United States is $26,362. If we assume that the annual salaries for Americans are normally distributed with a standard deviation equal to $6500, find the following:
  a. What percentage earn below $15,000?
  b. What percentage earn above $40,000?

6.56 According to the 1991 issue of *American Hospital Administration Hospital Statistics,* the average daily census total for 116 hospitals in Mississippi equals 10,872. Suppose the standard deviation of the daily census totals for these hospitals equals 1505 patients. If the daily census totals are normally distributed:
  a. What percentage of the days does the daily census total less than 8500 patients in these hospitals? Approximately how often should we expect this to occur?
  b. What percentage of the days does the daily census total exceed 12,500 patients in these hospitals? Approximately how often should we expect this to occur?

6.57 Use a computer or calculator to find the probability that one randomly selected value of x from a normal distribution, mean 584.2 and standard deviation 37.3, will have a value
  a. less than 525.   **b.** between 525 and 590.
  c. of at least 590.   **d.** Verify the results using Table 3.
  e. Explain any differences you may find.

MINITAB

Input 525 and 590 into C1; then continue with the cumulative probability commands on page 294, using 584.2 as μ, 37.3 as σ, and C2 as optional storage.

EXCEL

Input 525 and 590 into column A and activate the B1 cell; then continue with the cumulative probability commands on page 294, using 584.2 as μ and 37.3 as σ.

TI-83

Input 525 and 590 into L1; then continue with the cumulative probability commands on page 294 in L2, using 584.2 as μ, and 37.3 as σ.

6.58 a. Use a computer to generate your own abbreviated standard normal probability table (a short version of Table 3). Use z-values of 0.0 to 5.0 in intervals of 0.1.

b. How are the values obtained related to Table 3 entries? Make the necessary adjustment and store the results in a column.

c. Compare your results in (b) to the first column of Table 3. Comment on any differences you see.

MINITAB

a. Choose: `Calc > Make Patterned Data > Simple Set of Numbers`
    Enter: `Store patterned data in: C1`
             `From first value: 0`
             `To last value: 5`
             `In steps of: 0.1`

`Continue with the cumulative probability commands on page 294, using 0 as µ, 1 as σ, and C2 as optional storage.`

b. Choose: `Calc > Calculator`
    Enter: `Store result in variable: C3`
             `Expression: C2 − 0.5 > OK`
    Choose: `Manip > Display Data`
    Enter: `Columns to display: C1 C3`

EXCEL

a. Choose: `Tools > Data Analysis > Random Number Generation`
             `> OK`
    Enter: `Number of variables: 1`
             `Distribution: Patterned`
             `From: 0 to 5.1* in steps of 0.1`
             `(*stops before last value)`
             `repeat each number: 1 time`
             `repeating the sequence: 1 time`
             `Select: Output Range`
             `Enter:  (A1 or select cell) > OK`

`Continue with the cumulative probability commands on page 294, with activating cell B1 and using column A as x, 0 as µ and 1 as σ.`

b. `Activate cell C1, then continue with:`

    Choose: `Edit Formula =`
    Enter: `B1 − 0.5   > OK`
    Drag: `Bottom right corner of the C1 box down to give probabilities for the z-values`

6.59 Use a computer to compare a random sample to the population from which the sample was drawn. Consider the normal population with mean 100 and standard deviation 16.

a. List values of x from $\mu - 4\sigma$ to $\mu + 4\sigma$ in increments of half standard deviations and store them in a column.

b. Find the ordinate (y-value) corresponding to each abscissa (x-value) for the normal distribution curve for $N(100, 16)$ and store them in a column.

c. Graph the normal probability distribution curve for $N(100, 16)$.

d. Generate 100 random data values from the $N(100, 16)$ distribution and store them in a column.

e. Graph the histogram of the 100 data obtained in (d) using the numbers listed in (a) as class boundaries.

f. Calculate other helpful descriptive statistics of the 100 data and compare the data to the expected distribution. Comment on the similarities and the differences you see.

MINITAB

a. Use the Make Patterned Data commands in Exercise 6.58a, replacing the first value with 36, the last value with 164, and the steps with 8.

b. Choose: **Calc > Prob. Dist. > Normal**
Select: **Probability density**
Enter: Mean: **100**
Stand. dev.: **16**
Input column: **C1**
Optional Storage: **C2**

c. Use the PLOT commands on page 139 for the data in C1 and C2.

d. Use the generate RANDOM DATA commands on page 292 replacing n with 100, store in with C3, mean with 100 and standard deviation with 16.

e. Use the HISTOGRAM commands on page 56 for the data in C3, entering C1 as cutpoints.

f. Use the MEAN and STANDARD DEVIATION commands on pages 67 and 80 for the data in C3.

EXCEL

a. Use the RANDOM NUMBER GENERATION Patterned Distribution commands in Exercise 6.58a, replacing the first value with 36, the last value with 172, and the steps with 8.

b. Activate B1; then continue with:

Choose: **Paste function f_x > Statistical > NORMDIST > OK**
Enter: X: **(A1:A17 or select 'x value' cells)**
Mean: **100**
Standard dev.: **16**
Cumulative: **False** **> OK**
Drag: **Bottom right corner of the ordinate value box down to give other ordinates**

c. Use the CHART WIZARD XY(Scatter) commands on page 140 for the data in columns A and B.

d. Use the Normal RANDOM NUMBER GENERATION commands on page 292 replacing number of random numbers with 100, mean with 100, standard deviation with 16, and output range with C1.

e. Use the HISTOGRAM commands on page 57 with column C as the input range and column A as the bin range

f. Use the MEAN and STANDARD DEVIATION commands on pages 67 and 80 for the data in column C.

6.60 Suppose you were to generate several random samples, all the same size, all from the same normal probability distribution. Will they all be the same? How will they differ? By how much will they differ?

a. Use a computer or calculator to generate 10 different samples, all of size 100, all from the normal probability distribution of mean 200 and standard deviation 25.

b. Draw histograms of all 10 samples using the same class boundaries.

c. Calculate several descriptive statistics for all 10 samples, separately.

d. Comment on the similarities and the differences you see.

MINITAB

a. Use the generate RANDOM DATA commands on page 292 replacing n with 100, store in with C1-C10, mean with 200 and standard deviation with 25.

b. Use the HISTOGRAM commands on page 56 for the data in C1-C10, entering cutpoints from 100 to 300 in increments of 25 (100:300/25).

c. Use the DISPLAY DESCRIPTIVE STATISTICS command on page 97 for the data in C1-C10.

EXCEL

a. Use the Normal RANDOM NUMBER GENERATION commands on page 292 replacing number of variables with 10, number of random numbers with 100, mean with 200, and standard deviation with 25.

b. Use the RANDOM NUMBER GENERATION Patterned Distribution commands on page 297, replacing the first value with 100, the last value with 325, the steps with 25, and the output range with K1. Use the HISTOGRAM commands on page 57 for each of the columns A through J (input range) with column K as the bin range.

c. Use the DESCRIPTIVE STATISTICS commands on page 97 for the data in columns A through J.

TI-83

a. Use the 6:randNorm commands on page 292 replacing the mean with 200, the standard deviation with 25, and the number of trials with 100. Repeat 6 times using L1-L6 for storage.

b. Use the HISTOGRAM commands on page 58 for the data in L1-L6, entering WINDOW values: 100, 300, 25, −10, 60, 10, 1. Adjust with ZoomStat.

c. Use the 1-Var Stats command on page 97 for the data in L1-L6.

6.4   Notation

The z-score is used throughout statistics in a variety of ways; however, the relationship between the numerical value of z and the area under the **standard normal distribution** curve does not change. Since z will be used with great frequency, we want a convenient notation to identify the necessary information. The convention that we will use as an "algebraic name" for a specific z-score is $z(\alpha)$, where α represents the "area to the right" of the z being named.

ILLUSTRATION 6.15 ▼

$z(0.05)$ (read "z of 0.05") is the algebraic name for the z such that the area to the right and under the standard normal curve is exactly 0.05, as shown in Figure 6.6.

Figure 6.6
Area Associated with $z(0.05)$

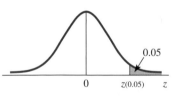

0.05

0        $z(0.05)$ z

Exercise 6.61
Draw a figure showing $z(0.15)$ on the standard normal curve.

ILLUSTRATION 6.16 ▼

$z_{(0.60)}$ [read "z of 0.60"] is that value of z such that 0.60 of the area lies to its right, as shown in Figure 6.7.

Figure 6.7
Area Associated with $z_{(0.60)}$

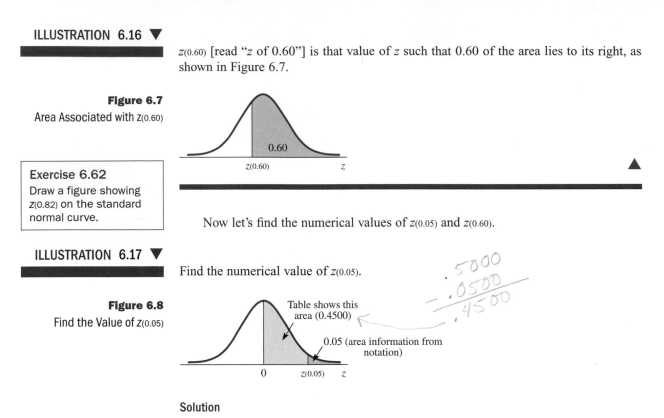

Exercise 6.62
Draw a figure showing $z_{(0.82)}$ on the standard normal curve.

Now let's find the numerical values of $z_{(0.05)}$ and $z_{(0.60)}$.

ILLUSTRATION 6.17 ▼

Find the numerical value of $z_{(0.05)}$.

Figure 6.8
Find the Value of $z_{(0.05)}$

Solution

We must convert the area information in the notation into information that can be used with Table 3 in Appendix B; see the areas shown in Figure 6.8.

When we look in Table 3, we look for an area as close as possible to 0.4500.

z	...	**0.04**		**0.05**	...
⋮					
1.6	...	0.4495	0.4500	0.4505	...
⋮					

Exercise 6.63
Find the value of $z_{(0.15)}$.

Therefore, $z_{(0.05)} = \mathbf{1.65}$. (Note: We will use the z that corresponds to the area closest in value. If the value happens to be exactly halfway between the table entries, always use the larger value of z.)

ILLUSTRATION 6.18 ▼

Find the value of $z_{(0.60)}$.

Solution

The value 0.60 is related to Table 3 by use of the area 0.1000, as shown in the following diagram. The closest values in Table 3 are 0.0987 and 0.1026.

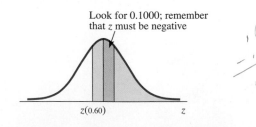

Exercise 6.64
Find the value of $z(0.82)$.

z	...	**0.05**		**0.06**	...
⋮					
0.2	...	0.0987	0.1000	0.1026	...
⋮					

Therefore, $z(0.60)$ is related to 0.25. Since $z(0.60)$ is below the mean, we conclude that $z(0.60) = -0.25$.   ▲

In later chapters we will use this notation on a regular basis. The values of z that will be used regularly come from one of the following situations: (1) the z-score such that there is a specified area in one tail of the normal distribution, or (2) the z-scores that bound a specified middle proportion of the normal distribution.

Illustration 6.17 showed that a commonly used one-tail situation; $z(0.05) = 1.65$ is located so that 0.05 of the area under the normal distribution curve is in the tail to the right.

ILLUSTRATION 6.19 ▼

Find $z(0.95)$.

Solution

$z(0.95)$ is located on the left-hand side of the normal distribution since the area to the right is 0.95. The area in the tail to the left then contains the other 0.05, as shown in Figure 6.9. Because of the symmetrical nature of the normal distribution, $z(0.95)$ is $-z(0.05)$, that is, $z(0.05)$ with its sign changed. Thus, $z(0.95) = -z(0.05) = -1.65$.

Figure 6.9
Area Associated with $z(0.95)$

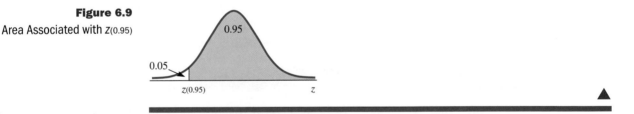

▲

When the middle proportion of a normal distribution is specified, we can still use the "area to the right" notation to identify the specific z-score involved.

ILLUSTRATION 6.20 ▼

Find the z-scores that bound the middle 0.95 of the normal distribution.

Solution

Given 0.95 as the area in the middle (Figure 6.10), the two tails must contain a total of 0.05. Therefore, each tail contains $\frac{1}{2}$ of 0.05, or 0.025, as shown in Figure 6.11.

Figure 6.10
Area Associated with
Middle 0.95

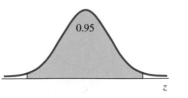

Figure 6.11

Finding z-Scores for Middle 0.95

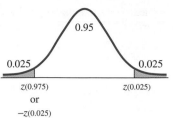

In order to find $z_{(0.025)}$ in Table 3, we must determine the area between the mean and $z_{(0.025)}$. It is $0.5000 - 0.0250 = 0.4750$, as shown in Figure 6.12.

Figure 6.12

Finding the Value of $z_{(0.025)}$

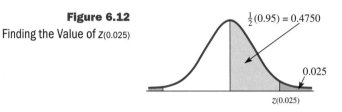

From Table 3,

z		$\cdots$	**0.06**	$\cdots$
$\vdots$				
1.9			0.4750	$\cdots$
$\vdots$				

Therefore, $z_{(0.025)} = 1.96$ and $z_{(0.975)} = -z_{(0.025)} = -1.96$. The middle 0.95 of the normal distribution is bounded by **−1.96** and **1.96**. ▲

EXERCISES • • • • • • • • • • • • • •

6.65 Using the $z_{(\alpha)}$ notation (identify the value of α used within the parentheses), name each of the standard normal variable z's shown in the following diagrams.

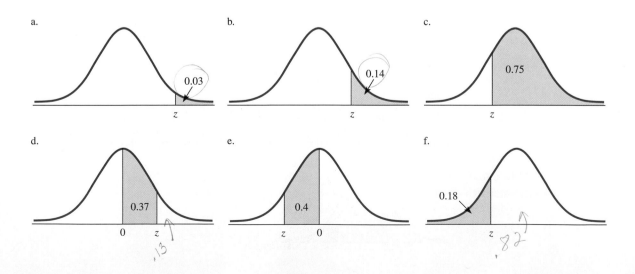

6.66 Using the $z_{(\alpha)}$ notation (identify the value of α used within the parentheses), name each of the standard normal variable z's shown in the following diagrams.

a.

b.

c.

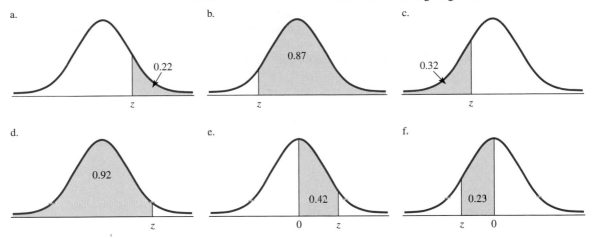

d.

e.

f.

6.67 We are often interested in finding the value of z that bounds a given area in the right-hand tail of the normal distribution, as shown in the accompanying figure. The notation $z(\alpha)$ represents the value of z such that $P(z > z(\alpha)) = \alpha$. Find the following:

   a. $z(0.025)$ **b.** $z(0.05)$ **c.** $z(0.01)$

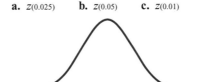

6.68 Use Table 3, Appendix B, to find the following values of z.
   a. $z(0.05)$ **b.** $z(0.01)$ **c.** $z(0.025)$ **d.** $z(0.975)$ **e.** $z(0.98)$

6.69 Complete the following charts of z-scores. The area A given in the tables is the area to the right under the normal distribution in the figures.
   a. z-scores associated with the right-hand tail: Given the area A, find $z_{(A)}$.

A	0.10	0.05	0.025	0.02	0.01	0.005
$z_{(A)}$						

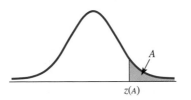

$z(A)$

   b. z-scores associated with the left-hand tail: Given the area B, find $z_{(B)}$.

B	0.995	0.99	0.98	0.975	0.95	0.90
$z_{(B)}$						

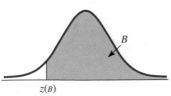

$z(B)$

6.70 **a.** Find the area under the normal curve for z between $z_{(0.95)}$ and $z_{(0.025)}$.

b. Find $z_{(0.025)} - z_{(0.95)}$.

6.71 The z notation, $z_{(\alpha)}$, combines two related concepts, the z-score and the area to the right, into a mathematical symbol. Identify the letter in each of the following as being a z-score or being an area, and then with the aid of a diagram explain what both the given number and the letter represent on the standard curve.

a. $z_{(A)} = 0.10$ **b.** $z_{(0.10)} = B$

c. $z_{(C)} = -0.05$ **d.** $-z_{(0.05)} = D$

6.72 Understanding the z notation, $z_{(\alpha)}$, requires us to know whether we have a z-score or an area. The following expressions use the z notation in a variety of ways, some typical and some not so typical. Find the value asked for in each of the following, and then with the aid of a diagram explain what your answer represents.

a. $z_{(0.08)}$

b. the area between $z_{(0.98)}$ and $z_{(0.02)}$

c. $z_{(1.00 - 0.01)}$

d. $z_{(0.025)} - z_{(0.975)}$

6.5   Normal Approximation of the Binomial

In Chapter 5 we introduced the **binomial distribution**. Recall that the binomial distribution is a probability distribution of the discrete random variable x, the number of successes observed in n repeated independent trials. We will now see how **binomial probabilities,** that is, probabilities associated with a binomial distribution, can be reasonably estimated by using the normal probability distribution.

Let's look first at a few specific binomial distributions. Figures 6.13a, 6.13b, and 6.13c show the probabilities of x for 0 to n for three situations: $n = 4$, $n = 8$, and $n = 24$. For each of these distributions, the probability of success for one trial is 0.5. Notice that as n becomes larger, the distribution appears more and more like the normal distribution.

To make the desired approximation, we need to take into account one major difference between the binomial and the normal probability distributions. The binomial random variable is **discrete,** whereas the normal random variable is **continuous.** Recall that Chapter 5 demonstrated that the probability assigned to a particular value of x should be shown on a diagram by means of a straight-line segment whose length represents the probability (as in Figure 6.13). Chapter 5 suggested, however, that we can also use a histogram in which the area of each bar is equal to the probability of x.

Let's look at the distribution of the binomial variable x, where $n = 14$ and $p = 0.5$. The probabilities for each x-value can be obtained from Table 2 in Appendix B. This distribution of x is shown in Figure 6.14. We see the very same distribution in Figure 6.15 in histogram form.

Let's examine $P(x = 4)$ for $n = 14$ and $p = 0.5$ to study the approximation technique. $P(x = 4)$ is equal to 0.061 (see Table 2 in Appendix B), the area of the bar above $x = 4$ in Figure 6.16. Area is the product of width and height. In this case the height is 0.061 and the width is 1.0; thus, the area is 0.061. Let's take a closer look at the width. For $x = 4$, the bar starts at 3.5 and ends at 4.5, so we are looking at an area bounded by $x = 3.5$ and $x = 4.5$. The addition and subtraction of 0.5 to the x-value is commonly called the **continuity correction factor.** It is our method of converting a discrete variable into a continuous variable.

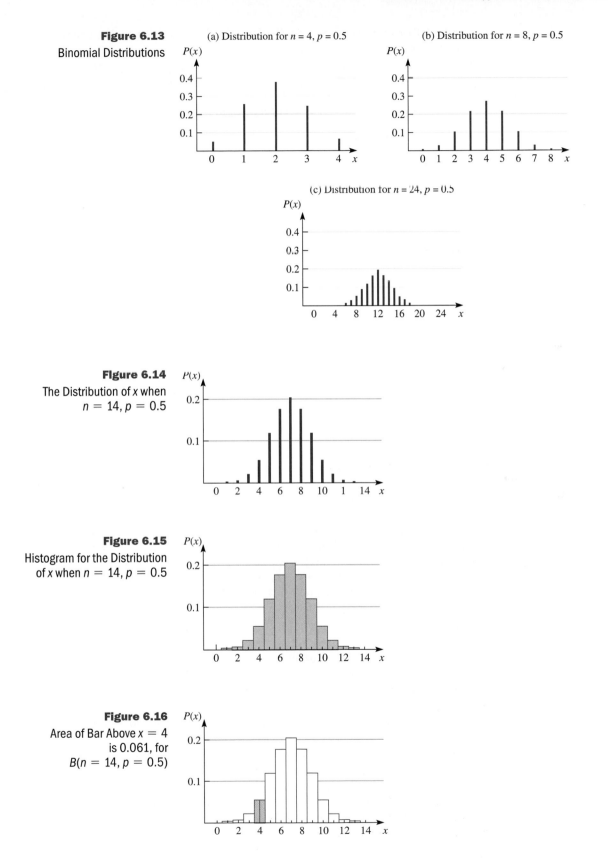

Figure 6.13
Binomial Distributions

(a) Distribution for $n = 4$, $p = 0.5$

(b) Distribution for $n = 8$, $p = 0.5$

(c) Distribution for $n = 24$, $p = 0.5$

Figure 6.14
The Distribution of x when
$n = 14$, $p = 0.5$

Figure 6.15
Histogram for the Distribution
of x when $n = 14$, $p = 0.5$

Figure 6.16
Area of Bar Above $x = 4$
is 0.061, for
$B(n = 14, p = 0.5)$

Now let's look at the normal distribution related to this situation. We will first need a normal distribution with a mean and a standard deviation equal to those of the binomial distribution we are discussing. Formulas (5.7) and (5.8) give us these values.

$$\mu = np = (14)(0.5) = \mathbf{7.0}$$

$$\sigma = \sqrt{npq} = \sqrt{(14)(0.5)(0.5)} = \sqrt{3.5} = \mathbf{1.87}$$

The probability that $x = 4$ is approximated by the area under the normal curve between $x = 3.5$ and $x = 4.5$, as shown in Figure 6.17. Figure 6.18 shows the entire distribution of the binomial variable x with a normal distribution of the same mean and standard deviation superimposed. Notice that the bars and the interval areas under the curve cover nearly the same area.

Figure 6.17

Probability That
$x = 4$ Is Approximated
by Shaded Area

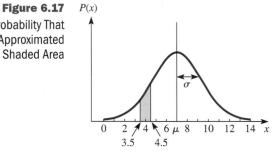

Figure 6.18

Normal Distribution
Superimposed over
Distribution for Binomial
Variable x

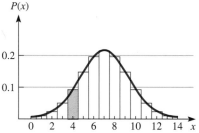

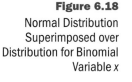

The probability that x is between 3.5 and 4.5 under this normal curve is found by using formula (6.3), Table 3, and the methods outlined in Section 6.3.

$$z = \frac{x - \mu}{\sigma}: \quad P(3.5 < x < 4.5) = P\left(\frac{3.5 - 7.0}{1.87} < z < \frac{4.5 - 7.0}{1.87}\right)$$

$$= P(-1.87 < z < -1.34)$$

$$= 0.4693 - 0.4099 = \mathbf{0.0594}$$

Since the binomial probability of 0.061 and the normal probability of 0.0594 are reasonably close in value, the normal probability distribution seems to be a reasonable approximation of the binomial distribution.

The normal approximation of the binomial distribution is also useful for values of p that are not close to 0.5. The binomial probability distributions shown in Figures 6.19 and 6.20 suggest that binomial probabilities can be approximated using the normal distribution. Notice that as n increases in size, the binomial distribution begins to look like the normal distribution. As the value of p moves away from 0.5, a larger n will be needed in order for the normal approximation to be reasonable. The "rule of thumb" is generally used as a guideline.

Figure 6.19

Binomial Distributions

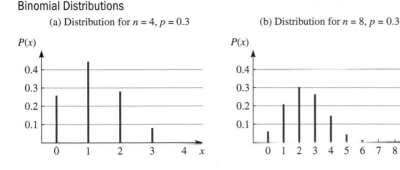

(a) Distribution for $n = 4$, $p = 0.3$

(b) Distribution for $n = 8$, $p = 0.3$

(c) Distribution for $n = 24$, $p = 0.3$

Figure 6.20

Binomial Distributions

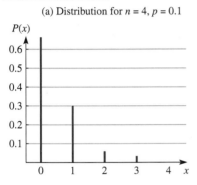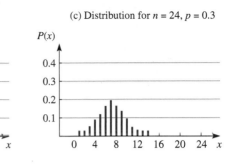

(a) Distribution for $n = 4$, $p = 0.1$

(b) Distribution for $n = 8$, $p = 0.1$

(c) Distribution for $n = 50$, $p = 0.1$

Exercise 6.73

Find the values np and nq for a binomial experiment with $n = 100$ and $p = 0.02$. Does this binomial distribution satisfy the rule for normal approximation? Explain.

RULE

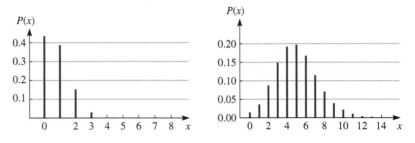

The normal distribution provides a reasonable approximation to a binomial probability distribution whenever the values of np and $n(1 - p)$ both equal or exceed 5.

By now you may be thinking, "So what? I will just use the binomial table and find the probabilities directly and avoid all the extra work." But consider for a moment the situation presented in Illustration 6.21.

ILLUSTRATION 6.21 ▼

An unnoticed mechanical failure has caused $\frac{1}{3}$ of a machine shop's production of 5000 rifle firing pins to be defective. What is the probability that an inspector will find no more than 3 defective firing pins in a random sample of 25?

Solution

In this illustration of a binomial experiment, x is the number of defectives found in the sample, $n = 25$, and $p = P(\text{defective}) = \frac{1}{3}$. To answer the question by using the binomial distribution, we will need to use the binomial probability function [formula (5.5)]:

$$P(x) = \binom{25}{x}\left(\frac{1}{3}\right)^x\left(\frac{2}{3}\right)^{25-x}, \text{ for } x = 0, 1, 2, \ldots, \text{ or } 25$$

We must calculate the values for $P(0)$, $P(1)$, $P(2)$, and $P(3)$, since they do not appear in Table 2. This is a very tedious job because of the size of the exponent. In situations such as this, we can use the normal approximation method.

Now let's find $P(x \le 3)$ by using the normal approximation method. We first need to find the mean and standard deviation of x [formulas (5.7) and (5.8)]:

$$\mu = np = (25)(\frac{1}{3}) = 8.333$$

$$\sigma = \sqrt{npq} = \sqrt{(25)(\frac{1}{3})(\frac{2}{3})} = \sqrt{5.55556} = 2.357$$

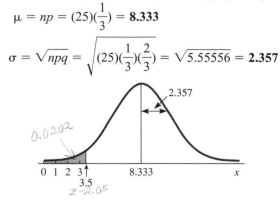

These values are shown in the figure above. The measure of the shaded area $(x < 3.5)$ represents the probability of $x = 0, 1, 2,$ or 3. Remember that $x = 3$, the discrete binomial variable, covers the continuous interval from 2.5 to 3.5.

$$P(x \text{ is no more than } 3) = P(x \le 3) \text{ (for discrete variable } x\text{)}$$

$$= P(x < 3.5) \text{ (using a continuous variable } x\text{)}$$

$$z = \frac{x - \mu}{\sigma}: \quad P(x < 3.5) = P\left(z < \frac{3.5 - 8.333}{2.357}\right) = P(z < -2.05)$$

$$= 0.5000 - 0.4798 = \mathbf{0.0202}$$

Thus, $P(\text{no more than three defectives})$ is approximately 0.02. ▲

Exercise 6.74
a. Calculate
$$P\left(x \le 3 \mid B(25, \frac{1}{3})\right).$$
(Hint: If you use a computer or calculator, use the commands on p. 258.)
b. How good was the normal approximation? Explain.

EXERCISES · · · · · · · · · · · · · ·

6.75 In which of the following binomial distributions does the normal distribution provide a reasonable approximation? Use computer commands to generate a graph of the distribution and compare the results to the "rule of thumb." State your conclusions.

a. $n = 10, p = 0.3$ b. $n = 100, p = 0.005$
c. $n = 500, p = 0.1$ d. $n = 50, p = 0.2$

MINITAB

Insert the specific n and p as needed in the procedure below.

Use the Make Patterned Data commands in Exercise 6.58a, replacing the first value with 0, the last value with n, and the steps with 1.

Use the Binomial Probability Distribution commands on page 258 using C2 as optional storage.

Use the PLOT commands on page 139 for the data in C1 and C2, replacing the data display with Project, then choosing Frame > Grid and entering Y for Direction.

EXCEL

Insert the specific n and p as needed in the following procedure.

Use the RANDOM NUMBER GENERATION Patterned Distribution commands in exercise 6.58a, replacing the first value with 0, the
last value with $n + 1$, the steps with 1, and the output range
with A1.

Activate cell B1 then use the Binomial Probability Distribution commands on page 258.

Use the Chart Wizard Column commands for the data in columns
A and B. Choosing the Series subcommand, input column B for
the Values and column A for the category (x) axis labels.
(Remove Series 1 if necessary.)

6.76 In order to see what happens when the normal approximation is improperly used, consider the binomial distribution with $n = 15$ and $p = 0.05$. Since $np = 0.75$, the rule of
thumb ($np > 5$ and $nq > 5$) is not satisfied. Using the binomial tables, find the probability
of one or fewer successes and compare this with the normal approximation.

6.77 Find the normal approximation for the binomial probability $P(x = 6)$, where $n = 12$ and
$p = 0.6$. Compare this to the value of $P(x = 6)$ obtained from Table 2.

6.78 Find the normal approximation for the binomial probability $P(x = 4, 5)$, where $n = 14$
and $p = 0.5$. Compare this to the value of $P(x = 4, 5)$ obtained from Table 2.

6.79 Find the normal approximation for the binomial probability $P(x \leq 8)$, where $n = 14$ and
$p = 0.4$. Compare this to the value of $P(x \leq 8)$ obtained from Table 2.

6.80 Find the normal approximation for the binomial probability $P(x \geq 9)$, where $n = 13$ and
$p = 0.7$. Compare this to the value of $P(x \geq 9)$ obtained from Table 2.

6.81 A drug manufacturer states that only 5% of the patients using a particular drug will experience side effects. Doctors at a large university hospital use the drug in treating 250 patients. What is the probability that 15 or fewer of the 250 patients experience side effects?

6.82 If 30% of all students entering a certain university drop out during or at the end of their
first year, what is the probability that more than 600 of this year's entering class of 1800
will drop out during or at the end of their first year?

6.83 A survey in the March 1994 issue of *Life* magazine indicated that 9 out of 10 Americans
pray frequently and earnestly, and almost all say God has answered their prayers. Assuming "9 out of 10" is accurate, use the normal approximation to the binomial to find the
probability that in a national survey of 1000 Americans, at least 925 will indicate that
they pray frequently and earnestly.

6.84 An article in *Life* magazine (June 1989) indicated that 60% of Americans have had a psychic experience. An example of a psychic experience is dreaming about an event before it
actually occurs. Some experts call psychic experiences precognitions, whereas others
write them off as pure coincidence. Suppose a national survey of 2000 Americans is conducted and each is asked whether or not they have had a psychic experience. Use the normal approximation to the binomial distribution to find the probability that over 1230 report such a phenomenon.

6.85 Not all NBA coaches who enjoyed lengthy careers were consistently putting together winning seasons with the teams they coached. For example, Bill Fitch, who coached for 25
seasons of professional basketball after starting his coaching career at the University of
Minnesota, won 944 games but lost 1106 while working with the Cavaliers, Celtics,
Rockets, Nets, and Clippers (*Sports Illustrated,* "Who Is Tim Floyd?", 8-3-98). If you
were to randomly select 60 box scores from the historical records of games in which Bill
Fitch coached one of the teams, what is the probability that less than half of them show
his team winning? To obtain your answer, use the normal approximation to the binomial
distribution.

6.86 According to the *Bureau of Justice Statistics Sourcebook of Criminal Justice Statistics
1992,* 4.5% of young adults reported using alcohol daily for the past 30 days. Use the
normal approximation to the binomial distribution to find the probability that, in a

Commands on pages 258 and 294 may be helpful

national poll of 1024 young adults, between 35 and 50 inclusive will indicate that they have used alcohol daily for the past 30 days.

a. Solve using normal approximation and Table 3.

b. Solve using a computer or calculator and the normal approximation method.

c. Solve using a computer or calculator and the binomial probability function.

6.87 According to the June 13, 1994, issue of *Time* magazine, the proportion of all workers who are union members equals 15.8%. Use the normal approximation to the binomial distribution to find the probability that, in a national survey of 2500 workers, at most 450 will be union members.

a. Solve using normal approximation and Table 3.

b. Solve using a computer or calculator and the normal approximation method.

Return to C H A P T E R   C A S E   S T U D Y

All normal probability distributions have the same shape and distribution relative to the mean and standard deviation. In this chapter we learned how to use the standard normal probability distribution to answer questions about all normal distributions. Let's return to the distribution of I.Q. scores discussed in the Chapter Case Study on page 277 and try out some of our new knowledge.

Exercise 6.88

a. Is the random variable, I.Q. score, a continuous variable?

b. The I.Q. score discussed seems to have a normal distribution. Find at least three bits of information about the I.Q. distribution indicating that the use of the standard normal distribution is appropriate.

c. What is the mean for this distribution of I.Q. scores?

d. Estimate the standard deviation for this distribution of I.Q. scores. Use at least two different tidbits of information from the article to obtain two separate estimates. Determine your answer.

e. What interval of I.Q. scores is classified as "superior"? What percentage of the adult population has "superior" intelligence?

f. What I.Q. score is two standard deviations above the mean? What is the classification of adults whose I.Q. score is more than two standard deviations above the mean?

g. What is the probability of randomly selecting one person from this population who is classified as "above normal"?

h. What proportion of the population is classified as "retarded"?

IN RETROSPECT

We have learned about the standard normal probability distribution, the most important family of continuous random variables. We have learned how to apply it to all other normal probability distributions and how to use it to estimate probabilities of binomial distributions. We have seen a wide variety of problems (variables) that have this normal distribution or are reasonably well approximated by it.

In the next chapter we will examine sampling distributions and learn how to use the standard normal probability to solve additional applications.

CHAPTER EXERCISES

6.89 According to Chebyshev's theorem, at least how much area is there under the standard normal distribution between $z = -2$ and $z = +2$? What is the actual area under the standard normal distribution between $z = -2$ and $z = +2$?

6.90 The middle 60% of a normally distributed population lies between what two standard scores?

6.91 Find the standard score z such that the area above the mean and below z under the normal curve is
   a. 0.3962   **b.** 0.4846   **c.** 0.3712

6.92 Find the standard score z such that the area below the mean and above z under the normal curve is
   a. 0.3212   **b.** 0.4788   **c.** 0.2700

6.93 Given that z is the standard normal variable, find the value of k such that
   a. $P(|z| > 1.68) = k$ **b.** $P(|z| < 2.15) = k$

6.94 Given that z is the standard normal variable, find the value of c such that
   a. $P(|z| > c) = 0.0384$ **b.** $P(|z| < c) = 0.8740$

6.95 Find the following values of z:
   a. $z_{(0.12)}$ **b.** $z_{(0.28)}$ **c.** $z_{(0.85)}$ **d.** $z_{(0.99)}$

6.96 Find the area under the normal curve that lies between the following pairs of z-values:
   a. $z = -3.00$ and $z = 3.00$ **b.** $z_{(0.975)}$ and $z_{(0.025)}$ **c.** $z_{(0.10)}$ and $z_{(0.01)}$

6.97 The length of life of a certain type of refrigerator is approximately normally distributed with a mean of 4.8 years and a standard deviation of 1.3 years.
   a. If this machine is guaranteed for 2 years, what is the probability that the machine you purchased will require replacement under the guarantee?
   b. What period of time should the manufacturer give as a guarantee if it is willing to replace only 0.5% of the machines?

6.98 An article in *USA Today* (4-4-91) quoted a study involving 3365 people in Minneapolis–St. Paul between 1980 and 1982 and another 4545 between 1985 and 1987. It found that the average cholesterol level for males was 200. The authors of the study say the results of their study are probably similar nationwide. Assume that the cholesterol values for males in the United States are normally distributed with a mean equal to 200 and a standard deviation equal to 25.
   a. What percentage have readings between 150 and 225?
   b. What percentage have readings that exceed 250?

6.99 A machine is programmed to fill 10-oz containers with a cleanser. However, the variability inherent in any machine causes the actual amounts of fill to vary. The distribution is normal with a standard deviation of 0.02 oz. What must the mean amount μ be in order that only 5% of the containers receive less than 10 oz?

6.100 In a large industrial complex, the maintenance department has been instructed to replace light bulbs before they burn out. It is known that the life of light bulbs is normally distributed with a mean life of 900 hours of use and a standard deviation of 75 hours. When should the light bulbs be replaced so that no more than 10% of them will burn out while in use?

6.101 Suppose that x has a binomial distribution with $n = 25$ and $p = 0.3$.
   a. Explain why the normal approximation is reasonable.
   b. Find the mean and standard deviation of the normal distribution that is used in the approximation.

6.102 Let x be a binomial random variable for $n = 30$ and $p = 0.1$.
   a. Explain why the normal approximation is not reasonable.
   b. Find the function used to calculate the probability of any x from $x = 0$ to $x = 30$.
   c. Use a computer or calculator to list the probability distribution.

6.103 **a.** Use a computer or calculator to list the binomial probabilities for the distribution where $n = 50$ and $p = 0.1$.
   b. Use the results from (a) and find $P(x \leq 6)$.
   c. Find the normal approximation for $P(x \leq 6)$, and compare the results with those in (b).

6.104 **a.** Use a computer or calculator to list both the probability distribution and the cumulative probability distribution for the binomial probability experiment with $n = 40$ and $p = 0.4$.
   b. Explain the relationship between the two distributions found in (a).
   c. If you could use only one of these lists when solving problems, which one would you prefer and why?

6.105 Consider the binomial experiment with $n = 300$ and $p = 0.2$.

> Use the cumulative probability commands.

   a. Set up, but do not evaluate, the probability expression for 75 or fewer successes in the 300 trials.
   b. Use a computer or calculator to find $P(x \leq 75)$ using the binomial probability function.
   c. Use a computer or calculator to find $P(x \leq 75)$ using the normal approximation.
   d. Compare the answers in (b) and (c).

6.106 The grades on an examination whose mean is 525 and whose standard deviation is 80 are normally distributed.
   a. Anyone who scores below 350 will be retested. What percentage does this represent?
   b. The top 12% are to receive a special commendation. What score must be surpassed to receive this special commendation?
   c. The interquartile range of a distribution is the difference between Q_1 and Q_3, $Q_3 - Q_1$. Find the interquartile range for the grades on this examination.
   d. Find the grade such that only 1 out of 500 will score above it.

6.107 A soft-drink vending machine can be regulated to dispense an average of μ oz of soft drink per glass.
   a. If the ounces dispensed per glass are normally distributed with a standard deviation of 0.2 oz, find the setting for μ that will allow 6-oz glass to hold (without overflowing) the amount dispensed 99% of the time.
   b. Use a computer or calculator to simulate drawing a sample of 40 glasses of soft drink from the machine [set using your answer to (a)].

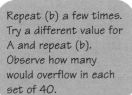

> Repeat (b) a few times. Try a different value for A and repeat (b). Observe how many would overflow in each set of 40.

MINITAB

Use the generate RANDOM DATA commands on page 292 replacing n with 40, store in with C1, the mean with the value calculated in (a), and standard deviation with 0.2.

Use the HISTOGRAM commands on page 56 for the data in C1, entering cutpoints from 5 to 6.2 in increments of 0.05 (5:6.2/0.05).

EXCEL

Use the Normal RANDOM NUMBER GENERATION commands on page 292 replacing n with 40, the mean with the value calculated in (a), the standard deviation with 0.2, and the output range with A1.

Use the RANDOM NUMBER GENERATION Patterned Distribution on page 297, replacing the first value with 5, the last value with 6.25, the steps with 0.05, and the output range with B1.

Use the HISTOGRAM commands with column A as the input
range and column B as the bin range.

TI-83

Use the 6:randNorm commands on page 292 replacing the
mean with the value calculated in (a), the standard deviation with 0.2, and the number of trials with 40. Store
in L1.

Use the HISTOGRAM commands on page 58 for the data in L1,
entering WINDOW VALUES: 5, 6.2, 0.05, 1, 10, 1, 1.

    c. What percentage of your sample would have overflowed the cup?
    d. Does your sample seem to indicate the setting for μ is going to work? Explain.

6.108 A company assets that 80% of the customers who purchase its special lawn mower will
have no repairs during the first two years of ownership. Your personal study has shown that
only 70 of the 100 in your sample lasted the two years without repair expenses. What is the
probability of your sample outcome or less if the actual expenses-free percentage is 80%?

6.109 A test-scoring machine is known to record an incorrect grade on 5% of the exams it
grades. Find, by the appropriate method, the probability that the machine records
    a. 3 wrong grades in a set of 5 exams.
    b. no more than 3 wrong grades in a set of 5 exams.
    c. no more than 3 wrong grades in a set of 15 exams.
    d. no more than 3 wrong grades in a set of 150 exams.

6.110 It is believed that 58% of married couples with children agree on methods of disciplin-
ing their children. Assuming this to be the case, what is the probability that in a random
survey of 200 married couples, we would find
    a. exactly 110 couples who agree?
    b. fewer than 110 couples who agree?
    c. more than 100 couples who agree?

6.111 Infant mortality rates are often used to assess quality of life and adequacy of health care.
The rate is based on the number of infant deaths that occur for every 1000 births. Prior
to President Clinton's trips to China and Russia in the summer of 1998, *Newsweek* magazine published a table showing the infant mortality rates of eight nations throughout the
world, including China and Russia, in 1996:

Nation	Infant Mortality
China	33
Germany	5
India	65
Japan	4
Mexico	32
Russia	17
S. Africa	49
United States	7

Source: Newsweek, "China by the Numbers: Portrait of a Nation," June 29, 1998.

Suppose the next 2000 births within each nation are tracked for the occurrence of infant
deaths.
    a. Construct a table showing the mean and standard deviation of the associated bino-
        mial distributions.
    b. In the final column of the table, find the probability that at least 70 infants from the
        samples within each nation will become casualties that contribute to the nation's
        mortality rate. Show all work.

6.112 If 60% of the registered voters plan to vote for Ralph Brown for mayor of a large city, what is the probability that less than half of the voters, in a poll of 200 registered voters, plan to vote for Ralph Brown?

6.113 A survey conducted by the Association of Executive Search Consultants revealed that 75% of all chief executive officers believe that corporations should have fast-track training programs installed to help develop especially talented employees. At the same time, the study found that only 47% of the companies actually have such programs operating at their companies. Average annual sales of the companies in the sample were $2.3 billion. (*Fortune*, "How to Tame the Fiercest Headhunter," 7-20-98.)

Suppose you randomly selected 50 of the questionnaires returned by the collection of CEOs. Use the normal approximation to the binomial distribution in order to find the probability that from within your collection:
  a. over 35 of the CEOs feel that corporations should have a fast-track program installed.
  b. less than 25 of the companies have a fast-track program in operation.
  c. between 30 and 40 of the CEOs feel that corporations should have a fast-track program installed.
  d. between 20 and 30 of the companies have a fast-track program in operation.

6.114 The following triangular distribution provides an approximation to the normal distribution. Line segment l_1 has the equation $y = \frac{1}{9}x + \frac{1}{3}$, and segment l_2 has the equation $y = -\frac{1}{9}x + \frac{1}{3}$.

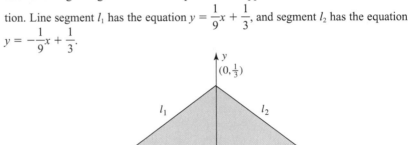

  a. Find the area under the entire triangular distribution.
  b. Find the area under the triangular distribution between 0 and 2.
  c. Find the area under the standard normal distribution between 0 and 2.
  d. Discuss the effectiveness of this "triangular" approximation.

VOCABULARY LIST

Be able to define each term. Pay special attention to the key terms, which are printed in **red.** In addition, describe in your own words, and give an example of, each term. Your examples should not be the ones given in class or in the textbook.

The bracketed numbers indicate the chapters in which the term first appeared, but you should define the terms again to show increased understanding. Page numbers indicate the first appearance of the term in Chapter 6.

area representation for probability (p. 279)
bell-shaped curve (p. 278)
binomial distribution [5] (p. 304)
binomial probability (p. 304)
continuity correction factor (p. 304)
continuous random variable (p. 278, 304)
discrete random variable [1, 5] (p. 278, 304)
normal approximation of binomial (p. 304)
normal curve (p. 278)

normal distribution (p. 278)
percentage (p. 279)
probability [4] (p. 279)
proportion (p. 279)
random variable [5] (p. 278)
standard normal distribution (p. 279, 286)
standard score [2] (p. 279, 286)
z-score [2] (p. 279, 286)

CHAPTER PRACTICE TEST

Part I: Knowing the Definitions

(1–10) Answer "True" if the statement is always true. If the statement is not always true, replace the words shown in bold with words that make the statement always true.

6.1 The normal probability distribution is symmetric about **zero.**

6.2 The total area under the curve of any normal distribution is **1.0**.

6.3 The theoretical probability that a particular value of a **continuous** random variable will occur is exactly zero.

6.4 The unit of measure for the standard score is the **same as the unit of measure of the data.**

6.5 All **normal** distributions have the same general probability function and distribution.

6.6 When using the notation $z_{(0.05)}$, the number in parentheses is the measure of the area to the **left** of the z-score.

6.7 Standard normal scores have a mean of **one** and a standard deviation of **zero.**

6.8 Probability distributions of **all** continuous random variables are normally distributed.

6.9 We are able to add and subtract the areas under the curve of a continuous distribution because these areas represent probabilities of **independent** events.

6.10 The most common distribution of a continuous random variable is the **binomial** probability.

Part II: Applying the Concepts

6.11 Find the following probabilities for z, the standard normal score:
  a. $P(0 < z < 2.42)$ **b.** $P(z < 1.38)$ **c.** $P(z < -1.27)$ **d.** $P(-1.35 < z < 2.72)$

6.12 Find the value of the z-score
  a. $P(z > ?) = 0.2643$ **b.** $P(z < ?) = 0.17$ **c.** $z_{(0.04)}$

6.13 Using the symbolic notation $z(\alpha)$, give the symbolic name for the z-score shown.

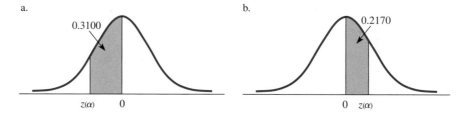

6.14 The lifetime of a flashlight battery is normally distributed about a mean of 35.6 hr with a standard deviation of 5.4 hr. Kevin selected one of these batteries at random and tested it. What is the probability that this one battery will last less than 40.0 hr?

6.15 The amount of time, x, spent commuting daily, one way, to college by students is believed to have a mean of 22 min with a standard deviation of 9 min. If the length of time spent commuting is approximately normally distributed, find the time, x, that separates the 25% who spend the most time commuting from the rest of the commuters.

6.16 Thousands of high school students take the SAT each year. The scores attained by the students in a certain city are approximately normally distributed with a mean of 490 and a standard deviation of 70. Find:
  a. the percentage of students who score between 600 and 700.
  b. the percentage of students who score less than 650.
  c. the 3rd quartile   **d.** the 15th percentile, P_{15} **e.** the 95th percentile, P_{95}

Part III: Understanding the Concepts

6.17 In 50 words, describe the standard normal distribution.

6.18 Describe the meaning of the symbol $z(\alpha)$.

6.19 Explain why the standard normal distribution, as computed in Table 3, Appendix B, can be used to find probabilities for all normal distributions.

7

Sample Variability

CHAPTER OUTLINE

7.1 Sampling Distributions
A distribution of values for a **sample statistic** is obtained by repeatedly sampling a population.

7.2 The Central Limit Theorem
The theorem describes the sampling distribution of **sample means.**

7.3 Application of the Central Limit Theorem
The behavior of the sample means is **predictable.**

CHAPTER CASE STUDY

Galluping Attitudes

Since its founding in 1935, the American Institute of Public Opinion, better known as the Gallup Poll, has put approximately 20,000 questions to more than 2 million people. One of the most interesting social changes to follow through the years is that of Americans' attitudes toward women at work and in politics. Herewith, an abbreviated reading of Gallup's progress report.

A WOMAN PRESIDENT

1937

Would you vote for a woman for President, if she qualified in every other respect?

		By Sex	Yes	No
Yes	34%	Men	27%	73%
No	66%	Women	41%	49%

1955

If the party whose candidate you most often support nominated a woman for President of the United States, would you vote for her if she seemed best qualified for the job?

		By Sex	Yes	No
Yes	52%	Men	47%	48%
No	44%	Women	57%	40%

1971

If your party nominated a woman for President, would you vote for her if she were qualified for the job?

		By Sex	Yes	No
Yes	66%	Men	65%	35%
No	29%	Women	67%	33%

1978

If your party nominated a woman for President, would you vote for her if she were qualified for the job?

		By Sex	Yes	No
Yes	76%	Men	76%	19%
No	19%	Women	77%	18%

1987

If your party nominated a woman for President, would you vote for her if she were qualified for the job?

		By Sex	Yes	No
Yes	82%	Men	81%	13%
No	12%	Women	83%	12%

2000   [What is your prediction?]

Would you vote for a woman for President, if she qualified in every other respect?

		By Sex	Yes	No
Yes	x %	Men	a %	b %
No	z %	Women	c %	d %

SELECTED NATIONAL TREND

Year	1937	1949	1955	1967	1967	1971	1975	1978	1983	1984	1987	2000
Yes	34%	48%	52%	57%	54%	66%	73%	76%	80%	78%	82%	xx %
No	66%	48%	44%	39%	39%	29%	23%	19%	16%	17%	12%	zz %

Source: Copyright by the *Gallup Report*. Reprinted by permission.

There are many reasons for repeatedly drawing samples from a population. "Galluping Attitudes" shows the Gallup Poll keeping track of America's attitude toward the possibility of having a woman President. In industry, products are constantly being sampled so that their quality can be monitored. However, in this chapter we talk about repeatedly drawing samples from a population for the purpose of better understanding the "behavior" of the sample mean. (See Exercise 7.37, p. 337.)

CHAPTER OBJECTIVES

In Chapters 1 and 2 we discussed how to obtain and describe a sample. The description of the sample data is accomplished by using three basic concepts: (1) measures of central tendency (the mean is the most popularly used sample statistic), (2) measures of dispersion (the standard deviation is most commonly used), and (3) kind of distribution (normal, skewed, rectangular, and so on). The question that seems to follow is, "What can be deduced about the statistical population from which the sample is taken?"

To put this query at a more practical level, suppose that we have just taken a sample of 25 rivets made for the construction of airplanes. The rivets were tested for shearing strength, and the force required to break each rivet was the response variable. The various descriptive measures (mean, standard deviation, type of distribution) can be found for this sample. However, it is not the sample itself that we are interested in. The rivets that were tested were destroyed during the test, so they can no longer be used in the construction of airplanes. What we are trying to determine, from this sample, is information about the total population, and we certainly cannot test every rivet that is produced (there are too many, and there would be none left for building airplanes). Therefore, we must somehow deduce information, that is, make inferences about the population based on the results observed in the sample.

Suppose that we take another sample of 25 rivets from the same supply and test them by the same procedure. Do you think that we would obtain the same sample mean from the second sample that we obtained from the first? the same standard deviation?

After considering these questions we might suspect that we need to investigate the variability in the sample statistics from sample to sample. Thus, we need to find (1) measures of central tendency for the sample statistics of importance, (2) measures of dispersion for the sample statistics, and (3) the pattern of variability (distribution) of the sample statistics. Once we have this information, we will be better able to predict (make inferences about) the population parameters.

The objective of this chapter is to study the measures and the patterns of variability for the distribution formed by repeatedly observing values of a sample mean. (Look back at your results for Exercise 6.60, p. 298. This exercise dealt with repeated samples taken from the same population. If you didn't complete that exercise before, it might be helpful to do so now.)

7.1   Sampling Distributions

To make inferences about a population, we need to discuss sample results a little more. A sample mean $\bar{x}$ is obtained from a sample. Do you expect that this value, $\bar{x}$, is exactly equal to the value of the population mean μ? Your answer should be "no." We do not expect that to happen, but we will be satisfied with our sample results if the sample mean is "close" to the value of the population mean. Let's consider a second question: If a second sample is taken, will the second sample have a mean equal to the population mean? Equal to the first sample mean? Again, no, we do not expect it to be equal to the population mean, nor do we expect the second sample mean to be a repeat of the first one. We do, however, again expect the values to be "close." (This argument should hold for any other sample statistic and its corresponding population value.)

The next questions should already have come to mind: What is "close"? How do we determine (and measure) this closeness? Just how would **repeated sample statistics** be distributed? To answer these questions we must look at a *sampling distribution*.

SAMPLING DISTRIBUTION OF A SAMPLE STATISTIC

The distribution of values for a sample statistic obtained from repeated samples, all of the same size and all drawn from the same population.

ILLUSTRATION 7.1 ▼

Let's consider a very small finite population to illustrate the concept of a sampling distribution: the set of single-digit even integers, $\{0, 2, 4, 6, 8\}$, and all possible samples of size 2. We will look at two different sampling distributions that might be formed: (1) the sampling distribution of sample means and (2) the sampling distribution of sample ranges.

First we need to list all possible samples of size 2; there are 25 possible samples:

$\{0, 0\}$	$\{2, 0\}$	$\{4, 0\}$	$\{6, 0\}$	$\{8, 0\}$
$\{0, 2\}$	$\{2, 2\}$	$\{4, 2\}$	$\{6, 2\}$	$\{8, 2\}$
$\{0, 4\}$	$\{2, 4\}$	$\{4, 4\}$	$\{6, 4\}$	$\{8, 4\}$
$\{0, 6\}$	$\{2, 6\}$	$\{4, 6\}$	$\{6, 6\}$	$\{8, 6\}$
$\{0, 8\}$	$\{2, 8\}$	$\{4, 8\}$	$\{6, 8\}$	$\{8, 8\}$

> Samples are drawn with replacement.

Each of these samples has a mean $\bar{x}$. These means are, respectively:

0	1	2	3	4
1	2	3	4	5
2	3	4	5	6
3	4	5	6	7
4	5	6	7	8

Exercise 7.1

Explain why the samples are equally likely; why $P(0) = 0.04$; and why $P(2) = 0.12$. (Illus. 7.1)

Each of these samples is equally likely, and thus each of the 25 sample means can be assigned a probability of $\frac{1}{25} = 0.04$. (Why? See Exercise 7.1.) The **sampling distribution for sample means** is shown in Table 7.1 as a **probability distribution** and shown in Figure 7.1 as a histogram.

Table 7.1   Sampling Distribution of Sample Means

$\bar{x}$	$P(\bar{x})$
0	0.04
1	0.08
2	0.12
3	0.16
4	0.20
5	0.16
6	0.12
7	0.08
8	0.04

Figure 7.1

Histogram: Sampling Distribution of Sample Means

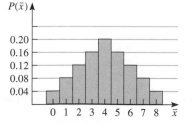

For the same set of all possible samples of size 2, let's find the sampling distribution for sample ranges. Each sample has a range R. The ranges are:

0	2	4	6	8
2	0	2	4	6
4	2	0	2	4
6	4	2	0	2
8	6	4	2	0

Again, each of these 25 sample ranges has a probability of 0.04. Table 7.2 shows the sampling distribution of sample ranges as a probability distribution, and Figure 7.2 shows the sampling distribution as a histogram.

Table 7.2 Sampling Distribution of Sample Ranges

R	$P(R)$
0	0.20
2	0.32
4	0.24
6	0.16
8	0.08

Figure 7.2
Histogram: Sampling Distribution of Sample Ranges

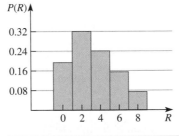

Illustration 7.1 is theoretical in nature and therefore expressed in probabilities. Since this population is quite small, it is easy to list all 25 possible samples of size 2 (a sample space) and assign probabilities. It is not always possible to do this.

Now, let's empirically (that is, by experimentation) investigate another sampling distribution.

ILLUSTRATION 7.2 ▼

Let's consider a population consisting of five equally likely integers: 1, 2, 3, 4, and 5. Let's observe a portion of the sampling distribution of sample means when 30 samples of size 5 are randomly selected. Figure 7.3 shows a histogram representation of the population.

Table 7.3 shows 30 samples and their means. The resulting sampling distribution, a **frequency distribution,** of sample means is shown in Figure 7.4. Notice that this distribution of sample means does not look like the population. Rather, it seems to display characteristics of a normal distribution; it is mounded and nearly symmetric about its mean (approximately 3.0).   ▲

Figure 7.3

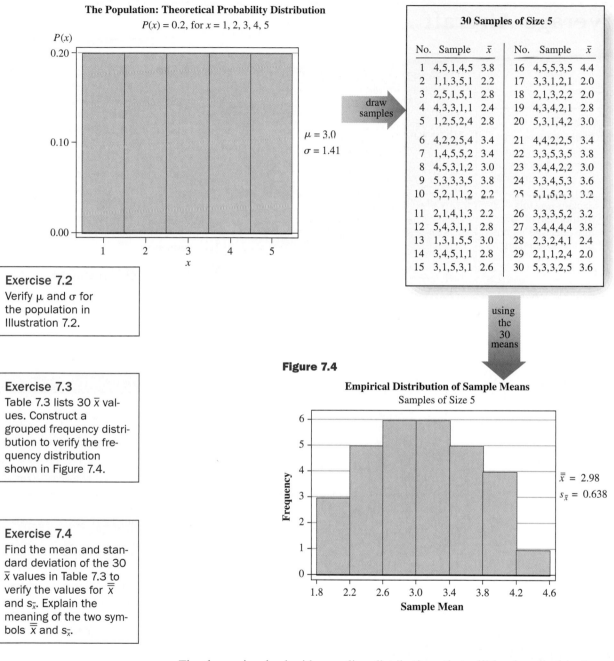

The Population: Theoretical Probability Distribution

$P(x) = 0.2$, for $x = 1, 2, 3, 4, 5$

$\mu = 3.0$
$\sigma = 1.41$

draw samples

Table 7.3

	30 Samples of Size 5				
No.	Sample	$\bar{x}$	No.	Sample	$\bar{x}$
1	4,5,1,4,5	3.8	16	4,5,5,3,5	4.4
2	1,1,3,5,1	2.2	17	3,3,1,2,1	2.0
3	2,5,1,5,1	2.8	18	2,1,3,2,2	2.0
4	4,3,3,1,1	2.4	19	4,3,4,2,1	2.8
5	1,2,5,2,4	2.8	20	5,3,1,4,2	3.0
6	4,2,2,5,4	3.4	21	4,4,2,2,5	3.4
7	1,4,5,5,2	3.4	22	3,3,5,3,5	3.8
8	4,5,3,1,2	3.0	23	3,4,4,2,2	3.0
9	5,3,3,3,5	3.8	24	3,3,4,5,3	3.6
10	5,2,1,1,2	2.2	25	5,1,5,2,3	3.2
11	2,1,4,1,3	2.2	26	3,3,3,5,2	3.2
12	5,4,3,1,1	2.8	27	3,4,4,4,4	3.8
13	1,3,1,5,5	3.0	28	2,3,2,4,1	2.4
14	3,4,5,1,1	2.8	29	2,1,1,2,4	2.0
15	3,1,5,3,1	2.6	30	5,3,3,2,5	3.6

using the 30 means

Figure 7.4

Empirical Distribution of Sample Means

Samples of Size 5

$\bar{\bar{x}} = 2.98$
$s_{\bar{x}} = 0.638$

Exercise 7.2

Verify μ and σ for the population in Illustration 7.2.

Exercise 7.3

Table 7.3 lists 30 $\bar{x}$ values. Construct a grouped frequency distribution to verify the frequency distribution shown in Figure 7.4.

Exercise 7.4

Find the mean and standard deviation of the 30 $\bar{x}$ values in Table 7.3 to verify the values for $\bar{\bar{x}}$ and $s_{\bar{x}}$. Explain the meaning of the two symbols $\bar{\bar{x}}$ and $s_{\bar{x}}$.

The theory involved with sampling distributions that will be described in the remainder of this chapter requires *random sampling*.

RANDOM SAMPLE

A sample obtained in such a way that each possible sample of fixed size n has an equal probability of being selected. (See p. 22.)

Case Study 7.1

Average Aircraft Age

Exercise 7.5

a. Construct a frequency histogram of average fleet age using class boundaries of 4.5, 6.5, ...
b. Explain why this distribution of "average fleet age" is not part of a sampling distribution.

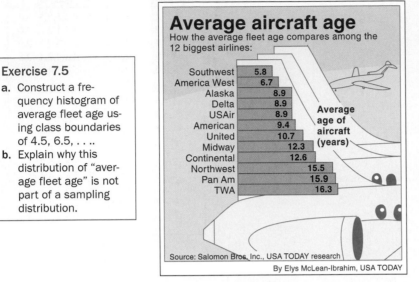

Average aircraft age

How the average fleet age compares among the 12 biggest airlines:

Airline	Average age of aircraft (years)
Southwest	5.8
America West	6.7
Alaska	8.9
Delta	8.9
USAir	8.9
American	9.4
United	10.7
Midway	12.3
Continental	12.6
Northwest	15.5
Pan Am	15.9
TWA	16.3

Source: Salomon Bros. Inc., USA TODAY research

By Elys McLean-Ibrahim, USA TODAY

USA Today's "Average aircraft age" shows the average age of the aircraft that make up the fleets of the 12 biggest airline companies. Each company reported their own average fleet age. The information is shown in the form of a bar graph so that each company's average fleet age can be easily visualized and compared to the others. Using the variable "average fleet age," a frequency distribution or a histogram could have been used to present the data; however, that distribution would not be part of a sampling distribution.

Source: Copyright 1991, USA TODAY.

Reprinted with permission.

EXERCISES • • • • • • • • • • • • • •

7.6   a. What is the sampling distribution of sample means?
   b. A sample of size 3 is taken from a population and the sample mean found. Describe how this sample mean is related to the sampling distribution of sample means.

7.7   Consider the set of odd single-digit integers {1, 3, 5, 7, 9}.
   a. Make a list of all samples of size 2 that can be drawn from this set of integers. (Sample with replacement; that is, the first number is drawn, observed, then replaced before the next drawing.)
   b. Construct the sampling distribution of sample means for samples of size 2 selected from this set.
   c. Construct the sampling distributions of sample ranges for samples of size 2.

7.8   Consider the set of even single-digit integers {0, 2, 4, 6, 8}.
   a. Make a list of all the possible samples of size 3 that can be drawn from this set of integers. (Sample with replacement; that is, the first number is drawn, observed, then replaced before the next drawing.)
   b. Construct the sampling distribution of the sample medians for samples of size 3.
   c. Construct the sampling distribution of the sample means for samples of size 3.

7.9   Using the telephone numbers listed in your local directory as your population, obtain randomly 20 samples of size 3. From each telephone number identified as a source, take the fourth, fifth, and sixth digits. (For example, for 245-8268, you would take the 8, the 2, and the 6 as your sample of size 3.)
   a. Calculate the mean of the 20 samples.
   b. Draw a histogram showing the 20 sample means. (Use classes −0.5 to 0.5, 0.5 to 1.5, 1.5 to 2.5, and so on.)
   c. Describe the distribution of $\bar{x}$'s that you see in (b) (shape of distribution, center, amount of dispersion).
   d. Draw 20 more samples and add the 20 new $\bar{x}$'s to the histogram in (b). Describe the distribution that seems to be developing.

7.10 Using a set of 5 dice, roll the dice and determine the mean number of dots showing on the five dice. Repeat the experiment until you have 25 sample means.
   a. Draw a dotplot showing the distribution of the 25 sample means. (See Illustration 7.2, p. 320.)
   b. Describe the distribution of $\bar{x}$'s in (a).
   c. Repeat the experiment to obtain 25 more sample means and add these 25 $\bar{x}$'s to your dotplot. Describe the distribution of 50 means.

7.11 From the table of random numbers in Appendix B, construct another table showing 20 sets of 5 randomly selected single-digit integers. Find the mean of each set, the grand mean, and compare this value to the theoretical population mean μ using the absolute difference and the % error. Show all work.

7.12 **a.** Simulate (using a computer or a random numbers table) the drawing of 100 samples, each size 5, from the uniform probability distribution of single-digit integers, 0 to 9.
   b. Find the mean for each sample.
   c. Construct a histogram of the sample means. (Use integer values as class marks.)
   d. Describe the sampling distribution shown in the histogram.

MINITAB

a. Use the Integer RANDOM DATA commands on page 182 replacing generate with 100, store in with C1-C5, minimum value with 0 and maximum value with 9.

b. Choose: **Calc > Row Statistics**
   Select: **Mean**
   Enter: Input variables: **C1-C5**
             Store result in: **C6**

c. Use the HISTOGRAM commands on page 56 for the data in C6, entering midpoints from 0 to 9 in increments of 1.

EXCEL

a. Input 0 through 9 into column A and corresponding 0.1 into column B; then continue with:

   Choose: **Tools > Data Analysis > Random Number Generation > OK**
   Enter: Number of Variables: **5**
             Number of Random Numbers: **100**
             Distribution: **Discrete**
             Value and Probability Input Range: **(A1:B10 or select cells)**
   Select: **Output Range**
   Enter: **(C1 or select cell)**

b. Activate cell H1.

   Choose: **Paste function, f_x > Statistical > AVERAGE > OK**
   Enter: Number1: **(C1:G1 or select cells)   > OK**
   Drag: **Bottom right corner of average value box down to give other averages**

c. Use the HISTOGRAM commands on page 57 with column H as the input range, column A as the bin range, and I1 as output range.

7.13 **a.** Use a computer to draw 200 random samples, each of size 10, from the normal probability distribution with a mean 100 and standard deviation 20.
   b. Find the mean for each sample.
   c. Construct a frequency histogram of the 200 sample means.
   d. Describe the sampling distribution shown in the histogram.

MINITAB

a. Use the Normal RANDOM DATA commands on page 292 replacing generate with 200, store in with C1-C10, mean with 100 and standard deviation with 20.

b. Choose: **Calc > Row Statistics**
   Select: **Mean**
   Enter: Input variables: **C1-C10**
          Store result in: **C11**

c. Use the HISTOGRAM commands on page 56 for the data in C11, entering cutpoints from 74.8 to 125.2 in increments of 6.3.

EXCEL

a. Use the Normal RANDOM NUMBER GENERATION commands on page 292 replacing number of variables with 10, number of random numbers with 200, mean with 100, and standard deviation with 20.

b. Activate cell K1.

   Choose: **Paste function, f_x > Statistical > AVERAGE > OK**
   Enter: **Number1: (A1:J1 or select cells) > OK**
   Drag: **Bottom right corner of average value box down to give other averages**

c. Use the RANDOM NUMBER GENERATION Patterned Distribution commands in Exercise 6.58(a) on page 297, replacing the first value with 74.8, the last value with 125.2, the steps with 6.3, and the output range with L1. Use the HISTOGRAM commands on page 57 with column K as the input range, column L as the bin range, and M1 as output range.

7.2 The Central Limit Theorem

On the preceding pages we discussed the sampling distributions of two statistics, sample means and sample ranges. There are many others that could be discussed; however, the only sampling distribution of concern to us at this time is the sampling distribution of sample means.

SAMPLING DISTRIBUTION OF SAMPLE MEANS

If all possible random samples, each of size n, are taken from any population with a mean μ and a standard deviation σ, the sampling distribution of sample means will

1. have a mean $\mu_{\bar{x}}$ equal to μ.
2. have a standard deviation $\sigma_{\bar{x}}$ equal to $\dfrac{\sigma}{\sqrt{n}}$.

Further, if the sampled population has a normal distribution , then the sampling distribution of $\bar{x}$ will also be normal for samples of all sizes.

Very useful information!

This is a very interesting two-part statement. The first part tells us about the relationship between the population mean and standard deviation, and the sampling distribution mean and standard deviation for all sampling distributions of sample means. The second part indicates that this information is not always useful. Stated differently it says that the mean value of only a few observations will be normally distributed when samples are drawn from a normally distributed population, but will not be normally distributed when the sampled population is uniform, skewed, or otherwise not normal. However, the central limit theorem gives us some additional and very important information about the sampling distribution of sample means.

CENTRAL LIMIT THEOREM

The sampling distribution of sample means will more closely resemble the normal as the sample size increases.

> Truly amazing; $\bar{x}$ is approximately normally distributed when n is large enough, no matter what shape the population is.

If the sampled distribution is normal, then the sampling distribution of sample means is normal, as stated above, and the Central Limit Theorem (CLT) does not apply. Interestingly, if the sampled population is not normal, the sampling distribution will still be approximately normally distributed under the right conditions. If the sampled distribution is nearly normal, the $\bar{x}$ distribution is approximately normal for fairly small n (possibly 15). When the sampled distribution lacks symmetry, n may have to be quite large (maybe 50 or more) before the normal distribution provides a satisfactory approximation.

STANDARD ERROR OF THE MEAN ($\sigma_{\bar{x}}$)

The standard deviation of the sampling distribution of sample means.

By combining the preceding information, we can describe the sampling distribution of $\bar{x}$ completely: (1) the location of the center (mean), (2) a measure of spread indicating how widely it is dispersed (standard deviation), and (3) an indication of how it is distributed.

1. $\mu_{\bar{x}} = \mu$; the mean of the sampling distribution ($\mu_{\bar{x}}$) is equal to the mean of the population (μ).

2. $\sigma_{\bar{x}} = \dfrac{\sigma}{\sqrt{n}}$; the standard error of the mean is equal to the standard deviation of the population (σ) divided by the square root of the sample size, n.

3. The distribution of sample means is normal when the parent population is normally distributed, and the CLT tells us the distribution of sample means becomes approximately normal (regardless of the shape of the parent population), when the sample size is large enough.

NOTE The n referred to is the size of each sample in the sampling distribution. (The number of repeated samples used in an empirical situation has no effect on the standard error.)

We do not show the proof for the above three facts in this text; however, their validity will be demonstrated by examining two illustrations. For the first illustration, let's consider a population for which the theoretical sampling distribution of all possible samples can be constructed.

ILLUSTRATION 7.3 ▼

Let's consider all possible samples of size 2 that could be drawn from a population that contains the three numbers 2, 4, and 6. First let's look at the population itself: Construct a histogram to picture its distribution, Figure 7.5; calculate the mean μ and the standard deviation σ, Table 7.4. (Remember: We must use the techniques from Chapter 5 for discrete probability distributions.)

Table 7.5 lists all the possible samples that could be drawn if samples of size 2 were to be drawn from this population. (One number is drawn, observed, and then returned to the population before the second number is drawn.) Table 7.5 also lists the means of

Figure 7.5

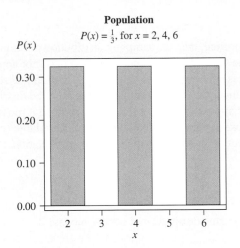

Population

$P(x) = \frac{1}{3}$, for $x = 2, 4, 6$

Table 7.5

All 9 possible samples of size 2

Sample	$\bar{x}$	Sample	$\bar{x}$	Sample	$\bar{x}$
2, 2	2	4, 2	3	6, 2	4
2, 4	3	4, 4	4	6, 4	5
2, 6	4	4, 6	5	6, 6	6

Table 7.6

Extensions Table for $\bar{x}$	$\bar{x}$	$P(\bar{x})$	$\bar{x}P(\bar{x})$	$\bar{x}^2P(\bar{x})$
	2	$\frac{1}{9}$	$\frac{2}{9}$	$\frac{4}{9}$
	3	$\frac{2}{9}$	$\frac{6}{9}$	$\frac{18}{9}$
	4	$\frac{3}{9}$	$\frac{12}{9}$	$\frac{48}{9}$
	5	$\frac{2}{9}$	$\frac{10}{9}$	$\frac{50}{9}$
	6	$\frac{1}{9}$	$\frac{6}{9}$	$\frac{36}{9}$
	Σ	$\frac{9}{9}$	$\frac{36}{9}$	$\frac{156}{9}$
		1.0	4.0	17.3$\bar{3}$

Table 7.4

Extensions Table for x	x	$P(x)$	$xP(x)$	$x^2P(x)$
	2	$\frac{1}{3}$	$\frac{2}{3}$	$\frac{4}{3}$
	4	$\frac{1}{3}$	$\frac{4}{3}$	$\frac{16}{3}$
	6	$\frac{1}{3}$	$\frac{6}{3}$	$\frac{36}{3}$
	Σ	$\frac{3}{3}$	$\frac{12}{3}$	$\frac{56}{3}$
		1.0	4.0	18.6$\bar{6}$

$\mu = 4.0$

$\sigma = \sqrt{18.6\bar{6} - (4.0)^2} = \sqrt{2.6\bar{6}} = 1.63$

 draw samples

$\mu_{\bar{x}} = 4.0$

$\sigma_{\bar{x}} = \sqrt{17.3\bar{3} - (4.0)^2} = \sqrt{1.3\bar{3}} = 1.15$

Figure 7.6

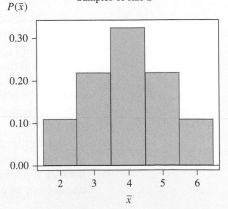

Sampling Distribution of Sample Means

Samples of size 2

these samples. The probability distribution for these means and the extensions are given in Table 7.6, along with the calculation of the mean and standard error of the mean for the sampling distribution. The histogram for the sampling distribution of sample means is shown in Figure 7.6.

Let's now check the truth of the three specific facts about the sampling distribution of sample means:

1. The mean $\mu_{\bar{x}}$ of the sampling distribution will equal the mean μ of the population: Both μ and $\mu_{\bar{x}}$ have the value **4.0**.

2. The standard error of the mean $\sigma_{\bar{x}}$ for the sampling distribution will equal the standard deviation σ of the population divided by the square root of the sample size n: $\sigma_{\bar{x}} = \mathbf{1.15}$ and $\sigma = 1.63$, $n = 2$, $\dfrac{\sigma}{\sqrt{n}} = \dfrac{1.63}{\sqrt{2}} = \mathbf{1.15}$; they are equal: $\sigma_{\bar{x}} = \dfrac{\sigma}{\sqrt{n}}$.

3. The distribution will become approximately normally distributed: The histogram (Figure 7.6) very strongly suggests normality. ▲

Illustration 7.3, a theoretical situation, suggests that all three facts appear to hold true. Do these three facts occur when actual data are collected? Let's look back at Illustration 7.2 (p. 320) and see if all three facts are supported by the empirical sampling distribution that occurred there. First, let's look at the population, the theoretical probability distribution from which the samples in Illustration 7.2 were taken. Figure 7.3 is the histogram showing the probability distribution for randomly selecting data from the population of equally likely integers 1, 2, 3, 4, 5. The population mean μ equals 3.0. The population standard deviation σ is $\sqrt{2}$, or 1.41. The population has a uniform distribution.

Now let's look at the empirical sampling distribution of the 30 sample means found in Illustration 7.2. Using the 30 values of $\bar{x}$ in Table 7.3, the observed mean of the $\bar{x}$'s, $\bar{\bar{x}}$, is 2.98 and the observed standard error of the mean, $s_{\bar{x}}$, is 0.638. The histogram of sampling distribution, Figure 7.4 (p. 321), appears to be mounded, approximately symmetrical, and centered near the value 3.0.

Now let's check the truth of the three specific properties.

1. $\mu_{\bar{x}}$ and μ will be equal.

   The mean of the population μ is 3.0, and the observed sampling distribution mean $\bar{\bar{x}}$ is 2.98; they are very close in value.

2. $\sigma_{\bar{x}}$ will equal $\dfrac{\sigma}{\sqrt{n}}$.

Exercise 7.14

If a population has a standard deviation σ of 25 units, what is the standard error of the mean if samples of size 16 are selected? samples of size 36? size 100?

   $\sigma = 1.41$ and $n = 5$; therefore, $\dfrac{\sigma}{\sqrt{n}} = \dfrac{1.41}{\sqrt{5}} = \mathbf{0.632}$; and $s_{\bar{x}} = \mathbf{0.638}$; they are very close in value. (Remember that we have taken only 30 samples, not all possible samples, of size 5.)

3. The sampling distribution of $\bar{x}$ will be approximately normally distributed.

   Even though the population has a rectangular distribution, the histogram in Figure 7.4 suggests that the $\bar{x}$ distribution has some of the properties of normality (mounded, symmetry).

Although Illustrations 7.2 and 7.3 do not constitute a proof, the evidence seems to strongly suggest that both statements, the sampling distribution of sample means and the central limit theorem, are true.

Having taken a look at these two specific illustrations, let's now look at four graphic illustrations that present the sampling distribution information and the CLT in a slightly different form. In each of these graphic illustrations there are four distributions. The first graph shows the distribution of the parent population, the distribution of the individual x-values. Each of the other three graphs shows a sampling distribution of sample means, $\bar{x}$'s, using three different sample sizes. In Figure 7.7 we have a uniform distribution, much like Figure 7.3 for the integer illustration, and the resulting distributions of sample means for samples of size 2, 5, and 30.

Figure 7.7

Uniform Distribution

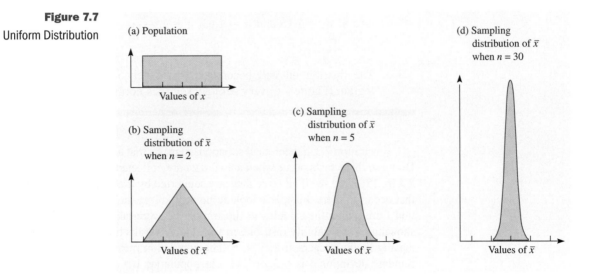

Figure 7.8 shows a U-shaped population and the three sampling distributions.

Figure 7.8

U-Shaped Distribution

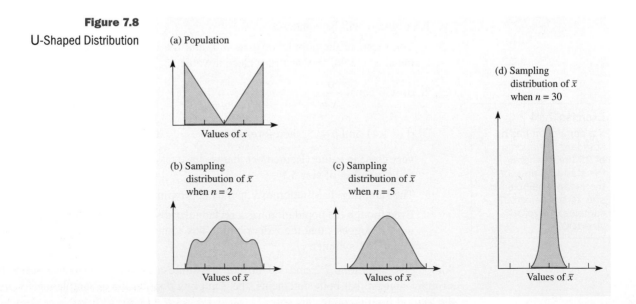

Figure 7.9 shows a **J-shaped population** and the three sampling distributions.

Figure 7.9
J-Shaped Distribution

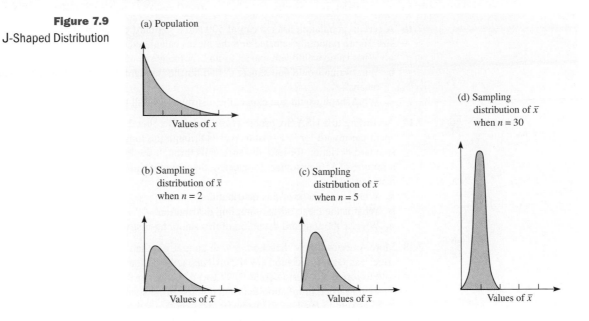

All three nonnormal distributions seem to verify the CLT; the sampling distributions of sample means appear to be approximately normal for all three when samples of size 30 were used. With the normal population (Figure 7.10), the sampling distributions for all sample sizes appear to be normal. Thus, you have seen an amazing phenomenon: No matter what the shape of a population, the sampling distribution of sample means either is normal or becomes approximately normal when n becomes sufficiently large.

Figure 7.10 shows a **normally distributed population** and the three sampling distributions.

Figure 7.10
Normal Distribution

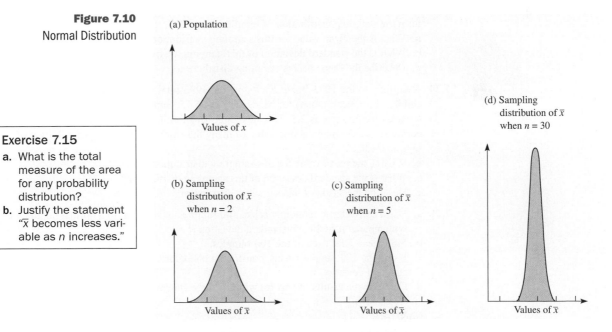

Exercise 7.15

a. What is the total measure of the area for any probability distribution?

b. Justify the statement "$\bar{x}$ becomes less variable as n increases."

You should notice one other point: The sample mean becomes less variable as the sample size increases. Notice that as n increases from 2 to 30, all the distributions become narrower and taller.

EXERCISES • • • • • • • • • • • • •

7.16 A certain population has a mean of 500 and a standard deviation of 30. Many samples of size 36 are randomly selected and the means calculated.
  a. What value would you expect to find for the mean of all these sample means?
  b. What value would you expect to find for the standard deviation of all these sample means?
  c. What shape would you expect the distribution of all these sample means to have?

7.17 According to a USA Snapshot® (*USA Today*, October 21–23, 1994), the average amount spent per month for long-distance calls through the long-distance carrier is $31.65. If the standard deviation for long-distance calls through the long-distance carrier is $12.25 and a sample of 150 customers is selected, the mean of this sample belongs to a sampling distribution.
  a. What is the shape of this distribution?
  b. What is the mean of this sampling distribution?
  c. What is the standard deviation of this sampling distribution?

7.18 More Americans heat their homes with natural gas than any other fuel. According to the American Gas Association (*1997 Gas Facts*), the national average price of natural gas sold to residential customers in 1997 was 62 cents per therm, about 18% less than it cost ten years earlier, in inflation-adjusted dollars. If the standard deviation for prices of natural gas sold to residential customers is 11 cents per therm and a random sample of 200 residential customers in 1997 is selected, the mean of this sample belongs to a sampling distribution.
  a. What is the shape of this sampling distribution?
  b. What is the mean of this sampling distribution?
  c. What is the standard deviation of this sampling distribution?

7.19 For the last 20 years the price that turkey farmers receive for turkeys has been relatively stable. According to *The World Almanac and Book of Facts 1998,* turkey farmers received an average of 43.3 cents per pound for their birds in 1996. Suppose the standard deviation of all prices received by turkey farmers is 7.5 cents per pound. The mean price received for a sample of 150 randomly selected turkey farmers in 1996 is one value of many that form the sampling distribution of sample means.
  a. What is the mean value for this sampling distribution?
  b. What is the standard deviation of this sampling distribution?
  c. Describe the shape of this sampling distribution.

7.20 According to the 1993 *World Factbook,* the 1993 total fertility rate (mean number of children born per woman) for Madagascar is 6.75. Suppose the standard deviation of the total fertility rate is 2.5. The mean number of children for a sample of 200 randomly selected women is one value of many that form the sampling distribution of sample means.
  a. What is the mean value for this sampling distribution?
  b. What is the standard deviation of this sampling distribution?
  c. Describe the shape of this sampling distribution.

7.21 a. Use a computer to randomly select 100 samples of size 6 from a normal population with a mean $\mu = 20$ and standard deviation $\sigma = 4.5$.
  b. Find mean $\bar{x}$ for each of the 100 samples.
  c. Using the 100 sample means, construct a histogram, find the mean $\bar{\bar{x}}$, and find the standard deviation $s_{\bar{x}}$.
  d. Compare the results of part (c) with the three statements made in the CLT.

MINITAB

a. Use the Normal RANDOM DATA commands on page 292 replacing generate with 100, store in with C1-C6, mean with 20 and standard deviation with 4.5.

b. Use the mean ROW STATISTICS commands on page 323 replacing input variables with C1-C6 and store result in with C7.

c. Use the HISTOGRAM commands on page 56 for the data in C7, entering cutpoints from 12.8 to 27.2 in increments of 1.8. Use the MEAN and STANDARD DEVIATION commands on pages 67 and 80 for the data in C7.

EXCEL

a. Use the Normal RANDOM NUMBER GENERATION commands on page 292 replacing number of variables with 6, number of random numbers with 100, mean with 20, standard deviation with 4.5, and output range with A1.

b. Activate cell G1.

Choose:	**Paste function, f_x > Statistical > AVERAGE > OK**
Enter:	**Number1: (A1:F1 or select cells) > OK**
Drag:	**Bottom right corner of average value box down to give other averages**

c. Use the RANDOM NUMBER GENERATION Patterned Distribution commands in Exercise 6.58(a) on page 297, replacing the first value with 12.8, the last value with 27.2, the steps with 1.8, and the output range with H1. Use the HISTOGRAM commands on page 57 with column G as the input range, column H as the bin range, and I1 as output range. Use the MEAN and STANDARD DEVIATION commands on pages 67 and 80 for the data in column G.

(If you use a computer, see Exercise 7.21)

7.22 **a.** Use a computer to randomly select 200 samples of size 24 from a normal population with a mean $\mu = 20$ and standard deviation $\sigma = 4.5$.

  b. Find mean $\bar{x}$ for each of the 200 samples.

  c. Using the 200 sample means, construct a histogram, find mean $\bar{\bar{x}}$, and find the standard deviation $s_{\bar{x}}$.

  d. Compare the results of part (c) with the three statements made in the CLT.

  e. Compare these results to the results obtained in Exercise 7.19. Specifically: What effect did the increase in sample size from 6 to 24 have? What effect did the increase from 100 samples to 200 samples have?

7.3   Application of the Central Limit Theorem

When the sampling distribution of sample means is normally distributed, or approximately normally distributed, we will be able to answer probability questions with the aid of the standard normal distribution, Table 3 of Appendix B.

ILLUSTRATION 7.4 ▼

Consider a normal population with $\mu = 100$ and $\sigma = 20$. If a sample of size 16 is selected at random, what is the probability that this sample will have a mean value between 90 and 110? That is, what is $P(90 < \bar{x} < 110)$?

Solution

Since the population is normally distributed, the sampling distribution of $\bar{x}$'s is normally distributed. To determine probabilities associated with a normal distribution, we will need to convert the statement $P(90 < \bar{x} < 110)$ to a probability statement involving the *z*-**score** in order to use Table 3 (Appendix B), the standard normal distribution table. The sampling distribution is shown in the following figure, with $P(90 < \bar{x} < 110)$

represented by the shaded area:

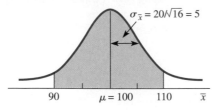

The formula for finding the z corresponding to a known value of $\bar{x}$ is

$$z = \frac{\bar{x} - \mu_{\bar{x}}}{\sigma_{\bar{x}}} \qquad (7.1)$$

The mean and standard error of the mean are: $\mu_{\bar{x}} = \mu$ and $\sigma_{\bar{x}} = \dfrac{\sigma}{\sqrt{n}}$. Therefore, we will rewrite formula (7.1) in terms of μ, σ, and n:

$$z = \frac{\bar{x} - \mu}{\sigma/\sqrt{n}} \qquad (7.2)$$

Returning to Illustration 7.4 and applying formula 7.2, we find:

the z-score for $\bar{x} = 90$: $\quad z = \dfrac{\bar{x} - \mu}{\sigma/\sqrt{n}} = \dfrac{90 - 100}{20/\sqrt{16}} = \dfrac{-10}{5} = \mathbf{-2.00}$

and

the z-score for $\bar{x} = 110$: $\quad z = \dfrac{\bar{x} - \mu}{\sigma/\sqrt{n}} = \dfrac{110 - 100}{20/\sqrt{16}} = \dfrac{10}{5} = \mathbf{2.00}$

Therefore,

$$P(90 < \bar{x} < 110) = P(-2.00 < z < 2.00) = 2(0.4772) = \mathbf{0.9544} \qquad \blacktriangle$$

> **Exercise 7.23**
>
> In Illustration 7.4, explain how 0.4772 was obtained and what it is.

Before we look at more illustrations, let's consider for a moment what is implied by saying that $\sigma_{\bar{x}} = \dfrac{\sigma}{\sqrt{n}}$. To demonstrate, let's suppose that $\sigma = 20$ and let's use a sampling distribution of samples of size 4. Now $\sigma_{\bar{x}}$ would be $20/\sqrt{4}$, or 10, and approximately 95% (0.9544) of all such sample means should be within the interval from 20 below to 20 above the population mean (within 2 standard deviations of the population mean). However, if the sample size were increased to 16, $\sigma_{\bar{x}}$ would become $20/\sqrt{16} = 5$ and approximately 95% of the sampling distribution would be within 10 units of the mean, and so on. As the sample size increases, the size of $\sigma_{\bar{x}}$ becomes smaller so that the distribution of sample means becomes much narrower. Figure 7.11 illustrates what happens to the distribution of $\bar{x}$'s as the size of the individual samples increases.

Figure 7.11

Distributions of Sample Means

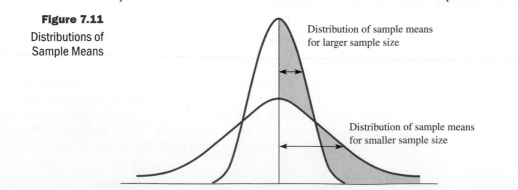

Recall that the area (probability) under the normal curve is always exactly one. So as the width of the curve narrows, the height will have to increase in order to maintain this area.

ILLUSTRATION 7.5 ▼

Kindergarten children have heights that are approximately normally distributed about a mean of 39 in. and a standard deviation of 2 in. A random sample of size 25 is taken and the mean $\bar{x}$ is calculated. What is the probability that this mean value will be between 38.5 and 40.0 inches?

Solution

We want to find: $P(38.5 < \bar{x} < 40.0)$. 38.5 and 40.0 are values of $\bar{x}$ and must be converted to z-scores (necessary for use of Table 3) as follows:

Using $z - \dfrac{\bar{x} - \mu}{\sigma/\sqrt{n}}$

Exercise 7.24

What is the probability that the sample of kindergarten children in Illustration 7.5 has a mean height less than 39.75 inches?

When $\bar{x} = 38.5$: $z = \dfrac{\bar{x} - \mu}{\sigma/\sqrt{n}} = \dfrac{38.5 - 39.0}{2/\sqrt{25}} = \dfrac{-0.5}{0.4} = \mathbf{-1.25}$

When $\bar{x} = 40.0$: $z = \dfrac{\bar{x} - \mu}{\sigma/\sqrt{n}} = \dfrac{40.0 - 39.0}{2/\sqrt{25}} = \dfrac{1.0}{0.4} = \mathbf{2.50}$

Therefore,

$$P(38.5 < \bar{x} < 40.0) = P(-1.25 < z < 2.50) = 0.3944 + 0.4938 = \mathbf{0.8882}$$

ILLUSTRATION 7.6 ▼

Referring to the heights of kindergarten children in Illustration 7.5, within what limits would the middle 90% of the sampling distribution of sample means for samples of size 100 fall?

Solution

The two tools we have to work with are formula (7.2) and Table 3. The formula relates the key values of the population to key values of the sampling distribution, and Table 3 relates area to z-scores. First, using Table 3, we find the middle 0.9000 is bounded by $z = \pm 1.65$.

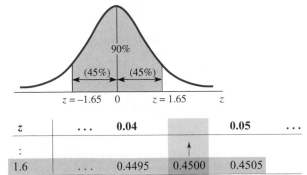

z	...	0.04		0.05	...
:					
1.6	...	0.4495	0.4500	0.4505	

Second, using formula (7.2), $z = \dfrac{\bar{x} - \mu}{\sigma/\sqrt{n}}$:

<table>
<tr><td>

Exercise 7.25

Referring to Illustration 7.5, what height would bound the lower 25% of all samples of size 25?

</td><td>

If $z = -1.65$: $-1.65 = \dfrac{\bar{x} - 39.0}{2/\sqrt{100}}$

$\bar{x} - 39 = (-1.65)(0.2)$

$\bar{x} = 39 - 0.33$

$= 38.67$

</td><td>

If $z = 1.65$: $1.65 = \dfrac{\bar{x} - 39.0}{2/\sqrt{100}}$

$\bar{x} - 39 = (1.65)(0.2)$

$\bar{x} = 39 + 0.33$

$= 39.33$

</td></tr>
</table>

Thus,

$$P(38.67 < \bar{x} < 39.33) = 0.90$$

Therefore, 38.67 in. and 39.33 in. are the limits that capture the middle 90% of the sample means.   ▲

EXERCISES   •   •   •   •   •   •   •   •   •   •   •   •   •

7.26 A random sample of size 36 is to be selected from a population that has a mean μ of 50 and a standard deviation σ of 10.

   a. This sample of 36 has a mean value of $\bar{x}$ that belongs to a sampling distribution. Find the shape of this sampling distribution.

   b. Find the mean of this sampling distribution.

   c. Find the standard error of this sampling distribution.

   d. What is the probability that this sample mean will be between 45 and 55?

   e. What is the probability that the sample mean will have a value greater than 48?

   f. What is the probability that the sample mean will be within 3 units of the mean?

7.27 Consider the approximately normal population of heights of male college students with mean μ of 69 in. and standard deviation σ of 4 in. A random sample of 16 heights is obtained.

   a. Describe the distribution of x, height of male college student.

   b. Find the proportion of male college students whose height is greater than 70 in.

   c. Describe the distribution of $\bar{x}$, the mean of samples of size 16.

   d. Find the mean and standard error of the $\bar{x}$ distribution.

   e. Find $P(\bar{x} > 70)$.

   f. Find $P(\bar{x} < 67)$.

7.28 The amount of fill (weight of contents) put into a glass jar of spaghetti sauce is normally distributed with mean $\mu = 850$ g and standard deviation $\sigma = 8$ g.

   a. Describe the distribution of x, the amount of fill per jar.

   b. Find the probability that one jar selected at random contains between 848 and 855 g.

   c. Describe the distribution of $\bar{x}$, the mean weight for a sample of 24 such jars of sauce.

   d. Find the probability that a random sample of 24 jars has a mean weight between 848 and 855 g.

7.29 The heights of the kindergarten children mentioned in Illustration 7.5 (p. 333) are approximately normally distributed with $\mu = 39$ and $\sigma = 2$.

   a. If an individual kindergarten child is selected at random, what is the probability that he or she has a height between 38 and 40 inches?

   b. A classroom of 30 of these children is used as a sample. What is the probability that the class mean $\bar{x}$ is between 38 and 40 inches?

   c. If an individual kindergarten child is selected at random, what is the probability that he or she is taller than 40 inches?

   d. A classroom of 30 of these kindergarten children is used as a sample. What is the probability that the class mean $\bar{x}$ is greater than 40 inches?

7.30 Compensations to chief executive officers of corporations vary substantially from one company to the next, both in amount and in form. A study published in *The Accounting Review* ("Accounting Transactions and CEO Cash Compensations," Vol. 73, No. 2) showed that the mean annual salary and bonus compensation to CEOs in 1997 was $634,961 with a standard deviation of $441,690.

    a. Assuming that annual salaries and bonus compensations are normally distributed, what is the probability that a randomly selected CEO received over $750,000 in 1997?

    b. Explain why the assumption of normally distributed may be very unlikely.

    c. A sample of 20 CEOs is taken and salaries and bonus compensations are reported. What is the probability that the sample mean salary and bonus compensation falls between $600,000 and $700,000?

    d. Explain why the standard normal distribution can be used to answer (c) with more confidence than (a).

7.31 According to the 1994 *World Almanac*, the average speed of winds in Honolulu, Hawaii, equals 11.4 miles per hour. Assume that wind speeds are approximately normally distributed with a standard deviation of 3.5 miles per hour.

    a. Find the probability that the wind speed on any one reading will exceed 13.5 miles per hour.

    b. Find the probability that the mean of a random sample of 9 readings exceeds 13.5 miles per hour.

    c. Do you think the assumption of normality is reasonable? Explain.

    d. What effect do you think the assumption of normality had on the answers to (a) and (b)? Explain.

7.32 According to the U.S. Dept. of Energy (*Monthly Energy Review,* June 1997), the average price of unleaded regular gasoline sold at service stations throughout the nation in 1996 was $1.23 per gallon. Assume that gasoline prices in general are normally distributed with a standard deviation of $.16 per gallon. A random sample of 45 stations in 1996 is selected and the pump prices for unleaded regular gasoline are recorded. Find the probability that the sample mean price:

    a. exceeds $1.28 per gallon.

    b. is less than $1.19 per gallon.

    c. is between $1.20 and $1.27 per gallon.

7.33 According to the *World Almanac and Book of Facts—1994,* the median weekly earnings of full-time wage and salary women, age 16 years or older in 1992, equals $381. Assume that the wages and salaries are normally distributed with $\sigma = \$85$.

    a. Find the probability that the mean weekly earnings of a sample of 250 such women is between $375 and $385, if the mean equals $381.

    b. Do you think the assumption of normality is reasonable? Explain.

    c. What effect do you think the assumption of normality about the x distribution had on the answer to (a)? Explain.

    d. Do you think the assumption of mean equals $381 is reasonable? Explain.

    e. What effect do you think the assumption about the value of the mean had on the answer to (a)? Explain.

7.34 According to the August 1994 issue of *Employment and Earnings,* the June 1994 average weekly earnings for employees in general automotive repair shops was $406.15. Suppose the standard deviation for the weekly earnings for such employees is $55.50. Assuming that this mean and standard deviation are the current values, find the following probabilities for the mean of a sample of 100 such employees.

    a. The probability that the mean of the sample is less than $400.

    b. The probability that the sample mean is between $400 and $410.

    c. The probability that the mean of the sample is greater than $415.

    d. Explain why the assumption of normality about the x distribution was not involved in the solution to (a), (b), and (c).

7.35 **a.** Find $P(4 < \bar{x} < 6)$ for a random sample of size 4 drawn from a normal population with mean $\mu = 5$ and standard deviation $\sigma = 2$.

b. Use a computer to randomly generate 100 samples, each of size 4, from a normal probability distribution with mean $\mu = 5$ and standard deviation $\sigma = 2$, and calculate the mean, $\bar{x}$, for each sample.

c. How many of the sample means in (b) have values between 4 and 6? What percentage is that?

d. Compare the answers to (a) and (c), and explain any differences that occurred.

MINITAB

a. Input the numbers 4 and 6 into C1. Use the CUMULATIVE NORMAL PROBABILITY DISTRIBUTION commands on page 294, replacing the mean with 5, the standard deviation with 1 (2/√4), the input column with C1, and the optional storage with C2. Find CDF(6) − CDF(4).

b. Use the Normal RANDOM DATA commands on page 292 replacing generate n with 100, store in with C3-C6, mean with 5, and standard deviation with 2. Use the mean ROW STATISTICS commands on page 323 replacing input variables with C3-C6 and store result in with C7.

c. Use the HISTOGRAM commands on page 56 for the data in C7, entering cutpoints from 1 to 9 in increments of 1, and selecting data labels.

EXCEL

a. Input the numbers 4 and 6 into column A. Activate cell B1. Use the CUMULATIVE NORMAL DISTRIBUTION commands on page 294, replacing x with the data values in A1:A2, the mean with 5, the standard deviation with 1 (2/√4), and cumulative with true. Find CDF(6) − CDF(4).

b. Use the Normal RANDOM NUMBER GENERATION commands on page 292 replacing number of variables with 4, number of random numbers with 100, mean with 5, standard deviation with 2, and output range with C1. Activate cell G1. Use the AVERAGE PASTE FUNCTION commands in Exercise 7.12(b) on page 323 replacing Number1 with C1:F1. Drag the G1 box down for other averages.

c. Use the RANDOM NUMBER GENERATION Patterned Distribution commands in Exercise 6.58(a) on page 297, replacing the first value with 0, the last value with 9, the steps with 1, and the output range with H1. Use the HISTOGRAM commands on page 57 with column G as the input range, column H as the bin range, and column I as the output range. (Use the frequencies from bins 5 and 6; they are upper limits.)

7.36 **a.** Find $P(46 < \bar{x} < 55)$ for a random sample size 16 drawn from a normal population with mean $\mu = 50$ and standard deviation $\sigma = 10$.

b. Use a computer to randomly generate 200 samples, each of size 16, from a normal probability distribution with mean $\mu = 50$ and standard deviation $\sigma = 10$, and calculate the mean, $\bar{x}$, for each sample.

c. How many of the sample means in (b) have values between 46 and 55? What percentage is that?

d. Compare the answers to (a) and (c), and explain any differences that occurred.

If you use a computer, see Exercise 7.35.

Return to C H A P T E R   C A S E   S T U D Y

The Gallup Poll has been surveying the public for many years. The results summarized in the Chapter Case Study list the findings for each of several years. When repeated sampling is used to track America's attitudes, the sample statistic reported (the percentage of yes responses) does not form a sampling distribution, but rather it forms a time series and demonstrates a trend. (Time series is a topic not covered in this text; however, many of its components are.) Complete the following questions to help recognize and understand the difference between repeated samples that belong to a sampling distribution and those that belong to a time series.

EXERCISE 7.37

  a. Plot a scatter diagram displaying the Selected National Trend information (p. 317), using the year as the input variable x and the percentage of yes responses as the output variable y.
  b. On the scatter diagram drawn in (a), plot the percentages of the no responses as a second output variable using the year number as the input variable x.
  c. Do you see what could be called a trend? Explain.
  d. Make a prediction for the percentage of Americans who would vote for a woman president. How did you use the case study information?
  e. Sampling distributions involve repeatedly sampling from the same population, but with a completely different purpose. Explain, in your own words, how a sampling distribution is different than the Chapter Case Study illustration.
  f. Repeated sampling, like that in the Chapter Case Study, is carried out for the purpose of "tracking" the statistic being studied. Describe, in your own words, the purpose of studying a statistic from repeated samples as a sampling distribution.

IN RETROSPECT

In Chapters 6 and 7 we have learned to use the standard normal probability distribution. We now have two formulas for calculating a z-score:

$$z = \frac{x - \mu}{\sigma} \qquad \text{and} \qquad z = \frac{\bar{x} - \mu}{\sigma/\sqrt{n}}$$

You must be careful to distinguish between these two formulas. The first gives the standard score when dealing with individual values from a normal distribution (x values). The second formula deals with a sample mean ($\bar{x}$-value). The key to distinguishing between the formulas is to decide whether the problem deals with an individual x or a sample mean, $\bar{x}$. If it deals with the individual values of x, we use the first formula, as presented in Chapter 6. If, on the other hand, the problem deals with a sample mean, $\bar{x}$, we use the second formula and proceed as illustrated in this chapter.

The basic purpose for considering what happens when a population is repeatedly sampled, as discussed in this chapter, is to form sampling distributions. The sampling distribution is then used to describe the variability that occurs from one sample to the next. Once this pattern of variability is known and understood for a specific sample statistic, we will be able to make predictions about the corresponding population parameter with a measure of how accurate the prediction is. The Central Limit Theorem helps describe the distribution for sample means. We will begin to make inferences about population means in Chapter 8.

There are other reasons for repeated sampling. Repeated samples are commonly used in the field of production control, in which samples are taken to determine whether a product is of the proper size or quantity. When the sample statistic does not fit the standards, a mechanical adjustment of the machinery is necessary. The adjustment is then followed by another sampling to be sure the production process is in control.

The "standard error of the ____" is the name used for the standard deviation of the sampling distribution for whatever statistic is named in the blank. In this chapter we have been concerned with the standard error of the mean. However, we could also work with the standard error of proportion, median, or any other statistic.

You should now be familiar with the concept of a sampling distribution and, in particular, with the sampling distribution of sample means. In Chapter 8 we will begin to make predictions about the values of population parameters.

CHAPTER EXERCISES

7.38 The diameters of Red Delicious apples in a certain orchard are normally distributed with a mean of 2.63 in. and a standard deviation of 0.25 in.

  a. What percentage of the apples in this orchard have diameters less than 2.25 in.?

  b. What percentage of the apples in this orchard are larger than 2.56 in.?

  A random sample of 100 apples is gathered and the mean diameter obtained is $\bar{x} = 2.56$.

  c. If another sample of size 100 is taken, what is the probability that its sample mean will be greater than 2.56 in.?

  d. Why is the z-score used in answering (a), (b), and (c)?

  e. Why is the formula for z used in (c) different from that used in (a) and (b)?

7.39 a. Find a value for e such that 95% of the apples in Exercise 7.38 are within e units of the mean 2.63. That is, find e such that $P(2.63 - e < x < 2.63 + e) = 0.95$.

  b. Find a value for E such that 95% of the samples of 100 apples taken from the orchard in Exercise 7.38 will have mean values within E units of the mean 2.63. That is, find E such that $P(2.63 - E < \bar{x} < 2.63 + E) = 0.95$.

7.40 Americans spend $2.8 billion per year on veterinary care, and the health care services offered to animals rival those provided to humans. The average dog owner in 1997 spent $275 per year on vet care. In addition, about 20% of all dog owners have indicated that they would be willing to spend over $5000 to extend their pet's life (*Life,* "Animal E.R.," July 1998). Assume that annual dog owner expenditure on vet care is normally distributed with a mean of $275 and a standard deviation of $95.

  a. What is the probability that a dog owner, randomly selected from the population, spent over $300 in vet care in 1997?

  b. Suppose a survey of 300 dog owners is conducted and each is asked to report the total of their vet care bills for the year. What is the probability that the mean annual expenditure of this sample falls between $293 and $305?

  c. Assume that life-extending offers are normally distributed with a standard deviation of $2000. What is the mean amount that dog owners would be willing to spend to extend their pet's life?

7.41 According to an article in *Newsweek* ("China by the Numbers: Portrait of a Nation," June 29, 1998), the rate of water pollution in China is more than twice that measured in the United States and over three times the amount measured for Japan. The mean emission of organic pollutants is 11.7 million pounds per day in China. Assume that water pollution in China is normally distributed throughout the year with a standard deviation of 2.8 million pounds of organic emissions per day.

  a. What is the probability that on any given day the water pollution in China exceeds 15 million pounds per day?

  b. If 20 days of water pollution readings in China are taken, what is the probability that the mean of this sample is less than 10 million pounds of organic emissions?

(continued)

7.42 According to an article in *Pharmaceutical News* (January 1991), a person age 65 or older will spend, on the average, $300 on personal-care products per year. If we assume that the amount spent on personal-care products by individuals 65 or older is normally distributed and has a standard deviation equal to $75, what is the probability that the mean amount spent by 25 randomly selected such individuals will fall between $250 and $350?

7.43 A shipment of steel bars will be accepted if the mean breaking strength of a random sample of 10 steel bars is greater than 250 pounds per square inch. In the past, the breaking strength of such bars has had a mean of 235 and a variance of 400.
  a. What is the probability, assuming that the breaking strengths are normally distributed, that one randomly selected steel bar will have a breaking strength in the range from 245 to 255?
  b. What is the probability that the shipment will be accepted?

7.44 A report in *Newsweek* (November 12, 1990) stated that the day-care cost per week in Boston is $109. If this figure is taken as the mean cost per week and if the standard deviation were known to be $20, find the probability that a sample of 50 day-care centers would show a mean cost of $100 or less per week.

7.45 A manufacturer of light bulbs says that its light bulbs have a mean life of 700 hr and a standard deviation of 120 hr. You purchased 144 of these bulbs with the idea that you would purchase more if the mean life of your sample were more than 680 hr. What is the probability that you will not buy again from this manufacturer?

7.46 A tire manufacturer claims (based on years of experience with its tires) that the mean mileage is 35,000 mi and the standard deviation is 5000 mi. A consumer agency randomly selects 100 of these tires and finds a sample mean of 31,000. Should the consumer agency doubt the manufacturer's claim?

7.47 For large samples, the sample sum ($\sum x$) has an approximately normal distribution. The mean of the sample sum is $n \cdot \mu$ and the standard deviation is $\sqrt{n} \cdot \sigma$. The distribution of savings per account for a savings and loan institution has a mean equal to $750 and a standard deviation equal to $25. For a sample of 50 such accounts, find the probability that the sum in the 50 accounts exceeds $38,000.

7.48 The baggage weights for passengers using a particular airline are normally distributed with a mean of 20 lb and a standard deviation of 4 lb. If the limit on total luggage weight is 2125 lb., what is the probability that the limit will be exceeded for 100 passengers?

7.49 A trucking firm delivers appliances for a large retail operation. The packages (or crates) have a mean weight of 300 lb. and a variance of 2500 lb.
  a. If a truck can carry 4000 lb. and 25 appliances need to be picked up, what is the probability that the 25 appliances will have an aggregate weight greater than the truck's capacity? (Assume that the 25 appliances represent a random sample.)
  b. If the truck has a capacity of 8000 lb., what is the probability that it will be able to carry the entire lot of 25 appliances?

7.50 A pop-music record firm wants the distribution of lengths of cuts on its records to have an average of 2 min 15 sec (135 sec) and a standard deviation of 10 sec so that disc jockeys will have plenty of time for commercials within each five-minute period. The population of times for cuts is approximately normally distributed with only a negligible skew to the right. You have just timed the cuts on a new release and have found that the 10 cuts average 140 sec.
  a. What percentage of the time will the average be 140 sec or longer, if the new release is randomly selected?
  b. If the music firm wants 10 cuts to average 140 sec less than 5% of the time, what must the population mean be, given that the standard deviation remains at 10 sec?

7.51 Let's simulate the sampling distribution related to the disc jockey's concern for "length of cut" in Exercise 7.50.
  a. Use a computer to randomly generate 50 samples, each of size 10, from a normal distribution with mean 135 and standard deviation 10. Find the "sample total" and the sample mean for each sample.

(continued)

b. Using the 50 sample means, construct a histogram and find their mean and standard deviation.

c. Using the 50 sample "totals," construct a histogram and find their mean and standard deviation.

d. Compare the results obtained in (b) and (c). Explain any similarities and any differences observed.

MINITAB

a. Use the Normal RANDOM DATA commands on page 292 replacing generate with 50, store in with C1-C10, mean with 135, and standard deviation with 10. Use the ROW STATISTICS commands on page 323 with selecting Sum, replacing input variables with C1-C10, and store result in with C11. Use the ROW STATISTICS commands again with selecting Mean and then replacing input variables with C1-C10 and store result in with C12.

b. Use the HISTOGRAM commands on page 56 for the data in C12, selecting midpoints and automatic. Use the MEAN and STANDARD DEVIATION commands on pages 67 and 80 for the data in C12.

c. Use the HISTOGRAM commands on page 56 for the data in C11, selecting midpoints and automatic. Use the MEAN and STANDARD DEVIATION commands on pages 67 and 80 for the data in C11.

d. Use the DISPLAY DESCRIPTIVE STATISTICS commands on page 97 for the data in C11 and C12.

EXCEL

a. Use the Normal RANDOM NUMBER GENERATION commands on page 292 replacing number of variables with 10, number of random numbers with 50, mean with 135, standard deviation with 10, and output range with A1.

Activate cell K1.

> Choose: **Paste function, f_x > All > SUM > OK**
> Enter: **Number1: (A1:J1 or select cells)**
> Drag: **Bottom right corner of sum value box down to give other sums**

Activate cell L1. Use the AVERAGE PASTE FUNCTION commands in Exercise 7.12b on page 323 replacing Number1 with A1:J1.

b. Use the RANDOM NUMBER GENERATION Patterned Distribution commands in Exercise 6.58(a) on page 297, replacing the first value with 125.4, the last value with 147.8, the steps with 3.2, and the output range with M1. Use the HISTOGRAM commands on page 57 with column L as the input range, column M as the bin range and N1 as the output range. Use the MEAN and STANDARD DEVIATION commands on pages 67 and 80 for the data in column L.

c. Use the RANDOM NUMBER GENERATION Patterned Distribution commands in Exercise 6.58(a) on page 297, replacing the first value with 1254, the last value with 1478, the

(continued)

steps with 32 and the output range with M20. Use the
HISTOGRAM commands on page 57 with column L as the
input range, cells M20-M26 as the bin range, and N20 as
output range. Use the MEAN and STANDARD DEVIATION commands on pages 67 and 80 for the data in column K.

d. Use the DESCRIPTIVE STATISTICS commands on page 97 for
the data in columns K and L.

7.52 **a.** Find the mean and standard deviation of x for a binomial probability distribution with $n = 16$ and $p = 0.5$.

b. Use a computer to construct the probability distribution and histogram for the binomial probability experiment with $n = 16$ and $p = 0.5$.

c. Use a computer to randomly generate 200 samples of size 25 from a binomial probability distribution with $n = 16$ and $p = 0.5$ and calculate the mean of each sample.

d. Construct a histogram and find the mean and standard deviation of the 200 sample means.

e. Compare the probability distribution of x found in (b) and the frequency distribution of $\bar{x}$ in (d). Does your information support the CLT? Explain.

MINITAB

b. Use the MAKE PATTERNED DATA commands in Exercise 6.58(a)
on page 297, replacing the first value with 0, the last
value with 16, and the steps with 1. Use the BINOMIAL
PROBABILITY DISTRIBUTIONS commands on page 258, replacing n with 16, p with 0.5, input column with C1, and
optional storage with C2. Use the PLOT commands on
page 139 replacing Y with C2, x with C1, and data display with area.

c. Use the BINOMIAL RANDOM DATA commands on page 267, replacing generate with 200, store in with C3-C27, number
of trials with 16, and probability with 0.5. Use the ROW
STATISTICS commands for a mean on page 323 replacing input variables with C3-C27 and store result in with C28.

d. Use the HISTOGRAM commands on page 56 for the data in
C28, selecting midpoints and automatic. Use the MEAN and
STANDARD DEVIATION commands on pages 67 and 80 for the
data in C28.

EXCEL

b. Input 0 through 16 into column A. Activate cell B1. Continue with the binomial probability commands on page
258, using $n = 16$ and $p = 0.5$. Drag B1 box down for other
probabilities. Activate columns A and B, then continue
with:

```
Choose:   Chart Wizard > Column > 1ˢᵗ picture > Next > Series
Choose:   Series 1 > Remove
Enter:    Category (x)axis labels: (A1:A17 or select
          'x value' cells)
Choose:   Next > Finish
```

c. Use the Binomial RANDOM NUMBER GENERATION commands from
Exercise 5.73 on page 268 replacing number of variables
with 25, number of random numbers with 200, p value with

(continued)

0.5, number of trials with 16, and output range with
C1. Activate cell AB1. Use the AVERAGE PASTE FUNCTION
commands in Exercise 7.12(b) on page 323 replacing Number 1 with C1:AA1. Drag AB1 box down for other averages.

    d. Use the RANDOM NUMBER GENERATION Patterned Distribution
commands in Exercise 6.58(a) on page 297, replacing the
first value with 6.8, the last value with 9.6, the steps
with 0.4, and the output range with AC1. Use the HISTOGRAM commands on page 57 with column AB as the input
range, column AC as the bin range, and AD1 as output
range. Use the MEAN and STANDARD DEVIATION commands on
pages 67 and 80 for the data in column BB.

7.53 **a.** Find the mean and standard deviation of x for a binomial probability distribution with $n = 200$ and $p = 0.3$.

> Use the commands in Exercise 7.52, making the necessary adjustments.

    b. Use a computer to construct the probability distribution and histogram for the random variable x of the binomial probability experiment with $n = 200$ and $p = 0.3$.

    c. Use a computer to randomly generate 200 samples of size 25 from a binomial probability distribution with $n = 200$ and $p = 0.3$ and calculate the mean $\bar{x}$ of each sample.

    d. Construct a histogram and find the mean and standard deviation of the 200 sample means.

    e. Compare the probability distribution of x found in (b) and the frequency distribution of $\bar{x}$ in (d). Does your information support the CLT? Explain.

7.54 A sample of 144 values is randomly selected from a population with mean, μ, equal to 45 and standard deviation, σ, equal to 18.

    a. Determine the interval (smallest value to largest value) within which you would expect a sample mean to lie.

    b. What is the amount of deviation from the mean for a sample mean of 46.3?

    c. What is the maximum deviation you have allowed for in your answer to (a)?

    d. How is this maximum deviation related to the standard error of the mean?

VOCABULARY LIST

Be able to define each term. Pay special attention to the key terms, which are printed in **red**. In addition, describe in your own words, and give an example of, each term. Your examples should not be ones given in class or in the textbook.

    The bracketed numbers indicate the chapter in which the term first appeared, but you should define the terms again to show increased understanding of their meaning. Page numbers indicate the first appearance of the term in Chapter 7.

Central Limit Theorem (p. 325)	sampling distribution (p. 319)
frequency distribution [2] (p. 320)	**sampling distribution of sample means**
probability distribution [5] (p. 319)	(p. 319, 324)
random sample [2] (p. 321)	**standard error of the mean** (p. 325)
repeated sampling (p. 318)	z-score [2, 6] (p. 331)

CHAPTER PRACTICE TEST

Part I: Knowing the Definitions

Answer "True" if the statement is always true. If the statement is not always true, replace the words shown in bold with words that make the statement always true.

7.1 A sampling distribution **is** a distribution listing all the sample statistics that describe a particular sample.

7.2 The histograms of **all** sampling distributions are symmetrically shaped.

7.3 The mean of the sampling distribution of $\bar{x}$'s is equal to the mean of the **sample.**

7.4 The standard error of the mean is the standard deviation of the population **from which the samples have been taken.**

7.5 The standard error of the mean **increases** as the sample size increases.

7.6 The shape of the distribution of sample means is always that of a **normal** distribution.

7.7 A **probability** distribution of a sample statistic is a distribution of all the values of that statistic that were obtained from all possible samples.

7.8 The Central Limit Theorem provides us with a description of the three characteristics of a sampling distribution of sample **medians.**

7.9 A **frequency** sample is obtained in such a way that all possible samples of a given size have an equal chance of being selected.

7.10 We **do not need** to repeatedly sample a population in order to use the concept of the sampling distribution.

Part II: Applying the Concepts

7.11 The lengths of the lake trout in Conesus Lake are believed to have a normal distribution with a mean length of 15.6 in. and a standard deviation of 3.8 in.
   a. Kevin is going fishing at Conesus Lake tomorrow. If he catches one lake trout, what is the probability that it is less than 15.0 in.?
   b. If Captain Brian's fishing boat takes 10 people fishing on Conesus Lake tomorrow and they catch a random sample of 16 lake trout, what is the probability that the mean length of their total catch is less than 15 in.?

7.12 Cigarette lighters manufactured by Easyvice Company are claimed to have a mean life of 20 months with a standard deviation of 6 months. The money-back guarantee allows you to return the lighter if it does not last at least 12 months from the date of purchase.
   a. If the length of life of these lighters is normally distributed, what percentage of the lighters will be returned to the company?
   b. If a random sample of 25 lighters is tested, what is the probability the sample mean will be more than 18 months?

7.13 Aluminum rivets produced by Rivets Forever, Inc., are believed to have shearing strengths that are distributed about a mean of 13.75 and have a standard deviation of 2.4. If this information is true and a sample of 64 such rivets is tested for shear strength, what is the probability that the mean strength will be between 13.6 and 14.2?

Part III: Understanding the Concepts

7.14 "Two heads are better than one." If that's true, then "how good would several heads be?" To find out, a statistics instructor drew a line across the chalkboard and asked her class to estimate its length to the nearest inch. She collected their estimates, which ranged from 33 to 61 in., and calculated the mean value. She reported that the mean was 42.25 in. She then measured the line and found it to be 41.75 in. long. Does this show that "several heads are better than one"? What statistical theory supports this occurrence? Explain how.

7.15 The sampling distribution of sample means is more than just a distribution of the mean values that occur when many samples are repeatedly taken from the same population. Describe what other specific condition must be met in order to have a sampling distribution of sample means.

7.16 Student A stated that "a sampling distribution of the standard deviation tell you how the standard deviation varies from sample to sample." Student B argues that "a population distribution tells you that." Who is right? Justify your answer.

7.17 Student A says that it is the "size of each sample used" and Student B says that it is the "number of samples used" that determines the spread of an empirical sampling distribution. Who is right? Justify your choice.

Working with Your Own Data

The Central Limit Theorem is very important to the development of the rest of this course. Its proof, which requires the use of calculus, is beyond the intended level of this course. However, the truth of the CLT can be demonstrated both theoretically and by experimentation. The following series of questions will help to verify the Central Limit Theorem both ways.

A The Population

Consider the theoretical population that contains the three numbers 0, 3, and 6 in equal proportions.

1. a. Construct the theoretical probability distribution for the drawing of a single number, with replacement, from this population.
   b. Draw a histogram of this probability distribution.
   c. Calculate the mean μ and the standard deviation σ for this population.

B The Sampling Distribution, Theoretically

Let's study the theoretical sampling distribution formed by the means of all possible samples of size 3 that can be drawn from the given population.

2. Construct a list showing all the possible samples of size 3 that could be drawn from this population. (There are 27 possibilities.)
3. Find the mean for each of the 27 possible samples listed in answer to question 2.
4. Construct the probability distribution (the theoretical sampling distribution of sample means) for these 27 sample means.
5. Construct a histogram for this sampling distribution of sample means.
6. Calculate the mean $\mu_{\bar{x}}$ and the standard error of the mean $\sigma_{\bar{x}}$ using the probability distribution found in B.4.
7. Show that the results found in answers 1(c), 5, and 6 support the three claims made by the Central Limit Theorem. Cite specific values to support your conclusions.

C The Sampling Distribution, Empirically

Let's now see whether the Central Limit Theorem can be verified empirically; that is, does it hold when the sampling distribution is formed by the sample means that result from several random samples?

8. Draw a random sample of size 3 from the given population. List your sample of three numbers and calculate the mean for this sample.

You may use a computer to generate your samples. You may take three identical "tags" numbered 0, 3, and 6, put them in a "hat," and draw your sample using replacement between each drawing. Or you may use dice; let 0 be represented by 1 and 2, let 3 be represented by 3 and 4, and 6 by 5 and 6. You may also use random numbers to simulate the drawing of your samples. Or you may draw your sample from the list of random samples at the bottom of the page. Describe the method you decide to use. (Ask your instructor for guidance.)

9. Repeat question 8 forty-nine (49) more times so that you have a total of fifty (50) sample means that have resulted from samples of size 3.
10. Construct a frequency distribution of the 50 sample means found in answering questions 8 and 9.
11. Construct a histogram of the frequency distribution of observed sample means.
12. Calculate the mean $\bar{\bar{x}}$ and standard deviation $s_{\bar{x}}$ of the frequency distribution formed by the 50 sample means.
13. Compare the observed values of $\bar{\bar{x}}$ and $s_{\bar{x}}$ with the values of $\mu_{\bar{x}}$ and $\sigma_{\bar{x}}$. Do they agree? Does the empirical distribution of $\bar{x}$ look like the theoretical one?

The following table contains 100 samples of size 3 that were generated randomly by computer:

6 3 0	0 3 0	6 6 0	3 3 6	6 6 3	6 3 3
0 0 3	3 0 6	3 3 0	3 6 6	0 3 0	6 6 3
6 6 6	0 3 0	6 3 6	0 6 3	6 0 3	6 3 3
6 0 0	3 0 6	6 3 3	3 3 0	3 3 0	3 3 3
3 3 3	3 0 0	6 6 6	3 3 6	0 0 6	0 6 3
6 6 6	0 0 6	3 3 0	0 6 6	0 0 3	6 6 3
0 0 6	0 0 6	6 6 6	6 3 6	6 6 0	3 0 0
3 6 6	6 3 0	3 6 3	3 0 0	3 3 6	0 6 0
3 0 0	0 3 6	6 3 3	6 0 6	3 3 6	6 0 3
0 3 6	3 6 3	6 6 3	6 6 0	3 3 3	3 0 0
6 3 0	6 6 0	0 3 0	6 6 0	3 6 6	0 3 6
6 3 3	0 3 0	6 6 0	6 6 3	6 6 0	3 0 3
3 6 3	3 6 0	0 0 6	0 3 3	3 6 6	0 3 6
0 6 0	6 0 0	0 6 0	0 6 6	0 3 3	0 3 6
3 3 6	3 3 3	3 3 6	6 3 6	3 3 3	3 6 6
6 3 3	3 0 0	3 0 6	6 0 3	3 6 6	6 0 3
0 3 3	6 3 0	0 3 6	0 3 6		

Three

Inferential Statistics

The Central Limit Theorem gave us some very important information about the sampling distribution of sample means. Specifically, it stated that in many realistic cases (when the random sample is large enough) a distribution of sample means is normally or approximately normally distributed about the mean of the population. With this information we were able to make probability statements about the likelihood of certain sample mean values occurring when samples are drawn from a population with a known mean and a known standard deviation. We are now ready to turn this situation around to the case in which the population mean is not known. We will draw one sample, calculate its mean value, and then make an inference about the value of the population mean based on the sample's mean value.

The objective of inferential statistics is to use the information contained in the sample data to increase our knowledge of the sampled population. In this part of the textbook we will learn about making two types of inferences: (1) estimating the value of a population parameter and (2) testing a hypothesis. Specifically, we will learn about making these two types of inferences for the mean μ of a normal population, for the standard deviation σ of a normal population, and for the probability parameter p of a binomial population.

William Gosset

WILLIAM GOSSET ("Student"), a British industrial statistician, was born in Canterbury, England, on June 13, 1876, to Frederick and Agnes (Vidal) Gosset. William's educational background included studies at Winchester College, New College, and Oxford University. Gosset was employed by the Arthur Guinness & Son Brewing Company as a brewer. In 1906 Gosset married Marjory Surtees Philpotts, and they became the parents of two daughters and a son. William Gosset died in Beaconsfield, England, on October 16, 1937.

Guinness liked their employees to use pen names if publishing papers, so in 1908 Gosset adopted the pen name "Student" under which he published what was probably his most noted contribution to statistics, "The Probable Error of a Mean." Guinness sent Gosset to work under Karl Pearson, at the University of London, and eventually he took charge of the new Guinness Brewery in London.

In his paper "The Probable Error of a Mean," Student set out to find the distribution of the amount of error in the sample mean, $(\bar{x} - \mu)$, when divided by s, where s was the estimate of σ from a sample of any known size. The probable error of a mean, $\bar{x}$, could be calculated for any size sample by using the distribution of $(\bar{x} - \mu)/(s/\sqrt{n})$. Even though he was well aware of the insufficiency of a small sample to determine the form of the distribution of $\bar{x}$, he chose the normal distribution for simplicity, stating his opinion: "It appears probable that the deviation from normality must be very severe to lead to serious error."

Student's t-distribution did not immediately gain popularity. In September 1922, even 14 years after its publication, Student wrote to Fisher: "I am sending you a copy of Student's Tables as you are the only man that's ever likely to use them!" Today, Student's t-distribution is widely used and respected in statistical research.

Chapter 8

Introduction to Statistical Inferences

CHAPTER CASE STUDY

Holiday Spending

Holiday spending is an annual "happening." The holiday spirit puts customers in a sharing and spending frame of mind and it puts the merchants in their best merchandizing frame of mind. Can you think of a better way to give a boost to year-end sales volume? The USA Snapshot® "Holiday home trimmings" that appeared in *USA Today*, December 8, 1998, describes one small part of the holiday spirit and spending that occurs annually.

USA SNAPSHOTS®
A look at statistics that shape our lives

Holiday home trimmings

American households spend an average of $59 per year on holiday decorations. Percent spending these amounts:

$0-$25 **36%**
$26-$49 **25%**
$50-$99 **21%**
$100 or more **17%**

Source: National Family Opinion Research for Department 56

By Anne R. Carey and Bob Laird, USA TODAY

One hundred fifty adult shoppers at a large shopping mall were asked "How much (to the nearest $25) do you anticipate your family will spend on holiday decorations this year?"

25	200	100	25	250	75	25	50	25	100	75	25	100	75	25
25	200	25	0	25	175	25	75	100	100	50	25	50	100	50
25	100	100	175	25	75	25	0	100	25	25	50	25	25	75
0	100	100	75	75	100	25	50	50	25	100	100	150	75	75
25	25	50	75	75	100	25	50	0	25	25	100	25	50	150
150	75	100	150	0	100	75	25	75	25	0	300	25	25	50
25	100	25	75	75	25	25	50	50	50	50	25	100	125	50
50	75	25	75	25	0	100	0	50	75	50	100	50	125	25
50	75	125	100	50	125	200	75	25	25	25	50	25	50	25
25	0	0	100	25	100	100	50	25	25	125	25	75	100	25

Use the above sample data to estimate the mean anticipated amount households living near this mall plan to spend on holiday decorations this year. Does the above sample suggest that the families who shop in this mall anticipate spending a different average amount than all Americans? Support your answer with statistical evidence found within this sample data. We will return to these questions. (See Exercise 8.121 on p. 404.)

CHAPTER OBJECTIVES

A random sample of 36 pieces of data yields a mean of 4.64. What can we deduce about the population from which the sample was taken? We will be asked to answer two types of questions:

1. What value or interval of values can we use to estimate the population mean?
2. Is the sample mean sufficiently different in value from the hypothesized mean value of 4.5 to justify arguing that 4.5 is not a correct value?

The first question requires us to make an estimate , whereas the second question requires us to make a decision about the population mean.

In this chapter we concentrate on learning about the basic concepts of estimation and hypothesis testing . We deal with questions about the population mean using two methods that assume the value of the population standard deviation is a known quantity. This assumption is seldom realized in real-life problems, but it will make our first look at the techniques of inference much simpler.

8.1   The Nature of Estimation

A company manufactures rivets for use in building aircraft. One characteristic of extreme importance is the "shearing strength" of each rivet. The company's engineers must monitor their production so that they are certain the shearing strength of their rivets meets the required specs. To accomplish this, they take a sample and determine the mean shearing strength of the sample. Based on this sample information, the company can estimate the mean shearing strength for all the rivets it is manufacturing.

A sample of 36 rivets is randomly selected, and each rivet is tested for shearing strength. The resulting sample mean is $\bar{x} = 924.23$ lb. Based on this sample, we say, "We believe the mean shearing strength of all such rivets is 924.23 lb."

NOTE 1 Shearing strength is the force required to break a material in a "cutting" action. Obviously, the manufacturer is not going to test all rivets since the test destroys each rivet tested. Therefore, samples are tested and the information about the sample must be used to make inferences about the population of all such rivets.

NOTE 2 Throughout Chapter 8 we will treat the standard deviation σ as a known, or given, quantity, and concentrate on learning the procedures for making statistical inferences about the population mean μ. Therefore, to continue the explanation of statistical inferences, we will assume $\sigma = 18$ for the specific rivet described in our example.

POINT ESTIMATE FOR A PARAMETER

A single number designed to estimate a quantitative parameter of a population; usually the value of the corresponding **sample statistic.**

That is, the sample mean, $\bar{x}$, is the point estimate (single number value) for the mean μ of the sampled population. For our rivet example, 924.23 is the point estimate for μ, the mean shearing strength of all rivets.

The quality of this point estimate should be questioned. Is the estimate exact? Is the estimate likely to be high? Or low? Would another sample yield the same result? Would another sample yield an estimate of nearly the same value? Or a value that is very different? How is "nearly the same" or "very different" measured? The quality of an estimation procedure (or method) is greatly enhanced if the sample statistic is both *less variable* and *unbiased*. The variability of a statistic is measured by the standard error of its sampling distribution. The sample mean can be made less variable by reducing its standard error, $\dfrac{\sigma}{\sqrt{n}}$. That requires using a larger sample, since as *n* increases, the standard error decreases.

> **Exercise 8.1**
>
> The use of a tremendously large sample does not solve the question of quality for an estimator. What problems do you anticipate with very large samples?

UNBIASED STATISTIC

A sample statistic whose sampling distribution has a mean value equal to the value of the population parameter being estimated. A statistic that is not unbiased is a **biased statistic.**

Figure 8.1 illustrates the concept of being unbiased and the effect of variability on the point estimate. The value *A* is the parameter being estimated, and the dots represent possible sample statistic values from the sampling distribution of the statistic. If *A* represents the true population mean μ, then the dots represent possible sample means from the $\overline{x}$ sampling distribution.

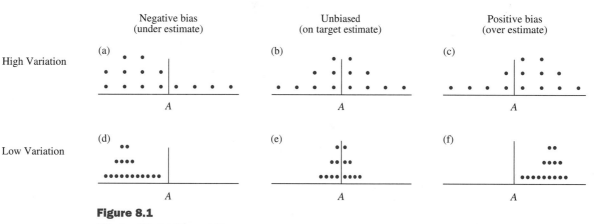

Figure 8.1

Effects of Variability and Bias

Figure 8.1(a), (c), (d), and (f) show biased statistics; (a) and (d) show sampling distributions whose mean values are less than the value of the parameter, while (c) and (f) show sampling distributions whose mean values are greater than the parameter. Figure 8.1(b) and (e) show sampling distributions that appear to have a mean value equal to the value of the parameter; therefore, they are unbiased. Figure 8.1(a), (b), and (c) show more variability, while (d), (e), and (f) show less variability in the sampling distributions. Diagram (e) represents the best situation, an estimator that is unbiased (on target) and with low variability (all values close to the target).

The sample mean, $\overline{x}$, is an unbiased statistic because the mean value of the sampling distribution of sample means, $\mu_{\overline{x}}$, is equal to the population mean, μ. (Recall that the sampling distribution of sample means has a mean $\mu_{\overline{x}} = \mu$.) Therefore, the sample statistic $\overline{x} = 924.23$ is an unbiased point estimate for the mean strength of all rivets being manufactured in our example.

Sample means vary in value and form a sampling distribution in which not all samples result in $\overline{x}$-values equal to the population mean. If that is the case, we should not expect this sample of 36 rivets to produce a point estimate (sample mean) that is exactly equal to the mean μ of the sampled population. We should, however, expect the point estimate to be fairly close in value to the population mean. The sampling distribution and the Central Limit Theorem provide the information needed to describe how close the point estimate, $\bar{x}$, is expected to be to the population mean, μ.

Recall that approximately 95% of a normal distribution is within two standard deviations of the mean and that the Central Limit Theorem describes the sampling distribution of sample means as being nearly normal when samples are large enough. Samples of size 36 from populations of variables like rivet strength are generally considered large enough. Therefore, we should anticipate 95% of all samples randomly selected from a population with unknown mean μ and standard deviation $\sigma = 18$ will have means $\bar{x}$ between

$$\mu - 2 \cdot \sigma_{\bar{x}} \quad \text{and} \quad \mu + 2 \cdot \sigma_{\bar{x}}$$

$$\mu - 2 \cdot \left(\frac{\sigma}{\sqrt{n}} \right) \quad \text{and} \quad \mu + 2 \cdot \left(\frac{\sigma}{\sqrt{n}} \right)$$

$$\mu - 2 \cdot \left(\frac{18}{\sqrt{36}} \right) \quad \text{and} \quad \mu + 2 \cdot \left(\frac{18}{\sqrt{36}} \right)$$

$$\mu - 6 \quad \text{and} \quad \mu + 6$$

Exercise 8.2

Explain why the standard error of sample means is 3 for the rivet example.

This suggests that 95% of all samples of size 36 randomly selected from the population of rivets should have a mean $\bar{x}$ between $\mu - 6$ and $\mu + 6$. Figure 8.2 shows the middle 95% of the distribution, the bounds of the interval covering the 95%, and the mean μ.

Figure 8.2

Sampling Distribution of $\bar{x}$'s, Unknown σ

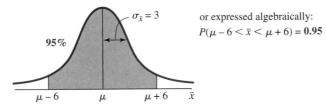

95%

$\sigma_{\bar{x}} = 3$

or expressed algebraically:
$P(\mu - 6 < \bar{x} < \mu + 6) = 0.95$

Now let's put all of this information together in the form of a *confidence interval*.

INTERVAL ESTIMATE

An interval bounded by two values and used to estimate the value of a population parameter. The values that bound this interval are statistics calculated from the sample that is being used as the basis for the estimation.

LEVEL OF CONFIDENCE $1 - \alpha$

The proportion of all interval estimates that include the parameter being estimated.

CONFIDENCE INTERVAL

An interval estimate with a specified level of confidence.

To construct the confidence interval, we will use the point estimate $\bar{x}$ as the central value of an interval in much the same way as we used the mean μ as the central value to find the interval that captures the middle 95% of the $\bar{x}$ distribution in Figure 8.2.

For our rivet example, we can find the bounds to an interval centered at $\bar{x}$:

$$\bar{x} - 2 \cdot \sigma_{\bar{x}} \quad \text{to} \quad \bar{x} + 2 \cdot \sigma_{\bar{x}}$$

$$924.23 - 6 \quad \text{to} \quad 924.23 + 6$$

The resulting interval is: 918.23 to 930.23

The level of confidence assigned to this interval is approximately 95%, or 0.95. The bounds of the interval are 2 multiples ($z = 2.0$) of the standard error from the sample mean, and by looking at Table 3 in Appendix B, we can more accurately determine the level of confidence as 0.9544. Putting all of this information together, we express the estimate as a confidence interval: **918.23 to 930.23** *is the 95.44% confidence interval for the mean shear strength of the rivets.* Or in an abbreviated form: **918.23 to 930.23,** *the 95.44% confidence interval for* μ.

> **Exercise 8.3**
>
> Verify that the level of confidence for a 2-standard-deviation interval is 95.44%.

Case Study   8.1

Gallup Report: Sampling Tolerances

> **Exercise 8.4**
>
> **a.** What is the "average of the repeated samplings"?
> **b.** What does "95 times out of 100" mean?
> **c.** If a value of 52 has a ±5 margin of error, how are these two numbers related to point estimate, interval width, and interval bounds?

We see survey results frequently in today's newspapers and magazines. A survey's results are often summarized using a statistic and the sampling error, with a footnote that says, "Margin of error: ±3." The statistic reported is the point estimate, and the margin of error is one-half of the interval width; therefore, 33 ± 3 represents the interval estimate 30 to 36 as Gallup describes in this article.

SAMPLING ERROR

In interpreting survey results, it should be borne in mind that all sample surveys are subject to sampling error, that is, the extent to which the results may differ from those that would be obtained if the whole population surveyed had been interviewed. The size of such sampling errors depends largely on the number of interviews.

A reported statistic is 33 and is subject to a sampling error of plus or minus 3. Another way of saying it is that very probably (95 times out of 100) the average of repeated samplings would be somewhere between 30 and 36.

Source: Copyright 1986 by Gallup Report. Reprinted by permission.

EXERCISES • • • • • • • • • • • • •

8.5 Identify each numerical value by "name" (mean, variance, etc.) and by symbol ($\bar{x}$, etc.):
  a. The mean height of 24 junior high school girls is 4'11".
  b. The standard deviation for I.Q. scores is 16.
  c. The variance among the test scores on last week's exam was 190.
  d. The mean height of all cadets who have ever entered West Point is 69 inches.

8.6 A random sample of the amount paid for taxi fare from downtown to the airport was obtained:

15	19	17	23	21	17	16	18	12	18	20	22	15	18	20

Use the data to find a point estimate for each of the following parameters.
  a. mean
  b. variance
  c. standard deviation

8.7 In each diagram below, I and II represent sampling distributions of two statistics that might be used to estimate a parameter. In each case, identify the statistic that you think would be the better estimator and describe why it is your choice.

a.   b.   c.

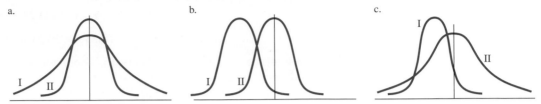

8.8 Suppose there are two statistics that will serve as an estimator for the same parameter. One of them is biased and the other unbiased.

   a. Everything else being equal, explain why you usually would prefer an unbiased estimator to a biased estimator.

   b. If a statistic is unbiased, does that ensure it is a good estimator? Why or why not? What other considerations must be taken into account?

   c. Describe a situation that might occur in which the biased statistic might be a better choice as an estimator than the unbiased statistic.

8.9 Being unbiased and having a small variability are two desirable characteristics of a statistic if it is going to be used as an estimator. Describe how the Central Limit Theorem addresses both of these properties when estimating the mean of a population.

8.10 Find the level of confidence assigned to an interval estimate of the mean formed using the interval

   a. $\bar{x} - 1.3 \cdot \sigma_{\bar{x}}$ to $\bar{x} + 1.3 \cdot \sigma_{\bar{x}}$ **b.** $\bar{x} - 1.65 \cdot \sigma_{\bar{x}}$ to $\bar{x} + 1.65 \cdot \sigma_{\bar{x}}$
   c. $\bar{x} - 1.96 \cdot \sigma_{\bar{x}}$ to $\bar{x} + 1.96 \cdot \sigma_{\bar{x}}$ **d.** $\bar{x} - 2.3 \cdot \sigma_{\bar{x}}$ to $\bar{x} + 2.3 \cdot \sigma_{\bar{x}}$
   e. $\bar{x} - 2.6 \cdot \sigma_{\bar{x}}$ to $\bar{x} + 2.6 \cdot \sigma_{\bar{x}}$

8.11 *Accounting Horizons* (June 1998) reported the results of a study concerning the KPMG Peat Marwick's "Research Opportunities in Auditing" (ROA) program that supported audit research projects with data from actual audits. A total of 174 projects were funded between 1977 and 1993. A sample of 25 of these projects revealed that 19 were valued at $17,320 each, and six were valued at $20,200 each. From the sample data, estimate the total value of the funding for all the projects.

8.12 A stamp dealer wishes to purchase a stamp collection that is believed to contain approximately 7000 individual stamps and approximately 4000 first day covers. Devise a plan that might be used to estimate the collection's worth.

8.2   Estimation of Mean μ (σ Known)

In Section 8.1 we surveyed the basic ideas of estimation : point estimate , interval estimate , level of confidence , and confidence interval . These basic ideas are interrelated and used throughout statistics when an inference calls for an estimate. In this section we formalize the interval estimation process as it applies to estimating the population mean μ based on a random sample under the restriction that the population standard deviation σ is a known value.

The sampling distribution of sample means and the Central Limit Theorem provide us with the needed information to ensure that the necessary *assumptions* are satisfied.

> **ASSUMPTIONS**
>
> The conditions that need to exist in order to correctly apply a statistical procedure.

NOTE The use of the word assumptions is somewhat a misnomer. It does not mean that we "assume" something to be the situation and continue, but that we must be sure the conditions expressed by the assumptions do exist before applying a particular statistical method.

> **THE ASSUMPTION FOR ESTIMATING MEAN μ USING A KNOWN σ**
>
> The sampling distribution of $\bar{x}$ has a normal distribution.

The information needed to ensure that this assumption (or condition) is satisfied is contained in the sampling distribution of sample means and in the Central Limit Theorem (see Chapter 7, p. 324 and 325).

> The sampling distribution for sample means $\bar{x}$ is distributed about a mean equal to μ with a standard error equal to $\sigma/\sqrt{n}$; and, (1) if the randomly sampled population is normally distributed, then $\bar{x}$ is normally distributed for all sample sizes, or (2) if the randomly sampled population is not normally distributed, then $\bar{x}$ is approximately normally distributed for sufficiently large sample sizes.

Therefore, we can satisfy the required assumption by either (1) knowing that the sampled population is normally distributed or (2) using a random sample containing a sufficiently large number of data. The first possibility is obvious. We either know enough about the population to know it is normally distributed or we don't. The second way to satisfy the assumption is by applying the CLT. Inspection of various graphic displays of the sample data should yield an indication of the type of distribution the population possesses. The CLT can be applied to smaller samples (say, $n = 15$ or larger) when the data provide a strong indication of a unimodal distribution that is approximately symmetric. If there is evidence of some skewness present in the data, then the sample size will need to be much larger (perhaps $n \geq 50$). If the data provide evidence of an extremely skewed or J-shaped distribution, the CLT will still apply if the sample is large enough. In extreme cases, "large enough" may be unrealistically or impracticably large.

> The help of a professional statistician should be sought when treating extremely skewed data.

WARNING There is no hard and fast rule defining "large enough"; the sample size that is "large enough" varies greatly according to the distribution of the population.

The $1 - \alpha$ confidence interval for the estimation of mean μ is found using the formula

$$\bar{x} - z_{(\alpha/2)} \cdot \frac{\sigma}{\sqrt{n}} \quad \text{to} \quad \bar{x} + z_{(\alpha/2)} \cdot \frac{\sigma}{\sqrt{n}} \tag{8.1}$$

The parts of the confidence interval formula are:

1. $\bar{x}$ is the point estimate and the center point of the confidence interval.
2. $z_{(\alpha/2)}$ is the **confidence coefficient.** It is the number of multiples of the standard error needed to formulate an interval estimate of the correct width to have a level of confidence of $1 - \alpha$. Figure 8.3 (p. 356) shows the relationship among: (a) the level of confidence (the middle portion of the distribution), (b) the $\alpha/2$ (the "area to the right" used with the critical-value notation), and (c) the confidence coefficient $z_{(\alpha/2)}$ (whose value is found using Table 4b of Appendix B).
3. $\sigma/\sqrt{n}$ is the **standard error of the mean,** the standard deviation of the sampling distribution of sample means.

Figure 8.3
Confidence Coefficient,
$z(\alpha/2)$.

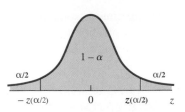

4. $z_{(\alpha/2)} \cdot \dfrac{\sigma}{\sqrt{n}}$ is one-half the width of the confidence interval (the product of the confidence coefficient and the standard error) and is called the **maximum error of estimate, E.**

5. $\bar{x} - z_{(\alpha/2)} \cdot \dfrac{\sigma}{\sqrt{n}}$ is called the **lower confidence limit** (LCL), and $\bar{x} + z_{(\alpha/2)} \cdot \dfrac{\sigma}{\sqrt{n}}$ is called the **upper confidence limit** (UCL) for the confidence interval.

The estimation procedure is organized into a five-step process that will take into account all of the above and produce both the point estimate and the confidence interval.

> Basically, the confidence interval is "point estimate ± maximum error."

THE CONFIDENCE INTERVAL: A FIVE-STEP PROCEDURE

STEP 1   The Set-Up:
Describe the population parameter of interest.

STEP 2   The Confidence Interval Criteria:
    a. Check the assumptions.
    b. Identify the probability distribution and the formula to be used.
    c. Determine the level of confidence, $1 - \alpha$.

STEP 3   The Sample Evidence:
Collect the sample information.

STEP 4   The Confidence Interval:
    a. Determine the confidence coefficient.
    b. Find the maximum error of estimate.
    c. Find the lower and upper confidence limits.

STEP 5   The Results:
State the confidence interval.

ILLUSTRATION 8.1 ▼

The student body at many community colleges is considered a "commuter population." The student activities' office wishes to answer the question "How far (one way) does the average community college student commute to college each day?" (Typically the "average student's commute distance" is meant to be the "mean distance" commuted by all students who commute.) A random sample of 100 commuting students was identified, and the one-way distance each commuted was obtained. The resulting sample mean distance was 10.22 miles. Estimate the mean one-way distance commuted by all commuting students using: (a) a point estimate and (b) a 95% confidence interval. (Use $\sigma = 6$ miles.)

Solution

(a) The point estimate for the mean one-way distance is **10.22** (the sample mean).

(b) 95% Confidence Interval

STEP 1 **The Set-Up:**
Describe the population parameter of interest:
The **mean** μ of the one-way distances commuted by all commuting community college students is the parameter of interest.

STEP 2 **The Confidence Interval Criteria:**
a. **Check the assumptions.**
σ is known. The variable "distance commuted" most likely has a skewed distribution since the vast majority of the students will commute between 0 and 25 miles, with fewer commuting more than 25 miles. A sample size of 100 should be large enough for the CLT to satisfy the assumption, the $\bar{x}$ **sampling distribution is approximately normal.**
b. **Identify the probability distribution and the formula to be used.**
The **standard normal distribution** (z) will be used to determine the confidence coefficient, and **Formula (8.1)** with $\sigma = 6$.
c. **State the level of confidence, $1 - \alpha$.**
The question asks for 95% confidence, or $1 - \alpha = 0.95$.

STEP 3 **The Sample Evidence:**
Collect the sample information.
The sample information is given in the statement of the problem:
$n = 100, \bar{x} = 10.22$.

STEP 4 **The Confidence Interval:**
a. **Determine the confidence coefficient.**
The confidence coefficient is found using Table 4B:

A Portion of Table 4B

α	$\ldots$	**0.05**
$z_{(\alpha/2)}$	$\ldots$	**1.96**
$1 - \alpha$	$\ldots$	**0.95**

Level of Confidence: $1 - \alpha = 0.95$ $\longrightarrow$ ($1 - \alpha$ row)

$\longrightarrow$ *Confidence Coefficient: $z_{(\alpha/2)} = 1.96$*

b. **Find the maximum error of estimate.**
Use the maximum error part of formula (8.1):
$$E = z_{(\alpha/2)} \cdot \frac{\sigma}{\sqrt{n}} = 1.96 \cdot \frac{6}{\sqrt{100}} = (1.96)(0.6) = \mathbf{1.176}$$

c. **Find the lower and upper confidence limits.**
Using the point estimate ($\bar{x}$) from step 3 and the maximum error (E) from step 4b, the confidence interval limits are:

$$\bar{x} - z_{(\alpha/2)} \cdot \frac{\sigma}{\sqrt{n}} \quad \text{to} \quad \bar{x} + z_{(\alpha/2)} \cdot \frac{\sigma}{\sqrt{n}}$$

$$10.22 - 1.176 \quad \text{to} \quad 10.22 + 1.176$$

$$9.044 \quad \text{to} \quad 11.396$$

$$\mathbf{9.04} \quad \text{to} \quad \mathbf{11.40}$$

Exercise 8.13

In your own words, describe the relationship between the point estimate, the level of confidence, the maximum error, and the confidence interval.

STEP 5 **The Results:**
State the confidence interval.
9.04 to 11.40, the 95% confidence interval for μ
That is, with 95% confidence we can say, "The mean one-way distance is between 9.04 and 11.40 miles."   ▲

Let's look at another illustration of the estimation procedure.

ILLUSTRATION 8.2 ▼

"Particle size" is an important property of latex paint and is monitored during production as part of the quality-control process. Thirteen particle-size measurements were taken using the Dwight P. Joyce Disc, and the sample mean was 3978.1 angstroms [where 1 angstrom $(1\text{Å}) = 10^{-8}$ cm]. The particle size, x, is normally distributed with a standard deviation $\sigma = 200$ angstroms. Find the 98% confidence interval for the mean particle size for this batch of paint.

Exercise 8.14

A machine produces parts whose lengths are normally distributed with $\sigma = 0.5$. A sample of ten parts has a mean length of 75.92.

a. Find the point estimate for μ.
b. Find the 98% confidence maximum error of estimate for μ.
c. Find the 98% confidence interval for μ.

Solution

STEP 1 **The Set-Up:**
Describe the population parameter of interest:
The **mean particle size, μ,** for the batch of paint from which the sample was drawn.

STEP 2 **The Confidence Interval Criteria:**
a. **Check the assumptions.**
σ is known. The variable "particle size" is normally distributed; therefore, the **sampling distribution of sample means is normal** for all sample sizes.
b. **Identify the probability distribution and the formula to be used.**
The **standard normal variable z,** using **formula (8.1)** and $\sigma = 200$.
c. **State the level of confidence, $1 - \alpha$.**
98%, or $1 - \alpha = 0.98$.

STEP 3 **The Sample Evidence:**
Collect the sample information. $n = 13$ and $\bar{x} = 3978.1$.

STEP 4 **The Confidence Interval:**
a. **Determine the confidence coefficient.**
The confidence coefficient is found using Table 4B: $z(\alpha/2) = z(0.01) = 2.33$.

A Portion of Table 4B

α	$\ldots$	0.02
$z(\alpha/2)$	$\ldots$	2.33
$1 - \alpha$	$\ldots$	0.98

Level of Confidence: $1 - \alpha = 0.98$ ⟶

⟶ *Confidence Coefficient: $z(\alpha/2) = 2.33$.*

b. **Find the maximum error of estimate.**
$$E = z(\alpha/2) \cdot \frac{\sigma}{\sqrt{n}} = 2.33 \cdot \frac{200}{\sqrt{13}} = (2.33)(55.47) = 129.2.$$

c. **Find the lower and upper confidence limits.**
Using the point estimate ($\bar{x}$) from step 3 and the maximum error (E) from step 4b, the confidence interval limits are:
$$\bar{x} - z(\alpha/2) \cdot \frac{\sigma}{\sqrt{n}} \quad \text{to} \quad \bar{x} + z(\alpha/2) \cdot \frac{\sigma}{\sqrt{n}}$$
$$3978.1 - 129.2 = 3848.9 \quad \text{to} \quad 3978.1 + 129.2 = 4107.3$$

STEP 5 **The Results:**
State the confidence interval.
3848.9 to 4107.3, the 98% confidence interval for μ
With 98% confidence we can say, "The mean particle size is between 3848.9 and 4107.3 angstroms." ▲

Let's take another look at the concept "level of confidence." It was defined to be the probability that the sample to be selected will produce interval bounds that contain the parameter.

ILLUSTRATION 8.3 ▼

Single-digit random numbers, like the ones in Table 1 (Appendix B), have a mean value μ = 4.5 and a standard deviation σ = 2.87 (see Exercise 5.35, p. 249). Draw a sample of 40 single-digit numbers from Table 1 and construct the 90% confidence interval for the mean. Does the resulting interval contain the expected value of μ, 4.5? If we were to select another sample of 40 single-digit numbers from Table 1, would the same results occur? What might happen if we were to select a total of 15 different samples and construct the 90% confidence interval related to each? Would the expected value for μ, namely 4.5, be contained in all of them? Should we expect all 15 confidence intervals to contain 4.5? Think about the definition of "level of confidence"; it says that in the

C a s e S t u d y 8.2

Rockies Snow Brings Little Water

When snow melts it becomes water, sometimes more water than at other times. This newspaper article compares the water content of snow from two areas in the United States that typically get about the same amount of snow annually. However, the water content is very different. There are several point estimates for the average included in the USA Today article.

Exercise 8.15

"Rockies snow brings little water" lists "14.36 inches" and "5.07 inches" as statistics and uses them as point estimates. Describe why these numbers are statistics and why they are also point estimates.

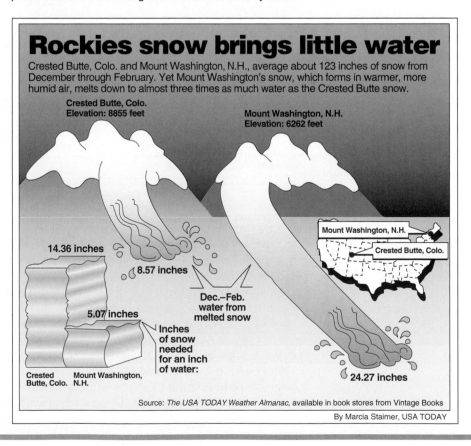

Rockies snow brings little water

Crested Butte, Colo. and Mount Washington, N.H., average about 123 inches of snow from December through February. Yet Mount Washington's snow, which forms in warmer, more humid air, melts down to almost three times as much water as the Crested Butte snow.

Crested Butte, Colo.
Elevation: 8855 feet

Mount Washington, N.H.
Elevation: 6262 feet

14.36 inches

8.57 inches

Dec.–Feb. water from melted snow

5.07 inches

Inches of snow needed for an inch of water:

Crested Butte, Colo.   Mount Washington, N.H.

Mount Washington, N.H.

Crested Butte, Colo.

24.27 inches

Source: *The USA TODAY Weather Almanac*, available in book stores from Vintage Books

By Marcia Staimer, USA TODAY

long run 90% of the samples will result in bounds that contain μ. In other words, 10% of the samples will not contain μ. Let's see what happens.

First we need to address the assumptions; if the assumptions are not satisfied, we cannot expect the 90% and the 10% to occur. We know: (1) the distribution of single-digit random numbers is rectangular (definitely not normal), (2) the distribution of single-digit random numbers is symmetric about their mean, (3) the $\bar{x}$ distribution for very small samples ($n = 5$) in Illustration 7.2 (page 321) displayed a distribution that appeared to be approximately normal, and (4) there should be no skewness involved. Therefore, it seems reasonable to assume that $n = 40$ is large enough to allow the CLT to apply.

The first random sample was drawn from Table 1 in Appendix B. The sample:

2	8	2	1	5	5	4	0	9	1
0	4	6	1	5	1	1	3	8	0
3	6	8	4	8	6	8	9	5	0
1	4	1	2	1	7	1	7	9	3

Summary of sample: $n = 40$, $\sum x = 159$, $\bar{x} = 3.975$
Resulting 90% confidence interval:

$$\bar{x} \pm z_{(\alpha/2)} \cdot \frac{\sigma}{\sqrt{n}}: \quad 3.975 \pm 1.65 \cdot \frac{2.87}{\sqrt{40}}$$

$$3.975 \pm (1.65)(0.454)$$

$$3.975 \pm 0.749$$

$$3.975 - 0.749 = \mathbf{3.23} \quad \text{to} \quad 3.975 + 0.749 = \mathbf{4.72}$$

3.23 to 4.72, the 90% confidence interval for μ

Figure 8.4 shows this interval estimate, its bounds, and the expected mean μ.

Figure 8.4

The 90% Confidence Interval

With 90% confidence, we think μ is somewhere within this interval

3.23 $\mu = 4.50$ 4.72 $\bar{x}$

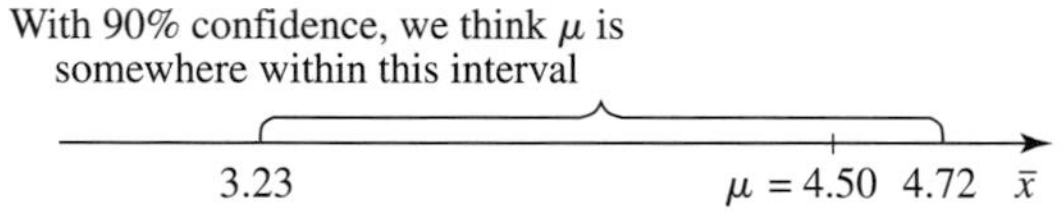

The expected value for the mean, 4.5, does fall within the bounds of the confidence interval for this sample. Let's now select 14 more random samples from Table 1 in Appendix B, each of size 40.

Table 8.1 lists the mean from the above sample and the means obtained from the 14 additional random samples of size 40. The 90% confidence intervals for the estimation of μ based on each of the 15 samples are listed in Table 8.1 and shown in Figure 8.5.

We see that 86.7% (13 of the 15) of the intervals contain μ and two of the 15 samples (sample 7 and sample 12) do not contain μ. The results here are "typical"; repeated experimentation might result in any number of intervals that contain 4.5. However, in the long run we should expect approximately $1 - \alpha = 0.90$ (or 90%) of the samples to result in bounds that contain 4.5 and approximately 10% that do not contain 4.5.

Table 8.1   Fifteen Samples of Size 40

Sample Number	Sample Mean $\bar{x}$	90% Confidence Interval Estimate for μ
1	3.975	3.23 to 4.72
2	4.64	3.89 to 5.39
3	4.56	3.81 to 5.31
4	3.96	3.21 to 4.71
5	5.12	4.37 to 5.87
6	4.24	3.49 to 4.99
7	3.44	2.69 to 4.19
8	4.60	3.85 to 5.35
9	4.08	3.33 to 4.83
10	5.20	4.45 to 5.95
11	4.88	4.13 to 5.63
12	5.36	4.61 to 6.11
13	4.18	3.43 to 4.93
14	4.90	4.15 to 5.65
15	4.48	3.73 to 5.23

Figure 8.5
Confidence Intervals from Table 8.1

The following commands will construct a 1 − α confidence interval for a mean μ with a given standard deviation σ.

Exercise 8.16

Using a computer or calculator, randomly generate a sample of 40 single-digit numbers and find the 90% confidence interval for μ. Repeat several times, observing whether or note 4.5 is in the interval each time. Describe your results.

(Use commands for generating integer data on page 182, then continue with confidence interval commands above.)

MINITAB

```
Input the data into C1; then continue with:

   Choose:   Stat > Basic Statistics > 1-Sample Z
   Enter:    Variables: C1
   Select:   Confidence interval
   Enter:    Level: 1 − α (ex.: 0.95 or 95.0)
             Sigma: σ
```

EXCEL

```
Input the data into column A; then continue with:

   Choose:   Tools > Data Analysis Plus
                   > Inference About A Mean(SIGMA Known) > OK
   Enter:    block coordinates: (A1:A20 or select cells)    > OK
   Choose:   Interval Estimate
   Enter:    Value of SIGMA: σ    > OK
             level of confidence: 1 − α (ex.: 0.95 or 95.0)
```

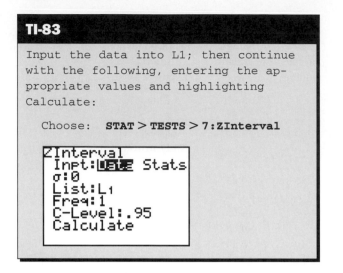

Sample Size

The confidence interval has two basic characteristics that determine its quality: its level of confidence and its width. It is preferred that the interval have a high level of confidence and be precise (narrow) at the same time. The higher the level of confidence, the more likely the interval is to contain the parameter, and the narrower the interval the more precise the estimation. However, these two properties seem to work against each other since it would seem that a narrower interval would tend to have a lower probability and a wider interval would be less precise. The maximum error part of the confidence interval formula specifies the relationship involved.

$$E = z(\alpha/2) \cdot \frac{\sigma}{\sqrt{n}} \qquad\qquad (8.2)$$

The components of this formula are: (a) maximum desired error E, one-half of the width of the confidence interval; (b) the confidence coefficient, $z(\alpha/2)$, which is determined by the level of confidence; (c) the sample size, n; and (d) the standard deviation, σ. The standard deviation σ is not a concern in this discussion because it is a constant (the standard deviation of a population does not change in value). That leaves three factors. Inspection of formula (8.2) indicates the following: Increasing the level of confidence will make the confidence coefficient larger and thereby require either the maximum error to increase or the sample size to increase; decreasing the maximum error will require the level of confidence to decrease or the sample size to increase; decreasing the sample size will force the maximum error to become larger or the level of confidence to decrease. We have a "three-way tug of war," as pictured in Figure 8.6 (p. 363). An increase or decrease to any one of the three factors has an effect on one or both of the other two factors. The statistician's job is to "balance" the level of confidence, the sample size, and the maximum error so that an acceptable interval results.

> When the denominator increases, the value of the fraction decreases.

Figure 8.6

The "Three-Way Tug of War" Between $1 - \alpha$, n, and E

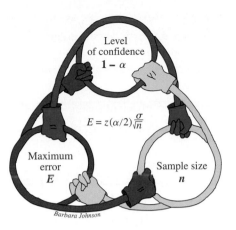

Barbara Johnson

Let's look at an illustration of this relationship in action.

ILLUSTRATION 8.4 ▼

Determine the size sample needed to estimate the mean weight of all second-grade boys if we want to be accurate within 1 lb. with 95% confidence. Assume a normal distribution and that the standard deviation of such weights is 3 lb.

Solution

> Instructions for using Table 4B are on pages 357 and 358.

The desired level of confidence determines the confidence coefficient: the confidence coefficient is found using Table 4B: $z(\alpha/2) = z(0.025) = $ **1.96.**

The desired maximum error is $E = 1.0$. Now we are ready to use the maximum error formula:

$$E = z_{(\alpha/2)} \frac{\sigma}{\sqrt{n}}: \quad 1.0 = 1.96 \cdot \frac{3}{\sqrt{n}}$$

$$\text{Solve for } n: \quad 1.0 = \frac{5.88}{\sqrt{n}}$$

$$\sqrt{n} = 5.88$$

$$n = (5.88)^2 = 34.57 = \mathbf{35}$$

> Complete Exercise 8.17 to see the effect that increasing the level of confidence has on sample size when the maximum error is kept the same.

Therefore, $n = 35$ is the sample size needed if you want a 95% confidence interval with a maximum error no larger than 1.0 lb.   ▲

NOTE When solving for the sample size n, it is customary to round up to the next larger integer, no matter what fraction (or decimal) results.

Using the maximum error formula (8.2) can be made a little easier by rewriting the formula into a form that expresses n in terms of the other values.

> **Exercise 8.17**
>
> Find the sample size needed to estimate μ of a normal population with σ = 3 to within 1 unit at the 98% level of confidence.

$$n = \left(\frac{z_{(\alpha/2)} \cdot \sigma}{E} \right)^2 \tag{8.3}$$

If the maximum error is expressed as a multiple of the standard deviation σ, the actual value of σ is not needed in order to calculate the sample size.

ILLUSTRATION 8.5 ▼

Find the sample size needed to estimate the population mean to within $\frac{1}{5}$ of a standard deviation with 99% confidence.

Solution

Determine the confidence coefficient (using Table 4b): $1 - \alpha = 0.99$, $z_{(\alpha/2)} = 2.58$.

The desired maximum error is $E = \frac{\sigma}{5}$. Now we are ready to use the sample size formula (8.3):

$$n = \left(\frac{z_{(\alpha/2)} \cdot \sigma}{E} \right)^2 : \quad n = \left(\frac{(2.58) \cdot (\sigma)}{\sigma/5} \right)^2 = \left(\frac{(2.58\sigma)(5)}{\sigma} \right)^2 = [(2.58)(5)]^2$$

$$n = (12.90)^2 = 166.41 = \mathbf{167}$$ ▲

EXERCISES · · · · · · · · · · · · · ·

8.18 Discuss the effect that each of the following have on the confidence interval.
a. point estimate   **b.** level of confidence
c. sample size   **d.** variability of the characteristic being measured

8.19 Discuss the conditions that must exist before we can estimate the population mean using the interval techniques of formula (8.1).

8.20 Determine the value of the confidence coefficient $z_{(\alpha/2)}$ for each situation described:
a. $1 - \alpha = 0.90$ **b.** $1 - \alpha = 0.95$
c. 98% confidence   **d.** 99% confidence

8.21 Given the information: the sampled population is normally distributed, $n = 16$, $\bar{x} = 28.7$, and $\sigma = 6$.
a. Find the 0.95 confidence interval for μ.
b. Are the assumptions satisfied? Explain why.

8.22 Given the information: the sampled population is normally distributed, $n = 55$, $\bar{x} = 78.2$, and $\sigma = 12$.
a. Find the 0.98 confidence interval for μ.
b. Are the assumptions met? Explain.

8.23 Given the information: $n = 86$, $\bar{x} = 128.5$, and $\sigma = 16.4$.
a. Find the 0.90 confidence interval for μ.
b. Are the assumptions satisfied? Explain why.

8.24 Given the information: $n = 22$, $\bar{x} = 72.3$, and $\sigma = 6.4$.
a. Find the 0.99 confidence interval for μ.
b. Are the assumptions satisfied? Explain why.

8.25 The Channel Tunnel train that connects England with France carries up to 650 passengers, and peak speeds of over 190 mph are occasionally obtained (*People*, "Match Point," 7-13-98). Recently Prince Harry paid $206 for his ticket to take the two-hour trip, whisking along at reportedly 186 mph. Assume the standard deviation is 19 mph in the course of all the journeys back and forth and that the train's speed is normally distributed. Suppose speed readings are made during the next 20 trips of the Channel Tunnel train, and the mean speed of these measurements is 184 mph.
a. What is the variable being studied?
b. Find the 90% confidence interval for the mean speed.
c. Find the 95% confidence interval for the mean speed.

8.26 A certain adjustment to a machine will change the length of the parts it is making but will not affect the standard deviation. The length of the parts is normally distributed, and the standard deviation is 0.5 mm. After an adjustment is made, a random sample is taken to determine the mean length of parts now being produced. The resulting lengths are:

| 75.3 | 76.0 | 75.0 | 77.0 | 75.4 | 76.3 | 77.0 | 74.9 | 76.5 | 75.8 |

    a. What is the parameter of interest?
    b. Find the point estimate for the mean length of all parts now being produced.
    c. Find the 0.99 confidence interval for μ.

8.27 A sample of 60 night school students' ages is obtained in order to estimate the mean age of night school students. $\bar{x} = 25.3$ years. The population variance is 16.
    a. Give a point estimate for μ.
    b. Find the 95% confidence interval for μ.
    c. Find the 99% confidence interval for μ.

8.28 The lengths of 200 fish caught in Cayuga Lake had a mean of 14.3 in. The population standard deviation is 2.5 in.
    a. Find the 90% confidence interval for the population mean length.
    b. Find the 98% confidence interval for the population mean length.

8.29 An article titled "A Comparison of the Effects of Constant Co-operative Grouping versus Variable Co-operative Grouping on Mathematics Achievement Among Seventh Grade Students" (*International Journal of Mathematics Education in Science and Technology,* Vol. 24, No. 5, 1993) gives the mean percentile score on the California Achievement Test (CAT) for 20 students to be 55.20. Assume the population of CAT scores is normally distributed and that σ = 19.5.
    a. Make a point estimate for the mean of the population the sample represents.
    b. Find the maximum error of estimate for a level of confidence equal to 95%.
    c. Construct a 95% confidence interval for the population mean.
    d. Explain the meaning of each of the above answers.

8.30 About 67% of married adults say they consult with their spouse before spending $352, the average of the amount for which married adults say they consult (Yankelovich Partner for Lutheran Brotherhood).
    a. Based on the above information, what can you conclude about the variable, amount spent before consulting with spouse? What is the $352?

A survey of 500 married adults was taken from a nearby neighborhood and gave a sample mean of $289.75.
    b. Construct a 0.98 confidence interval for the mean amount for all married adults. Use σ = $600.
    c. Based on the above answers, what can you conclude about the mean amount before consulting with spouse for adults in the sampled neighborhood compared to the general population?

8.31 According to a USA Snapshot® (*USA Today,* 11-3-94), the annual teaching income for ski instructors in the Rocky Mountain and Sierra areas is $5600. (Assume σ = $1000.)
    a. If this figure is based on a survey of 15 instructors and if the annual incomes are normally distributed, find a 90% confidence interval for μ, the mean annual teaching income for all ski instructors in the Rocky Mountain and Sierra areas.
    b. If the distribution of annual incomes is not normally distributed, what effect do you think that would have on the interval answer in part (a)? Explain.

8.32 How large a sample should be taken if the population mean is to be estimated with 99% confidence to within $75? The population has a standard deviation of $900.

8.33 A high-tech company wants to estimate the mean number of years of college education its employees have completed. A good estimate of the standard deviation for the number of years of college is 1.0. How large a sample needs to be taken to estimate μ to within 0.5 of a year with 99% confidence?

8.34 By measuring the amount of time it takes a component of a product to move from one work station to the next, an engineer has estimated that the standard deviation is 5 sec.
    a. How many measurements should be made in order to be 95% certain that the maximum error of estimation will not exceed 1 sec?
    b. What sample size is required for a maximum error of 2 sec?

8.35 The new mini-laptop computers can deliver as much computing power as machines ten times their size, but they weigh in at less than three pounds. Experts predict they will soon replace their traditional 6- to 8-pound older laptop brothers as the computer industry continues its relentless quest for smallness (*Fortune*, "Time to Dump Heavyweights," 5-25-98). If the standard deviation of the weight of all mini-laptops is 0.4 pounds, how large a sample would be needed to estimate the population mean weight if the maximum error of estimate is to be 0.15 pounds with 95% confidence?

8.36 According to a USA Snapshot® (*USA Today*, 10-7-98), adults visit a public library an average of seven times per year. A random sample of the adults in a metropolitan area is to be commissioned by the library's planning board. They are interested in estimating the mean number of visits made to the library by the adults of their community. How large will the sample need to be if they wish to estimate the mean within 0.3 of one standard deviation with 0.98 confidence?

8.3   The Nature of Hypothesis Testing

We all make decisions every day of our lives. Some of these decisions are of major importance; others are seemingly insignificant. All decisions follow the same basic pattern. We weigh the alternatives; then, based on our beliefs and preferences, and whatever evidence is available, we arrive at a decision and take the appropriate action. The statistical hypothesis test follows much the same process, except that it involves statistical information. In this section, we will develop many of the concepts and attitudes of the hypothesis test while looking at several decision-making situations without any statistics.

A friend is having a party (Super Bowl party, home-from-college party, you know the situation, any excuse will do), and you have been invited. You must make a decision: attend or not attend. Simple decision; well maybe, except that you want to go only if you can be convinced the party is going to be more fun than your friend's typical party; further, you definitely do not want to go if it's going to be just another dud of a party. You have taken the position that "the party will be a dud" and you will not go unless you become convinced otherwise. Your friend assures you, "Guaranteed, the party will be a great time!" Do you go or not?

The decision-making process starts by identifying **something of concern** and then formulating **two hypotheses** about it.

HYPOTHESIS

A statement that something is true.

Your friend's statement, "The party will be a great time," is a hypothesis. Your position, "The party will be a dud," is also a hypothesis.

STATISTICAL HYPOTHESIS TEST

A process by which a decision is made between two opposing hypotheses. The two opposing hypotheses are formulated so that each hypothesis is the negation of the other. (That way one of them is always true, and the other one is always false.) Then one hypothesis is tested, hoping that it can be shown to be a very improbable occurrence, thereby implying the other hypothesis is likely the truth.

The two hypotheses involved in making a decision are known as the *null hypothesis* and the *alternative hypothesis.*

NULL HYPOTHESIS,* H_o

The hypothesis we will test. Generally this is a statement that a population parameter has a specific value. The null hypothesis is so named because it is the "starting point" for the investigation. (The phrase "there is no difference" is often used in its interpretation.)

ALTERNATIVE HYPOTHESIS, H_a

A statement about the same population parameter that is used in the null hypothesis. Generally this is a statement that specifies the population parameter has a value different, in some way, from the value given in the null hypothesis. The rejection of the null hypothesis will imply the likely truth of this alternative hypothesis.

With regard to your friend's party, the two opposing viewpoints or hypotheses would be: "The party will be a great time" and "The party will be a dud." Which statement becomes the null hypothesis, and which becomes the alternative hypothesis?

Determining the statement of the null hypothesis and the statement of the alternative hypothesis is a very important step. The *basic idea* of the hypothesis test is for the evidence to have a chance to "disprove" the null hypothesis. The null hypothesis is the statement that the evidence might disprove. *Your concern* (belief or desired outcome), as the person doing the testing, is expressed in the alternative hypothesis. As the person making the decision, you believe that the evidence will demonstrate the feasibility of your "theory" by demonstrating the *unlikelihood* of the truth of the null hypothesis. The alternative hypothesis is sometimes referred to as the *research hypothesis* since it represents what the researcher hopes will be found to be "true." (If so, he or she will get a paper out of the research.)

Since the "evidence" (who's going to the party, what is going to be served, and so on) can only demonstrate the unlikeliness of the party being a dud, your initial position, "The party will be a dud," becomes the null hypothesis. Your friend's claim, "The party will be a great time," then becomes the alternative hypothesis.

$$H_o: \text{``Party will be a dud''} \quad \text{vs.} \quad H_a: \text{``Party will be a great time''}$$

ILLUSTRATION 8.6 ▼

You are testing a new design for airbags used in automobiles, and you are concerned that they might not open properly. State the null and alternative hypotheses.

---

*We use the notation H_o for the null hypothesis to contrast it with H_a for the alternative hypothesis. Other texts may use H_0 (subscript zero) in place of H_o and H_1 in place of H_a.

Solution

The two opposing possibilities are "bags open properly" or "bags do not open properly." Testing can only produce evidence that discredits the hypothesis "the bags open properly." Therefore, the null hypothesis is "they do open properly" and the alternative hypothesis is "they do not open properly." ▲

The alternative hypothesis can be the statement the experimenter wants to show to be true.

ILLUSTRATION 8.7 ▼

An engineer wishes to show that the new formula that was just developed results in a quicker-drying paint. State the null and alternative hypotheses.

Solution

The two opposing possibilities are "does dry quicker" and "does not dry quicker." Since the engineer wishes to show "does dry quicker," the alternative hypothesis is "paint made with the new formula does dry quicker" and the null hypothesis is "paint made with the new formula does not dry quicker." ▲

Occasionally it might be reasonable to hope that the evidence does not lead to a rejection of the null hypothesis. Such is the case in Illustration 8.8.

ILLUSTRATION 8.8 ▼

You suspect that a brand-name detergent outperforms the store's brand of detergent, and you wish to test the two detergents because you would prefer to buy the cheaper store brand. State the null and alternative hypotheses.

Solution

Your suspicion, "the brand-name detergent outperforms the store brand" is the reason for the test and therefore becomes the alternative hypothesis.

> H_o: "There is no difference in detergent performance."

> H_a: "The brand-name detergent performs better than the store brand."

However, as a consumer, you are hoping not to reject the null hypothesis for budgetary reasons. ▲

Exercise 8.37

You are testing a new detonating system for explosives and you are concerned that the system is not reliable. State the null and alternative hypotheses.

Before returning to our example about the party, we need to look at the four possible outcomes that could result from the null hypothesis being either true or false and the decision being either to "reject H_o" or to "fail to reject H_o." Table 8.2 shows these four possible outcomes. A **type A correct decision** occurs when the null hypothesis is true, and we decide in its favor. A **type B correct decision** occurs when the null hypothesis is false, and the decision is in opposition to the null hypothesis. A **type I error** will be committed when a true null hypothesis is rejected, that is, when the null hypothesis is true but we decided against it. A **type II error** is committed when we decide in favor of a null hypothesis that is actually false.

Case Study 8.3

Evaluation of Teaching Techniques

Abstract: This study tests the effect of homework collection and quizzes on exam scores.

The hypothesis for this study is that an instructor can improve a student's performance (exam scores) through influencing the student's perceived effort-reward probability. An instructor accomplishes this by assigning tasks (teaching techniques) which are a part of a student's grade and are perceived by the student as a means of improving his or her grade in the class. The student is motivated to increase effort to complete those tasks which should also improve understanding of course material. The expected final result is improved exam scores.

The null hypothesis for this study is:

H_o: Teaching techniques have no significant effect on students' exam scores. . . .

Source: David R. Vruwink and Janon R. Otto, *The Accounting Review,* Vol. LXII, No. 2, April 1987, Reprinted by permission.

Table 8.2   Four Possible Outcomes in a Hypothesis Test

	Null Hypothesis	
Decision	**True**	**False**
Fail to reject H_o	Type A correct decision	Type II error
Reject H_o	Type I error	Type B correct decision

ILLUSTRATION 8.9 ▼

Describe the four possible outcomes and the resulting actions that would occur for the hypothesis test in Illustration 8.8.

Solution

Recall: H_o: "There is no difference in detergent performance"
$\quad\quad\quad\quad$ H_a: "The brand-name detergent performs better than the store brand"

	Null Hypothesis Is True	**Null Hypothesis Is False**
Fail to Reject H_o	*Type A Correct Decision* **Truth of situation:** There is no difference between the detergents. **Conclusion:** It was determined that there was no difference. **Action:** The consumer bought the cheaper detergent, saving money and getting the same results.	*Type II Error* **Truth of situation:** The brand-name detergent is better. **Conclusion:** It was determined that there was no difference. **Action:** The consumer bought the cheaper detergent, saving money and getting inferior results.
Reject H_o	*Type I Error* **Truth of situation:** There is no difference between the detergents. **Conclusion:** It was determined that the brand-name detergent was better. **Action:** The consumer bought the brand-name detergent, spending extra money to attain no better results.	*Type B Correct Decision* **Truth of situation:** The brand-name detergent is better. **Conclusion:** It was determined that the brand-name detergent was better. **Action:** The consumer bought the brand-name detergent, spending more and getting better results.

▲

(See Note on p. 370)

Exercise 8.40

Describe how the type II error in the party example represents a "lost opportunity."

NOTE The type II error often results in what represents a "lost opportunity"; lost in this situation is the chance to use a product that yields better results.

When a decision is made, it would be nice to always make the correct decision. This, however, is not possible in statistics, since we make our decisions on the basis of sample information. The best we can hope for is to control the probability with which an error occurs. The probability assigned to the type I error is α (called **"alpha"**; α is the first letter of the Greek alphabet). The probability of the type II error is β (called **"beta"**; β is the second letter of the Greek alphabet). See Table 8.3.

Exercise 8.41

Explain why α is not always the probability of rejecting the null hypothesis.

Table 8.3 Probability with Which Decisions Occur

Error in Decision	Type	Probability	Correct Decision	Type	Probability
Rejection of a true H_o	I	α	Failure to reject a true H_o	A	$1 - \alpha$
Failure to reject a false H_o	II	β	Rejection of a false H_o	B	$1 - \beta$

Exercise 8.42

Explain how assigning a small probability to an error controls the likelihood of its occurrence.

To control these errors we will assign a small probability to each of them. The most frequently used probability values for α and β are 0.01 and 0.05. The probability assigned to each error depends on its seriousness. The more serious the error, the less willing we are to have it occur, and therefore a smaller probability will be assigned. α and β are each probabilities of errors, each under separate conditions, and they cannot be combined. Therefore, we cannot determine a single probability for making an incorrect decision. Likewise, the two correct decisions are distinctly separate and each has its own probability; $1 - \alpha$ is the probability of a correct decision when the null hypothesis is true, and $1 - \beta$ is the probability of a correct decision when the null hypothesis is false. $1 - \beta$ is called the power of the statistical test since it is the measure of the ability of a hypothesis test to reject a false null hypothesis, a very important characteristic.

REMEMBER Regardless of the outcome of a hypothesis test, you are never certain that a correct decision has been reached.

Let's look back at the two possible errors in decision that could occur in Illustration 8.9 while testing the detergents. Most people would become upset if they found out they were spending extra money for a detergent that performed no better than the cheaper brand. Likewise, most people would become upset if they found out they could have been buying a better detergent. Evaluating the relative seriousness of these errors requires knowing whether this is your personal laundry or a professional laundry business, and how much extra the brand-name detergent costs, and so on.

There is an interrelationship between the probability of the type I error (α), the probability of the type II error (β), and the sample size (n). This relationship is very much like the relationship between level of confidence, maximum error, and sample size discussed on pages 362 and 363. Figure 8.7 shows the "three-way tug of war" among α, β, and n. If any one of the three is increased or decreased, it has an effect on one or both of the others.

If α is reduced, then either β must increase or n must be increased; if β is decreased, then either α increases or n must be increased; if n is decreased, then either α increases or β increases. The choices for α, β, and n are definitely not arbitrary. At this time in our study of statistics, α will be given in the statement of the problem, as will the sample size n. Further discussion on the role of β, the P(type II error), is left for another time.

Figure 8.7

The "Three-Way Tug of War"
Between α, β, and *n*

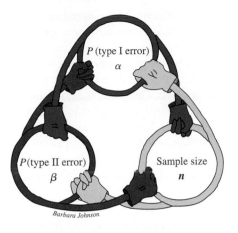

Barbara Johnson

LEVEL OF SIGNIFICANCE α

The probability of committing the type I error.

Establishing the level of significance can be thought of as a "managerial decision." Typically, someone in charge determines the level of probability with which they are willing to risk a type I error.

At this point in the hypothesis test procedure, the evidence is collected, summarized, and the value of a *test statistic* is calculated.

TEST STATISTIC

A random variable whose value is calculated from the sample data and is used in making the decision "fail to reject H_o" or "reject H_o."

The value of the calculated test statistic is used in conjunction with a decision rule to determine either "reject H_o " or "fail to reject H_o." This **decision rule** must be established prior to collecting the data and specifics how you will reach the decision.

Back to your friend's party: You have to weigh the history of your friend giving parties, the time and place, others going, and so on, against your own criteria and then make your decision. As a result of the decision about the null hypothesis ("the party will be a dud"), you will take the appropriate action; you will either go to or not go to the party.

To complete a hypothesis test, you will need to write a conclusion that carefully describes the meaning of the decision relative to the intent of the hypothesis test.

THE CONCLUSION

   a. If the decision is "reject H_o," then the conclusion should be worded something like, "There is sufficient evidence at the α level of significance to show that . . . (the meaning of the alternative hypothesis)."

   b. If the decision is "fail to reject H_o," then the conclusion should be worded something like, "There is not sufficient evidence at the α level of significance to show that . . . (the meaning of the alternative hypothesis)."

When writing the decision and the conclusion, remember that (1) the decision is about H_o and (2) the conclusion is a statement about whether or not the contention of H_a was upheld. This is consistent with the "attitude" of the whole hypothesis test procedure. The null hypothesis is the statement that is "on trial," and therefore the decision must be about it. The contention of the alternative hypothesis is the thought that brought about the need for a decision. Therefore, the question that led to the alternative hypothesis must be answered when the conclusion is written.

We must always remember that when the decision is made, nothing has been proved. Both decisions can lead to errors: "Fail to reject H_o" could be a type II error (the lack of sufficient evidence has led to great parties being missed more than once), and "reject H_o" could be a type I error (more than one person has decided to go to a party that was a dud).

EXERCISES

8.43 State the null and alternative hypotheses for each of the following:
  a. You are investigating a complaint that "special delivery mail takes too much time" to be delivered.
  b. You want to show that people find the new design for a recliner chair more comfortable than the old design.
  c. You are trying to show that cigarette smoke has an effect on the quality of a person's life.
  d. You are testing a new formula for hair conditioner hoping to show it is effective on "split ends."

8.44 When a parachute is inspected, the inspector is looking for anything that might indicate the parachute might not open.
  a. State the null and alternative hypotheses.
  b. Describe the four possible outcomes that can result depending on the truth of the null hypothesis and the decision reached.
  c. Describe the seriousness of the two possible errors.

8.45 When a medic at the scene of a serious accident inspects each victim, she administers the appropriate medical assistance to all victims, unless she is certain the victim is dead.
  a. State the null and alternative hypotheses.
  b. Describe the four possible outcomes that can result depending on the truth of the null hypothesis and the decision reached.
  c. Describe the seriousness of the two possible errors.

8.46 A supplier of highway materials claims he can supply an asphalt mixture that will make roads paved with his materials less slippery when wet. A general contractor who builds roads wishes to test the supplier's claim. The null hypothesis is "roads paved with this asphalt mixture are no less slippery than roads paved with other asphalt." The alternative hypothesis is "roads paved with this asphalt mixture are less slippery than roads paved with other asphalt."
  a. Describe the meaning of the two possible types of errors that can occur in the decision when this hypothesis test is completed.
  b. Describe how the null hypothesis, as stated above, is a "starting point" for the decision to be made about the asphalt.

8.47 Describe the action that would result in a type I error and a type II error if each of the following null hypotheses were tested. (Remember, the alternative hypothesis is the negation of the null hypothesis.)
  a. H_o: The majority of Americans favor laws against assault weapons.
  b. H_o: This fast-food menu is not low salt.
  c. H_o: This building must not be demolished.
  d. H_o: There is no waste in government spending.

8.48 Describe the action that would result in a correct decision type A and a correct decision type B, if each of the null hypotheses in Exercise 8.47 were tested.

8.49 **a.** If the null hypothesis is true, what decision error could be made?
b. If the null hypothesis is false, what decision error could be made?
c. If the decision "reject H_o" is made, what decision error could have been made?
d. If the decision "fail to reject H_o" is made, what decision error could have been made?

8.50 The director of an advertising agency is concerned with the effectiveness of a television commercial.
a. What null hypothesis is she testing if she commits a type I error when she erroneously says that the commercial is effective?
b. What null hypothesis is she testing if she commits a type II error when she erroneously says that the commercial is effective?

8.51 **a.** If α is assigned the value 0.001, what are we saying about the type I error?
b. If α is assigned the value 0.05, what are we saying about the type I error?
c. If α is assigned the value 0.10, what are we saying about the type I error?

8.52 **a.** If β is assigned the value 0.001, what are we saying about the type II error?
b. If β is assigned the value 0.05, what are we saying about the type II error?
c. If β is assigned the value 0.10, what are we saying about the type II error?

8.53 **a.** If the null hypothesis is true, the probability of a decision error is identified by what name?
b. If the null hypothesis is false, the probability of a decision error is identified by what name?

8.54 Suppose that a hypothesis test is to be carried out by using α = 0.05. What is the probability of committing a type I error?

8.55 The conclusion is part of the hypothesis test that communicates the findings of the test to the reader. As such, it needs special attention so that the reader receives an accurate picture of the findings.
a. Carefully describe the "attitude" of the statistician and the statement of the conclusion when the decision is "reject H_o."
b. Carefully describe the "attitude" and the statement of the conclusion when the decision is "fail to reject H_o."

8.56 Find the power of a test when the probability of the type II error is
a. 0.01   **b.** 0.05   **c.** 0.10

8.57 A normally distributed population is known to have a standard deviation of 5, but its mean is in question. It has been argued to be either μ = 80 or μ = 90, and the following hypothesis test has been devised to settle the argument. The null hypothesis, H_o: μ = 80, will be tested using one randomly selected data and comparing it to the critical value 86. If the data is greater than or equal to 86, the null hypothesis will be rejected.
a. Find α, the probability of the type I error.
b. Find β, the probability of the type II error.

8.58 Suppose the argument in Exercise 8.57 was to be settled using a sample of size 4; find α and β.

8.4   Hypothesis Test of Mean μ (σ Known): A Probability-Value Approach

In Section 8.3 we surveyed the concepts and much of the reasoning behind a hypothesis test while looking at nonstatistical illustrations. In this section we are going to formalize the hypothesis test procedure as it applies to statements concerning the mean μ of a population under the restriction that σ, the population standard deviation, is a known value.

THE ASSUMPTION FOR HYPOTHESIS TESTS ABOUT MEAN μ USING A KNOWN σ

The sampling distribution of $\bar{x}$ has a normal distribution.

The information needed to ensure that this assumption is satisfied is contained in the sampling distribution of sample means and in the Central Limit Theorem (see Chapter 7, p. 324, 325).

The sampling distribution for sample means $\bar{x}$ is distributed about a mean equal to μ with a standard error equal to $\sigma/\sqrt{n}$; and, (1) if the randomly sampled population is normally distributed, then $\bar{x}$ is normally distributed for all sample sizes, or (2) if the randomly sampled population is not normally distributed, then $\bar{x}$ is approximately normally distributed for sufficiently large sample sizes.

The hypothesis test is a well-organized, step-by-step procedure used to make a decision. Two different formats are commonly used for hypothesis testing. The *probability-value approach*, or simply *p-value approach*, is the hypothesis test process that has gained popularity in recent years, largely as a result of the convenience and the "number crunching" ability of the computer. This approach is organized as a five-step procedure.

THE PROBABILITY-VALUE HYPOTHESIS TEST: A FIVE-STEP PROCEDURE

STEP 1 **The Set-Up:**
  a. Describe the population parameter of interest.
  b. State the null hypothesis (H_o) and the alternative hypothesis (H_a).

STEP 2 **The Hypothesis Test Criteria:**
  a. Check the assumptions.
  b. Identify the probability distribution and the test statistic formula to be used.
  c. Determine the level of significance, α.

STEP 3 **The Sample Evidence:**
  a. Collect the sample information.
  b. Calculate the value of the test statistic.

STEP 4 **The Probability Distribution:**
  a. Calculate the *p*-value for the test statistic.
  b. Determine whether or not the *p*-value is smaller than α.

STEP 5 **The Results:**
  a. State the decision about H_o.
  b. State a conclusion about H_a.

A commercial aircraft manufacturer buys rivets for use in assembling airliners. Each rivet supplier that wants to sell rivets to the aircraft manufacturer must demonstrate that its rivets meet the required specifications. One of the specs is: "The mean shearing strength of all such rivets, μ, is at least 925 lb." Each time the aircraft manufacturer buys rivets, it is concerned that the mean strength might be less than the 925-lb specification.

> Think about the consequences of using weak rivets.

NOTE 1 Each individual rivet has a shearing strength, which is determined by measuring the force required to shear ("break") the rivet. Clearly, not all the rivets can be tested. Therefore, a sample of rivets will be tested, and a decision about the mean strength of all the untested rivets will be based on the mean from those sampled and tested.

NOTE 2 We will use $\sigma = 18$ for our rivet example.

STEP 1 **The Set-Up:**

a. **Describe the population parameter of interest.**
   The population parameter of interest is the **mean μ, the mean shearing strength** (or mean force required to shear) of the rivets being considered for purchase.

b. **State the null hypothesis (H_o) and the alternative hypothesis (H_a).**

The null hypothesis and the alternative hypothesis are formulated by inspecting the problem or statement to be investigated and first formulating two opposing statements concerning the mean μ. For our example, these two opposing statements are: (A) "The mean shearing strength is less than 925" ($\mu < 925$, the aircraft manufacturer's concern), and (B) "The mean shearing strength is at least 925" ($\mu = 925$, the rivet supplier's claim and the aircraft manufacturer's spec).

NOTE The trichotomy law from algebra states that two numerical values must be related in exactly one of three possible relationships: $<$, $=$, or $>$. All three of these possibilities must be accounted for between the two opposing hypotheses in order for the two hypotheses to be negations of each other. The three possible combinations of signs and hypotheses are shown in Table 8.4. Recall that the null hypothesis assigns a specific value to the parameter in question and therefore "equals" will always be part of the null hypothesis.

Table 8.4   The Three Possible Statements of Null and Alternative Hypotheses

Null Hypothesis	Alternative Hypothesis
1. greater than or equal to ($\geq$)	less than ($<$)
2. less than or equal to ($\leq$)	greater than ($>$)
3. equal to ($=$)	not equal to ($\neq$)

The parameter of interest, the population mean μ, is related to the value 925.
Statement (A) becomes the alternative hypothesis:

$$H_a: \mu < 925 \text{ (the mean is less than 925)}$$

This statement represents the aircraft manufacturer's concern and says "the rivets do not meet the required specs."
Statement (B) becomes the null hypothesis:

$$H_o: \mu = 925 \ (\geq) \text{ (the mean is at least 925)}$$

This hypothesis represents the negation of the aircraft manufacturer's concern and states that "the rivets do meet the required specs."

NOTE The null hypothesis will be written with just the equal sign ("a value is assigned"). When "equal" is paired with "less than" or paired with "greater than," the combined symbol is written beside the null hypothesis as a reminder that all three signs have been accounted for in these two opposing statements.

Exercise 8.59

In this example, the aircraft builder, the buyer of the rivets, is concerned that the rivets might not meet the mean-strength spec. State the aircraft manufacturer's null and alternative hypotheses.

More specific instructions on page 367.

Before continuing with our example, let's look at three illustrations that demonstrate formulating the statistical null and alternative hypotheses involving the population mean μ. Illustrations 8.10 and 8.11 each demonstrate a "one-tailed" alternative hypothesis.

ILLUSTRATION 8.10 ▼

Suppose the EPA was suing the city of Rochester for noncompliance with carbon monoxide standards. Specifically, the EPA would want to show that the mean level of carbon monoxide in downtown Rochester's air is dangerously high, higher than 4.9 parts per million. State the null and alternative hypotheses.

Solution

To state the two hypotheses, we first need to identify the population parameter in question: the "mean level of carbon monoxide pollution in Rochester." The parameter μ is being compared to the value 4.9 parts per million, the specific value of interest. The EPA is questioning the value of μ and wishes to show it to be higher than 4.9 (that is, $\mu > 4.9$). The three possible relationships (1) $\mu < 4.9$, (2) $\mu = 4.9$, or (3) $\mu > 4.9$ must be arranged to form two opposing statements: One states the EPA's position, "the mean level is higher than 4.9 ($\mu > 4.9$)," and the other states the negation, "the mean level is not higher than 4.9 ($\mu \leq 4.9$)." One of these two statements will become the null hypothesis H_o, and the other will become the alternative hypothesis H_a.

RECALL There are two rules for forming the hypotheses: (1) The null hypothesis states that the parameter in question has a specified value ("H_o must contain the equal sign"), and (2) the EPA's contention becomes the alternative hypothesis ("higher than"). Both rules indicate:

$$H_o: \mu = 4.9 \ (\leq) \quad \text{and} \quad H_a: \mu > 4.9$$

ILLUSTRATION 8.11 ▼

An engineer wants to show that applications of paint made with the new formula dry and are ready for the next coat in a mean time of less than 30 minutes. State the null and alternative hypotheses for this test situation.

Solution

The parameter of interest is the mean drying time per application, and 30 minutes is the specified value. $\mu < 30$ corresponds to "the mean time is less than 30," whereas $\mu \geq 30$ corresponds to the negation, "the mean time is not less than 30." Therefore, the hypotheses are

$$H_o: \mu = 30 \ (\geq) \quad \text{and} \quad H_a: \mu < 30 \qquad ▲$$

Illustration 8.12 demonstrates a "two-tailed" alternative hypothesis.

ILLUSTRATION 8.12 ▼

Job satisfaction is very important when it comes to getting workers to produce. A standard job-satisfaction questionnaire was administered by union officers to a sample of assembly-line workers in a large plant in hopes of showing that the assembly workers' mean score on this questionnaire would be different from the established mean of 68. State the null and alternative hypotheses.

Solution

Either the mean job satisfaction score is different from 68 ($\mu \neq 68$) or the mean is equal to 68 ($\mu = 68$). Therefore,

$$H_o: \mu = 68 \quad \text{and} \quad H_a: \mu \neq 68$$

▲

Exercise 8.60

Professor Hart does not believe a statement he heard: "The mean weight of college women is 54.4 kg." State the null and alternative hypotheses he would use to challenge this statement.

NOTE 1 The alternative hypothesis is referred to as being "two-tailed" when H_a is "not equal."

NOTE 2 When "less than" is combined with "greater than," they become "not equal to."

The viewpoint of the experimenter greatly affects the way the hypotheses are formed. Generally, the experimenter is trying to show that the parameter value is different from the value specified. Thus, the experimenter is often hoping to be able to reject the null hypothesis so that the experimenter's theory has been substantiated. Illustrations 8.10, 8.11, and 8.12 also represent the three possible arrangements for the $<$, $=$, and $>$ relationships between the parameter μ and a specified value.

Table 8.5 lists some additional common phrases used in claims and indicates their negations and the hypothesis where each phrase will be used. Again, notice that "equals" is always in the null hypothesis. Also notice that the negation of "less than" is "greater than or equal to." Think of negation as "all the others" from the set of three signs.

Table 8.5   Common Phrases and Their Negations

H_o: ($\geq$)	H_a: ($<$)	H_o: ($\leq$)	H_a: ($>$)	H_o: ($=$)	H_a: ($\neq$)
at least	less than	at most	more than	is	is not
no less than	less than	no more than	more than	not different from	different from
not less than	less than	not greater than	greater than	same as	not same as

Exercise 8.61

State the null and alternative hypotheses used to test each of the following claims.

a. The mean reaction time is greater than 1.25 seconds.
b. The mean score on that qualifying exam is less than 335.
c. The mean selling price of homes in the area is not $230,000.

Once the null and alternative hypotheses are established, we will work under the assumption that the null hypothesis is a true statement until there is sufficient evidence to reject it. This situation might be compared to a courtroom trial, where the accused is assumed to be innocent (H_o: Defendant is innocent vs. H_a: Defendant is not innocent) until sufficient evidence has been presented to show that innocence is totally unbelievable ("beyond reasonable doubt"). At the conclusion of the hypothesis test, we will make one of two possible decisions. We will decide in opposition to the null hypothesis and say that we "reject H_o" (this corresponds to "conviction" of the accused in a trial); or we will decide in agreement with the null hypothesis and say that we "fail to reject H_o" (this corresponds to "fail to convict" or an "acquittal" of the accused in a trial).

Let's return to the rivet example we interrupted on page 375 and continue with Step 2. Recall:

$$H_o: \mu = 925 \ (\geq) \ (\text{at least 925}) \quad H_a: \mu < 925 \ (\text{less than 925})$$

STEP 2   The Hypothesis Test Criteria:

    a. Check the assumptions.

        σ is known. Variables like shearing strength typically have a mounded distribution; therefore, a sample of size 50 should be large enough for the

CLT to apply and insure that the **sampling distribution of sample means will be approximately normally distributed.**

b. **Identify the probability distribution and the test statistic to be used.**
The standard normal probability distribution is used since $\bar{x}$ is expected to have an approximately normal distribution.

For a hypothesis test of μ, we will want to compare the value of the sample mean to the value of the population mean as stated in the null hypothesis. This comparison is accomplished using formula (8.4).

$$z \star = \frac{\bar{x} - \mu}{\sigma/\sqrt{n}} \qquad\qquad (8.4)$$

The resulting calculated value is identified as $z \star$ ("z star") since it is expected to have a standard normal distribution when the null hypothesis is true and the assumptions have been satisfied. The $\star$ ("star") is to remind us that this is the calculated value of the test statistic.

The test statistic to be used is

$$z \star = \frac{\bar{x} - \mu}{\sigma/\sqrt{n}} \text{ with } \sigma = 18.$$

c. **Determine the level of significance, α.** Setting α was described as a managerial decision in Section 8.3. To see what is involved in determining α, the probability of the type I error, for our rivet example, we start by identifying the four possible outcomes, their meaning, and the action related to each.

The type I error occurs when a true null hypothesis is rejected. This would occur when the manufacturer tested rivets that in truth did meet the specs and rejected them. Undoubtedly this would lead to the rivets not being purchased even though the manufacturer unknowingly did meet the specs. In order for the manager to set a level of significance, related information is needed; namely, how soon is the new supply of rivets needed? If they are needed tomorrow and this is the only vendor with an available supply, waiting a week to find acceptable rivets could be very expensive; therefore, rejecting good rivets could be considered a serious error. On the other hand, if the rivets are not needed until next month, then this error may not be very serious. Only the manager will know all the ramifications, and therefore the manager's input is important here.

After much consideration, the manager assigns the level of significance: $\alpha = 0.05$

STEP 3   The Sample Evidence:
a. **Collect the sample information.**

We are ready for the data. The sample must be a random sample drawn from the population whose mean μ is being questioned.

A random sample of 50 rivets is selected, each rivet is tested, and the sample mean shearing strength is calculated.

$$\bar{x} = 921.18 \quad \text{and} \quad n = 50$$

b. **Calculate the value of the test statistic.**

The sample evidence ($\bar{x}$ and n found in step 3a) is next converted into the **calculated value of the test statistic, $z \star$,** using formula (8.4).

(μ is 925 from H_o; $\sigma = 18$ is the known quantity, given in Note 2, p. 350.)

$$z \star = \frac{\bar{x} - \mu}{\sigma/\sqrt{n}}: \quad z \star = \frac{921.18 - 925.0}{18/\sqrt{50}} = \frac{-3.82}{2.5456} = -1.50$$

Do Exercise 8.62 before continuing.

Exercise 8.62
Identify the four possible outcomes and describe the situation involved with each outcome with regard to the aircraft manufacturer's testing and buying rivets. Which is the more serious error; the type I or the type II? Explain.

There is more to this scenario, but hopefully you get the idea.

α will be assigned in the statement of exercises.

Exercise 8.63
Calculate the test statistic $z \star$, given H_o: $\mu = 56$, $\sigma = 7$, $\bar{x} = 54.3$, and $n = 36$.

STEP 4   **The Probability Distribution:**
   a. Calculate the *p*-value.

PROBABILITY-VALUE, OR *p*-VALUE

The probability that the test statistic could be the value it is or a more extreme value (in the direction of the alternative hypothesis) when the null hypothesis is true. (*Note:* The symbol **P** will be used to represent the *p*-value, especially in algebraic situations.)

Draw a sketch of the standard normal distribution and locate z ★ (found in step 3b) on it. To identify the area that represents the *p*-value, look at the sign in the alternative hypothesis.

> For this test, the alternative hypothesis indicates that we are interested in that part of the sampling distribution which is *"less than"* z ★. Therefore, the *p*-value is the area that lies to the *left* of z ★. Shade this area.

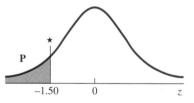

To find the *p*-value, you may use any one of three methods:

Complete instructions for using Table 3 are on page 280.

Method 1   Use Table 3 in Appendix B to determine the tabled area related to $z = 1.50$, then calculate the *p*-value by subtracting from 0.5000:

$$\text{The } p\text{-value} = P(z < z ★) = P(z < -1.50)$$
$$= P(z > 1.50) = 0.5000 - 0.4332 = \mathbf{0.0668}$$

or

Let the tables do the work the computer will typically do.

Method 2   Use Table 5 in Appendix B and the symmetry property: Table 5 is set up to allow you to read the *p*-value directly from the table.

> Since $P(z < -1.50) = P(z > 1.50)$, simply locate $z ★ = 1.50$ on Table 5 and read the *p*-value.

$$P(z < -1.50) = \mathbf{0.0668}$$

Instructions for using this command are on page 294.

or

Method 3   Use the cumulative probability function on a computer or calculator to find the *p*-value.

$$P(z < -1.50) = \mathbf{0.0668}$$

Try it! See if you get the same answer?

   b. Determine whether or not the *p*-value is smaller than α.
   The *p*-value (0.0668) **is not smaller** than α (0.05).

Exercise 8.64

a. Calculate the *p*-value, given H_a: $μ < 45$ and $z ★ = -2.3$.

b. Calculate the *p*-value, given H_a: $μ > 58$ and $z ★ = 1.8$.

STEP 5   **The Results:**
   a. State the decision about H_o.

Is the *p*-value small enough to indicate that the sample evidence is highly unlikely in the event that the null hypothesis is true? In order to make the decision, we need to know the *decision rule*.

Exercise 8.65

a. What decision is reached when the *p*-value is greater than α?

b. What decision is reached when α is greater than the *p*-value?

DECISION RULE

a. If the *p*-value is *less than* or *equal to* the level of significance α, then the decision must be **reject H₀**.

b. If the *p*-value is *greater than* the level of significance α, then the decision must be **fail to reject H₀**.

Decision about H₀   "Fail to reject H₀."

b. Write the conclusion about Hₐ.

There is not sufficient evidence at the 0.05 level of significance to show that the mean shearing strength of the rivets is less than 925.

> Specific information about writing the conclusion is on page 371.

"We failed to convict" the null hypothesis. In other words, a sample mean as small as 921.18 is likely to occur (as defined by α) when the true population mean value is 925.0 and $\bar{x}$ is normally distributed. The resulting action by the manager would be to buy the rivets.

NOTE When the decision reached is "fail to reject H₀ " (or "accept H₀," as many will improperly say), it simply means "for the lack of better information, act as if the null hypothesis is true."

Table 8.6   Finding *p*-values

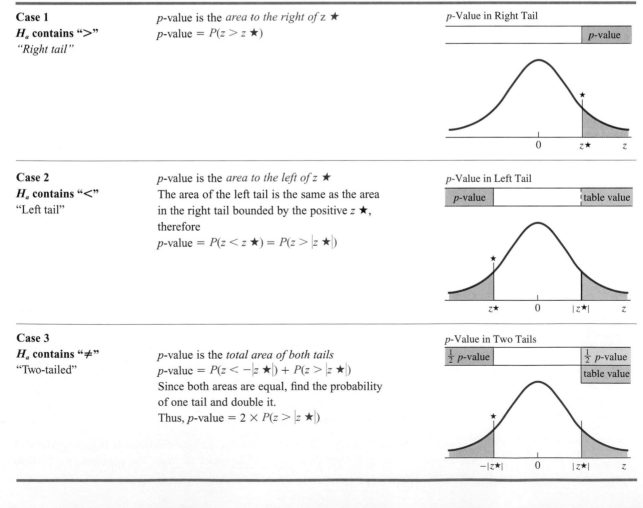

Case 1
Hₐ contains ">"
"Right tail"

p-value is the *area to the right of z ★*
p-value = $P(z > z \star)$

p-Value in Right Tail

Case 2
Hₐ contains "<"
"Left tail"

p-value is the *area to the left of z ★*
The area of the left tail is the same as the area in the right tail bounded by the positive z ★, therefore
p-value = $P(z < z \star) = P(z > |z \star|)$

p-Value in Left Tail

Case 3
Hₐ contains "≠"
"Two-tailed"

p-value is the *total area of both tails*
p-value = $P(z < -|z \star|) + P(z > |z \star|)$
Since both areas are equal, find the probability of one tail and double it.
Thus, *p*-value = $2 \times P(z > |z \star|)$

p-Value in Two Tails

Before looking at another example, let's look at the procedures for finding the p-value. The p-value is represented by the area, under the curve of the probability distribution for the test statistic, that is more extreme than the calculated value of the test statistic. There are three separate cases, and the direction (or sign) of the alternative hypothesis is the key. Table 8.6 outlines the procedure for all three cases.

Let's look at an illustration involving the two-tailed procedure.

ILLUSTRATION 8.13 ▼

Exercise 8.66

Find the test statistic z ★ and the p-value for each of the following situations.

a. H_o: μ = 22.5,
   H_a: μ > 22.5,
   $\bar{x}$ = 24.5, σ = 6,
   n = 36
b. H_o: μ = 200,
   H_a: μ < 200,
   $\bar{x}$ = 192.5, σ = 40,
   n = 50
c. H_o: μ = 12.4,
   H_a: μ ≠ 12.4,
   $\bar{x}$ = 11.52, σ = 2.2,
   n = 16

Many of the large companies in a certain city have for years used the Kelley Employment Agency for testing prospective employees. The employment selection test used has historically resulted in scores normally distributed about a mean of 82 and a standard deviation of 8. The Brown Agency has developed a new test that is quicker and easier to administer and therefore less expensive. Brown claims that its test results are the same as those obtained on the Kelley test. Many of the companies are considering a change from the Kelley Agency to the Brown Agency in order to cut costs. However, they are unwilling to make the change if the Brown test results have a different mean value. An independent testing firm tested 36 prospective employees. A sample mean of 79 resulted. Determine the p-value associated with this hypothesis test. (Assume σ = 8.)

Solution

STEP 1   The Set-Up:
   a. **Describe the population parameter of interest.**
      The population mean μ, **the mean of all test scores** using the Brown Agency test.
   b. **State the null hypothesis (H_o) and the alternative hypothesis (H_a).**
      The hypotheses: The Brown Agency's test results "will be different" (the concern) if the mean test score is not equal to 82. They "will be the same" if the mean is equal to 82.
      Therefore,   **H_o: μ = 82** (test results have the same mean)
                    H_a: μ ≠ 82 (test results have a different mean)

STEP 2   The Hypothesis Test Criteria:
   a. **Check the assumptions.**
      σ is known. If the Brown test scores are distributed the same as the Kelley scores, they will be normally distributed and the **sampling distribution will be normal** for all sample sizes.
   b. **Identify the probability distribution and the test statistic to be used.**
      The **standard normal probability distribution** and the test statistic

      $$z \star = \frac{\bar{x} - \mu}{\sigma/\sqrt{n}} \text{ will be used with } \sigma = 8.$$

   c. **Determine the level of significance, α.**

The level of significance is omitted when a question asks for the p-value and not a decision.

STEP 3   The Sample Evidence:
   a. **Collect the sample information.**
      Sample information: **n = 36, $\bar{x}$ = 79.**
   b. **Calculate the value of the test statistic.**
      (μ is 82 from H_o; σ = 8 is the known quantity)

      $$z \star = \frac{\bar{x} - \mu}{\sigma/\sqrt{n}}: \quad z \star = \frac{79 - 82}{8/\sqrt{36}} = \frac{-3}{1.3333} = \textbf{-2.25}$$

STEP 4 **The Probability Distribution:**
   a. Calculate the *p*-value.

Since the alternative hypothesis indicates a two-tailed test, we must find the probability associated with both tails. The *p*-value is found by doubling the area of one tail. (See Table 8.6, p. 380.)

Since $z\star = -2.25$, the value of $|z\star| = 2.25$.

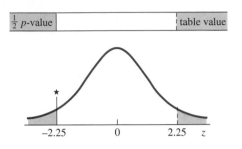

The *p*-value $= 2 \times P(z > |z\star|) = 2 \times P(z > 2.25)$

Using Table 3: *p*-value $= 2 \times P(z > 2.25) = 2 \times (0.5000 - 0.4878) = 2(0.0122) = \mathbf{0.0244}$

or

Using Table 5: *p*-value $= 2 \times P(z > 2.25) = 2(0.0122) = \mathbf{0.0244}$

or

Use the cumulative probability function on a computer or calculator:
p-value $= 2 \times P(z < -2.25) = 2(0.0122) = \mathbf{0.0244}.$

   b. Determine whether or not the *p*-value is smaller than α.
   A comparison is not possible; no α value was given in statement of question.

STEP 5 **The Results:**
   The *p*-value for this hypothesis test is 0.0244. Each individual company now will make a decision whether to (a) continue to use Kelley's services or (b) change to the Brown Agency. Each will need to establish the level of significance that best fits its own situation and then make a decision using the decision rule described previously. ▲

See instructions on page 379.

See instructions on page 379.

See instructions on pages 294 and 379.

The idea of the *p*-value is to express the degree of belief in the null hypothesis:

(a) When the *p*-value is minuscule (something like 0.0003), the null hypothesis would be rejected by everybody because the sample results are very unlikely for a true H_o;

(b) When the *p*-value is fairly small (like 0.012), the evidence against H_o is quite strong and H_o will be rejected by many;

(c) When the *p*-value begins to get larger (say, 0.02 to 0.08), there is too much probability that data like the sample involved could have occurred even if H_o were true, and the rejection of H_o is not an easy decision; and

(d) When the *p*-value gets large (like 0.15 or more), the data are not at all unlikely if the H_o is true, and no one will reject H_o.

The advantages of the *p*-value approach are: (1) The results of the test procedure are expressed in terms of a continuous probability scale from 0.0 to 1.0, rather than simply on a "reject" or "fail to reject" basis. (2) A *p*-value can be reported and the user of the information can decide on the strength of the evidence as it applies to his or her own situation. (3) Computers can do all the calculations and report the *p*-value, thus eliminating the need for tables.

The disadvantage of the *p*-value approach is the tendency for people to put off determining the level of significance. This should not be allowed to happen, as it is then possible for someone to set the level of significance after the fact, leaving open the possibility that his or her "preferred" decision will result. This is probably only important when the reported *p*-value falls in the "hard choice" range (say, 0.02 to 0.08), case (c) as described above.

> Do your opponents show you their poker hands before you bet?

ILLUSTRATION 8.14 ▼

According to the results of Exercise 5.35, page 249, the mean of single-digit random numbers is 4.5 and the standard deviation is σ = 2.87. Draw a random sample of 40 single-digit numbers from Table 1 in Appendix B and test the hypothesis "the mean of the single-digit numbers in Table 1 is 4.5." Use α = 0.10.

Solution

STEP 1   The Set-Up:
    a. Describe the population parameter of interest.
        The population parameter of interest is the mean μ of the population of single-digit numbers in Table 1 of Appendix B
    b. State the null hypothesis (H_o) and the alternative hypothesis (H_a).
        H_o: μ = 4.5 (mean is 4.5)
        H_a: μ ≠ 4.5 (mean is not 4.5)

STEP 2   The Hypothesis Test Criteria:
    a. Check the assumptions.
        σ is known. Samples of size 40 should be large enough to satisfy the CLT; see discussion of this on page 359.
    b. Identify the probability distribution and the test statistic to be used.
        The standard normal probability distribution, the test statistic is

$$z \star = \frac{\bar{x} - \mu}{\sigma/\sqrt{n}}, \sigma = 2.87.$$

    c. Determine the level of significance, α.
        α = 0.10 (given in the statement of the problem).

STEP 3   The Sample Evidence:
    a. Collect the sample information.
        The following random sample was drawn from Table 1.

2	8	2	1	5	5	4	0	9	1
0	4	6	1	5	1	1	3	8	0
3	6	8	4	8	6	8	9	5	0
1	4	1	2	1	7	1	7	9	3

        From the sample: $\bar{x}$ = 3.975, *n* = 40.
    b. Calculate the value of the test statistic.
        Using formula (8.4), and μ is 4.5 from H_o; σ = 2.87,

$$z \star = \frac{\bar{x} - \mu}{\sigma/\sqrt{n}}: \quad z \star = \frac{3.975 - 4.50}{2.87/\sqrt{40}} = \frac{-0.525}{0.454} = -1.156 = \mathbf{-1.16}$$

STEP 4 **The Probability Distribution:**
   a. Calculate the *p*-value.

Since the alternative hypothesis indicates a two-tailed test, we must find the probability associated with both tails. The *p*-value is found by doubling the area of one tail.

Since $z\bigstar = -1.16$, the value of $|z\bigstar| = 1.16$.

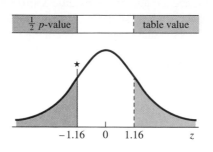

The *p*-value $= 2 \times P(z > |z\bigstar|)$

$$\mathbf{P} = 2 \times P(z > 1.16) = 2 \times (0.5000 - 0.3770) = 2(0.1230) = \mathbf{0.2460}$$

   b. Determine whether or not the *p*-value is smaller than α.
   The *p*-value (0.2460) is greater than α (0.10).

STEP 5 **The Results:**
   a. State the decision about H_o.
   "Fail to reject H_o."
   b. Write a conclusion about H_a.
   The observed sample mean is not significantly different from 4.5 at the 0.10 level of significance. ▲

> **Exercise 8.67**
> Calculate the *p*-value, given H_a: $\mu \neq 245$ and $z\bigstar = 1.1$.

Suppose that we were to take another sample of size 40 from Table 1. Would we obtain the same results? Suppose that we took a third sample and a fourth. What results might we expect? What does the *p*-value in Illustration 8.14 measure? Table 8.7 lists (a) the means obtained from 50 different random samples of size 40 that were taken from Table 1 in Appendix B, (b) the 50 values of $z\bigstar$ corresponding to the 50 $\bar{x}$'s, and (c) their 50 corresponding *p*-values. Figure 8.8 shows a histogram of the 50 $z\bigstar$-values.

Table 8.7
a. The Means of 50 Random Samples Taken from Table 1 in Appendix B

3.850	5.075	4.375	4.675	5.200	4.250	3.775	4.075	5.800	4.975
4.225	4.125	4.350	4.925	5.100	4.175	4.300	4.400	4.775	4.525
4.225	5.075	4.325	5.025	4.725	4.600	4.525	4.800	4.550	3.875
4.750	4.675	4.700	4.400	5.150	4.725	4.350	3.950	4.300	4.725
4.975	4.325	4.700	4.325	4.175	3.800	3.775	4.525	5.375	4.225

b. The $z\bigstar$-Values Corresponding to the 50 Means

−1.432	1.267	−0.275	0.386	1.543	−0.551	−1.598	−0.937	2.865	1.047
−0.606	−0.826	−0.331	0.937	1.322	−0.716	−0.441	−0.220	0.606	0.055
−0.606	1.267	−0.386	1.157	0.496	0.220	0.055	0.661	0.110	−1.377
0.551	0.386	0.441	−0.220	1.432	0.496	−0.331	−1.212	−0.441	0.496
1.047	−0.386	0.441	−0.386	−0.716	−1.543	−1.598	0.055	1.928	−0.606

c. The p-Values Corresponding to the 50 Means

0.152	0.205	0.783	0.700	0.123	0.582	0.110	0.349	0.004	0.295
0.545	0.409	0.741	0.349	0.186	0.474	0.659	0.826	0.545	0.956
0.545	0.205	0.700	0.247	0.620	0.826	0.956	0.509	0.912	0.168
0.582	0.700	0.659	0.826	0.152	0.620	0.741	0.226	0.659	0.620
0.295	0.700	0.659	0.700	0.474	0.123	0.110	0.956	0.054	0.545

Figure 8.8

The 50 Values of $z \star$ from Table 8.7

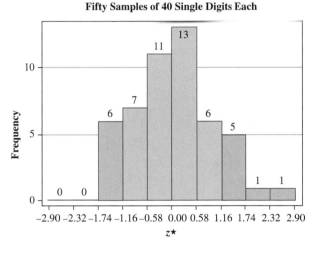

Fifty Samples of 40 Single Digits Each

Exercise 8.68

Describe in your own words what the p-value measures.

The p-value approach was "made" for the computer!

The histogram shows that 6 values of $z \star$ were less than -1.16 and 7 values were greater than 1.16. That means 13 of the 50 samples, or 26%, have mean values more extreme than the mean ($\bar{x} = 3.975$) in Illustration 8.14. This observed relative frequency of 0.26 represents an empirical look at the p-value. Notice that the empirical value for the p-value (0.26) is very similar to the calculated p-value of 0.2460. Check the list of p-values; do you find 13 of the 50 p-values are less than 0.2460? Which samples resulted in a $|z \star| = 1.16$? Which samples resulted in a p-value greater than 0.2460? How do they compare?

The following commands will complete a hypothesis test for the mean μ for a given standard deviation σ.

Exercise 8.69

Use a computer or calculator to select 40 random single-digit numbers. Find the sample mean, $z \star$, and p-value. Repeat several times as in Table 8.7. Describe your findings.

(Use commands for generating integer data on page 182, then continue with these hypothesis test commands.)

MINITAB

```
Input the data into C1; then continue
with:

   Choose:  Stat > Basic Statistics
            > 1-Sample Z
   Enter:   Variables: C1
   Select:  Test mean
   Enter:   μ
   Choose:  Alternative: less than or
            not equal or greater than
   Enter:   Sigma: σ
```

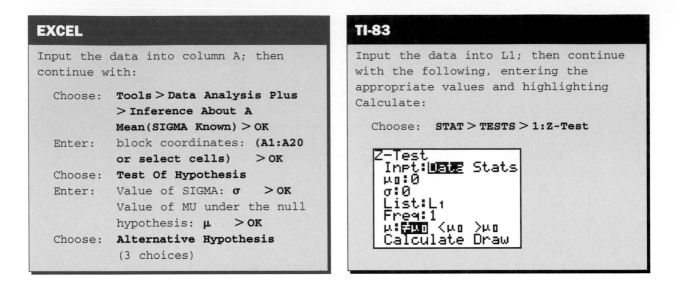

EXCEL

Input the data into column A; then continue with:

Choose: **Tools > Data Analysis Plus > Inference About A Mean(SIGMA Known) > OK**

Enter: **block coordinates: (A1:A20 or select cells)   > OK**

Choose: **Test Of Hypothesis**

Enter: Value of SIGMA: **σ**   **> OK**
          Value of MU under the null hypothesis: **μ   > OK**

Choose: **Alternative Hypothesis**
          (3 choices)

TI-83

Input the data into L1; then continue with the following, entering the appropriate values and highlighting Calculate:

Choose: **STAT > TESTS > 1:Z-Test**

```
Z-Test
 Inpt:DATA Stats
 μ0:0
 σ:0
 List:L₁
 Freq:1
 μ:≠μ0 <μ0 >μ0
Calculate Draw
```

Exercise 8.70

Describe how MINITAB found each of the six numerical values it reported as results.

The MINITAB solution to the rivet example, used in this section (pp. 374–375, 377–380), is shown below.

```
Z-Test

TEST OF MU = 925.00 VS MU L.T. 925.00
THE ASSUMED SIGMA = 18.0

      N      MEAN     STDEV    SE MEAN       Z     P VALUE
     50    921.18     17.58      2.546    -1.50     0.0668
```

When the computer is used, all that is left for you is to make the decision and to write the conclusion.

EXERCISES • • • • • • • • • • • • • •

8.71 State the null hypothesis H_o and the alternative hypothesis H_a that would be used for a hypothesis test related to each of the following statements.
   a. The mean age of the students enrolled in evening classes at a certain college is greater than 26 years.
   b. The mean weight of packages shipped on Air Express during the past month was less than 36.7 lb.
   c. The mean life of fluorescent light bulbs is at least 1600 hr.
   d. The mean weight of college football players is no more than 210 lb.
   e. The mean strength of welds by a new process is different from 570 lb. per unit area, the mean strength of welds by the old process.

8.72 A manufacturer wishes to test the hypothesis that "by changing the formula of their toothpaste it will give its users improved protection." The null hypothesis represents the idea that "the change will not improve the protection," and the alternative represents "the change will improve the protection." Describe the meaning of the two possible types of errors that can occur in the decision when this test of the hypothesis is conducted.

8.73 Suppose we want to test the hypothesis that the mean hourly charge for automobile repairs is at least $60 per hour at the repair shops in a nearby city. Explain the conditions that would exist if we make an error in decision by committing a
  a. type I error.   **b.** type II error.

8.74 Describe how the null hypothesis, as stated in Illustration 8.11 (p. 376), is a "starting point" for the decision to be made about the drying time for paint made with the new formula.

8.75 Assume that z is the test statistic and calculate the value of $z \bigstar$ for each of the following:
  a. H_o: $\mu = 10$, $\sigma = 3$, $n = 40$, $\bar{x} = 10.6$
  b. H_o: $\mu = 120$, $\sigma = 23$, $n = 25$, $\bar{x} = 126.2$
  c. H_o: $\mu = 18.2$, $\sigma = 3.7$, $n = 140$, $\bar{x} = 18.93$
  d. H_o: $\mu = 81$, $\sigma = 13.3$, $n = 50$, $\bar{x} = 79.6$

8.76 Assume that z is the test statistic and calculate the value of $z \bigstar$ for each of the following:
  a. H_o: $\mu = 51$, $\sigma = 4.5$, $n = 40$, $\bar{x} = 49.6$
  b. H_o: $\mu = 20$, $\sigma = 4.3$, $n = 75$, $\bar{x} = 21.2$
  c. H_o: $\mu = 138.5$, $\sigma = 3.7$, $n = 14$, $\bar{x} = 142.93$
  d. H_o: $\mu = 815$, $\sigma = 43.3$, $n = 60$, $\bar{x} = 799.6$

8.77 There are only two possible decisions as a result of a hypothesis test.
  a. State the two possible decisions.
  b. Describe the conditions that will lead to each of the two decisions identified in (a).

8.78 For each of the following pairs of values, state the decision that will occur and state why.
  a. p-value $= 0.014$, $\alpha = 0.02$
  b. p-value $= 0.118$, $\alpha = 0.05$
  c. p-value $= 0.048$, $\alpha = 0.05$
  d. p-value $= 0.064$, $\alpha = 0.10$

8.79 The calculated p-value for a hypothesis test is p-value $= 0.084$. What decision about the null hypothesis would occur if
  a. the hypothesis test is completed at the 0.05 level of significance?
  b. the hypothesis test is completed at the 0.10 level of significance?

8.80 **a.** A one-tailed hypothesis test is to be completed at the 0.05 level of significance. What calculated values of p will cause a rejection of H_o?
  b. A two-tailed hypothesis test is to be completed at the 0.02 level of significance. What calculated values of p will cause a "fail to reject H_o" decision?

8.81 Calculate the p-value for each of the following:
  a. H_o: $\mu = 10$, H_a: $\mu > 10$, $z \bigstar = 1.48$
  b. H_o: $\mu = 105$, H_a: $\mu < 105$, $z \bigstar = -0.85$
  c. H_o: $\mu = 13.4$, H_a: $\mu \neq 13.4$, $z \bigstar = 1.17$
  d. H_o: $\mu = 8.56$, H_a: $\mu < 8.56$, $z \bigstar = -2.11$
  e. H_o: $\mu = 110$, H_a: $\mu \neq 110$, $z \bigstar = -0.93$

8.82 Calculate the p-value for each of the following:
  a. H_o: $\mu = 20$, H_a: $\mu < 20$ ($\bar{x} = 17.8$, $\sigma = 9$, $n = 36$)
  b. H_o: $\mu = 78.5$, H_a: $\mu > 78.5$ ($\bar{x} = 79.8$, $\sigma = 15$, $n = 100$)
  c. H_o: $\mu = 1.587$, H_a: $\mu \neq 1.587$ ($\bar{x} = 1.602$, $\sigma = 0.15$, $n = 50$)

8.83 Find the value of $z \bigstar$ for each of the following:
  a. H_o: $\mu = 35$ versus H_a: $\mu > 35$ when p-value $= 0.0582$
  b. H_o: $\mu = 35$ versus H_a: $\mu < 35$ when p-value $= 0.0166$
  c. H_a: $\mu = 35$ versus H_a: $\mu \neq 35$ when p-value $= 0.0042$

8.84 The null hypothesis, H_o: $\mu = 48$, was tested against the alternative hypothesis, H_a: $\mu > 48$. A sample of 75 resulted in a calculated p-value of 0.102. If $\sigma = 3.5$, find the value of the sample mean, $\bar{x}$.

8.85 The following computer output was used to complete a hypothesis test.

```
TEST OF MU = 525.00 VS MU L.T.  525.00
THE ASSUMED SIGMA = 60.0

       N    MEAN    STDEV   SE MEAN     Z    P VALUE
      38   512.14   64.78    9.733    -1.32   0.093
```

    a. State the null and alternative hypotheses.
    b. If the test is completed using $\alpha = 0.05$, what decision and conclusion are reached?
    c. Verify the value of the standard error of the mean.

8.86 The following computer output was used to complete a hypothesis test.

```
TEST OF MU = 6.250 VS MU N.E.  6.250
THE ASSUMED SIGMA = 1.40

       N    MEAN    STDEV   SE MEAN     Z    P VALUE
      78   6.596   1.273    0.1585    2.18   0.029
```

    a. State the null and alternative hypotheses.
    b. If the test is completed using $\alpha = 0.05$, what decision and conclusion are reached?
    c. Verify the value of the standard error of the mean.
    d. Find the values for $\sum x$ and $\sum x^2$.

8.87 An article titled "Comparisons of Mathematical Competencies and Attitudes of Elementary Education Majors with Established Norms of a General College Population" (*School Science and Mathematics,* Vol. 93, No. 3, March 1993) reported the mean score on a test of mathematical competency for 165 elementary education majors to be 32.63. Test the null hypothesis that μ, the mean for the population of elementary education majors, is 35.70 (the established norm of the general college population) versus the alternative that $\mu < 35.70$. Assume that $\sigma = 6.73$.
    a. Describe the parameter of interest.   **b.** State the null and alternative hypotheses.
    c. Calculate the value for $z \star$ and find the *p*-value.
    d. State your decision and conclusion using $\alpha = 0.001$.

8.88 When the workers for a major automobile manufacturer go on strike, there are repercussions to the rest of the economy, and, in particular, the dealers who sell the cars and trucks feel the pinch. Dealers like to maintain a 2-month supply to give their customers adequate selection, but when the manufacturer cannot deliver the vehicles, inventories dwindle. The June 1997 days' supply of Chevrolet S-10 pickup trucks was 106 vehicles, but shortly after the United Auto Workers strike in Flint, Michigan, the June 1998 days' supply of these trucks had dropped to a mean of 38 and a standard deviation of 16 (*Newsweek,* "Big, Empty Lots," 7-27-98). Suppose one month after the strike was settled, 150 dealers are sampled and S-10 inventories are taken that yield a mean days' supply of 41 trucks. Based on this new evidence, complete the hypothesis test of H_0: $\mu = 38$ vs. H_a: $\mu > 38$ at the 0.02 level of significance using the probability-value approach.
    a. Define the parameter.   **b.** State the null and alternative hypotheses.
    c. Specify the hypothesis test criteria.   **d.** Present the sample evidence.
    e. Find the probability distribution information.
    f. Determine the results.

8.89 Who says that the more you spend on a wristwatch the more accurately the watch will run? Some say that nowadays you can buy a quartz watch for under $25 that runs just as accurately as watches costing four times as much. Suppose the average accuracy for all

(continued)

watches being sold today, irrespective of price, is within 19.8 seconds per month with a standard deviation of 9.1 seconds. A random sample of 36 quartz watches priced less than $25 is taken, and their accuracy check reveals a sample mean error of 22.7 seconds per month. Based on this evidence, complete the hypothesis test of H_0: $\mu = 20$ vs. H_a: $\mu > 20$ at the 0.05 level of significance using the probability-value approach.

   a. Define the parameter.   **b.** State the null and alternative hypotheses.
   c. Specify the hypothesis test criteria.   **d.** Present the sample evidence.
   e. Find the probability distribution information.
   f. Determine the results.

8.90 According to an article in *Good Housekeeping* (February 1991) a 128-lb woman who walks for 30 minutes four times a week at a steady, 4 mi./hr pace can lose up to 10 pounds over a span of a year. Suppose 50 women with weights between 125 and 130 lb performed the four walks per week for a year and at the end of the year the average weight loss for the 50 was 9.1 lb. Assuming that the standard deviation, σ, is 5, complete the hypothesis test of H_o: $\mu = 10.0$ vs. H_a: $\mu \neq 10.0$ at the 0.05 level of significance using the *p*-value approach.

8.5   Hypothesis Test of Mean μ (σ Known): A Classical Approach

In Section 8.3 we surveyed the concepts and much of the reasoning behind a hypothesis test while looking at nonstatistical illustrations. In this section we are going to formalize the hypothesis test procedure as it applies to statements concerning the mean μ of a population under the restriction that σ, the population standard deviation, is a known value.

THE ASSUMPTION FOR HYPOTHESIS TESTS ABOUT MEAN μ USING A KNOWN σ

The sampling distribution of $\bar{x}$ has a normal distribution.

The information needed to ensure that this assumption is satisfied is contained in the sampling distribution of sample means and in the Central Limit Theorem (see Chapter 7, p. 324, 325).

> The sampling distribution for sample means $\bar{x}$ is distributed about a mean equal to μ with a standard error equal to $\sigma/\sqrt{n}$; and, (1) if the randomly sampled population is normally distributed, then $\bar{x}$ is normally distributed for all sample sizes, or (2) if the randomly sampled population is not normally distributed, then $\bar{x}$ is approximately normally distributed for sufficiently large sample sizes.

The hypothesis test is a well-organized, step-by-step procedure used to make a decision. Two different formats are commonly used for hypothesis testing. The *classical approach* is the hypothesis test process that has enjoyed popularity for many years. This approach is organized as a five-step procedure .

THE CLASSICAL HYPOTHESIS TEST: A FIVE-STEP PROCEDURE

STEP 1 The Set-up:
   a. Describe the population parameter of interest.
   b. State the null hypothesis (H_o) and the alternative hypothesis (H_a).

STEP 2   **The Hypothesis Test Criteria:**
   a. Check the assumptions.
   b. Identify the probability distribution and the test statistic to be used.
   c. Determine the level of significance, α.

STEP 3   **The Sample Evidence:**
   a. Collect the sample information.
   b. Calculate the value of the test statistic.

STEP 4   **The Probability Distribution:**
   a. Determine the critical region and critical value(s).
   b. Determine whether or not the calculated test statistic is in the critical region.

STEP 5   **The Results:**
   a. State the decision about H_o.
   b. State the conclusion about H_a.

> Using weak rivets could have terrible consequences.

Exercise 8.91
In this example, the aircraft builder, the buyer of the rivets, is concerned that the rivets might not meet the mean-strength spec. State the aircraft manufacturer's null and alternative hypotheses.

> More specific instructions on page 367.

A commercial aircraft manufacturer buys rivets for use in assembling airliners. Each rivet supplier who wants to sell rivets to the aircraft manufacturer must demonstrate that its rivets meet the required specifications. One of the specs is: "The mean shearing strength of all such rivets, μ, is at least 925 lb." Each time the aircraft manufacturer buys rivets, it is concerned that the mean strength might be less than the 925-lb specification.

NOTE 1 Each individual rivet has a shearing strength, which is determined by measuring the force required to shear ("break") the rivet. Clearly, not all the rivets can be tested. Therefore, a sample of rivets will be tested, and a decision about the mean strength of all the untested rivets will be based on the mean from those sampled and tested.

NOTE 2 We will use $\sigma = 18$ for our rivet example.

STEP 1   **The Set-Up:**
   a. **Describe the population parameter of interest.**
      The population parameter of interest is the mean μ, the mean shearing strength (or mean force required to shear) of the rivets being considered for purchase.
   b. **State the null hypothesis (H_o) and the alternative hypothesis (H_a).**

The null hypothesis and the alternative hypothesis are formulated by inspecting the problem or statement to be investigated and first formulating two opposing statements concerning the mean μ. For our example, these two opposing statements are: (A) "The mean shearing strength is less than 925" ($\mu < 925$, the aircraft manufacturer's concern); and (B) "The mean shearing strength is at least 925" ($\mu = 925$, the rivet supplier's claim and the aircraft manufacturer's spec).

NOTE The trichotomy law from algebra states that two numerical values must be related in exactly one of three possible relationships: $<$, $=$, or $>$. All three of these possibilities must be accounted for between the two opposing hypotheses in order for the two hypotheses to be negations of each other. The three possible combinations of signs and hypotheses are shown in Table 8.8. Recall that the null hypothesis assigns a specific value to the parameter in question and therefore "equals" will always be part of the null hypothesis.

Table 8.8   The Three Possible Statements of Null and Alternative Hypotheses

Null Hypothesis	Alternative Hypothesis
1. greater than or equal to ($\geq$)	less than ($<$)
2. less than or equal to ($\leq$)	greater than ($>$)
3. equal to ($=$)	not equal to ($\neq$)

The parameter of interest, the population mean μ, is related to the value 925.
   Statement (A) becomes the alternative hypothesis:

$$H_a: \mu < 925 \text{ (the mean is less than 925)}$$

This statement represents the aircraft manufacturer's concern and states "the rivets do not meet the required specs."
   Statement (B) becomes the null hypothesis:

$$H_o: \mu = 925 \ (\geq) \text{ (the mean is at least 925)}$$

This hypothesis represents the negation of the aircraft manufacturer's concern and states that "the rivets do meet the required specs."

NOTE The null hypothesis will be written with just the equal sign ("a value is assigned"). When "equal" is paired with "less than" or paired with "greater than," the combined symbol is written beside the null hypothesis as a reminder that all three signs have been accounted for in these two opposing statements.

   Before continuing with our example, let's look at three illustrations that demonstrate formulating the statistical null and alternative hypotheses involving population mean μ. Illustrations 8.15 and 8.16 each demonstrate a "one-tailed" alternative hypothesis .

ILLUSTRATION 8.15 ▼

A consumer advocate group would like to disprove a car manufacturer's claim that a specific model will average 24 miles per gallon of gasoline. Specifically, the group would like to show that the mean miles per gallon is considerably lower than 24 miles per gallon. State the null and alternative hypotheses.

Solution

To state the two hypotheses, we first need to identify the population parameter in question: the "mean mileage attained by this car model." The parameter μ is being compared to the value 24 miles per gallon, the specific value of interest. The advocates are questioning the value of μ and wish to show it to be lower than 24 (that is, $\mu < 24$). There are three possible relationships: (1) $\mu < 24$, (2) $\mu = 24$, or (3) $\mu > 24$. These three statements must be arranged to form two opposing statements: one that states what the advocates are trying to show, "the mean level is lower than 24 ($\mu < 24$)," whereas the "negation" is "the mean level is not lower than 24 ($\mu \geq 24$)." One of these two statements will become the null hypothesis, and the other will become the alternative hypothesis.

RECALL There are two rules for forming the hypotheses: (1) The null hypothesis states that the parameter in question has a specified value ("H_o must contain the equal sign"), and (2) the consumer advocate group's contention becomes the alternative hypothesis ("lower than"). Both rules indicate:

$$H_o: \mu = 24 \ (\geq) \quad \text{and} \quad H_a: \mu < 24$$

ILLUSTRATION 8.16 ▼

Suppose the EPA was suing a large manufacturing company for not meeting federal emissions guidelines. Specifically, the EPA is claiming that the mean amount of sulfur dioxide in the air is dangerously high, higher than 0.09 parts per million. State the null and alternative hypotheses for this test situation.

Solution

The parameter of interest is the mean amount of sulfur dioxide in the air, and 0.09 parts per million is the specified value. $\mu > 0.09$ corresponds to "the mean amount is greater than 0.09," whereas $\mu \leq 0.09$ corresponds to the negation, "the mean time is not greater than 0.09." Therefore, the hypotheses are

$$H_o: \mu = 0.09 \ (\leq) \quad \text{and} \quad H_a: \mu > 0.09 \qquad \blacktriangle$$

Illustration 8.17 demonstrates a "two-tailed" alternative hypothesis .

ILLUSTRATION 8.17 ▼

Job satisfaction is very important when it comes to getting workers to produce. A standard job satisfaction questionnaire was administered by union officers to a sample of assembly line workers in a large plant in hope of showing that the assembly workers' mean score on this questionnaire would be different from the established mean of 68. State the null and alternative hypotheses.

Solution

Either the **mean** job satisfaction score **is different from** 68 ($\mu \neq$ **68**) or the mean score **is equal to** 68 ($\mu =$ **68**). Therefore,

$$H_o: \mu = 68 \quad \text{and} \quad H_a: \mu \neq 68 \qquad \blacktriangle$$

NOTE 1 When "less than" is combined with "greater than," they become "not equal to."

NOTE 2 The alternative hypothesis is referred to as being "two-tailed" when H_a is "not equal."

The viewpoint of the experimenter greatly affects the way the hypotheses are formed. Generally, the experimenter is trying to show that the parameter value is different from the value specified. Thus, the experimenter is often hoping to be able to reject the null hypothesis so that the experimenter's theory has been substantiated. Illustrations 8.15, 8.16, and 8.17 also represent the three possible arrangements for the $<$, $=$, and $>$ relationships between the parameter μ and a specified value.

Table 8.9 lists some additional common phrases used in claims and indicates the phrase of its negation and the hypothesis where each phrase will be used. Again, notice that "equals" is always in the null hypothesis. Also notice that the negation of "less than" is "not less than," which is equivalent to "greater than or equal to." Think of negation of "one sign" as the "other two signs combined."

Exercise 8.92
Professor Hart does not believe the statement "the mean distance commuted daily by the nonresident students at our college is no more than 9 miles." State the null and alternative hypotheses he would use to challenge this statement.

Table 8.9 Common Phrases and Their Negations

$H_o: (\geq)$	$H_a: (<)$	$H_o: (\leq)$	$H_a: (>)$	$H_o: (=)$	$H_a: (\neq)$
at least	less than	at most	more than	is	is not
no less than	less than	no more than	more than	not different from	different from
not less than	less than	not greater than	greater than	same as	not same as

Exercise 8.93

State the null and alternative hypotheses used to test each of the following claims.

a. The mean reaction time is less than 1.25 seconds.
b. The mean score on that qualifying exam is different from 335.
c. The mean selling price of homes in the area is no more than $230,000.

Once the null and alternative hypotheses are established, we will work under the assumption that the null hypothesis is a true statement until there is sufficient evidence to reject it. This situation might be compared to a courtroom trial, where the accused is assumed to be innocent (H_o: Defendant is innocent vs. H_a: Defendant is not innocent) until sufficient evidence has been presented to show that innocence is totally unbelievable ("beyond reasonable doubt"). At the conclusion of the hypothesis test, we will make one of two possible decisions. We will decide in opposition to the null hypothesis and say that we "reject H_o" (this corresponds to "conviction" of the accused in a trial); or we will decide in agreement with the null hypothesis and say that we "fail to reject H_o" (this corresponds to "fail to convict" or an "acquittal" of the accused in a trial).

Let's return to the rivet example we interrupted on page 390 and continue with step 2. Recall:

$$H_o: \mu = 925 \ (\geq) \ (\text{at least } 925) \qquad H_a: \mu < 925 \ (\text{less than } 925)$$

STEP 2   The Hypothesis Test Criteria:
 a. **Check the assumptions.**
   σ is known. Variables like shearing strength typically have a mounded distribution; therefore, a sample of size 50 should be large enough for the CLT to satisfy the assumption, **the sampling distribution of sample means is approximately normally distributed.**
 b. **Identify the probability distribution and the test statistic to be used.**
   The standard normal probability distribution is used since $\bar{x}$ is expected to have a normal or approximately normal distribution.

For a hypothesis test of μ, we will want to compare the value of the sample mean to the value of the population mean as stated in the null hypothesis. This comparison is accomplished using formula (8.4).

$$z \star = \frac{\bar{x} - \mu}{\sigma/\sqrt{n}} \qquad (8.4)$$

The resulting calculated value is identified as $z \star$ ("z star") since it is expected to have a standard normal distribution when the null hypothesis is true and the assumptions have been satisfied. The $\star$ ("star") is to remind us that this is the calculated value of the test statistic.

The test statistic to be used is

$$z \star = \frac{\bar{x} - \mu}{\sigma/\sqrt{n}}$$

Do Exercise 8.94 before continuing.

 c. **Determine the level of significance, α.**

Setting α was described as a managerial decision in Section 8.3. To see what is involved in determining α, the probability of the type I error, for our rivet example, we start by identifying the four possible outcomes, their meaning, and the action related to each.

Exercise 8.94

Identify the four possible outcomes and describe the situation involved with each outcome with regard to the aircraft manufacturer's testing and buying rivets. Which is the more serious error; the type I or the type II? Explain.

The type I error occurs when a true null hypothesis is rejected. This would occur when the manufacturer tested rivets that in truth did meet the specs, and rejected them. Undoubtedly this would lead to the rivets not being purchased even though the manufacturer unknowingly did meet the specs. In order for the manager to set a level of significance, related information is needed; namely, how soon is the new supply of rivets needed? If they are needed tomorrow and this is the only vendor with an available supply, waiting a week to find acceptable rivets could be very expensive; therefore, rejecting good rivets could be considered a serious error. On the other hand, if the rivets are

> There is more to this scenario, but hopefully you get the idea.

> α will be assigned in the statement of exercises.

not needed until next month, then this error may not be very serious. Only the manager will know all the ramifications, and therefore the manager's input is important here.

After much consideration, the manager assigns the level of significance:

$\alpha = 0.05$

STEP 3   The Sample Evidence:
   a. Collect the sample information.

We are ready for the data. The sample must be a random sample drawn from the population whose mean μ is being questioned.

> A random sample of 50 rivets is selected, each rivet is tested, and the sample mean shearing strength is calculated.
>
> $$\bar{x} = 921.18 \quad \text{and} \quad n = 50$$

   b. Calculate the value of the test statistic.

The sample evidence ($\bar{x}$ and n found in step 3a) is next converted into the **calculated value of the test statistic, $z \star$,** using formula (8.4).

> (μ is 925 from H_o; $\sigma = 18$ is the known quantity, given in Note 2, p. 350.)
>
> $$z \star = \frac{\bar{x} - \mu}{\sigma/\sqrt{n}} \ : \quad z \star = \frac{921.18 - 925.0}{18/\sqrt{50}} = \frac{-3.82}{2.5456} = -1.50$$

STEP 4   The Probability Distribution:
   a. Determine the critical region and critical value(s).

The standard normal variable z is our test statistic for this hypothesis test; therefore, draw a sketch of the standard normal distribution, label the scale as z, and locate its mean value, 0.

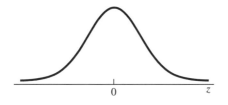

CRITICAL REGION

The set of values for the test statistic that will cause us to reject the null hypothesis. The set of values that are not in the critical region is called the **noncritical region** (sometimes called the *acceptance region*).

Recall that we are working under the assumption that the null hypothesis is true. Thus, we are assuming that the mean shearing strength of all rivets in the sampled population is 925. If this is the case, then when we select a random sample of 50 rivets, we can expect this sample mean, $\bar{x}$, to be part of a normal distribution that is centered at 925 and to have a standard error of $\sigma/\sqrt{n} = 18/\sqrt{50}$, or approximately 2.55. Approximately 95% of the sample mean values will be greater than 920.8 (a value 1.65 standard errors below the mean, $925 - (1.65)(2.55) = 920.8$). Thus, if H_o is true and $\mu = 925$, then we expect $\bar{x}$ to be greater than 920.8 approximately 95% of the time and less than 920.8 only 5% of the time.

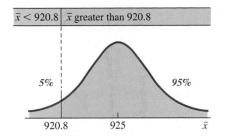

If, however, the value of $\bar{x}$ that we obtain from our sample is less than 920.8, say 919.5, we will have to make a choice. It could be that either: (A) such an $\bar{x}$-value (919.5) is a member of the sampling distribution with mean 925 although it has a very low probability of occurrence (less than 0.05), or (B) $\bar{x} = 919.5$ is a member of a sampling distribution whose mean is less than 925, which would make it a value that is more likely to occur.

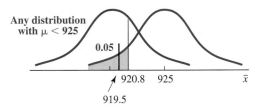

In statistics, we "bet" on the "more probable to occur" and consider the second choice (B) to be the right one. Thus, the left-hand tail of the z distribution becomes the critical region. And the level of significance α becomes the measure of its area.

CRITICAL VALUE(S)

The "first" or "boundary" value(s) of the critical region(s).

> Shading will be used to identify the critical region.

> Information about the critical value notation, $z(\alpha)$, is on pages 299–302.

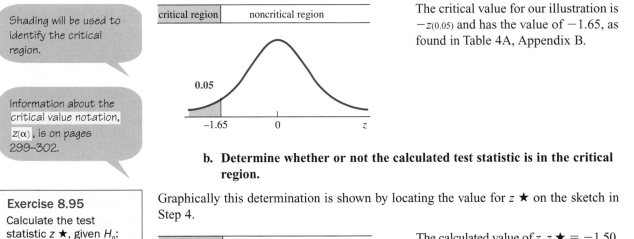

The critical value for our illustration is $-z_{(0.05)}$ and has the value of -1.65, as found in Table 4A, Appendix B.

b. Determine whether or not the calculated test statistic is in the critical region.

Graphically this determination is shown by locating the value for $z \bigstar$ on the sketch in Step 4.

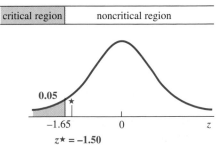

The calculated value of z, $z \bigstar = -1.50$, is **not in the critical region** (it is in the unshaded portion of the figure).

> **Exercise 8.95**
>
> Calculate the test statistic $z \bigstar$, given H_o: $\mu = 356$, $\sigma = 17$, $\bar{x} = 354.3$, and $n = 120$.

STEP 5 The Result:
   a. State the decision about H_o.

Exercise 8.96

a. What decision is reached when the test statistic falls in the critical region?
b. What decision is reached when the test statistic falls in the noncritical region?

In order to make the decision, we need to know the *decision rule.*

DECISION RULE

a. If the test statistic falls **within the critical region,** then the decision must be **reject** H_o. (The critical value is part of the critical region.)
b. If the test statistic is **not in the critical region,** then the decision must be **fail to reject** H_o.

Specific information about writing the conclusion is on page 371.

Decision about H_o "Fail to reject H_o."

   b. Write the conclusion about H_a.
      There is not sufficient evidence at the 0.05 level of significance to show that the rivets have a mean shearing strength less than 925.

"We failed to convict" the null hypothesis. In other words, a sample mean as small as 921.18 is likely to occur (as defined by α) when the true population mean value is 925.0. Therefore, the resulting action would be to buy the rivets.

Before we look at another illustration, let's summarize briefly some of the details we have seen thus far:

1. The null hypothesis specifies a particular value of a population parameter.
2. The alternative hypothesis can take three forms. Each form dictates a specific location of the critical region(s), as shown in the following table.
3. For many hypothesis tests, the sign in the alternative hypothesis "points" in the direction in which the critical region is located. [Think of the not equal to sign ($\neq$) as being both less than ($<$) and greater than ($>$), thus pointing in both directions.]

Exercise 8.97

Find the critical region and value(s) for H_a: $\mu < 19$, and $\alpha = 0.01$.

	Sign in the Alternative Hypothesis		
	$<$	$\neq$	$>$
Critical Region	One region Left side **One-tailed test**	Two regions Half on each side **Two-tailed test**	One region Right side **One-tailed test**

Exercise 8.98

Find the critical region and value(s) for H_a: $\mu > 34$, and $\alpha = 0.02$.

The value assigned to α is called the significance level of the hypothesis test. Alpha cannot be interpreted to be anything other than the risk (or probability) of rejecting the null hypothesis when it is actually true. We will seldom be able to determine whether the null hypothesis is true or false; we will decide only to "reject H_o" or to "fail to reject H_o." The relative frequency with which we reject a true hypothesis is α, but we will never know the relative frequency with which we make an error in decision. The two ideas are quite different; that is, a type I error and an error in decision are two different things altogether. Remember, there are two types of errors: type I and type II.

Let's look at another hypothesis test, one involving the two-tailed procedure.

ILLUSTRATION 8.18 ▼

It has been claimed that the mean weight of women students at a college is 54.4 kg. Professor Hart does not believe the statement and sets out to show the mean weight is not 54.4 kg. To test the claim he collects a random sample of 100 weights from among the women students. A sample mean of 53.75 kg results. Is this sufficient evidence for Professor Hart to reject the statement? Use $\alpha = 0.05$ and $\sigma = 5.4$ kg.

Exercise 8.99

How many pounds is 54.4 kilograms?

Solution

STEP 1 **The Set-Up:**
  a. **Describe the population parameter of interest.**
  The population parameter of interest is the mean μ, the mean weight of all women students at the college.
  b. **State the null hypothesis (H_o) and the alternative hypothesis (H_a).**
  The mean weight **is equal** to 54.4 kg, or the mean weight **is not equal** to 54.4 kg.

$$H_o: \mu = \mathbf{54.4} \text{ (mean weight is 54.4)}$$

$$H_a: \mu \neq \mathbf{54.4} \text{ (mean weight is not 54.4)}$$
$$\text{(Remember: } \neq \text{ is } < \text{ and } > \text{ together.)}$$

STEP 2 **The Hypothesis Test Criteria:**
  a. **Check the assumptions.**
  σ is known. Weights of an adult group of women are generally approximately normally distributed; therefore, a sample of $n = 100$ is large enough to allow the CLT to apply.
  b. **Identify the probability distribution and the test statistic to be used.**
  The **standard normal probability distribution** and the test statistic

$$z \bigstar = \frac{\bar{x} - \mu}{\sigma / \sqrt{n}} \text{ will be used. } \sigma = \mathbf{5.4.}$$

  c. **Determine the level of significance, α.**
  $\alpha = \mathbf{0.05}$ (given in the statement of problem).

STEP 3 **The Sample Evidence:**
  a. **Collect the sample information.**
  $\bar{x} = \mathbf{53.75}$ and $n = \mathbf{100}$
  b. **Calculate the value of the test statistic.**
  Using formula (8.4), information from $H_o: \mu = 54.4$ and $\sigma = 5.4$ (known).

$$z \bigstar = \frac{\bar{x} - \mu}{\sigma / \sqrt{n}}: \quad z \bigstar = \frac{53.75 - 54.4}{5.4 / \sqrt{100}} = \frac{-0.65}{0.54} = -1.204 = \mathbf{-1.20}$$

STEP 4 **The Probability Distribution:**
  a. **Determine the critical region and critical value(s).**

Information about the critical value notation is on pages 299–302.

Instructions for using Table 4B are on page 357.

The critical region is both the left tail and the right tail because both smaller and larger values of the sample mean suggest the null hypothesis is wrong. The level of significance will be split in half, with 0.025 being the measure of each tail. The critical values are found in Table 4B, Appendix B: $\pm z_{(0.025)} = \pm 1.96$

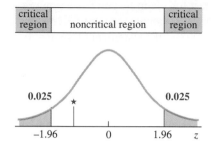

b. **Make the comparison: determine whether or not the calculated test statistic is in the critical region.**
The calculated value of z, $z \star = -1.20$, is **not in the critical region** (shown in red on figure above).

STEP 5 The Results:

a. **State the decision about H_o.**
"Fail to reject H_o."

b. **Write the conclusion about H_a.**
There is not sufficient evidence at the 0.05 level of significance to show that the women students have a mean weight different than the 54.4 kg claimed. In other words, there is no statistical evidence to support Professor Hart's contentions.

ILLUSTRATION 8.19 ▼

The student body at many community colleges is considered a "commuter population." The following question was asked of the Student Affairs Office at one such college: "How far (one way) does the average community college student commute to college daily?" The office answered: "No more than 9.0 mi." The inquirer was not convinced of the truth of this and decided to test the statement. She took a random sample of 50 students and found a mean commuting distance of 10.22 mi. Test the hypothesis stated above at a significance level of $\alpha = 0.05$, using $\sigma = 5$ mi.

Solution

STEP 1 The Set-Up:

a. **Describe the population parameter of interest.**
The population parameter of interest is the mean μ, **the mean one-way distance** traveled by all commuting students.

b. **State the null hypothesis (H_o) and the alternative hypothesis (H_a).**
The hypotheses: The claim "no more than 9.0 mi." implies that the three possible relationships should be grouped "no more than 9.0" ($\leq$) versus "more than 9.0" ($>$).

$$H_o: \mu = 9.0 \ (\leq) \ (\text{no more than 9.0 mi.})$$

$$H_a: \mu > 9.0 \ (\text{more than 9.0 mi.})$$

STEP 2 **The Hypothesis Test Criteria:**
    a. Check the assumptions.
        σ is known. Commuting distance is a mounded and skewed distribution, but with a sample size $n = 50$, the CLT will hold; therefore, the assumptions are satisfied.
    b. Identify the probability distribution and the test statistic to be used.
        The **standard normal probability distribution** and the test statistic

$$z \star = \frac{\bar{x} - \mu}{\sigma/\sqrt{n}} \text{ will be used. } \sigma = 5.$$

    c. Determine the level of significance, α.
        $\alpha = 0.05$ (given in the statement of problem).

STEP 3 **The Sample Evidence:**
    a. Collect the sample information.
        $\bar{x} = 10.22$ and $n = 50$

    b. Calculate the value of the test statistic.
        Using formula (8.4), information from H_o: $\mu = 9.0$ and $\sigma = 5.0$ was given.

$$z \star = \frac{\bar{x} - \mu}{\sigma/\sqrt{n}}: \quad z \star = \frac{10.22 - 9.0}{5.0/\sqrt{50}} = \frac{1.22}{0.707} = \mathbf{1.73}$$

STEP 4 **The Probability Distribution:**
    a. Determine the critical region and critical value(s).
        The critical region is the right tail because the alternative hypothesis is "greater than." The critical value is $z_{(0.05)} = 1.65$.

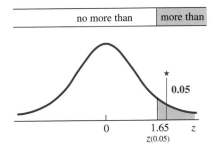

    b. Determine whether or not the calculated test statistic is in the critical region.
        The calculated value of z, $z \star = 1.73$, is **in the critical region** (shown in red on figure above).

STEP 5 **The Results:**
    a. State the decision about H_o.
        "Reject H_o."
    b. Write the conclusion about H_a.
        There is sufficient evidence at the 0.05 level of significance to conclude that the average commuting community college student probably travels more than 9.0 miles one way to college.

ILLUSTRATION 8.20 ▼

According to the results of Exercise 5.35, page 249, the mean of single-digit random numbers is 4.5 and the standard deviation is $\sigma = 2.87$. Draw a random sample of 40 single-digit numbers from Table 1 in Appendix B and test the hypothesis, "the mean of the single-digit numbers in Table 1 is 4.5." Use $\alpha = 0.10$.

Solution

STEP 1   The Set-Up:

    a. Describe the population parameter of interest.

    The parameter of interest is the mean μ of the population of single-digit numbers in Table 1 of Appendix B.

    b. State the null hypothesis (H_o) and the alternative hypothesis (H_a).

$$H_o: \mu = 4.5 \text{ (mean is 4.5)}$$
$$H_a: \mu \neq 4.5 \text{ (mean is not 4.5)}$$

STEP 2   The Test Criteria:

    a. Check the assumptions.

    σ is known. Samples of size 40 should be large enough to satisfy the CLT; see discussion of this on page 359.

    b. Identify the probability distribution and the test statistic to be used.

    The standard normal probability distribution and the test statistic

$$z \bigstar = \frac{\bar{x} - \mu}{\sigma/\sqrt{n}} \text{ will be used. } \sigma = 2.87.$$

    c. Determine the level of significance, α.

    $\alpha = 0.10$ (given in the statement of problem).

STEP 3   The Sample Evidence:

    a. Collect the sample information.

    The following random sample was drawn from Table 1.

2	8	2	1	5	5	4	0	9	1
0	4	6	1	5	1	1	3	8	0
3	6	8	4	8	6	8	9	5	0
1	4	1	2	1	7	1	7	9	3

    the sample statistics: $\bar{x} = 3.975$, $n = 40$.

    b. Calculate the value of the test statistic.

    Using formula (8.4), information from H_o: $\mu = 4.5$ and $\sigma = 2.87$.

$$z \bigstar = \frac{\bar{x} - \mu}{\sigma/\sqrt{n}}: \quad z \bigstar = \frac{3.975 - 4.50}{2.87/\sqrt{40}} = \frac{-0.525}{0.454} = -1.156 = -1.16$$

STEP 4   The Probability Distribution:

    a. Determine the critical region and critical value(s).

    A two-tailed critical region will be used, and 0.05 will be the area of each tail. The critical values are $\pm z_{(0.05)} = \pm 1.65$.

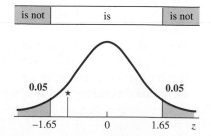

b. **Determine whether or not the calculated test statistic is in the critical region.**
The calculated value of z, $z \star = -1.20$, is **not in the critical region** (shown in red on figure above).

STEP 5   **The Result:**
a. **State the decision about H_o.**
Fail to reject H_o.
b. **Write the conclusion about H_a.**
The observed sample mean is not significantly different from 4.5 at the 0.10 significance level.   ▲

Suppose that we were to take another sample of size 40 from the table of random digits. Would we obtain the same results? Suppose that we took a third sample or a fourth? What results might we expect? What is the level of significance? Yes, its value is 0.10, but what does it measure? Table 8.10 lists the means obtained from 20 different random samples of size 40 that were taken from Table 1 in Appendix B. The calculated value of $z \star$ that corresponds to each $\bar{x}$ and the decision each would dictate are also listed. The 20 calculated z-scores are shown in Figure 8.9. Note that 3 of the 20 samples (or 15%) caused us to reject the null hypothesis, even though we know the null hypothesis is true for this situation. Can you explain this?

Table 8.10   Twenty Random Samples of Size 40 Taken from Table 1 in Appendix B

Sample Number	Sample Mean ($\bar{x}$)	Calculated z ($z \star$)	Decision Reached
1	4.62	+0.26	Fail to reject H_o
2	4.55	+0.11	Fail to reject H_o
3	4.08	−0.93	Fail to reject H_o
4	5.00	+1.10	Fail to reject H_o
5	4.30	−0.44	Fail to reject H_o
6	3.65	−1.87	Reject H_o
7	4.60	+0.22	Fail to reject H_o
8	4.15	−0.77	Fail to reject H_o
9	5.05	+1.21	Fail to reject H_o
10	4.80	+0.66	Fail to reject H_o
11	4.70	+0.44	Fail to reject H_o
12	4.88	+0.83	Fail to reject H_o
13	4.45	−0.11	Fail to reject H_o
14	3.93	−1.27	Fail to reject H_o
15	5.28	+1.71	Reject H_o
16	4.20	−0.66	Fail to reject H_o
17	3.48	−2.26	Reject H_o
18	4.78	+0.61	Fail to reject H_o
19	4.28	−0.50	Fail to reject H_o
20	4.23	−0.61	Fail to reject H_o

Exercise 8.100

Use a computer or calculator to select 40 random single-digit numbers. Repeat it several times as in Table 8.10. Find the sample mean and $z \star$. Describe your findings after several tries.

(Use commands for generating integer data on page 182, then continue with the hypothesis test commands on page 385.)

REMEMBER α is the probability that we "reject H_o" when it is actually a true statement. Therefore, we can anticipate that a type I error will occur α of the time when testing a true null hypothesis. In the above empirical situation, we observed a 15% rejection rate. If we were to repeat this experiment many times, the proportion of samples that would lead to a rejection would vary, but the observed relative frequency of rejection should be approximately α or 10%.

Figure 8.9
z-Scores from Table 8.10

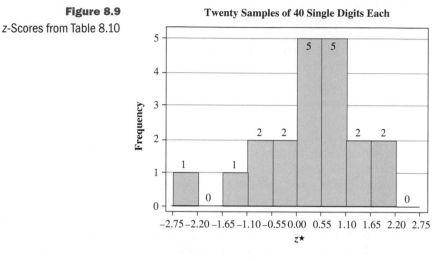

EXERCISES • • • • • • • • • • • • • •

8.101 State the null hypothesis, H_o, and the alternative hypothesis, H_a, that would be used for a hypothesis test related to each of the following statements.
   a. The mean age of the youths who hang out at the mall is less than 16 years.
   b. The mean height of professional basketball players is more than 6 ft 6 in.
   c. The mean elevation drop for ski trails at eastern ski centers is at least 285 ft.
   d. The mean diameter of the rivets is no more than 0.375 in.
   e. The mean cholesterol level of male college students is different from 200 units.

8.102 Suppose you want to test the hypothesis that "the mean salt content of frozen 'lite' dinners is more than 350 mg per serving." An average of 350 mg is an acceptable amount of salt per serving; therefore, you use it as the standard. The null hypothesis is "the average content is not more than 350 mg" ($\mu = 350$). The alternative hypothesis is "the average content is more than 350 mg" ($\mu > 350$).
   a. Describe the conditions that would exist if your decision results in a type I error.
   b. Describe the conditions that would exist if your decision results in a type II error.

8.103 Suppose you wanted to test the hypothesis that the mean minimum home service call charge for plumbers is at most $85 in your area. Explain the conditions that would exist if you make an error in decision by committing a
   a. type I error.   **b.** type II error.

8.104 Describe how the null hypothesis in Illustration 8.19 is a "starting point" for the decision to be made about the mean one-way distance.

8.105   a. What is the critical region?   **b.** What is the critical value?

8.106 Since the size of the type I error can always be made smaller by reducing the size of the critical region, why don't we always choose critical regions that make α extremely small?

8.107 Determine the critical region and critical values for z that would be used to test the null hypothesis at the given level of significance, as described in each of the following:
   a. H_o: $\mu = 20$, H_a: $\mu \neq 20$, $\alpha = 0.10$
   b. H_o: $\mu = 24$ ($\leq$), H_a: $\mu > 24$, $\alpha = 0.01$
   c. H_o: $\mu = 10.5$ ($\geq$), H_a: $\mu < 10.5$, $\alpha = 0.05$
   d. H_o: $\mu = 35$, H_a: $\mu \neq 35$, $\alpha = 0.01$

8.108 Determine the critical region and the critical values used to test the following null hypotheses.
  a. H_o: $\mu = 55$ $(\geq)$, H_a: $\mu < 55$, $\alpha = 0.02$
  b. H_o: $\mu = -86$ $(\geq)$, H_a: $\mu < -86$, $\alpha = 0.01$
  c. H_o: $\mu = 107$, H_a: $\mu \neq 107$, $\alpha = 0.05$
  d. H_o: $\mu = 17.4$ $(\leq)$, H_a: $\mu > 17.4$, $\alpha = 0.10$

8.109 The null hypothesis, H_o: $\mu = 250$, was tested against the alternative hypothesis, H_a: $\mu < 250$. A sample of $n = 85$ resulted in a calculated test statistic of $z \bigstar = -1.18$. If $\sigma = 22.6$, find the value of the sample mean, $\bar{x}$. Find the sum of the sample data, $\sum x$.

8.110 Find the value of $\bar{x}$ for each of the following:
  a. H_o: $\mu = 580$, $z \bigstar = 2.10$, $\sigma = 26$, $n = 55$
  b. H_o: $\mu = 75$, $z \bigstar = -0.87$, $\sigma = 9.2$, $n = 35$

8.111 The calculated value of the test statistic is actually the number of standard errors that the sample mean differs from the hypothesized value of μ in the null hypothesis. Suppose that the null hypothesis is H_o: $\mu = 4.5$, σ is known to be 1.0, and a sample of size 100 results in $\bar{x} = 4.8$.
  a. How many standard errors is $\bar{x}$ above 4.5?
  b. If the alternative hypothesis is H_a: $\mu > 4.5$ and $\alpha = 0.01$, would you reject H_o?

8.112 Consider the hypothesis test where the hypotheses are H_o: $\mu = 26.4$ and H_a: $\mu < 26.4$. A sample of size 64 is randomly selected and yields a sample mean of 23.6.
  a. If it is known that $\sigma = 12$, how many standard errors below $\mu = 26.4$ is the sample mean, $\bar{x} = 23.6$?
  b. If $\alpha = 0.05$, would you reject H_o? Explain.

8.113 There are only two possible decisions as a result of a hypothesis test.
  a. State the two possible decisions.
  b. Describe the conditions that will lead to each of the two decisions identified in (a).

8.114 **a.** What proportion of the probability distribution is in the critical region, provided the null hypothesis is correct?
  b. What error could be made if the test statistic falls in the critical region?
  c. What proportion of the probability distribution is in the noncritical region, provided the null hypothesis is not correct?
  d. What error could be made if the test statistic falls in the noncritical region?

8.115 The following computer output was used to complete a hypothesis test.

```
TEST OF MU = 15.0000 VS MU N.E.  15.0000
THE ASSUMED SIGMA = 0.50

      N        MEAN       STDEV      SE MEAN        Z
      30      15.6333     0.4270     0.0913        6.94
```

  a. State the null and alternative hypotheses.
  b. If the test is completed using $\alpha = 0.01$, what decision and conclusion are reached?
  c. Verify the value of the standard error of the mean.

8.116 The following computer output was used to complete a hypothesis test.

```
TEST OF MU = 72.00 VS MU G.T.  72.00
THE ASSUMED SIGMA = 12.0

      N        MEAN       STDEV      SE MEAN        Z
      36       75.2       11.87       2.00         1.60
```

(continued)

a. State the null and alternative hypotheses.
b. If the test is completed using $\alpha = 0.05$, what decision and conclusion are reached?
c. Verify the value of the standard error of the mean.

8.117 The gestation period (the elapsed time between conception and birth) of gray squirrels measured in captivity is listed as 44 days as estimated by the author of *Walker's Mammals of the World* (Johns Hopkins University Press, 1991). It is recognized that the potential life span of animals is rarely attained in nature, but the gestation period could be either shorter or longer.

Suppose the gestation period of a sample of 81 squirrels living in the wild is measured using the latest techniques available, and the mean length of time is found to be 42.5 days. Test the hypothesis that squirrels living in the wild have the same gestation period as those in captivity at the 0.05 level of significance. Assume that $\sigma = 5$ days. Use the classical approach.

a. Define the parameter.   **b.** State the null and alternative hypotheses.
c. Specify the hypothesis test criteria.   **d.** Present the sample evidence.
e. Find the probability distribution information.
f. Determine the results.

8.118 The dollar value of a college education is often measured by comparing the average annual pay for workers who were graduated from college to the pay earned by those who never attended college after receiving a high school diploma. A USA Snapshot® (*USA Today,* 9-11-98) listed Census Bureau data that showed average earnings in 1996 of male workers who possessed a bachelor's degree as $53,102 and that of male high school graduates with no college experience as $34,034. Suppose a random sample of 150 male college graduates with bachelor's degrees is taken in 1998 to check for any possible salary increase and a sample mean of $53,500 is obtained. Test the hypothesis that no increase in average salary occurred during the 2-year period at the 0.05 level of significance. Assume that $\sigma = \$3900$ per year. Use the classical approach.

a. Define the parameter.   **b.** State the null and alternative hypotheses.
c. Specify the hypothesis test criteria.   **d.** Present the sample evidence.
e. Find the probability distribution information.
f. Determine the results.

8.119 The manager at Air Express feels that the weights of packages shipped recently are less than in the past. Records show that in the past packages have had a mean weight of 36.7 lb. and a standard deviation of 14.2 lb. A random sample of last month's shipping records yielded a mean weight of 32.1 lb. for 64 packages. Is this sufficient evidence to reject the null hypothesis in favor of the manager's claim? Use $\alpha = 0.01$.

8.120 A fire insurance company felt that the mean distance from a home to the nearest fire department in a suburb of Chicago was at least 4.7 mi. It set its fire insurance rates accordingly. Members of the community set out to show that the mean distance was less than 4.7 mi. This, they felt, would convince the insurance company to lower its rates. They randomly identified 64 homes and measured the distance to the nearest fire department for each. The resulting sample mean was 4.4. If $\sigma = 2.4$ mi., does the sample show sufficient evidence to support the community's claim at the $\alpha = 0.05$ level of significance?

Return to CHAPTER CASE STUDY

The USA Snapshot® "Holiday home trimmings" presented information about all American households (page 349). One hundred fifty adult shoppers at a large shopping mall were asked "How much (to the nearest $25) do you anticipate your family will spend on holiday decorations this year?"

25	200	100	25	250	75	25	50	25	100	75	25	100	75	25
25	200	25	0	25	175	25	75	100	100	50	25	50	100	50
25	100	100	175	25	75	25	0	100	25	25	50	25	25	75
0	100	100	75	75	100	25	50	50	25	100	100	150	75	75
25	25	50	75	75	100	25	50	0	25	25	100	25	50	150
150	75	100	150	0	100	75	25	75	25	0	300	25	25	50
25	100	25	75	75	25	25	50	50	50	50	25	100	125	50
50	75	25	75	25	0	100	0	50	75	50	100	50	125	25
50	75	125	100	50	125	200	75	25	25	25	50	25	50	25
25	0	0	100	25	100	100	50	25	25	125	25	75	100	25

EXERCISE 8.121

Use the above sample data to describe the anticipated amount households living near this mall plan to spend on holiday decorations this year.

  a. Describe the sample data using several numerical statistics and at least one graph.
  b. Estimate the mean anticipated amount households living near this mall plan to spend on holiday decorations this year. Use 95% level of confidence and assume $\sigma = 70$.
  c. Does the above sample suggest that the families who shop in this mall anticipate spending a different average amount than all Americans according to "Holiday home trimmings"? Use $\alpha = 0.05$.
  d. Are the assumptions for the confidence interval and hypothesis test methods satisfied? Explain.

IN RETROSPECT

Two forms of inference were studied in this chapter: estimation and hypothesis testing. They may be, and often are, used separately. It would, however, seem natural for the rejection of a null hypothesis to be followed by a confidence interval. (If the value claimed is wrong, we will often want an estimate of the true value.)

These two forms of inference are quite different but they are related. There is a certain amount of crossover between the use of the two inferences. For example, suppose that you had sampled and calculated a 90% confidence interval for the mean of a population. The interval was 10.5 to 15.6. Following this someone claims that the true mean is 15.2. Your confidence interval can be compared to this claim. If the claimed value falls within your interval estimate, you would fail to reject the null hypothesis that $\mu = 15.2$ at a 10% level of significance in a two-tailed test. If the claimed value (say, 16.0) falls outside the interval, you would then reject the null hypothesis that $\mu = 16.0$ at $\alpha = 0.10$ in a two-tailed test. If a one-tailed test is required, or if you prefer a different value of α, a separate hypothesis test must be used.

Many users of statistics (especially those marketing a product) will claim that their statistical results prove that their product is superior. But remember, the hypothesis test does not prove or disprove anything. The decision reached in a hypothesis test has probabilities associated with the four various situations. If "fail to reject H_o" is the decision, it is possible that an error has occurred. Further, if "reject H_o" is the decision reached, it is possible for this to be an error. Both errors have probabilities greater than zero.

In this chapter we have restricted our discussion of inferences to the mean of a population for which the standard deviation is known. In Chapters 9 and 10, we will discuss inferences about the population mean and remove the restriction about the known value for standard deviation. We will also look at inferences about the parameters proportion, variance and standard deviation.

CHAPTER EXERCISES

8.122 A sample of 64 measurements is taken from a continuous population, and the sample mean is found to be 32.0. The standard deviation of the population is known to be 2.4. An interval estimation is to be made of the mean with a level of confidence of 90%. State or calculate the following items.

    a. $\bar{x}$ **b.** μ **c.** n **d.** $1 - \alpha$ **e.** $z_{(\alpha/2)}$ **f.** $\sigma_{\bar{x}}$

    g. E (maximum error of estimate)   **h.** upper confidence limit

    i. lower confidence limit

8.123 Suppose that a confidence interval is assigned a level of confidence of $1 - \alpha = 95\%$. How is the 95% used in constructing the confidence interval? If $1 - \alpha$ was changed to 90%, what effect would this have on the confidence interval?

8.124 Suppose a hypothesis test is assigned a level of significance of $\alpha = 0.01$.

    a. How is the 0.01 used in completing the hypothesis test?

    b. If α is changed to 0.05, what effect would this have on the test procedure?

8.125 The expected mean of a continuous population is 100, and its standard deviation is 12. A sample of 50 measurements gives a sample mean of 96. Using a 0.01 level of significance, a test is to be made to decide between "the population mean is 100" or "the population mean is different than 100." State or find each of the following:

    a. H_o **b.** H_a **c.** α **d.** μ (based on H_o)   **e.** $\bar{x}$

    f. σ **g.** $\sigma_{\bar{x}}$ **h.** $z \star$, z-score for $\bar{x}$ **i.** p-value   **j.** decision

    k. Sketch the standard normal curve and locate $z \star$, p-value, and α.

8.126 The expected mean of a continuous population is 200, and its standard deviation is 15. A sample of 80 measurements gives a sample mean of 205. Using a 0.01 level of significance, a test is to be made to decide between "the population mean is 200" or "the population mean is different than 200." State or find each of the following:

    a. H_o **b.** H_a **c.** α **d.** $z_{(\alpha/2)}$ **e.** μ (based on H_o)

    f. $\bar{x}$ **g.** σ **h.** $\sigma_{\bar{x}}$ **i.** $z \star$, z-score for $\bar{x}$ **j.** decision

    k. Sketch the standard normal curve and locate $\alpha/2$, $z_{(\alpha/2)}$, the critical region, and $z \star$.

8.127 From a population of unknown mean μ and a standard deviation $\sigma = 5.0$, a sample of $n = 100$ is selected and the sample mean 40.6 is found. Compare the concepts of estimation and hypothesis testing by answering the following:

    a. Determine the 95% confidence interval for μ.

    b. Complete the hypothesis test involving H_a: $\mu \neq 40$ using the p-value approach and $\alpha = 0.05$.

    c. Complete the hypothesis test involving H_a: $\mu \neq 40$ using the classical approach and $\alpha = 0.05$.

    d. On one sketch of the standard normal curve, locate the interval representing the confidence interval from (a); the $z \star$, p-value, and α from (b); and the $z \star$ and critical regions from (c). Describe the relationship between these three separate procedures.

8.128 From a population of unknown mean μ and a standard deviation $\sigma = 5.0$, a sample of $n = 100$ is selected and the sample mean 41.5 is found. Compare the concepts of estimation and hypothesis testing by answering the following:

    a. Determine the 95% confidence interval for μ.

    b. Complete the hypothesis test involving H_a: $\mu \neq 40$ using the p-value approach and $\alpha = 0.05$.

    c. Complete the hypothesis test involving H_a: $\mu \neq 40$ using the classical approach and $\alpha = 0.05$.

    d. On one sketch of the standard normal curve, locate the interval representing the confidence interval from (a); the $z \star$, p-value, and α from (b); and the $z \star$ and critical regions from (c). Describe the relationship between these three separate procedures.

8.129 From a population of unknown mean μ and a standard deviation $\sigma = 5.0$, a sample of $n = 100$ is selected and the sample mean 40.9 is found. Compare the concepts of estimation and hypothesis testing by answering the following: *(continued)*

a. Determine the 95% confidence interval for μ.

b. Complete the hypothesis test involving H_a: $\mu > 40$ using the p-value approach and $\alpha = 0.05$.

c. Complete the hypothesis test involving H_a: $\mu > 40$ using the classical approach and $\alpha = 0.05$.

d. On one sketch of the standard normal curve, locate the interval representing the confidence interval from (a); the $z \star$, p-value, and α from (b); and the $z \star$ and critical regions from (c). Describe the relationship between these three separate procedures.

8.130 The standard deviation of a normally distributed population is equal to 10. A sample size of 25 is selected, and its mean is found to be 95.

a. Find an 80% confidence interval for μ.

b. If the sample size were 100, what would be the 80% confidence interval?

c. If the sample size were 25 but the standard deviation were 5 (instead of 10), what would be the 80% confidence interval?

8.131 The weights of full boxes of a certain kind of cereal are normally distributed with a standard deviation of 0.27 oz. A sample of 18 randomly selected boxes produced a mean weight of 9.87 oz.

a. Find the 95% confidence interval for the true mean weight of a box of this cereal.

b. Find the 99% confidence interval for the true mean weight of a box of this cereal.

c. What effect did the increase in the level of confidence have on the width of the confidence interval?

8.132 Waiting times (in hours) at a popular restaurant are believed to be approximately normally distributed with a variance of 2.25 hr during busy periods.

a. A sample of 20 customers revealed a mean waiting time of 1.52 hr. Construct the 95% confidence interval for the population mean.

b. Suppose that the mean of 1.52 hr had resulted from a sample of 32 customers. Find the 95% confidence interval.

c. What effect does a larger sample size have on the confidence interval?

8.133 A random sample of the scores of 100 applicants for clerk-typist positions at a large insurance company showed a mean score of 72.6. The preparer of the test maintained that qualified applicants should average 75.0.

a. Determine the 99% confidence interval for the mean score of all applicants at the insurance company. Assume that the standard deviation of test scores is 10.5.

b. Can the insurance company conclude that it is getting qualified applicants (as measured by this test)?

8.134 Over the years, there has been a general trend that farms have been increasing in size, whereas the rural population has continually declined. On the other hand, the National Agricultural Statistics Service of the U.S. Department of Agriculture (1998) indicated that the average acreage per farm was 470 acres in both 1996 and 1997 among the 2.06 million or so farms in the nation. Total farm acreage declined slightly, but so did the number of farms. Therefore, the average farm size did not change. A random sample of 100 farms in rural America in 1998 reveals an average of 495 acres. Find the 95% confidence interval for the mean farm size in 1998. Assume the standard deviation is 190 acres per farm.

8.135 An article titled "Evaluation of a Self-Efficacy Scale for Preoperative Patients" (*AORN Journal,* Vol. 60, No. 1, July 1994) describes a 32-item rating scale used to determine efficacy (measure of effectiveness) expectations as well as outcome expectations. The 32-item scale was administered to 200 preoperative patients. The mean efficacy expectation score for the ambulating item equaled 4.00. Construct the 0.95 confidence interval for the mean of all such preoperative patients. Use $\sigma = 0.94$.

8.136 An automobile manufacturer wants to estimate the mean gasoline mileage that its customers will obtain with its new compact model. How many sample runs must be performed in order that the estimate be accurate to within 0.3 mpg at 95% confidence? (Assume that $\sigma = 1.5$.)

(continued)

8.137 A fish hatchery manager wants to estimate the mean length of her 3-year-old hatchery-raised trout. She wants to make a 99% confidence interval accurate to within $\frac{1}{3}$ of a standard deviation. How large a sample does she need to take?

8.138 We are interested in estimating the mean life of a new product. How large a sample do we need to take in order to estimate the mean to within $\frac{1}{10}$ of a standard deviation with 90% confidence?

8.139 According to an article in *Health* magazine (March 1991), supplementation with potassium reduced the blood pressure readings in a group of mild hypertensive patients from an average of $\frac{158}{100}$ to $\frac{143}{84.5}$. Consider a study involving the use of potassium supplementation to reduce the systolic blood pressure for mild hypertensive patients. Suppose 75 patients with mild hypertension were placed on potassium for 6 weeks. The response measured was the systolic reading at the beginning of the study minus the systolic reading at the end of the study. The mean drop in the systolic reading was 12.5 units. Assume the population standard deviation, σ, to be 7.5 units. Calculate the value of the test statistic $z \star$ and the p-value for testing H_o: $\mu = 10.0$ vs. H_a: $\mu > 10.0$.

8.140 The college bookstore tells prospective students that the average cost of its textbooks is $32 per book with a standard deviation of $4.50. The engineering science students think that the average cost of their books is higher than the average for all students. To test the bookstore's claim against their alternative, the engineering students collect a random sample of size 45.
   a. If they use $\alpha = 0.05$, what is the critical value of the test statistic?
   b. The engineering students' sample data are summarized by $n = 45$ and $\sum x = 1470.25$. Is this sufficient evidence to support their contention?

8.141 A manufacturing process produces ball bearings with diameters having a normal distribution and a standard deviation of $\sigma = 0.04$ cm. Ball bearings that have diameters that are too small, or too large, are undesirable. To test the null hypothesis that $\mu = 0.50$ cm., a sample of 25 is randomly selected and the sample mean is found to be 0.51.
   a. Design a null and alternative hypothesis such that rejection of the null hypothesis will imply that the ball bearings are undesirable.
   b. Using the decision rule established in (a), what is the p-value for the sample results?
   c. If the decision rule in (a) is used with $\alpha = 0.02$, what is the critical value for the test statistic?

8.142 A rope manufacturer, after conducting a large number of tests over a long period of time, has found that the rope has a mean breaking strength of 300 lb. and a standard deviation of 24 lb. Assume that these values are μ and σ. It is believed that by using a recently developed high-speed process, the mean breaking strength has been decreased.
   a. Design a null and alternative hypothesis such that rejection of the null hypothesis will imply that the mean breaking strength has decreased.
   b. Using the decision rule established in (a), what is the p-value associated with rejecting the null hypothesis when 45 tests result in a sample mean of 295?
   c. If the decision rule in (a) is used with $\alpha = 0.01$, what is the critical value for the test statistic and what value of $\bar{x}$ corresponds to it if a sample of size 45 is used?

8.143 Worker honey bees leave the hive on a regular basis and travel to flowers and other sources of pollen and nectar before returning to the hive to deliver their cargo. The process is repeated several times each day in order to feed younger bees and support the hive's production of honey and wax. The worker bee can carry an average of 0.0113 gm of pollen and nectar per trip, with a standard deviation of 0.0063 gm. Fuzzy Drone is entering the honey and beeswax business with a new strain of Italian bees that are reportedly capable of carrying larger loads of pollen and nectar than the typical honey bee. After installing three hives, Fuzzy isolated 200 bees before and after their return trip and carefully weighed their cargoes. The sample mean weight of the pollen and nec-

(continued)

tar was 0.0124 gm. Can Fuzzy's bees carry a greater load of pollen and nectar than the rest of the honey bee population? Complete the appropriate hypothesis test at the 0.01 level of significance.

a. Solve using the *p*-value approach.   **b.** Solve using the classical approach.

8.144 For a population of humans to sustain itself, there must be an average of just over 2 births for each woman of reproductive age. This fertility rate varies substantially from within one nation to the next. Peter Drucker recently pointed out in *Fortune* magazine (9-28-98) that the average for Japan has dropped to 1.5 births for each woman of reproductive age, which could reduce Japan's population from 135 million in 1997 to 50 million by the end of the 21st century.

Suppose a random sample of 300 Japanese women of reproductive age is taken in 1998, and the sample mean fertility rate is measured at 1.45. Assume the standard deviation is 0.75. Did the rate decline? Complete the appropriate hypothesis test at the 0.05 level of significance.

a. Solve using the *p*-value approach.   **b.** Solve using the classical approach.

8.145 In a large supermarket the customer's waiting time to check out is approximately normally distributed with a standard deviation of 2.5 min. A sample of 24 customer waiting times produced a mean of 10.6 min. Is this evidence sufficient to reject the supermarket's claim that its customer checkout time averages no more than 9 min? Complete this hypothesis test using the 0.02 level of significance.

a. Solve using the *p*-value approach.   **b.** Solve using the classical approach.

8.146 At a very large firm, the clerk-typists were sampled to see whether the salaries differed among departments for workers in similar categories. In a sample of 50 of the firm's accounting clerks, the average annual salary was $16,010. The firm's personnel office insists that the average salary paid to all clerk-typists in the firm is $15,650 and that the standard deviation is $1800. At the 0.05 level of significance, can we conclude that the accounting clerks receive, on the average, a different salary from that of the clerk-typists?

a. Solve using the *p*-value approach.   **b.** Solve using the classical approach.

8.147 Jack Williams is vice president of marketing for one of the largest natural gas companies in the nation. During the past 4 years, he has watched two major factors erode away the profits and sales of the company. First, the average price of crude oil has been virtually flat, and many of his industrial customers are burning heavy oil rather than natural gas to fire their furnaces, irrespective of added smoke stack emissions. Second, both residential and commercial customers are still pursuing energy conservation techniques (e.g., adding extra insulation, installing clock-drive thermostats, and sealing cracks around doors and windows to eliminate cold air infiltration). In 1997 residential customers bought an average of 129.2 mcf of natural gas from Jack's company ($\sigma = 18$ mcf), based on internal company billing records, but environmentalists have claimed that conservation is cutting fuel consumption up to 3% per year. Jack has commissioned you to conduct a spot check now to see if any change in annual usage has transpired before his next meeting with the officers of the corporation. A sample of 300 customers selected randomly from the billing records reveals an average of 127.1 mcf during the past 12 months. Did consumption decline?

a. Complete the appropriate hypothesis test at the 0.01 level of significance using the probability-value approach so that you can properly advise Jack prior to his meeting.

b. Since you are Jack's assistant, why is it best for you to use the *p*-value approach?

 8.148 A manufacturer of automobile tires believes it has developed a new rubber compound that has superior wearing qualities. It produced a test run of tires made with this new compound and had them road tested. The data recorded was the amount of tread wear per 10,000 miles. In the past, the mean amount of tread wear per 10,000 miles, for tires of this quality, has been 0.0625 inches.

The null hypothesis to be tested here is "the mean amount of wear on the tires made with the new compound is the same mean amount of wear with the old compound,

(continued)

0.0625 inches per 10,000 miles," H_o: $\mu = 0.0625$. Three possible alternative hypotheses could be used: (1) H_a: $\mu < 0.0625$, (2) H_a: $\mu \neq 0.0625$, (3) H_a: $\mu > 0.0625$.

a. Explain the meaning of each of these three alternatives

b. Which one of the possible alternative hypotheses should the manufacturer use if it hopes to conclude that "use of the new compound does yield superior wear"?

8.149 All drugs must be approved by the Food and Drug Administration (FDA) before they can be marketed by a drug company. The FDA must weigh the error of marketing an ineffective drug, with the usual risks of side effects, against the consequences of not allowing an effective drug to be sold. Suppose, using standard medical treatment, that the mortality rate (r) of a certain disease is known to be A. A manufacturer submits for approval a drug that is supposed to treat this disease. The FDA sets up the hypothesis to test the morality rate for the drug as (1) H_o: $r = A$, H_a: $r < A$, $\alpha = 0.005$; or (2) H_o: $r = A$, H_a: $r > A$, $\alpha = 0.005$.

a. If $A = 0.95$, which test do you think the FDA should use? Explain.

b. If $A = 0.05$, which test do you think the FDA should use? Explain.

8.150 The drug manufacturer in Exercise 8.149 has a different viewpoint on the matter. It wants to market the new drug starting as soon as possible so that it can beat its competitors to the marketplace and make lots of money. Its position is, "Market the drug unless the drug is totally ineffective."

a. How would the drug company set up the alternative hypothesis if it were doing the testing? H_a: $r < A$, H_a: $r \neq A$, or H_a: $r > A$? Explain why.

b. Does the mortality rate ($A = 0.95$ or $A = 0.05$) of the existing treatment affect the alternative? Explain.

8.151 The following computer output shows a simulated sample of size 25 randomly generated from a normal population with $\mu = 130$ and $\sigma = 10$. A confidence interval command was then used to set a 95% confidence interval for μ.

```
116.187 119.832 121.782 122.320 141.436 129.197 119.172
120.713 135.765 131.153 122.307 126.155 137.545 141.154
123.405 143.331 121.767 109.742 140.524 150.600 121.655
127.992 136.434 139.768 125.594

  N    MEAN    STDEV    SE MEAN    95.0 PERCENT C.I.
 25   129.02   10.18     2.00      (125.10, 132.95)
```

a. State the confidence interval that resulted.

b. Verify the values reported for the standard error of mean and the interval bounds.

8.152 Use a computer and generate 50 random samples, each of size $n = 25$, from a normal probability distribution with $\mu = 130$ and standard deviation $\sigma = 10$.

a. Calculate the 95% confidence interval based on each sample mean.

b. What proportion of these confidence intervals contain $\mu = 130$?

c. Explain what the proportion found in (b) represents.

8.153 The following computer output shows a simulated sample of size 28 randomly generated from a normal population with $\mu = 18$ and $\sigma = 4$. Computer commands were then used to complete a hypothesis test for $\mu = 18$ against a two-tailed alternative.

a. State the alternative hypothesis, the decision, and the conclusion that resulted.

b. Verify the values reported for the standard error of mean, $z \star$, and the p-value.

```
18.7734 21.4352 15.5438 20.2764 23.2434 15.7222 13.9368
14.4112 15.7403 19.0970 19.0032 20.0688 12.2466 10.4158
 8.9755 18.0094 20.0112 23.2721 16.6458 24.6146 17.8078
16.5922 16.1385 12.3115 12.5674 18.9141 22.9315 13.3658
```

(continued)

```
TEST OF MU = 18.000 VS MU N.E. 18.000
THE ASSUMED SIGMA = 4.00

      N    MEAN   STDEV   SE MEAN    Z     P VALUE
     28   17.217   4.053   0.756   -1.04    0.30
```

8.154 Use a computer and generate 50 random samples, each of size $n = 28$, from a normal probability distribution with $\mu = 18$ and standard deviation $\sigma = 4$.

   a. Calculate the $z \star$ corresponding to each sample mean.

   b. In regard to the p-value approach, find the proportion of 50 $z \star$-values that are "more extreme" than the $z = -1.04$ that occurred in Exercise 8.153 (H_a: $\mu \neq 18$). Explain what this proportion represents.

   c. In regard to the classical approach, find the critical values for a two-tailed test using $\alpha = 0.10$; find the proportion of 50 $z \star$-values that fall in the critical region. Explain what this proportion represents.

8.155 Use a computer and generate 50 random samples, each of size $n = 28$, from a normal probability distribution with $\mu = 19$ and standard deviation $\sigma = 4$.

   a. Calculate the $z \star$ corresponding to each sample mean that would result when testing the null hypothesis $\mu = 18$.

   b. In regard to the p-value approach, find the proportion of 50 $z \star$-values that are "more extreme" than the $z = -1.04$ that occurred in Exercise 8.153 (H_a: $\mu \neq 18$). Explain what this proportion represents.

   c. In regard to the classical approach, find the critical values for a two-tailed test using $\alpha = 0.10$; find the proportion of 50 $z \star$-values that fall in the noncritical region. Explain what this proportion represents.

VOCABULARY LIST

Be able to define each term. Pay special attention to the key terms, which are printed in **red.** In addition, describe in your own words, and give an example of, each term. Your examples should not be ones given in class or in the textbook.

The bracketed numbers indicate the chapters in which the terms first appeared, but you should define the terms again to show increased understanding of their meaning. Page numbers indicate the first appearance of the term in Chapter 8.

alpha (α) (p. 370)
alternative hypothesis (pp. 367, 375, 391)
assumptions (pp. 354, 355, 374, 389)
beta (b) (p. 370)
biased statistics (p. 351)
calculated value ($z \star$) (pp. 378, 394)
conclusion (pp. 371, 380, 396)
confidence coefficient (p. 355)
confidence interval (p. 352)
confidence interval procedure (p. 356)
critical region (p. 394)
critical value (p. 395)
decision rule (pp. 371, 380, 396)
estimation (p. 350)
hypothesis (p. 366)
hypothesis test (p. 366)
hypothesis test, classic procedure (p. 389)
hypothesis test, p-value procedure (p. 374)
interval estimate (p. 352)
level of confidence (p. 352)
level of significance (p. 371, 378, 393)

lower confidence limit (p. 356)
maximum error of estimate (p. 356)
noncritical region (p. 394)
null hypothesis (pp. 367, 375, 391)
parameter [1] (pp. 350, 356, 367, 374, 389)
point estimate (p. 350)
p-value (p. 379)
sample size (p. 362)
sample statistic [1, 2] (p. 350)
standard error of mean [7] (p. 355)
test criteria (pp. 377, 394)
test statistic (pp. 371, 378, 393)
type A correct decision (p. 368)
type B correct decision (p. 368)
type I error (p. 368)
type II error (p. 368)
unbiased statistic (p. 351)
upper confidence limit (p. 356)
$z_{(\alpha)}$ [6] (pp. 355, 395)
$z \star$ ("z star") (pp. 378, 393)

CHAPTER PRACTICE TEST

Part I: Knowing the Definitions

Answer "True" if the statement is always true. If the statement is not always true, replace the words shown in bold with words that make the statement always true.

8.1 **Beta** is the probability of a type I error.

8.2 $1 - \alpha$ is known as the level of significance of a hypothesis test.

8.3 The standard error of the mean is the standard deviation of the **sample selected.**

8.4 The maximum error of estimate is controlled by three factors: **level of confidence, sample size,** and **standard deviation.**

8.5 Alpha is the measure of the area under the curve of the standard score that lies in the **rejection region** for H_o.

8.6 The risk of making a **type I error** is directly controlled in a hypothesis test by establishing a level for α.

8.7 Failing to reject the null hypothesis when it is false is a **correct decision.**

8.8 If the noncritical region in a hypothesis test is made wider (assuming σ and n remain fixed), α becomes larger.

8.9 Rejection of a null hypothesis that is false is a **type II error.**

8.10 To conclude that the mean is higher (or lower) than a claimed value, the value of the test statistic must fall in the **acceptance region.**

Part II: Applying the Concepts

Answer all questions, showing all formulas, substitutions, and work.

8.11 An unhappy post office customer is disturbed with the waiting time to buy stamps. Upon registering his complaint he was told, "The average waiting time in the past has been about 4 min with a standard deviation of 2 min." The customer collected a sample of $n = 45$ customers and found the mean wait was 5.3 min. Find the 95% confidence interval for the mean waiting time.

8.12 State the null (H_o) and the alternative (H_a) hypotheses that would be used to test each of the following claims.
   a. The mean weight of professional football players is more than 245 lb.
   b. The mean monthly amount of rainfall in Monroe County is less than 4.5 in.
   c. The mean weight of the baseball bats used by major league players is not equal to 35 oz.

8.13 Determine the level of significance, test statistic, critical region, and critical value(s) that would be used in completing each hypothesis test using $\alpha = 0.05$.
   a. H_o: $\mu = 43$ **b.** H_o: $\mu = 0.80$ **c.** H_o: $\mu = 95$
   H_a: $\mu < 43$ H_a: $\mu > 0.80$ H_a: $\mu \neq 95$
   (given $\sigma = 6$)   (given $\sigma = 0.13$)   (given $\sigma = 12$)

8.14 Find each of the following:
   a. $z(0.05)$ **b.** $z(0.01)$ **c.** $z(0.12)$

8.15 In the past, the grapefruits grown in a particular orchard have had a mean diameter of 5.50 in. and a standard deviation of 0.6 in. The owner believes this year's crop is larger than in the past. He collected a random sample of 100 and found a sample mean diameter of 5.65 in.
   a. Find the value of the test statistic, $z \star$, that corresponds to $\overline{x} = 5.65$.
   b. Calculate the p-value for the owner's hypothesis.

8.16 A manufacturer of light bulbs claims that its light bulbs have a mean life of 1520 hr with a standard deviation of 85 hr. A random sample of 40 such bulbs is selected for testing. If the sample produces a mean value of 1498.3 hr, is there sufficient evidence to claim that the mean life is less than the manufacturer claimed? Use $\alpha = 0.01$.

Part III: Understanding the Concepts

8.17 Sugar Creek Convenience Stores has commissioned a statistical firm to survey its customers in order to estimate the mean amount spent per customer. From previous records the standard deviation is believed to be $\sigma = \$5$. In their proposal to Sugar Creek, the statistics firm states that they plan to base the estimate for the mean amount spent on a sample of size 100 and use the 95% confidence level. Sugar Creek's president has suggested that the sample size be increased to 400. If nothing else changes, what effect will this increase in the sample size have on
a. the point estimate for the mean?
b. the maximum error of estimation?
c. the confidence interval?

The CEO wants the level of confidence increased to 99%. If nothing else changes, what effect will this increase in level of confidence have on
d. the point estimate for the mean?
e. the maximum error of estimation?
f. the confidence interval?

8.18 The noise level in a hospital may be a critical factor influencing a patient's rate of recovery. Suppose for the sake of discussion that a research commission has recommended a maximum mean noise level of 30 db with a standard deviation of 10 db. The staff of a hospital intends to sample one of its wards to determine if the noise level is significantly higher than the recommended level. The following hypothesis test will be completed.

$$H_o: \mu = 30 \ (\leq) \text{ vs. } H_a: \mu > 30, \ \alpha = 0.05$$

a. Identify the correct interpretation for each hypothesis with regard to the recommendation and justify your choice.
   H_o: (a) noise level is not significantly higher than the recommended level, or
      (b) noise level is significantly higher than the recommended level
   H_a: (a) noise level is not significantly higher than the recommended level, or
      (b) noise level is significantly higher than the recommended level
b. Which statement below best describes the type I error?
   (1) Decision reached was, noise level is within level recommended, when in fact it actually was within.
   (2) Decision reached was, noise level is within level recommended, when in fact it actually exceeded it.
   (3) Decision reached was, noise level exceeds the level recommended, when in fact it actually was within.
   (4) Decision reached was, noise level is within level recommended, when in fact it actually exceeded it.
c. Which statement in (b) best describes the type II error?
d. If α were changed from 0.05 to 0.01, identify and justify the effect (increases, decreases, or remains the same) on each of the following: $P(\text{type I error})$, $P(\text{type II error})$.

8.19 The alternative hypothesis is sometimes called the "research hypothesis." The conclusion is a statement written about the alternative hypothesis. Explain why these two statements are compatible.

Chapter 9

Inferences Involving One Population

CHAPTER CASE STUDY

Paying Their Own Way

Many students work full-time or part-time, during the academic year or during the summer, or some combination, to earn part of (maybe even all of) their college expenses. Do you work to pay for some part of your college expenses? Do your friends work to pay for part of their college expenses? How much did you or your friends each earn last month? The USA Snapshot®, "Working their way through," that appeared in *USA Today*, March 17, 1998, describes the average monthly earnings of American college students.

USA SNAPSHOTS®

A look at statistics that shape your finances

Working their way through

Two-thirds of college students say they have a job.
Average monthly earnings:

Do not have a job 33%
Less than $100 4%
$100-$199 10%
$200-$299 13%
$300-$399 12%
$400-$499 11%
$500 or more 17%

Source: Campus Concepts   By Cindy Hall and Jerry Mosemak, USA TODAY

The information on the graph seems to suggest that the average amount earned each month by college students is approximately $350. What are the statistical procedures used for estimating the mean amount earned per month by working college students?

Listed below is the amount earned last month by each in a sample of 35 college students.

0	0	105	0	313	453	769	415	244	0	333	0
0	362	276	158	409	0	0	534	449	281	37	338
240	0	0	0	142	0	519	356	280	161	0	

Use this sample data to estimate the mean amount earned by a college student per month. Find two kinds of estimates, a point estimate and a confidence interval without knowledge of the population standard deviation, σ. These are among the methods to be learned in this chapter. We will return to these questions at the end of the chapter. (See Exercise 9.123, p. 463.)

CHAPTER OBJECTIVES

In Chapter 8 we learned about two forms of statistical inference: confidence intervals and hypothesis testing. The study of these two types of inference was restricted to the unlikely circumstance that the population parameter σ was known so that we could focus our attention on learning the basic procedures. The standard deviation or variance of a population is seldom known in a real-world application, so now we will remove that restriction and find out how both types of inference about the population mean μ are treated in a more realistic manner. We will also learn how to perform both types of inference when our concern is about the population parameter p, the binomial probability of success, and σ, the population standard deviation.

Figure 9.1

Do I Use the z-Statistic or the t-Statistic?

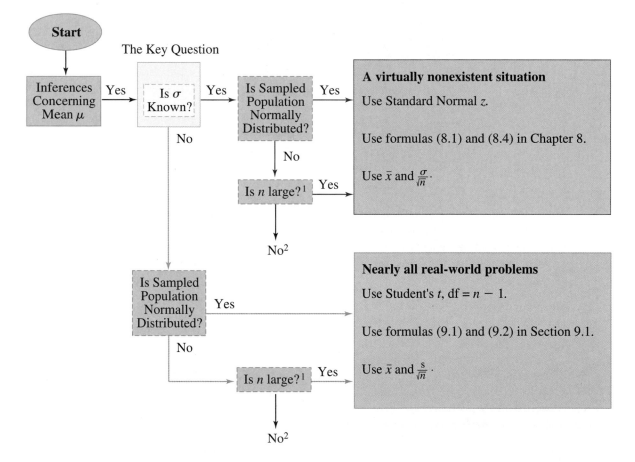

1. Is n large? – Samples as small as $n = 15$ or 20 may be considered large enough for the central limit theorem to hold if the sample data a unimodal, nearly symmetrical, short-tailed, and without outliers. Samples that are not symmetrical require larger sample sizes, with 50 sufficing, except for extremely skewed samples. See discussion on page 355.
2. Requires the use of a nonparametric technique; see Chapter 14.

9.1   Inferences About Mean μ (σ Unknown)

Inferences about the population mean μ are based on the sample mean $\bar{x}$ and information obtained from the sampling distribution of sample means. Recall the sampling distribution of sample means has a mean μ and a **standard error** of $\sigma/\sqrt{n}$ for all samples of size n, and it is normally distributed when the sampled population has a normal distribution or approximately normal when the **sample size** is sufficiently large. This means the test statistic $z \bigstar = \dfrac{\bar{x} - \mu}{\sigma/\sqrt{n}}$ has a standard normal distribution. However, when σ **is unknown,** the standard error $\sigma/\sqrt{n}$ is also unknown. Therefore, the sample standard deviation s will be used as the point estimate for σ. As a result, an estimated standard error of the mean, $s/\sqrt{n}$, will be used and our test statistic will become $\dfrac{\bar{x} - \mu}{s/\sqrt{n}}$.

When a **known** σ is being used to make an inference about mean μ, a sample provides one value for use in the formulas; that one value is $\bar{x}$. When the sample standard deviation s is also used, the sample provides two values: the sample mean $\bar{x}$ and the estimated standard error $s/\sqrt{n}$. As a result, the z-statistic will be replaced with a statistic that accounts for the use of an estimated standard error. This new statistic is known as the **Student's t-statistic.**

In 1908, W. S. Gosset, an Irish brewery employee, published a paper about this t-distribution under the pseudonym "Student." In deriving the t-distribution, Gosset assumed that the samples were taken from normal populations. Although this might seem to be quite restrictive, satisfactory results are obtained when selecting large samples from many nonnormal populations.

Figure 9.1 presents a diagrammatic organization for the inferences about the population mean as discussed in Chapter 8 and in this first section of Chapter 9. Two situations exist: σ is known, or σ is unknown. As stated before, σ is almost never a known quantity in real-world problems; therefore, the standard error will almost always be estimated using $s/\sqrt{n}$. The use of an estimated standard error of the mean requires the use of the t-distribution. Almost all real-world inferences about the population mean will be completed using the Student's t-statistic.

The t-distribution has the following properties (also see Figure 9.2, p. 418):

PROPERTIES OF THE t-DISTRIBUTION (df $>$ 2)[1]

1. t is distributed with a mean of 0.
2. t is distributed symmetrically about its mean.
3. t is distributed so as to form a family of distributions, a separate distribution for each different number of degrees of freedom (df $\geq$ 1).
4. The t-distribution approaches the **standard normal distribution** as the number of degrees of freedom increases.
5. t is distributed with a variance greater than 1, but as the degrees of freedom increases, the variance approaches 1.
6. t is distributed so as to be less peaked at the mean and thicker at the tails than the normal distribution.

---

1. Not all of the properties hold for df $= 1$ and df $= 2$. Since we will not encounter situations where df $= 1, 2$, these special cases are not discussed further.

Figure 9.2

Student's *t*-Distributions

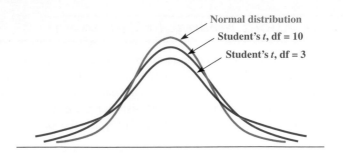

DEGREES OF FREEDOM, df

A **parameter** that identifies each different distribution of the Student's *t*-distribution. For the methods presented in this chapter, the value of df will be the sample size minus 1, df = $n - 1$.

The number of degrees of freedom associated with s^2 is the divisor $(n - 1)$ used to define sample variance s^2 [formula (2.6), p. 77]. That is, df = $n - 1$. Sample variance is the mean of the squared deviations. The number of degrees of freedom is the "number of unrelated deviations" available for use in estimating σ^2. Recall that the sum of the deviations, $\sum(x - \bar{x})$, must be zero. From a sample of size n, only the first $n - 1$ of these deviations has freedom of value. That is, the last, or nth, value of $(x - \bar{x})$ must make the sum of the n deviations total exactly zero. As a result, variance is said to average $n - 1$ unrelated squared deviation values, and this number, $n - 1$, is named "degrees of freedom."

Although there is a separate *t*-distribution for each degrees of freedom, df = 1, df = 2, . . ., df = 20, . . ., df = 40, and so on, only certain key **critical values of *t*** will be necessary for our work. Consequently, the table for Student's *t*-distribution (Table 6 in Appendix B) is a table of critical values rather than a complete table, such as Table 3 is for the standard normal distribution for *z*. As you look at Table 6, you will note that the left side of the table is identified by "df," degrees of freedom. This left-hand column starts at 3 at the top and lists consecutive df values to 30, then jumps to 35, . . ., to "df > 100" at the bottom. As previously stated, as the degrees of freedom increases, the *t*-distribution approaches the characteristics of the standard normal *z*-distribution. Once df is "greater than 100," the critical values of the *t*-distribution are the same as the corresponding critical values of the standard normal distribution as shown in Table 4A in Appendix B.

The critical values of the Student's *t*-distribution that are to be used for both constructing a confidence interval and for hypothesis testing will be obtained from Table 6 in Appendix B. To obtain the value of *t*, you will need to know two identifying values: (1) df, the number of degrees of freedom (identifying the distribution of interest); and (2) α, the area under the curve to the right of the right-hand critical value. A notation much like that used with *z* will be used to identify a critical value. $t_{(df, \alpha)}$, read as "t of df, α," is the symbol for the value of *t* with df degrees of freedom and an area of α in the right-hand tail, as shown in Figure 9.3.

Exercise 9.1

Make a list of four numbers that total "zero." How many numbers were you able to pick without restriction? Explain how this demonstrates "degrees of freedom."

Exercise 9.2

Explain the relationship between the critical values found in the bottom row of Table 6 and the critical values of *z* given in Table 4.

Figure 9.3

t-Distribution Showing $t_{(df, \alpha)}$

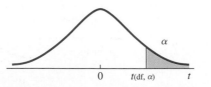

ILLUSTRATION 9.1 ▼

Find the value of $t_{(10, 0.05)}$ (see the diagram).

Solution

There are 10 degrees of freedom, and 0.05 is the area to the right. In Table 6, Appendix B, we look for the row df = 10 and the column marked "Amount of α in one-tail", α = 0.05. At their intersection, we see that $t_{(10, 0.05)}$ = **1.81.**

Exercise 9.3

Find the value of:

a. $t_{(12, 0.01)}$
b. $t_{(22, 0.025)}$

Portion of Table 6

df	...	Amount of α in one-tail	...	
		0.05		
⋮				
10		1.81 ──────⟶		$t_{(10, 0.05)}$ = **1.81**

▲

For the values of t on the left side of the mean, we can use one of two notations. The t-value shown in Figure 9.4 could be named $t_{(df, 0.95)}$, since the area to the right of it is 0.95, or it could be identified by $-t_{(df, 0.05)}$, since the t-distribution is symmetric about its mean, zero.

Figure 9.4
t-Value on Left Side

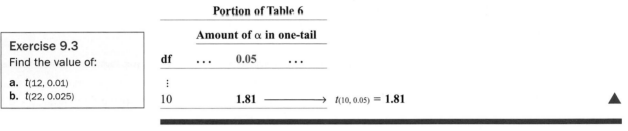

ILLUSTRATION 9.2 ▼

Find the value of $t_{(15, 0.95)}$.

Exercise 9.4

Find the value of:

a. $t_{(18, 0.90)}$
b. $t_{(9, 0.99)}$

Solution

There are 15 degrees of freedom. In Table 6 we look for the column marked α = 0.05 (one-tail) and its intersection with row df = 15. The table gives us $t_{(15, 0.05)}$ = 1.75; therefore, $t_{(15, 0.95)}$ = $-t_{(15, 0.05)}$ = -1.75 (the value is negative because it is to the left of the mean; see the following figure).

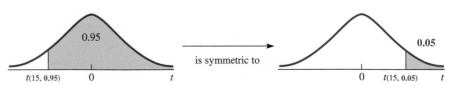

ILLUSTRATION 9.3 ▼

Find the values of the *t*-distribution that bound the middle 0.90 of the area under the curve for the distribution with df = 17.

Solution

The middle 0.90 leaves 0.05 for the area of each tail. The value of *t* that bounds the right-hand tail is $t(17, 0.05) = \mathbf{1.74}$, as found in Table 6. The value bounding the left-hand tail is -1.74, since the *t*-distribution is symmetric about its mean, zero.

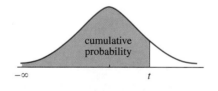

Exercise 9.5
Find the values of *t* that bound the middle 0.95 of the distribution for df = 12.

▲

If the df needed is not listed in the left-hand column of Table 6, then use the next smaller value of df that is listed. For example, $t(72, 0.05)$ will be estimated using $t(70, 0.05) = 1.67$.

Most computer software packages or statistical calculators will calculate the area related to a specified *t*-value. The accompanying figure shows the relationship between the cumulative probability distribution and a specific *t*-value for a *t*-distribution with df degrees of freedom.

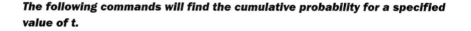

The following commands will find the cumulative probability for a specified value of t.

Exercise 9.6
Use a computer/calculator to find the area to the left of $t = -2.12$ with df = 18. Draw a sketch showing the question with the answer.

MINITAB	
Choose:	**Calc > Probability Distributions > t**
Select:	**Cumulative Probability**
Enter:	Degrees of freedom: **df**
Select:	**Input constant***
Enter:	**t-value** (ex. 1.74)

*Select Input column if several *t*-values are stored in C1. Use C2 for optional storage. If the area in the right tail is needed, subtract the calculated probability from 1.

The following commands will find the probability in one or two tails for a given t-value.

EXCEL

If several *t*-values (nonnegative) are to be used, input the values into column A and activate B1; then continue with:

Choose: **Paste function f_x > Statistical > TDIST > OK**
Enter: *X*: **individual *t*-value or (A1:A5 or select 't-value' cells)***
         Deg_freedom: **df**
         Tails: **1** or **2** (one- or two-tailed distributions)
Drag*: **Bottom right corner of the B1 cell down to give other probabilities**

To find the probability within the two tails or the cumulative probability for a one tail, subtract the calculated probability from 1.

The following commands will find the cumulative probability for a specified t-value.

TI-83

Choose: **2ⁿᵈ > DISTR > 5:tcdf(**
Enter: **−1EE99, *t*-value, df)**

NOTE To find the probability between two *t*-values, enter the two values in place of −1EE99 and *t*-value.

If the area in the right tail is needed, subtract the calculated probability from 1.

Exercise 9.7
Use a computer/calculator to find the area to the right of $t = 1.12$ with df = 15. Draw a sketch showing the question with the answer.

We are now ready to make inferences about the population mean μ using the sample standard deviation. As mentioned earlier, use of the *t*-distribution has a condition.

THE ASSUMPTION FOR INFERENCES ABOUT MEAN μ WHEN σ IS UNKNOWN

The sampled population is normally distributed.

Confidence Interval Procedure

The procedure used to make confidence intervals using the sample standard deviation is very similar to that used when σ is known (see pp. 354–358). The difference is the use of the Student's *t* in place of the standard normal *z* and the use of *s*, the sample standard deviation, as an estimate of σ. The Central Limit Theorem implies that this technique can also be applied to nonnormal populations when the sample size is sufficiently large.

The formula for the $1 - \alpha$ confidence interval for μ is

$$\bar{x} - t(\text{df}, \alpha/2) \cdot \frac{s}{\sqrt{n}} \quad \text{to} \quad \bar{x} + t(\text{df}, \alpha/2) \cdot \frac{s}{\sqrt{n}}, \text{ where df} = n - 1 \qquad \textbf{(9.1)}$$

ILLUSTRATION 9.4 ▼

A random sample of size 20 is taken from the weights of babies born at Northside Hospital during the year 1994. A mean of 6.87 lb and a standard deviation of 1.76 lb were found for the sample. Estimate, with 95% confidence, the mean weight of all babies born in this hospital in 1994. Based on past information, it is assumed that weights of newborns are normally distributed.

Solution

> The five-step confidence interval procedure is on page 356.

STEP 1 **The Set-Up:**
Describe the population parameter of interest.
μ, **the mean weight** of newborns at Northside Hospital.

STEP 2 **The Confidence Interval Criteria:**
a. **Check the assumptions.**
σ is unknown and past information indicates that the **sampled population is normal.**
b. **Identify the probability distribution and formula to be used.**
The Student's *t*-**distribution** will be used with **Formula (9.1).**
c. **Determine the level of confidence, $1 - \alpha$.**
$1 - \alpha = 0.95$.

STEP 3 **The Sample Evidence:**
Collect the sample information.

$n = \textbf{20}, \bar{x} = \textbf{6.87},$ and $s = \textbf{1.76}.$

STEP 4 **The Confidence Interval:**
a. **Determine the confidence coefficients.**

> Recall, confidence intervals are two-tailed situations.

Since $1 - \alpha = 0.95$, $\alpha = 0.05$; and therefore $\alpha/2 = 0.025$. Also, since $n = 20$, df $= 19$. From Table 6, at the intersection of row df $= 19$ and one-tailed column $\alpha = 0.025$, we find $t(\text{df}, \alpha/2) = t(19, 0.025) = \textbf{2.09}$. See the figure.

> df is used to find the confidence coefficient in Table 6, while *n* is used in the formula.

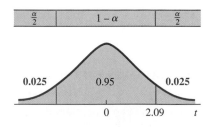

Information about the confidence coefficient and using Table 6 are on pages 418-420.
b. **Find the maximum error of estimate**

$$E = t(\text{df}, \alpha/2) \cdot \frac{s}{\sqrt{n}}:$$

$$E = t(19, 0.025) \cdot \frac{s}{\sqrt{n}} = 2.09 \cdot \frac{1.76}{\sqrt{20}} = (2.09)(0.394) = \textbf{0.82}$$

c. **Find the lower and upper confidence limits.**

$$\bar{x} - E \quad \text{to} \quad \bar{x} + E$$

$$6.87 - 0.82 \quad \text{to} \quad 6.87 + 0.82$$

$$\textbf{6.05} \quad \textbf{to} \quad \textbf{7.69}$$

Exercise 9.8
Construct a 95% confidence interval estimate for the mean μ using the sample information $n = 24$, $\bar{x} = 16.7$, and $s = 2.6$.

STEP 5   **The Results:**
State the confidence interval.
6.05 to 7.69, the 95% confidence interval for μ.
That is, with 95% confidence we estimate the mean weight to be between 6.05 and 7.69 lb.   ▲

The following commands will construct a $1 - \alpha$ confidence interval for a mean μ when the population standard deviation is unknown.

MINITAB

Input the data into C1; then continue with:

    Choose: **Stat > Basic Statistics > 1-Sample *t***

    Enter: Variables: **C1**
    Select: **Confidence interval**
    Enter: Level: **1 $-$ α** (ex. 95.0)

EXCEL

Input the data into column A; then continue with:

    Choose: **Tools > Data Analysis Plus > Inference About A Mean(SIGMA Unknown) > OK**

    Enter: **block coordinates: (A1:A20 or select cells) > OK**

    Choose: **Interval Estimate**
    Enter: level of confidence: **1 $-$ α** (ex. 0.95 or 95.0)

Exercise 9.9
Use a computer/calculator to construct a 0.98 confidence interval using the sample data:

6 7 12 9 10 8 5 9 7 9 6 5

TI-83

Input the data into L1; then continue with the following, entering the appropriate values and highlight Calculate:

    Choose: **STAT > TESTS > 8:Tinterval**

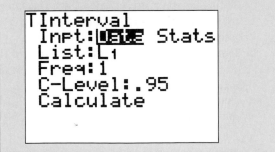

The solution to Illustration 9.4 looks like this when solved on MINITAB:

Compare the MINITAB output to the solution of Illustration 9.4.

```
Confidence Interval

  Variable    N    Mean   StDev   SE Mean      95.0% C.I.
        C1   20   6.870   1.760     0.394   (6.047, 7.693)
```

Hypothesis-Testing Procedure

The t-statistic is used to complete a hypothesis test about the population mean μ in much the same manner as z was used in Chapter 8. In hypothesis-testing situations, we will use formula (9.2) to calculate the value of the **test statistic $t \bigstar$.**

$$t \bigstar = \frac{\bar{x} - \mu}{s/\sqrt{n}}, \text{ with df} = n - 1 \qquad (9.2)$$

The **calculated t** is the number of estimated standard errors $\bar{x}$ is from the hypothesized mean μ. As with confidence intervals, the Central Limit Theorem indicates that the t-distribution can also be applied to nonnormal populations when the **sample size** is sufficiently large.

ILLUSTRATION 9.5 ▼

Let's return to the hypothesis of Illustration 8.10 (p. 376) where the EPA wanted to show that "the mean carbon monoxide level of air pollution is higher than 4.9." Does a random sample of 22 readings (sample results: $\bar{x} = 5.1$ and $s = 1.17$) present sufficient evidence to support the EPA's claim? Use $\alpha = 0.05$. Previous studies have indicated that such readings have an approximately normal distribution.

Solution

The five-step p-value hypothesis test procedure is on page 314.

STEP 1 **The Set-Up:**
  a. **Describe the population parameter of interest.**
     The parameter of interest is μ, **the mean pollution level** of air in downtown Rochester.
  b. **State the null hypothesis (H_o) and the alternative hypothesis (H_a).**

     H_o: $\mu = 4.9$ ($\leq$) (no higher than)

     H_a: $\mu > 4.9$ (higher than)

Procedures for writing H_o and H_a are on pages 375-377.

STEP 2 **The Hypothesis Test Criteria:**
  a. **Check the assumptions.**
     The assumptions are satisfied since **the sampled population is approximately normal** and the sample size is large enough for the CLT to apply (see p. 416). σ **is unknown.**
  b. **Identify the probability distribution and the test statistic to be used.**
     The t-distribution with df $= n - 1 = 21$, and the test statistic is $t \bigstar$, Formula (9.2).
  c. **Determine the level of significance, α**
     $\alpha = 0.05$.

STEP 3 **The Sample Evidence:**
  a. **Collect the sample information.**
     $n = 22$, $\bar{x} = 5.1$, and $s = 1.17$.

Exercise 9.10

Calculate the value of $t\,\bigstar$ for the hypothesis test: H_o: $\mu = 32$, H_a: $\mu > 32$, $n = 16$, $\bar{x} = 32.93$, $s = 3.1$.

b. Calculate the value of the test statistic.

Using Formula (9.2),

$$t\,\bigstar = \frac{\bar{x} - \mu}{s/\sqrt{n}}: \quad t\,\bigstar = \frac{5.1 - 4.9}{1.17/\sqrt{22}} = \frac{0.20}{0.2494} = 0.8018 = \mathbf{0.80}$$

STEP 4   The Probability Distribution:

USING THE *p*-VALUE PROCEDURE:

a. Calculate the *p*-value.

Use the right-hand tail since the H_a expresses concern for values related to "higher than."

$\mathbf{P} = P(t > 0.80$, with df = 21) as shown on the figure.

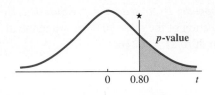

To find the *p*-value, use one of three methods:

1. Use Table 6 (Appendix B) to place bounds on the *p*-value: **0.10 < P < 0.25**
2. Use Table 7 (Appendix B) to read the value directly from the table: **P = 0.216**
3. Use a computer or calculator to calculate the *p*-value: **P = 0.2163**

Specific details follow this illustration.

b. Determine whether or not the *p*-value is smaller than α.

The *p*-value **is not smaller** than the level of significance, α

OR

USING THE CLASSICAL PROCEDURE:

a. Determine the critical region and critical value(s).

The critical region is the right-hand tail since the H_a expresses concern for values related to "higher than." The critical value is found at the intersection of the df = 21 row and the one-tailed 0.05 column of Table 6:

$$t(21, 0.05) = \mathbf{1.72.}$$

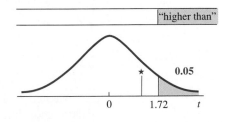

Specific instructions are on page 419.

b. Determine whether or not the calculated test statistic is in the critical region.

$t\,\bigstar$ **is not in** the critical region, as shown in **red** on the figure above.

Exercise 9.11

Find the value of **P** and state the decision for the hypothesis test in Exercise 9.10 using $\alpha = 0.05$.

STEP 5   The Results:

a. State the decision about H_o.

Fail to reject H_o.

b. State the conclusion about H_a.

At the 0.05 level of significance, the EPA does not have sufficient evidence to show that "the mean carbon monoxide level is higher than 4.9."

▲

Exercise 9.12

Find the critical region and the critical value, and state the decision for the hypothesis test in Exercise 9.10 using $\alpha = 0.05$.

Calculating the *p*-value when using the *t*-distribution:

Method 1   *Using Table 6 in Appendix B to "place bounds" on the p-value:* By inspecting the df = 21 row of Table 6, you can determine an interval within which the *p*-value lies. Locate the $t\,\bigstar$ along the row labeled df = 21. If $t\,\bigstar$ is not listed, locate the two table values it falls between, and read the bounds for the *p*-value from the top of the table. In this case, $t\,\bigstar = 0.80$ is between 0.686 and 1.32; therefore, **P** is between 0.10 and 0.25. Use the one-tailed heading since H_a is one-tailed in this illustration. (Use the two-tailed heading when H_a is two-tailed.)

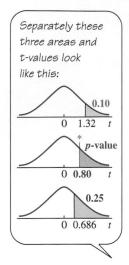

Separately these three areas and *t*-values look like this:

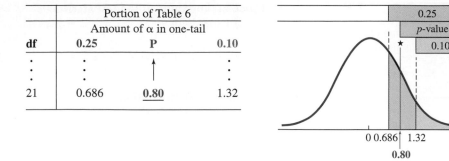

	Portion of Table 6		
	Amount of α in one-tail		
df	**0.25**	**P**	**0.10**
⋮	⋮	↑	⋮
21	0.686	**0.80**	1.32

The 0.686 entry in the table tells us that $P(t > 0.686) = 0.25$, as shown on the above figure in *purple*. The 1.32 entry in the table tells us that $P(t > 1.32) = 0.10$, as shown in *orange*. You can see that the p-value **P** (shown in *red*) is between 0.10 and 0.25. Therefore, $0.10 < P < 0.25$, and we say that 0.10 and 0.25 are the "bounds" for the p-value.

Method 2   *Using Table 7 in Appendix B to read the p-value directly from the table:* Locate the p-value at the intersection of the $t ★ = 0.80$ row and the df = 21 column. The p-value for $t ★ = 0.80$ with df = 21 is **0.216.**

	Portion of Table 7		
$t ★$	**df**	**...**	**21**
⋮			↓
0.80			0.216 ⟶ **P** = $P(t > 0.80$, with df = 21) = **0.216**

Method 3   If you are doing the hypothesis test with the aid of a computer or calculator, most likely it will calculate the p-value for you or you may use the cumulative probability distribution commands described on page 420.

Let's look at a two-tailed hypothesis-testing situation.

ILLUSTRATION 9.6 ▼

On a popular self-image test, which results in normally distributed scores, the mean score for public-assistance recipients is expected to be 65. A random sample of 28 public-assistance recipients in Emerson County is given the test. They achieve a mean score of 62.1, and their scores have a standard deviation of 5.83. Do the Emerson County public-assistance recipients test differently, on the average, than what is expected, at the 0.02 level of significance?

Solution

STEP 1   **The Set-Up:**
   a. **Describe the population parameter of interest.**
      The parameter of interest is, μ, **the mean self-image test score** for all Emerson County public-assistance recipients.
   b. **State the null hypothesis (H_o) and the alternative hypothesis (H_a).**

   H_o: $\mu = 65$ (mean is 65)

   H_a: $\mu \neq 65$ (mean is different from 65)

STEP 2   **The Hypothesis Test Criteria:**
   a. **Check the assumptions.**
      The **test is expected to produce normally distributed scores;** therefore, the assumption has been satisfied. σ **is unknown.**

b. **Identify the probability distribution and the test statistic to be used.**
The t-distribution with df $= n - 1 = 27$, and the test statistic is $t \star$, Formula (9.2).

c. **Determine the level of significance, α**
$\alpha = 0.02$ (given in statement of problem).

STEP 3 **The Sample Evidence:**
a. **Collect the sample information.**
$n = 28$, $\bar{x} = 62.1$, and $s = 5.83$.

b. **Calculate the value of the test statistic.**
Using Formula (9.2),

$$t \star = \frac{\bar{x} - \mu}{s/\sqrt{n}}: \quad t \star = \frac{62.1 - 65.0}{5.83/\sqrt{28}} = \frac{-2.9}{1.1018} = -2.632 = -\mathbf{2.63}$$

STEP 4 **The Probability Distribution:**

USING THE p-VALUE PROCEDURE:

a. **Calculate the p-value.**
Use both tails since the H_a expresses concern for values related to "different than."
$\mathbf{P} = P(t < -2.63) + P(t > 2.63) = 2 \cdot P(t > 2.63)$, with df $= 27$ as shown on the figure.

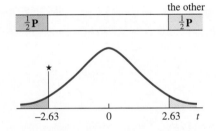

the other
$\frac{1}{2}\mathbf{P}$ $\frac{1}{2}\mathbf{P}$

-2.63 0 2.63 t

To find the p-value, use one of three methods:
1. Use Table 6 (Appendix B) to place bounds on the p-value.
   $\mathbf{0.01 < P < 0.02}$
2. Use Table 7 (Appendix B) to place bounds on the p-value:
   $\mathbf{0.012 < P < 0.016}$
3. Use a computer or calculator to calculate the p-value.
   $\mathbf{P = 0.0140}$
Specific details follow this illustration.

b. **Determine whether or not the p-value is smaller than α**
The p-value **is smaller than** the level of significance, α.

OR

USING THE CLASSICAL PROCEDURE:

a. **Determine the critical region and critical value(s).**
The critical region is both tails since the H_a expresses concern for values related to "different than." The critical value is found at the intersection of the df $= 27$ row and the one-tailed 0.01 column of Table 6.

$t(27, 0.01) = \mathbf{2.47}$

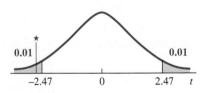

0.01 0.01

-2.47 0 2.47 t

b. **Determine whether or not the calculated test statistic is in the critical region.**
$t \star$ **is in** the critical region, as shown in **red** on the figure above.

STEP 5 **The Results:**
a. **State the decision about H_o.**
Reject H_o.
b. **State the conclusion about H_a.**
At the 0.02 level of significance, we do have sufficient evidence to conclude that the Emerson County assistance recipients test significantly differently, on the average, than the expected 65. ▲

Exercise 9.13
Calculate the value of $t \star$ for the following hypothesis test:
H_o: $\mu = 73$, H_a: $\mu \neq 73$, $\alpha = 0.05$, $n = 12$, $\bar{x} = 71.46$, $s = 4.1$.

Exercise 9.14
Find the critical region and the critical values for the hypothesis test in Exercise 9.13; state the decision using $\alpha = 0.05$.

Calculating the p-value when using the t-distribution:

Exercise 9.15

Use Table 6 or Table 7 to find the value of **P** for the hypothesis test in Exercise 9.13; state the decision using $\alpha = 0.05$.

Method 1   Using Table 6, find 2.63 between two entries in the df = 27 row, read the bounds for **P** from the two-tail heading at the top of the table; **0.01 < P < 0.02.**

Method 2   Using Table 7: Generally, bounds found using Table 7 will be narrower than the bounds found using Table 6. The table below shows you how to read the bounds from Table 7; find $t \star = 2.63$ between two rows and df = 27 between two columns; locate the four intersections of these columns and rows; the value of $\frac{1}{2}$**P** is bounded by the upper-left and the lower-right of these table entries.

Portion of Table 7

Degrees of Freedom

$t \star$	25	27	29
⋮			⋮
2.6	0.008		0.007
2.63		$\frac{1}{2}$**P**	$0.006 < \frac{1}{2}P < 0.008$
2.7	0.006		0.006 / **0.012 < P < 0.016**

Exercise 9.16

Use a computer/calculator to find the *p*-value for Exercise 9.13.

Method 3   If you are doing the hypothesis test with the aid of a computer or calculator, most likely it will calculate the *p*-value for you (do not double). Or you may use the cumulative probability distribution commands described on page 420.

The following commands will complete a hypothesis test for the mean μ when the population standard deviation is unknown.

Exercise 9.17

Use a computer/ calculator to complete the hypothesis test:
$H_o: \mu = 52$, $H_a: \mu < 52$, $\alpha = 0.01$ using the data:

45 47 46 58 59 49
46 54 53 52 47 41.

MINITAB

```
Input the data into C1; then continue with:

    Choose:   Stat > Basic Statistics > 1-Sample t
    Enter:    Variables: C1
    Select:   Test mean
    Enter:    μ
    Choose:   Alternative: less than or not equal or greater
              than
```

EXCEL

```
Input the data into column A; then
continue with:

    Choose:   Tools > Data Analysis Plus
              > Inference About A
              Mean(SIGMA Unknown) > OK
    Enter:    block coordinates: (A1:A20
              or select cells)    > OK
    Choose:   Test of Hypothesis
    Enter:    Value of MU under the null
              hypothesis: μ    > OK
    Choose:   Alternative Hypothesis
              (3 choices)
```

TI-83

```
Input the data into L1; then continue
with the following, entering the appro-
priate values and highlighting Calculate:

    Choose:   STAT > TESTS > 2:T-Test
```

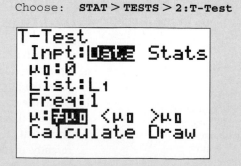

Compare MINITAB results to the solution found in Illustration 9.6.

The MINITAB solution to Illustration 9.6:

```
T-Test of the Mean
Test of mu = 65 vs mu ≠ 65

  Variable    N    Mean   StDev   SE Mean      T   P-Value
        C1   28    62.1    5.83     1.102    2.63    0.0140
```

Case Study 9.1

Mother's Use of Personal Pronouns When Talking with Toddlers

The calculated t-value and the probability value for five different hypothesis tests are given in the following article. The expression t(44) = 1.92 means t ★ = 1.92 with df = 44 and is significant with p-value < 0.05. Can you verify the p-values? Explain.

Exercise 9.18

a. Verify that $t(44) = 1.92$ is significant at the 0.05 level.
b. Verify that $t(44) = 3.41$ is significant at the 0.01 level.
c. Explain why $t(44) = 1.81$, $p < .10$, makes sense only if the hypothesis test is two-tailed. If the test is one-tailed, what level would be reported?

ABSTRACT. The verbal interaction of 2-year-old children (N = 46; 16 girls, 30 boys) and their mothers was audiotaped, transcribed, and analyzed for the use of personal pronouns, the total number of utterances, the child's mean length of utterance, and the mother's responsiveness to her child's utterances. Mothers' use of the personal pronoun "we" was significantly related to their children's performance on the Stanford-Binet at age 5 and the Wechsler Intelligence Scale for Children at age 8. Mothers' use of "we" in social-vocal interchange, indicating a system for establishing a shared relationship with the child, was closely connected with their verbal responsiveness to their children. The total amount of maternal talking, the number of personal pronouns used by mothers, and their verbal responsiveness to their children were not related to mothers' social class or years of education.

Mothers tended to use more first person singular pronouns (I and me), $t(44) = 1.81$, $p < .10$, and used significantly more first person plural pronouns (we), $t(44) = 1.92$, $p < .05$, with female children than with male children. The mothers also were more verbally responsive to their female children, $t(44) = 2.0$, $p < .06$.

In general, mothers talked more to their first-born children, $t(44) = 3.41$, $p < .001$, and were more responsive to their first-born children, $t(44) = 3.71$, $p < .001$. Yet the proportion of personal pronouns used when speaking to first-born children was not different from that used when speaking to later born children.

Source: Dan R. Laks, Leila Beckwith, Sarale E. Cohen, *The Journal of Genetic Psychology*, 151(1), 25–32, 1990. Reprinted with permission of the Helen Dwight Reid Educational Foundation. Published by Heldref Publications, 4000 Albemarie St., N.W., Washington, D.C., 20016. Copyright (c) 1990.

EXERCISES • • • • • • • • •

9.19 Find these critical values using Table 6 in Appendix B:
  a. $t(25, 0.05)$
  b. $t(10, 0.10)$
  c. $t(15, 0.01)$
  d. $t(21, 0.025)$
  e. $t(21, 0.95)$
  f. $t(26, 0.975)$
  g. $t(27, 0.99)$
  h. $t(60, 0.025)$

9.20 Using the notation of Exercise 9.19, name and find the following critical values of t:

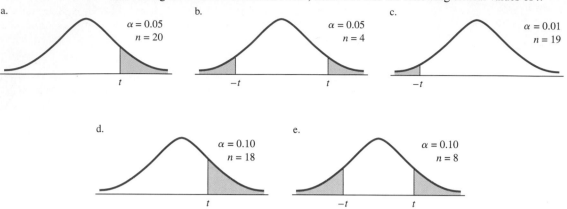

a. $\alpha = 0.05$, $n = 20$ b. $\alpha = 0.05$, $n = 4$ c. $\alpha = 0.01$, $n = 19$ d. $\alpha = 0.10$, $n = 18$ e. $\alpha = 0.10$, $n = 8$

9.21 Ninety percent of the Student's t-distribution lies between $t = -1.89$ and $t = 1.89$ for how many degrees of freedom?

9.22 Ninety percent of the Student's t-distribution lies to the right of $t = -1.37$ for how many degrees of freedom?

9.23 **a.** Find the first percentile of the Student's t-distribution with 24 degrees of freedom.
  b. Find the 95th percentile of the Student's t-distribution with 24 degrees of freedom.
  c. Find the first quartile of the Student's t-distribution with 24 degrees of freedom.

9.24 Find the percent of the Student's t-distribution that lies between the following values.
  a. df $= 12$ and t ranges from -1.36 to 2.68
  b. df $= 15$ and t ranges from -1.75 to 2.95

9.25 **a.** State two ways in which the standard normal distribution and the Student's t-distribution are alike.
  b. State two ways in which they are different.

9.26 The variance for each of the Student's t-distributions is equal to df/(df $- 2$). Find the standard deviation for a Student's t-distribution with each of the following degrees of freedom.
  a. 10   **b.** 20   **c.** 30
  d. Explain how this verifies Property 5 of the t-distributions listed on p. 417.

9.27 The data from a study reported in "White-Collar Crime and Criminal Careers: Some Preliminary Findings" (*Crime and Delinquency,* July 1990) indicate that white-collar criminals are likely to be older and to show a lower frequency of offending than street criminals. For example, the mean age of onset of offending for those convicted of antitrust offenses was 54 years with $n = 35$. If the standard deviation is estimated to be 7.5 years, set a 90% confidence interval on the true mean age.

9.28 The Robertson square drive screw was invented in 1908, but it has only gained in popularity with American woodworkers and home craftsmen within the last ten years. The advantages of square drives over conventional screws is indeed remarkable—most notably greater strength, increased holding power, and reduced driving resistance and "cam-out." Strength test results published in McFeely's 1998 catalog (*McFeely's Square Drive Screws,* Cat. 98D, Fall 1998) revealed that #8 square drive screws fail only after an average of 48 inch-pounds of torque is applied, nearly 50% more strength than the more common slotted- or Phillips-head wood screw.

Suppose an independent testing laboratory selects 22 square drive screws, selected randomly from a box of 1000 screws, and obtains a mean failure torque of 48.2 inch-pounds and a standard deviation of 5.1 inch-pounds. Estimate with 95% confidence the mean failure torque of the #8 wood screws based on the study by the independent laboratory. Specify the population parameter of interest, the criteria, the sample evidence, and the interval limits.

9.29 There seems to be no end to how large the signing bonuses professional athletes can obtain when they start their careers. When the Indianapolis Colts gave Peyton Manning $11.6 million and the San Diego Chargers awarded Ryan Leaf $11.25 million as signing bonuses in 1998, both these amounts exceeded what the 1989 first-round draft pick, Troy Aikman, earned in his first five National Football League seasons combined while playing for the Dallas Cowboys (*Sports Illustrated,* 8-10-98).

Suppose a sample of 18 new NFL players report their signing bonuses at the start of the 1998 season, and the results show a mean of $3.81 million and a standard deviation of $1.7 million.
   a. Estimate with 95% confidence the mean signing bonus based on the report. (Specify the population parameter of interest, the criteria, the sample evidence, and the interval limits.)
   b. Discuss how this situation does or does not satisfy the assumptions for the inference.

9.30 While doing an article on the high cost of college education, a reporter took a random sample of the cost of textbooks for a semester. The random variable x is the cost of one book. Her sample data can be summarized by $n = 41$, $\sum x = 550.22$, and $\sum (x - \bar{x})^2 = 1617.984$.
   a. Find the sample mean, $\bar{x}$.
   b. Find the sample standard deviation, s.
   c. Find the 90% confidence interval to estimate the true mean textbook cost for the semester based on this sample.

9.31 Ten randomly selected shut-ins were each asked to list how many hours of television they watched per week. The results are

82   66   90   84   75   88   80   94   110   91

Determine the 90% confidence interval estimate for the mean number of hours of television watched per week by shut-ins. Assume the number of hours is normally distributed.

9.32 The addition of a new accelerator is claimed to decrease the drying time of latex paint by more than 4%. Several test samples were conducted with the following percentage decrease in drying time.

5.2   6.4   3.8   6.3   4.1   2.8   3.2   4.7

If we assume that the percentage decrease in drying time is normally distributed:
   a. Find the 95% confidence interval for the true mean decrease in the drying time based on this sample. (The sample mean and standard deviation were found in answering Exercise 2.151).
   b. Did the interval estimate reached in (a) result in the same conclusion as you expressed in answering part (c) of Exercise 2.151 for these same data?

9.33 The pulse rates for 13 adult women were

83   58   70   56   76   64   80   76   70   97   68   78   108

Verify the results shown on the last line of the MINITAB output below.

```
MTB > TINTERVAL 90 PERCENT CONFIDENCE INTERVAL FOR
DATA IN C1

           N    MEAN   STDEV   SE MEAN   90.0 PERCENT C.I.
   C1    13   75.69   14.54    4.03      (68.50, 82.88)
```

9.34 The weights of the drained fruit found in 21 randomly selected cans of peaches packed by Sunny Fruit Cannery were (in ounces)

11.0   11.6   10.9   12.0   11.5   12.0   11.2   10.5   12.2   11.8   12.1
11.6   11.7   11.6   11.2   12.0   11.4   10.8   11.8   10.9   11.4

(continued)

Using a computer or a calculator,
  a. Calculate the sample mean and standard deviation.
  b. Assume normality and construct the 98% confidence interval for the estimate of the mean weight of drained peaches per can.

9.35 State the null hypothesis, H_o, and the alternative hypothesis, H_a, that would be used to test each of the following claims.
  a. The mean weight of honey bees is at least 11 g.
  b. The mean age of patients at Memorial Hospital is no more than 54 years.
  c. The mean amount of salt in granola snack bars is different from 75 mg.

9.36 Determine the p-value for the following hypothesis tests involving the Student's t-distribution with 10 degrees of freedom.
  a. H_o: $\mu = 15.5$, H_a: $\mu < 15.5$, $t\star = -2.01$
  b. H_o: $\mu = 15.5$, H_a: $\mu > 15.5$, $t\star = 2.01$
  c. H_o: $\mu = 15.5$, H_a: $\mu \neq 15.5$, $t\star = 2.01$
  d. H_o: $\mu = 15.5$, H_a: $\mu \neq 15.5$, $t\star = -2.01$

9.37 Determine the critical region and critical value(s) that would be used in the classical approach to test the null hypotheses below.
  a. H_o: $\mu = 10$, H_a: $\mu \neq 10$ ($\alpha = 0.05$, $n = 15$)
  b. H_o: $\mu = 37.2$, H_a: $\mu > 37.2$ ($\alpha = 0.01$, $n = 25$)
  c. H_o: $\mu = -20.5$, H_a: $\mu < -20.5$ ($\alpha = 0.05$, $n = 18$)
  d. H_o: $\mu = 32.0$, H_a: $\mu > 32.0$ ($\alpha = 0.01$, $n = 42$)

9.38 Compare the p-value and classical approaches to hypothesis testing by comparing the p-value and decision of the p-value approach to the critical values and decision of the classical approach for each of the following situations. Use $\alpha = 0.05$.
  a. H_o: $\mu = 128$, H_a: $\mu \neq 128$, $n = 15$, $t\star = 1.60$
  b. H_o: $\mu = 18$, H_a: $\mu > 18$, $n = 25$, $t\star = 2.16$
  c. H_o: $\mu = 38$, H_a: $\mu < 38$, $n = 45$, $t\star = -1.73$
  d. Compare the results of the two techniques for each case.

9.39 A student group maintains that the average student must travel for at least 25 minutes in order to reach college each day. The college admissions office obtained a random sample of 31 one-way travel times from students. The sample had a mean of 19.4 min and a standard deviation of 9.6 min. Does the admissions office have sufficient evidence to reject the students' claim? Use $\alpha = 0.01$.
  a. Solve using the p-value approach.   **b.** Solve using the classical approach.

9.40 Homes in a nearby college town have a mean value of $88,950. It is assumed that homes in the vicinity of the college have a higher value. To test this theory, a random sample of 12 homes is chosen from the college area. Their mean valuation is $92,460 and the standard deviation is $5200. Complete a hypothesis test using $\alpha = 0.05$. Assume prices are normally distributed.
  a. Solve using the p-value approach.   **b.** Solve using the classical approach.

9.41 An article in the *American Journal of Public Health* (March 1994) describes a large study involving 20,143 individuals. The article states that the mean percentage intake of kilocalories from fat was 38.4% with a range from 6% to 71.6%. A small sample study was conducted at a university hospital to determine if the mean intake of patients at that hospital was different from 38.4%. A sample of 15 patients had a mean intake of 40.5% with a standard deviation equal to 7.5%. Assuming that the sample is from a normally distributed population, test the hypothesis of "different from" using a level of significance equal to 0.05.
  a. What evidence do you have that the assumption of normality is reasonable? Explain.
  b. Complete the test using the p-value approach. Include $t\star$, p-value, and your conclusion.
  c. Complete the test using the classical approach. Include the critical values, $t\star$, and your conclusion.

9.42 Consumers enjoy the deep selection of merchandise made possible by specialty stores that sacrifice breadth for greater depth. Consider stores that carry only Levi Strauss pants.

(continued)

The company reports that a fully stocked Levi's store carries 130 ready-to-wear pairs of jeans for any given waist and inseam, and the company is phasing in two more lines of pants (Personal Pair and Original Spin) that it claims will eventually quadruple that number (*Fortune*, 9-28-98).

Suppose a random sample of 24 Levi stores is sampled two months after the phase-in process has been launched, and inventories are taken at each of the stores in the sample for all sizes of jeans. The sample mean number of choices for any given size is 141.3, and the standard deviation is 36.2. Does this sample of stores carry a greater selection of jeans, on the average, than what is expected, at the 0.01 level of significance?

a. Solve using the *p*-value approach.   **b.** Solve using the classical approach.

9.43 It is claimed that the students at a certain university will score an average of 35 on a given test. Is the claim reasonable if a random sample of test scores from this university yields 33, 42, 38, 37, 30, 42? Complete a hypothesis test using $\alpha = 0.05$. Assume test results are normally distributed.

a. Solve using the *p*-value approach.   **b.** Solve using the classical approach.

9.44 Gasoline pumped from a supplier's pipeline is supposed to have an octane rating of 87.5. On 13 consecutive days a sample was taken and analyzed with the following results.

| 88.6 | 86.4 | 87.2 | 88.4 | 87.2 | 87.6 | 86.8 | 86.1 | 87.4 | 87.3 | 86.4 | 86.6 | 87.1 |

a. If the octane ratings have a normal distribution, is there sufficient evidence to show that these octane readings were taken from gasoline with a mean octane significantly less than 87.5 at the 0.05 level? (The sample mean and standard deviation were found in answering Exercise 2.152.)

b. Did the statistical decision reached in (a) result in the same conclusion as you expressed in answering part (c) of Exercise 2.152 for this same data?

9.45 In order to test the null hypothesis "the mean weight for adult males equals 160 lb" against the alternative "the mean weight for adult males exceeds 160 lb," the weights of 16 males were determined with the following results.

| 173 | 178 | 145 | 146 | 157 | 175 | 173 | 137 |
| 152 | 171 | 163 | 170 | 135 | 159 | 199 | 131 |

Assume normality and verify the results shown on the following MINITAB analysis by calculating the values yourself.

```
TEST OF MU = 160.00 VS MU G.T.  160.00

              N     MEAN      STDEV    SE MEAN      T     P VALUE
      C1     16    160.25     18.49     4.62      0.05     0.48
```

9.46 "Obesity raises heart-attack risk" according to a study published in the March 1990 issue of the *New England Journal of Medicine*. "Those about 15 to 25 percent above desirable weight had twice the heart disease rate." Suppose the data listed below are the percentages above desired weight for a sample of patients involved in a similar study.

18.3	19.7	22.1	19.2	17.5	12.7	22.0	17.2	21.1	16.2	15.4
19.9	21.5	19.8	22.5	16.5	13.0	22.1	27.7	17.9	22.2	19.7
18.1	22.4	17.3	13.3	22.1	16.3	21.9	16.9	15.4	19.3

Use a computer or calculator to test the null hypothesis, $\mu = 18\%$, versus the alternative hypothesis, $\mu \neq 18\%$. Use $\alpha = 0.05$.

9.47 Use a computer or calculator to complete the calculations and the hypothesis test for this exercise. Delco Products, a division of General Motors, produces commutators designed to be 18.810 mm in overall length. (A commutator is a device used in the electrical

(continued)

Many of these data sets are on your StatSource CD.

system of an automobile.) The following sample of 35 commutators was taken while monitoring the manufacturing process.

18.802	18.810	18.780	18.757	18.824	18.827	18.825
18.809	18.794	18.787	18.844	18.824	18.829	18.817
18.785	18.747	18.802	18.826	18.810	18.802	18.780
18.830	18.874	18.836	18.758	18.813	18.844	18.861
18.824	18.835	18.794	18.853	18.823	18.863	18.808

Source: With permission of Delco Products Division, GMC.

Is there sufficient evidence to reject the claim that these parts meet the design requirements "mean length is 18.810" at the $\alpha = 0.01$ level of significance?

9.48 How important is the assumption, "the sampled population is normally distributed" to the use of the Student's *t*-distribution? Using a computer, simulate drawing 100 samples of size 10 from each of three different types of population distributions; namely, a normal, a uniform, and an exponential. First generate 1000 data from the population and construct a histogram to see what the population looks like. Then generate 100 samples of size 10 from the same population; each row represents a sample. Calculate the mean and standard deviation for each of the 100 samples. Calculate $t \star$ for each of the 100 samples. Construct histograms of the 100 sample means and the 100 $t \star$- values. (Additional details can be found in the Statistical Tutor.)

For the samples from the normal population:
a. Does the *x*-bar distribution appear to be normal? Find percentages for intervals and compare to the normal distribution.
b. Does the distribution of $t \star$ appear to have a *t*-distribution with df = 9? Find percentages for intervals and compare them to the *t*-distribution.

For the samples from the rectangular or uniform population:
c. Does the *x*-bar distribution appear to be normal? Find percentages for intervals and compare them to the normal distribution.
d. Does the distribution of $t \star$ appear to have a *t*-distribution with df = 9? Find percentages for intervals and compare them to the *t*-distribution.

For the samples from the skewed (exponential) population:
e. Does the *x*-bar distribution appear to be normal? Find percentages for intervals and compare them to the normal distribution.
f. Does the distribution of $t \star$ appear to have a *t*-distribution with df = 9? Find percentages for intervals and compare them to the *t*-distribution.

In summary:
g. In each of the preceding three situations, the sampling distribution for *x*-bar appears to be slightly different than the distribution of $t \star$. Explain why.
h. Does the normality condition appear to be necessary in order for the calculated test statistic $t \star$ to have a Student's *t*-distribution? Explain.

9.2 Inferences About the Binomial Probability of Success

Perhaps the most common inference of all is an inference involving the **binomial parameter *p*,** the "probability of success." Yes, every one of us uses this inference, even if only very casually. There are thousands of examples of situations in which we are concerned about something either "happening" or "not happening." There are only two possible outcomes of concern, and that is the fundamental property of a **binomial experiment.** The other needed ingredient is for multiple independent trials to exist. Asking 5 people whether they are "for" or "against" some issue can create 5 independent trials; if 200 people are asked the same question, 200 independent trials may be involved; if 30 items are inspected to see if each "exhibits a particular property" or "not," there will be 30 repeated trials; these are the makings of a binomial inference.

The binomial parameter p is defined to be the probability of success on a single trial in a binomial experiment. We define p', the **observed** or **sample binomial probability,** to be

Complete details about binomial experimentation can be found on pages 253–256.

$$p' = \frac{x}{n} \qquad (9.3)$$

where the **random variable** x represents the number of successes that occur in a sample consisting of n trials. Also recall: The mean and standard deviation of the binomial random variable x are found by using formula (5.7), $\mu = np$, and formula (5.8), $\sigma = \sqrt{npq}$, where $q = 1 - p$. The distribution of x is considered to be approximately normal if n is larger than 20 and if np and nq are both larger than 5. This commonly accepted rule of thumb allows us to use the **standard normal distribution** to estimate probabilities for the binomial random variable x, the number of successes in n trials, and to make inferences concerning the binomial parameter p, the probability of success while performing an individual trial.

Generally, it is easier and more meaningful to work with the distribution of p' (the observed probability of occurrence) than with x (the number of occurrences). Consequently, we will convert formulas (5.7) and (5.8) from the units of x (integers) to units of proportions (percentages expressed as decimals) by dividing each formula by n, as shown in Table 9.1.

Table 9.1   Formulas (9.4) and (9.5)

	Variable	Mean	Standard Deviation
	x	$\mu_x = np$ **(5.7)**	$\sigma_x = \sqrt{npq}$ **(5.8)**
to change x to p', divide by n	$\dfrac{x}{n}$	$\dfrac{np}{n}$	$\dfrac{\sqrt{npq}}{n}$
	p'	$\mu_{p'} = p$ **(9.4)**	$\sigma_{p'} = \sqrt{\dfrac{pq}{n}}$ **(9.5)**

Exercise 9.49
a. Does it seem reasonable that the mean of the sampling distribution of observed values of p' should be p, the true proportion? Explain.
b. Explain why p' is an unbiased estimator for the population p.

Recall that $\mu_{p'} = p$ and that the sample statistic p' is an **unbiased estimator for p.** Therefore, the information about the sampling distribution of p' is summarized as follows:

If a sample of size n is randomly selected from a large population with $p = P(\text{success})$, then the sampling distribution of p' has

1. a mean $\mu_{p'}$ equal to p,
2. a standard error $\sigma_{p'}$ equal to $\sqrt{\dfrac{pq}{n}}$, and
3. an approximately normal distribution if n is sufficiently large.

Exercise 9.50
Show that $\dfrac{\sqrt{npq}}{n}$ simplifies to $\sqrt{\dfrac{pq}{n}}$.

In practice, use of the following guidelines will ensure normality :

1. The sample size is greater than 20.
2. The products np and nq are both larger than 5.
3. The sample consists of less than 10% of the population.

> ## THE ASSUMPTIONS FOR INFERENCES ABOUT THE BINOMIAL PARAMETER p
>
> The n random observations forming the sample are selected independently from a population that is not changing during the sampling.

Confidence Interval Procedure

> The standard deviation of a sampling distribution is called "standard error."

Inferences concerning the population binomial parameter p, P(success), are made using procedures that closely parallel the inference procedures for the population mean μ. When we estimate the **population proportion p,** we will base our estimations on the **unbiased sample statistic p'.** The point estimate, p', becomes the center of the confidence interval, and the maximum error of estimate is a multiple of the **standard error.** The **level of confidence** determines the confidence coefficient, the number of multiples of the standard error.

$$p' - z(\alpha/2) \cdot \sqrt{\frac{p'q'}{n}} \quad \text{to} \quad p' + z(\alpha/2) \cdot \sqrt{\frac{p'q'}{n}} \qquad (9.6)$$

where $p' = \dfrac{x}{n}$ and $q' = 1 - p'$. Notice that the standard error, $\sqrt{\dfrac{pq}{n}}$, has been replaced by $\sqrt{\dfrac{p'q'}{n}}$. Since we are estimating p, we do not know its value and therefore must use the best replacement available. That replacement is p', the observed value or the point estimate for p. This replacement will cause little change in the standard error or the width of our confidence interval provided n is sufficiently large.

ILLUSTRATION 9.7 ▼

While talking about the cars that fellow students drive, several statements were made about types, ages, makes, colors, and so on. Dana decided he wanted to estimate the proportion of convertibles students drove, so he randomly identified 200 cars in the student parking lot, of which he found 17 to be convertibles. Find the 90% confidence interval for the proportion of convertibles driven by students.

Solution

> The five-step confidence interval procedure is on page 356.

STEP 1 **The Set-Up:**
Describe the population parameter of interest.
The parameter of interest is p, **the proportion (percentage) of convertibles** driven by students.

STEP 2 **The Confidence Interval Criteria:**
a. Check the assumptions.
The sample was randomly selected, and each subject's response was independent of those of the others surveyed.
b. Identify the probability distribution and formula to be used.
The standard normal distribution will be used with formula (9.6) as the test statistic. p' is expected to be approximately normal since $n = 200$ is greater than 20 and both np [approximated by $np' = 200(17/200) = 17$] and nq [approximated by $nq' = 200(183/200) = 183$] are larger than 5.
c. Determine the level of confidence, $1 - \alpha$.
$1 - \alpha = 0.90$.

STEP 3   **The Sample Evidence:**
Collect the sample information.
$n = 200$ cars were identified, and $x = 17$ were convertibles.
$$p' = \frac{x}{n} = \frac{17}{200} = 0.085.$$

STEP 4   **The Confidence Interval:**
a. **Determine the confidence coefficients.**
This is the z-score $z_{(\alpha/2)}$, "z of one-half of alpha" identifying the number of standard errors needed to attain the level of confidence and is found using Table 4 in Appendix B, $z_{(\alpha/2)} = z_{(0.05)} = 1.65$ (see diagram)

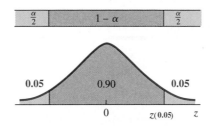

Specific instructions are on pages 355-356.

b. **Find the maximum error of estimate.**
Using the maximum error part of formula (9.6),
$$E = z_{(\alpha/2)} \cdot \sqrt{\frac{p'q'}{n}} = 1.65 \cdot \sqrt{\frac{(0.085)(0.915)}{200}}$$
$$= (1.65)\sqrt{0.000389} = (1.65)(0.020) = \mathbf{0.033}$$

c. **Find the lower and upper confidence limits.**
$$p' - E \quad \text{to} \quad p' + E$$
$$0.085 - 0.033 \quad \text{to} \quad 0.085 + 0.033$$
$$\mathbf{0.052} \quad \text{to} \quad \mathbf{0.118}$$

STEP 5   **The Results:**
State the confidence interval.
0.052 to 0.118, 90% confidence interval for $p = P$(drives convertible).
That is, the true proportion of students who drive convertibles is between 0.052 and 0.118, with 90% confidence. ▲

> **Exercise 9.51**
>
> Another sample is taken to estimate the proportion of convertibles (Illustration 9.7). Results: $n = 400$, $x = 92$. Find:
>
> a. the estimate for the standard error.
> b. the 95% confidence interval.

The following commands will construct a $1 - \alpha$ confidence interval for a proportion p.

EXCEL

```
Input the data into column A using 0's for failures (or no's) and 1's for suc-
cesses (or yes's); then continue with:

  Choose:  Tools > Data Analysis Plus > Inference About A Proportion > OK
  Enter:   block coordinates: (A2:A20 or select cells)   > OK
                       (do not include column name)
           Code for success: 1   > OK
  Choose:  Interval Estimate
  Enter:   level of confidence: 1 − α (ex. 0.95 or 95.0)
```

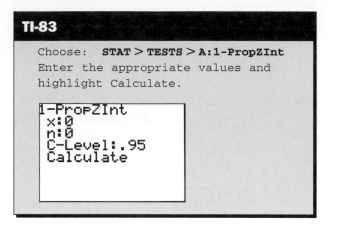

Sample Size

By using the maximum-error part of the confidence interval formula, it is possible to determine the **size of the sample** that must be taken in order to estimate p with a desired accuracy. The **maximum error of estimate for a proportion** is

$$E = z_{(\alpha/2)} \cdot \sqrt{\frac{pq}{n}} \qquad \textbf{(9.7)}$$

In order to determine the sample size from this formula, we must decide on the quality we want for our final confidence interval. This quality is measured in two ways: the level of confidence and the preciseness (narrowness) of the interval. The level of confidence we establish will in turn determine the confidence coefficient, $z_{(\alpha/2)}$. The desired preciseness will determine the maximum error of estimate, E. (Remember that we are estimating p, the binomial probability; therefore, E will typically be expressed in hundredths.)

Remember: $q = 1 - p$.

For ease of use, formula (9.7) can be solved for n as follows:

$$n = \frac{[z_{(\alpha/2)}]^2 \cdot p^* \cdot q^*}{E^2} \qquad \textbf{(9.8)}$$

where p^* and q^* are provisional values of p and q used for planning.

By inspecting formula (9.8), we can observe that there are three components determining the sample size: (1) level of confidence ($1 - \alpha$, which in turn determines the confidence coefficient), (2) the provisional value of p (p^* determines the value of q^*), and (3) the maximum error, E. An increase or decrease in one of these three components affects the sample size. If the level of confidence is increased or decreased (while the other components are held constant), then the sample size will increase or decrease, respectively. If the product of p^* and q^* is increased or decreased (other components held constant), then the sample size will increase or decrease, respectively. (The product $p^* \cdot q^*$ is largest when $p^* = 0.5$ and decreases as the value of p^* becomes further from 0.5.) An increase or decrease in the desired maximum error will have the opposite effect on the sample size since E appears in the denominator of the formula. If no provisional values for p and q are available, then use $p^* = 0.5$ and $q^* = 0.5$. Using $p^* = 0.5$ is safe because it gives the largest sample size of any possible value of p. Using $p^* = 0.5$ works reasonably well when the true value is "near 0.5" (say, between 0.3 and 0.7); however, as p gets nearer to either 0 or 1, a sizable overestimate in sample size will occur.

ILLUSTRATION 9.8 ▼

Determine the sample size that is required to estimate the true proportion of blue-eyed community college students if you want your estimate to be within 0.02 with 90% confidence.

Solution

STEP 1   The level of confidence is $1 - \alpha = 0.90$; therefore, the confidence coefficient is $z(\alpha/2) = z(0.05) = 1.65$ from Table 4 in Appendix B; see diagram.

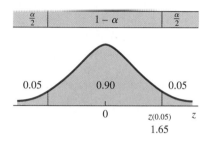

Exercise 9.52

Find the sample size n needed for a 95% interval estimate in Illustration 9.8.

STEP 2   The desired maximum error is $E = 0.02$.

STEP 3   Since no estimate was given for p, use $p^* = 0.5$, and $q^* = 1 - p^* = 0.5$.

STEP 4   Use formula (9.8) to find n:

$$n = \frac{[z(\alpha/2)]^2 \cdot p^* \cdot q^*}{E^2} : \quad n = \frac{(1.65)^2 \cdot (0.5) \cdot (0.5)}{(0.02)^2} = \frac{0.680625}{0.0004} = 1701.56 = \mathbf{1702}$$

ILLUSTRATION 9.9 ▼

An automobile manufacturer purchases bolts from a supplier who claims the bolts to be approximately 5% defective. Determine the sample size that will be required to estimate the true proportion of defective bolts if we want our estimate to be within ±0.02 with 90% confidence.

Solution

STEP 1   The level of confidence is $1 - \alpha = 0.90$; the confidence coefficient is $z(\alpha/2) = z(0.05) = 1.65$

STEP 2   The desired maximum error is $E = 0.02$.

STEP 3   Since there is an estimate for p (supplier's claim is "5% defective"), use $p^* = 0.05$ and $q^* = 1 - p^* = 0.95$.

STEP 4   Use formula (9.8) to find n:

$$n = \frac{[z(\alpha/2)]^2 \cdot p^* \cdot q^*}{E^2} :$$

Always round sample size up to the next larger integer no matter how small the decimal.

$$n = \frac{(1.65)^2 \cdot (0.05) \cdot (0.95)}{(0.02)^2} = \frac{0.12931875}{0.0004} = 323.3 = \mathbf{324} \qquad ▲$$

Notice the difference in the sample size required in Illustrations 9.8 and 9.9. The only mathematical difference between the problems is the value that was used for p^*.

Exercise 9.53

Find n for a 90% confidence interval for p with $E = 0.02$ using an estimate of $p = 0.25$.

In Illustration 9.8 we used $p^* = 0.5$, and in Illustration 9.9 we used $p^* = 0.05$. Recall that the use of the provisional value $p^* = 0.5$ gives the maximum sample size. As you can see, it will be an advantage to have an indication of the value expected for p, especially as p becomes increasingly further from 0.5.

Hypothesis-Testing Procedure

When the binomial parameter p is to be tested using a hypothesis-testing procedure, we will use a test statistic that represents the difference between the observed proportion and the hypothesized proportion, divided by the standard error. This test statistic is assumed to be normally distributed when the null hypothesis is true, when the assumptions for the test have been satisfied, and when n is sufficiently large ($n > 20$, $np > 5$, and $nq > 5$).

The value of the **test statistic $z \bigstar$** is calculated using formula (9.9):

p' is from sample; p is from H_o and $q = 1 - p$

$$ z \bigstar = \frac{p' - p}{\sqrt{\dfrac{pq}{n}}} \quad \text{where } p' = \frac{x}{n} \tag{9.9} $$

ILLUSTRATION 9.10 ▼

Many people sleep-in on the weekends to make up for "short nights" during the work-week. The Better Sleep Council reports that 61% of us get more than seven hours of sleep per night on the weekend. A random sample of 350 adults found that 235 had more than seven hours each night last weekend. At the 0.05 level of significance, does this evidence show that more than 61% get seven or more hours per night on the weekend?

Solution

STEP 1 **The Set-Up:**
  a. **Describe the population parameter of interest.**
    p, **the proportion** of adults who get more than seven hours of sleep per night on weekends.
  b. **State the null hypothesis (H_o) and the alternative hypothesis (H_a).**

  H_o: $p = P(7+ \text{ hours of sleep}) = 0.61$ ($\leq$) (no more than 61%)

  H_a: $p > 0.61$ (more than 61%)

STEP 2 **The Hypothesis Test Criteria:**
  a. **Check the assumptions.**
    The random sample of 350 adults was independently surveyed.
  b. **Identify the probability distribution and the test statistic to be used.**
    The **standard normal z** will be used with **formula (9.9).** Since $n = 350$ is larger than 20 and both $np = (350)(0.61) = 213.5$ and $nq = (350)(0.39) = 136.5$ are larger than 5, p' is expected to be approximately normally distributed.
  c. **Determine the level of significance, α.**
    $\alpha = 0.05$.

STEP 3 **The Sample Evidence:**
  a. **Collect the sample information.**

  $$ n = 350 \text{ and } x = 235; \ p' = \frac{x}{n} = \frac{235}{350} = 0.671. $$

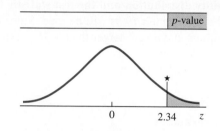

Exercise 9.54

Calculate the test statistic $z \star$ used in testing:
H_o: $p = 0.70$ vs.
H_a: $p > 0.70$, with
the sample $n = 300$
and $x = 224$.

b. Calculate the value of the test statistic.

Using formula (9.9),

$$z \star = \frac{p' - p}{\sqrt{\dfrac{pq}{n}}}; z \star = \frac{0.671 - 0.61}{\sqrt{\dfrac{(0.61)(0.39)}{350}}} = \frac{0.061}{\sqrt{0.0006797}} = \frac{0.061}{0.0261} = \textbf{2.34}$$

STEP 4   The Probability Distribution:

---

p-VALUE:

a. Calculate the *p*-value.

Use the right-hand tail since the H_a expresses concern for values related to "more than."

$$\textbf{P} = p\text{-value} = P(z > 2.34) \text{ as shown in the figure.}$$

To find the *p*-value, use one of three methods:

1. Use Table 3 (Appendix B) to calculate *p*-value:
   $\textbf{P} = 0.5000 - 0.4904 = \textbf{0.0096}$
2. Use Table 5 (Appendix B) to place bounds on the *p*-value:
   $\textbf{0.0094} < \textbf{P} < \textbf{0.0107}$
3. Use a computer or calculator to calculate *p*-value:
   $\textbf{P} = \textbf{0.0096}$

For specific instructions, see below.

b. Determine whether or not the *p*-value is smaller than α.
The *p*-value **is smaller** than α.

OR   CLASSICAL:

a. Determine the critical region and critical value(s).

The critical region is the right-hand tail since the H_a expresses concern for values related to "more than." The critical value is obtained from Table 4A: $z_{(0.05)} = \textbf{1.65.}$

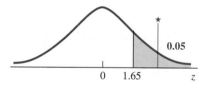

Specific instructions for finding critical values are on page 394–395.

b. Determine whether or not the calculated test statistic is in the critical region.

$z \star$ **is in** the critical region, as shown in red in the figure above.

Exercise 9.55

Find the value of **P** for the hypothesis test in Exercise 9.54; state the decision using α = 0.05.

STEP 5   The Results:

a. State the decision about H_o.
   Reject H_o.

b. State the conclusion about H_a.
   There is sufficient reason to conclude that the proportion of adults in the sampled population who are getting more than seven hours of sleep nightly on weekends is significantly higher than 61% at the 0.05 level of significance.

Exercise 9.56

Find the critical region and the critical values for the hypothesis test in Exercise 9.54; state the decision using α = 0.05.

Method 3   If you are doing the hypothesis test with the aid of a computer or calculator, most likely it will calculate the *p*-value for you, or you may use the cumulative probability distribution commands described on page 294.

ILLUSTRATION 9.11 ▼

While talking about the cars that fellow students drive (Illustration 9.7, p. 436), Tom made the claim that 15% of the students drive convertibles. Jody finds this hard to believe and she wants to check the validity of Tom's claim, using Dana's random sample.

At a level of significance of 0.10, is there sufficient evidence to reject Tom's claim if there were 17 convertibles in his sample of 200 cars?

Solution

STEP 1   The Set-Up:

    a. Describe the population parameter of interest.
    $p = P$(student drives convertible).

    b. State the null hypothesis (H_o) and the alternative hypothesis (H_a).

      H_o: $p = 0.15$ (15% do drive convertibles)

      H_a: $p \neq 0.15$ (the percent is different from 15)

STEP 2   The Hypothesis Test Criteria:

    a. Check the assumptions.
    The sample was randomly selected, and each subject's response was independent of other responses.

    b. Identify the probability distribution and the test statistic to be used.
    The standard normal z and formula (9.9) will be used. Since $n = 200$ is larger than 20 and both np and nq are larger than 5, p' is expected to be approximately normally distributed.

    c. Determine the level of significance, α
    $\alpha = 0.10$.

STEP 3   The Sample Evidence:

    a. Collect the sample information.

$$n = 200 \text{ and } x = 17; \quad p' = \frac{x}{n} = \frac{17}{200} = 0.085$$

    b. Calculate the value of the test statistic.
    Using formula (9.9),

$$z\bigstar = \frac{p' - p}{\sqrt{\dfrac{pq}{n}}}: \quad z\bigstar = \frac{0.085 - 0.150}{\sqrt{\dfrac{(0.15)(0.85)}{200}}} = \frac{-0.065}{\sqrt{0.00064}} = \frac{-0.065}{0.022525} = \mathbf{-2.57}$$

STEP 4   The Probability Distribution:

p-VALUE:		CLASSICAL:
a. Calculate the *p*-value. Use both tails since the H_a expresses concern for values related to "different from."	**OR**	**a. Determine the critical region and critical value(s).** The critical region is two-tailed since the H_a expresses concern for values related to "different from." The critical value is obtained from Table 4B: $z_{(0.05)} = \mathbf{1.65}$.

$$\mathbf{P} = p\text{-value} = P(z < -2.57) + P(z > 2.57)$$
$$= 2 \times P(z > 2.57) \text{ as shown in the figure.}$$

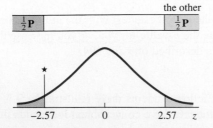

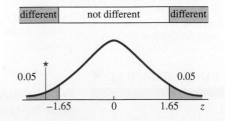

p-VALUE:

To find the *p*-value, use one of three methods:
1. Use Table 3 (Appendix B):
   $P = 2 \times (0.5000 - 0.4949) = \mathbf{0.0102}$
2. Use Table 5 (Appendix B) to place bounds on the *p*-value:
   $0.0047 < \frac{1}{2}P < 0.0054$
   $\mathbf{0.0094 < P < 0.0108}$
3. Use a computer or calculator to calculate the *p*-value:
   $\mathbf{P = 0.0102}$

For specific instructions, see page 379.

b. Determine whether or not the *p*-value is smaller than α.
The *p*-value **is smaller** than α.

CLASSICAL:

OR For specific instructions, see page 394–395.

b. Determine whether or not the calculated test statistic is in the critical region.

$z \star$ **is in** the critical region, as shown in **red** in the figure above.

STEP 5 The Results:
a. State the decision about H_o.
   Reject H_o.
b. State the conclusion about H_a.
   There is sufficient evidence to reject Tom's claim and conclude that the percentage of students who drive convertibles is significantly different from 15% at the 0.10 level of significance. ▲

The following commands will complete a hypothesis test for a proportion p.

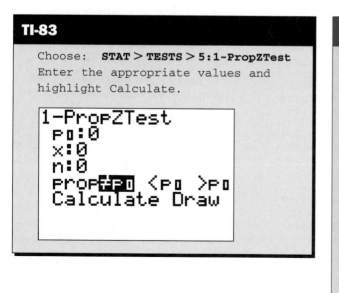

TI-83

Choose: **STAT > TESTS > 5:1-PropZTest**
Enter the appropriate values and highlight Calculate.

```
1-PropZTest
 P0:0
 x:0
 n:0
 prop≠P0 <P0 >P0
 Calculate Draw
```

EXCEL

Input the data into column A using 0's for failures (or no's) and 1's for successes (or yes's); then continue with:

Choose: **Tools > Data Analysis Plus > Inference About A Proportion > OK**
Enter: block coordinates:
         (A2:A20 or select cells) > OK
         (do not include column name)
         Code for success: **1** > OK
Choose: **Test of Hypothesis**
Enter: value of *p* under the null
         hypothesis: **p** > OK
Choose: Alternative: **less than** or **not equal** or **greater than**

There is a relationship between confidence intervals and two-tailed hypothesis tests when the level of confidence and the level of significance add up to 1. The confidence coefficients and the critical values are the same, meaning the width of the confidence

Exercise 9.57

Show that the hypothesis test completed as Illustration 9.11 was unnecessary since the confidence interval had already been completed in Illustration 9.7.

interval and the width of the noncritical region are the same. The point estimate is the center of the confidence interval, and the hypothesized mean is the center of the noncritical region. Therefore, if the hypothesized value of p is contained in the confidence interval, then the test statistic will be in the noncritical region (see Figure 9.5). Further, if the hypothesized probability p does not fall within the confidence interval, then the test statistic will be in the critical region (see Figure 9.6). This comparison should be used only when the hypothesis test is two-tailed and when the same value of α is involved for both procedures.

Case Study 9.2

The Methods Behind the Polling Madness

The "Sampling error" graph shows the margin of error that results from using samples of various sizes. It specifically points out that a sample of size $n = 751$ yields a 3.6% (or 0.036) sampling error. This sampling error is also called the "maximum error" of estimate and is a multiple of the standard error of proportion.

The "Applying sampling error" chart shows how the estimates we read as headlines can be interpreted and also helps explain why the various polls seem to report what appear at first glance to be inconsistent findings. Notice that when the sampling error of 3.6% is applied to the 44%, the proportion of voters favoring Clinton on that day could have been any percentage from 40.4% to 47.6% and for Bush the percentage could have been any value from 29.4% to 36.6%; thus, the margin Clinton had that day could have been anywhere from 3.8% to 18.2%. If you had taken your own random sample of 751 voters on that day, do you think your sample results would have been the same 33% and 44%?

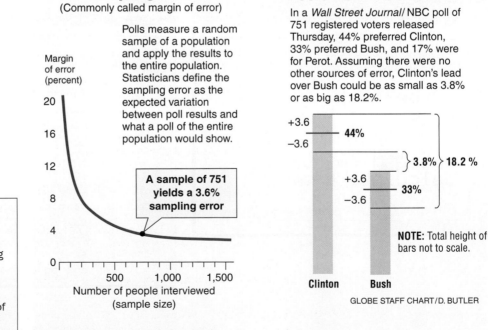

Sampling error
(Commonly called margin of error)

Polls measure a random sample of a population and apply the results to the entire population. Statisticians define the sampling error as the expected variation between poll results and what a poll of the entire population would show.

A sample of 751 yields a 3.6% sampling error

Margin of error (percent)

Number of people interviewed (sample size)

Applying sampling error

In a *Wall Street Journal*/ NBC poll of 751 registered voters released Thursday, 44% preferred Clinton, 33% preferred Bush, and 17% were for Perot. Assuming there were no other sources of error, Clinton's lead over Bush could be as small as 3.8% or as big as 18.2%.

NOTE: Total height of bars not to scale.

Clinton   Bush

GLOBE STAFF CHART / D. BUTLER

Source: The Boston Globe, November 2, 1992.

Exercise 9.58

a. Find the standard error of proportion when $n = 751$ using $p^* = 0.5$.
b. What level of confidence is related to the sampling error of 0.036 reported in Case Study 9.2?

Figure 9.5

Confidence Interval Contains p

Figure 9.6

Confidence Interval Does Not Contain p

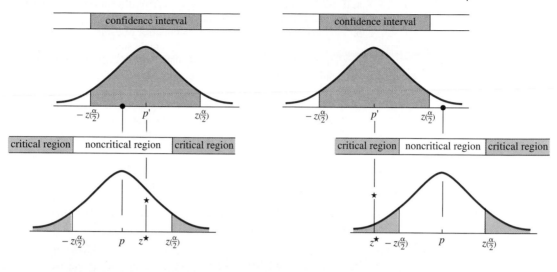

EXERCISES • • • • • • • • • • • • • •

9.59 Forty-five (45) of the 150 elements in a random sample are classified as "success."
  a. Explain why x and n are assigned the values 45 and 150, respectively.
  b. Determine the value of p'. Explain how p' is found and the meaning of p'.

9.60 For each of the following situations, find p'.
  a. $x = 24$ and $n = 250$
  b. $x = 640$ and $n = 2050$
  c. 892 of 1280 responded "yes"

9.61 **a.** What is the relationship between $p = P(\text{success})$ and $q = P(\text{failure})$? Explain.
  b. Explain why the relationship between p and q can be expressed by the formula $q = 1 - p$.
  c. If $p = 0.6$, what is the value of q?
  d. If the value of $q' = 0.273$, what is the value of p'?

9.62 Find (1) α, (2) area of one-tail, and (3) the confidence coefficients of z that are used with each of the following levels of confidence.
  a. $1 - \alpha = 0.90$ **b.** $1 - \alpha = 0.95$ **c.** $1 - \alpha = 0.98$ **d.** $1 - \alpha = 0.99$

9.63 A telephone survey was conducted to estimate the proportion of households with a personal computer. Of the 350 households surveyed, 75 had a personal computer.
  a. Give a point estimate for the proportion in the population who have a personal computer.
  b. Give the maximum error of estimate with 95% confidence.

9.64 A bank randomly selected 250 checking-account customers and found that 110 of them also had savings accounts at this same bank. Construct a 95% confidence interval for the true proportion of checking-account customers who also have savings accounts.

9.65 In a sample of 60 randomly selected students, only 22 favored the amount being budgeted for next year's intramural and interscholastic sports. Construct the 99% confidence interval for the proportion of all students who support the proposed budget amount.

9.66 *USA Today* (2-2-95) reported on a poll of 750 children and teenagers, aged 10 to 16. Two of the findings reported were: "4 out of 5 kids say entertainment television should teach youngsters right from wrong," and "2 out of 3 say shows like *The Simpsons* and *Married . . . With Children* encourage kids to disrespect their parents." The

(continued)

poll (margin of error: plus/minus 3%) was sponsored by Children Now, an advocacy group.

a. Find the point estimate, the maximum error of estimate, and the 95% confidence interval that results from the "4 out of 5" sample summary.

b. Find the point estimate, the maximum error of estimate, and the 95% confidence interval that results from the "2 out of 3" sample summary.

c. Compare the maximum error calculated in parts (a) and (b) to the margin of error mentioned in the article. What level of confidence do you believe the margin of error "plus/minus 3%" represents? Explain.

d. In Case Study 9.2, the sampling error for a sample of 751 is reported as being 3.6%. The article mentioned above reports that a sample of size 750 has a margin of error of 3%. Explain why these two polls have different sampling errors. Assume both are using 95% confidence.

9.67 An article titled "Why Don't Women Buy CDs?" appeared in the September 1994 issue of *Music* magazine. Yehuda Shapiro, marketing director of Virgin Retail Europe, found that across Europe 40% of his customers who buy classical records are women. Determine a 90% confidence interval for the true value of p if the 40% estimate is based on 1000 randomly selected buyers.

9.68 Adverse drug reactions to legally prescribed medicines are among the leading causes of death in the United States. According to a recent study published in the *Journal of the American Medical Association,* just over 100,000 people die due to complications resulting from legally prescribed drugs, whereas 5,000 to 10,000 die from using illicit drugs (*People*, 9-7-98).

You decide to investigate drug-related deaths in your city by monitoring the next 250 incidences. From among these cases, 223 were caused by legally prescribed drugs; the rest were the result of illicit drug use. Find the 95% confidence interval for the proportion of drug-related deaths that are caused by legally prescribed drugs.

9.69 "Parents should spank children when they think it is necessary, said 51% of adult respondents to a survey—though most child-development experts say spanking is not appropriate. The survey of 7225 adults . . . was co-sponsored by *Working Mother* magazine and Epcot Center at Walt Disney World." This statement appeared in the Rochester *Democrat & Chronicle* (12-20-90). Find the 99% confidence maximum error of estimate for the parameter p, P(should spank when necessary), for the adult population.

9.70 In a survey of 12,000 adults aged 19 to 74, National Cancer Institute researchers found that 9% in the survey ate at least the recommended two servings of fruit or juice and three servings of vegetables per day (*Ladies Home Journal,* April 1991). Use this information to determine a 95% confidence interval for the true proportion in the population who follow the recommendation.

9.71 Construct 90% confidence intervals for the binomial parameter p for each of the following pairs of values. Write your answers on the chart.

Observed Proportion $p' = x/n$	Sample Size	Lower Limit	Upper Limit
a. $p' = 0.3$	$n = 30$		
b. $p' = 0.7$	$n = 30$		
c. $p' = 0.5$	$n = 10$		
d. $p' = 0.5$	$n = 100$		
e. $p' = 0.5$	$n = 1000$		

f. Explain the relationship between answers (a) and (b).

g. Explain the relationship between answers (c), (d), and (e).

9.72 a. Calculate the maximum error of estimate for p for the 95% confidence interval for each of the situations listed in the table that follows.

Approximate Value of p

Sample Size n	0.1	0.3	0.5	0.7	0.9
100					
500					
1000					
1500					

   b. Explain the relationship between answers in columns 0.1 and 0.9; 0.3 and 0.7.
   c. Explain the relationship between answers in column 0.5 and the values that can be
      read from the sampling-error graph shown in Case Study 9.2 on page 444.

9.73 Karl Pearson once tossed a coin 24,000 times and recorded 12,012 heads.
   a. Calculate the point estimate for $p = P(\text{head})$ based on Pearson's results.
   b. Determine the standard error of proportion.
   c. Determine the 95% confidence interval estimate for $p = P(\text{head})$.
   d. It must have taken Mr. Pearson many hours to toss a coin 24,000 times. You can simu-
      late 24,000 coin tosses using the computer and calculator commands listed below.
      (*Note:* A Bernoulli experiment is like a "single" trial binomial experiment. That is,
      one toss of a coin is one Bernoulli experiment with $p = 0.5$; and 24,000 tosses of a
      coin either is a binomial experiment with $n = 24{,}000$ or is 24,000 Bernoulli experi-
      ments. Code: $0 = $ tail, $1 = $ head. The sum of the 1's will be the number of heads in
      the 24,000 tosses.)

```
MINITAB: Choose Calc > Random Data > Bernoulli, then enter
0.5 for the probability. Sum the data and divide by
24,000.
```

```
EXCEL: Choose Tools > Data Analysis > Random Number Genera-
tion > Bernoulli, then enter 0.5 for p. Sum the data and
divide by 24,000.
```

```
TI-83: Choose MATH > PRB > 5:randInt, then enter 0, 1, num-
ber of trials. The maximum number of elements (trials) in
a list is 999. (slow process for large n's) Sum the data
and divide by n.
```

   e. How do your simulated results compare to Pearson's?
   f. Use these commands and generate another set of 24,000 coin tosses. Compare these
      results to those above. Explain what you can conclude from these results.

9.74 The "rule of thumb," stated on page 435, indicated that we would expect the sampling
     distribution of p' to be approximately normal when "$n > 20$ and both np and nq were
     greater than 5." What happens when these guidelines are not followed?
   a. Use the following set of computer/calculator commands to show you what happens. Try
      $n = 15$ and $p = 0.1$. Do the distributions look normal? Explain what causes the "gaps."
      Why do the histograms look alike? Try some different combinations of n and p:

```
MINITAB: Choose Calc > Random Data > Binomial to simulate
1000 trials for an n of 15 and a p of 0.5. Divide each
generated value by n, forming a column of sample p's.
Calculate a z-value for each sample p by using z = (p'-p)/
√p(1 − p)/n. Construct a histogram for the sample p's and
another histogram for the z's.
```

```
EXCEL: Choose Tools > Data Analysis > Random Number Genera-
tion > Binomial to simulate a 1000 trials for an n of 15
and a p of 0.5. Divide each generated value by n, forming
```

a column of sample p's. Calculate a z-value for each sample p by using $z = (p'-p)/\sqrt{p(1-p)/n}$. Construct a histogram for the sample p's and another histogram for the z's.

TI-83: Choose MATH > PRB > 7:randBin, then enter n, p, number of trials. The maximum number of elements (trials) in a list is 999. (slow process for large n's) Divide each generated value by n, forming a list of sample p's. Calculate a z-value for each sample p by using $z = (p'-p)/\sqrt{p(1-p)/n}$. Construct a histogram for the sample p's and another histogram for the z's.

b. Try $n = 15$ and $p = 0.01$.
c. Try $n = 50$ and $p = 0.03$.
d. Try $n = 20$ and $p = 0.2$.
e. Try $n = 20$ and $p = 0.8$.
f. What happens when the rule of thumb is not followed?

9.75 According to the June 1994 issue of *Bicycling,* only 16% of all bicyclists own helmets. You wish to conduct a survey in your city to determine what percent of the bicyclists own helmets. Use the national figure of 16% for your initial estimate of p.
a. Find the sample size if you want your estimate to be within 0.02 with 90% confidence.
b. Find the sample size if you want your estimate to be within 0.04 with 90% confidence.
c. Find the sample size if you want your estimate to be within 0.02 with 98% confidence.
d. What effect does changing the maximum error have on the sample size? Explain.
e. What effect does changing the level of confidence have on the sample size? Explain.

9.76 A bank believes that approximately $\frac{2}{5}$ of its checking-account customers have used at least one other service provided by the bank within the last six months. How large a sample will be needed to estimate the true proportion to within 5% at the 98% level of confidence?

9.77 Paul Polger, a meteorologist with the National Weather Service, says that weathermen now accurately predict 82% of extreme weather events, up from 60% a decade ago. In fact, Polger claims, "We're doing as well in a two-day forecast as we did in a one-day forecast twenty years ago" (*Life,* August 1998).

You wish to conduct a study of extreme weather forecast accuracy by comparing local forecasts with actual weather conditions occurring in your city.
a. What is the best estimate available for the probability of accuracy in predicting extreme weather events?
b. Find the sample size if you want your estimate to be within 0.02 with 90% confidence.
c. Find the sample size if you want your estimate to be within 0.04 with 95% confidence.
d. Find the sample size if you want your estimate to be within 0.06 with 99% confidence.
e. If the level of confidence remains constant, what happens to the required sample size if you wish to double the maximum error of your estimate?

9.78 According to the May 1990 issue of *Good Housekeeping,* only about 14% of lung cancer patients survive for five years after diagnosis. Suppose you wanted to see if this survival rate were still true. How large a sample would you need to take to estimate the true proportion surviving for five years after diagnosis to within 1% with 95% confidence? (Use the 14% as the value of p.)

9.79 State the null hypothesis, H_o, and the alternative hypothesis, H_a, that would be used to test these claims:
a. More than 60% of all students at our college work part-time jobs during the academic year.
b. The probability of our team winning tonight is less than 0.50.

(continued)

    c. No more than one-third of cigarette smokers are interested in quitting.

    d. At least 50% of all parents believe in spanking their children when appropriate.

    e. A majority of the voters will vote for the school budget this year.

    f. At least three-quarters of the trees in our county were seriously damaged by the storm.

    g. The results show the coin was not tossed fairly.

    h. The single-digit numbers generated by the computer do not seem to be equally likely with regard to being odd or even.

9.80 Determine the test criteria that would be used to test the following hypotheses when z is used as the test statistic and the classical approach is used.

    a. H_o: $p = 0.5$ and H_a: $p > 0.5$, with $\alpha = 0.05$

    b. H_o: $p = 0.5$ and H_a: $p \neq 0.5$, with $\alpha = 0.05$

    c. H_o: $p = 0.4$ and H_a: $p < 0.4$, with $\alpha = 0.10$

    d. H_o: $p = 0.7$ and H_a: $p > 0.7$, with $\alpha = 0.01$

9.81 Determine the p-value for each of the following hypothesis-testing situations.

    a. H_o: $p = 0.5$, H_a: $p \neq 0.5$, $z \bigstar = 1.48$

    b. H_o: $p = 0.7$, H_a: $p \neq 0.7$, $z \bigstar = -2.26$

    c. H_o: $p = 0.4$, H_a: $p > 0.4$, $z \bigstar = 0.98$

    d. H_o: $p = 0.2$, H_a: $p < 0.2$, $z \bigstar = -1.59$

9.82 The binomial random variable, x, may be used as the test statistic when testing hypotheses about the binomial parameter, p, when n is small (say, 15 or less). Use Table 2 in Appendix B and determine the p-value for each of the following situations.

    a. H_o: $p = 0.5$, H_a: $p \neq 0.5$, where $n = 15$ and $x = 12$

    b. H_o: $p = 0.8$, H_a: $p \neq 0.8$, where $n = 12$ and $x = 4$

    c. H_o: $p = 0.3$, H_a: $p > 0.3$, where $n = 14$ and $x = 7$

    d. H_o: $p = 0.9$, H_a: $p < 0.9$, where $n = 13$ and $x = 9$

9.83 The binomial random variable, x, may be used as the test statistic when testing hypotheses about the binomial parameter, p. When n is small (say, 15 or less), Table 2 in Appendix B provides the probabilities for each value of x separately, thereby making it unnecessary to estimate probabilities of the discrete binomial random variable with the continuous standard normal variable z. Use Table 2 and determine the value of α for each of the following:

    a. H_o: $p = 0.5$ and H_a: $p > 0.5$, where $n = 15$ and the critical region is $x = 12, 13, 14, 15$

    b. H_o: $p = 0.3$ and H_a: $p < 0.3$, where $n = 12$ and the critical region is $x = 0, 1$

    c. H_o: $p = 0.6$ and H_a: $p \neq 0.6$, where $n = 10$ and the critical region is $x = 0, 1, 2, 3, 9, 10$

    d. H_o: $p = 0.05$ and H_a: $p > 0.05$, where $n = 14$ and the critical region is $x = 4, 5, 6, 7, \ldots, 14$

9.84 Use Table 2 in Appendix B and determine the critical region used in testing each of the following hypotheses. (*Note:* Since x is discrete, choose critical regions that do not exceed the value of α given.)

    a. H_o: $p = 0.5$ and H_a: $p > 0.5$, where $n = 15$ and $\alpha = 0.05$

    b. H_o: $p = 0.5$ and H_a: $p \neq 0.5$, where $n = 14$ and $\alpha = 0.05$

    c. H_o: $p = 0.4$ and H_a: $p < 0.4$, where $n = 10$ and $\alpha = 0.10$

    d. H_o: $p = 0.7$ and H_a: $p > 0.7$, where $n = 13$ and $\alpha = 0.01$

9.85 You are testing the hypothesis $p = 0.7$ and have decided to reject this hypothesis if after 15 trials you observe 14 or more successes.

    a. If the null hypothesis is true and you observe 13 successes, then which of the following will you do: (1) correctly fail to reject H_o? (2) correctly reject H_o? (3) commit a type I error? (4) commit a type II error?

    b. Find the significance level of your test.

    c. If the true probability of success is $\dfrac{1}{2}$ and you observe 13 successes, then which of the following will you do: (1) correctly fail to reject H_o? (2) correctly reject H_o? (3) commit a type I error? (4) commit a type II error?

    d. Calculate the p-value for your hypothesis test after 13 successes are observed.

9.86 You are testing the null hypothesis $p = 0.4$ and will reject this hypothesis if $z \star$ is less than -2.05.

    a. If the null hypothesis is true and you observe $z \star$ equal to -2.12, then which of the following results: (1) correctly fail to reject H_o? (2) correctly reject H_o? (3) commit a type I error? (4) commit a type II error?

    b. What is the significance level for this test?

    c. What is the p-value for $z \star = -2.12$?

9.87 An insurance company states that 90% of its claims are settled within 30 days. A consumer group selected a random sample of 75 of the company's claims to test this statement. If the consumer group found that 55 of the claims were settled within 30 days, do they have sufficient reason to support their contention that fewer than 90% of the claims are settled within 30 days? Use $\alpha = 0.05$.

    a. Solve using the p-value approach.   **b.** Solve using the classical approach.

9.88 The popularity of personal watercraft (PWCs, also known as jet skis) continues to increase, despite the apparent danger associated with their use. In fact, a sample of 54 reported watercraft accidents to the Game and Parks Commission in the state of Nebraska revealed in 1997 that 85% of them involved PWCs even though only 8% of the motorized boats registered in the state are PWCs. (*Nebraskaland,* June 1998).

Suppose the national average proportion of watercraft accidents in 1997 involving PWCs was 78%. Does the watercraft accident rate for PWCs in the state of Nebraska exceed the nation as a whole? Use a 0.01 level of significance.

    a. Solve using the p-value approach.   **b.** Solve using the classical approach.

9.89 A politician claims that she will receive 60% of the vote in an upcoming election. The results of a properly designed random sample of 100 voters showed that 50 of those sampled will vote for her. Is it likely that her assertion is correct at the 0.05 level of significance?

    a. Solve using the p-value approach.   **b.** Solve using the classical approach.

9.90 The full-time student body of a college is composed of 50% males and 50% females. Does a random sample of students (30 male, 20 female) from an introductory chemistry course show sufficient evidence to reject the hypothesis that the proportion of male and female students who take this course is the same as that of the whole student body? Use $\alpha = 0.05$.

    a. Solve using the p-value approach.   **b.** Solve using the classical approach.

9.91 The first baby ever conceived through in vitro fertilization (IVF) was born in England in 1978. In the 20 years that followed, 10 million women have received such care for infertility. The procedure, which can cost upward of $12,000, has experienced an average success rate of 22.5% nationwide, but that rate is increasing with advances in technology (*Family Circle,* 9-15-98).

Suppose a recent study of 200 women attempting to overcome infertility using the IVF procedure shows that 61 were actually successful in becoming pregnant. Do the results show a greater success rate for the sample than expected based on the historical success rate? Use $\alpha = 0.05$.

    a. Solve using the p-value approach.   **b.** Solve using the classical approach.

9.92 The article "Making Up for Lost Time" (*U.S. News & World Report,* July 30, 1990) reported that more than half of the country's workers aged 45 to 64 want to quit work before they reach age 65. Suppose you conduct a survey of 1000 randomly chosen workers in order to test $H_o: p = 0.5$ versus $H_a: p < 0.5$, where p represents the proportion who want to quit before they reach age 65. Four hundred sixty of the 1000 sampled want to quit work before age 65. Use $\alpha = 0.01$.

    a. Calculate the value of the test statistic.

    b. Solve using the p-value approach.

    c. Solve using the classical approach.

9.3   Inferences About Variance and Standard Deviation

Problems often arise that require us to make inferences about variability . For example, a soft-drink bottling company has a machine that fills 16-oz bottles. It needs to control the standard deviation σ (or variance σ^2) in the amount of soft drink, x, put into each bottle. The mean amount placed in each bottle is important, but a correct mean amount does not ensure that the filling machine is working correctly. If the variance is too large, there will be many bottles that are overfilled and many that are underfilled. Thus, the bottling company will want to maintain as small a standard deviation (or variance) as possible.

When discussing inferences about the spread of data, it is customary to talk about variance instead of the standard deviation, because the techniques (the formulas used) employ the sample variance rather than the standard deviation. However, remember that the standard deviation is the positive square root of the variance; thus, to talk about the variance of a population is comparable to talking about the standard deviation.

Inferences about the variance of a normally distributed population use the **chi-square, χ^2, distributions.** ("ki-square"; that's "ki" as in kite: χ is the Greek lowercase letter chi.) The chi-square distributions , like the Student t-distributions, are a family of probability distributions, each one being identified by the **parameter,** number of **degrees of freedom.** In order to use the chi-square distribution, we must be aware of its properties (see Figure 9.7).

PROPERTIES OF THE CHI-SQUARE DISTRIBUTION

1. χ^2 is nonnegative in value; it is zero or positively valued.
2. χ^2 is not symmetrical; it is skewed to the right.
3. χ^2 is distributed so as to form a family of distributions, a separate distribution for each different number of degrees of freedom .

Figure 9.7
Various Chi-Square Distributions

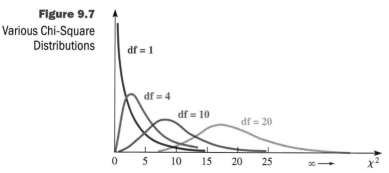

The **critical values for chi-square** are obtained from Table 8 in Appendix B . Each critical value is identified by two pieces of information: degrees of freedom (df) and the area under the curve to the right of the critical value being sought. Thus, χ^2(df, α) (read "chi-square of df, alpha" is the symbol used to identify the critical value of chi-square with df degrees of freedom and with α area to the right, as shown in Figure 9.8 (p. 452). Since the chi-square distribution is not symmetrical, the critical values associated with right and left tails are given separately in Table 8.

Figure 9.8
Chi-Square Distribution
Showing $\chi^2(df, \alpha)$

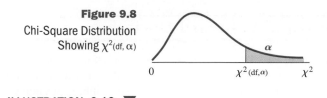

ILLUSTRATION 9.12 ▼

Find $\chi^2(20, 0.05)$.

Solution

See the figure. Use Table 8 in Appendix B to find the value of $\chi^2(20, 0.05)$ at the intersection of row df = 20 and column α = 0.05, as shown in the table below:

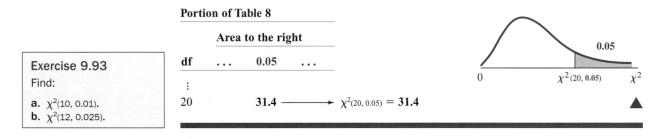

Portion of Table 8

	Area to the right		
df	...	**0.05**	...
⋮			
20		**31.4**	

$\chi^2(20, 0.05)$ = **31.4**

Exercise 9.93
Find:

a. $\chi^2(10, 0.01)$.
b. $\chi^2(12, 0.025)$.

NOTE When df > 2, the mean value of the chi-square distribution is df. The mean is located to the right of the mode (the value where the curve reaches its high point) and just to the right of the median (the value that splits the distribution, 50% on either side). By locating the zero at the left extreme and the value of df on your sketch of the χ^2 distribution, you will establish an approximate scale so that other values can be located in their respective positions. See Figure 9.9.

Figure 9.9
Location of Mean, Median, and Mode for χ^2 Distribution

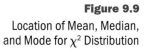

ILLUSTRATION 9.13 ▼

Find $\chi^2(14, 0.90)$.

Solution

See the figure. Use Table 8 in Appendix B to find the value of $\chi^2(14, 0.90)$ at the intersection of row df = 14 and column α = 0.90, as shown in the table below:

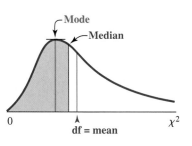

Portion of Table 8

	Area to the right		
df	...	**0.90**	...
⋮			
20		**7.79**	

$\chi^2(14, 0.90)$ = **7.79.**

Exercise 9.94
Find:

a. $\chi^2(10, 0.95)$.
b. $\chi^2(22, 0.995)$.

Most computer software packages or statistical calculators will calculate the area related to a specified χ^2-value. The accompanying figure shows the relationship between the cumulative probability distribution and a specific χ^2-value for a χ^2-distribution with df degrees of freedom.

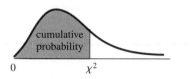

0   χ^2

The following commands will find the cumulative probability for a specified value of χ^2.

MINITAB

Choose: **Calc > Probability Distribu-tions > Chi-Square**

Select: **Cumulative Probability**

Enter: Degrees of freedom: **df**

Select: **Input constant***

Enter: χ^2**-value** (ex. 47.25)

*Select Input column if several χ^2-values are stored in C1. Use C2 for optional storage. If the area in the right tail is needed, subtract the calculated probability from 1.

TI-83

Choose: **2ⁿᵈ > DISTR > 7: χ^2cdf (**

Enter: **0, χ^2-value, df)**

If the area in the right tail is needed, subtract the calculated probability from 1.

The following commands will find the probability in one tail for a given χ^2-value (i.e., area to the right). Only for values in the left tail, is subtraction from one necessary.

Exercise 9.95

Use a computer/calculator to find the area (a) to the left, and (b) to the right of the calculated $\chi^2 \star = 20.2$ with df = 15.

EXCEL

If several χ^2-values are to be used, input the values into column A and activate B1; then continue with:

Choose: **Paste function f_x > Statistical > CHIDIST > OK**

Enter: X: **individual χ^2-value or (A1:A5 or select 'χ^2-value' cells)**

         Deg_freedom: **df**   **> OK**

Drag: **Bottom right corner of the B1 cell down to give other probabilities**

We are ready to use chi-square to make inferences about the population variance or standard deviation.

> ### THE ASSUMPTIONS FOR INFERENCES ABOUT THE VARIANCE σ^2 OR STANDARD DEVIATION σ
>
> The sampled population is normally distributed.

The t procedures for inferences about the mean (Section 9.1) were based on the assumption of normality, but they are generally very useful even when the sampled population is nonnormal, especially for larger samples. However, the same is not true about the inference procedures for standard deviation. The statistical procedures for standard deviation are very sensitive to nonnormal distributions (skewness, in particular), making it very difficult to determine whether an apparently significant result is the result of the sample evidence or a violation of the assumptions. Therefore, the only inference procedure to be presented here will be the hypothesis test for the standard deviation of a normal population.

The **test statistic** that will be used in testing hypotheses about the population variance or standard deviation is obtained by using the formula

$$\chi^2 \bigstar = \frac{(n-1)s^2}{\sigma^2}, \quad \text{with df} = n - 1. \tag{9.10}$$

When random samples are drawn from a normal population of a known variance σ^2, the quantity $\dfrac{(n-1)s^2}{\sigma^2}$ possesses a probability distribution that is known as the chi-square distribution with $n - 1$ degrees of freedom.

Hypothesis-Testing Procedure

Let's return to the illustration about the bottling company that wishes to detect when the variability in the amount of soft drink placed in each bottle gets out of control. A variance of 0.0004 is considered acceptable, and the company will want to adjust the bottle-filling machine when the variance, σ^2, becomes larger than this value. The decision will be made using the hypothesis-testing procedure.

ILLUSTRATION 9.14 ▼

The soft-drink bottling company wants to control the variability in the amount of fill by not allowing the variance to exceed 0.0004. Does a sample of size 28 with a variance of 0.0007 indicate that the bottling process is out of control (with regard to variance) at the 0.05 level of significance?

Solution

STEP 1 **The Set-Up:**
   a. **Describe the population parameter of interest.**
      The **variance σ^2** for the amount of fill of a soft drink during a bottling process.
   b. **State the null hypothesis (H_o) and the alternative hypothesis (H_a).**
      The null hypothesis is "the variance is no larger than the specified value 0.0004"; the alternative hypothesis is "the variance is larger than 0.0004."

H_o: $\sigma^2 = 0.0004$ ($\leq$) (variance is not larger than 0.0004)

H_a: $\sigma^2 > 0.0004$ (variance is larger than 0.0004)

STEP 2   **The Hypothesis Test Criteria:**
 a. **Check the assumptions.**
     The amount of fill put into a bottle is generally normally distributed. By checking the distribution of the sample, we could verify this.
 b. **Identify the probability distribution and the test statistic to be used.**
     The chi-square distribution will be used and formula (9.10), with df $= n - 1 = 28 - 1 = 27$.
 c. **Determine the level of significance, α.**
     $\alpha = 0.05$.

Exercise 9.96

Find the test statistic for the hypothesis test: H_o: $\sigma^2 = 532$ vs. H_a: $\sigma^2 > 532$ using sample information $n = 18$ and $s^2 = 785$.

STEP 3   **The Sample Evidence:**
 a. **Collect the sample information.**
     $n = 28$ and $s^2 = 0.0007$.
 b. **Calculate the value of the test statistic.**
     Using formula (9.10),

$$\chi^2 \bigstar = \frac{(n-1)s^2}{\sigma^2}: \quad \chi^2 \bigstar = \frac{(28-1)(0.0007)}{0.0004} = \frac{(27)(0.0007)}{0.0004} = \mathbf{47.25}$$

STEP 4   **The Probability Distribution:**

USING THE *p*-VALUE PROCEDURE:

 a. **Calculate the *p*-value.**
 Use the right-hand tail since the H_a expresses concern for values related to "larger than."
 $\mathbf{P} = P(\chi^2 > 47.25$, with df $= 27)$ as shown in the figure.

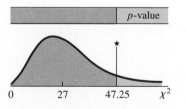

To find the *p*-value, use one of two methods:
1. Use Table 8 (Appendix B) to place bounds on the *p*-value:
   0.005 < P < 0.01
2. Use a computer or calculator to calculate the *p*-value:
   P = 0.0093

Specific instructions follow this illustration.

 b. **Determine whether or not the *p*-value is smaller than α.**
 The *p*-value **is smaller** than the level of significance, α (0.05).

OR

USING THE CLASSICAL PROCEDURE:

 a. **Determine the critical region and critical value(s).**
 The critical region is the right-hand tail since the H_a expresses concern for values related to "larger than." The critical value is obtained from Table 8, at the intersection of row df $= 27$ and column $\alpha = 0.05$:

$$\chi^2(27, 0.05) = \mathbf{40.1}$$

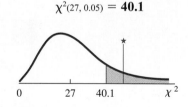

For specific instructions, see page 452.

 b. **Determine whether or not the calculated test statistic is in the critical region.**
 $\chi^2 \bigstar$ **is in** the critical region, as shown in red in the figure above.

STEP 5   **The Results:**
 a. **State the decision about H_o.**
     Reject H_o.
 b. **State the conclusion about H_a.**
     At the 0.05 level of significance, we conclude that the bottling process is out of control with regard to the variance.   ▲

Exercise 9.97

Complete the hypothesis test in Exercise 9.96 using the *p*-value method and $\alpha = 0.01$.

Exercise 9.98

Complete the hypothesis test in Exercise 9.96 using the classical method and $\alpha = 0.01$.

Calculating the *p*-value when using the χ^2-distribution:

Method 1 *Using Table 8 in Appendix B to "place bounds" on the p-value:* By inspecting the df = 27 row of Table 8, you can determine an interval within which the *p*-value lies. Locate the $\chi^2 \star$ along the row labeled df = 27. If $\chi^2 \star$ is not listed, locate the two values $\chi^2 \star$ falls between, and then read the bounds for the *p*-value from the top of the table. In this case, $\chi^2 \star = 47.25$ is between 47.0 and 49.6; therefore, P is between 0.005 and 0.01.

Portion of Table 8

df	...	0.01	P	0.005	
					0.005 < P < 0.01
$\vdots$		$\uparrow$		$\uparrow$	
27		47.0	**47.25**	49.6	

with heading **α in Right-hand tail**

Method 2 *Using a computer or calculator:* use the χ^2 probability or χ^2 cumulative probability distribution commands on page 453 to find the *p*-value associated with $\chi^2 \star = 47.25$.

ILLUSTRATION 9.15 ▼

Find the *p*-value for the following hypothesis test.

$$H_o: \sigma^2 = 12$$

$$H_a: \sigma^2 < 12 \quad \text{with df} = 15 \text{ and } \chi^2 \star = 7.88.$$

Solution

Since the concern is for "smaller" values (alternative hypothesis is "less than"), the *p*-value is the area to the left of $\chi^2 \star = 7.88$ as shown in the figure.

$$\mathbf{P} = P(\chi^2 < 7.88, \text{ with df} = 15).$$

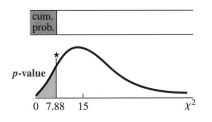

To find the *p*-value, use one of two methods:

Method 1 *Use Table 8 (Appendix B) to place bounds on the p-value:* inspect the df = 15 row locating $\chi^2 \star = 7.88$, the $\chi^2 \star$ value is between entries, the interval that bounds **P** is read from the left-hand tail heading at the top of the table.

Portion of Table 8

df	...	0.05	P	0.10	
					0.05 < P < 0.10
$\vdots$		$\uparrow$		$\uparrow$	
15	...	7.26	**7.88**	8.55	

with heading **α in Left-hand tail**

Method 2   *Using a computer or calculator:* use the χ^2 probability or χ^2 cumulative probability distribution commands on page 453 to find the *p*-value associated with $\chi^2 \bigstar = 7.88$.

$$p\text{-value} = P(\chi^2 < 7.88 | df = 15) = \mathbf{0.0715.}$$

ILLUSTRATION 9.16 ▼

A photographic chemical is claimed by its manufacturer to have a shelf life that is normally distributed about a mean of 180 days with a standard deviation of no more than 10 days. As a user of this chemical, Fast Photo is concerned that the standard deviation might be different from 10 days; otherwise, they will buy a larger quantity while the chemical is part of a special promotion. Twelve samples were randomly selected and tested, with a standard deviation of 14 days resulting. At the 0.05 level of significance, does this sample present sufficient evidence to show the standard deviation is different from 10 days?

Solution

STEP 1   **The Set-Up:**

      a. Describe the population parameter of interest.
          The **standard deviation** σ for the shelf life of the chemical.

      b. State the null hypothesis (H_o) and the alternative hypothesis (H_a).
          The null hypothesis is "the standard deviation is the specified value 10 days"; the alternative hypothesis is "the standard deviation is different from 10 days."

          H_o: $\sigma = 10$ (standard deviation is 10)

          H_a: $\sigma \neq 10$ (standard deviation is different from 10)

STEP 2   **The Hypothesis Test Criteria:**

      a. Check the assumptions.
          The manufacturer claims "shelf life" is normally distributed; this could be verified by checking the distribution of the sample.

      b. Identify the probability distribution and the test statistic to be used.
          The chi-square distribution will be used and formula (9.10), with $df = n - 1 = 12 - 1 = 11$.

      c. Determine the level of significance, α.
          $\alpha = 0.05$.

STEP 3   **The Sample Evidence:**

      a. Collect the sample information.
          $n = 12$ and $s = 14$.

      b. Calculate the value of the test statistic.
          Using formula (9.10),

> **Exercise 9.99**
>
> Find the test statistic for the hypothesis test:
> H_o: $\sigma^2 = 52$ vs.
> H_a: $\sigma^2 \neq 52$ using sample information $n = 41$ and $s^2 = 78.2$.

$$\chi^2 \bigstar = \frac{(n-1)s^2}{\sigma^2}: \quad \chi^2 \bigstar = \frac{(12-1)(14)^2}{(10)^2} = \frac{2156}{100} = \mathbf{21.56}$$

STEP 4 The Probability Distribution:

USING THE p-VALUE PROCEDURE:

a. Calculate the p-value.
Since the concern is for values "different from" 10, the p-value is the area of both tails. The area of each tail represents $\frac{1}{2}$**P**. Since $\chi^2 \star = 21.56$ is in the right tail, the area of the right tail is $\frac{1}{2}$**P**. $\frac{1}{2}$**P** $= P(\chi^2 > 21.56$, with df $= 11$) as shown in the figure

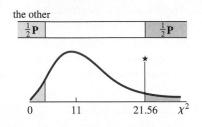

To find $\frac{1}{2}$**P**, use one of two methods:
1. Use Table 8 (Appendix B) to place bounds on the $\frac{1}{2}$**P**:
   0.025 $< \frac{1}{2}$P $<$ 0.05
   Double both bounds to find the bounds for **P**.
   $2 \times (0.025 < \frac{1}{2}$**P** $< 0.05)$ becomes **0.05 $<$ P $<$ 0.10**
2. Use a computer or calculator to find $\frac{1}{2}$**P**: $\frac{1}{2}$**P** $= 0.0280$; therefore **P $=$ 0.0560**
   Specific instructions follow this illustration.

b. Determine whether or not the p-value is smaller than α.
The p-value **is not smaller** than the level of significance, α (0.05).

OR

USING THE CLASSICAL PROCEDURE:

a. Determine the critical region and critical value(s).
The critical region is split into two equal parts since the H_a expresses concern for values related to "different from." The critical values are obtained from Table 8, at the intersections of row df $= 11$ with columns $\alpha = 0.975$ and 0.025 (area to right):

$$\chi^2(11, 0.975) = \textbf{3.82 and}$$
$$\chi^2(11, 0.025) = \textbf{21.9.}$$

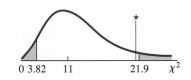

For specific instructions, see page 452.

b. Determine whether or not the calculated test statistic is in the critical region.
$\chi^2 \star$ is not in the critical region; see figure above.

Exercise 9.100
Complete the hypothesis test in Exercise 9.99 using the classical method and $\alpha = 0.05$.

STEP 5 The Results:
    a. State the decision about H_o.
        Fail to reject H_o.
    b. State the conclusion about H_a.
        There is not sufficient evidence at the 0.05 significance level to conclude that the shelf life of this chemical does have a standard deviation different from 10 days. Therefore, Fast Photo should purchase the chemical.

Calculating the p-value when using the χ^2-distribution:
    Method 1 *Use Table 8 (Appendix B) to place bounds on the p-value:* inspect the df $= 11$ row locating $\chi^2 \star = 21.56$, notice that 21.56 is between two table entries. The bounds for $\frac{1}{2}$**P** are read from the right-hand tail heading at the top of the table.

Portion of Table 8

df	...	0.05	$\frac{1}{2}$P	0.025
$\vdots$		$\uparrow$		$\uparrow$
11		19.7	**21.56**	21.9

$\longrightarrow$ 0.025 $< \frac{1}{2}$P $<$ 0.05

Exercise 9.101
Complete the hypothesis test in Exercise 9.99 using the p-value method and $\alpha = 0.05$.

    Double both bounds to find the bounds for **P**: $2 \times (0.025 < \frac{1}{2}$**P** $< 0.05)$ becomes **0.05 $<$ P $<$ 0.10.**

Exercise 9.102

Use a computer/calcula-tor to find the p-value for the hypothesis test: $H_o: \sigma^2 = 7$ vs. $H_a: \sigma^2 \neq 7$, if $\chi^2 \star = 6.87$ for a sample of $n = 15$.

Method 2   Using a computer or calculator: use the χ^2 probability or χ^2 cumulative probability distribution commands on page 453 to find the *p*-value associated with $\chi^2 \star = 21.56$. Remember to double the probability.   ▲

NOTE When sample data are skewed, just one outlier can greatly affect the standard deviation. It is very important, especially when using small samples, that the sampled population be normal; otherwise, the procedures are not reliable.

EXERCISES • • • • • • • • • • • • • •

9.103 **a.** Calculate the standard deviation for each set: **A:** 5, 6, 7, 7, 8, 10; **B:** 5, 6, 7, 7, 8, 15.
　　　b. What effect did the largest value changing from 10 to 15 have on the standard deviation?
　　　c. Why do you think 15 might be called an outlier?

9.104 Find these critical values by using Table 8 of Appendix B.
　　　a. $\chi^2(18, 0.01)$　　**b.** $\chi^2(16, 0.025)$　　**c.** $\chi^2(8, 0.10)$
　　　d. $\chi^2(28, 0.01)$　　**e.** $\chi^2(22, 0.95)$　　**f.** $\chi^2(10, 0.975)$
　　　g. $\chi^2(50, 0.90)$　　**h.** $\chi^2(24, 0.99)$

9.105 Using the notation of Exercise 9.104, name and find the critical values of χ^2.

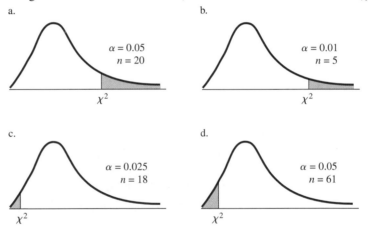

9.106 Using the notation of Exercise 9.104, name and find the critical values of χ^2.

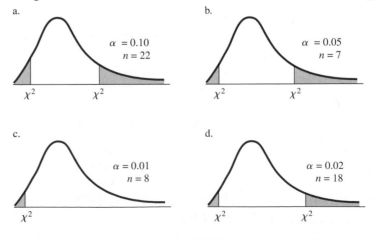

9.107 **a.** What value of chi-square for 5 degrees of freedom subdivides the area under the distribution curve such that 5% is to the right and 95% is to the left?
  b. What is the value of the 95th percentile for the chi-square distribution with 5 degrees of freedom?
  c. What is the value of the 90th percentile for the chi-square distribution with 5 degrees of freedom?

9.108 **a.** The central 90% of the chi-square distribution with 11 degrees of freedom lies between what values?
  b. The central 95% of the chi-square distribution with 11 degrees of freedom lies between what values?
  c. The central 99% of the chi-square distribution with 11 degrees of freedom lies between what values?

9.109 For a chi-square distribution having 12 degrees of freedom, find the area under the curve for chi-square values ranging from 3.57 to 21.0.

9.110 For a chi-square distribution having 35 degrees of freedom, find the area under the curve between $\chi^2{(35, 0.96)}$ and $\chi^2{(35, 0.15)}$.

9.111 State the null hypothesis, H_o, and the alternative hypothesis, H_a, that would be used to test these claims:
  a. The standard deviation has increased from its previous value of 24.
  b. The standard deviation is no larger than 0.5 oz.
  c. The standard deviation is not equal to 10.
  d. The variance is no less than 18.
  e. The variance is different from the value of 0.025, the value called for in the specs.
  f. The variance has increased from 34.5.

9.112 Calculate the value for the test statistic, $\chi^2 \bigstar$, for each of these situations:
  a. H_o: $\sigma^2 = 20$, $n = 15$, $s^2 = 17.8$
  b. H_o: $\sigma^2 = 30$, $n = 18$, $s = 5.7$
  c. H_o: $\sigma = 42$, $n = 25$, $s = 37.8$
  d. H_o: $\sigma = 12$, $n = 37$, $s^2 = 163$

9.113 Calculate the p-value for each of the following hypothesis tests.
  a. H_a: $\sigma^2 \neq 20$, $n = 15$, $\chi^2 \bigstar = 27.8$
  b. H_a: $\sigma^2 > 30$, $n = 18$, $\chi^2 \bigstar = 33.4$
  c. H_a: $\sigma^2 \neq 42$, df $= 25$, $\chi^2 \bigstar = 37.9$
  d. H_a: $\sigma^2 < 12$, df $= 40$, $\chi^2 \bigstar = 26.3$

9.114 Determine the critical region and critical value(s) that would be used to test the following using the classical approach:
  a. H_o: $\sigma = 0.5$ and H_a: $\sigma > 0.5$, with $n = 18$ and $\alpha = 0.05$
  b. H_o: $\sigma^2 = 8.5$ and H_a: $\sigma^2 < 8.5$, with $n = 15$ and $\alpha = 0.01$
  c. H_o: $\sigma = 20.3$ and H_a: $\sigma \neq 20.3$, with $n = 10$ and $\alpha = 0.10$
  d. H_o: $\sigma^2 = 0.05$ and H_a: $\sigma^2 \neq 0.05$, with $n = 8$ and $\alpha = 0.02$
  e. H_o: $\sigma = 0.5$ and H_a: $\sigma < 0.5$, with $n = 12$ and $\alpha = 0.10$

9.115 A random sample of 51 observations was selected from a normally distributed population. The sample means was $\bar{x} = 98.2$, and the sample variance was $s^2 = 37.5$. Does this sample show sufficient reason to conclude that the population standard deviation is not equal to 8 at the 0.05 level of significance?

9.116 In the past the standard deviation of weights of certain 32.0-oz packages filled by a machine was 0.25 oz. A random sample of 20 packages showed a standard deviation of 0.35 oz. Is the apparent increase in variability significant at the 0.10 level of significance? Assume package weight is normally distributed.
  a. Solve using the p-value approach.
  b. Solve using the classical approach.

9.117 In the United States, 36% of all people have a medically treatable foot problem. Among patients requiring surgery, 80 to 90% are women. The problem is compounded by mail-order sales of shoes that are the wrong size or do not fit, but people wear the shoes anyway rather than send them back. Although many experts have blamed high-heel shoes for most of the troubles, a recent study of 368 women complaining of foot problems showed that 88 percent were wearing shoes that were too small (*Ladies Home Journal,* "My Aching Feet", June 1998).

Suppose the standard deviation of shoe sizes for all manufacturers is 0.32. A separate study is conducted of 27 mail order sellers of women's shoes and a standard deviation of 0.51 is obtained from the sample. Do mail-order distributors sell shoes to women that vary more in size than shoe manufacturers in general, at the 0.01 level of significance?

   a. What role does the assumption of normality play in this solution? Explain.

   b. Describe how you might attempt to determine whether or not it is realistic to assume that shoe sizes are normally distributed.

   c. Solve using the *p*-value approach.

   d. Solve using the classical approach.

9.118 A commercial farmer harvests his entire field of a vegetable crop at one time. Therefore, he would like to plant a variety of green beans that mature all at one time (small standard deviation between maturity times of individual plants). A seed company has developed a new hybrid strain of green beans that it believes to be better for the commercial farmer. The maturity time of the standard variety has an average of 50 days and a standard deviation of 2.1 days. A random sample of 30 plants of the new hybrid showed a standard deviation of 1.65 days. Does this sample show a significant lowering of the standard deviation at the 0.05 level of significance? Assume that maturity time is normally distributed.

   a. Solve using the *p*-value approach.

   b. Solve using the classical approach.

9.119 A car manufacturer claims that the miles per gallon for a certain model has a mean equal to 40.5 mi. with a standard deviation equal to 3.5 mi. Use the following data, obtained from a random sample of 15 such cars, to test the hypothesis that the standard deviation differs from 3.5. Use $\alpha = 0.05$. Assume normality.

37.0	38.0	42.5	45.0	34.0	32.0	36.0	35.5
38.0	42.5	40.0	42.5	36.0	30.0	37.5	

   a. Solve using the *p*-value approach.

   b. Solve using the classical approach.

9.120 Farm real estate values in rural America fluctuate substantially from state to state and county to county, thus making it difficult for buyers purchasing land or landowners to know precisely what their property is actually worth. For example, the average value of ranchland in Missouri in 1998 was $548 per acre, whereas the same average in three nearby states (Kansas, Nebraska, and Oklahoma) was over $200 less (*Regional Economic Digest,* "Survey of Agricultural Credit Conditions," First Quarter 1998).

This discrepancy could be caused by an exaggerated variability in the value of ranchland acreage in the state of Missouri. Assume that the combined four-state region yielded a standard deviation of $85 per acre. Suppose a sample of 31 landowners in Missouri who recently sold their property was taken and a sample standard deviation of $125 per acre resulted. Is the variability in ranchland value in Missouri, at the 0.05 level of significance, greater than the variability for the region as a whole?

   a. Solve using the *p*-value approach.

   b. Solve using the classical approach.

9.121 The chi-square distribution was described on page 451 as a family of distributions. Let's investigate these distributions and observe some of their properties.

a. Use the MINITAB commands below and generate several large random samples of data from various chi-square distributions. Use df values of 1, 2, 3, 5, 10, 20, and 80 (others if you wish).

```
Choose:  Calc > Random Data > ChiSquare
Enter:   Generate: 1000 rows of data
         Store in column(s): C1
         Degrees of freedom: df

Use Stat > Basic Statistic > Display Descriptive Statis-
tics to calculate the mean and median of the data in
C1. Use Graph > Histogram to construct a histogram of
the data in C1.
```

b. What appears to be the relationship between the mean of the sample and the number of degrees of freedom?

c. How do the values of the mean, median, and mode appear to be related? Do your results agree with the information on page 452?

d. Have the computer generate samples for two additional degrees of freedom df = 120 and 150. Describe how these distributions seem to be changing as df increases.

9.122 How important is the assumption "the sampled population is normally distributed" for the use of the chi-square distributions? Use a computer and the two sets of MINITAB commands that can be found in the *Statistical Tutor* to simulate drawing 200 samples of size 10 from each of two different types of population distributions. The first commands will generate 2000 data and construct a histogram so that you can see what the population looks like. The next commands will generate 200 samples of size 10 from the same population; each row represents a sample. The following commands will calculate the standard deviation and χ^2 ★for each of the 200 samples. The last commands construct histograms of the 200 sample standard deviations and the 200 χ^2 ★-values. (Additional details can be found in the *Statistical Tutor*.)

For the samples from the normal population:

a. Does the sampling distribution of sample standard deviations appear to be normal? Describe the distribution.

b. Does the χ^2-distribution appear to have a chi-square distribution with df = 9? Find percentages for intervals (less than 2, less than 4, . . ., more than 15, more than 20, etc.), and compare them to the percentages expected as estimated using Table 8, "Critical Values of the Chi-Square Distribution," in Appendix B.

For the samples from the skewed population:

c. Does the sampling distribution of sample standard deviations appear to be normal? Describe the distribution.

d. Does the χ^2-distribution appear to have a chi-square distribution with df = 9? Find percentages for intervals (less than 2, less than 4, . . ., more than 15, more than 20, etc.), and compare them to the percentages expected as estimated using Table 8.

In summary:

e. Does the normality condition appear to be necessary in order for the calculated test statistic χ^2 ★ to have a χ^2-distribution? Explain.

Return to C H A P T E R   C A S E   S T U D Y

Many students work full-time or part-time. How much did you or your friends each earn last month? You are prepared to give a full answer to this question now. You learned how to describe data in Chapter 2 and how to make inferences in Chapter 9. (See Chapter Case Study, p. 415.)

Listed below is the amount earned last month by each student in a sample of 35 college students.

0	0	105	0	313	153	769	415	244	0	333	0
0	362	276	158	409	0	0	534	449	281	37	338
240	0	0	0	142	0	519	356	280	161	0	

Exercise 9.123

Use these sample data to describe the amount earned by working college students.
a. Describe the population of interest.
b. How many of the students in the sample above are working?
c. Describe the variable, amount earned by a working college student last month, using one graph, one measure of central tendency, and one measure of dispersion.
d. Find evidence to show that the assumptions for use of Student's *t*-distribution have been satisfied.
e. Estimate the mean amount earned by a college student per month using a point estimate and a 95% confidence interval.
f. The Statistical Snapshot® in the Chapter Case Study (p. 415) suggests the average amount earned each month by college students is approximately $350. Does the sample show sufficient reason to reject that claim? Use $\alpha = 0.05$.

IN RETROSPECT

We have been studying inferences, both confidence intervals and hypothesis tests, for the three basic population parameters (mean μ, proportion p, and standard deviation σ) of a single population. Most inferences about a single population are concerned with one of these three parameters. Figure 9.10 presents a visual organization of the techniques studied throughout Chapters 8 and 9 along with the key questions that you must ask yourself as you are deciding which test statistic and which formula to use.

In this chapter we also used the maximum error of estimate, formula (9.7), to determine the size of sample required to make estimations about the population proportion with the desired accuracy. Case Study 9.2 presents a graph showing the "margin of error" (maximum error of estimate) for sample sizes up to 1500. These are for 95% confidence interval estimates as you found out in Exercise 9.58. By combining a point estimate with its corresponding maximum error of estimate (sampling error), we can construct an interval estimate based on the information reported. Most polls and surveys use the 95% confidence level, even though they do not report the 95%.

In the next chapter we will discuss inferences about two populations whose respective means, proportions and standard deviations are to be compared.

Figure 9.10

Choosing the Right Inference Technique

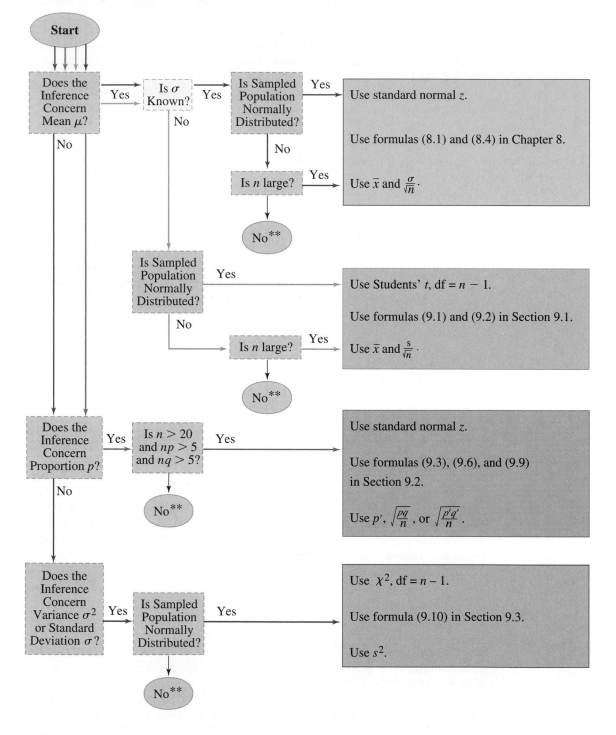

No** means that a nonparametric technique (normal distribution not required) is used: see Chaper 14.

CHAPTER EXERCISES

9.124 A natural gas utility is considering a contract for purchasing tires for its fleet of service trucks. The decision will be based on expected mileage. For a sample of 100 tires tested, the mean mileage was 36,000 and the standard deviation was 2000 miles. Estimate the mean mileage that the utility should expect from these tires using a 96% confidence interval.

9.125 One of the objectives of a large medical study was to estimate the mean physician fee for cataract removal. For 25 randomly selected cases the mean fee was found to be $1550 with a standard deviation of $125. Set a 99% confidence interval on μ, the mean fee for all physicians. Assume fees are normally distributed.

9.126 Oranges are selected at random from a large shipment that just arrived. The sample was taken to estimate the size (circumference, in inches) of the oranges. The sample data are summarized as follows: $n = 100$, $\sum x = 878.2$, and $\sum(x - \bar{x})^2 = 49.91$.
a. Determine the sample mean and standard deviation.
b. What is the point estimate for μ, the mean circumference of all oranges in the shipment?
c. Find the 95% confidence interval for μ.

9.127 Many U.S. housewives wouldn't think of leaving home to go shopping without their stash of coupons, as reported in a 1998 issue of *Family Circle* (9-15-98). In fact, couponing has been a popular practice for over 30 years. The amount that people save at the supermarket by redeeming coupons varies substantially; some shoppers routinely save $50 or more per trip, whereas others save little if anything. Couponing has also been criticized for generating sales of frivolous products and overstocking of items that ordinarily would have remained on the shelf, and it takes longer to check out at the cash register. On the other hand, coupon queens tend to be more educated and living in higher-income households.

Suppose the mean of all coupon sales at supermarkets in the United States is $10. A random sample of 25 shoppers with annual household incomes exceeding $75,000 is taken and reveals a mean redemption of $15 and a standard deviation of $7. Do shoppers from the higher-income group redeem coupons worth more than those redeemed by the rest of the nation? Use $\alpha = 0.01$.

9.128 It has been suggested that abnormal male children tend to occur more in children born to older-than-average parents. Case histories of 20 abnormal males were obtained, and the ages of the 20 mothers were

| 31 | 21 | 29 | 28 | 34 | 45 | 21 | 41 | 27 | 31 |
| 43 | 21 | 39 | 38 | 32 | 28 | 37 | 28 | 16 | 39 |

The mean age at which mothers in the general population give birth is 28.0 years.
a. Calculate the sample mean and standard deviation.
b. Does the sample give sufficient evidence to support the claim that abnormal male children have older-than-average mothers? Use $\alpha = 0.05$. Assume ages have a normal distribution.

9.129 The water pollution readings at State Park Beach seem to be lower than last year. A sample of 12 readings was randomly selected from the records of this year's daily readings:

| 3.5 | 3.9 | 2.8 | 3.1 | 3.1 | 3.4 | 4.8 | 3.2 | 2.5 | 3.5 | 4.4 | 3.1 |

Does this sample provide sufficient evidence to conclude that the mean of this year's pollution readings is significantly lower than last year's mean of 3.8 at the 0.05 level? Assume that all such readings have a normal distribution.

9.130 In a large cherry orchard the average yield has been 4.35 tons per acre for the last several years. A new fertilizer was tested on 15 randomly selected one-acre plots. The yields from these plots follow:

| 3.56 | 5.00 | 4.88 | 4.93 | 3.92 | 4.25 | 5.12 | 5.13 | 4.79 | 4.45 |
| 5.35 | 4.81 | 3.48 | 4.45 | 4.72 |

At the 0.05 level of significance, do we have sufficient evidence to claim that there was a significant increase in production? Assume yield per acre is normally distributed.

9.131 Home schooling became legal in all 50 states in 1993, thus allowing parents to take charge of their kids' education from kindergarten to college. Researchers estimate that as many as 1.5 million children and teenagers in 1998 were being taught primarily by their mothers and fathers, about five times as many as there were ten years earlier. This number is rather remarkable considering that the number of two-income households also rose during the same period. (Who's staying at home teaching the kids?) The average ACT score for a home-schooled student in 1998 was 23, whereas the average for traditionally schooled students was 21 (*Newsweek,* 10-5-98).

Suppose a recent survey of 22 home schoolers in your state revealed a mean ACT score of 23.2 and a standard deviation of 4.1. Do the ACT scores of home schoolers in your state exceed the scores for the traditionally schooled? Use the 0.05 level of significance.

9.132 A manufacturer of television sets claims that the maintenance expenditures for its product will average no more than $50 during the first year following the expiration of the warranty. A consumer group has asked you to substantiate or discredit the claim. The results of a random sample of 50 owners of such television sets showed that the mean expenditure was $61.60 and the standard deviation was $32.46. At the 0.01 level of significance, should you conclude that the producer's claim is true or not likely to be true?

9.133 The *LEXIS,* a national law journal, found from a survey conducted on April 6–7, 1991, that nearly two-thirds of the 800 people surveyed said doctors should not be prosecuted for helping people with terminal illnesses commit suicide. The poll carries a margin of error of plus or minus 3.5%.
   a. Describe how this survey of 800 people fits the properties of a binomial experiment. Specifically identify: n, a trial, success, p, and x.
   b. Exactly what is the "two-thirds" reported? How was it obtained? Is it a parameter or a statistic?
   c. Calculate the 95% confidence maximum error of estimate for the population proportion of all people who believe doctors should not be prosecuted.
   d. How is the maximum error, found in (c), related to the 3.5% mentioned in the survey report?

9.134 The marketing research department of an instant-coffee company conducted a survey of married men to determine the proportion of married men who preferred their brand. Twenty of the 100 in the random sample preferred the company's brand. Use a 95% confidence interval to estimate the proportion of all married men who prefer this company's brand of instant coffee. Interpret your answer.

9.135 A company is drafting an advertising campaign that will involve endorsements by noted athletes. In order for the campaign to succeed, the endorser must be both highly respected and easily recognized. A random sample of 100 prospective customers are shown photos of various athletes. If the customer recognizes an athlete, then the customer is asked whether he or she respects the athlete. In the case of a top woman golfer, 16 of the 100 respondents recognized her picture and indicated that they also respected her. At the 95% level of confidence, what is the true proportion with which this woman golfer is both recognized and respected?

9.136 A local auto dealership advertises that 90% of customers whose autos were serviced by their service department are pleased with the results. As a researcher, you take exception to this statement because you are aware that many people are reluctant to express dissatisfaction even if they are not pleased. A research experiment was set up in which those in the sample had received service by this dealer within the past two weeks. During the interview, the individuals were led to believe that the interviewer was new in town and was considering taking his car to this dealer's service department. Of the 60 sampled, 14 said that they were dissatisfied and would not recommend the department.
   a. Estimate the proportion of dissatisfied customers using a 95% confidence interval.
   b. Given your answer to (a), what can be concluded about the dealer's claim?

9.137 In obtaining the sample size to estimate a proportion, the formula $n = \dfrac{[z(\alpha/2)]^2 pq}{E^2}$ is used.

If a reasonable estimate of p is not available, then it is suggested that $p = 0.5$ be used because this will give the maximum value for n. Calculate the value of $pq = p(1 - p)$ for $p = 0.1, 0.2, 0.3, \ldots, 0.8, 0.9$ in order to obtain some idea about the behavior of the quantity pq.

9.138 The so-called "glass ceiling effect" and numerous other reasons have prevented women from reaching the top of the corporate employment ladder compared to men. *Fortune* reported on 10/12/98 that women constitute 11% of corporate directors within the list of *Fortune* 500 companies, even though women represent a much higher percentage (40%) of the total work force employed in management positions in America. The percentage of female corporate directors and officers, however, has been rising steadily, and power appears to be shifting to people who are not in traditional corporate America.

You wish to conduct a study to estimate the percentage of female corporate directors in the companies with headquarters in your state. Assume the population proportion is 11% as reported by *Fortune*. What sample size must you use if you want your estimate to be within:
a. 0.03 with 90% confidence?
b. 0.06 with 95% confidence?
c. 0.09 with 99% confidence?

9.139 *Prevention* magazine reported in its latest survey that 64% of adult Americans, or 98 million people, were overweight. The telephone survey of 1254 randomly selected adults was conducted November 8–29, 1990, and had a margin of error of three percentage points.
a. Calculate the maximum error of estimate for 0.95 confidence with $p' = 0.64$.
b. How is the margin of error of three percentage points related to answer (a)?
c. How large a sample would be needed to reduce the maximum error to 0.02 with 95% confidence?

9.140 "Two of five Americans believe the country should rely on nuclear power more than other energy sources for energy in the 1990s, according to a poll released yesterday. . . . The telephone poll, taken April 10–11, has a margin of error of plus or minus 3 points." This statement appeared in the Rochester *Democrat & Chronicle* on April 21, 1991. Forty percent, plus or minus three points, sounds like a confidence interval.
a. What is another name for the "margin of error of plus or minus 3 points"?
b. If we assume a 95% level of confidence, how large a sample was needed for a maximum error of 0.03?

9.141 A machine is considered to be operating in an acceptable manner if it produces 0.5% or fewer defective parts. It is not performing in an acceptable manner if more than 0.5% of its production is defective. The hypothesis H_o: $p = 0.005$ is tested against the hypothesis H_a: $p > 0.005$ by taking a random sample of 50 parts produced by the machine. The null hypothesis is rejected if two or more defective parts are found in the sample. Find the probability of the type I error.

9.142 You are interested in comparing the hypothesis $p = 0.8$ against the alternative $p < 0.8$. In 100 trials you observe 73 successes. Calculate the p-value associated with this result.

9.143 An instructor asks each of the 54 members of his class to write down "at random" one of the numbers $1, 2, 3, \ldots, 13, 14, 15$. Since the instructor believes that students like gambling, he considers that 7 and 11 are lucky numbers. He counts the number of students, x, who selected 7 or 11. How large must x be before the hypothesis of randomness can be rejected at the 0.05 level?

9.144 Today's newspapers and magazines often report the findings of survey polls about various aspects of life. *The American Gender Evolution* (November 1990) reports "62% of the men believe both partners should earn a living." Other bits of information given in the article are: "telephone survey of 1201 adults" and "has a sampling error of $\pm 3\%$." Relate this information to the statistical inferences you have been studying in this chapter.

a. Is a percentage of people a population parameter, and if so, how is it related to any of the parameters that we have studied?

b. Based on the information given, find the 95% confidence interval for the true proportion of men who believe both partners should earn a living.

c. Explain how the terms "point estimate," "level of confidence," "maximum error of estimate," and "confidence interval" relate to the values reported in the article and to your answers in (b).

9.145 In order to test the hypothesis that the standard deviation on a standard test is 12, a sample of 40 randomly selected students was tested. The sample variance was found to be 155. Does this sample provide sufficient evidence to show that the standard deviation differs from 12 at the 0.05 level of significance?

9.146 Bright-Lite claims its 60-watt light bulb burns with a length of life that is approximately normally distributed with a standard deviation of 81 hours. A sample of 101 bulbs had a variance of 8075. Is this sufficient evidence to reject Bright-Lite's claim in favor of the alternative, "the standard deviation is larger than 81 hours," at the 0.05 level of significance?

9.147 A production process is considered out of control if the produced parts have a mean length different from 27.5 mm or a standard deviation that is greater than 0.5 mm. A sample of 30 parts yields a sample mean of 27.63 mm and a sample standard deviation of 0.87 mm. If we assume part length is a normally distributed variable, does this sample indicate that the process should be adjusted in order to correct the standard deviation of the product? Use $\alpha = 0.05$.

9.148 Julie Jackson operates a franchised restaurant that specializes in soft ice cream cones and sundaes. Recently she received a letter from corporate headquarters warning her that her shop was in danger of losing its franchise because the average sales per customer had dropped "substantially below the average for the rest of the corporation." The statement may be true, but Julie is convinced that such a statement is completely invalid to justify threatening a closing. The variation in sales at her restaurant is bound to be larger than most, primarily because she serves more children, elderly, and single adults rather than large families who run up big bills at the other restaurants. Therefore, her average ticket is likely to be smaller and exhibit greater variability. To prove her point, Julie obtained the sales records from the whole company and found that the standard deviation was $2.45 per sales ticket. She then conducted a study of the last 71 sales tickets at her store and found a standard deviation of $2.95 per ticket. Is the variability in sales at Julie's franchise, at the 0.05 level of significance, greater than the variability for the company?

VOCABULARY LIST

Be able to define each term. Pay special attention to the key terms, which are printed in **red.** In addition, describe in your own words, and give an example of, each term. Your examples should not be ones given in class or in the textbook.

The bracketed numbers indicate the chapter(s) in which the term appeared previously, but you should define the terms again to show increased understanding of their meaning. Page numbers indicate the first appearance of the term in Chapter 9.

assumptions [8] (p. 421, 436, 454)
binomial experiment [5] (p. 434)
calculated value [8] (p. 425, 441, 455)
chi-square (p. 451)
conclusion [8] (p. 425, 441, 455)
confidence interval [8] (p. 421, 436)

critical region [8] (p. 425, 441, 455)
critical value [8] (p. 418, 451)
decision [8] (p. 425, 441, 455)
degrees of freedom (p. 418, 451)
hypothesis test [8] (p. 424, 440, 454)
inference [8] (p. 416, 434, 451)

CHAPTER PRACTICE TEST

Part I: Knowing the Definitions

Answer "True" if the statement is always true. If the statement is not always true, replace the words shown in bold with words that make the statement always true.

9.1 The Student's t-distributions have an approximately normal distribution but are **more** dispersed than the standard normal distribution.

9.2 The **chi-square** distribution is used for inferences about the mean when the σ is unknown.

9.3 The **Student's** t-distribution is used for all inferences about a population's variance.

9.4 If the test statistic falls in the critical region, the null hypothesis has **been proven true.**

9.5 When the test statistic is t and the number of degrees of freedom gets very large, the critical value of t is very close to that of the **standard normal** z.

9.6 When making inferences about one mean when the value of σ is not known, the **z-score** is the test statistic.

9.7 The chi-square distribution is a skewed distribution whose mean value is **2** for df > 2.

9.8 Often the concern with testing the variance (or standard deviation) is to keep its size under control or relatively small. Therefore, many of the hypothesis tests with chi-square will be **one-tailed.**

9.9 $\sqrt{npq}$ is the standard error of proportion.

9.10 The sampling distribution of p' is approximately distributed as a **Student's** t-distribution.

Part II: Applying the Concepts

Answer all questions, showing all formulas, substitutions, and work.

9.11 Find each of the following:
   a. $z(0.02)$
   b. $t(18, 0.95)$
   c. $\chi^2(25, 0.95)$

9.12 A random sample of 25 data was selected from a normally distributed population for the purpose of estimating the population mean, μ. The sample statistics are $n = 25$, $\bar{x} = 28.6$, $s = 3.50$.
   a. Find the point estimate for μ.
   b. Find the maximum error of estimate for the 0.95 confidence interval estimate.
   c. Find the lower confidence limit (LCL) and the upper confidence limit (UCL) for the 0.95 confidence interval estimate for μ.

9.13 Thousands of area elementary school students were recently given a nationwide standardized exam testing their composition skills. If 64 of a random sample of 100 students passed this exam, construct the 0.98 confidence interval estimate for the true proportion of all area students who passed the exam.

9.14 State the null (H_o) and the alternative (H_a) hypotheses that would be used to test each of the following claims.

 a. The mean weight of professional basketball players is no more than 225 lb.

 b. Approximately 40% of MCC's daytime students own their own car.

 c. The standard deviation for the monthly amounts of rainfall in Monroe County is less than 3.7 in.

9.15 Determine the level of significance, test statistic, critical region, and critical values(s) that would be used in completing each hypothesis test using the classical approach with an $\alpha = 0.05$.

 a. H_o: $\mu = 43$ vs. H_a: $\mu < 43$,   (given $\sigma = 6$)

 b. H_o: $\mu = 95$ vs. H_a: $\mu \neq 95$,   (σ unknown, $n = 22$)

 c. H_o: $p = 0.80$ vs. H_a: $p > 0.80$

 d. H_o: $\sigma = 12$ vs. H_a: $\sigma \neq 12$ ($n = 28$)

9.16 The manufacturer of a new model car, called Orion, claims the typical Orion will average 26 mpg of gasoline. An independent consumer group is somewhat skeptical of this claim and thinks the mean gas mileage is less than the 26 claimed. A sample of 24 randomly selected Orions produced the following results:

Sample statistics: mean = 24.15, st. dev. = 4.87

At the 0.05 level of significance, does the consumer group have sufficient evidence to refute the manufacturer's claim?

9.17 A coffee machine is supposed to dispense 6 fluid ounces of coffee into a paper cup. In reality, the amount dispensed varies from cup to cup. However, if the machine is operating properly, the standard deviation of the amounts dispensed should be 0.1 or less of an ounce. A random sample of 15 cups produced a standard deviation of 0.13 oz. Does this represent sufficient evidence, at the 0.10 level of significance, to conclude that "the machine is not operating properly"?

9.18 An unhappy customer is disturbed with the waiting time at the post office when buying stamps. Upon registering his complaint, he was told, "You wait more than one minute for service no more than half of the time when only buying stamps." Not believing this to be the case, the customer collected some data from people who had just purchased only stamps.

Sample statistics: $n = 60$, $x = n$ (wait more than one minute) $= 35$

At the 0.02 level of significance, does our unhappy customer have sufficient evidence to refute the post office's claim?

Part III: Understanding the Concepts

9.19 Student B says the range of a set of data may be used to obtain a crude estimate for the standard deviation of a population. Student A is not sure. How will student B correctly explain how and under what circumstances his statement is true?

9.20 Is it (a) the null hypothesis or (b) the alternative hypothesis that the researcher usually believes to be true? Explain.

9.21 When you reject a null hypothesis, student A says that you are expressing disbelief in the value of the parameter as claimed in the null hypothesis. Whereas student B says that you are expressing the belief that the sample statistic came from a population other than the one related to the parameter claimed in the null hypothesis. Who is correct? Explain.

9.22 "The Student t-distribution must be used when making inferences about the population mean, μ, when the population standard deviation, σ, is not known" is a true statement. Student A states that the z-score plays a role sometimes when using the t-distribution. Explain the conditions that exist and the role played by z that make student A's statement correct.

9.23 Student A says that the percentage of the sample means that fall outside the critical values of the sampling distribution determined by a true null hypothesis is the p-value for the test. Student B says that the percentage student A is describing is the level of significance. Who is correct? Explain.

9.24 Student A carries out a study in which he is willing to run a 1% risk of making a type I error. He rejects the null hypothesis and claims that his statistic is significant at the 99% level of confidence. Student B argues that student A's claim is not properly worded. Who is correct? Explain.

9.25 Student A claims that when you employ a 95% confidence interval to determine an estimation, you do not know for sure whether your inference is correct (the parameter is contained within the interval) or not. Student B claims that you do; you have shown that the parameter cannot be less than the lower limit or greater than the upper limit of the interval. Who is right? Explain.

9.26 Student A says that the best way to improve a confidence interval estimate is to increase the level of confidence. Student B argues that using a high confidence level does not really improve the desirability of the resulting interval estimate. Who is right? Explain.

Chapter 10

Inferences Involving Two Populations

CHAPTER CASE STUDY

Who Knows the American Flag?

You probably assume that everybody knows the answer to simple historical questions like: How many stars are there on the American flag? How many stripes are there on the American flag? What do the stars represent? The stripes represent? Interestingly enough not everyone knows the answers to these questions and it will most likely be a shock when you find out how large the percentage of adults is that do not know. The USA Snapshot® shown below appeared in *USA Today*, 1998, and lists the percentage of adults by gender and by region of the United States who knew the number of stars.

USA SNAPSHOTS®

A look at statistics that shape the nation

Stars on Old Glory

Asked how many stars there are on the flag, more men (76%) correctly said 50 than women (65%). Adults who knew by region:

Region	Percent
South	71%
West	70%
Northeast	68%
Midwest	64%

Source: Maritz AmeriPoll

By Cindy Hall and Gary Visgaitis, USA TODAY

Two hundred adults in Erie County, New York were asked how many stars there are on the American flag. The table below shows the numbers of adults belonging to each category. The sample results were tallied twice, once by gender and again by residence of the adult answering the question.

	Men	Women	City	Urban	Rural
n(Knew)	72	72	57	58	31
n(Didn't know)	22	34	25	14	15

Does this information seem to be consistent with the results found in the Maritz AmeriPoll? Is there a significant difference between the percentage of men and the percentage of women who answered the question correctly? Is there a difference between the percentage of city and the percentage of urban adults who answered the question correctly? Statistical methods for comparing two populations will be studied in this chapter, and we will return to this case study at the end of the chapter. (See Exercise 10.107, p. 525)

CHAPTER OBJECTIVES

In Chapters 8 and 9 you were introduced to the basic concepts of estimation and hypothesis testing in connection with inferences about one population and the parameters mean, proportion and variance (or standard deviation). In this chapter we continue to investigate the inferences about these same three parameters, but we will now use these parameters as a basis for comparing two populations.

When sampling two different populations, two types of samples are used: independent samples and dependent samples. After learning about the two types of samples in Section 10.1, we will learn the techniques for using the mean of the paired differences, the difference between two means, the difference between two proportions, and the ratio of two variances for comparing two populations. Since we will be studying several different situations, Figure 10.1 is offered as a "road map" to help you organize these various inference techniques.

Figure 10.1

"Road Map" to Two Population Inferences

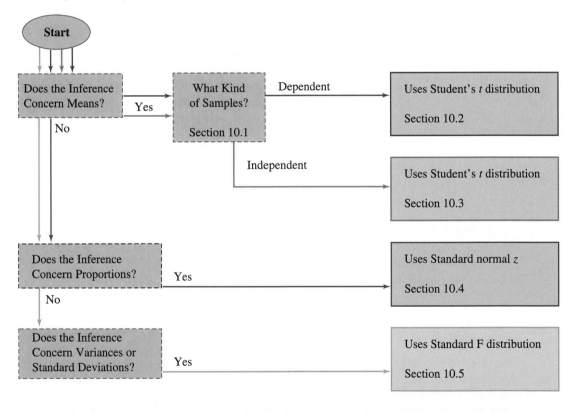

10.1 Independent and Dependent Samples

In this chapter we are going to study the procedures for making inferences about two populations. When comparing two populations, we need two samples, one from each population. Two basic kinds of samples can be used: independent and dependent. The dependence or independence of two samples is determined by the sources used for the data. A **source** can be a person, an object, or anything that yields a piece of data. If the same set of sources or related sets are used to obtain the data representing both populations, we have **dependent samples.** If two unrelated sets of sources are used, one set from each population, we have **independent samples.** The following illustrations should amplify these ideas.

ILLUSTRATION 10.1 ▼

A test will be conducted to see whether the participants in a physical fitness class actually improve in their level of fitness. It is anticipated that approximately 500 people will sign up for this course. The instructor decides that she will give 50 of the participants a set of tests before the course begins (a pretest) and then will give another set of tests to 50 participants at the end of the course (a posttest). The following two sampling procedures are proposed:

> Plan A: Randomly select 50 participants from the list of those enrolled and give them the pretest. At the end of the course, make another or second random selection of size 50 and give them the posttest.

> Plan B: Randomly select 50 participants and give them the pretest; give the same set of 50 the posttest upon completion of the course.

Plan A illustrates independent sampling; the sources (the class participants) used for each sample (pretest and posttest) were selected separately. Plan B illustrates dependent sampling; the sources used for both samples (pretest and posttest) are the same.   ▲

> **Exercise 10.1**
> Describe how you could select two independent samples from among your classmates to compare the heights of female and male students.

Typically, when both a pretest and a posttest are used, the same subjects are used in the study. Thus, pretest versus posttest (before versus after) studies are usually dependent samples.

ILLUSTRATION 10.2 ▼

A test is being designed to compare the wearing quality of two brands of automobile tires. The automobiles will be selected and equipped with the new tires and then driven under "normal" conditions for one month. Then a measurement will be taken to determine how much wear took place. Two plans are proposed:

> Plan C: A sample of cars will be selected randomly and equipped with brand A and driven for the month, and another sample of cars will be selected and equipped with brand B and driven for the month.

> Plan D: A sample of cars will be selected randomly, equipped with one tire of brand A and one tire of brand B (the other two tires are not part of the test), and driven for the month.

In this illustration we might suspect that many other factors must be taken into account when testing automobile tires—such as age, weight, and mechanical condition of the car; driving habits of drivers; location of the tire on the car; and where and how much the car is driven. However, at this time we are trying only to illustrate dependent and independent samples. Plan C is independent (unrelated sources), and plan D is dependent (common sources).   ▲

> **Exercise 10.2**
> Describe how you could select two dependent samples from among your classmates to compare their heights as they entered high school to when they entered college.

Independent and dependent samples each have their advantages; these will be emphasized later. Both methods of sampling are often used.

Case Study 10.1

Exploring the Traits of Twins

Studies that involve identical twins are a natural for the dependent sampling technique discussed in this section.

A NEW STUDY SHOWS THAT KEY CHARACTERISTICS MAY BE INHERITED

Like many identical twins reared apart, Jim Lewis and Jim Springer found they had been leading eerily similar lives. Separated four weeks after birth in 1940, the Jim twins grew up 45 miles apart in Ohio and were reunited in 1979. Eventually they discovered that both drove the same model blue Chevrolet, chain-smoked Salems, chewed their fingernails, and owned dogs named Toy. Each had spent a good deal of time vacationing at the same three-block strip of beach in Florida. More important, when tested for such personality traits as flexibility, self-control, and sociability, the twins responded almost exactly alike.

> **Exercise 10.3**
> Explain why studies involving identical twins result in dependent samples of data.

The project is considered the most comprehensive of its kind. The Minnesota researchers report the results of six-day tests of their subjects, including 44 pairs of identical twins who were brought up apart. Well-being, alienation, aggression, and the shunning of risk or danger were found to owe as much or more to nature as to nurture. Of eleven key traits or clusters of traits analyzed in the study, researchers estimated that a high of 61 percent of what they call "society potency" (a tendency toward leadership or dominance) is inherited, while "social closeness" (the need for intimacy, comfort and help) was lowest, at 33 percent.

Source: Copyright 1987 by Time Inc. All rights reserved. Reprinted by permission of *TIME.*

EXERCISES • • • • • • • • • • • •

10.4 In trying to estimate the amount of growth that took place in the trees planted by the County Parks Commission recently, 36 trees were randomly selected from the 4000 planted. The heights of these trees were measured and recorded. One year later another set of 42 trees was randomly selected and measured. Do the two sets of data (36 heights, 42 heights) represent dependent or independent samples? Explain.

10.5 Twenty people were selected to participate in a psychology experiment. They answered a short multiple-choice quiz about their attitudes on a particular subject and then viewed a 45-minute film. The following day the same 20 people were asked to answer a follow-up questionnaire about their attitudes. At the completion of the experiment, the experimenter will have two sets of scores. Do these two samples represent dependent or independent samples? Explain.

10.6 An experiment is designed to study the effect diet has on the uric acid level. Twenty white rats are used for the study. Ten rats are randomly selected and given a junk-food diet. The other ten receive a high-fiber, low-fat diet. Uric acid levels of the two groups are determined. Do the resulting sets of data represent dependent or independent samples? Explain.

10.7 Two different types of disc centrifuges are used to measure the particle size in latex paint. A gallon of paint is randomly selected, and ten specimens are taken from it for testing on each of the centrifuges. There will be two sets of data, ten data each, as a result of the testing. Do the two sets of data represent dependent or independent samples? Explain.

10.8 An insurance company is concerned that garage A charges more for repair work than garage B charges. It plans to send 25 cars to each garage and obtain separate estimates for the repairs needed for each car.
  a. How can the company do this and obtain independent samples? Explain in detail.
  b. How can the company do this and obtain dependent samples? Explain in detail.

10.9 A study is being designed to determine the reasons why adults choose to follow a healthy diet plan. One thousand men and 1000 women will be surveyed. Upon completion, the

(continued)

reasons that men choose a healthy diet will be compared to the reasons that women choose a healthy diet.

    a. How can the data be collected if independent samples are to be obtained? Explain in detail.

    b. How can the data be collected if dependent samples are to be obtained? Explain in detail.

 10.10 Suppose that 400 students in a certain college are taking elementary statistics this semester. Two samples of size 25 are needed in order to test some precourse skill against the same skill after the students complete the course.

    a. Describe how you would obtain your samples if you were to use dependent samples.

    b. Describe how you would obtain your samples if you were to use independent samples.

10.2 Inferences Concerning the Mean Difference Using Two Dependent Samples

The procedures for comparing two population means are based on the relationship between two sets of sample data, one sample from each population. When dependent samples are involved, the data are thought of as "paired data." The data may be paired as a result of the data being obtained from certain "before" and "after" studies; from pairs of identical twins as in Case Study 10.1; from a "common" source, as with the amounts of tire wear for each brand from a car in plan D of Illustration 10.2; or from matching two subjects of similar traits to form "matched pairs." The pairs of data values are compared directly to one another by using the difference in their numerical values. The resulting difference is called a **paired difference.**

$$\text{Paired difference: } d = x_1 - x_2 \qquad \textbf{(10.1)}$$

    The concept of using paired data this way has a built-in ability to remove the effect of otherwise uncontrolled factors. The tire-wear problem (Illustration 10.2, p. 476) is an excellent example of such additional factors. The wearing ability of a tire is greatly affected by a multitude of factors: the size, weight, age, and condition of the car, the driving habits of the driver, the number of miles driven, the condition and types of roads driven on, the quality of the material used to make the tire, and so on. We have created paired data by placing one tire from each brand on a car. Since one tire of each brand will be tested under the same conditions, same car, same driver, and so on, the extraneous causes of wear are neutralized.

    A test was conducted to compare the wearing quality of the tires produced by two tire companies using plan D, as described in Illustration 10.2. All the aforementioned factors will have an equal effect on both brands of tires, car by car. Table 10.1 gives the amount of wear (in thousandths of an inch) that resulted from the test. One tire of each brand was placed on each of six test cars. The position (left or right side, front or back) was determined with the aid of a random numbers table.

Table 10.1   Amount of Tire Wear

Car	1	2	3	4	5	6
Brand A	125	64	94	38	90	106
Brand B	133	65	103	37	102	115

Since the various cars, drivers, and conditions are the same for each tire of a paired set of data, it would make sense to use a third variable, the paired difference *d*.

Exercise 10.11
Find the paired differences, $d = A - B$, for this set of data:

Pairs	1	2	3	4	5
Sample A	3	6	1	4	7
Sample B	2	5	1	2	8

Our two dependent samples of data will be combined into one set of d values, where $d = B - A$.

Car	1	2	3	4	5	6
$d = B - A$	8	1	9	−1	12	9

The sample statistics that are needed will be the mean of the sample differences, $\bar{d}$

$$\bar{d} = \frac{\sum d}{n} \tag{10.2}$$

and the standard deviation of the sample differences, s_d

Formulas (10.2) and (10.3) are adaptations of formulas (2.1) and (2.10).

$$s_d = \sqrt{\frac{\sum d^2 - \left[\dfrac{(\sum d)^2}{n}\right]}{n - 1}} \tag{10.3}$$

ILLUSTRATION 10.3 ▼

Find the mean and standard deviation of the paired differences in Table 10.1.

Exercise 10.12
Find the mean $\bar{d}$ and the standard deviation s_d of the paired differences in Exercise 10.11.

Solution

The summary of data: $n = 6$, $\sum d = 38$, and $\sum d^2 = 372$.
  The mean:

n = number of dependent pairs

$$\bar{d} = \frac{\sum d}{n}: \quad \bar{d} = \frac{38}{6} = 6.333 = \mathbf{6.3}$$

The standard deviation:

$$s_d = \sqrt{\frac{\sum d^2 - \left[\dfrac{(\sum d)^2}{n}\right]}{n - 1}}: \quad s_d = \sqrt{\frac{372 - \left[\dfrac{(38)^2}{6}\right]}{6 - 1}} = \sqrt{26.27} = 5.13 = \mathbf{5.1}$$

The difference between the two population means, when dependent samples are used (often called **"dependent means"**), is equivalent to the **mean of the paired differences.** Therefore, when an inference is to be made about the difference of two means and paired differences are being used, the inference will in fact be about the mean of the paired differences. The sample mean of the paired differences will be used as the point estimate for these inferences.

  In order to make inferences about the mean of all possible paired differences μ_d, we need to know about the sampling distribution of $\bar{d}$.

When paired observations are randomly selected from normal populations, the paired difference, $d = x_1 - x_2$, will be normally distributed about a mean μ_d with a standard deviation of σ_d.

This is another situation in which the t-test for one mean is applied; namely, we wish to make inferences about an unknown mean (μ_d) where the random variable (d) involved has an approximately normal distribution with an unknown standard deviation (σ_d).

Inferences about the mean of all possible paired differences μ_d are based on samples of n dependent pairs of data and the **t-distribution** with $n - 1$ degrees of freedom, under the following assumption.

THE ASSUMPTION FOR INFERENCES ABOUT THE MEAN OF PAIRED DIFFERENCES μ_d

The paired data are randomly selected from normally distributed populations.

Confidence Interval Procedure

Formula (10.4) is an adaptation of formula (9.1).

The $1 - \alpha$ **confidence interval for estimating the mean difference** μ_d is found using this formula:

$$\bar{d} - t_{(df, \alpha/2)} \cdot \frac{s_d}{\sqrt{n}} \quad \text{to} \quad \bar{d} + t_{(df, \alpha/2)} \cdot \frac{s_d}{\sqrt{n}}, \qquad \text{where df} = n - 1 \qquad \textbf{(10.4)}$$

ILLUSTRATION 10.4 ▼

Construct the 95% confidence interval for the mean difference in the paired data on tire wear, as reported in Table 10.1. The sample information is $n = 6$ pieces of paired data, $\bar{d} = 6.3$, and $s_d = 5.1$ (calculated in Illustration 10.3). Assume the amount of wear is approximately normally distributed for both brands of tires.

Solution

STEP 1   The Set-Up:
   Describe the population parameter of interest.
   μ_d, **the mean difference** in the amount of wear that occurred between the two brands of tires.

STEP 2   The Confidence Interval Criteria:
   a. Check the assumptions.
      Both sampled populations are approximately normal.
   b. Identify the probability distribution and the formula to be used.
      The t-distribution with df = $6 - 1 = 5$ and formula (10.4) will be used.
   c. State the level of confidence, $1 - \alpha$.
      $1 - \alpha = \textbf{0.95}$.

STEP 3   The Sample Evidence:
   Collect the sample information.
   $n = 6, \bar{d} = 6.3$, and $s_d = 5.1$

STEP 4   The Confidence Interval:
   a. Determine the confidence coefficient.

Exercise 10.13

Find $t_{(15, 0.025)}$. Describe the role this number plays in the confidence interval.

   This is a two-tailed situation with $\dfrac{\alpha}{2} = 0.025$ in one tail; from Table 6 in Appendix B, $t_{(df, \alpha/2)} = t_{(5, 0.025)} = \textbf{2.57.}$

   For specific instructions about confidence coefficients and Table 6, see page 422.
   b. Find the maximum error of estimate.
      Using the maximum error part of formula (10.4),

$$E = t_{(df, \alpha/2)} \cdot \frac{s_d}{\sqrt{n}}: \quad E = 2.57 \cdot \left(\frac{5.1}{\sqrt{6}}\right) = (2.57)(2.082) = 5.351 = \textbf{5.4}$$

Exercise 10.14

a. Find the 95% confidence interval for μ_d given: $n = 26$, $d = 6.3$, and $s_d = 5.1$.

b. Compare your interval to the interval found in Illustration 10.4.

c. **Find the lower and upper confidence limits.**

$$\bar{d} \pm E$$

$$6.3 \pm 5.4$$

$$6.3 - 5.4 = \mathbf{0.9} \quad \text{to} \quad 6.3 + 5.4 = \mathbf{11.7}$$

STEP 5   **The Results:**

State the confidence interval.

0.9 to 11.7, the 95% confidence interval for μ_d.

That is, with 95% confidence we can say that the mean difference in the amount of wear is between 0.9 and 11.7 thousandths of an inch.   ▲

NOTE This confidence interval is quite wide, due, in part, to the small sample size. Recall from the Central Limit Theorem that as the sample size increases, the standard error (estimated by $\frac{s_d}{\sqrt{n}}$) decreases.

The following commands will construct a $1 - \alpha$ confidence interval for mean μ_d with unknown standard deviation for two dependent sets of sample data.

MINITAB

```
Input the paired data into C1 and C2, then continue with:

  Choose:   Calc > Calculator
  Enter:    Store result in variable: C3
            Expression: C1 - C2*  > OK
  Choose:   Stat > Basic Statistics > 1-Sample t
  Enter:    C3
  Select:   Confidence interval
  Enter:    Level: 1 - α (ex. 95.0)
*Enter the expression in the order that is needed: C1 - C2 or C2 - C1.
```

EXCEL

```
Input the paired data into columns A and B, activate C1 or
C2 (depending on whether column headings are used or not),
then continue with:

  Choose:   Edit Formula (=)
  Enter:    A2 - B2* (if column headings are used)
  Drag:     Bottom right corner of C2 down to give other
            differences
  Choose:   Tools > Data Analysis Plus
            > Inference About A Mean(SIGMA Unknown) > OK
  Enter:    block coordinates: (C2:C20 or select cells)    > OK
  Choose:   Interval Estimate
  Enter:    level of confidence: 1 - α (ex. 0.95 or 95.0)
*Enter the expression in the order that is needed: A2 - B2 or B2 - A2.
```

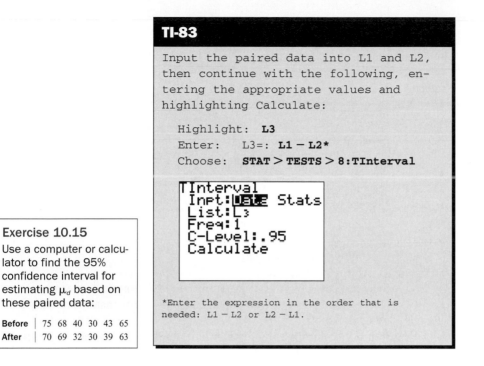

Exercise 10.15

Use a computer or calculator to find the 95% confidence interval for estimating μ_d based on these paired data:

Before	75	68	40	30	43	65
After	70	69	32	30	39	63

The solution to Illustration 10.4 looks like this when solved on MINITAB:

> MINITAB is shown; however EXCEL and TI outputs are comparable.

```
Confidence Intervals

    Variable    N    Mean    StDev    SE Mean    95.0% C.I.
        C3      6    6.33    5.13      2.09      (0.95, 11.71)
```

Hypothesis Testing Procedure

When testing a null **hypothesis about the mean difference,** the test statistic used will be the difference between the sample mean $\bar{d}$ and the hypothesized value of μ_d, divided by the estimated **standard error.** This statistic is assumed to have a t-distribution when the null hypothesis is true and the assumptions for the test are satisfied. The value of the **test statistic $t \star$** is calculated using formula (10.5):

> Formula (10.5) is an adaptation of formula (9.2).

$$t \star = \frac{\bar{d} - \mu_d}{s_d/\sqrt{n}}, \text{ where df} = n - 1 \tag{10.5}$$

NOTE A hypothesized mean difference, μ_d, can be any specified value. The most common value specified is zero; however, the difference can be nonzero.

ILLUSTRATION 10.5 ▼

In a study dealing with high blood pressure and the drugs used to aid in controlling it, the effect of calcium channel blockers on pulse rate was one of many specific concerns. Twenty-six patients were randomly selected from a large pool of potential subjects, and their pulse rates were established. A calcium channel blocker was administered to each patient for a fixed period of time, and then each patient's pulse rate was again determined. The two resulting sets of data appeared to have an approximately normal distribution, and resulting statistics were $\overline{d} = 1.07$ and $s_d = 1.74$ ($d = $ before $-$ after). Does the resulting sample information provide sufficient evidence to show that this calcium channel blocker did lower the pulse rate? Use $\alpha = 0.05$.

Solution

> "Lower rate" means "after" is less than "before" and "before − after" is positive.

STEP 1 **The Set-Up:**
    a. Describe the population parameter of interest.
        μ_d, **the mean difference** (reduction) in pulse rate from before to after having used the calcium channel blocker for the time period of the test.
    b. State the null hypothesis (H_o) and the alternative hypothesis (H_a).

$$H_o: \mu_d = 0 \ (\leq) \ \text{(did not lower rate)} \quad \text{Remember: } d = \text{before} - \text{after.}$$

$$H_a: \mu_d > 0 \ \text{(did lower rate)}$$

STEP 2 **The Hypothesis Test Criteria:**
    a. Check the assumptions.
        Since the data in **both sets are approximately normal,** it seems reasonable to assume that the two populations are approximately normally distributed.
    b. Identify the probability distribution and the test statistic to be used.
        The **t-distribution** with df $= n - 1 = 25$ and **test statistic is $t \bigstar$,** formula (10.5).
    c. Determine the level of significance, α.
        $\alpha = 0.05$.

STEP 3 **The Sample Evidence:**
    a. Collect the sample information.

$$n = 26, \overline{d} = 1.07, \text{ and } s_d = 1.74.$$

    b. Calculate the value of the test statistic.

$$t \bigstar = \frac{\overline{d} - \mu_d}{s_d/\sqrt{n}}: \quad t \bigstar = \frac{1.07 - 0.0}{1.74/\sqrt{26}} = \frac{1.07}{0.34} = \mathbf{3.14}$$

STEP 4 **The Probability Distribution:**

p-VALUE:

a. Calculate the p-value.
Use the right-hand tail since the H_a expresses concern for values related to "greater than."
$\mathbf{P} = P(t > 3.14, \text{ with df} = 25)$ as shown in the figure.

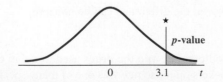

To find the p-value, use one of the three methods:
1. Use Table 6 (Appendix B) to place bounds on the p-value:
    P < 0.005

OR   **CLASSICAL:**

a. Determine the critical region and critical value(s).
The critical region is the right-hand tail since the H_a expresses concern for values related to "greater than." The critical value is obtained from Table 6:
$t_{(25, 0.05)} = \mathbf{1.71.}$

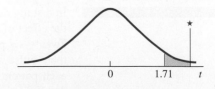

p-VALUE:

2. Use Table 7 (Appendix B) to read the value directly from the table: **P = 0.002**
3. Use a computer or calculator to calculate the *p*-value:
    P = 0.0022
    Specific instructions are on page 425, 426.

b. Determine whether or not the *p*-value is smaller than α.
The *p*-value **is smaller** than the level of significance, α.

CLASSICAL:

OR   Specific instructions are on page 419.

b. Determine whether or not the cal-culated test statistic is in the criti-cal region.
t ★ **is in** the critical region, as shown in red in the figure above.

Exercise 10.16

Complete the hypothesis test with alternative hy-pothesis $\mu_d > 0$ based on the paired data listed below and $d = B - A$. Use $\alpha = 0.05$. Assume normality.

A 700 830 860 1080 930
B 720 820 890 1100 960

a. Use the *p*-value approach.
b. Use the classical approach.

STEP 5   The Results:

a. **State the decision about H_o.**
    Reject H_o.

b. **State the conclusion about H_a.**
    At the 0.05 level of significance, we can conclude that the calcium chan-nel blocker does lower the pulse rate.   ▲

"Statistical significance" does not always have the same meaning when the "prac-tical" application of the results is considered. In the preceding detailed hypothesis test, the results showed a statistical significance with a *p*-value of 0.002, that is, 2 chances in 1000. However, a more practical question might be, "Is lowering the pulse rate by this small average amount, estimated to be 1.07 beats per minute, worth the risks involving possible side effects of this medication?" Actually the whole issue is far more reaching than just this one issue of pulse rate.

The following commands will complete a hypothesis test for the mean μ_d with un-known standard deviation for two dependent sets of sample data.

```
MINITAB

Input the paired data into C1 and C2,
then continue with:

    Choose:  Calc > Calculator
    Enter:   Store result in variable:
             C3
             Expression: C1 − C2*    > OK
    Choose:  Stat > Basic Statistics >
             1-Sample t
    Enter:   Variables: C3
    Select:  Test mean
    Enter:   μ_d
    Choose:  Alternative: less than or
             not equal or greater than
    *Enter the expression in the order that is
    needed: C1 − C2 or C2 − C1.
```

TI-83

Input the paired data into L1 and L2, then continue with the following, entering the appropriate values, and highlighting Calculate:

Highlight: **L3**
Enter: L3=: **L1 − L2***
Choose: **STAT > TESTS > 2:T-Test. . .**

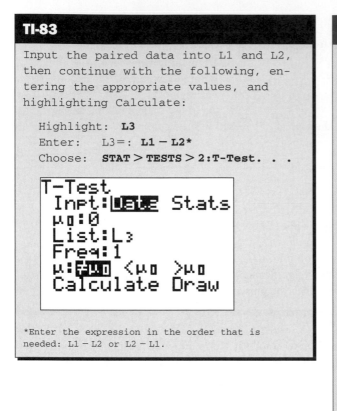

*Enter the expression in the order that is needed: L1 − L2 or L2 − L1.

EXCEL

Input the paired data into columns A and B, then continue with:

Choose: **Tools > Data Analysis > t-Test: Paired Two Sample for Means > OK**
Enter: Variable 1 Range: **(A1:A20 or select cells)**
          Variable 2 Range: **(B1:B20 or select cells)**
             (subtracts: Var1 − Var2)
          Hypothesized Mean Difference: μ_d (usually 0)
Select: **Labels** (if necessary)
Enter: α (ex. 0.05)
Select: **Output Range**
Enter: **(C1 or select cell)**

Use Format > Column > AutoFit Selection to make the output more readable. The output shows p-values and critical values for one- and two-tailed tests.

The hypothesis test may also be done by first subtracting the two columns and then using the inference about a mean (sigma unknown) commands on page 428 on the differences.

Exercise 10.17

Use a computer or calculator to complete the hypothesis test with alternative hypothesis $\mu_d < 0$ based on the paired data listed below and $d = M - N$. Use $\alpha = 0.02$. Assume normality.

M	58	78	45	38	49	62
N	62	86	42	39	47	68

The solution to Illustration 10.5 looks like this when solved on MINITAB:

```
T-Test of the Mean
Test of mu = 0.00 vs mu > 0.00

      Variable    N    Mean    StDev    SE Mean    T      P-Value
         C3      26    1.07    1.74      0.34     3.14    0.002
```

ILLUSTRATION 10.6 ▼

Suppose the sample data in Table 10.1 (p. 477) had been collected with the hope of showing that "the two brands do not wear equally." Do the data provide sufficient evidence for us to conclude that the two brands show unequal wear, at the 0.05 level of significance? Assume the amount of wear is approximately normally distributed for both brands of tires.

Solution

STEP 1 **The Set-Up:**
  a. Describe the population parameter of interest.
     μ_d, **the mean difference** in the amount of wear between the two brands.

b. State the null hypothesis (H_o) and the alternative hypothesis (H_a).

H_o: $\mu_d = 0$ (no difference) Remember: $d = B - A$.

H_a: $\mu_d \neq 0$ (difference)

STEP 2   The Hypothesis Test Criteria:
   a. Check the assumptions.
   The assumption of normality is included in the statement of the problem.
   b. Identify the probability distribution and the test statistic to be used.

   The *t*-distribution with df $= n - 1 = 6 - 1 = 5$, and $t \star = \dfrac{\bar{d} - \mu_d}{s_d/\sqrt{n}}$.

   c. Determine the level of significance, α.
   $\alpha = 0.05$.

STEP 3   The Sample Evidence:
   a. Collect the sample information.
   $n = 6$, $\bar{d} = 6.3$, and $s_d = 5.1$.
   b. Calculate the value of the test statistic.

   $$t \star = \frac{\bar{d} - \mu_d}{s_d/\sqrt{n}}: \quad t \star = \frac{6.3 - 0.0}{5.1/\sqrt{6}} = \frac{6.3}{2.08} = 3.03$$

STEP 4   The Probability Distribution:

Exercise 10.18

Complete the hypothesis test with alternative hypothesis $\mu_d \neq 0$ based on the paired data listed below and $d = O - Y$. Use $\alpha = 0.01$. Assume normality.

| **Oldest** | 199 | 162 | 174 | 159 | 173 |
| **Youngest** | 194 | 162 | 167 | 156 | 176 |

a. Use the *p*-value approach.
b. Use the classical approach.

---

p-**VALUE:**

a. Calculate the *p*-value.
Use both tails since the H_a expresses concern for values related to "different from."
$\mathbf{P} = p\text{-value} = P(t < -3.03) + P(t > 3.03)$
$= 2 \times \mathrm{P}(|t| > 3.03)$ as shown in the figure.

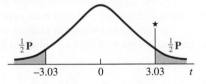

To find the *p*-value, you have three options:
1. Use Table 6 (Appendix B): $\mathbf{0.02 < P < 0.005}$
2. Use Table 7 (Appendix B) to place bounds on the *p*-value:
   $\mathbf{0.026 < P < 0.030}$
3. Use a computer or calculator to calculate the *p*-value:
   $\mathbf{P} = 2 \times 0.0145 = \mathbf{0.0290}$

Specific instructions, see page 425, 426.

b. Determine whether or not the *p*-value is smaller than α.
The *p*-value **is smaller** than α.

OR

CLASSICAL:

a. Determine the critical region and critical value(s).
The critical region is two-tailed since the H_a expresses concern for values related to "different from." The critical value is obtained from Table 6: $t_{(5, 0.025)} = \mathbf{2.57}$.

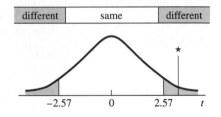

For specific instructions, see page 419.

b. Determine whether or not the calculated test statistic is in the critical region.
$t \star$ is in the critical region, as shown in red in the figure above.

STEP 5   The Results:
   a. State the decision about H_o.
   Reject H_o.
   b. State the conclusion about H_a.
   There is a significant difference in the mean amount of wear at the 0.05 level of significance.   ▲

EXERCISES • • • • • • • • • • • • • • •

10.19 Salt-free diets are often prescribed to people with high blood pressure. The following data were obtained from an experiment designed to estimate the reduction in diastolic blood pressure as a result of following a salt-free diet for two weeks. Assume diastolic readings to be normally distributed.

Before	93	106	87	92	102	95	88	110
After	92	102	89	92	101	96	88	105

a. What is the point estimate for the mean reduction in the diastolic reading after two weeks on this diet?

b. Find the 98% confidence interval for the mean reduction.

10.20 All students who enroll in a certain memory course are given a pretest before the course begins. A random sample of ten students who finished the course were given a posttest, and their scores are listed below.

Student	1	2	3	4	5	6	7	8	9	10
Before	93	86	72	54	92	65	80	81	62	73
After	98	92	80	62	91	78	89	78	71	80

MINITAB was used to find the 95% confidence interval for the mean improvement in memory resulting from taking the memory course, as measured by the difference in test scores (d = after − before). Verify the results shown on the output by calculating the values yourself.

```
Confidence Intervals

    Variable    N    Mean    StDev    SE Mean    95% C.I.
          C3   10    6.10     4.79       1.52    (2.67, 9.53)
```

10.21 An experiment was designed to estimate the mean difference in weight gain for pigs fed ration A as compared to those fed ration B. Eight pairs of pigs were used. The pigs within each pair were littermates. The rations were assigned at random to the two animals within each pair. The gains (in pounds) after 45 days are shown in the following table.

Litter	1	2	3	4	5	6	7	8
Ration A	65	37	40	47	49	65	53	59
Ration B	58	39	31	45	47	55	59	51

Assuming weight gain is normal, find the 95% confidence interval estimate for the mean of the differences μ_d, where d = ration A − ration B.

10.22 A sociologist is studying the effects of a certain motion picture film on the attitudes of black men toward white men. Twelve black men were randomly selected and asked to fill out a questionnaire before and after viewing the film. The scores received by the 12 men are shown in the following table. Assuming the questionnaire scores are normal, construct a 95% confidence interval for the mean shift in score that takes place when this film is viewed.

Before	10	13	18	12	9	8	14	12	17	20	7	11
After	5	9	13	17	4	5	11	14	13	18	7	12

10.23 Two men, A and B, who usually commute to work together, decide to conduct an experiment to see whether one route is faster than the other. The men feel that their driving

(continued)

habits are approximately the same, and therefore they decide on the following procedure. Each morning for two weeks A will drive to work on one route and B will use the
other route. On the first morning, A will toss a coin. If heads appear, he will use route I;
if tails appear, he will use route II. On the second morning, B will toss the coin: heads,
route I; tails, route II. The times, recorded to the nearest minute, are shown in the following table. Assume commute times are normal and estimate the mean of the differences with a 95% confidence interval.

					Day					
Route	M	Tu	W	Th	F	M	Tu	W	Th	F
I	29	26	25	25	25	24	26	26	30	31
II	25	26	25	25	24	23	27	25	29	30

10.24 When the Dow Jones industrial average dropped 512 points on August 31, 1998, it was the
largest point drop since the 554-point drop that hit the stock market on October 27, 1987.
Prior to that drop, some experts had predicted a 9500 Dow before the end of the year, but
after the August slide, opinions changed. The consensus by many analysts seemed to be
that the market would not bounce back in time to even approach the level predicted by earlier estimates. The bull was tired. (*Fortune,* "Requiem for the Bull," 9-28-98)

A random sample of 18 closing stock prices from the New York Stock Exchange was
taken on August 25, 1998 (about one week prior to the "crash"), and again using the
same stocks on September 15, 1998. Both dates are on a Tuesday. Results are shown in
the table below:

Stock	August 25	September 15	Stock	August 25	September 15
1	$10^1/_2$	11	10	$7^3/_4$	$8^1/_2$
2	$11^1/_4$	9	11	$14^1/_2$	$13^1/_2$
3	$23^3/_4$	$21^1/_4$	12	$16^1/_4$	$17^1/_4$
4	14	$11^1/_2$	13	$65^1/_2$	$63^1/_2$
5	$12^1/_2$	$7^1/_4$	14	$53^1/_4$	$56^1/_4$
6	$19^3/_4$	$18^1/_2$	15	$43^1/_2$	$41^1/_2$
7	$27^1/_4$	$22^1/_2$	16	71	$70^1/_4$
8	32	$24^1/_4$	17	$32^1/_2$	$34^1/_2$
9	$56^1/_2$	$59^1/_4$	18	$17^1/_4$	$16^1/_2$

a. On the basis of these data, construct a 90% confidence interval for the mean change
from August 25 to September 15.
b. Can you conclude that the stock market was still suffering from the August 31 slide
by September 15 or had it recovered? Explain.

10.25 State the null hypothesis, H_o, and the alternative hypothesis, H_a, that would be used to
test these claims:
a. The mean of the differences between the posttest and the pretest scores is greater
than 10.
b. The mean weight gain, due to the change in diet for the laboratory animals, is at least
10 oz.
c. The mean weight loss experienced by people on a new diet plan was no less than 12 lb.
d. As a result of a special training session, it is believed that the mean of the difference
in performance scores will not be zero.

10.26 Determine the test criteria that would be used with the classical approach to test the following hypotheses when t is used as the test statistic.
a. H_o: $\mu_d = 0$ and H_a: $\mu_d > 0$, with $n = 15$ and $\alpha = 0.05$
b. H_o: $\mu_d = 0$ and H_a: $\mu_d \neq 0$, with $n = 25$ and $\alpha = 0.05$
c. H_o: $\mu_d = 1.45$ and H_a: $\mu_d < 1.45$, with $n = 12$ and $\alpha = 0.10$
d. H_o: $\mu_d = 0.75$ and H_a: $\mu_d > 0.75$, with $n = 18$ and $\alpha = 0.01$

10.27 The corrosive effects of various soils on coated and uncoated steel pipe was tested by using a dependent sampling plan. The data collected are summarized by $n = 40$, $\sum d = 220$, $\sum d^2 = 6222$, where d is the amount of corrosion on the coated portion subtracted from the amount of corrosion on the uncoated portion. Does this sample provide sufficient reason to conclude that the coating is beneficial? Use $\alpha = 0.01$.
a. Solve using the p-value approach.　　**b.** Solve using the classical approach.

10.28 To test the effect of a physical fitness course on one's physical ability, the number of sit-ups that a person could do in one minute, both before and after the course, was recorded. Ten randomly selected participants scored as shown in the following table. Can you conclude that a significant amount of improvement took place? Use $\alpha = 0.01$.

Before	29	22	25	29	26	24	31	46	34	28
After	30	26	25	35	33	36	32	54	50	43

a. Solve using the p-value approach.　　**b.** Solve using the classical approach.

10.29 A group of ten recently diagnosed diabetics was tested to determine whether an educational program was effective in increasing their knowledge of diabetes. They were given a test, before and after the educational program, concerning self-care aspects of diabetes. The scores on the test were as follows:

Patient	1	2	3	4	5	6	7	8	9	10
Before	75	62	67	70	55	59	60	64	72	59
After	77	65	68	72	62	61	60	67	75	68

The following MINITAB output may be used to determine whether the scores improved as a result of the program. Verify the values shown on the output [mean difference (MEAN), standard deviation (STDEV), standard error of the difference (SE MEAN), $t \star$ (T), and p-value] by calculating the values yourself.

```
TEST OF MU = 0.000 VS MU G.T. 0.000

          N      MEAN     STDEV    SE MEAN        T     P VALUE
   C3    10     3.200     2.741      0.867     3.69      0.0025
```

10.30 As metal parts experience wear, the metal is displaced. The table lists displacement measurements (mm) on metal parts that have undergone durability cycling for the equivalent of 100,000-plus miles. The first column is the serial number for the part, the second column lists the before test (BT) displacement measurement of the new part, the third column lists the end of test (EOT) measurements, and the fourth column lists the change (i.e., wear) in the parts.

	Displacement (mm)		
Serial Number	BT	EOT	Difference
1	4.609	4.604	−0.005
2	5.227	5.208	−0.019
3	5.255	5.193	−0.062
4	4.622	4.601	−0.021
5	4.630	4.589	−0.041
6	5.207	5.188	−0.019
7	5.239	5.198	−0.041
8	4.605	4.596	−0.009
9	4.622	4.576	−0.046
10	4.753	4.736	−0.017

(continued)

11	5.226	5.218	−0.008
12	5.094	5.057	−0.037
13	4.702	4.683	−0.019
14	5.152	5.111	−0.041

Source: Problem data provided by AC Rochester Division, General Motors, Rochester, NY

a. Does it seem right that all the difference values are negative? Explain.

Use a computer or calculator to complete the following:
b. Find the sample mean, variance, and standard deviation for the before test (BT) data.
c. Find the sample mean, variance, and standard deviation for the end of test (EOT) data.
d. Find the sample mean, variance, and standard deviation for the difference data.
e. How is the sample mean difference related to the means of BT and EOT? Are the variances and standard deviations related in the same way?
f. At the 0.01 level, do the data show that a significant amount of wear took place?

 10.31 The amount of general anesthetic a patient should receive prior to surgery has received considerable public attention. According to the American Society of Anesthesiologists, every year about 40,000 (some researchers have put the figure closer to 200,000) of the 28 million patients who undergo general anesthesia experience limited awareness during surgery because of resistance to the medication or from too little dosage. Patients commonly report overhearing doctors conversing with nurses and assistants during operations. (*People,* "Wake-Up Call," 6-29-98).

Suppose a study is conducted using 20 patients who are having eye surgery performed on both eyes, two weeks separating the treatments on each eye. Ten of the patients are given a lighter dose of general anesthetic prior to surgery on the first eye and the other ten patients are given a heavier dose. The following week, the procedure is reversed. Two days after each surgery is performed, the patients are asked to rate the amount of pain and discomfort they experienced on a scale from 0 (none) to 10 (unbearable). Results are shown below:

Subject	Light Dosage	Heavy Dosage	Subject	Light Dosage	Heavy Dosage
1	4	3	11	6	7
2	6	5	12	7	5
3	5	6	13	10	7
4	8	4	14	3	2
5	4	5	15	1	0
6	9	6	16	5	6
7	3	2	17	6	3
8	7	8	18	8	5
9	8	5	19	4	2
10	9	7	20	2	0

Can you conclude that the heavier dosage of anesthetic resulted in the patients experiencing lower pain and discomfort after the eye surgery? Use the 0.01 level of significance.
a. Solve using the *p*-value approach.   **b.** Solve using the classical approach.

 10.32 An article titled, "Influencing Diet and Health Through Project LEAN" (*Journal of Nutrition Education,* July/August 1994) compared 28 individuals with borderline-high or high cholesterol levels before and after a nutrition education session. The participants' cholesterol levels were significantly lowered, and the *p*-value was reported to be less than 0.001. A similar study involving 10 subjects was performed with the following cholesterol reading results: *(continued)*

Subject	1	2	3	4	5	6	7	8	9	10
Presession	295	279	250	235	255	290	310	260	275	240
Postsession	265	266	245	240	230	230	235	250	250	215

Let d = presession cholesterol − postsession cholesterol.

Test the null hypothesis that the mean difference equals zero versus the alternative that the mean difference is positive at $\alpha = 0.05$. Assume normality.

a. Solve using the p-value approach.   **b.** Solve using the classical approach.

10.3 Inferences Concerning the Difference Between Means Using Two Independent Samples

> Why is $\bar{x}_1 - \bar{x}_2$ an unbiased estimator of $\mu_1 - \mu_2$?

When comparing the means of two populations, we typically consider the difference between their means, $\mu_1 - \mu_2$ (often called **"independent means"**). The inferences about $\mu_1 - \mu_2$ will be based on the difference between the observed sample means, $\bar{x}_1 - \bar{x}_2$. This observed difference, $\bar{x}_1 - \bar{x}_2$, belongs to a sampling distribution, the characteristics of which are described in the following statement.

> Even though this is the standard error for the differences (−) between two means, the variances under the radical in the formula are added (+).

If independent samples of sizes n_1 and n_2 are drawn randomly from large populations with means μ_1 and μ_2 and variances σ_1^2 and σ_2^2, respectively, the sampling distribution of $\bar{x}_1 - \bar{x}_2$, the difference between the sample means, has

1. a mean, $\mu_{\bar{x}_1 - \bar{x}_2} = \mu_1 - \mu_2$

2. a **standard error**, $\sigma_{\bar{x}_1 - \bar{x}_2} = \sqrt{\left(\dfrac{\sigma_1^2}{n_1}\right) + \left(\dfrac{\sigma_2^2}{n_2}\right)}$ **(10.6)**

If both populations have normal distributions, then the sampling distribution of $\bar{x}_1 - \bar{x}_2$ will also be normally distributed.

Exercise 10.33

Two independent random samples resulted in the following:

Sample 1: $n_1 = 12$, $s_1^2 = 190$
Sample 2: $n_2 = 18$, $s_2^2 = 150$

Find the estimate for the standard error for the difference between two means.

The preceding statement is true for all sample sizes given that the populations involved are normal and the population variances σ_1^2, and σ_2^2 are known quantities. However, as with inferences about one mean, the variance of a population is generally an unknown quantity. Therefore, it will be necessary to estimate the standard error by replacing the variances, σ_1^2 and σ_2^2, in formula (10.6) with the best estimates available, namely, the sample variances, s_1^2 and s_2^2. The *estimated standard error* will be found using formula (10.7).

$$\text{Estimated standard error} = \sqrt{\left(\frac{s_1^2}{n_1}\right) + \left(\frac{s_2^2}{n_2}\right)} \qquad \textbf{(10.7)}$$

Inferences about the difference between two population means $\mu_1 - \mu_2$ will be based on the following assumptions.

THE ASSUMPTIONS FOR INFERENCES ABOUT THE DIFFERENCE BETWEEN TWO MEANS, $\mu_1 - \mu_2$:

The samples are randomly selected from normally distributed populations, and the samples are selected in an independent manner.

NO ASSUMPTIONS ARE MADE ABOUT THE POPULATION VARIANCES.

Since the samples provide the information for determining the standard error, the *t*-**distribution** will be used as the test statistic. The inferences are divided into two cases.

Case 1: The *t*-distribution will be used, and the number of degrees of freedom will be calculated.

Case 2: The *t*-distribution will be used, and the number of degrees of freedom will be approximated.

Case 1 will occur when you are completing the inferences *using a computer/ statistical calculator and the statistical software/program calculates the number of degrees of freedom* for you. The calculated value for df is a function of both sample sizes and their relative sizes, and both sample variances and their relative sizes. The value of df will be a number between the smaller of $df_1 = n_1 - 1$ or $df_2 = n_2 - 1$, and the sum of the degrees of freedom, $df_1 + df_2 = [(n_1 - 1) + (n_2 - 1)] = n_1 + n_2 - 2$.

Case 2 will occur when you are completing the inference *without the aid of a computer or calculator and their statistical software.* Use of the *t*-distribution with the smaller of $df_1 = n_1 - 1$ or $df_2 = n_2 - 1$ degrees of freedom will give conservative results. Because of this approximation, the true level of confidence for an interval estimate will be slightly higher than the reported level of confidence; or the true *p*-value and the true level of significance for a hypothesis test will be slightly less than reported. The gap between these reported values and the true values will be quite small, unless the sample sizes are quite small and unequal or the sample variances are very different in value. The gap will decrease as the samples increase in size or as the sample variances are more alike in value.

Since the only difference between the two cases is the number of degrees of freedom used to identify the *t*-distribution involved, we will study case 2 first.

NOTE $A > B$ ("A is greater than B") is equivalent to $B < A$ ("B is less than A"). When the difference between A and B is being discussed, it is customary to express the difference as "larger subtract smaller" so that the resulting difference is positive, $A - B > 0$. To express the difference as "smaller subtract larger" results in $B - A < 0$ ("the difference is negative") and is at best "clever," and is usually unnecessarily confusing. Therefore, it is recommended that the difference be expressed "larger subtract smaller."

> **Exercise 10.34**
>
> Two independent random samples of sizes 18 and 24 were obtained to make inferences about the difference between two means. What is the number of degrees of freedom? Discuss both cases.

> Would you say the difference between 5 and 8 is −3? How would you express the difference? Explain.

Confidence Interval Procedure

We will use formula (10.8) for calculating the endpoints of the **1 − α confidence interval.**

$$(\bar{x}_1 - \bar{x}_2) - t_{(df,\, \alpha/2)} \cdot \sqrt{\left(\frac{s_1^2}{n_1}\right) + \left(\frac{s_2^2}{n_2}\right)} \quad \text{to} \quad (\bar{x}_1 - \bar{x}_2) + t_{(df,\, \alpha/2)} \cdot \sqrt{\left(\frac{s_1^2}{n_1}\right) + \left(\frac{s_2^2}{n_2}\right)} \quad \textbf{(10.8)}$$

where df equals the smaller of df_1 or df_2 when calculating the confidence interval without the aid of a computer and its statistical software.

ILLUSTRATION 10.7 ▼

The heights (measured in inches) of 20 randomly selected women and 30 randomly selected men were independently obtained from the student body of a certain college in order to estimate the difference in their mean heights. The sample information is given in Table 10.2. Assume that heights are approximately normally distributed for both populations.

Table 10.2 Sample Information for Illustration 10.7

Sample	Number	Mean	Standard Deviation
Female (f)	20	63.8	2.18
Male (m)	30	69.8	1.92

Find the 95% confidence interval for the difference between the mean heights, $\mu_m - \mu_f$.

Solution

STEP 1 **The Set-Up:**
Describe the population parameter of interest.
$\mu_m - \mu_f$, **the difference between the mean height** of male students and the mean height of female students.

STEP 2 **The Confidence Interval Criteria:**
a. **Check the assumptions.**
Both populations are approximately normal, and the samples were random and **independently selected.**
b. **Identify the probability distribution and the formula to be used.**
The *t*-distribution with df = 19; the smaller of $n_m - 1 = 30 - 1 = 29$ or $n_f - 1 = 20 - 1 = 19$, and formula (10.8).
c. **State the level of confidence, $1 - \alpha$.**
$1 - \alpha = 0.95$.

STEP 3 **The Sample Evidence:**
Collect the sample information.
See Table 10.2 above.

STEP 4 **The Confidence Interval:**
a. **Determine the confidence coefficient.**
We have a two-tailed situation with $\alpha/2 = 0.025$ in one tail and df = 19. From Table 6 in Appendix B, $t_{(df, \alpha/2)} = t_{(19, 0.025)} = 2.09$. See the figure.

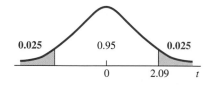

0.025   0.95   0.025

0   2.09   *t*

The exercise box on left

Exercise 10.35
Find the 90% confidence interval for the difference between two means based on this information about two samples. Assume independent samples from normal populations.

Sample	Number	Mean	St. Dev.
1	20	35	22
2	15	30	16

See page 422 for instructions for using Table 6.
b. **Find the maximum error of estimate.**
Using the maximum error part of formula (10.8),

$$E = t_{(df, \alpha/2)} \cdot \sqrt{\left(\frac{s_m^2}{n_m}\right) + \left(\frac{s_f^2}{n_f}\right)}:$$

$$E = 2.09 \cdot \sqrt{\left(\frac{1.92^2}{30}\right) + \left(\frac{2.18^2}{20}\right)} = (2.09)(0.60) = \mathbf{1.25}$$

c. **Find the lower and upper confidence limits.**

$$(\bar{x}_m - \bar{x}_f) \pm E$$

$$6.00 \pm 1.25$$

$$6.00 - 1.25 = \mathbf{4.75} \quad \text{to} \quad 6.00 + 1.25 = \mathbf{7.25}$$

STEP 5   **The Results:**
State the confidence interval.
4.75 to 7.25, the 95% confidence interval for $\mu_m - \mu_f$
That is, with 95% confidence, we can say that the difference between the mean heights of male and female students is between 4.75 and 7.25 inches.

▲

Hypothesis-Testing Procedure

When testing a null **hypothesis about the difference between two population means,** the test statistic used will be the difference between the observed difference of sample means and the hypothesized difference of the population means, divided by the estimated standard error. The test statistic is assumed to have approximately a t-distribution when the null hypothesis is true and the normality assumption has been satisfied. The calculated value of the **test statistic** is found using formula (10.9).

$$t \bigstar = \frac{(\bar{x}_1 - \bar{x}_2) - (\mu_1 - \mu_2)}{\sqrt{\left(\frac{s_1^2}{n_1}\right) + \left(\frac{s_2^2}{n_2}\right)}}, \tag{10.9}$$

with df equal to the smaller of df_1 or df_2 when calculating $t \bigstar$ *without* the aid of a computer/statistical calculator and its statistical software/programs.

NOTE A hypothesized difference between the two population means $\mu_1 - \mu_2$ can be any specified value. The most common value specified is zero; however, the difference can be nonzero.

ILLUSTRATION 10.8 ▼

Suppose that we are interested in comparing the academic success of college students who belong to fraternal organizations with the academic success of those who do not belong to fraternal organizations. The reason for the comparison centers on the recent concern that the fraternity members, on the average, are achieving at a lower academic level than nonfraternal students achieve. (Cumulative grade-point average is used to measure academic success.) Random samples of size 40 are taken from each population. The sample results are listed below in Table 10.3.

Table 10.3   Sample Information for Illustration 10.8

Sample	Number	Mean	Standard Deviation
Fraternity members (*f*)	40	2.03	0.68
Nonmembers (*n*)	40	2.21	0.59

Complete a hypothesis test using $\alpha = 0.05$. Assume that grade-point averages for both groups are approximately normally distributed.

Solution

STEP 1 **The Set-Up:**

a. **Describe the population parameter of interest.**

$\mu_n - \mu_f$, **the difference between the mean grade-point average** for the nonfraternity members and that for the fraternity members.

b. **State the null hypothesis (H_o) and the alternative hypothesis (H_a).**

$H_o: \mu_n - \mu_f = 0 \ (\leq)$ (fraternity averages are no lower)

$H_a: \mu_n - \mu_f > 0$ (fraternity averages are lower)

> Remember: "Larger subtract smaller" results in a positive difference.

STEP 2 **The Hypothesis Test Criteria:**

a. **Check the assumptions.**

Both populations are approximately normal, and random samples were selected. Since the two populations are separate, **the samples are independent.**

b. **Identify the probability distribution and the test statistic to be used.**

The **t-distribution** with df = the smaller of df_n or df_f; since both n's are 40, df = 40 − 1 = **39**, and $t \star$ calculated using formula (10.9).

c. **Determine the level of significance, α.**

$\alpha = 0.05.$

STEP 3 **The Sample Evidence:**

a. **Collect the sample information.**

See Table 10.3, page 493.

b. **Calculate the value of the test statistic.**

$$t \star = \frac{(\bar{x}_n - \bar{x}_f) - (\mu_n - \mu_f)}{\sqrt{\left(\frac{s_n^2}{n_n}\right) + \left(\frac{s_f^2}{n_f}\right)}}: \quad t \star = \frac{(2.21 - 2.03) - (0.00)}{\sqrt{\left(\frac{0.59^2}{40}\right) + \left(\frac{0.68^2}{40}\right)}}$$

$$t \star = \frac{0.18}{\sqrt{0.00870 + 0.01156}} = \frac{0.18}{0.1423} = \mathbf{1.26}$$

Exercise 10.36

Find the value of $t \star$ for the difference between two means based on this information about two samples:

Sample	Number	Mean	St. Dev.
1	18	38.2	14.2
2	25	43.1	10.6

STEP 4 **The Probability Distribution:**

p-VALUE:

a. **Calculate the *p*-value.**

Use the right-hand tail since the H_a expresses concern for values related to "greater than."

$\mathbf{P} = P(t > 1.26,$ with df $= 39)$ as shown in the figure.

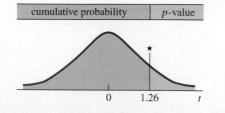

| cumulative probability | *p*-value |

OR

CLASSICAL:

a. **Determine the critical region and critical value(s).**

The critical region is the right-hand tail since the H_a expresses concern for values related to "greater than." The critical value is obtained from Table 6: $t_{(39, 0.05)} = \mathbf{1.69}.$

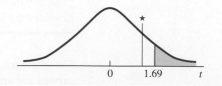

***p*-VALUE:**

To find the *p*-value, use one of three methods:
1. Use Table 6 (Appendix B) to place bounds on the *p*-value:
   0.10 < P < 0.25
2. Use Table 7 (Appendix B) to place bounds on the *p*-value:
   0.100 < P < 0.119
3. Use a computer or calculator to calculate the *p*-value:
   P = 0.1076

Specific details follow this illustration

b. Determine whether or not the *p*-value is smaller than α.
The *p*-value **is not smaller** than α.

CLASSICAL:

OR   See page 419 for information about critical values.

b. Determine whether or not the calculated test statistic is in the critical region.
t ★ **is not in** the critical region, as shown in **red** in the figure above.

STEP 5   The Results:
   a. **State the decision about** H_o**.**
      Fail to reject H_o.
   b. **State the conclusion about** H_a**.**
      At the 0.05 level of significance, the claim that the fraternity members achieve at a level lower than nonmembers is not supported by the sample data.

> When df is not in table, use the next smaller df-value.

To find the *p*-value for Illustration 10.8, use one of three methods:
   Method 1   *Using Table 6:* find 1.26 between two entries in the df = 39 row, read the bounds for **P** from the one-tail heading at the top of the table;   **0.10 < P < 0.25**
   Method 2   *Using Table 7:* find *t* ★ = 1.26 between two rows and df = 39 between two columns, read the bounds for $P(t > 1.26|df = 39)$;   **0.100 < P < 0.119**
   Method 3   If you are doing the hypothesis test with the aid of a computer or calculator, most likely it will calculate the *p*-value for you (see page 491) or you may use the cumulative probability distribution commands described in Chapter 9 on page 420, 421.

ILLUSTRATION 10.9 ▼

Many students have complained that the soft-drink vending machine A (in the student recreation room) dispenses a different amount of drink than machine B (in the faculty lounge). To test this belief, a student randomly sampled several servings from each machine and carefully measured them, with the results as shown in Table 10.4.

Table 10.4   Sample Information for Illustration 10.9

Machine	Number	Mean	Standard Deviation
A	10	5.38	1.59
B	12	5.92	0.83

Does this evidence support the hypothesis that the mean amount dispensed by A is different from the amount dispensed by B? Assume the amounts dispensed by both machines are normally distributed and complete the test using α = 0.10.

Solution

STEP 1   The Set-Up:
   a. **Describe the population parameter of interest.**
      $\mu_B - \mu_A$**, the difference between the mean amount dispensed** by machine B and the mean amount dispensed by machine A.

Remember: "Larger subtract smaller" results in a positive difference.

b. State the null hypothesis (H_o) and the alternative hypothesis (H_a).

H_o: $\mu_B - \mu_A = 0$ (A dispenses the same amount as B)

H_a: $\mu_B - \mu_A \neq 0$ (A dispenses a different amount than B)

STEP 2 **The Hypothesis Test Criteria:**

a. **Check the assumptions.**

Both populations are assumed to be approximately normal, and the samples were random and independently selected.

b. **Identify the probability distribution and the test statistic to be used.**

The *t*-distribution with df = the smaller of $n_A - 1 = 10 - 1 = 9$ or $n_B - 1 = 12 - 1 = 11$; df = 9, and $t\bigstar$ calculated using formula (10.9).

c. **Determine the level of significance, α.**

$\alpha = 0.10$.

STEP 3 **The Sample Evidence:**

a. **Collect the sample information.**

See Table 10.4.

b. **Calculate the value of the test statistic.**

$$t\bigstar = \frac{(\bar{x}_B - \bar{x}_A) - (\mu_B - \mu_A)}{\sqrt{\left(\frac{s_B^2}{n_B}\right) + \left(\frac{s_A^2}{n_A}\right)}}; \quad t\bigstar = \frac{(5.92 - 5.38) - (0.00)}{\sqrt{\left(\frac{0.83^2}{12}\right) + \left(\frac{1.59^2}{10}\right)}}$$

$$t\bigstar = \frac{0.54}{\sqrt{0.0574 + 0.2528}} = \frac{0.54}{0.557} = 0.97$$

STEP 4 **The Probability Distribution:**

p-VALUE:

a. Calculate the *p*-value.

Use both tails since the H_a expresses concern for values related to "different from."

$\mathbf{P} = p\text{-value} = P(t < -0.97) + P(t > 0.97)$
$= 2 \times P(|t| > 0.97 | df = 9)$

See figure below.

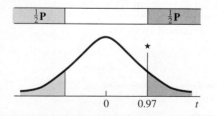

To find the *p*-value, you have three options:

1. Use Table 6 (Appendix B): **0.20 < P < 0.50**
2. Use Table 7 (Appendix B) to place bounds on the *p*-value:
   0.340 < P < 0.394
3. Use a computer or calculator to calculate the *p*-value:
   P = 2 × 0.1787 = 0.3574

For specific instructions, see below.

b. Determine whether or not the *p*-value is smaller than α.
The *p*-value **is not smaller** than α.

OR

CLASSICAL:

a. Determine the critical region and critical value(s).

The critical region is two-tailed since the H_a expresses concern for values related to "different from." The right-hand critical value is obtained from Table 6: $t_{(9, 0.05)} = \mathbf{1.83}$.

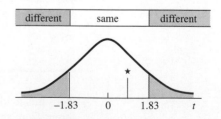

For specific instructions, see page 419.

b. Determine whether or not the calculated test statistic is in the critical region.

$t\bigstar$ **is not in** the critical region as shown in red in the figure above.

Exercise 10.37

Suppose the calculated $t\star$ had been 1.80 in Illustration 10.9. Using df = 9 or using df = 20 results in different answers. Explain how the word "conservative" (p. 491) applies here.

STEP 5   The Results:
   a. State the decision about H_o.
      Fail to reject H_o.
   b. State the conclusion about H_a.

The evidence is not sufficient to show that machine A dispenses a different amount of soft drink than machine B, at the 0.10 level of significance. Thus, for lack of evidence we will proceed as though the two machines dispense, on average, the same amount.   ▲

To find the p-value for Illustration 10.9, use one of three methods.

   Method 1   *Using Table 6:* find 0.97 between two entries in the df = 9 row, read the bounds for **P** from the two-tail heading at the top of the table;   **0.20 < P < 0.50**

   Method 2   *Using Table 7:* find $t\star = 0.97$ between two rows and df = 9 between two columns, read the bounds for $P(t > 0.97|\text{df} = 9)$; $0.170 < \frac{1}{2}\mathbf{P} < 0.197$, therefore **0.340 < P < 0.394**

   Method 3   If you are doing the hypothesis test with the aid of a computer or calculator, most likely it will calculate the p-value (do not double) for you (see page 491) or you may use the cumulative probability distribution commands described in Chapter 9 on page 420, 421.

   Most computer/calculator statistical packages will complete the inferences for the difference between two means by calculating the number of degrees of freedom.

Exercise 10.38

For the hypothesis test involving H_a:
$\mu_B - \mu_A \neq 0$ with df = 18 and $t\star = 1.3$.

a. Find the p-value.
b. Find the critical values given $\alpha = 0.05$.

The following commands will complete a hypothesis test for the difference between two population means with unknown standard deviation given two independent sets of sample data.

MINITAB

MINITAB's TWOSAMPLE command performs both the confidence interval and the hypothesis test at the same time.

Input the two independent sets of data into C1 and C2, then continue with:

```
Choose:  Stat > Basic Statistics
         > 2-Sample t
Select:  Samples in different columns
Enter:   First: C1
         Second: C2
Choose:  Alternative: less than or
         not equal or greater than
Enter:   Confidence level: 1 − α (ex.
         0.95 or 95.0)
Select:  Assume equal variances (if
         known)
```

EXCEL

Input the two independent sets of data into columns A and B, then continue with:

```
Choose:  Tools > Data Analysis
         > t-Test: Two-Sample
         Assuming Unequal Vari-
         ances > OK
Enter:   Variable 1 Range: (A1:A20
         or select cells)
         Variable 2 Range: (B1:B20
         or select cells)
         Hypothesized Mean Differ-
         ence: μ_B − μ_A (usually 0)
Select:  Labels (if necessary)
Enter:   α (ex. 0.05)
Select:  Output Range
Enter:   (C1 or select cell)
```

Use Format > Column > AutoFit Selection to make the output more readable. The output shows p-values and critical values for one- and two-tailed tests.

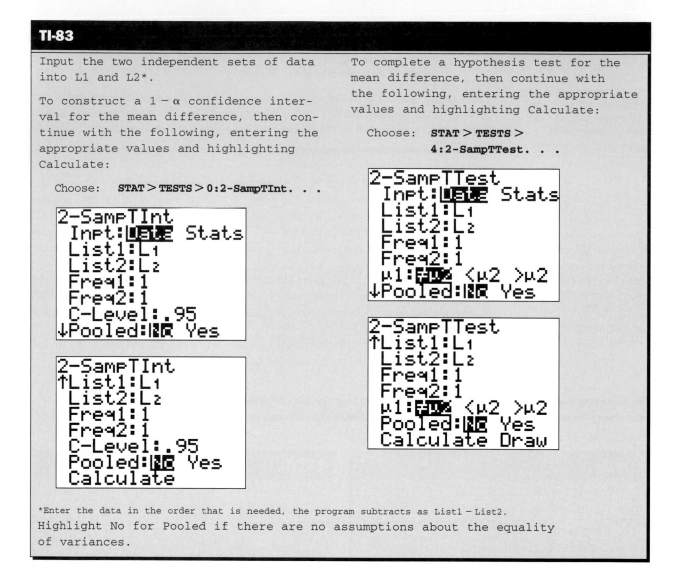

Illustration 10.8 (pp. 493–495) was solved using MINITAB. With 40 cumulative grade-point averages for nonmembers in C1 and 40 averages for fraternity members in C2, the above commands resulted in the output shown below. Compare these results to the solution of Illustration 10.8. Notice the difference in **P** and df values. Explain.

```
Two Sample T-Test and Confidence Interval
Twosample T for C1 vs C2

          N     Mean     StDev     SE Mean
C1       40     2.21     0.59      0.09
C2       40     2.03     0.68      0.11

95% C.I. for mu C1 − mu C2: (−0.10, 0.46)
T-Test mu C1 = mu C2 (vs >): T = 1.26 P = 0.106 DF = 75
```

Case Study   10.2

An Empirical Study of Faculty Evaluation Systems: Business Faculty Perceptions

Hsin-Min Tong and Allen L. Bures report a whole series of t-test results from comparing various factors of perceptions between AACSB accredited schools and non-AACSB accredited schools. For each t-test they report a two-tailed probability (p-value).

Table 1 shows how respondents from American Assembly of Collegiate Schools of Business (AACSB) accredited and non-AACSB-accredited schools rated the importance of the ten faculty evaluation factors. Clearly AACSB-accredited schools and non-AACSB-accredited schools are different, with the AACSB-accredited institutions placing greater weight on research and publication, and with the nonaccredited institutions putting more emphasis on classroom teaching, campus committee work, student advising, public service, advisor to student organizations, and consultation (business, government). There is no significant difference between AACSB-accredited and non-AACSB-accredited schools in the importance given to a faculty member's activity in professional societies.

Exercise 10.39

a. Explain why the samples discussed in this article are independent samples.
b. Many of the reported p-values are 0.000, meaning that the p-value is less than 0.0005. Explain what is implied by these p-values.
c. The p-value for "activity in professional societies" is 0.394. Estimate the calculated $t \star$ -value. Explain what is implied by this p-value.
d. Explain why there is reason to question the use of the two-sample t-test as studied in this chapter to analyze the data Tong and Bures collected. (*Hint:* It has to do with assumptions.)

Table 1   *t*-Test Results of Importance of Faculty Evaluation Factors, AACSB-Accredited vs. Non-AACSB-Accredited Schools

Factor	AACSB-Accredited Schools ($n = 176$)	Non-AACSB-Accredited Schools ($n = 74$)	Two-Tailed *t*-Test Probability
Articles in professional journals	4.49	2.87	0.000
Classroom teaching	3.34	4.31	0.000
Books as author or editor	3.23	2.45	0.000
Papers at professional meetings	3.09	2.65	0.001
Activity in professional societies	2.55	2.65	0.394
Campus committee work	2.25	3.16	0.000
Student advising	1.80	3.36	0.000
Public service	1.98	2.47	0.000
Advisor to student organizations	1.71	2.30	0.000
Consultation (business, government)	1.73	2.30	0.001

Note: A five-point scale with 1 = "not at all important" to 5 = "extremely important" was used in this analysis.

Source: Hsin-Min Tong and Allen L. Bures in *Journal of Education for Business*, April 1987. Reprinted with permission of the Helen Dwight Reid Educational Foundation. Published by Heldref Publications, 4000 Albemarle St., N.W., Washington, D.C. 20016. Copyright (c) 1987.

EXERCISES • • • • • • • • • • • • • •

10.40 Find the confidence coefficient, $t_{(df, \alpha/2)}$, that would be used to find the maximum error for each of the following situations when estimating the difference between two means, $\mu_1 - \mu_2$.

a. $1 - \alpha = 0.95$, $n_1 = 25$, $n_2 = 15$ b. $1 - \alpha = 0.98$, $n_1 = 43$, $n_2 = 32$
c. $1 - \alpha = 0.99$, $n_1 = 19$, $n_2 = 45$

10.41 An experiment was conducted to compare the mean absorptions of two drugs in specimens of muscle tissue. Seventy-two tissue specimens were randomly divided into two equal groups. Each group was tested with one of the two drugs. The sample results were $\bar{x}_A = 7.9$, $\bar{x}_B = 8.5$, $s_A = 0.11$, and $s_B = 0.10$. Assume both populations are normal. Construct the 98% confidence interval for the difference in the mean absorption rates.

10.42 Experimentation with a new rocket nozzle has led to two slightly different designs. The following data summaries resulted from testing these two designs.

	n	$\sum x$	$\sum x^2$
Design 1	36	278.4	2163.76
Design 2	42	310.8	2332.26

Determine the 99% confidence interval for the difference in the means for these two rocket nozzles.

10.43 The two independent samples shown in the following table were obtained in order to estimate the difference between the two population means. Construct the 98% confidence interval.

Sample A	6	7	7	6	6	5	6	8	5	4
Sample B	7	2	4	3	3	5	4	6	4	2

10.44 A study was designed to estimate the difference in diastolic blood pressures between men and women. MINITAB was used to construct a 99% confidence interval for the difference between the means based on the following sample data.

Males	76	76	74	70	80	68	90	70	90	72
	76	80	68	72	96	80				

Females	76	70	82	90	68	60	62	68	80	74
	60	62	72							

```
Two sample T for Males vs Females

                   N         Mean        StDev        SE Mean
      Males       16         77.37        8.35          2.1
    Females       13         71.08        9.22          2.6

99% C.I. for mu males - mu females:  (-2.9, 15.5)
```

Verify the results [the two sample means and standard deviations, and the confidence interval bounds] by calculating the values yourself.

10.45 A study comparing attitudes toward death was conducted in which organ donors (individuals who had signed organ donor cards) were compared with nondonors. The study is reported in the journal *Death Studies* (Vol. 14, No. 3, 1990). Templer's Death Anxiety Scale (DAS) was administered to both groups. On this scale, high scores indicate high anxiety concerning death. The results were reported as follows.

	n	Mean	St. Dev.
Organ Donors	25	5.36	2.91
Nonorgan Donors	69	7.62	3.45

Construct the 95% confidence interval for the difference between the means, $\mu_{non} - \mu_{donor}$.

10.46 "Is the length of a steel bar affected by the heat treatment technique used?" This was the question being tested when the following data were collected.

Heat Treatment	Lengths (to the nearest inch)										
1	156	159	151	153	157	159	155	155	151	152	158
	154	156	156	157	155	156	159	153	157	157	159
	158	155	159	152	150	154	156	156	157	160	
2	154	156	150	151	156	155	153	154	149	150	150
	151	154	155	155	154	154	156	150	151	156	154
	153	154	149	150	150	151	154	148	155	158	

a. Find the means and standard deviations for the two sets of data.
b. Find evidence about the sample data (both graphic and numeric) that supports the assumption of normality for the two sampled populations.
c. Find the 95% confidence interval for $\mu_1 - \mu_2$.

10.47 A USA Snapshot® (*USA Today*, 10-12-94) reported the longest average workweeks for nonsupervisory employees in private industry to be in mining (45.4 hours) and manufacturing (42.3 hours). The same article reported the shortest average workweeks to be in retail trade (29 hours) and services (32.4 hours). A study conducted in Missouri found the following results for a similar study.

Industry	n	Average Hours/Week	Standard Deviation
Mining	15	47.5	5.5
Manufacturing	10	43.5	4.9

Set a 95% confidence interval on the difference in average length of workweek between mining and manufacturing. Assume normality for the sampled populations and that the samples were selected randomly.

10.48 An article titled "Stages of Change for Reducing Dietary Fat to 30% of Energy or Less" (*Journal of the American Dietetic Association*, October 1994) measured the energy from fat (expressed as a percent) for two different groups. Sample 1 was a random sample of 614 adults who responded to mailed questionnaires, and sample 2 was a convenience sample of 130 faculty, staff, and graduate students. The following table gives the percent of energy from fat for the two groups.

Group	n	Mean	Standard Deviation
1	614	35.0	6.3
2	130	32.0	9.1

a. Construct the 95% confidence interval for $\mu_1 - \mu_2$.
b. Do these samples satisfy the assumptions for this confidence interval? Explain.

10.49 State the null and alternative hypotheses that would be used to test the following claims.
a. There is a difference between the mean age of employees at two different large companies.
b. The mean of population 1 is greater than the mean of population 2.
c. The difference between the means of the two populations is more than 20 pounds.
d. The mean of population A is less than 50 more than the mean of population B.

10.50 Calculate the estimate for the standard error of difference between two independent means for each of the following cases.
a. $s_1^2 = 12$, $s_2^2 = 15$, $n_1 = 16$,   and $n_2 = 21$
b. $s_1^2 = 0.054$, $s_2^2 = 0.087$, $n_1 = 8$,   and $n_2 = 10$
c. $s_1 = 2.8$, $s_2 = 6.4$, $n_1 = 16$,   and $n_2 = 21$

10.51 Determine the p-value for the following hypothesis tests for the difference between two means with population variances unknown.
  a. $H_a: \mu_1 - \mu_2 > 0$, $n_1 = 6$, $n_2 = 10$, $t \star = 1.3$
  b. $H_a: \mu_1 - \mu_2 < 0$, $n_1 = 16$, $n_2 = 9$, $t \star = -2.8$
  c. $H_a: \mu_1 - \mu_2 \neq 0$, $n_1 = 26$, $n_2 = 16$, $t \star = 1.8$
  d. $H_a: \mu_1 - \mu_2 \neq 5$, $n_1 = 26$, $n_2 = 35$, $t \star = -1.8$

10.52 Determine the critical values that would be used for the following hypothesis tests (using the classical approach) about the difference between two means with population variances unknown.
  a. $H_a: \mu_1 - \mu_2 \neq 0$, $n_1 = 26$, $n_2 = 16$, $\alpha = 0.05$
  b. $H_a: \mu_1 - \mu_2 < 0$, $n_1 = 36$, $n_2 = 27$, $\alpha = 0.01$
  c. $H_a: \mu_1 - \mu_2 > 0$, $n_1 = 8$, $n_2 = 11$, $\alpha = 0.10$
  d. $H_a: \mu_1 - \mu_2 \neq 10$, $n_1 = 14$, $n_2 = 15$, $\alpha = 0.05$

10.53 A study was designed to compare the attitudes of two groups of nursing students toward computers. Group 1 had previously taken a statistical methods course that involved significant computer interaction through the use of statistical packages. Group 2 had taken a statistical methods course that did not use computers. The students' attitudes were measured by administering the Computer Anxiety Index (CAIN). The results were as follows:

Group 1 (with computers): $n = 10$ $\bar{x} = 60.3$ $s = 7.5$

Group 2 (without computers): $n = 15$ $\bar{x} = 67.2$ $s = 2.1$

Do the data show that the mean score for those with computer experience was significantly less than the mean score for those without computer experience? Use $\alpha = 0.05$.

10.54 If a random sample of 18 homes south of Center Street in Provo has a mean selling price of $125,000 and a standard deviation of $2400, and a random sample of 18 homes north of Center Street has a mean selling price of $127,000 and a standard deviation of $4800 can you conclude that there is a significant difference between the selling price of homes in these two areas of Provo at the 0.05 level?
  a. Solve using the p-value approach.:   **b.** Solve using the classical approach.

10.55 The computer age has allowed teachers to use electronic tutorials in order to motivate their students to learn. *Issues in Accounting Education* (May 1998) published the results of a 1998 study revealing that an electronic tutorial, along with intentionally induced peer pressure, was effective in enhancing pre-class preparations and in improving class attendance, test scores, and course evaluations when used by students studying tax accounting.

Suppose a similar study is conducted at your school using an electronic study guide (ESG) as a tutor for students in accounting principles. For one course section, the students were required to use a new ESG computer program that generated and scored chapter review quizzes and practice examinations, presented textbook chapter reviews, and tracked progress. Students could use the computer to build, take, and score their own simulated tests and review materials at their own pace before they took their formal in-class quizzes and exams composed of different questions. The same instructor taught the other course section, used the same textbook, and gave the same daily assignments, but he did not require the students to use the ESG. Identical tests were administered to both sections, and the mean scores of all tests and assignments at the end of the year were tabulated:

Section	n	Mean Score	Std. Dev.
ESG (1)	38	79.6	6.9
No ESG (2)	36	72.8	7.6

Do these results show that the mean scores of tests and assignments for students taking accounting principles with an ESG to help them are significantly greater than those not using an ESG? Use a 0.01 level of significance.
  a. Solve using the p-value approach.   **b.** Solve using the classical approach.

10.56 The purchasing department for a regional supermarket chain is considering two sources from which to purchase 10-lb bags of potatoes. A random sample taken from each source shows the following results.

	Idaho Supers	Idaho Best
Number of Bags Weighed	100	100
Mean Weight	10.2 lb	10.4 lb
Sample Variance	0.36	0.25

At the 0.05 level of significance, is there a difference between the mean weights of the 10-lb bags of potatoes?
  a. Solve using the *p*-value approach.   **b.** Solve using the classical approach.

10.57 MINITAB was used to complete a *t*-test of the difference between the two means using the following two independent samples.

Sample 1	33.7	21.6	32.1	38.2	33.2	35.9	34.1	39.8
	23.5	21.2	23.3	18.9	30.3			
Sample 2	28.0	59.9	22.3	43.3	43.6	24.1	6.9	14.1
	30.2	3.1	13.9	19.7	16.6	13.8	62.1	28.1

```
Two sample T for sample 1 vs sample 2

            N      Mean     StDev    SE Mean
 sample1    13     29.68     7.07      2.0
 sample2    16     26.9     17.4       4.4

T-Test mu sample1 = mu sample2 (vs not =):
T = 0.59 P = 0.56 DF = 20
```

  a. Verify the results (two sample means and standard deviations, and the calculated *t* ★) by calculating the values yourself.
  b. Use Table 7 in Appendix B to verify the *p*-value based on the calculated df.
  c. Find the *p*-value using the smaller number of degrees of freedom. Compare the two *p*-values.

10.58 A study was conducted to assess the safety and efficiency of receiving nitroglycerin from a transdermal system (i.e., a patch worn on the skin), which intermittently delivers the medication, versus oral medication (pills). Twenty patients who suffer from angina (chest pain) due to physical effort were enrolled in trials. All received patches, some ($n = 8$) contained nitroglycerin, the others ($n = 12$) contained a placebo. Suppose the resulting "time to angina" data were summarized:

	Active	Placebo	Difference	SE	*p*-Value[a]
			Mean Time to Angina (sec)		
Day 1 AM	320.00	287.00	33.00	9.68	0.0029
Day 7 PM	314.00	285.25	28.75	13.74	0.0500

[a]For treatment difference.

  a. Determine the value of *t* for the difference between two independent means given the difference and the standard error (SE) for the day 1 AM data.
  b. Verify the *p*-value.
  c. Determine the value of *t* for the difference between two independent means given the difference and the standard error (SE) for the day 7 PM data.
  d. Verify the *p*-value.

10.59 Twenty laboratory mice were randomly divided into two groups of ten. Each group was fed according to a prescribed diet. At the end of three weeks, the weight gained by each animal was recorded. Do the data in the following table justify the conclusion that the mean weight gained on diet B was greater than the mean weight gained on diet A, at the $\alpha = 0.05$ level of significance?

Diet A	5	14	7	9	11	7	13	14	12	8
Diet B	5	21	16	23	4	16	13	19	9	21

   a. Solve using the *p*-value approach.   **b.** Solve using the classical approach.

10.60 The quality of latex paint is monitored by measuring different characteristics of the paint. One characteristic of interest is the particle size. Two different types of disc centrifuges (JLDC, Joyce Loebl Disc Centrifuge, and the DPJ, Dwight P. Joyce disc) are used to measure the particle size. It is thought that these two methods yield different measurements. Thirteen readings were taken from the same batch of latex paint using both the JLDC and the DPJ discs.

JLDC	4714	4601	4696	4896	4905	4870	4987	5144	3962	4006	4561	4626	4924
DPJ	4295	4271	4326	4530	4618	4779	4752	4744	3764	3797	4401	4339	4700

Source: With permission of SCM Corporation.

Assuming particle size is normally distributed:
   a. Determine whether there is a significant difference between the readings at the 0.10 level of significance.
   b. What is your estimate for the difference between the two readings?

10.61 The material used in making parts affects not only how long the part lasts but also how difficult it is to disassemble to repair. The following measurements are for screw torque removal for a specific screw after several operations of use. The first column lists the part number, the second column lists the screw torque removal measurements for assemblies made with material A, and the third column lists the screw torque removal measurements for assemblies made with material B. Assume torque measurements are normally distributed.

Removal Torque (NM, Newton-meters)

Part Number	1	2	3	4	5	6	7	8	9	10	11	12	13	14	15
Material A	16	14	13	17	18	15	17	16	14	16	15	17	14	16	15
Material B	11	14	13	13	10	15	14	12	11	14	13	12	11	13	12

Source: Problem data provided by AC Rochester Division, General Motors, Rochester, NY.

   a. Find the sample mean, variance, and standard deviation for the material A data.
   b. Find the sample mean, variance, and standard deviation for the material B data.
   c. At the 0.01 level, do these data show a significant difference in the mean torque required to remove the screws from the two different materials?

10.62 Some 20 million Americans visit chiropractors annually, and the number of practitioners has nearly doubled in the last two decades, to 55,000 in the United States, according to the American Chiropractic Association. The *New England Journal of Medicine* (Fall 1998) released a report in the fall of 1998 showing results of a study that compared chiropractic spinal manipulation (CSM) with physical therapy for treating acute lower back pain. After two years of treatment, CSM was found to be no more effective at either reducing missed work or preventing a relapse.

Suppose a similar study of 60 patients is made by dividing the sample into two groups. One group is given CSM and the other receives physical therapy for one year. During the period, the number of missed days at work due to lower back pain is measured to reveal the table of results: *(continued)*

Group	n	Mean	Std. Dev.
CSM (1)	32	10.6	4.8
Therapy (2)	28	12.5	6.3

Do these results show that the mean number of missed days of work for people suffering from acute back pain is significantly less for those receiving CSM than for those undergoing physical therapy? Use a 0.01 level of significance.

a. Solve using the p-value approach.

b. Solve using the classical approach.

10.63 Use a computer to demonstrate the truth of the statement describing the sampling distribution of $\bar{x}_1 - \bar{x}_2$. Use two theoretical normal populations: $N_1(100, 20)$ and $N_2(120, 20)$.

See the Statistical Tutor for additional information about commands.

a. To get acquainted with the two theoretical populations, randomly select a very large sample from each. Generate 2000 data values; calculate the mean and standard deviation; and construct a histogram using class boundaries that are multiples of one-half of a standard deviation (10) starting at the mean for each population.

b. If samples of size 8 are randomly selected from each population, what do you expect the distribution of $\bar{x}_1 - \bar{x}_2$ to be like (shape of distribution, mean, standard error)?

c. Randomly draw a sample of size 8 from each population, and find the mean of each sample. Find the difference between the sample means. Repeat 99 more times.

d. The set of 100 $(\bar{x}_1 - \bar{x}_2)$ values form an empirical sampling distribution of $\bar{x}_1 - \bar{x}_2$. Describe the empirical distribution: shape (histogram), mean, and standard error. (Use class boundaries that are multiples of the standard error from mean for easy comparison to the expected.)

e. Using the information found above, verify the statement about the $\bar{x}_1 - \bar{x}_2$ sampling distribution made on page 490.

f. Repeat the experiment a few times and compare the results.

10.64 One of the reasons for being conservative when determining the number of degrees of freedom to use with the t-distribution was the possibility that the population variances might be unequal. Extremely different values cause a lowering in the number of df used. Repeat Exercise 10.63 using theoretical normal distributions of $N(100, 9)$ and $N(120, 27)$ and both sample sizes of 8. Check all three properties of the sampling distribution: normality, its mean value, and its standard error. Describe in detail what you discover. Do you think we should be concerned about the choice of df? Explain.

10.65 Unbalanced sample sizes is a factor in determining the number of degrees of freedom for inferences about the difference between two means. Repeat Exercise 10.63 using theoretical normal distributions of $N(100, 20)$ and $N(120, 20)$ and sample sizes of 5 and 20. Check all three properties of the sampling distribution: normality, its mean value, and its standard error. Describe in detail what you discover. Do you think we should be concerned when using unbalanced sample sizes? Explain.

10.66 One part of the assumptions for the two-sample t-test is the "sampled populations are to be normally distributed." What happens when they are not normally distributed? Repeat Exercise 10.63 using two theoretical populations that are not normal and using samples of size 10. The exponential distribution uses a continuous random variable, it has a J-shaped distribution, and its mean and standard deviation are the same value. Use the two exponential distributions with means of 50 and 80: Exp(50) and Exp(80). Check all three properties of the sampling distribution: normality, its mean value, and its standard error. Describe in detail what you discover. Do you think we should be concerned when sampling nonnormal populations? Explain.

10.4 Inferences Concerning the Difference Between Proportions Using Two Independent Samples

The 3 "p" words (proportion, percentage, probability) are all the **binomial parameter p,** "P(success)."

We are often interested in making statistical comparisons between the **proportions, percentages,** or **probabilities** associated with two populations. The following questions ask for such comparisons: Is the proportion of homeowners who favor a certain tax proposal different from the proportion of renters who favor it? Did a larger percentage of this semester's class than of last semester's class pass statistics? Is the probability of a Democratic candidate winning in New York greater than the probability of a Republican candidate winning in Texas? Do students' opinions about the new code of conduct differ from those of the faculty? You have probably asked similar questions.

Binomial experiments are completely defined on page 253.

RECALL The properties of a **binomial experiment** are as follows:

1. The observed probability is $p' = x/n$, where x is the number of observed successes in n trials.
2. $q' = 1 - p'$.
3. p is the probability of success on an individual trial in a binomial probability experiment of n repeated independent trials.

In this section, we will compare two population proportions using the difference between the observed proportions, $p_1' - p_2'$, of two independent samples. The observed difference, $p_1' - p_2'$, belongs to a sampling distribution, the characteristics of which are described in the following statement.

Exercise 10.67

Only 75 of the 250 people interviewed were able to name the vice president of the United States. Find the values for x, n, p', and q'.

If independent samples of sizes n_1 and n_2 are drawn randomly from large populations with $p_1 = P_1$ (success) and $p_2 = P_2$ (success), respectively, the sampling distribution of $p_1' - p_2'$ has these properties:

1. a mean $\mu_{p_1' - p_2'} = p_1 - p_2$

2. a **standard error** $\sigma_{p_1' - p_2'} = \sqrt{\left(\dfrac{p_1 q_1}{n_1}\right) + \left(\dfrac{p_2 q_2}{n_2}\right)}$

3. an approximately normal distribution if n_1 and n_2 are sufficiently large

(10.10)

In practice, use the following guidelines to ensure normality.

1. The sample sizes are both larger than 20.
2. The products $n_1 p_1$, $n_1 q_1$, $n_2 p_2$, and $n_2 q_2$ are all larger than 5.
3. The samples consist of less than 10% of their respective populations.

NOTE p_1 and p_2 are unknown; therefore, the products mentioned in guideline (2) above will be estimated by $n_1 p_1'$, $n_1 q_1'$, $n_2 p_2'$, and $n_2 q_2'$.

Inferences about the difference between two population proportions, $p_1 - p_2$, will be based on the following assumptions.

Exercise 10.68

If $n_1 = 40$, $p_1' = 0.9$, $n_2 = 50$, and $p_1' = 0.9$:

a. Find the estimated values for both np's and both nq's.
b. Would this situation satisfy the guidelines for approximately normal? Explain.

ASSUMPTIONS FOR INFERENCES ABOUT THE DIFFERENCE BETWEEN TWO PROPORTIONS $p_1 - p_2$

The n_1 random observations and the n_2 random observations forming the two samples are selected independently from two populations that are not changing during the sampling.

Confidence-Interval Procedure

When we estimate the difference between two proportions $p_1 - p_2$, we will base our estimates on the unbiased sample statistic $p_1' - p_2'$. The point estimate, $p_1' - p_2'$, becomes the center of the confidence interval and the confidence limits are found using formula (10.11).

$$(p_1' - p_2') - z(\alpha/2) \cdot \sqrt{\left(\frac{p_1' q_1'}{n_1}\right) + \left(\frac{p_2' q_2'}{n_2}\right)} \quad \text{to} \quad (p_1' - p_2') + z(\alpha/2) \cdot \sqrt{\left(\frac{p_1' q_1'}{n_1}\right) + \left(\frac{p_2' q_2'}{n_2}\right)} \quad \textbf{(10.11)}$$

ILLUSTRATION 10.10 ▼

In studying his campaign plans, Mr. Morris wishes to estimate the difference between men's and women's views regarding his appeal as a candidate. He asks his campaign manager to take two random independent samples and find the 99% confidence interval for the difference. A sample of 1000 voters was taken from each population, with 388 men and 459 women favoring Mr. Morris.

Solution

The campaign manager determined the confidence interval as follows:

STEP 1   **The Set-Up:**
Describe the population parameter of interest.
$p_w - p_m$, **the difference between the proportion** of men voters and the proportion of women voters who plan to vote for Mr. Morris.

> It is customary to place the larger value first; that way, the point estimate for the difference is a positive value.

STEP 2   **The Confidence Interval Criteria:**
a. Check the assumptions.
The samples are randomly and independently selected.
b. Identify the probability distribution and the formula to be used.
The **standard normal distribution** [Populations are large (all voters); the sample sizes are larger than 20; the estimated values for $n_m p_m$, $n_m q_m$, $n_w p_w$, and $n_w q_w$ are all larger than 5. Therefore, the sampling distribution of $p_w' - p_m'$ should have an approximately normal distribution] and $z \star$ will be calculated using formula (10.11).
c. State the level of confidence, $1 - \alpha$.
$1 - \alpha = 0.99$.

STEP 3   **The Sample Evidence:**
Collect the sample information.

Sample: $n_m = 1000$, $x_m = 388$, $n_w = 1000$, $x_w = 459$.

$$p_m' = \frac{x_m}{n_m} = \frac{388}{1000} = \textbf{0.388}; \quad \text{and} \quad q_m' = 1 - 0.388 = \textbf{0.612}$$

$$p_w' = \frac{x_w}{n_w} = \frac{459}{1000} = \textbf{0.459}; \quad \text{and} \quad q_w' = 1 - 0.459 = \textbf{0.541}$$

STEP 4   **The Confidence Interval:**
a. Determine the confidence coefficient.

Two-tailed situation, with $\dfrac{\alpha}{2}$ in each tail: From Table 4B,

$z(\alpha/2) = z(0.005) = 2.58.$

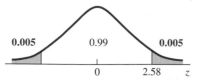

Instructions for using Table 4B are on page 355, 356.

b. **Find the maximum error of estimate.**

Using the maximum error part of formula (10.11),

$$E = z_{(\alpha/2)} \cdot \sqrt{\left(\frac{p'_w q'_w}{n_w}\right) + \left(\frac{p'_m q'_m}{n_m}\right)}:$$

$$E = 2.58 \cdot \sqrt{\left(\frac{(0.459)(0.541)}{1000}\right) + \left(\frac{(0.388)(0.612)}{1000}\right)}$$

$$= 2.58\sqrt{0.000248 + 0.000237} = (2.58)(0.022) = \textbf{0.057}$$

c. **Find the lower and upper confidence limits.**

$$(p'_w - p'_m) \pm E$$

$$0.071 \pm 0.057$$

$$0.071 - 0.057 = \textbf{0.014} \quad \text{to} \quad 0.071 + 0.057 = \textbf{0.128}$$

Exercise 10.69

Find the 95% confidence interval for $p_A - p_B$.

Sample	n	x
A	125	45
B	150	48

STEP 5 **The Results:**

State the confidence interval.

0.014 to 0.128, the 99% confidence interval for $p_w - p_m$.

With 99% confidence, we can say that there is a difference of from 1.4% to 12.8% in Mr. Morris's voter appeal. That is, a larger proportion of women than men favor Mr. Morris, and the difference in proportion is between 1.4% and 12.8%. ▲

The following commands will construct a 1 − α confidence interval for the difference between two proportions $p_1 - p_2$, for two independent sets of sample data.

EXCEL

```
Input the data for the first sample
into column A using 0's for failures
(or no's) and 1's for successes (or
yes's), then repeat the same procedure
for the second sample in column B;
then continue with:
```

Choose: **Tools > Data Analysis**
Plus > Inference About
Two Proportions > OK

Enter: block coordinates: **(A2:B20**
or select cells) > OK
(do not include column name)
Code for success: **1** > OK

Choose: **Interval Estimate**

Enter: level of confidence: **1 − α**
(ex. 0.95 or 95.0)

TI-83

Choose: **STAT > TESTS > B:2-PropZInt**
Enter the appropriate values and
highlight Calculate

```
2-PropZInt
 x1:0
 n1:0
 x2:0
 n2:0
 C-Level:.95
 Calculate
```

Hypothesis-Testing Procedure

When the null **hypothesis, "there is no difference between two proportions,"** is being tested, the **test statistic** will be the difference between the observed proportions divided by the **standard error** and it is found using formula (10.12).

$$z \star = \frac{(p_1' - p_2')}{\sqrt{pq\left[\left(\frac{1}{n_1}\right) + \left(\frac{1}{n_2}\right)\right]}} \tag{10.12}$$

NOTES

1. The null hypothesis is $p_1 = p_2$, or $p_1 - p_2 = 0$ ("difference is zero").
2. Nonzero differences between proportions are not discussed in this section.
3. The numerator of formula (10.12) could be written as $(p_1' - p_2') - (p_1 - p_2)$, but since the null hypothesis is assumed to be true during the test, $p_1 - p_2 = 0$. By substitution the numerator becomes simply $p_1' - p_2'$.
4. Since the null hypothesis is $p_1 = p_2$, the standard error of $p_1' - p_2'$, $\sqrt{\left(\frac{p_1 q_1}{n_1}\right) + \left(\frac{p_2 q_2}{n_2}\right)}$, can be written as $\sqrt{pq\left[\left(\frac{1}{n_1}\right) + \left(\frac{1}{n_2}\right)\right]}$, where $p = p_1 = p_2$ and $q = 1 - p$.
5. When the null hypothesis states $p_1 = p_2$ and does not specify the values of either p_1 or p_2, the two sets of sample data will be pooled to obtain the estimate for p. This **pooled observed probability** (known as p_p') is the total number of successes divided by the total number of observations with the two samples combined; and it is found using formula (10.13).

$$p_p' = \frac{x_1 + x_2}{n_1 + n_2} \tag{10.13}$$

and q_p' is its complement,

$$q_p' = 1 - p_p' \tag{10.14}$$

When the pooled estimate, p_p', is being used, formula (10.12) becomes formula (10.15).

$$z \star = \frac{(p_1' - p_2')}{\sqrt{(p_p')(q_p')\left[\left(\frac{1}{n_1}\right) + \left(\frac{1}{n_2}\right)\right]}} \tag{10.15}$$

Exercise 10.70

Find the values of p_p' and q_p' for these samples:

Sample	x	n
E	15	250
R	25	275

ILLUSTRATION 10.11 ▼

A salesman for a new manufacturer of cellular phones claims not only that they cost the retailer less but also that the percentage of defective cellular phones found among his products will be no higher than the percentage of defectives found in a competitor's line. To test this statement, the retailer took random samples of each manufacturer's product. The sample summaries are given in Table 10.5. Can we reject the salesman's claim at the 0.05 level of significance?

Table 10.5   Cellular Phone Sample Information

Product	Number Defective	Number Checked
Salesman's	15	150
Competitor's	6	150

Solution

STEP 1 **The Set-Up:**

a. **Describe the population parameter of interest.**

$p_s - p_c$, **the difference between the proportion** of defectives of the salesman's product and the proportion of defectives in the competitor's product.

b. **State the null hypothesis (H_o) and the alternative hypothesis (H_a).**

The concern of the retailer is that the salesman's less expensive product may be of a poorer quality, meaning a greater proportion of defectives. By using the difference, the "suspected larger proportion minus the smaller proportion," the alternative hypothesis is "the difference is positive (greater than zero)."

$$H_o: p_s - p_c = 0$$

($\leq$) (salesman's product defective rate is no higher than competitor's)

$$H_a: p_s - p_c > 0$$

(salesman's product defective rate is higher than competitor's)

STEP 2 **The Hypothesis Test Criteria:**

a. **Check the assumptions.**

Random samples were selected from the product of two different manufacturers.

b. **Identify the probability distribution and the test statistic to be used.**

The standard normal distribution [Populations are very large (all cellular phones produced); the samples are larger than 20; the estimated products $n_s p_s'$, $n_s q_s'$, $n_c p_c'$, and $n_c q_c'$ are all larger than 5. Therefore, the sampling distribution should have an approximately normal distribution.] And $z \star$ will be calculated using formula (10.15).

c. **Determine the level of significance, α. $\alpha = 0.05$.**

STEP 3 **The Sample Evidence:**

a. **Collect the sample information.**

$$p_s = \frac{x_s}{n_s} = \frac{15}{150} = 0.10; \quad p_c = \frac{x_c}{n_c} = \frac{6}{150} = 0.04$$

$$p_p' = \frac{x_1 + x_2}{n_1 + n_2} = \frac{15 + 6}{150 + 150} = \frac{21}{300} = 0.07;$$

$$q_p' = 1 - p_p' = 1 - 0.07 = 0.93$$

b. **Calculate the value of the test statistic.**

$$z \star = \frac{(p_s' - p_c')}{\sqrt{\left(p_p'\right)\left(q_p'\right)\left[\left(\dfrac{1}{n_s}\right) + \left(\dfrac{1}{n_c}\right)\right]}} :$$

$$z \star = \frac{(0.10 - 0.04)}{\sqrt{(0.07)(0.93)\left[\left(\dfrac{1}{150}\right) + \left(\dfrac{1}{150}\right)\right]}}$$

$$z \star = \frac{0.06}{\sqrt{0.000868}} = \frac{0.06}{0.02946} = 2.04$$

Exercise 10.71

Find the value of $z \star$ that would be used to test the difference between the proportions, given the following:

Sample	n	x
G	380	323
H	420	332

STEP 4   **The Probability Distribution:**

p-**VALUE:**   **OR**   **CLASSICAL:**

a. Calculate the *p*-value.

Use the right-hand tail since the H_a expresses concern for values related to "higher than."

$\mathbf{P} = $ *p*-value $= P(z > 2.04)$ as shown in the figure.

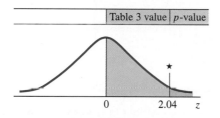

To find the *p*-value, you have three options:

1. Use Table 3 (Appendix B) to calculate *p*-value:
   $\mathbf{P} = 0.5000 - 0.4793 = \mathbf{0.0207}$
2. Use Table 5 (Appendix B) to place bounds on the *p*-value:
   $\mathbf{0.0202 < P < 0.0228}$
3. Use a computer or calculator: $\mathbf{P = 0.0207}$

Specific instructions follow this illustration.

b. Determine whether or not the *p*-value is smaller than α.

The *p*-value **is smaller** than α.

a. Determine the critical region and critical value(s).

The critical region is the right-hand tail since the H_a expresses concern for values related to "higher than." The critical value is obtained from Table 4A: $z_{(0.05)} = \mathbf{1.65}.$

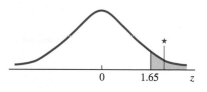

For specific instructions, see page 394, 395.

b. Determine whether or not the calculated test statistic is in the critical region.

$z \star$ **is in** the critical region, as shown in red in figure above.

STEP 5   **The Results:**

a. State the decision about H_o.

Reject H_o.

b. State the conclusion about H_a.

At the 0.05 level of significance, there is sufficient evidence to reject the salesman's claim; the proportion of his company's cellular phones that are defective is higher than the proportion of his competitor's cellular phones that are defective.   ▲

Exercise 10.72

Find the *p*-value for the test with alternative hypothesis $p_E < p_R$ using the data in Exercise 10.70.

The following commands will complete a hypothesis test for the difference between two proportions $p_1 - p_2$, for two independent sets of sample data.

EXCEL

```
Input the data for the first sample into column A using 0's for failures (or no's)
and 1's for successes (or yes's), then repeat the same procedure for the second
sample in column B; then continue with:

  Choose:   Tools > Data Analysis Plus > Inference About Two Proportions > OK
  Enter:    block coordinates: (A2:B20 or select cells)   > OK
                               (do not include column name)
            Code for success: 1   > OK
  Choose:   Test of Hypothesis (Case 1)
  Choose:   Alternative: HA: P1-P2 not equal to zero or
                         HA: P1-P2 greater than zero or
                         HA: P1-P2 less than zero
```

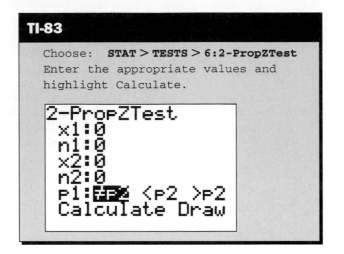

Case Study 10.3

Smokers Need More Time in Recovery Room

The following report says that a study involving 327 patients in Long Branch, New Jersey, showed that 38% of the nonsmokers spent less than one hour in recovery after surgery, compared to 23% of smokers. Further, 19% of the smokers spent more than two hours, compared to 7% of the nonsmokers.

Smokers take longer to recover from anesthesia after surgery than nonsmokers do, new research shows.

A study reported at the American Society of Anesthesiologists meeting in Las Vegas found that smokers were nearly three times more likely than nonsmokers to spend two or more hours in the recovery room. "It's a very strong difference," says researcher Dr. David Handlin, of Monmouth Medical Center, Long Branch, NJ.

Of the 327 patients studied:

* 38% of nonsmokers spent less than one hour in the recovery room, vs. 23% of smokers.
* 19% of smokers spent more than two hours in recovery, vs. 7% of nonsmokers.

Longer recovery room stays demand more nursing care and drive up health care costs, says Handlin. Other studies have shown that smokers tend to have a higher risk of complications from surgery. Handlin says this is the first to show that smokers take longer to recover from anesthesia.

Longer recovery stays may be due to mild lung disease or a lowered ability of the blood to carry oxygen, says Handlin. Some experts say quitting smoking weeks before surgery may cut recovery time.

Source: Copyright 1990, *USA TODAY*. Reprinted with permission.

> ### Exercise 10.73
>
> "38% of the nonsmokers spent less than one hour in recovery vs. 23% of smokers" was reported in "Smokers Need More Time in Recovery Room." Assuming the patients were equally divided between the two samples, find the *p*-value associated with the hypothesis test using this information. Do you believe this information is significant? Explain.

EXERCISES • • • • • • • • • • • • • • • •

10.74 In a random sample of 40 brown-haired individuals, 22 indicated that they used hair coloring. In another random sample of 40 blonde individuals, 26 indicated that they used hair coloring. Use a 92% confidence interval to estimate the difference in the proportion of these groups that use hair coloring.

10.75 An article titled "Nurse Executive Turnover" (*Nursing Economics,* January/February 1993) compared two groups of nurse executives. One group had participated in a unique

(continued)

program for nurse executives called the Wharton Fellows Program, and the other group had not participated in the program. Eighty-seven of 341 Wharton Fellows had experienced one change in position, and 9 of 40 non-Wharton Fellows had experienced one change in position. Find a 99% confidence interval for the difference in population proportions.

10.76 In a survey of 300 people from city A, 128 preferred New Spring soap to all other brands of deodorant soap. In city B, 149 of 400 people preferred New Spring. Find the 98% confidence interval for the difference in the two proportions.

10.77 The proportions of defective parts produced by two machines were compared, and the following data were collected.

$$\text{Machine 1: } n = 150; \text{ number of defective parts} = 12$$

$$\text{Machine 2: } n = 150: \text{ number of defective parts} = 6$$

Determine a 90% confidence interval for $p_1 - p_2$.

10.78 Show that the standard error of $p'_1 - p'_2$, which is $\sqrt{\left(\dfrac{p_1 q_1}{n_1}\right) + \left(\dfrac{p_2 q_2}{n_2}\right)}$, reduces to $\sqrt{pq\left[\left(\dfrac{1}{n_1}\right) + \left(\dfrac{1}{n_2}\right)\right]}$, when $p_1 = p_2 = p$.

10.79 State the null hypothesis, H_o, and the alternative hypothesis, H_a, that would be used to test these claims:
  a. There is no difference between the proportions of men and women who will vote for the incumbent in next month's election.
  b. The percentage of boys who cut classes is greater than the percentage of girls who cut classes.
  c. The percentage of college students who drive old cars is higher than the percentage of noncollege people of the same age who drive old cars.

10.80 Determine the critical region and critical value(s) that would be used to test (classical procedure) the following hypotheses when z is used as the test statistic.
  a. $H_o: p_1 = p_2$ vs. $H_a: p_1 > p_2$,   with $\alpha = 0.05$
  b. $H_o: p_A = p_B$ vs. $H_a: p_A \neq p_B$,   with $\alpha = 0.05$
  c. $H_o: p_1 - p_2 = 0$ vs. $H_a: p_1 - p_2 < 0$,   with $\alpha = 0.04$
  d. $H_o: p_m - p_f = 0$ vs. $H_a: p_m - p_f > 0$,   with $\alpha = 0.01$

10.81 A 1998 study of the Y2K problem investigated consumer opinions over what should be done to handle the situation and who should be responsible for monitoring the progress. In response to the question, "Who should monitor the report on progress in solving the Y2K problem?", 34% of the respondents surveyed felt that it was the government's responsibility (*Newsweek*, 10-5-98).

Suppose you believe that differences in opinion exist between rural and city dwellers on whether the government should monitor the Y2K problem. A study of 250 heads of households in the city and 200 rural heads of households are asked the above question. You find that 100 of the city dwellers and 64 of the rural dwellers believed that it was the government's responsibility. Is there a significant difference in the opinions of the two groups? Use $\alpha = 0.05$.
  a. Solve using the *p*-value approach.   **b.** Solve using the classical approach.

10.82 In a survey of working parents (both parents working), one of the questions asked was "Have you refused a job, promotion, or transfer because it would mean less time with your family?" Two hundred men and 200 women were asked this question. Twenty-nine percent of the men and 24% of the women responded "yes." Based on this survey, can we conclude that there is a difference in the proportion of men and women responding "yes" at the 0.05 level of significance?

10.83 Two randomly selected groups of citizens were exposed to different media campaigns that dealt with the image of a political candidate. One week later the citizen groups

(*continued*)

were surveyed to see whether they would vote for the candidate. The results were as follows:

	Exposed to Conservative Image	Exposed to Moderate Image
Number in Sample	100	100
Proportion for the Candidate	0.40	0.50

Is there sufficient evidence to show a difference in the effectiveness of the two image campaigns at the 0.05 level of significance?
a. Solve using the p-value approach.   b. Solve using the classical approach.

10.84 U.S. military active duty personnel and their dependents are provided with free medical care. A survey was conducted to compare obstetrical care between military and civilian (pay for medical care) families ("Use of Obstetrical Care Compared Among Military and Civilian Families," *Public Health Reports,* May/June 1989). The number of women who began prenatal care by the second trimester were reported as follows:

	Military	Civilian
Prenatal Care Began by 2nd Trimester	358	6786
Total Sample	407	7363

Is there a significant difference between the proportion of military and civilian women who begin prenatal care by the second trimester? Use $\alpha = 0.02$.

10.85 The July 28, 1990, issue of *Science News* reported that smoking boosts death risk for diabetics. Suppose as a follow-up study we investigated the smoking rates for male and female diabetics and obtained the following data.

Gender	n	Number Who Smoke
Male	500	215
Female	500	170

a. Test the research hypothesis that the smoking rate (proportion of smokers) is higher for males than for females. Calculate the p-value.
b. What decision and conclusion would be reached at the 0.05 level of significance?

10.86 The guidelines to ensure the sampling distribution of $p'_1 - p'_2$ is normal include several conditions about the size of several values. The two binomial distributions $B(100, 0.3)$ and $B(100, 0.4)$ satisfy all of those guidelines.
a. Verify that $B(100, 0.3)$ and $B(100, 0.4)$ satisfy all guidelines.
b. Use a computer to randomly generate 200 random samples from each of the binomial populations. Find the observed proportion for each sample and the value of the 200 differences between two proportions.
c. Describe the observed sampling distribution using both graphic and numerical statistics.
d. Does the empirical sampling distribution appear to have an approximately normal distribution? Explain

See the Statistical Tutor for additional information about commands.

10.5 Inferences Concerning the Ratio of Variances Using Two Independent Samples

When comparing two populations, it is natural that we compare their two most fundamental distribution characteristics, their "center" and their "spread," by comparing their means and standard deviations. We have learned, in two of the previous sections, how to use the t-distribution to make inferences comparing two population means using either dependent or independent samples. These procedures were intended for uses with normal populations but work quite well even when the populations are not exactly normally distributed.

The next logical step in comparing two populations is to compare their standard deviations, the most often used measure of spread. However, sampling distributions dealing with sample standard deviations (or variances) are very sensitive to slight departures from the assumptions. Therefore, the only inference procedure to be presented here will be the **hypothesis test for the equality of standard deviations (or variances)** for two normal populations.

The soft-drink bottling company discussed in Section 9.3 (page 451) is trying to decide whether to install a modern, high-speed bottling machine. There are, of course, many concerns in making this decision, and one of them is the concern that the increased speed may result in an increased variability in the amount of fill placed in each bottle; such an increase would not be acceptable. To this concern, the manufacturer of the new system responded that the variance in fills will be no larger with the new machine than with the old. (The new system will fill several bottles in the same amount of time as the old system fills one bottle—this is the reason why the change is being considered.) A test is set up to statistically test the bottling company's concern "standard deviation of new machine is greater than standard deviation of old" against the manufacturer's claim "standard deviation of new is no greater than standard deviation of old."

ILLUSTRATION 10.12 ▼

State the null and alternative hypotheses to be used for comparing the variances of the two soft-drink bottling machines.

Solution

There are several equivalent ways to express the null and alternative hypotheses, but since the test procedure uses the ratio of variances, the recommended convention is to express the null and alternative hypotheses as ratios of the population variances. Further, it is recommended that the "larger" or "expected to be larger" variance be used as the numerator. The concern of the soft-drink company is that the new modern machine (m) will result in a larger standard deviation in the amount of fill than its present machine (p); $\sigma_m > \sigma_p$, or equivalently, $\sigma_m^2 > \sigma_p^2$, that becomes $\dfrac{\sigma_m^2}{\sigma_p^2} > 1$. We want to test the manufacturer's claim (the null hypothesis) against the company's concern (the alternative hypothesis).

> **Exercise 10.87**
> Explain why the inequality $\sigma_m^2 > \sigma_p^2$ is equivalent to $\dfrac{\sigma_m^2}{\sigma_p^2} > 1$.

$$H_o: \frac{\sigma_m^2}{\sigma_p^2} = 1 \ (m \text{ is no more variable})$$

$$H_a: \frac{\sigma_m^2}{\sigma_p^2} > 1 \ (m \text{ is more variable})$$

Inferences about the ratio of variances for two normally distributed populations use the **F-distribution.** The F-distribution, similar to the Student's t-distribution and the χ^2-distribution, is a family of probability distributions. Each F-distribution is identified by two numbers of degrees of freedom, one for each of the two samples involved. ▲

Before continuing with the details of the hypothesis test procedure, let's learn about the F-distribution.

> **PROPERTIES OF THE F-DISTRIBUTION**
> 1. F is nonnegative in value; it is zero or positively valued.
> 2. F is nonsymmetrical; it is skewed to the right.
> 3. F is distributed so as to form a family of distributions; there is a separate distribution for each pair of numbers of degrees of freedom.

For inferences discussed in this section, the number of degrees of freedom for each sample is $df_1 = n_1 - 1$ and $df_2 = n_2 - 1$. Each different combination of degrees of freedom results in a different F-distribution; and each F-distribution looks approximately like the distribution shown in Figure 10.2.

Figure 10.2
F Distribution

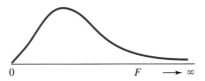

The critical values for the F-distribution are identified using three values:

df_n, the degrees of freedom associated with the sample whose variance is in the numerator of the calculated F,

df_d, the degrees of freedom associated with the sample whose variance is in the denominator, and

α, the area under the distribution curve to the right of the critical value being sought.

Therefore, the symbolic name for a critical value of F will be $F_{(df_n, df_d, \alpha)}$, as shown in Figure 10.3.

Figure 10.3
A Critical Value of F

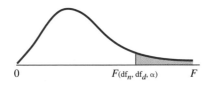

Since it takes three values to identify a single critical value of F, making tables for F is not as simple as with previously studied distributions. The tables presented in this textbook are organized so as to have a different table for each different value of α, the "area to the right." Table 9a in Appendix B shows the critical values for $F_{(df_n, df_d, \alpha)}$, when $\alpha = 0.05$; Table 9b gives the critical values when $\alpha = 0.025$; Table 9c gives the values when $\alpha = 0.01$.

ILLUSTRATION 10.13 ▼

Find $F_{(5, 8, 0.05)}$, the critical F-value for samples of size 6 and size 9 with 5% of the area in the right-hand tail.

Solution

Using Table 9a ($\alpha = 0.05$); find the intersection of column df = 5 (for numerator) and row df = 8 (for denominator), read the value: $F_{(5, 8, 0.05)} = $ **3.69**. See the accompanying partial table.

Portion of Table 9a ($\alpha = 0.05$)

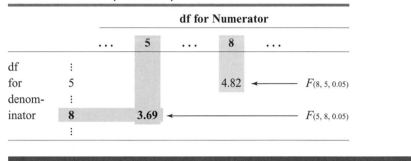

Exercise 10.88
Find the values of $F_{(12, 24, 0.01)}$ and $F_{(24, 12, 0.01)}$.

		df for Numerator		
	...	**5**	...	**8**	...

df for denom-inator				
5				4.82 ← $F_{(8, 5, 0.05)}$
8		3.69 ← $F_{(5, 8, 0.05)}$		

Notice that $F_{(8, 5, 0.05)}$ is 4.82. The degrees of freedom associated with the numerator and with the denominator must be kept in the correct order; 3.69 is quite different from 4.82. Check some other pairs to verify that interchanging the degrees of freedom numbers will result in different F-values.

The following commands will find the cumulative probability for a specified value of F.

MINITAB

Choose: **Calc > Probability Distributions > F**
Select: **Cumulative Probability**
Enter: Numerator degrees of freedom: **df$_n$**
           Denominator degrees of freedom: **df$_d$**
Select: **Input constant***
Enter: **F-value** (ex. 1.74)

*Select Input column if several F-values are stored in C1. Use C2 for optional storage. If the area in the right tail is needed, subtract the calculated probability from 1.

TI-83

Choose: **2nd > DISTR > 9:Fcdf(**
Enter: **0, F-value, df$_n$, df$_d$)**

Note: To find the probability between two F-values, enter the two values in place of 0 and F-value.

If the area in the right tail is needed, subtract the calculated probability from 1.

The following commands will find the probability in the right tail for a given F-value.

EXCEL

```
If several F-values are to be used, input the values into column A and activate
B1; then continue with:

    Choose:   Paste function fx > Statistical > FDIST > OK
    Enter:    X: individual F-value or (A1:A5 or select
              'F-value' cells)*
              Deg_freedom 1: dfn
              Deg_freedom 2: dfd
    Drag*:    Bottom right corner of the B1 cell down to give other probabilities
To find the probability for the left tail (the cumulative probability up to the F-
value), subtract the calculated probability from 1.
```

We are ready to use F to complete a hypothesis test about the ratio of two population variances.

THE ASSUMPTIONS FOR INFERENCES ABOUT THE RATIO OF TWO VARIANCES

The samples are randomly selected from normally distributed populations, and the two samples are selected in an independent manner.

ILLUSTRATION 10.14 ▼

Recall that our soft-drink bottling company was to make a decision about the equality of the variances of amounts of fill between its present machine and a modern high-speed outfit. Does the sample information in Table 10.6 present sufficient evidence to reject the null hypothesis (the manufacturer's claim) that the modern high-speed bottle-filling machine fills bottles with no more variance than the company's present machine? Assume the amount of fill is normally distributed for both machines, and complete the test using $\alpha = 0.01$.

Table 10.6 Sample Information for Illustration 10.14

Sample	n	s^2
Present machine (p)	22	0.0008
Modern high-speed machine (m)	25	0.0018

Solution

STEP 1 **The Set-Up:**

a. **Describe the population parameter of interest.**

$\dfrac{\sigma_m^2}{\sigma_p^2}$, **the ratio of the variances** in the amount of fill placed in bottles, the modern machine to the company's present machine.

b. **State the null hypothesis (H_o) and the alternative hypothesis (H_a).**

The hypotheses were established in Illustration 10.12 (see p. 515):

$$H_o: \frac{\sigma_m^2}{\sigma_p^2} = 1 \ (m \text{ is no more variable}) \quad \text{vs.} \quad H_a: \frac{\sigma_m^2}{\sigma_p^2} > 1 \ (m \text{ is more variable})$$

Exercise 10.89

Express the H_o and H_a of Illustration 10.14 equivalently in terms of standard deviations.

NOTE When the "expected to be largest" variance is in the numerator for a one-tail test, the alternative hypothesis will state, "the ratio of the variances is greater than 1."

STEP 2   **The Hypothesis Test Criteria:**
   a. Check the assumptions.
        The assumptions: The **sampled populations are normally distributed** (given in statement of problem) and the **samples are independently selected** (drawn from two separate populations).
   b. Identify the probability distribution and the test statistic to be used.
        The **F-distribution** with the ratio of the sample variances, formula (10.16):

$$F \star = \frac{s_m^2}{s_p^2}, \text{ with } df_m = n_m - 1 \text{ and } df_p = n_p - 1 \qquad \textbf{(10.16)}$$

The sample variances are assigned to the numerator and denominator in the order established by the null and alternative hypotheses for one-tail tests. The calculated ratio, $F \star$, will have an F-distribution with $df_n = n_n - 1$ (numerator) and $df_d = n_d - 1$ (denominator) when the assumptions and the null hypothesis are true.

   c. Determine the level of significance, α.
        $\alpha = 0.01$.

STEP 3   **The Sample Evidence:**
   a. Collect the sample information.
        See Table 10.6 on page 518.
   b. Calculate the value of the test statistic.
        Using Formula (10.16),

> **Exercise 10.90**
> Calculate $F \star$ given $s_1 = 3.2$ and $s_2 = 2.6$.

$$F \star = \frac{s_m^2}{s_p^2}: \qquad F \star = \frac{0.0018}{0.0008} = \textbf{2.25}$$

The number of degrees of freedom for the numerator is, $df_n = 24 \, (25 - 1)$ since the sample from the modern high-speed machine is associated with the numerator, as specified by the null hypothesis. $df_d = 21$ since the sample associated with the denominator has size 22.

STEP 4   **The Probability Distribution:**

p-**VALUE:**   **CLASSICAL:**

a. Calculate the *p*-value.   OR   **a. Determine the critical region and**
Use the right-hand tail since the H_a expresses concern for   **critical value(s).**
values related to "more than."   The critical region is the right-hand tail
$\mathbf{P} = P(F > 2.25$, with $df_n = 24$ and $df_d = 21)$ as shown in   since the H_a expresses concern for values
the figure.   related to "more than." $df_n = 24$ and $df_d = 21$. The critical value is obtained from
                                                     Table 9(c): $F_{(24, \, 21, \, 0.01)} = \textbf{2.80}$.

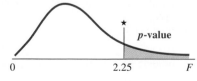

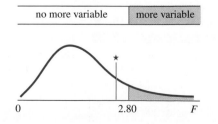

To find the *p*-value, you have two options:
1. Use Tables 9(a) and (b) (Appendix B) to place bounds on the *p*-value: **0.025 < P < 0.05**

p-VALUE:

2. Use a computer or calculator to calculate the *p*-value:
   P = 0.0323

Specific instructions follow this illustration.

b. Determine whether or not the *p*-value is smaller than α.

The *p*-value **is not smaller** than the level of significance, $\alpha(0.01)$.

CLASSICAL:

OR   For additional instructions, see page 517.

b. Determine whether or not the calculated test statistic is in the critical region.

$F \bigstar$ **is not in** the critical region, as shown in **red** in the figure above.

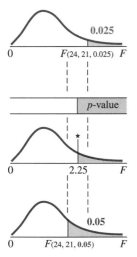

STEP 5   The Results:

a. State the decision about H_o.
   Fail to reject H_o.

b. State the conclusion about H_a.
   At the 0.01 level of significance, the samples do not present sufficient evidence to indicate an increase in variance.

Calculating the p-value when using the F-distribution:

   Method 1   *Using Table 9 in Appendix B to "place bounds" on the* p-*value:* Using Tables 9a, b, and c in Appendix B to estimate the *p*-value is very limited. However, for Illustration 10.14, the *p*-value can be estimated. By inspecting Tables 9a and 9b, you will find that $F_{(24, 21, 0.025)} = 2.37$ and $F_{(24, 21, 0.05)} = 2.05$. $F \bigstar = 2.25$ is between the values 2.37 and 2.05; therefore, the *p*-value is between 0.025 and 0.05; **0.025 < P < 0.05.**

   Method 2   If you are doing the hypothesis test with the aid of a computer or calculator, most likely it will calculate the *p*-value for you or you may use the cumulative probability distribution commands described on page 517, 518.   ▲

Exercise 10.91

Find the critical value for the hypothesis test with H_a: $\sigma_1 > \sigma_2$, with $n_1 = 7$, $n_2 = 10$, using the classical approach and $\alpha = 0.05$.

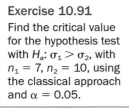

α must still be split between the two tails for a two-tailed H_a.

ILLUSTRATION 10.15 ▼

   The tables of critical values for the *F*-distribution give only the right-hand critical values. This will not be a problem since the right-hand critical value is the only critical value that will be needed. You can adjust the numerator-denominator order so that all the "activity" is in the right-hand tail. There are two cases: one-tailed tests and two-tailed tests.

One-tailed tests: Arrange the null and alternative hypotheses so that the alternative is always "greater than." The $F \bigstar$-value is calculated using the same order as specified in the null hypothesis (as in Illustration 10.14; also, see Illustration 10.15).

Two-tailed tests: When the value of $F \bigstar$ is calculated, always use the sample with the largest variance for the "numerator"; this will make $F \bigstar$ larger than 1 and place it in the right-hand tail of the distribution. Thus, you will need only the critical value for the right-hand tail (see Illustration 10.16).

Reorganize the following alternative hypothesis so that the critical region will be the right-hand tail.

$$H_a: \sigma_1^2 < \sigma_2^2 \quad \text{or} \quad \frac{\sigma_1^2}{\sigma_2^2} < 1 \quad \text{(Population 1 is less variable)}$$

Solution

Reverse the direction of the inequality, and reverse the roles of the numerator and denominator.

$$H_a: \sigma_2^2 > \sigma_1^2 \quad \text{or} \quad \frac{\sigma_2^2}{\sigma_1^2} > 1 \quad \text{(Population 2 is more variable)}$$

The calculated test statistic $F \bigstar$ will be $\dfrac{s_2^2}{s_1^2}$.

ILLUSTRATION 10.16 ▼

Find $F \bigstar$ and the critical values for the following hypothesis test so that only the right-hand critical value is needed. Use $\alpha = 0.05$ and the sample information $n_1 = 10$, $n_2 = 8$, $s_1 = 5.4$, $s_2 = 3.8$.

> **Exercise 10.92**
> What would be the value of $F \bigstar$ in Illustration 10.16 if $F \bigstar = \dfrac{s_2^2}{s_1^2}$ were used? Why is it less than 1?

$$H_o: \sigma_2^2 = \sigma_1^2 \quad \text{or} \quad \frac{\sigma_2^2}{\sigma_1^2} = 1 \qquad \text{vs.} \qquad H_a: \sigma_2^2 \neq \sigma_1^2 \quad \text{or} \quad \frac{\sigma_2^2}{\sigma_1^2} \neq 1$$

Solution

When the alternative hypothesis is two-tailed ($\neq$), the calculated $F \bigstar$ can be either $F \bigstar = \dfrac{s_1^2}{s_2^2}$ or $F \bigstar = \dfrac{s_2^2}{s_1^2}$. The choice is ours; we only need to make sure that we keep the df_n and df_d in the correct order. We make the choice by looking at the sample information and using the sample with the larger standard deviation or variance as the sample for the numerator. Therefore, in this illustration,

$$F \bigstar = \frac{s_1^2}{s_2^2} = \frac{5.4^2}{3.8^2} = \frac{29.16}{14.44} = \mathbf{2.02}$$

The critical values for this test are: left tail, $F(9, 7, 0.975)$, and right tail, $F(9, 7, 0.025)$, as shown in the figure.

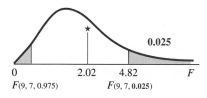

Since we chose the sample with the larger standard deviation (or variance) for the numerator, the value of $F \bigstar$ will be greater than 1 and will be in the right-hand tail; therefore, only the right-hand critical value is needed. (All critical values for left-hand tails will be values between 0 and 1.) ▲

Exercise 10.93

When a hypothesis test is two-tailed and EXCEL is used to calculate the p-value, what additional step must be taken?

The following commands will complete a hypothesis test for the ratio between two population variances, $\frac{\sigma_1^2}{\sigma_2^2}$, *for two independent sets of sample data.*

TI-83

Input the data for the numerator (larger spread) into L1 and the data for the denominator (smaller spread) in L2, then continue with the following, entering the appropriate values and highlighting Calculate:

Choose: **STAT > TESTS >**
              D:2-SampFTest. . .

```
2-SampFTest
 Inpt:Data Stats
 List1:L1
 List2:L2
 Freq1:1
 Freq2:1
 σ1:≠σ2 <σ2 >σ2
 Calculate Draw
```

EXCEL

Input the data for the numerator (larger spread) into column A and the data for the denominator (smaller spread) in column B, then continue with:

Choose: **Tools > Data Analysis > F-Test: Two-Sample for Variances > OK**

Enter: **Variable 1 Range: (A1:A20 or select cells)**
           Variable 2 Range: (B1:B20 or select cells)

Select: **Labels** (if necessary)

Enter: **α** (ex. 0.05)

Select: **Output Range**

Enter: **(C1 or select cell)**

Use Format > Column > AutoFit Selection to make the output more readable. The output shows the p-value and critical value for a one-tailed test.

Case Study   10.4

Personality Characteristics of Police Academy Applicants:

Comparisons Across Subgroups and with Other Populations

Exercise 10.94

a. What null and alternative hypotheses did Carpenter and Raza test?

b. What does "$p <$ 0.005" mean?

c. Use a computer or calculator to calculate the p-value for $F(237, 305) = 1.36$.

Bruce N. Carpenter and Susan M. Raza concluded that "police applicants are somewhat more like each other than are those in the normative population" when the F-test of homogeneity of variance resulted in a p-value of less than 0.005. Homogeneity means that the group's scores are less variable than the scores for the normative population.

To determine whether police applicants are a more homogeneous group than the normative population, the *F*-test of homogeneity of variance was used. With the exception of scales F, K, and 6, where the differences are nonsignificant, the results indicate that the police applicants form a somewhat more homogeneous group than the normative population [$F(237,305) = 1.36, p < .005$]. Thus, police applicants are somewhat more like each other than are individuals in the normative population.

Source: Reproduced from the *Journal of Police Science and Administration,* Vol. 15, no. 1, pp. 10–17, with permission of the International Association of Chiefs of Police, PO Box 6010, 13 Firstfield Road, Gaithersburg, MD 20878.

EXERCISES • • • • • • • • • • • • • •

10.95 State the null hypothesis, H_o, and the alternative hypothesis, H_a, that would be used to test the following claims.
   a. The variances of populations A and B are not equal.
   b. The standard deviation of population I is larger than the standard deviation of population II.
   c. The ratio of the variances for populations A and B is different from 1.
   d. The variability within population C is less than the variability within population D.

10.96 Using the $F_{(df_1,\, df_2,\, \alpha)}$ notation name each of the critical values shown on the following figures.

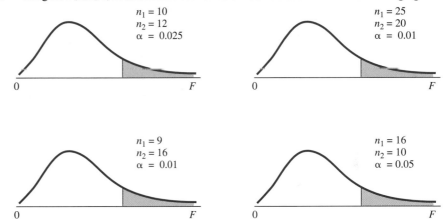

10.97 Find the following critical values for F from Tables 9a, 9b, and 9c in Appendix B.
   a. $F_{(24,\, 12,\, 0.05)}$ **b.** $F_{(30,\, 40,\, 0.01)}$ **c.** $F_{(12,\, 10,\, 0.05)}$
   d. $F_{(5,\, 20,\, 0.01)}$ **e.** $F_{(15,\, 18,\, 0.025)}$ **f.** $F_{(15,\, 9,\, 0.025)}$
   g. $F_{(40,\, 30,\, 0.05)}$ **h.** F_w

10.98 Determine the critical region and critical value(s) that would be used to test the following hypotheses using the classical model when $F\,\bigstar$ is used as the test statistic.
   a. $H_o\colon \sigma_1^2 = \sigma_2^2$ vs. $H_a\colon \sigma_1^2 > \sigma_2^2$, with $n_1 = 10$, $n_2 = 16$, and $\alpha = 0.05$
   b. $H_o\colon \dfrac{\sigma_1^2}{\sigma_2^2} = 1$ vs. $H_a\colon \dfrac{\sigma_1^2}{\sigma_2^2} \neq 1$, with $n_1 = 25$, $n_2 = 31$, and $\alpha = 0.05$
   c. $H_o\colon \dfrac{\sigma_1^2}{\sigma_2^2} = 1$ vs. $H_a\colon \dfrac{\sigma_1^2}{\sigma_2^2} > 1$, with $n_1 = 10$, $n_2 = 10$, and $\alpha = 0.01$
   d. $H_o\colon \sigma_1 = \sigma_2$ vs. $H_a\colon \sigma_1 < \sigma_2$, with $n_1 = 25$, $n_2 = 16$, and $\alpha = 0.01$

10.99 **a.** Two independent samples, each of size 3, are drawn from a normally distributed population. Find the probability that one of the sample variances is at least 19 times larger than the other one.
   b. Two independent samples, each of size 6, are drawn from a normally distributed population. Find the probability that one of the sample variances is no more than 11 times larger than the other one.

10.100 The quality of the end product is somewhat determined by the quality of the materials used. Textile mills monitor the tensile strength of the fibers used in weaving their yard goods. The following data are tensile strengths of cotton fibers from two suppliers.

Supplier A	78	82	85	83	77	84	90	82	93	82
	80	82	77	80	80					
Supplier B	76	79	83	78	72	73	69	80	74	77
	78	78	73	76	78	79				

(continued)

Calculate the observed value of F, $F \star$ for comparing the variances of these two sets of data.

10.101 A bakery is considering buying one of two gas ovens. The bakery requires that the temperature remain constant during a baking operation. A study was conducted to measure the variance in temperature of the ovens during the baking process. The variance in temperature before the thermostat restarted the flame for the Monarch oven was 2.4 for 16 measurements. The variance for the Kraft oven was 3.2 for 12 measurements. Does this information provide sufficient reason to conclude that there is a difference in the variances for the two ovens? Assume measurements are normally distributed and use a 0.01 level of significance.

10.102 A study was conducted to determine whether or not there was equal variability in male and female systolic blood pressures. Random samples of 16 men and 13 women were used to test the experimenter's claim that the variances were unequal. MINITAB was used to calculate the standard deviations, $F \star$ and the p-value.

Men	120	120	118	112	120	114	130	114	124	125
	130	100	120	108	112	122				

Women	122	102	118	126	108	130	104	116	102	122
	120	118	130							

```
Standard deviation of Men = 7.8864
Standard deviation of Women = 9.9176

  F*        cumprob     p-value
  1.58144   0.801026    0.198974
```

Verify these results by calculating the values yourself.

10.103 A study in *Pediatric Emergency Care* (June 1994) compared the injury severity between younger and older children. One measure reported was the Injury Severity Score (ISS). The standard deviation of ISS scores for 37 children eight years or younger was 23.9, and the standard deviation for 36 children older than eight years was 6.8. Assume that ISS scores are normally distributed for both age groups.
   a. At the 0.01 level of significance, is there sufficient reason to conclude that the standard deviation of ISS scores for younger children is larger than the standard deviation of ISS scores for older children?
   b. Use a computer or a calculator to calculate the p-value for this hypothesis test.

10.104 Television viewing time appears to be quite different from one age group to the next. According to Nielson Media Research, adults in 1997 over age 54 watched TV for an average of 11.2 hours per week between 8:00 and 11:00 p.m. On the other hand, adults between the ages of 25 and 54 averaged 8.5 hours of viewing time per week. Media experts claim that such data is suspect to interpretation because younger adults are more unsettled and exhibit a greater variety of life styles than older adults, many of whom are retired and display more stability. (*The World Almanac and Book of Facts 1998.*)

A recent study is conducted to check the results of the national survey, and the sampling data is shown in the table below:

Viewing Age Group	n	**Weekly Mean** Viewing Time (hrs)	**Standard** Deviation
25 to 54 (Y)	27	8.8	2.9
55 and over (O)	28	11.6	1.6

(continued)

Test the hypothesis of equal variances against the alternative hypothesis that younger adults have a greater variance. Use the 0.05 level of significance.

a. Solve using the *p*-value approach.   **b.** Solve using the classical approach.

10.105 Use a computer to demonstrate the truth of the theory presented in this section.

See the *Statistical Tutor* for additional information about commands.

   a. The underlying assumptions are "the populations are normally distributed" and while conducting a hypothesis test for the equality of two standard deviations, it is assumed that the standard deviations are equal. Generate very large samples of two theoretical populations: $N(100, 20)$ and $N(120, 20)$. Find graphic and numerical evidence that the populations satisfy the assumptions.

   b. Randomly select 100 samples, each of size 8, from both populations and find the standard deviation of each sample.

   c. Using the first sample drawn from each population as a pair, calculate the $F \star$-statistic. Repeat for all samples. Describe the sampling distribution of the 100 $F \star$-values using both graphic and numerical statistics.

   d. Generate the probability distribution for $F(7, 7)$, and compare it to the observed distribution of $F \star$. Do the two graphs agree? Explain.

10.106 It was stated in this section that the *F*-test was very sensitive to minor digressions from the assumptions. Repeat Exercise 10.105 using $N(100, 20)$ and $N(120, 30)$. Notice that the only change from Exercise 10.105 is the seemingly slight increase in the standard deviation of the second population. Answer the same questions using the same kind of information and you will see very different results.

Return to C H A P T E R   C A S E   S T U D Y

One of the most commonly seen applications of statistics is the poll percentages, reported in the news, of people who say, think, or do some specific thing. So who does "know the American flag"? Let's apply some of the statistical methods learned in this chapter by returning to the Chapter Case Study (p. 473) and answering the questions asked there.

EXERCISE 10.107

Two hundred adults in Erie County, New York, were asked how many stars there are on the American flag. The table below shows the number of adults belonging to each category. The sample results were tallied twice, by gender and by residence of adult answering question.

	Men	Women	City	Urban	Rural
n(Knew)	72	72	57	58	31
n(Didn't know)	22	34	25	14	15

   a. Is there a significant difference between the percentage of men and the percentage of women who answered the question correctly? Use $\alpha = 0.05$.

   b. Is there a difference between the percentage of city and the percentage of urban adults who answered the question correctly? Use $\alpha = 0.05$.

   c. Does this information seem to be consistent with the results found in the Maritz AmeriPoll? Give a detailed answer using statistical methods studied in this chapter.

IN RETROSPECT

In this chapter we began the comparisons of two populations by first distinguishing between independent and dependent samples, which are statistically very important and useful sampling procedures. We then proceeded to examine the inferences concerning the comparison of means, proportions, and variances for two populations.

The use of confidence intervals and hypothesis tests can sometimes be interchanged; that is, a confidence interval can be used in place of a hypothesis test. For example, Illustration 10.10 (p. 507) called for a confidence interval. Now suppose that Mr. Morris asked: "Is there a difference in my voter appeal to men voters as opposed to women voters?" To answer his question, you would not need to complete a hypothesis test if you chose to test at $\alpha = 0.01$ using a two-tailed test. "No difference" would mean a difference of zero, which is not included in the interval from 0.014 to 0.128 (the interval determined in Illustration 10.10). Therefore, a null hypothesis of "no difference" would be rejected, thereby substantiating the conclusion that a significant difference exists in voter appeal between the two groups.

We are always making comparisons between two groups. We compare means and we compare proportions. In this chapter we have learned how to statistically compare two populations by making inferences about their means, proportions, or variances. For convenience, Table 10.7 identifies the formulas to use when making inferences about comparisons between two populations.

In Chapters 8 through 10 we have learned how to use confidence intervals and hypothesis tests to answer questions about means, proportions, and standard deviations for one or two populations. From here we can expand our techniques to include inferences about more than two populations as well as inferences of different types.

Table 10.7   Formulas to Use for Inferences Involving Two Populations

Situations	Test Statistic	Confidence Interval	Hypothesis Test
		Formula to Be Used	
Difference between two means			
Using dependent samples	t	Formula (10.4) (p. 479)	Formula (10.5) (p. 481)
Using independent samples	t	Formula (10.8) (p. 491)	Formula (10.9) (p. 493)
Difference between two proportions	z	Formula (10.11) (p. 507)	Formula (10.12) (p. 509)
Difference between two variances	F		Formula (10.16) (p. 519)

CHAPTER EXERCISES

10.108 The diastolic blood pressures for 15 patients were determined using two techniques: the standard method used by medical personnel and a method using an electronic device with a digital readout. The results were as follows:

Patient	1	2	3	4	5	6	7	8	9	10	11	12	13	14	15
Standard method	72	80	88	80	80	75	92	77	80	65	69	96	77	75	60
Digital method	70	76	87	77	81	75	90	75	82	64	72	95	80	70	61

Assuming blood pressure is normally distributed, determine the 90% confidence interval for the mean difference in the two readings, where d = standard method − digital readout.

10.109 Using a 95% confidence interval, estimate the difference in I.Q. between the oldest and the youngest members (brothers and sisters) of a family based on the following random sample of I.Q.s. Assume normality.

Oldest	145	133	116	128	85	100	105	150	97	110	120	130
Youngest	131	119	103	93	108	100	111	130	135	113	108	125

10.110 We want to know which of two types of filters should be used. A test was designed in which the strength of a signal could be varied from zero to the point where the operator first detects the image. At this point, the intensity setting is recorded. The lower the setting, the better. Twenty operators were asked to make one reading for each filter.

Operator	1	2	3	4	5	6	7	8	9	10
Filter1	96	83	97	93	99	95	97	91	100	92
Filter2	92	84	92	90	93	91	92	90	93	90

Operator	11	12	13	14	15	16	17	18	19	20
Filter1	88	89	85	94	90	92	91	78	77	93
Filter2	88	89	86	91	89	90	90	80	80	90

Assuming the intensity readings are normally distributed, estimate the mean difference between the two readings using a 90% confidence interval.

10.111 Ten new recruits participated in a rifle-shooting competition at the end of their first day at training camp. The same ten competed again at the end of a full week of training and practice. Their resulting scores are shown in the following table.

Time of Competition	Recruit									
	1	2	3	4	5	6	7	8	9	10
First day	72	29	62	60	68	59	61	73	38	48
One week later	75	43	63	63	61	72	73	82	47	43

Does this set of ten pairs of data show that there was a significant amount of improvement in the recruits' shooting abilities during the week? Use $\alpha = 0.05$.

10.112 Twelve automobiles were selected at random to test two new mixtures of unleaded gasoline. Each car was given a measured allotment of the first mixture, x, and driven; then the distance traveled was recorded. The second mixture, y, was immediately tested in the same manner. The order in which the x and y mixtures were tested was also randomly assigned. The results are given in the following table.

Mixture	Car											
	1	2	3	4	5	6	7	8	9	10	11	12
x	7.9	5.6	9.2	6.7	8.1	7.3	8.1	5.4	6.9	6.1	7.1	8.1
y	7.7	6.1	8.9	7.1	7.9	6.7	8.2	5.0	6.2	5.7	6.2	7.5

Can you conclude that there is no real difference in mileage obtained by these two gasoline mixtures at the 0.10 level of significance? Assume mileage is normal.
a. Solve using the p-value approach.   **b.** Solve using the classical approach.

10.113 A test that measures math anxiety was given to 50 male and 50 female students. The results were as follows:

$$\text{Males:} \quad \bar{x} = 70.5, \quad s = 13.2$$

$$\text{Females:} \quad \bar{x} = 75.7, \quad s = 13.6$$

Construct a 95% confidence interval for the difference between the mean anxiety scores.

10.114 The same achievement test is given to soldiers selected at random from two units. The scores they attained are summarized as follows:

$$\text{Unit 1:} \quad n_1 = 70 \quad \bar{x}_1 = 73.2 \quad s_1 = 6.1$$

$$\text{Unit 2:} \quad n_2 = 60 \quad \bar{x}_2 = 70.5 \quad s_2 = 5.5$$

Construct a 90% confidence interval for the difference in the mean level of the two units.

10.115 Ten soldiers were selected at random from each of two companies to participate in a rifle-shooting competition. Their scores are shown in the following table.

| Company A | 72 | 29 | 62 | 60 | 68 | 59 | 61 | 73 | 38 | 48 |
| Company B | 75 | 43 | 63 | 63 | 61 | 72 | 73 | 82 | 47 | 43 |

Construct a 95% confidence interval for the difference between the mean scores for the two companies.

10.116 The performance on an achievement test in a beginning computer science course was administered to two groups. One group had a previous computer science course in high school; the other group did not. The test results are below. Assuming test scores are normal, construct a 98% confidence interval for the difference between the two means.

| Group 1 (had high school course) | 17 | 18 | 27 | 19 | 24 | 36 | 27 | 26 | 35 | 22 | 18 | 29 |
| | 29 | 26 | 33 | | | | | | | | | |

| Group 2 (no high school course) | 19 | 25 | 28 | 27 | 21 | 24 | 18 | 14 | 28 | 21 | 22 | 20 |
| | 21 | 14 | 29 | 28 | 25 | 17 | 20 | 28 | 31 | 27 | | |

10.117 Two methods were used to study the latent heat of ice fusion. Both method A (an electrical method) and method B (a method of mixtures) were conducted with the specimens cooled to $-0.72°C$. The data in the following table represent the change in total heat from $-0.72°C$ to water at $0°C$ in calories per gram of mass.

| Method A | 79.98 | 80.04 | 80.02 | 80.04 | 80.03 | 80.03 | 80.04 | 79.97 | 80.05 | 80.03 |
| | 80.02 | 80.00 | 80.02 | | | | | | | |

| Method B | 80.02 | 79.94 | 79.98 | 79.97 | 79.97 | 80.03 | 79.95 | 79.97 | | |

Construct a 95% confidence interval for the difference between the means.

10.118 A test concerning some of the fundamental facts about AIDS was administered to two groups, one consisting of college graduates and the other consisting of high school graduates. A summary of the test results follows:

College graduates: $n = 75$ $\bar{x} = 77.5$ $s = 6.2$

High school graduates: $n = 75$ $\bar{x} = 50.4$ $s = 9.4$

Do these data show that the college graduates, on the average, score significantly higher on the test? Use $\alpha = 0.05$.

10.119 George Johnson is the head coach of a college football team that trains and competes at home on artificial turf. George is concerned that the 40-yard sprint time recorded by his players and others increases substantially when running on natural turf as opposed to artificial turf. If so, there is little comparison between his players' speed and those of his opponents whenever his team plays on grass. George's next opponent plays on grass, so he surveyed all the opponents' starters in the next game and obtained their best 40-yard sprint times. He then compared them to the best times turned in by his own players on artificial turf. The results are shown in the table below:

Player Group	n	Mean (sec)	Std. Dev.
Artificial Turf	22	4.85	0.31
Grass	22	4.96	0.42

Do Coach Johnson's players have a lower mean sprint time? Test at the 0.05 level of significance to advise Coach Johnson.
a. Solve using the *p*-value approach.   **b.** Solve using the classical approach.

10.120 To compare the merits of two short-range rockets, 8 of the first kind and 10 of the second kind are fired at a target. The first kind has a mean target error of 36 ft and a standard deviation of 15 ft, while the second kind has a mean target error of 52 ft and a standard deviation of 18 ft. Does this indicate that the second kind of rocket is less accurate than the first? Use $\alpha = 0.01$ and assume normal distribution for target error.

10.121 The following data were collected concerning waist sizes of men and women. Do these data present sufficient evidence to conclude that men have larger mean waist sizes than women at the 0.05 level of significance? Assume waist sizes are normally distributed.

Men	33	33	30	34	34	40	35	35	32
	34	32	35	32	32	34	36	30	38
Women	22	29	27	24	28	28			
	27	26	27	26	25				

10.122 A group of 17 students participated in an evaluation of a special training session that claimed to improve memory. The students were randomly assigned to two groups: group A, the test group, and group B, the control group. All 17 students were tested for the ability to remember certain material. Group A was given the special training; group B was not. After one month both groups were tested again, with the results as shown in the following table. Do these data support the alternative hypothesis that the special training is effective at the $\alpha = 0.01$ level of significance?

Time of Test	Group A Students									Group B Students							
	1	2	3	4	5	6	7	8	9	10	11	12	13	14	15	16	17
Before	23	22	20	21	23	18	17	20	23	22	20	23	17	21	19	20	20
After	28	29	26	23	31	25	22	26	26	23	25	26	18	21	17	18	20

10.123 A survey was conducted to determine the proportion of Democrats as well as Republicans who support a "get tough" policy in South America. The results of the survey were as follows:

Democrats: $n = 250$ number in support $= 120$

Republicans: $n = 200$ number in support $= 105$

Construct the 98% confidence interval for the difference between the proportions of support.

10.124 According to a report in *Science News* (Vol. 137, No. 8), the percentage of seniors who used an illicit drug during the previous month was 19.7% in 1989. The figure was 21.3% in 1988. The annual survey of 17,000 seniors is conducted by researchers at the University of Michigan in Ann Arbor.
  a. Set a 95% confidence interval on the true decrease in usage.
  b. Does the interval found in part (a) suggest that there has been a significant decrease in the usage of illicit drugs by seniors? Explain.

10.125 Of a random sample of 100 stocks on the New York Stock Exchange, 32 made a gain today. A random sample of 100 stocks on the American Stock Exchange showed 27 stocks making a gain.
  a. Construct a 99% confidence interval, estimating the difference in the proportion of stocks making a gain.
  b. Does the answer to (a) suggest that there is a significant difference between the proportions of stocks making gains on the two stock exchanges?

10.126 A consumer group compared the reliability of two comparable microcomputers from two different manufacturers. The proportion requiring service within the first year after purchase was determined for samples from each of two manufacturers.

Manufacturer	Sample Size	Proportion Needing Service
1	75	0.15
2	75	0.09

Find a 0.95 confidence interval for $p_1 - p_2$.

10.127   According to "Holiday blues," the USA Snapshot® reported in the December 24, 1997 *USA Today,* 65% of adult men and 55% of adult women have suffered from holiday depression.

Give details to support each of your answers.
a. If these statistics came from samples of 100 men and 100 women, is the difference significant?
b. If these statistics came from samples of 150 men and 150 women, is the difference significant?
c. If these statistics came from samples of 200 men and 200 women, is the difference significant?

10.128   In determining the "goodness" of a test question, a teacher will often compare the percentage of better students who answer it correctly to the percentage of the poorer students who answer it correctly. One expects that the better students will answer the question correctly more frequently than the poorer students. On the last test, 35 of the students with the top 60 grades and 27 with the bottom 60 answered a certain question correctly. Did the students with the top grades do significantly better on this question? Use $\alpha = 0.05$.
a. Solve using the *p*-value approach.   **b.** Solve using the classical approach.

10.129   The April 4, 1991, issue of *USA Today* reported results from the *New England Journal of Medicine.* In a study of 987 deaths in southern California, the average right-hander died at age 75 and the average left-hander died at age 66. In addition, it was found that 7.9% of the lefties died from accident-related injuries, excluding vehicles, versus 1.5% for the right-handers; and 5.3% of the left-handers died while driving vehicles versus 1.4% of the right-handers.

Suppose you examine 1000 randomly selected death certificates of which 100 were left-handers and 900 were right-handers. If you found that 5 of the left-handers and 18 of the right-handers died while driving a vehicle, would you have evidence to show that the proportion of left-handers who die at the wheel is significantly higher than the proportion of right-handers? Calculate the *p*-value and interpret its meaning.

10.130   PC users are often victimized by hardware problems. A 1998 study (*PC World,* "Which PC Makers Can You Trust?", November 1998) revealed that hardware problems reported to manufacturers could not be fixed by one in three owners of personal computers. Home PC owners fared even worse than those with work PCs, facing longer waits for service and getting even fewer problems resolved. Relatively few owners gave service technicians high marks for having adequate knowledge or for exerting sincere efforts to help solve the problems with the hardware.

Suppose a study is conducted to compare the service provided by manufacturers to both home PC owners and work PC owners. Of 220 home PC owners who had trouble, 98 reported that their problem was not resolved satisfactorily. When the same question was asked to 180 work PC owners who experienced difficulty, 52 reported that the problem was not resolved. Did the home PC owners experience more problems that could not be solved with help from the manufacturer? Use the 0.05 level of significance to answer the question.
a. Solve using the *p*-value approach.   **b.** Solve using the classical approach.

10.131   Who wins disputed cases whenever there is a change made to the tax laws, the taxpayer or the Internal Revenue Service? The latest trend indicates that the burden of proof in all court cases has shifted from the taxpayers to the IRS, which tax experts predict could set off more intrusive questioning. Of the accountants, lawyers, and other tax professionals surveyed by RIA Group, a tax-information publisher, 55% expect at least a slight increase in taxpayer wins. (*Fortune,* "Tax Reform?", 9-28-98).

Suppose samples of 175 accountants and 165 lawyers are asked, "Do you expect taxpayers to win more court cases because of the new burden of proof rules?" Of those surveyed, 101 accountants replied yes, and 84 lawyers said yes. Do the two expert groups differ in their opinions? Use a 0.01 level of significance to answer the question.
a. Solve using the *p*-value approach.   **b.** Solve using the classical approach.

10.132 "It's a draw, according to two Australian researchers. By age 25, up to 29% of all men and up to 34% of all women have some gray hair, but this difference is so small that it's considered insignificant" ("Silver Threads Among the Gold: Who'll Find Them First, a Man or a Woman?" *Family Circle,* June 26, 1990). If 1000 men and 1000 women were involved in this research, would the 5% difference mentioned be significant at the 0.01 level? Explain, include details to support your answer.

10.133 A manufacturer designed an experiment to compare the difference between men and women with respect to the times they require to assemble a product. Fifteen men and 15 women were tested to determine the time they required, on the average, to assemble the product. The time required by the men had a standard deviation of 4.5 min, and the times required by the women had a standard deviation of 2.8 min. Do these data show that the amount of time needed by men is more variable than the time needed by the women? Use $\alpha = 0.05$ and assume the time are approximately normally distributed.
a. Solve using the *p*-value approach. **b.** Solve using the classical approach.

10.134 A soft-drink distributor is considering two new models of dispensing machines. Both the Harvard Company machine and the Fizzit machine can be adjusted to fill the cups to a certain mean amount. However, the variation in the amount dispensed from cup to cup is a primary concern. Ten cups dispensed from the Harvard machine showed a variance of 0.065, whereas 15 cups dispensed from the Fizzit machine showed a variance of 0.033. The factory representative from the Harvard Company maintains that his machine had no more variability than the Fizzit machine. Assume the amount dispensed is normally distributed.

At the 0.05 level of significance, does the sample refute the representative's assertion?
a. Solve using the *p*-value approach. **b.** Solve using the classical approach.

10.135 Mindy Fernandez is in charge of production at the new sport utility vehicle (SUV) assembly plant that just opened in her town. Lately she has been concerned that the wheel lug bolts don't match the chrome lug nuts close enough to keep the assembly of the wheels operating smoothly. Workers are complaining that cross-threading is happening so often that threads are being stripped by the air wrenches and that torque settings also have to be adjusted downward to prevent stripped threads even if the parts match up. In an effort to determine whether the fault lies with the lug nuts or the bolts, Mindy has decided to ask the quality control department to test a random sample of 60 lug nuts and 40 bolts to see if the variances in threads are the same for both parts. The report from the technician indicated that the thread variance of the sampled lug nuts was 0.00213 and that the thread variance for the sampled bolts was 0.00166. What can Mindy conclude about the equality of the variances at the 0.05 level of significance?
a. Solve using the *p*-value approach. **b.** Solve using the classical approach.

VOCABULARY LIST

Be able to define each term. Pay special attention to the key terms, which are printed in **red.** In addition, describe in your own words, and give an example of, each term. Your examples should not be ones given in class or in the textbook.

The bracketed numbers indicate the chapters in which the term previously appeared, but you should define the terms again to show increased understanding of their meaning. Page numbers indicate the first appearance of the term in Chapter 10.

binomial experiment [5, 9] (p. 506)
confidence interval [8, 9] (pp. 479, 491, 507)
dependent means (p. 478)
dependent samples (p. 475)
F-distribution (p. 515)
F-statistic (p. 519)
hypothesis test [8, 9] (pp. 481, 493, 509, 515)
independent means (p. 490)
independent samples (p. 475)
mean difference (p. 478)
paired difference (p. 477)

percentage [5] (p. 506)
probability [5] (p. 506)
proportion [5] (p. 506)
pooled observed probability (p. 509)
p-value [8, 9] (pp. 482, 494, 511, 519)
source (of data) (p. 474)
standard error [8, 9] (pp. 481, 490, 506, 509)
t-distribution [9] (pp. 479, 491)
test statistic (pp. 481, 493, 509, 519)
t-statistic [9] (pp. 481, 493)
z-statistic [8, 9] (p. 509)

CHAPTER PRACTICE TEST

Part I: Knowing the Definitions

Answer "True" if the statement is always true. If the statement is not always true, replace the words shown in bold with words that make the statement always true.

10.1 When the means of two unrelated samples are used to compare two populations, we are dealing with **two dependent means.**

10.2 The use of **paired data (dependent means)** often allows for the control of unmeasurable or confounding variables because each pair is subjected to these confounding effects equally.

10.3 The **chi-square distribution** is used for making inferences about the ratio of the variances of two populations.

10.4 The **z-distribution** is used when two dependent means are to be compared.

10.5 In comparing two independent means when the σ's are unknown, we need to use the **standard normal** distribution.

10.6 The **standard normal score** is used for all inferences concerning population proportions.

10.7 The F-distribution is a **symmetric** distribution.

10.8 The number of degrees of freedom for the critical value of t is equal to **the smaller of $n_1 - 1$ or $n_2 - 1$** when making inferences about the difference between two independent means for the case when the degrees of freedom is estimated.

10.9 In constructing a confidence interval for the mean difference in paired data, the interval **increases** in width when the sample size is increased.

10.10 A **pooled estimate** for any statistic in a problem dealing with two populations is a value arrived at by combining the two separate sample statistics so as to achieve the best possible point estimate.

Part II: Applying the Concepts

Answer all questions, showing all formulas, substitutions, and work.

10.11 State the null (H_o) and the alternative (H_a) hypotheses that would be used to test each of these claims:

a. There is no significant difference in the mean batting averages for the baseball players of the two major leagues.

b. The standard deviation for the monthly amounts of rainfall in Monroe County is less than the standard deviation for the monthly amounts of rainfall in Orange County.

c. There is a significant difference between the percentage of male and female college students who own their own car.

d. There is a significant increase in performance as indicated by the mean difference in pretest and posttest scores.

10.12 Determine the test statistic, critical region, and critical value(s) that would be used in completing each hypothesis test using the classical procedure with $\alpha = 0.05$.

a. $H_o: p_1 - p_2 = 0$
$H_a: p_1 - p_2 \neq 0$

b. $H_o: \mu_d = 12$
$H_a: \mu_d \neq 12$
($n = 28$)

c. $H_o: \mu_1 - \mu_2 = 17$
$H_a: \mu_1 - \mu_2 > 17$
($n_1 = 8, n_2 = 10$)

d. $H_o: \mu_1 - \mu_2 = 37$
$H_a: \mu_1 - \mu_2 < 37$
($n_1 = 38, n_2 = 50$)

e. $H_o: \sigma_m^2 = \sigma_p^2$
$H_a: \sigma_m^2 > \sigma_p^2$
($n_m = 16, n_p = 25$)

10.13 Find each of the following:

a. $z(0.02)$ **b.** $t(15, 0.025)$ **c.** $F_{(24, 12, 0.05)}$ **d.** $F_{(12, 24, 0.05)}$

e. $z(0.04)$ **f.** $t(38, 0.05)$ **g.** $t(23, 0.99)$ **h.** $z(0.90)$

10.14 Twenty college freshmen were randomly divided into two groups. The members of one group were assigned to a statistics section using programmed materials only. Members of the other group were assigned to a section in which the professor lectured. At the end of the semester, all were given the same final exam. The results were as follows:

Programmed	76	60	85	58	91	44	82	64	79	88
Lecture	81	62	87	70	86	77	90	63	85	83

At the 5% level of significance, do these data show sufficient evidence to conclude that on the average the students in the lecture sections performed significantly better on the final exam?

10.15 The weights of eight people before they stopped smoking and five weeks after they stopped smoking are as follows:

Person	1	2	3	4	5	6	7	8
Before	148	176	153	116	129	128	120	132
After	154	179	151	121	130	136	125	128

At the 0.05 level of significance, does this sample present enough evidence to justify the conclusion that weight increases if one quits smoking?

10.16 In a nationwide sample of 600 school-age boys and 500 school-age girls, 288 boys and 175 girls admitted to having committed a destruction-of-property offense. Use these sample data and construct a 95% confidence interval for the difference between the proportion of boys and girls who have committed this offense.

Part III: Understanding the Concepts

10.17 To compare the accuracy of two short-range missiles, eight of the first kind and ten of the second kind are fired at a target. Let x be the distance by which the missile missed the target. Do these two sets of data (eight distances and ten distances) represent dependent or independent samples? Explain.

10.18 Let's assume that there are 400 students in our college taking elementary statistics this semester. Describe how you could obtain two dependent samples of size 20 from these students in order to test some precourse skill against the same skill after completing the course. Be very specific.

10.19 Student A says he "doesn't see what all the fuss about the difference between independent and dependent means is all about; the results are almost the same regardless of the method used." Professor C suggests, "Student A should compare the procedures a bit more carefully." Help student A discover that there is a substantial difference between the procedures.

10.20 Suppose you are testing H_o: $\mu_d = 0$ vs. H_a: $\mu_d < 0$ and the sample paired differences are all negative. Does this mean there is sufficient evidence to reject the null hypothesis? How can it not be significant? Explain.

10.21 Truancy is very disruptive to the educational system. A group of high school teachers and counselors have developed a group counseling program that they hope will help improve the truancy situation in their school. They have selected the 80 students in their school with the worst truancy records and have randomly assigned half of them to the group counseling program. At the end of the school year, the 80 students will be rated with regard to their truancy. When the scores have been collected, they will be turned over to you for evaluation. Explain what you will do to complete the study.

10.22 You wish to estimate and compare the proportion of Catholic families whose children attend a private school to the proportion of non-Catholic families whose children attend private schools. How would you go about estimating the two proportions and the difference between them?

Working with Your Own Data

History contains many stories about consumers and various products they purchase. An exhibit at the Boston Museum of Science tells such a mathematician–baker story. A man named Poincare bought one loaf of bread daily from his local baker; a loaf that was supposed to weigh 1 kilogram. After a year of weighing and recording the weight of each loaf, Poincare found a normal distribution with mean 950 grams. The police were called and the baker was told to behave himself; however, a year later Poincare reported that the baker had not reformed and the police confronted the baker again. The baker questioned, "How could Poincare have known that we always gave him the largest loaf?" Poincare then showed the police the second year of his record, a bell-shaped curve with a mean of 950 grams but truncated on the left side.

As consumers, we all purchase many bottled, boxed, canned, or packaged products. Seldom, if ever, do any of us question whether or not the content amount stated on the container is fulfilled. Here are a few typical content listings found on various containers we purchase:

28 FL OZ (1 PT 12 OZ)	750 ml
5 FL OZ (148 ml)	32 FL OZ (1 QT) 0.951
NET WT 10 OZ 283 GRAMS	NET WT $3^3/_4$ OZ 106 g—48 tea bags
140 1-PLY NAPKINS	77 SQ FT—92 TWO-PLY SHEETS—11 × 11 IN.

Have you ever thought, "I wonder if I am getting the amount that I am paying for?" And if this thought did cross your mind, did you attempt to check the validity of the content claim? The following article appeared in *The Times Union* of Rochester, New York, in 1972.

Milk Firm Accused of Short Measure*

The processing manager of Dairylea Cooperative, Inc., has been named in a warrant charging that the cooperative is distributing cartons of milk in the Rochester area containing less than the quantity represented.

. . . an investigator found shortages in four quarts of Dairylea milk purchased Friday. Asst. Dist. Atty. Howard R. Relin, who issued the warrant, said the shortages ranged from $1\frac{1}{8}$ to $1\frac{1}{4}$ ounces per quart. A quart of milk contains 32 fluid ounces.

. . . the state Agriculture and Markets Law . . . provides that a seller of a commodity shall not sell or deliver less of the commodity than the quantity represented to be sold.

. . . the purpose of the law under which . . . the dairy is charged is to ensure honest, accurate, and fair dealing with the public. There is no requirement that intent to violate the law be proved, he said.

*From *The Times-Union,* Rochester, N.Y., Feb. 16, 1972.

This situation poses a very interesting legal problem: There is no need to show intent to "short the customer." If caught, the violators are fined automatically and the fines are often quite severe.

A

A High-Speed Filling Operation

A high-speed piston-type machine used to fill cans with hot tomato juice was sold to a canning company. The guarantee stated that the machine would fill 48-oz cans with a mean amount of 49.5 oz, a standard deviation of 0.072 oz, and a maximum spread of 0.282 oz while operating at a rate of filling 150 to 170 cans per minute. On August 12, 1994, a sample of 42 cans was gathered and the following weights were recorded. The weights, measured to the nearest $\frac{1}{8}$ oz, are recorded as variations from 49.5 oz.

$-\frac{1}{8}$	0	$-\frac{1}{8}$	0	0	0	$-\frac{1}{8}$	0	0	0	0	0	$-\frac{1}{8}$	0
$\frac{1}{8}$	0	$\frac{1}{8}$	0	$\frac{1}{8}$	0	0	0	$-\frac{1}{8}$	0	0	0	0	$-\frac{1}{8}$
0	0	0	0	0	0	0	0	0	0	0	0	0	0

1. Calculate the mean $\bar{x}$, the standard deviation s, and the range of the sample data.
2. Construct a histogram picturing the sample data.
3. Does the amount of fill differ from the prescribed 49.5 oz at the α 0.05 level? Test the hypothesis that $\mu = 49.5$ against an appropriate alternative.
4. Does the amount of variation, as measured by the range, satisfy the guarantee?
5. Assuming that the filling machine continues to fill cans with an amount of tomato juice that is distributed normally and the mean and standard deviation are equal to the values found in question 1: What is the probability that a randomly selected can will contain less than the 48 oz claimed on the label?
6. If the amount of fill per can is normally distributed and the standard deviation can be maintained, find the setting for the mean value that would allow only 1 can in every 10,000 to contain less than 48 oz.

B

Your Own Investigation

Select a packaged product that has a quantity of fill per package that you can and would like to investigate.
1. Describe your selected product, including the quantity per package, and describe how you plan to obtain your data.
2. Collect your sample of data. (Consult your instructor for advice on size of sample.)
3. Calculate the mean $\bar{x}$ and standard deviation s for your sample data.
4. Construct a histogram or stem-and-leaf diagram picturing the sample data.
5. Does the mean amount of fill meet the amount prescribed on the label? Test using $\alpha = 0.05$.
6. Assume that the item you selected is filled continually. The amount of fill is normally distributed, and the mean and standard deviation are equal to the values found in question 3. What is the probability that one randomly selected package contains less than the prescribed amount?

More Inferential Statistics

In Part Three we studied the two inferential statistical techniques: confidence intervals and hypothesis tests, for one and two populations, and the three parameters: mean, μ; proportion, p; and standard deviation, σ. In Part Four we will learn how to use the confidence interval and the hypothesis test techniques in other situations. Some of these situations will be extensions of previous methods. (For example, analysis of variance will be used to deal with more than two populations when the question involves the mean, and the binomial experiment will be expanded to a multinomial experiment.) Other situations will be alternatives to methods previously studied (such as the nonparametric techniques) or will deal with new inferences (such as the nonparametric methods and the regression and correlation inference techniques).

Sir Ronald A. Fisher

SIR RONALD A. FISHER, British statistician, was born in London on February 17, 1890. In 1912, after earning a B.A. from Gonville and Caius College in Cambridge, Fisher worked as a statistician for an investment company until 1915, at which time he became a public school teacher. In 1917 he married Ruth Eillean Gralton Guiness with whom he had eight children. Fisher received his M.A. in 1920 and his Sc.D. in 1926. Many years later, he retired to Australia, where he died at the age of 72 on July 29, 1962.

In 1919 Fisher was hired by Rothamstead Experimental Station to do statistical work with its plant breeding experiments. It was there that he pioneered the applications of statistical procedures to the design of scientific experiments. During his employment at Rothamstead, Fisher introduced the principle of randomization and originated the concept of the analysis of variance (an even more important achievement). In 1925 Fisher wrote "Statistical Methods for Research," a work that remained in print for more than 50 years. Fisher has played a major role in the development and application of statistics.

Chapter

11

Applications of Chi-Square

CHAPTER OUTLINE

11.1 Chi-Square Statistic
The chi-square distribution will be used to test hypotheses concerning **enumerated data.**

11.2 Inferences Concerning Multinomial Experiments
A multinomial experiment differs from a binomial experiment in that **each trial has many outcomes** rather than two outcomes.

11.3 Inferences Concerning Contingency Tables
Contingency tables are tabular representations of frequency counts for data in a **two-way classification.**

CHAPTER CASE STUDY

Cooling Your Mouth After a Great Hot Taste

If you like hot foods, you probably have a preferred way to "cool" your mouth after eating a delicious spicy favorite. Some of the more common methods used by people are drinking water, milk, soda or beer, or eating bread or other food. There are even a few who prefer not to cool their mouth on such occasions and therefore do nothing to cool their mouth. The USA Snapshot® shown below appeared in *USA Today*, March 6, 1998 and shows the top six ways adults say they cool their mouths after hot sauce.

USA SNAPSHOTS®
A look at statistics that shape our lives

Some like it hot

The 10th annual National Fiery Foods Show, a festival of chilis and hot sauces, opens Friday in Albuquerque. Top six ways adults say they cool their mouth after hot sauce:

Method	Percentage
Water	43%
Bread	19%
Milk	15%
Beer	7%
Soda	7%
Don't	6%

Source: Cholula Hot Sauce   By Anne R. Carey and Suzy Parker, USA TODAY

Two hundred adults professing to love hot spicy food were asked to name their favorite way to cool their mouth after eating food with hot sauce. Below is the summary of the resulting sample.

Method	Water	Milk	Soda	Beer	Bread	Other	Nothing
Number	73	35	20	19	29	11	13

How similar is the distribution in this sample to the distribution of percentages in the Snapshot? Does the sample show a distribution that is significantly different from the distribution shown in the "Some like it hot" graph? We will learn about statistical methods that can be used to answer these questions, and we will return to this case study at the end of the chapter. (See Exercise 11.37, p. 566)

CHAPTER OBJECTIVES

Previously we discussed and used the chi-square distribution to test the value of the variance (or standard deviation) of a single population. The chi-square distribution may also be used for tests in other types of situations. In this chapter we are going to investigate two tests: a multinomial experiment and the contingency table. These two types of tests are to compare experimental results with expected results in order to determine (1) preferences, (2) independence, and (3) homogeneity.

The information that we will use in these techniques will be **enumerative;** that is, the data will be placed in categories and counted. These counts become the enumerative information used.

11.1   Chi-Square Statistic

There are many problems for which the data are categorized and the results shown by way of counts. For example, a set of final exam scores can be displayed as a frequency distribution. These frequency numbers are counts, the number of data that fall in each cell. A survey asks voters whether they are registered as Republican, Democrat, or other, and whether or not they support a particular candidate. The results are usually displayed on a chart that shows the number of voters for each possible category. There were numerous illustrations of this way of presenting data throughout the previous ten chapters.

Suppose that we have a number of **cells** into which n observations have been sorted. (The term cell is synonymous with the term class; the terms class and frequency were defined and first used in earlier chapters. Before continuing, a brief review of Sections 2.1, 2.2, and 3.1 might be beneficial.) The **observed frequencies** in each cell are denoted by $O_1, O_2, O_3, \ldots, O_k$ (see Table 11.1). Note that the sum of all observed frequencies is equal to

$$O_1 + O_2 + \cdots + O_k = n,$$

where n is the sample size. What we would like to do is to compare the observed frequencies with some **expected,** or theoretical **frequencies,** denoted by $E_1, E_2, E_3, \ldots, E_k$ (see Table 11.1), for each of these cells. Also, the sum of these expected frequencies must be exactly

$$E_1 + E_2 + \cdots + E_k = n$$

Table 11.1   Observed Frequencies

	\multicolumn{6}{c}{k **Categories**}					
	1st	**2nd**	**3rd**	**...**	**kth**	**Total**
Observed Frequencies	O_1	O_2	O_3	$\cdots$	O_k	n
Expected Frequencies	E_1	E_2	E_3	$\cdots$	E_k	n

We will then decide whether the observed frequencies seem to agree or disagree with the expected frequencies. This will be accomplished by a **hypothesis test** using **chi-square,** χ^2 ("ki-square"; that's "ki" as in kite; χ is the Greek lowercase letter chi).

The calculated value of the test statistic will be $\chi^2 \bigstar$,

$$\chi^2 \bigstar = \sum_{\text{all cells}} \frac{(O - E)^2}{E} \tag{11.1}$$

This calculated value for chi-square is the sum of several nonnegative numbers, one from each cell (or category). The numerator of each term in the formula for $\chi^2 \bigstar$ is the square of the difference between the values of the observed and the expected frequencies. The closer together these values are, the smaller the value of $(O - E)^2$; the farther apart, the larger the value $(O - E)^2$. The denominator for each cell puts the size of the numerator into perspective. That is, a difference $(O - E)$ of 10 resulting from frequencies of 110 (O) and 100 (E) seems quite different from a difference of 10 resulting from 15 (O) and 5 (E).

These ideas suggest that small values of chi-square indicate agreement between the two sets of frequencies, whereas larger values indicate disagreement. Therefore, it is customary for these tests to be one-tailed, with the critical region on the right.

In repeated sampling, the calculated value of $\chi^2 \bigstar$ [formula (11.1)] will have a sampling distribution that can be approximated by the chi-square probability distribution when n is large. This approximation is generally considered adequate when all the expected frequencies are greater than or equal to 5. Recall that the chi-square distributions, like the Student t-distributions, are a family of probability distributions, each one being identified by the parameter, number of **degrees of freedom,** df. The appropriate value of df will be described with each specific test. In order to use the chi-square distribution, we must be aware of its properties; these properties are outlined in Section 9.3 on page 451 (also, see Figure 9.7). The critical values for chi-square are obtained from Table 8 in Appendix B. (Specific instructions are in Section 9.3, p. 452.)

> **Exercise 11.1**
> Find:
>
> **a.** $\chi^2(10, 0.01)$
> **b.** $\chi^2(12, 0.025)$
> **c.** $\chi^2(10, 0.95)$
> **d.** $\chi^2(22, 0.995)$

ASSUMPTION FOR USING CHI-SQUARE TO MAKE INFERENCES BASED ON ENUMERATIVE DATA

The sample information is obtained using a random sample drawn from a population in which each individual is classified according to the categorical variable(s) involved in the test.

A categorical variable is a variable that classifies or categorizes each individual into exactly one of several cells or classes; these cells or classes are all-inclusive and mutually exclusive. The resulting face from a rolled die is a categorical variable: The list of outcomes {1, 2, 3, 4, 5, 6} form a set of all-inclusive and mutually exclusive categories.

In this chapter we permit a certain amount of "liberalization" with respect to the null hypothesis and its testing. In previous chapters the null hypothesis has always been a statement about a population parameter (μ, σ, or p). However, there are other types of hypotheses that can be tested, such as "this die is fair" or "height and weight of individuals are independent." Notice that these hypotheses are not claims about a parameter, although sometimes they could be stated with parameter values specified.

Suppose that I claim that "this die is fair," $p = P(\text{any one number}) = \frac{1}{6}$, and you want to test it. What would you do? Was your answer something like "Roll this die many times, recording the results"? Suppose that you decide to roll the die 60 times. If the die is fair, what do you expect will happen? Each number (1, 2, . . ., 6) should appear approximately $\frac{1}{6}$ of the time (that is, 10 times). If it happens that approximately 10 of each number occur, you will certainly accept the claim of fairness ($p = \frac{1}{6}$, for each value).

If it happens that the die seems to favor some numbers, you will reject the claim. (The test statistic $\chi^2 \star$ will have a large value in this case, as we will soon see.)

11.2   Find these critical values by using Table 8 of Appendix B.
    a. $\chi^2(18, 0.01)$ **b.** $\chi^2(16, 0.025)$ **c.** $\chi^2(40, 0.10)$ **d.** $\chi^2(45, 0.01)$
    Using the notation seen in (a)–(d), name and find the critical values of χ^2.

e.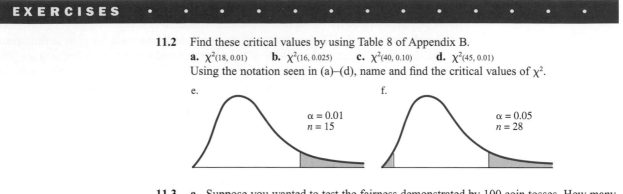

$\alpha = 0.01$
$n = 15$

f.

$\alpha = 0.05$
$n = 28$

11.3   **a.** Suppose you wanted to test the fairness demonstrated by 100 coin tosses. How many heads would you expect to result? tails?
    b. Suppose you were testing the fairness of a die by rolling it 200 times. What is the expected frequency for each number?

11.2   Inferences Concerning Multinomial Experiments

The preceding die problem is a good illustration of a **multinomial experiment.** Let's consider this problem again. Suppose that we want to test this die (at $\alpha = 0.05$) and decide whether to fail to reject or reject the claim "this die is fair." (The probability of each number is $\frac{1}{6}$.) The die is rolled from a cup onto a smooth flat surface 60 times, with the observed frequencies shown in the following table.

Number	1	2	3	4	5	6
Observed freq.	7	12	10	12	8	11

The null hypothesis that the die is fair is assumed to be true. This allows us to calculate the expected frequencies. If the die was fair, we certainly would expect 10 occurrences for each number.

Now let's calculate an observed value of χ^2. These calculations are shown in Table 11.2. The calculated value is $\chi^2 \star = 2.2$.

Table 11.2   Computations for Calculating χ^2

Number	Observed (O)	Expected (E)	$O - E$	$(O - E)^2$	$\dfrac{(O - E)^2}{E}$
1	7	10	−3	9	0.9
2	12	10	2	4	0.4
3	10	10	0	0	0.0
4	12	10	2	4	0.4
5	8	10	−2	4	0.4
6	11	10	1	1	0.1
Total	60	60	0 (ck)		**2.2**

NOTE $\sum(O - E)$ must equal zero since $\sum O = \sum E = n$. You can use this fact as a check, as shown in Table 11.2.

Before continuing, let's set up the hypothesis-testing procedure.

STEP 1   **The Set-Up:**
    a. **Describe the population parameter of interest.**
    The probability with which each face lands on top; $P(1)$, $P(2)$, $P(3)$, $P(4)$, $P(5)$, $P(6)$.
    b. **State the null hypothesis (H_o) and the alternative hypothesis (H_a).**

    H_o: **The die is fair** (each $p = \dfrac{1}{6}$).

    H_a: **The die is not fair** (at least one p is different from the others).

STEP 2   **The Hypothesis Test Criteria:**
    a. **Check the assumptions.**
    The data were collected in a random manner and each outcome is one of the six numbers.

In a multinomial experiment, df $= k - 1$, where k is the number of cells.

    b. **Identify the probability distribution and the test statistic to be used.**
    The **chi-square distribution** will be used and **formula (11.1),** with df $= k - 1 = 6 - 1 = \mathbf{5}$.
    c. **Determine the level of significance, α.**
    $\alpha = \mathbf{0.05}$.

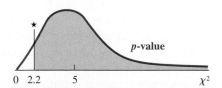
Computer and calculator commands to find the probability associated with a specified chi-square value can be found in Chapter 9 on page 453

STEP 3   **The Sample Evidence:**
    a. **Collect the sample information.**
    Sample information is in Table 11.2.
    b. **Calculate the value of the test statistic.** Using formula (11.1),

$$\chi^2 \star = \sum_{\text{all cells}} \frac{(O - E)^2}{E}: \quad \chi^2 \star = \mathbf{2.2} \text{ (calculations shown in Table 11.2)}$$

STEP 4   **The Probability Distribution:**

USING THE p-VALUE PROCEDURE:

a. **Calculate the p-value.**
Use the right-hand tail since "larger" values of chi-square disagree with null hypothesis:
$\mathbf{P} = P(\chi^2 > \mathbf{2.2} \mid \mathbf{df} = \mathbf{5})$ as shown in the figure.

OR

USING THE CLASSICAL PROCEDURE:

a. **Determine the critical region and critical value(s).**
The critical region is the right-hand tail since "larger" values of chi-square disagree with null hypothesis. The critical value is obtained from Table 8, at the intersection of row df $= 5$ and column $\alpha = 0.05$:

$$\chi^2_{(5,\,0.05)} = \mathbf{11.1.}$$

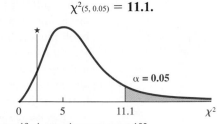

For specific instructions, see page 452.

To find the p-value, you have two options:
1. Use Table 8 (Appendix B) to place bounds on the p-value: $\mathbf{0.75 < P < 0.90}$
2. Use a computer or calculator to calculate the p-value: $\mathbf{P = 0.821}$

For specific instructions, see page 456.

b. **Determine whether or not the p-value is smaller than α.**
The p-value **is not smaller** than the level of significance, α

b. **Determine whether or not the calculated test statistic is in the critical region.**
$\chi^2 \star$ **is not in** the critical region, as shown in red in the figure above.

STEP 5 **The Results:**

    a. State the decision about H_o.

       Fail to reject H_o.

    b. State the conclusion about H_a.

       At the 0.05 level of significance, the observed frequencies are not signifi-
       cantly different from those expected of a fair die.

Before we look at other illustrations, we must define the term multinomial experiment and we need to state the guidelines for completing the chi-square test for it.

MULTINOMIAL EXPERIMENT

An experiment with the following characteristics:

1. It consists of n repeated (identical) independent trials .
2. The outcome of each trial fits into exactly one of k possible cells.
3. There is a probability associated with each particular cell, and these individual probabilities remain constant during the experiment. (It must be true that $p_1 + p_2 + \cdots + p_k = 1$.)
4. The experiment will result in a set of k observed frequencies, $O_1, O_2, \ldots, O_k$, where each O_i is the number of times a trial outcome falls into that particular cell. (It must be the case that $O_1 + O_2 + \cdots + O_k = n$.)

The die example meets the definition of a multinomial experiment because it has all four of the characteristics described in the definition.

1. The die was rolled n (60) times in an identical fashion, and these trials were independent of each other. (The result of each trial was unaffected by the results of other trials.)
2. Each time the die was rolled, one of six numbers resulted, and each number was associated with a cell.
3. The probability associated with each cell was $\frac{1}{6}$, and this was constant from trial to trial. (Six values of $\frac{1}{6}$ sum to 1.0.)
4. When the experiment was complete, we had a list of six frequencies (7, 12, 10, 12, 8, and 11) that summed to 60, indicating that each of the outcomes was taken into account.

The testing procedure for multinomial experiments is very similar to the testing procedure described in previous chapters. The biggest change comes with the statement of the null hypothesis. It may be a verbal statement, such as in the die example: "This die is fair." Often the alternative to the null hypothesis is not stated. However, in this book the alternative hypothesis will be shown, since it aids in organizing and understanding the problem. However, it will not be used to determine the location of the critical region, as was the case in previous chapters. For multinomial experiments we will always use a one-tailed critical region, and it will be the right-hand tail of the χ^2 distribution, because larger deviations (positive or negative) from the expected values lead to an increase in the calculated $\chi^2 \bigstar$ value.

The critical value will be determined by the level of significance assigned (α) and the number of degrees of freedom. The number of degrees of freedom (df) will be 1 less than the number of cells (k) into which the data are divided:

$$\text{df} = k - 1 \tag{11.2}$$

Each expected frequency, E_i, will be determined by multiplying the total number of trials n by the corresponding probability (p_i) for that cell. That is,

$$E_i = n \cdot p_i \tag{11.3}$$

One guideline should be met to ensure a good approximation to the chi-square distribution: Each expected frequency should be at least 5 (that is, each $E_i \geq 5$). Sometimes it is possible to combine "smaller" cells to meet this guideline. If this guideline cannot be met, corrective measures to ensure a good approximation should be used. These corrective measures are not covered in this book but are discussed in many other sources.

ILLUSTRATION 11.1 ▼

College students have regularly insisted on freedom of choice when registering for courses. This semester there were seven sections of a particular mathematics course. They were scheduled to meet at various times with a variety of instructors. Table 11.3 shows the number of students who selected each of the seven sections. Do the data indicate that the students had a preference for certain sections, or do they indicate that each section was equally likely to be chosen?

Table 11.3   Data for Illustration 11.1

| | Section | | | | | | | |
	1	2	3	4	5	6	7	Total
Number of Students	18	12	25	23	8	19	14	119

Solution

If no preference were shown in the selection of sections, we would expect the 119 students to be equally distributed among the seven classes. Thus, if no preference were the case, then we would expect 17 students to register for each section. The test is completed as shown in steps 1–5, at the 5% level of significance.

STEP 1   The Set-Up:
   a. Describe the population parameter of interest.
      Preference of each section, the probability that a particular section is selected at registration.
   b. State the null hypothesis (H_o) and the alternative hypothesis (H_a).

      H_o: **There was no preference shown** (equally distributed).

      H_a: **There was a preference shown** (not equally distributed).

STEP 2   The Hypothesis Test Criteria:
   a. Check the assumptions.
      The 119 students represent a random sample of the population of all students who register for this particular course. Since there were no new regulations related to the selection of courses, this or others, and registration seemed to proceed in its usual patterns, there is no reason to believe this is other than a random sample.
   b. Identify the probability distribution and the test statistic to be used.
      The **chi-square distribution** will be used and **formula (11.1)**, with df = **6.**
   c. Determine the level of significance, α.
      $\alpha = 0.05.$

STEP 3 **The Sample Evidence:**
Sample information is in Table 11.3.
b. Calculate the value of the test statistic.
Using formula (11.1),

$$\chi^2 \bigstar = \sum_{\text{all cells}} \frac{(O-E)^2}{E} : \chi^2 \bigstar = \frac{(18-17)^2}{17} + \frac{(12-17)^2}{17} + \frac{(25-17)^2}{17}$$

$$+ \frac{(23-17)^2}{17} + \frac{(8-17)^2}{17} + \frac{(19-17)^2}{17} + \frac{(14-17)^2}{17}$$

$$\chi^2 \bigstar = \frac{(1)^2 + (-5)^2 + (8)^2 + (6)^2 + (-9)^2 + (2)^2 + (-3)^2}{17}$$

$$\chi^2 \bigstar = \frac{1+25+64+36+81+4+9}{17} = \frac{220}{17} = 12.9411$$

$$\chi^2 \bigstar = \mathbf{12.94}$$

STEP 4 **The Probability Distribution:**

USING THE *p*-VALUE PROCEDURE:

a. Calculate the *p*-value.
Use the right-hand tail since "larger" values of chi-square disagree with null hypothesis:
P = $P(\chi^2 > 12.94|df = 6)$ as shown in the figure.

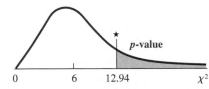

To find the *p*-value, you have two options:
1. Use Table 8 (Appendix B) to place bounds on the *p*-value: **0.025 < P < 0.05**
2. Use a computer or calculator to calculate the *p*-value: **P = 0.044**

For specific instructions, see page 456.

b. Determine whether or not the *p*-value is smaller than α.
The *p*-value **is smaller** than the level of significance, α

OR

USING THE CLASSICAL PROCEDURE:

a. Determine the critical region and critical value(s).
The critical region is the right-hand tail since "larger" values of chi-square disagree with the null hypothesis. The critical value is obtained from Table 8, at the intersection of row df = 6 and column α = 0.05:
$$\chi^2_{(6,\,0.05)} = \mathbf{12.6.}$$

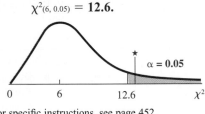

For specific instructions, see page 452.

b. Determine whether or not the calculated test statistic is in the critical region.
$\chi^2 \bigstar$ **is in** the critical region, as shown in **red** in the figure above.

STEP 5 **The Results:**
a. State the decision about H_o.
Reject H_o.
b. State the conclusion about H_a.
At the 0.05 level of significance, there does seem to be a preference shown. We cannot determine, from the given information, what the preference is. It could be teacher preference, time preference, or a case of schedule conflict.
Conclusions must be worded carefully to avoid suggesting conclusions that we cannot support.

Not all multinomial experiments result in equal expected frequencies, as we will see in Illustration 11.2.

ILLUSTRATION 11.2 ▼

The Mendelian theory of inheritance claims that the frequencies of round and yellow, wrinkled and yellow, round and green, and wrinkled and green, peas will occur in the ratio 9:3:3:1 when two specific varieties of peas are crossed. In testing this theory, Mendel obtained frequencies of 315, 101, 108, and 32, respectively. Do these sample data provide sufficient evidence to reject this theory at the 0.05 level of significance?

Solution

STEP 1 **The Set-Up:**
   a. Describe the population parameter of interest.
      The proportions: P(round and yellow), P(wrinkled and yellow), P(round and green), P(wrinkled and green).
   b. State the null hypothesis (H_o) and the alternative hypothesis (H_a).

   H_o: **9:3:3:1 is the ratio of inheritance.**

   H_a: **9:3:3:1 is not the ratio of inheritance.**

STEP 2 **The Hypothesis Test Criteria:**
   a. Check the assumptions.
      We will assume that Mendel's results form a random sample.
   b. Identify the probability distribution and the test statistic to be used.
      The **chi-square distribution** will be used and **formula (11.1),** with df = **3.**

C a s e   S t u d y   11.1

Why We Rearrange Furniture

Exercise 11.4

Verify that "Why we re-arrange furniture" is a multinomial experiment.

a. What is one trial?
b. What is the variable?
c. What are the many levels of results from each trial?

The data collected from a survey of 300 adults and used to create "Why we rearrange furniture" is actually that of a multinomial experiment.

USA SNAPSHOTS®
A look at statistics that shape our lives

Why we rearrange furniture

Other
Purchasing new furniture — 14%
Bored with arrangement — 36%
15%
16% / 19%
Redecorating
Moving to new residence

Source: Southwestern Bell Freedom Phone survey of 300 adults
By Marcy E. Mullins, USA TODAY

Source: Copyright 1990, USA TODAY. Reprinted with permission.

c. **Determine the level of significance, α.**
   $\alpha = 0.05$.

STEP 3 **The Sample Evidence:**
a. **Collect the sample information.**
   The observed frequencies were: **315, 101, 108, and 32**
   $n = \sum O_i = 315 + 101 + 108 + 32 = 556$
b. **Calculate the value of the test statistic.**

The ratio $9:3:3:1$ indicates probabilities of $\dfrac{9}{16}, \dfrac{3}{16}, \dfrac{3}{16}$, and $\dfrac{1}{16}$. There-fore, the expected frequencies are $\dfrac{9n}{16}, \dfrac{3n}{16}, \dfrac{3n}{16}$, and $\dfrac{1n}{16}$, which are 312.75, 104.25, 104.25, and 34.75.

The computations for calculating $\chi^2 \star$ are shown in Table 11.4.

> Mendel's data are so good, many researchers think he fudged the counts.

> **Exercise 11.5**
> Explain how $9:3:3:1$ becomes $\dfrac{9}{16}, \dfrac{3}{16}, \dfrac{3}{16}$, and $\dfrac{1}{16}$.

Table 11.4   Computations Needed to Calculate $\chi^2 \star$

O	E	$O - E$	$\dfrac{(O - E)^2}{E}$
315	312.75	2.25	0.0162
101	104.25	−3.25	0.1013
108	104.25	3.75	0.1349
32	34.75	−2.75	0.2176
556	556.00	0 (ck)	0.4700

$$\longrightarrow \quad \chi^2 \star = \sum_{\text{all cells}} \frac{(O - E)^2}{E} = 0.47.$$

> **Exercise 11.6**
> Explain how 312.75, 2.25, and 0.0162 were obtained in Table 11.4.

Case Study 11.2

Why People Volunteer

> **Exercise 11.7**
> Why is the information shown in "Why people volunteer" not that of a multinomial experiment? Be specific.

"Why people volunteer" displays the results of surveying adults about why they say they volunteer their time to help various causes.

USA SNAPSHOTS®
A look at statistics that shape the nation

Why people volunteer
Adults say they volunteer to:

Help people	87%
Make community a better place	72%
Be with people they enjoy	56%
Be with people with same ideals	51%
Learn about issue/problem	41%
Fulfill civic duty	39%

Note: Could name more than one

Source: American Association of Retired Persons   By Cindy Hall and Genevieve Lynn, USA TODAY

Source: Copyright 1998, USA TODAY. Reprinted with permission.

STEP 4 **The Probability Distribution:**

USING THE *p*-VALUE PROCEDURE:

a. Calculate the *p*-value.
Use the right-hand tail since "larger" values of chi-square disagree with null hypothesis:
$P = P(\chi^2 > 0.47 \mid df = 3)$ as shown in the figure.

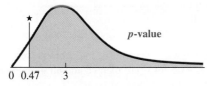

To find the *p*-value, you have two options:
1. Use Table 8 (Appendix B) to place bounds on the *p*-value: $0.90 < P < 0.95$
2. Use a computer or calculator to calculate the *p*-value:
   P = 0.925

For specific instructions, see page 456.

b. Determine whether or not the *p*-value is smaller than α.
The *p*-value **is not smaller** than the level of significance, α.

OR

USING THE CLASSICAL PROCEDURE:

a. Determine the critical region and critical value(s).
The critical region is the right-hand tail since "larger" values of chi-square disagree with the null hypothesis. The critical value is obtained from Table 8, at the intersection of row df = 3 and column α = 0.05:
$$\chi^2_{(3,\,0.05)} = \textbf{7.82.}$$

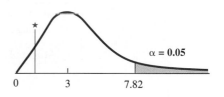

For specific instructions, see page 452.

b. Determine whether or not the calculated test statistic is in the critical region.
$\chi^2 \bigstar$ **is not in** the critical region, as shown in **red** in the figure above.

STEP 5   **The Results:**
   a. State the decision about H_o.
      Fail to reject H_o.
   b. State the conclusion about H_a.
      At the 0.05 level of significance, there is not sufficient evidence to reject Mendel's theory.

EXERCISES

11.8   State the null hypothesis H_o and the alternative hypothesis H_a that would be used to test the following statements.
   a. The five numbers, 1, 2, 3, 4, and 5, are equally likely to be drawn.
   b. That multiple-choice question has a history of students selecting answers in the ratio of $2:3:2:1$.
   c. The poll will show a distribution of 16%, 38%, 41%, and 5% for the possible ratings of excellent, good, fair, and poor on that issue.

11.9   Determine the *p*-value for the following hypothesis test involving the χ^2-distribution with 12 degrees of freedom.
   a. $H_o: P(1) = P(2) = P(3) = P(4) = 0.25$, with $\chi^2 \bigstar = 12.25$.
   b. $H_o: P(\text{I}) = 0.25, P(\text{II}) = 0.40, P(\text{III}) = 0.35$, with $\chi^2 \bigstar = 5.98$.

11.10   Determine the critical value and critical region that would be used in the classical approach to test the null hypothesis for each of the following multinomial experiments.
   a. $H_o: P(1) = P(2) = P(3) = P(4) = 0.25$, with α = 0.05.
   b. $H_o: P(\text{I}) = 0.25, P(\text{II}) = 0.40, P(\text{III}) = 0.35$, with α = 0.01.

11.11 A manufacturer of floor polish conducted a consumer-preference experiment to determine which of five different floor polishes was the most appealing in appearance. A sample of 100 consumers viewed five patches of flooring that had each received one of the five polishes. Each consumer indicated the patch he or she preferred. The lighting and background were approximately the same for all patches. The results were as follows:

Polish	A	B	C	D	E	Total
Frequency	27	17	15	22	19	100

   a. State the hypothesis for "no preference" in statistical terminology.
   b. What test statistic will be used in testing this null hypothesis?
   c. Complete the hypothesis test using $\alpha = 0.10$.
      (1) Solve using the p-value approach.   (2) Solve using the classical approach.

11.12 A certain type of flower seed will produce magenta, chartreuse, and ochre flowers in the ratio $6:3:1$ (one flower per seed). A total of 100 seeds are planted and all germinate, yielding the following results.

Magenta	Chartreuse	Ochre
52	36	12

   a. If the null hypothesis $(6:3:1)$ is true, what is the expected number of magenta flowers?
   b. How many degrees of freedom are associated with chi-square?
   c. Complete the hypothesis test using $\alpha = 0.10$.
      (1) Solve using the p-value approach.   (2) Solve using the classical approach.

11.13 A nationwide survey of consumers in 1998 (*Newsweek,* "Don't Bank on It," 8-24-98) was conducted to determine the level of uncertainty surrounding the Y2K problem. CIO Communications, Inc. asked heads of households, "What are you planning to do with your money if the Y2K problem isn't solved by mid-1999?" The results are shown in the table below:

Plans for Money (National)	Percent Responding
Hide it (mattress stuffing)	25
Deposit it in several banks	11
Deposit it in one bank	16
Don't know	48
Total	100

Suppose a local follow-up survey is conducted using 300 respondents from your city who answer exactly the same question. Results from the follow-up study are tabulated below:

Plans for Money (Local)	Number Responding
Hide it (mattress stuffing)	72
Deposit it in several banks	32
Deposit it in one bank	47
Don't know	149
Total	300

Does the distribution of responses differ from the distribution obtained from the nationwide survey? Test at the 0.05 level of significance.
   a. Solve using the p-value approach.   **b.** Solve using the classical approach.

11.14 A large supermarket carries four qualities of ground beef. Customers are believed to purchase these four qualities with probabilities of 0.10, 0.30, 0.35, and 0.25, respectively, from the least to most expensive. A sample of 500 purchases resulted in sales of 46, 162, 191, and 101 of the respective qualities. Does this sample contradict the expected proportions? Use $\alpha = 0.05$. Use a computer to complete this exercise.
   a. Solve using the *p*-value approach.   **b.** Solve using the classical approach.

11.15 One of the major benefits of e-mail is its ability to communicate rapidly without getting a busy signal or no answer—two major criticisms of voice telephone calls. But does it succeed in helping to solve the problems people have trying to run computer hardware? A study in 1998 polled the opinions of consumers who tried to use e-mail for help by posting a message online to their PC manufacturer or authorized representative (*PC World,* "PC World's 1998 Reliability and Service Survey," November 1998). Results are shown in the following table:

Result of Online Query	Percent
Never got a response	14
Got a response, but it didn't help	30
Response helped, but didn't solve problem	34
Response solved problem	22

As marketing manager for a large PC manufacturer, you decide to conduct a survey of your customers using your e-mail records to compare against the published results. In order to make a fair comparison, you elect to use the same questionnaire and examine returns from 500 customers who attempted to use e-mail for help from your technical support staff. The results are shown below:

Result of Online Query	Number Responding
Never got a response	35
Got a response, but it didn't help	102
Response helped, but didn't solve problem	125
Response solved problem	238
Total	500

Does the distribution of responses differ from the distribution obtained from the published survey? Test at the 0.01 level of significance.
   a. Solve using the *p*-value approach.   **b.** Solve using the classical approach.

11.16 The 1993 edition of the *Digest of Educational Statistics* gives the following distribution of the ages of persons 18 and over who hold a bachelor's or higher degree.

Age	18–24	25–34	35–44	45–54	55–64	65 or over
Percent	5	29	30	16	10	10

A survey of 500 randomly chosen persons age 18 and over who hold a bachelor's or higher degree in Alaska gave the following distribution.

Age	18–24	25–34	35–44	45–54	55–64	65 or over
Number	30	150	155	75	35	55

Test the null hypothesis that the age distribution is the same in Alaska as it is nationally at a level of significance equal to 0.05. Include in your answer the calculated $\chi^2 \star$, the *p*-value or critical value, decision, and conclusion.

11.17 A program for generating random numbers on a computer is to be tested. The program is instructed to generate 100 single-digit integers between 0 and 9. The frequencies of the observed integers were as follows:

Integer	0	1	2	3	4	5	6	7	8	9
Frequency	11	8	7	7	10	10	8	11	14	14

At the 0.05 level of significance, is there sufficient reason to believe that the integers are not being generated uniformly?
a. Solve using the *p*-value approach.   **b.** Solve using the classical approach.

11.18 *Nursing Magazine* (March 1991) reported results of a survey of over 1800 nurses across the country concerning job satisfaction and retention. Nurses from magnet hospitals (hospitals that successfully attract and retain nurses) describe the staffing situation in their units as follows:

Staffing Situation	Percent
1. Desperately short of help—patient care has suffered	12%
2. Short, but patient care hasn't suffered	32%
3. Adequate	38%
4. More than adequate	12%
5. Excellent	6%

A survey of 500 nurses from nonmagnet hospitals gave the following responses to the staffing situation.

Staffing Situation	1	2	3	4	5
Number	165	140	125	50	20

Do the data indicate that the nurses from the nonmagnet hospitals have a different distribution of opinions? Use $\alpha = 0.05$.
a. Solve using the *p*-value approach.   **b.** Solve using the classical approach.

11.19 Why is this chi-square test typically a one-tail test with the critical region in the right tail?
a. What kind of value would result if the observed frequencies and the expected frequencies were very close in value? Explain how you would interpret this situation.
b. Suppose you had to roll a die 60 times as an experiment to test the fairness of the die as discussed in the example on page 542; but instead of rolling the die yourself, you paid your little brother $1 to roll it 60 times and keep a tally of the numbers. He agreed to perform this deed for you and ran off to his room with the die, returning in a few minutes with his resulting frequencies. He demanded his $1. You, of course, pay him before he hands over his results, which are: 10, 10, 10, 10, 10, and 10. The observed results are exactly what you had "expected." Right? Explain your reactions. What value of $\chi^2\star$ will result? What do you think happened? What do you demand of your little brother and why? What possible role might the left tail have in the hypothesis test?
c. Why is the left tail not typically of concern?

 11.20 According to the September 20, 1994, issue of *Family Circle,* 39% of Americans own guns for hunting, 30% for protection, 19% for both hunting and protection, and 12% for other reasons. A survey in Memphis, Tennessee, of 2000 individuals gave the following results. *(continued)*

Why Do You Own a Gun?	Number Responding
Hunting	740
Protection	600
Both	360
Other	300

a. Test the null hypothesis that the distribution of reasons for owning a gun is the same in Memphis as it is nationally as reported by *Family Circle.* Use a level of significance equal to 0.05.

b. What caused the calculated value of $\chi^2 \star$ to be so large? Does it seem right that one cell should have this much effect on the results? How could this test be completed differently (hopefully, more meaningfully) so that the results might not be affected as they were in (a)? Be specific.

11.3   Inferences Concerning Contingency Tables

A **contingency table** is an arrangement of data into a two-way classification . The data are sorted into cells , and the number of data in each cell is reported. The contingency table involves two factors (or variables), and the usual question concerning such tables is whether the data indicate that the two variables are independent or dependent (see pp. 131–134).

There are two different tests that use the contingency table format. The first one we will look at is the test of independence.

Test of Independence

To illustrate a test of independence, let's consider a random sample showing the gender of liberal arts college students and their favorite academic area.

ILLUSTRATION 11.3 ▼

Each person in a group of 300 students was identified as male or female and then asked whether he or she preferred taking liberal arts courses in the area of math–science, social science, or humanities. Table 11.5 is a contingency table that shows the frequencies found for these categories. Does this sample present sufficient evidence to reject the null hypothesis "preference for math–science, social science, or humanities is independent of the gender of a college student"? Complete the **hypothesis test** using the 0.05 level of significance.

Table 11.5   Contingency Table Showing Sample Results for Illustration 11.3

	Favorite Subject Area			
Gender	Math–Science (MS)	Social Science (SS)	Humanities (H)	Total
Male (*m*)	37	41	44	122
Female (*f*)	35	72	71	178
Total	72	113	115	300

Solution

STEP 1 **The Set-Up:**
  a. **Describe the population parameter of interest.**
     The independence of variables "gender" and "favorite course" require us to discuss the probability of the various answers and the effect that answers of one variable have on the probability of answers related to the other variable. Independence, as defined in Chapter 4, requires $P(MS|M) = P(MS|F) = P(MS)$; that is, gender has no effect on the probability of a person's choice being math–science, and so on.
  b. **State the null hypothesis (H_o) and the alternative hypothesis (H_a).**

     H_o: Preference for math–science, social science, or humanities
         is independent of the gender of a college student.

     H_a: Subject preference **is not independent** of the gender of the student.

STEP 2 **The Hypothesis Test Criteria:**
  a. **Check the assumptions.**
     The sample information is obtained using random samples drawn from two separate populations in which each individual is classified according to their favorite subject area.
  b. **Identify the probability distribution and the test statistic to be used.**
     In the case of contingency tables, the number of degrees of freedom is exactly the same number as the number of cells in the table that may be filled in freely when you are given the **marginal totals**. The totals in this illustration are shown in the following table.

			122
			178
72	113	115	300

Given these totals, you can fill in only two cells before the others are all determined. (The totals must, of course, be the same.) For example, once we pick two arbitrary values (say, 50 and 60) for the first two cells of the first row, the other four cell values are fixed (see the following table).

50	60	C	122
D	E	F	178
72	113	115	300

They have to be $C = 12$, $D = 22$, $E = 53$, and $F = 103$. Otherwise, the totals will not be correct. Therefore, for this problem there are two free choices. Each free choice corresponds to 1 degree of freedom. Hence the number of degrees of freedom for our example is 2 (df = 2).
     The **chi-square distribution** will be used and **formula (11.1)**, with df = **2.**
  c. **Determine the level of significance, α.**
     $\alpha = 0.05.$

STEP 3 **The Sample Evidence:**
  a. **Collect the sample information.**
     Sample results are listed in Table 11.5.

b. **Calculate the value of the test statistic.**

Before the calculated value of chi-square can be found, we need to determine the **expected values,** E, for each cell. To do this we must recall the null hypothesis, which asserts that these factors are independent. Therefore, we would expect the values to be distributed in proportion to the marginal totals. There are 122 males; we would expect them to be distributed among MS, SS, and H proportionally to the 72, 113, and 115 totals. Thus, the expected cell counts for males are

$$\frac{72}{300} \cdot 122 \qquad \frac{113}{300} \cdot 122 \qquad \frac{115}{300} \cdot 122$$

Similarly, we would expect for the females

$$\frac{72}{300} \cdot 178 \qquad \frac{113}{300} \cdot 178 \qquad \frac{115}{300} \cdot 178$$

Thus, the expected values are as shown in Table 11.6. Always check the marginal totals for the expected values against the marginal totals for the observed values.

Table 11.6   Expected Values

	MS	SS	H	Total
Male	29.28	45.95	46.77	122.00
Female	42.72	67.05	68.23	178.00
Total	72.00	113.00	115.00	300.00

▲

NOTE   We can think of the computation of the expected values in a second way. Recall that we assume the null hypothesis to be true until there is evidence to reject it. Having made this assumption in our example, in effect we are saying that the event that a student picked at random is male and the event that a student picked at random prefers math–science courses are independent. Our point estimate for the probability that a student is male is 122/300, and the point estimate for the probability that the student prefers math–science courses is 72/300. Therefore, the probability that both events occur is the product of the probabilities. [Refer to formula (4-8a), p. 210.] Thus, (122/300) · (72/300) is the probability of a selected student being male and preferring math–science. Therefore, the number of students out of 300 that are expected to be male and prefer math–science is found by multiplying the probability (or proportion) by the total number of students (300). Thus, the expected number of males who prefer math–science is (122/300)(72/300)(300) = (122/300)(72) = 29.28. The other expected values can be determined in the same manner.

Typically the contingency table is written so that it contains all this information (see Table 11.7).

Exercise 11.21

Find the expected value for the cell shown.

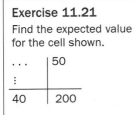

Table 11.7   Contingency Table Showing Sample Results and Expected Values

Gender of Student	Favorite Subject Area			Total
	Math–Science (MS)	Social Science (SS)	Humanities (H)	
Male	37 (29.28)	41 (45.95)	44 (46.77)	122
Female	35 (42.72)	72 (67.05)	71 (68.23)	178
Total	72	113	115	300

The calculated chi-square is

$$\chi^2 \bigstar = \sum_{\text{all cells}} \frac{(O - E)^2}{E}: \quad \chi^2 \bigstar = \frac{(37 - 29.28)^2}{29.28} + \frac{(41 - 45.95)^2}{45.95} + \frac{(44 - 46.77)^2}{46.77}$$

$$+ \frac{(35 - 42.72)^2}{42.72} + \frac{(72 - 67.05)^2}{67.05} + \frac{(71 - 68.23)^2}{68.23}$$

$$\chi^2 \bigstar = 2.035 + 0.533 + 0.164 + 1.395 + 0.365 + 0.112$$

$$\chi^2 \bigstar = \mathbf{4.604}$$

STEP 4 **The Probability Distribution:**

USING THE *p*-VALUE PROCEDURE:

a. Calculate the *p*-value.
Use the right-hand tail since "larger" values of chi-square disagree with the null hypothesis:
$\mathbf{P} = P(\chi^2 > \mathbf{4.604} | \mathbf{df = 2})$ as shown in the figure.

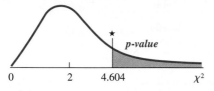

To find the *p*-value, you have two options:
1. Use Table 8 (Appendix B) to place bounds on the *p*-value: **0.10 < P < 0.25**
2. Use a computer or calculator to calculate the *p*-value: **P = 0.1001**

For specific instructions, see page 456.

b. Determine whether or not the *p*-value is smaller than α.
The *p*-value **is not smaller** than α.

OR

USING THE CLASSICAL PROCEDURE:

a. Determine the critical region and critical value(s).
The critical region is the right-hand tail since "larger" values of chi-square disagree with the null hypothesis. The critical value is obtained from Table 8, at the intersection of row df = 2 and column α = 0.05:
$$\chi^2_{(2, \, 0.05)} = \mathbf{5.99}.$$

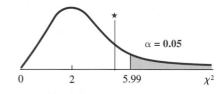

For specific instructions, see page 452.

b. Determine whether or not the calculated test statistic is in the critical region.
$\chi^2 \bigstar$ **is not in** the critical region, as shown in red in the figure above.

STEP 5 **The Results:**

a. State the decision about H_o.
Fail to reject H_o.

b. State the conclusion about H_a.
At the 0.05 level of significance, the evidence does not allow us to reject the idea of independence between the gender of a student and the student's preferred academic subject area. ▲

In general, the **$r \times c$ contingency table** (*r* is the number of **rows**; *c* is the number of **columns**) will be used to test the independence of the row factor and the column factor. The number of **degrees of freedom** will be determined by

$$df = (r - 1) \cdot (c - 1), \tag{11.4}$$

where r and c are both greater than 1. (This value for df should agree with the number of cells counted according to the general description on p. 554).

The **expected frequencies** for an $r \times c$ contingency table will be found by means of the formulas in each cell in Table 11.8, where n = grand total. In general, the expected frequency at the intersection of the ith row and the jth column is given by

$$E_{i,j} = \frac{Row\ total \times Column\ total}{Grand\ total} = \frac{R_i \times C_j}{n} \qquad (11.5)$$

Table 11.8   Expected Frequencies for an $r \times c$ Contingency Table

Rows	\multicolumn{6}{c}{Columns}	Total
	1	2	$\cdots$	jth Column	$\cdots$	c	
1	$\dfrac{R_1 \times C_1}{n}$	$\dfrac{R_1 \times C_2}{n}$	$\cdots$	$\dfrac{R_1 \times C_j}{n}$	$\cdots$	$\dfrac{R_1 \times C_c}{n}$	R_1
2	$\dfrac{R_2 \times C_1}{n}$						R_2
$\vdots$	$\vdots$			$\vdots$			$\vdots$
ith Row	$\dfrac{R_j \times C_1}{n}$			$\dfrac{R_i \times C_j}{n}$			R_i
$\vdots$	$\vdots$			$\vdots$			$\vdots$
r							
Total	C_1	C_2	$\cdots$	C_j	$\cdots$	$\cdots$	n

We should again observe the previously mentioned guideline: Each $E_{i,j}$ should be at least 5.

NOTE   The notation used in Table 11.8 and formula (11.5) may be unfamiliar to you. For convenience in referring to cells or entries in a table, $E_{i,j}$ can be used to denote the entry in the ith row and the jth column. That is, the first letter in the subscript corresponds to the row number and the second letter corresponds to the column number. Thus, $E_{1,2}$ is the entry in the first row, second column, and $E_{2,1}$ is the entry in the second row, first column. Referring to Table 11.6 (p. 555), $E_{1,2}$ for that table is 45.95 and $E_{2,1}$ is 42.72. The notation used in Table 11.8 is interpreted in a similar manner; that is, R_1 corresponds to the total from row 1, and C_1 corresponds to the total from column 1.

Exercise 11.22

Identify these values from Table 11.7.
a. C_2 b. R_1
c. n d. $E_{2,3}$

Test of Homogeneity

The second type of contingency table problem is called a test of homogeneity. This test is used when one of the two variables is controlled by the experimenter so that the row (or column) totals are predetermined.

For example, suppose that we were to poll registered voters about a piece of legislation proposed by the governor. In the poll, 200 urban, 200 suburban, and 100 rural residents will be randomly selected and asked whether they favor or oppose the governor's proposal. That is, a simple random sample is taken for each of these three groups. A total of 500 voters are to be polled. But notice that it has been predetermined (before the sample is taken) just how many are to fall within each row category, as shown in Table 11.9, and each category is sampled separately.

Table 11.9 Registered Voter Poll with Predetermined Row Totals

Type of Residence	Favor	Oppose	Total
Urban			200
Suburban			200
Rural			100
Total			500

In a test of this nature, we are actually testing the hypothesis "the distribution of proportions within the rows is the same for all rows." That is, the distribution of proportions in row 1 is the same as in row 2, is the same as in row 3, and so on. The alternative to this is "the distribution of proportions within the rows is not the same for all rows." This type of example may be thought of as a comparison of several multinomial experiments.

Beyond this conceptual difference, the actual testing for independence and homogeneity with contingency tables is the same. Let's demonstrate this **hypothesis test** by completing the polling illustration.

ILLUSTRATION 11.4 ▼

Each person in a random sample of 500 registered voters (200 urban, 200 suburban, and 100 rural residents) was asked his or her opinion about the governor's proposed legislation. Does the sample evidence shown in Table 11.10 support the hypothesis that "voters within the different residence groups have different opinions about the governor's proposal"? Use $\alpha = 0.05$.

Table 11.10 Sample Results for Illustration 11.4

Type of Residence	Favor	Oppose	Total
Urban	143	57	200
Suburban	98	102	200
Rural	13	87	100
Total	254	246	500

Solution

STEP 1 **The Set-Up:**
    a. Describe the population parameter of interest.

        The proportion of voters who favor or oppose (that is, the proportion of urban voters who favor, the proportion of suburban voters who favor, the proportion of rural voters who favor, and the proportion of all three groups, separately, who oppose).

b. State the null hypothesis (H_o) and the alternative hypothesis (H_a).

H_o: The proportion of voters favoring the proposed legislation is the same in all three groups.

H_a: The proportion of voters favoring the proposed legislation is not the same in all three groups. (That is, in at least one group the proportions are different from the others.)

STEP 2   The Hypothesis Test Criteria:

a. Check the assumptions.

The sample information is obtained using three random samples drawn from three separate populations in which each individual is classified according to their opinion.

b. Identify the probability distribution and the test statistic to be used.

The **chi-square distribution** will be used and **formula (11.1)**, with
$$df = (r - 1)(c - 1) = (3 - 1)(2 - 1) = 2$$

c. Determine the level of significance, α.
$\alpha = 0.05.$

Case Study   11.3

Westerners Bake Their Spuds

Exercise 11.23

Refer to Case Study 11.3.

a. Express the percentage of Americans who prefer "baked" to "other" vs. region as a 2×4 contingency table.

b. Explain why the following question could be tested using the chi-square statistic: "Is the preference for baked the same in all four regions of the USA?"

c. Explain why this is a test of homogeneity.

The USA Snapshot® "Westerners bake their spuds" reports the percentage of Americans that prefer to eat baked potatoes by region as well as for the whole country. If the actual number of people in each category were given, we would have a contingency table and we would be able to complete a hypothesis test about the homogeneity of the four regions.

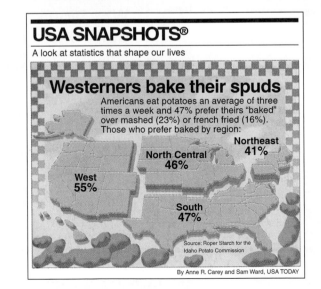

USA SNAPSHOTS®

A look at statistics that shape our lives

Westerners bake their spuds

Americans eat potatoes an average of three times a week and 47% prefer theirs "baked" over mashed (23%) or french fried (16%). Those who prefer baked by region:

Northeast 41%

North Central 46%

West 55%

South 47%

Source: Roper Starch for the Idaho Potato Commission

By Anne R. Carey and Sam Ward, USA TODAY

Case Study 11.4

Why We Don't Exercise

Exercise 11.24

The USA Snapshots® "Why we don't exercise" and "Why women don't exercise" in Case Study 11.4 show the responses given by 1000 men and 1000 women who do not exercise.

a. Express the information on these circle graphs as a contingency table showing relative frequencies. Use two columns (M, W) and six rows (the six different responses).

b. Change the relative frequencies to frequencies and construct another 6 × 2 contingency table.

c. Complete a hypothesis test to determine if there is a significant difference between the distribution of responses given by men and women. Use α = 0.05.

The USA Snapshots® below show circle graphs that represent relative frequency distributions for six categories of responses by men and women separately. Does the distribution of responses given by men appear to be significantly different from the distribution of responses given by women? This question calls for a test of homogeneity. Can the sample information given here be expressed on a contingency table?

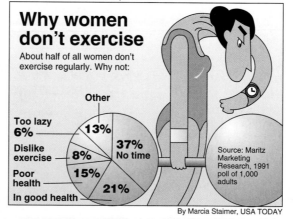

Source: Copyright 1991, USA TODAY. Reprinted with permission.

STEP 3 **The Sample Evidence:**
  a. **Collect the sample information.**
     Sample results are listed in Table 11.10.
  b. **Calculate the value of the test statistic.**
     The expected values are found by using formula (11.5) (p. 557) and are given in Table 11.11.

Table 11.11 **Sample Results and Expected Values**

Type of Residence	Governor's Proposal		
	Favor	**Oppose**	**Total**
Urban	143 (101.6)	57 (98.4)	200
Suburban	98 (101.6)	102 (98.4)	200
Rural	13 (50.8)	87 (49.2)	100
Total	254	246	500

NOTE Each expected value is used twice in the calculation of $\chi^2 \star$; therefore, it is a good idea to keep extra decimal places while doing the calculations.
   The calculated chi-square is

$$\chi^2 \star = \sum_{\text{all cells}} \frac{(O - E)^2}{E} : \chi^2 \star = \frac{(143 - 101.6)^2}{101.6} + \frac{(57 - 98.4)^2}{98.4} + \frac{(98 - 101.6)^2}{101.6}$$

$$+ \frac{(102 - 98.4)^2}{98.4} + \frac{(13 - 50.8)^2}{50.8} + \frac{(87 - 49.2)^2}{49.2}$$

$$\chi^2 \star = 16.87 + 17.42 + 0.13 + 0.13 + 28.13 + 29.04$$

$$\chi^2 \star = \mathbf{91.72}$$

STEP 4 **The Probability Distribution:**

USING THE *p*-VALUE PROCEDURE:

a. Calculate the *p*-value.
Use the right-hand tail since "larger" values of chi-square disagree with the null hypothesis:
$\mathbf{P} = P(\chi^2 > \mathbf{91.72} | \mathbf{df = 2})$ as shown in the figure.

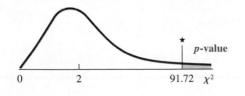

OR   **USING THE CLASSICAL PROCEDURE:**

a. Determine the critical region and critical value(s).
The critical region is the right-hand tail since "larger" values of chi-square disagree with the null hypothesis. The critical value is obtained from Table 8, at the intersection of row df = 2 and column $\alpha = 0.05$:

$$\chi^2_{(2,\,0.05)} = \mathbf{5.99.}$$

USING THE *p*-VALUE PROCEDURE:

To find the *p*-value, you have two options:
1. Use Table 8 (Appendix B) to place bounds on the *p*-value: **P< 0.005**
2. Use a computer or calculator to calculate the *p*-value: **P = 0.000+**

For specific instructions, see page 456.

b. Determine whether or not the *p*-value is smaller than α.

The *p*-value **is smaller** than α.

USING THE CLASSICAL PROCEDURE:

OR

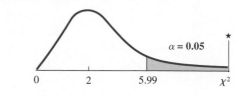

For specific instructions, see page 452.

b. Determine whether or not the calculated test statistic is in the critical region.

χ^2 ★ **is in** the critical region, as shown in **red** in the figure above.

STEP 5   The Results:

a. State the decision about H_o.
Reject H_o.

b. State the conclusion about H_a.
The three groups of voters do not all have the same proportions favoring the proposed legislation.

The following commands will complete a hypothesis test of independence or homogeneity.

MINITAB

Input each column of observed frequencies from the contingency table into C1, C2, . . .; then continue with:

Choose: **Stat > Tables > Chi-Square Test**
Enter: Columns containing the table: **C1 C2. . .**

TI-83

Input the observed frequencies from the $r \times c$ contingency table into an $r \times c$ matrix A. Set up matrix B as an empty $r \times c$ matrix for the expected frequencies;

Choose: **MATRX > EDIT > 1:[A]**
Enter: **r > ENTER > c > ENTER**
           Each observed frequency with an ENTER afterwards
Choose: **MATRX > EDIT > 2[B]**
Enter: **r > ENTER > c > ENTER**
then continue with:

Choose: **STAT > TESTS > C:χ^2 – Test. . .**
Enter: Observed: **[A]** or wherever the contingency table is located
           Expected: **[B]** place for expected frequencies
Highlight: **Calculate   > ENTER**

EXCEL

Input each column of observed frequencies from the contingency table into columns A, B, . . .; then continue with:

Choose: **Tools > Data Analysis Plus > CHI-Square Test of a Contingency Table > OK**
Enter: block coordinates: **(A1:B4 or select cells**–do not include headings)

The solution to Illustration 11.4 looks like this when solved on Minitab:

```
With data in columns C1 and C2 as shown:
  COMPUTER SOLUTION MINITAB Printout for Illustration 11.4:
            C1        C2
           143.      57.
            98.     102.
            13.      87.

  Expected counts are printed below observed counts

          favor     oppose     Total
    1       143        57        200
          101.60     98.40

    2        98       102        200
          101.60     98.40

    3        13        87        100
           50.80     49.20

  Total    254       246        500

  Chi-Sq = 16.870 + 17.418 +
              0.128 + 0.132 + 28.127 + 29.041 = 91.715
  DF = 2,  P-value = 0.000
```

EXERCISES

11.25 State the null hypothesis H_o and the alternative hypothesis H_a that would be used to test the following statements.
  a. The voters expressed preferences that were not independent of their party affiliations.
  b. The distribution of opinions is the same for all three communities.
  c. The proportion of "yes" responses was the same for all categories surveyed.

11.26 The "test of independence" and the "test of homogeneity" are completed in identical fashion, using the contingency table to display and organize the calculations. Explain how these two hypothesis tests differ.

11.27 A random sample of 500 married men was taken; each person was cross-classified as to the size community that he was presently residing in and the size community that he was reared in. The results are shown in the following table.

Size Community Reared In	Size of Community Residing In			Total
	Under 10,000	**10,000 to 49,999**	**50,000 or over**	**Total**
Under 10,000	24	45	45	114
10,000 to 49,999	18	64	70	152
50,000 or over	21	54	159	234
Total	63	163	274	500

Does this sample contradict the claim of independence at the 0.01 level of significance?
  a. Solve using the *p*-value approach.   **b.** Solve using the classical approach.

11.28 A survey of randomly selected travelers who visited the service station restrooms of a large U.S. petroleum distributor showed the following results:

Gender of Respondent	Quality of Restroom Facilities			Totals
	Above Average	**Average**	**Below Average**	
Female	7	24	28	59
Male	8	26	7	41
Totals	15	50	35	100

Using $\alpha = 0.05$, does the sample present sufficient evidence to reject the hypothesis "Quality of responses is independent of the gender of the respondent"?
a. Solve using the *p*-value approach.   **b.** Solve using the classical approach.

11.29 A survey of employees at an insurance firm was concerned with worker–supervisor relationships. One statement for evaluation was "I am not sure what my supervisor expects." The results of the survey are found in the following contingency table.

Years of Employment	I Am Not Sure What My Supervisor Expects		
	True	**Not True**	**Totals**
Less than 1 year	18	13	31
1 to 3 years	20	8	28
3 to 10 years	28	9	37
10 years or more	26	8	34
Totals	92	38	130

Can we reject the hypothesis that "the responses to the statement and the years of employment are independent" at the 0.10 level of significance?
a. Solve using the *p*-value approach.   **b.** Solve using the classical approach.

11.30 The following table is from the July 1993 publication of *Vital and Health Statistics* from the Centers for Disease Control and Prevention/National Center for Health Statistics. The individuals in the following table have an eye irritation, or a nose irritation, or a throat irritation. They have only one of the three.

Type of Irritation	Age (years)			
	18–29	**30–44**	**45–64**	**65 and over**
Eye	440	567	349	59
Nose	924	1311	794	102
Throat	253	311	157	19

Is there sufficient evidence to reject the hypothesis that the type of ENT irritation is independent of the age group at a level of significance equal to 0.05.
a. Solve using the *p*-value approach.   **b.** Solve using the classical approach.

11.31 The manager of an assembly process wants to determine whether the number of defective articles manufactured depends on the day of the week the articles are produced. She collected the following information.

Day of Week	Mon.	Tues.	Wed.	Thurs.	Fri.
Nondefective	85	90	95	95	90
Defective	15	10	5	5	10

Is there sufficient evidence to reject the hypothesis that the number of defective articles is independent of the day of the week on which they are produced? Use $\alpha = 0.05$.
a. Solve using the *p*-value approach.   **b.** Solve using the classical approach.

11.32 A study of the Harvard Business School conducted in 1998 (*Fortune,* "Tales of the Trailblazers," 10-12-98) concentrated on the career paths and lifestyles of the women who

(continued)

were graduates of the program. The focus was on the class of 1973 and the class of 1983. The class of 1973 was the first to include "a solid number of women." But did things change ten years later? Consider the following table:

Harvard Business

School Class	Men	Women	Total
1973	742	34	776
1983	538	189	727
Total	1,280	223	1,503

At the 0.01 level of significance, did the distribution of men and women who completed the program significantly change between 1973 and 1983?
a. Solve using the *p*-value approach.   **b.** Solve using the classical approach.

11.33 Students use many kinds of criteria when selecting courses. "Teacher who is a very easy grader" is often one criterion. Three teachers are scheduled to teach statistics next semester. A sample of previous grade distributions for these three teachers is shown below.

	Professor		
Grades	**#1**	**#2**	**#3**
A	12	11	27
B	16	29	25
C	35	30	15
Other	27	40	23

At the 0.01 level of significance, is there sufficient evidence to conclude "The distribution of grades is not the same for all three professors"?
a. Solve using the *p*-value approach.   **b.** Solve using the classical approach.
c. Which professor is the easiest grader? Explain, citing specific supporting evidence.

11.34 Fear of darkness is a common emotion. The following data were obtained by asking 200 individuals in each age group whether they had serious fears of darkness. At $\alpha = 0.01$, do we have sufficient evidence to reject the hypothesis that "the same proportion of each age group has serious fears of darkness"? (*Hint:* The contingency table must account for all 1000 people.)

Age Group	Elementary	Jr. High	Sr. High	College	Adult
No. Who Fear Darkness	83	72	49	36	114

a. Solve using the *p*-value approach.   **b.** Solve using the classical approach.

11.35 Every two years the Josephson Institute of Ethics conducts a survey of American teenage ethics. One of the objectives of the study is to determine the propensity of teenagers to lie, cheat, and steal. The 1998 report (Josephson Institute of Ethics, "Survey of American Teen-age Ethics," October 1998) split the sample of 20,829 respondents into two major student groups for comparison—high school and middle school. The students admitted to performing the various acts during the past year. The following table summarizes two portions of the results:

Cheated During a Test in School	Yes	No	Total
Middle school	5,437	4,632	10,069
High school	7,532	3,228	10,760
Total	12,969	7,860	20,829

Lied to a parent	Yes	No	Total
Middle school	8,861	1,208	10,069
High school	9,899	861	10,760
Total	18,760	2,069	20,829

(continued)

Does the sample evidence show that students in high school and middle school have a different tendency to: (1) cheat during a test in school? (2) lie to at least one of their parents? Use the 0.01 level of significance in each case.

a. Solve using the *p*-value approach.   **b.** Solve using the classical approach.

11.36 Case Study 11.3 (p. 559) reports percentages describing people's preferences with regards to how potatoes are prepared. Do you believe there is a significant difference between the four regions of America with regards to the percentage that prefer baked? Notice that the article does not mention the sample size.

a. Assume the percentages reported were based on four samples of size 100 from each region and calculate $\chi^2 \star$ and its *p*-value.

b. Repeat (a) using sample sizes of 200 and 300.

c. Are the four percentages, reported in the USA Snapshot®, who prefer baked potatoes significantly different? Describe in detail the circumstances for which they are significantly different.

Return to CHAPTER CASE STUDY

EXERCISE 11.37

Two hundred adults professing to love hot spicy food were asked to name their favorite way to cool their mouth after eating food with hot sauce. Below is the summary of the resulting sample.

Method	Water	Milk	Soda	Beer	Bread	Other	Nothing
Number	73	35	20	19	29	11	13

How similar is the distribution in this sample to the distribution of percentages in the Snapshot?

    a. What information was collected from each adult in the sample?

    b. Define the population and the variable involved in the above sample.

    c. Construct the graph that you believe best displays the distribution of these data.

    d. Does the sample show a distribution that is significantly different from the distribution shown in the "Some like it hot" graph (p. 539)? Use $\alpha = 0.05$.

    e. Write a paragraph (50+ words) describing why the statistical method used in part (c) is appropriate for this set of data.

    f. Write a paragraph (50+ words) describing the meaning of the assumptions and the results of the above statistical procedure.

IN RETROSPECT

In this chapter we have been concerned with tests of hypotheses using chi-square, the cell probabilities associated with the multinomial experiment, and the simple contingency table. In each case the basic assumptions are that a large number of observations have been made and that the resulting test statistic, $\sum \dfrac{(O - E)^2}{E}$, is approximately distributed as chi-square. In general, if *n* is large and the minimum allowable expected cell size is 5, this assumption is satisfied.

The contingency table can be used to test independence or homogeneity. The test for homogeneity and the test for independence look very similar and, in fact, are carried out in exactly the same way. The concepts being tested, however, same distributions and independence, are quite different. The two tests are easily distinguished from one another, for the test of homogeneity has predetermined marginal totals in one direction in the table. That is, before the data are collected, the experimenter determines how many subjects will be observed in each category. The only predetermined number in the test of independence is the grand total.

A few words of caution: The correct number of degrees of freedom is critical if the test results are to be meaningful. The degrees of freedom determine, in part, the critical region, and its size is important. Like other tests of hypothesis, failure to reject H_o does not mean outright acceptance of the null hypothesis.

CHAPTER EXERCISES

11.38 The psychology department at a certain college claims that the grades in its introductory course are distributed as follows: 10% A's, 20% B's, 40% C's, 20% D's, and 10% F's. In a poll of 200 randomly selected students who had completed this course, it was found that 16 had received A's, 43 B's, 65 C's, 48 D's, and 28 F's. Does this sample contradict the department's claim at the 0.05 level?
a. Solve using the p-value approach.   **b.** Solve using the classical approach.

11.39 When interbreeding two strains of roses, we expect the hybrid to appear in three genetic classes in the ratio $1:3:4$. If the results of an experiment yield 80 hybrids of the first type, 340 of the second type, and 380 of the third type, do we have sufficient evidence to reject the hypothesized genetic ratio at the 0.05 level of significance?
a. Solve using the p-value approach.   **b.** Solve using the classical approach.

11.40 A sample of 200 individuals are tested for their blood type, and the results are used to test the hypothesized distribution of blood types:

Blood Type	A	B	O	AB
Percent	0.41	0.09	0.46	0.04

The observed results were as follows:

Blood Type	A	B	O	AB
Number	75	20	95	10

At the 0.05 level of significance, is there sufficient evidence to show that the stated distribution is incorrect?

11.41 As reported in *USA Today*, 10-6-97, about 8.9 million families sent students to college this year and over half live away from home. Where students live:

Parent or guardian's home	46%
Campus housing	26%
Off-campus rental	18%
Own off-campus housing	9%
Other arrangements	2%

Note: Exceeds 100% due to rounding error.

A random sample of 1000 college students resulted in the following information.

Parent or guardian's home	484
Campus housing	230
Off-campus rental	168
Own off-campus housing	96
Other arrangements	22

(continued)

Is the distribution of this sample significantly different than the distribution reported in the newspaper? Use $\alpha = 0.05$. (To adjust for the rounding error, subtract 2 from each expected frequency.)

a. Solve using the *p*-value approach.　　**b.** Solve using the classical approach.

11.42 How often do you review your pay stub to check that the correct taxes are being withheld? On 6-8-98, *USA Today* reported that American adults check as follows:

Always	53%
Most of the time	12%
Occasionally	14%
Never	10%
Don't get a paycheck	10%
Not sure	1%

A random sample of 650 workers resulted in the following frequencies:

Always	342
Most of the time	94
Occasionally	68
Never	73
Don't get a paycheck	60
Not sure	13

Is the distribution of this sample significantly different than the distribution reported in the newspaper? Use $\alpha = 0.05$.

a. Solve using the *p*-value approach.　　**b.** Solve using the classical approach.

11.43 Most golfers are probably happy to play 18 holes of golf whenever they get a chance to play. In 1998, Ben Winter, a club professional, played 306 holes in one day at a charity golf marathon in Stevens, Pennsylvania. A nationwide survey conducted by *Golf* magazine ("18 is Not Enough," September 1998) over the Internet revealed the following frequency distribution of the most number of holes ever played by the respondents in one day:

Most Holes Played in One Day	Percent
18	5
19 to 27	12
28 to 36	28
37 to 45	20
46 to 54	18
55 or more	17

Suppose one of your local public golf courses asks the next 200 golfers who tee off to answer the same question. The following table summarizes their responses:

Most Holes Played in One Day	Number
18	12
19 to 27	35
28 to 36	60
37 to 45	44
46 to 54	35
55 or more	14

Does the distribution of "marathon golfers" at your public course differ from the distribution compiled by *Golf* magazine using responses polled on the Internet? Test at the 0.01 level of significance.

a. Solve using the *p*-value approach.　　**b.** Solve using the classical approach.

11.44 The weights (x) of 300 adult males were determined and used to test the hypothesis that the weights were normally distributed with a mean of 160 lb and a standard deviation of 15 lb. The data were grouped into the following classes:

Weight (x)	Observed Frequency
$x < 130$	7
$130 \leq x < 145$	38
$145 \leq x < 160$	100
$160 \leq x < 175$	102
$175 \leq x < 190$	40
190 and over	13

Using the normal tables, the percentages for the classes are 2.28%, 13.59%, 34.13%, 34.13%, 13.59%, and 2.28%, respectively. Do the observed data show significant reason to discredit the hypothesis that the weights are normally distributed with a mean of 160 lb and a standard deviation of 15 lb? Use $\alpha = 0.05$.
a. Verify the percentages for the classes.
b. Solve using the p-value approach.   c. Solve using the classical approach.

11.45 An article titled "Human Papillomavirus Infection and Its Relationship to Recent and Distant Sexual Partners" (*Obstetrics & Gynecology,* November 1994) gave the following results concerning age and the percent who were HPV-positive among the 290 participants in the study.

Age	N	HPV-Positive (%)
≤ 20	27	40.7
21–25	81	37.0
26–30	108	31.5
31–35	74	24.3

Complete the test of the hypothesis that the same proportion of each age group is HPV-positive for the population this sample represents. Use $\alpha = 0.05$.
a. Solve using the p-value approach.   b. Solve using the classical approach.

11.46 The following table shows the number of reported crimes committed last year in the inner part of a large city. The crimes were classified according to type of crime and district of the inner city where it occurred. Do these data show sufficient evidence to reject the hypothesis that the type of crime and the district in which it occurred are independent? Use $\alpha = 0.01$.

District	Crime				
	Robbery	Assault	Burglary	Larceny	Stolen Vehicle
1	54	331	227	1090	41
2	42	274	220	488	71
3	50	306	206	422	83
4	48	184	148	480	42
5	31	102	94	596	56
6	10	53	92	236	45

a. Solve using the p-value approach.   b. Solve using the classical approach.

11.47 Based on the results of a survey questionnaire, 400 individuals were classified as either politically conservative, moderate, or liberal. In addition, each was classified by age, as shown in the following table.

(continued)

	Age Group			
	20–35	**36–50**	**Over 50**	**Totals**
Conservative	20	40	20	80
Moderate	80	85	45	210
Liberal	40	25	45	110
Totals	140	150	110	400

Is there sufficient evidence to reject the hypothesis that "political preference is independent of age"? Use $\alpha = 0.01$.
a. Solve using the *p*-value approach.   **b.** Solve using the classical approach.

 11.48 "Cramped quarters" is a common complaint by airline travelers. A random sample of 100 business travelers and 150 leisure travelers were asked where they would "most like more space." The resulting answers are summarized in the table below.

Place	**Business**	**Leisure**
Overhead space on plane	15	9
Hotel room	29	49
Leg room on plane	91	66
Rental car size	10	20
Other	5	6

Does this sample information present sufficient evidence to conclude that the business traveler and the leisure traveler differ in where they would most like additional space? Use $\alpha = 0.05$.
a. Solve using the *p*-value approach.   **b.** Solve using the classical approach.

11.49 Four brands of popcorn were tested for popping. One hundred kernels of each brand were popped, and the number of kernels not popped was recorded in each test (see the following table). Can we reject the null hypothesis that all four brands pop equally? Test at $\alpha = 0.05$.

Brand	**A**	**B**	**C**	**D**
No. Not Popped	14	8	11	15

a. Solve using the *p*-value approach.   **b.**   Solve using the classical approach.

11.50 An average of two players per boys' or girls' high school basketball team is injured during a season. The table below shows the distribution of injuries for a random sample of 1000 girls and 1000 boys taken from the 1996–1997 season records of all reported injuries.

Injury	**Girls**	**Boys**
Ankle/foot	360	383
Hip/thigh/leg	166	147
Knee	130	103
Forearm/wrist/hand	112	115
Face/scalp	88	122
All others	144	130

Does this sample information present sufficient evidence to conclude that the distribution of injuries is different for girls than boys? Use $\alpha = 0.05$.
a. Solve using the *p*-value approach.   **b.** Solve using the classical approach.

11.51 Does test failure reduce academic aspirations and thereby contribute to the decision to drop out of school? These were the concerns of a study titled "Standards and School Dropouts: A National Study of Tests Required for High School Graduation." The table reports the responses of 283 students selected from schools with low graduation rates to the question "Do tests required for graduation discourage some students from staying in school?"

	Urban	Suburban	Rural	Total
Yes	57	27	47	131
No	23	16	12	51
Unsure	45	25	31	101
Total	125	68	90	283

Source: American Journal of Education (November 1989), University of Chicago Press. ©1989 by The University of Chicago. All Rights Reserved.

Does there appear to be a relationship at the 0.05 level of significance between a student's response and the school's location?

11.52 Last year's work record for absenteeism in each of four categories for 100 randomly selected employees is compiled in the following table. Do these data provide sufficient evidence to reject the hypothesis that the rate of absenteeism is the same for all categories of employees? Use $\alpha = 0.01$ and 240 workdays for the year.

	Married Male	Single Male	Married Female	Single Female
Number of Employees	40	14	16	30
Days Absent	180	110	75	135

a. Solve using the *p*-value approach.   **b.** Solve using the classical approach.

11.53 If you were to roll a die 600 times, how different from 100 could the observed frequencies for each face be before the results would become significantly different from equally likely at the 0.05 level?

11.54 Using the students at your college or a population of your choice, collect your own set of "favorite hot sauce coolant" data and repeat Exercise 11.37 (a–f).

11.55 Consider the following set of data.

	Response		
	Yes	No	Total
Group 1	75	25	100
Group 2	70	30	100
Total	145	55	200

a. Compute the value of the test statistic $z \bigstar$ that would be used to test the null hypothesis that $p_1 = p_2$, where p_1 and p_2 are the proportions of "yes" responses in the respective groups.
b. Compute the value of the test statistic $\chi^2 \bigstar$ that would be used to test the hypothesis that "response is independent of group."
c. Show that $\chi^2 \bigstar = (z \bigstar)^2$.

VOCABULARY LIST

Be able to define each term. Pay special attention to the key terms, which are printed in **red**. In addition, describe in your own words, and give an example of, each term. Your examples should not be ones given in class or in the textbook.

The bracketed numbers indicate the chapters in which the term previously appeared, but you should define the terms again to show increased understanding of their meaning. Page numbers indicate the first appearance of the term in Chapter 11.

cell (p. 540)
chi-square [9] (p. 540)
column [3] (p. 557)
contingency table (p. 553)
degrees of freedom [9, 10] (pp. 541, 557)
enumerative data (p. 540)
expected frequency (pp. 540, 557)
expected value (p. 555)

homogeneity (p. 558)
hypothesis test [8, 9, 10] (pp. 540, 553, 558)
independence [4] (p. 553)
marginal totals [3] (p. 554)
multinomial experiment (pp. 542, 544)
observed frequency [2, 4] (p. 540)
$r \times c$ contingency table (p. 557)
rows [3] (p. 557)

CHAPTER PRACTICE TEST

Part I: Knowing the Definitions

Answer "True" if the statement is always true. If the statement is not always true, replace the words shown in bold with words that make the statement always true.

11.1 The number of degrees of freedom for a test of a multinomial experiment is **equal to** the number of cells in the experimental data.

11.2 The **expected frequency** in a chi-square test is found by multiplying the hypothesized probability of a cell by the number of pieces of data in the sample.

11.3 The **observed** frequency of a cell should not be allowed to be smaller than 5 when a chi-square test is being conducted.

11.4 In the **multinomial experiment** we have $(r - 1)$ times $(c - 1)$ degrees of freedom (r is the number of rows, and c is the number of columns).

11.5 A multinomial experiment consists of the ***n* identical, independent trials.**

11.6 A **multinomial experiment** arranges the data into a two-way classification such that the totals in one direction are predetermined.

11.7 The charts for both the multinomial experiment and the contingency table **must** be set in such a way that each piece of data will fall into exactly one of the categories.

11.8 The test statistic $\sum \dfrac{(O - E)^2}{E}$ has a distribution that is **approximately normal.**

11.9 The data used in a chi-square multinomial test are always **enumerative** in nature.

11.10 The null hypothesis being tested by a test of **homogeneity** is that the distribution of proportions is the same for each of the subpopulations.

Part II: Applying the Concepts

Answer all questions. Show formulas, substitutions, and work.

11.11 State the null and alternative hypotheses that would be used to test each of the following claims:
  a. The single-digit numerals generated by a certain random number generator were not equally likely.

(continued)

      b. The results of the last election in our city suggest that the votes cast were not independent of the voter's registered party.

      c. The distributions of types of crimes committed against society are the same in the four largest U.S. cities.

11.12 Find each of the following:

    a. $\chi^2(12, 0.975)$ **b.** $\chi^2(17, 0.005)$

11.13 Three hundred consumers were each asked to identify which one of three different items they found to be the most appealing. The table shows the number that preferred each item.

Item	1	2	3
Number	85	103	112

Do these data present sufficient evidence at the 0.05 level of significance to indicate that the three items are not equally preferred?

11.14 To study the effect of the type of soil on the amount of growth attained by a new hybrid plant, saplings were planted in three different types of soil and their subsequent amounts of growth classified into three categories:

Growth	Soil Type		
	Clay	Sand	Loam
Poor	16	8	14
Average	31	16	21
Good	18	36	25
Total	65	60	60

Does the quality of growth appear to be distributed differently for the tested soil types at the 0.05 level?

    a. State the null and alternative hypotheses.

    b. Find the expected value for the cell containing 36.

    c. Calculate the value of chi-square for these data.

    d. Find the p-value.

    e. Find the test criteria [level of significance, test statistic, its distribution, critical region, and critical value(s)].

    f. State the decision and the conclusion for this hypothesis test.

Part III: Understanding the Concepts

11.15 Explain how a multinomial experiment and a binomial experiment are similar and also how they are different.

11.16 Explain the distinction between a test for independence and a test for homogeneity.

11.17 Student A says that the tests for independence and homogeneity are the same, and student B says that they are not at all alike because they are tests of different concepts. Both students are partially right and partially wrong. Explain.

11.18 You are interpreting the results of an opinion poll on the role of recycling in your town. A random sample of 400 people were asked to respond strongly in favor, slightly in favor, neutral, slightly against, or strongly against on each of several questions. There are four key questions that concern you, and you plan to analyze their results.

    a. How do you calculate the expected probabilities for each answer?

    b. How would you decide if the four questions were answered the same?

Chapter 12

Analysis of Variance

CHAPTER CASE STUDY

Time Spent Reading the Newspaper

How much time did you spend reading the newspaper yesterday? How much time did your parents spend reading the newspaper yesterday? Does everybody spend the same amount of time? What was the mean amount of time spent reading the newspaper yesterday by people of your age? What was the mean amount of time spent reading the newspaper yesterday by people of your parents' age? Do you think that a person's age will have any effect on the amount of time spent reading the newspaper yesterday? The USA Snapshot® "Speed readers" that appeared in *USA Today,* July 14, 1998, seems to suggest that older adults spend more time reading the newspaper than do younger adults.

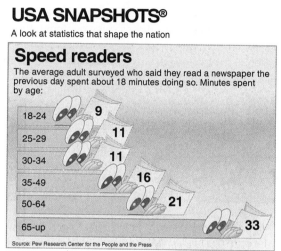

USA SNAPSHOTS®
A look at statistics that shape the nation

Speed readers
The average adult surveyed who said they read a newspaper the previous day spent about 18 minutes doing so. Minutes spent by age:

Age	Minutes
18-24	9
25-29	11
30-34	11
35-49	16
50-64	21
65-up	33

Source: Pew Research Center for the People and the Press

By Anne R. Carey and Web Bryant, USA TODAY

Twenty-five newspaper readers were asked how much time, estimated to the nearest minute, they spent reading yesterday's newspaper. Their times are listed by age category of the reader.

	Age of Adult					
	18–24	25–29	30–34	35–49	50–64	65–up
x,	10	13	14	20	20	8
time spent	5	8	7	12	12	30
reading	15	20	20	5	18	35
newspaper	4	10		28	15	20
yesterday				15	30	

Does the sample show that a person's age has an effect on the amount of time spent daily reading the newspaper? Does the sample show that a person's age has an effect on the average amount of time spent daily reading the newspaper? Are these different questions? We will learn about statistical methods that can be used to answer these questions in this chapter, and we will return to this case study at the end of the chapter.
(See Exercise 12.24, p. 595)

CHAPTER OBJECTIVES

Previously, we have tested hypotheses comparing two means. In this chapter we are concerned with testing a hypothesis comparing several means . The **analysis of variance technique (ANOVA,** pronounced AN-NO-VA), which we are about to explore, will be used to test a hypothesis about several means, for example,

$$H_o: \mu_1 = \mu_2 = \mu_3 = \mu_4 = \mu_5$$

By using our former technique for hypotheses about two means, we could test several hypotheses if each stated a comparison of two means. For example, we could test

$H_1: \mu_1 = \mu_2$	$H_2: \mu_1 = \mu_3$	$H_3: \mu_1 = \mu_4$	$H_4: \mu_1 = \mu_5$	$H_5: \mu_2 = \mu_3$
$H_6: \mu_2 = \mu_4$	$H_7: \mu_2 = \mu_5$	$H_8: \mu_3 = \mu_4$	$H_9: \mu_3 = \mu_5$	$H_{10}: \mu_4 = \mu_5$

In order to test the null hypothesis, H_o, that all five means are equal, we would have to test each of these ten hypotheses using our former technique for two means. Rejection of any one of the ten hypotheses about two means would cause us to reject the null hypotheses that all five means are equal. If we failed to reject all ten hypotheses about the means, we would fail to reject the main null hypothesis. Suppose we tested a null hypothesis that dealt with several means by testing all the possible pairs of two means; the overall type I error rate would become much larger than the value of α associated with a single test. The ANOVA techniques allow us to test the null hypothesis (all means are equal) against the alternative hypothesis (at least one mean value is different) with a specified value of α.

In this chapter we introduce ANOVA. ANOVA experiments can be very complex, depending on the situation. We will restrict our discussion to the most basic experimental design, the single-factor ANOVA .

12.1   Introduction to the Analysis of Variance Technique

We will begin our discussion of the analysis of variance technique by looking at an illustration.

ILLUSTRATION 12.1 ▼

The temperature at which a plant is maintained is believed to affect the rate of production in the plant. The data in Table 12.1 are the number, x, of units produced in one hour for randomly selected one-hour periods when the production process in the plant was operating at each of three temperature levels . The data values resulting from repeated samplings are called **replicates.** Four replicates, or data values, were obtained for two of the temperatures and five were obtained for the third temperature. Do these data suggest that temperature has a significant effect on the production level at the 0.05 level?

Table 12.1 Sample Results for Illustration 12.1

	Temperature Levels		
	Sample from 68°F ($i = 1$)	Sample from 72°F ($i = 2$)	Sample from 76°F ($i = 3$)
	10	7	3
	12	6	3
	10	7	5
	9	8	4
		7	
Column	$C_1 = 41$	$C_2 = 35$	$C_3 = 15$
Totals	$\bar{x}_1 = 10.25$	$\bar{x}_2 = 7.0$	$\bar{x}_3 = 3.75$

The level of production is measured by the mean value; $\bar{x}_i$ indicates the observed production mean at level i, where $i = 1$, 2, and 3 corresponds to temperatures of 68°F, 72°F, and 76°F, respectively. There is a certain amount of variation among these means. Since sample means do not necessarily repeat when repeated samples are taken from a population, some variation can be expected, even if all three population means are equal. We will next pursue the question "Is this variation among the $\bar{x}$'s due to chance, or is it due to the effect that temperature has on the production rate?"

Solution

STEP 1 **The Set-Up:**

a. **Describe the population parameter of interest.**

The "mean" at each **level of the test factor** is of interest: the mean production rate at 68°F μ_{68}, the mean production rate at 72°F μ_{72}, and the mean production rate at 76°F μ_{76}. The factor being tested, "plant temperature," has three levels 68°F, 72°F, or 76°F.

b. **State the null hypothesis (H_o) and the alternative hypothesis (H_a).**

$$H_o: \mu_{68} = \mu_{72} = \mu_{76}$$

That is, the true production mean is the same at each temperature level tested. In other words, the temperature does not have a significant effect on the production rate. The alternative to the null hypothesis is

$$H_a: \text{Not all temperature level means are equal}$$

Thus, we will want to reject the null hypothesis if the data show that one or more of the means are significantly different from the others.

STEP 2 **The Test Criteria:**

a. **Check the assumptions.**

The data were randomly collected and are independent of each other. The effects due to chance and untested factors are assumed to be normally distributed. (See pages 583 and 584 for further discussion.)

b. **Identify the probability distribution and the test statistic to be used.**

We will make the decision to reject H_o or fail to reject H_o by using the F distribution and an F-test statistic. (See page 578.)

c. **Determine the level of significance, α.**

$\alpha = 0.05$ (given in the statement of the problem).

Exercise 12.1

Draw a dotplot of the data in Table 12.1. Represent the data using integers 1, 2, and 3 to indicate the level of test factor the data are from. Do you see a "difference" between the levels?

Exercise 12.2

Each department at the large industrial plant is rated weekly. State the hypotheses used to test "the mean weekly ratings are the same in three departments."

STEP 3 **The Sample Evidence:**
    a. **Collect the sample information.**
       The sample data are listed in Table 12.1, page 577.
    b. **Calculate the value of the test statistic.**

Recall from Chapter 10 that the calculated value of F is the ratio of two variances. The analysis of variance procedure will separate the variation among the entire set of data into two categories. To accomplish this separation, we first work with the numerator of the fraction used to define **sample variance,** formula (2.6):

$$s^2 = \frac{\sum(x - \bar{x})^2}{n - 1}$$

The numerator of this fraction is called the **sum of squares:**

$$\text{sum of squares} = \sum(x - \bar{x})^2 \tag{12.1}$$

We calculate the **total sum of squares, SS(total),** for the total set of data by using a formula that is equivalent to formula (12.1) but does not require the use of $\bar{x}$. This equivalent formula is

$$\text{SS(total)} = \sum(x^2) - \frac{(\sum x)^2}{n} \tag{12.2}$$

Now we can find the SS(total) for our illustration by using formula (12.2):

First:
$$\sum(x^2) = 10^2 + 12^2 + 10^2 + 9^2 + 7^2 + 6^2 + 7^2 + 8^2 + 7^2 + 3^2 + 3^2 + 5^2 + 4^2 = 731$$
$$\sum x = 10 + 12 + 10 + 9 + 7 + 6 + 7 + 8 + 7 + 3 + 3 + 5 + 4 = 91$$

Then using formula 12.2:

$$\text{SS(total)} = \sum(x^2) - \frac{(\sum x)^2}{n}: \quad \text{SS(total)} = 731 - \frac{(91)^2}{13} = 731 - 637 = \mathbf{94}$$

Next, 94, the SS(total), must be separated into two parts: the sum of squares due to temperature levels, SS(temperature), and the sum of squares due to experimental error of replication, SS(error). This splitting is often called **partitioning,** since SS(temperature) + SS(error) = SS(total); that is, in our illustration SS(temperature) + SS(error) = 94. The sum of squares, **SS(factor)** [SS(temperature) for our illustration] that measures the **variation between the factor levels** (temperatures) is found by using formula (12.3):

$$\text{SS(factor)} = \left(\frac{C_1^2}{k_1} + \frac{C_2^2}{k_2} + \frac{C_3^2}{k_3} + \cdots\right) - \frac{(\sum x)^2}{n} \tag{12.3}$$

where C_i represents the column total, k_i represents the number of replicates at each level of the factor, and n represents the total sample size ($n = \sum k_i$).

NOTE The data have been arranged so that each column represents a different level of the factor being tested.

    Now we can find the SS(temperature) for our illustration by using formula (12.3):

$$\text{SS(factor)} = \left(\frac{C_1^2}{k_1} + \frac{C_2^2}{k_2} + \frac{C_3^2}{k_3} + \cdots\right) - \frac{(\sum x)^2}{n}:$$

$$\text{SS(temperature)} = \left(\frac{41^2}{4} + \frac{35^2}{5} + \frac{15^2}{4}\right) - \frac{(91)^2}{13}$$

$$= (420.25 + 245.00 + 56.25) - 637.0 = 721.5 - 637.0 = \mathbf{84.5}$$

The sum of squares **SS(error)** that measures the **variation within the rows** is found by using formula (12.4):

$$SS(\text{error}) = \sum(x^2) - \left(\frac{C_1^2}{k_1} + \frac{C_2^2}{k_2} + \frac{C_3^2}{k_3} + \cdots \right) \qquad (12.4)$$

The SS(error) for our illustration can now be found by using formula (12.4):

$$SS(\text{error}) = \sum(x^2) - \left(\frac{C_1^2}{k_1} + \frac{C_2^2}{k_2} + \frac{C_3^2}{k_3} + \cdots \right):$$

First: $\sum(x^2) = 731$ (found previously)

$$\left(\frac{C_1^2}{k_1} + \frac{C_2^2}{k_2} + \frac{C_3^2}{k_3} + \cdots \right) = 721.5 \text{ (found previously)}$$

Then using formula 12.4

$$SS(\text{error}) = \sum(x^2) - \left(\frac{C_1^2}{k_1} + \frac{C_2^2}{k_2} + \frac{C_3^2}{k_3} + \cdots \right): \quad SS(\text{error}) = 731.0 - 721.5 = \mathbf{9.5}$$

NOTE   SS(total) = SS(factor) + SS(error). Inspection of formulas (12.2), (12.3), and (12.4) will verify this.

For convenience we will use an ANOVA table to record the sums of squares and to organize the rest of the calculations. The format of an ANOVA table is shown in Table 12.2.

Table 12.2   Format for ANOVA Table

Source	df	SS	MS
Factor		84.5	
Error		9.5	
Total		94.0	

We have calculated the three sums of squares for our illustration. The **degrees of freedom,** df, associated with each of the three sources are determined as follows:

1. df(factor) is one less than the number of levels (columns) for which the factor is tested:

$$df(\text{factor}) = c - 1 \qquad (12.5)$$

where c represents the number of levels for which the factor is being tested (number of columns on the data table).

2. df(total) is one less than the total number of data:

$$df(\text{total}) = n - 1 \qquad (12.6)$$

where n represents the number of data in the total sample (that is, $n = k_1 + k_2 + k_3 + \cdots$, where k_i is the number of replicates at each level tested).

3. df(error) is the sum of the degrees of freedom for all the levels tested (columns in the data table). Each column has $k_i - 1$ degrees of freedom; therefore,

$$df(\text{error}) = (k_1 - 1) + (k_2 - 1) + (k_3 - 1) + \cdots$$

or

$$df(\text{error}) = n - c \qquad (12.7)$$

The degrees of freedom for our illustration are

$$df(\text{temperature}) = c - 1 = 3 - 1 = \mathbf{2}$$

$$df(\text{total}) = n - 1 = 13 - 1 = \mathbf{12}$$

$$df(\text{error}) = n - c = 13 - 3 = \mathbf{10}$$

The sums of squares and the degrees of freedom must check. That is,

$$SS(\text{factor}) + SS(\text{error}) = SS(\text{total}) \qquad \mathbf{(12.8)}$$

$$df(\text{factor}) + df(\text{error}) = df(\text{total}) \qquad \mathbf{(12.9)}$$

The **mean square** for the factor being tested, the **MS(factor),** and for error, the **MS(error),** will be obtained by dividing the sum-of-squares value by the corresponding number of degrees of freedom. That is,

$$MS(\text{factor}) = \frac{SS(\text{factor})}{df(\text{factor})} \qquad \mathbf{(12.10)}$$

$$MS(\text{error}) = \frac{SS(\text{error})}{df(\text{error})} \qquad \mathbf{(12.11)}$$

The mean squares for our illustration are

$$MS(\text{temperature}) = \frac{SS(\text{temperature})}{df(\text{temperature})} = \frac{84.5}{2} = \mathbf{42.25}$$

$$MS(\text{error}) = \frac{SS(\text{error})}{df(\text{error})} = \frac{9.5}{10} = \mathbf{0.95}$$

The complete ANOVA table appears as shown in Table 12.3.

Table 12.3 ANOVA Table for Illustration 12.1

Source	df	SS	MS
Temperature	2	84.5	42.25
Error	10	9.5	0.95
Total	12	94.0	

The hypothesis test is now completed using the two mean squares as the measures of variance. The calculated value of the **test statistic, $F\bigstar$,** is found by dividing the MS(factor) by the MS(error):

$$F\bigstar = \frac{MS(\text{factor})}{MS(\text{error})} \qquad \mathbf{(12.12)}$$

The calculated value of F for our illustration is found by using formula (12.12):

$$F\bigstar = \frac{MS(\text{factor})}{MS(\text{error})} : \quad F\bigstar = \frac{MS(\text{temperature})}{MS(\text{error})} = \frac{42.25}{0.95} = \mathbf{44.47}$$

NOTE Since the calculated value of F, $F\bigstar$, is found by dividing MS(temperature) by MS(error), the number of degrees of freedom for the numerator is $df(\text{temperature}) = 2$ and the number of degrees of freedom for the denominator is $df(\text{error}) = 10$.

STEP 4 The Probability Distribution:

p-VALUE PROCEDURE:

a. Calculate the *p*-value.
Use the right-hand tail since larger values of $F \star$ indicate "not all equal" as expressed by H_a.
$\mathbf{P} = P(F > 44.47 \,|\, df_n = 2, df_d = 10)$ as shown in the figure.

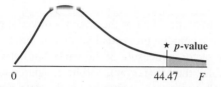

To find the *p*-value, you have two options:
1. Use Tables 9(c) (Appendix B) to place bounds on the *p*-value: **P < 0.01**
2. Use a computer or calculator to calculate the *p*-value: **P = 0.00001**

For additional instructions, see page 520.

b. Determine whether or not the *p*-value is smaller than α.
The *p*-value **is smaller** than the level of significance, α (0.05).

OR

CLASSICAL PROCEDURE:

a. Determine the critical region and critical value(s)
The critical region is the right-hand tail since large values of $F \star$ indicate "not all equal" as expressed by H_a. $df_n = 2$ and $df_d = 10$. The critical value is obtained from Table 9(a):
$F_{(2, 10, 0.05)} = \mathbf{4.10.}$

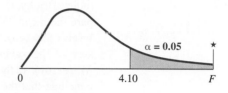

For additional instructions, see page 517.

b. Determine whether or not the calculated test statistic is in the critical region.
$F \star$ **is in** the critical region, as shown in **red** in the figure above.

Exercise 12.3
Use a computer or calculator to find the *p*-value when $F \star = 4.572$, $df_{factor} = 5$, and $df_{error} = 22$.

STEP 5   The Results:
a. State the decision about H_o.
Reject H_o.
b. State the conclusion about H_a.
At least one of the room temperatures does have a significant effect on the production rate. The differences in the mean production rates at the tested temperature levels were found to be significant.

The mean at 68°F is certainly different from the mean at 76°F, since the sample means for these levels are the largest and smallest, respectively. Whether any other pairs of means are significantly different cannot be determined from the ANOVA procedure alone.   ▲

In this section, we have seen how the ANOVA technique separated the variance among the sample data into two measures of variance: (1) MS(factor), the measure of variance between the levels being tested, and (2) MS(error), the measure of variance within the levels being tested. Then these measures of variance can be compared. For our illustration, the between-level variance was found to be significantly larger than the within-level variance (experimental error). This led us to the conclusion that temperature did have a significant effect on the variable *x*, the number of units of production completed per hour.

In the next section we will use several illustrations to demonstrate the logic of the analysis of variance technique.

12.2 The Logic Behind ANOVA

Many experiments are conducted to determine the effect that different levels of some test factor have on a **response variable.** The test factor may be temperature (as in Illustration 12.1), the manufacturer of a product, the day of the week, or any number of other things. In this chapter we are investigating the single-factor analysis of variance. Basically, the design for the single-factor ANOVA is to obtain independent random samples at each of the several levels of the factor being tested. We will then make a statistical decision concerning the effect that the levels of the test factors have on the response (observed) variable.

Illustrations 12.2 and 12.3 demonstrate the logic of the analysis of variance technique. Briefly, the reasoning behind the technique proceeds like this: In order to compare the means of the levels of the test factor, a measure of the **variation between the levels** (between columns on the data table), the **MS(factor),** will be compared to a measure of the **variation within the levels** (within the columns on the data table), the **MS(error).** If the MS(factor) is significantly larger than the MS(error), then we will conclude that the means for each of the factor levels being tested are not all the same. This implies that the factor being tested does have a significant effect on the response variable. If, however, the MS(factor) is not significantly larger than the MS(error), we will not be able to reject the null hypothesis that all means are equal.

ILLUSTRATION 12.2 ▼

Do the data in Table 12.4 show sufficient evidence to conclude that there is a difference in the three population means μ_F, μ_G, and μ_H?

Table 12.4 Sample Results for Illustration 12.2

	Factor Levels	
Sample from Level F	**Sample from Level G**	**Sample from Level H**
3	5	8
2	6	7
3	5	7
4	5	8
$C_F = 12$	$C_G = 21$	$C_H = 30$
$\bar{x}_F = 3.00$	$\bar{x}_G = 5.25$	$\bar{x}_H = 7.50$

Figure 12.1 shows the relative relationship among the three samples. A quick look at the figure suggests that the three sample means are different from each other, implying that the sampled populations have different mean values. These three samples demonstrate

Figure 12.1
Data from Table 12.4

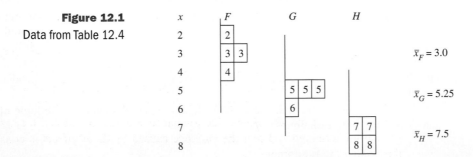

relatively little within-sample variation, although there is a relatively large amount of between-sample variation.   ▲

Let's look at another illustration.

ILLUSTRATION 12.3 ▼

Do the data in Table 12.5 show sufficient evidence to conclude that there is a difference in the three population means μ_J, μ_K, and μ_L?

Table 12.5   Sample Results for Illustration 12.3

| | Factor Levels | |
Sample from Level J	Sample from Level K	Sample from Level L
3	5	6
8	4	2
6	3	7
4	7	5
$C_J = 21$	$C_K = 19$	$C_L = 20$
$\bar{x}_J = 5.25$	$\bar{x}_K = 4.75$	$\bar{x}_L = 5.00$

Figure 12.2 shows the relative relationship among the three samples. A quick look at the figure does not suggest that the three sample means are different from each other.

Figure 12.2
Data from Table 12.5

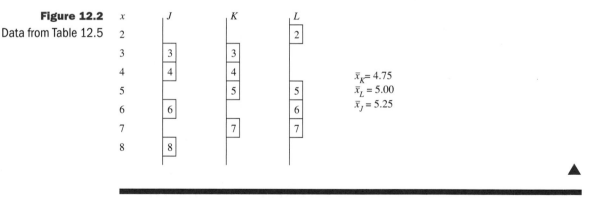

There is little between-sample variation for these three samples (that is, the sample means are relatively close in value), whereas the within-sample variation is relatively large (that is, the data values within each sample cover a relatively wide range of values).

To complete a hypothesis test for analysis of variance, we must agree on some ground rules, or assumptions . In this chapter we will use the following three basic assumptions:

1. Our goal is to investigate the effect that various levels of the factor under test have on the response variable. Typically, we want to find the level that yields the most advantageous values of the response variable. This, of course, means that we probably will want to reject the null hypothesis in favor of the alternative. Then a follow-up study could determine the "best" level of the factor.

2. We must assume that the effects due to chance and due to untested factors are normally distributed and that the variance caused by these effects is constant throughout the experiment.

3. We must assume independence among all observations of the experiment. (Recall that independence means the results of one observation of the experiment do not affect the results of any other observation.) We will usually conduct the tests in a **randomized** order to ensure independence. This technique also helps to avoid data contamination.

12.3   Applications of Single-Factor ANOVA

Before continuing our ANOVA discussion, let's identify the notation, particularly the subscripts that are used (see Table 12.6). Notice that each piece of data has two subscripts; the first subscript indicates the column number (test factor level), and the second subscript identifies the replicate (row) number. The column totals, C_i, are listed across the bottom of the table. The grand total, T, is equal to the sum of all x's and is found by adding the column totals. Row totals can be used as a cross-check but serve no other purpose.

A **mathematical model** (equation) is often used to express a particular situation. In Chapter 3 we used a mathematical model to help explain the relationship between the values of bivariate data. The equation $\hat{y} = b_0 + b_1 x$ served as the model when we believed that a straight-line relationship existed. The probability functions studied in Chapter 5 are also examples of mathematical models. For the single-factor ANOVA, the mathematical model, formula (12.13), is an expression of the composition of each piece of data entered in our data table.

Case Study   12.1

High School Homework Time

Exercise 12.4
Referring to "High school homework time,"

a. Does the category appear to have an effect on the average amount of study time? Explain.

b. The Snapshot shows seven categories. Explain why these seven categories cannot be used to organize data for one-way. How could these categories be used?

This USA Snapshot® reports that the average amount of time spent weekly by high school students is down to 6.1 hours from 6.6 hours as reported in 1996–1997. Does it appear that the category to which the student belongs has an effect on the average weekly study time?.

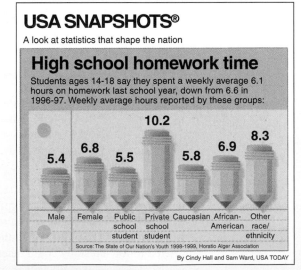

Source: Copyright 1998, USA TODAY. Reprinted with permission.

Exercise 12.5

Consider the following table for a single-factor ANOVA. Find the following:

a. $x_{1,2}$ b. $x_{2,1}$ c. C_1
d. Σx e. $\Sigma (C_i)^2$

	Level of Factor		
Replicates	1	2	3
1	3	2	7
2	0	5	4
3	1	4	5

Table 12.6   Notation Used in ANOVA

	Factor Levels				
Replicates	Sample from Level 1	Sample from Level 2	Sample from Level 3	...	Sample from Level C
$k = 1$	$x_{1,1}$	$x_{2,1}$	$x_{3,1}$		$x_{c,1}$
$k = 2$	$x_{1,2}$	$x_{2,2}$	$x_{3,2}$		$x_{c,2}$
$k = 3$	$x_{1,3}$	$x_{2,3}$	$x_{3,3}$		$x_{c,3}$
$\vdots$					
Column Totals	C_1	C_2	C_3	...	C_c	T

$$T = \text{grand total} = \text{sum of all } x\text{'s} = \Sigma x = \Sigma C_i$$

$$x_{c,k} = \mu + F_c + \epsilon_{k(c)} \tag{12.13}$$

We interpret each term of this model as follows:

1. μ is the mean value for all the data without respect to the test factor.

2. F_c is the effect that the factor being tested has on the response variable at each different level c.

3. $\epsilon_{k(c)}$ (ϵ is the lowercase Greek letter epsilon) is the **experimental error** that occurs among the k replicates in each of the c columns.

Let's look at another hypothesis test concerning an analysis of variance.

Case Study   12.2

A Time for Giving

Exercise 12.6

The amount of money donated varies from adult to adult within each age group, just as it varies from person to person for all ages. "A time for giving" suggests that the average amount donated is affected by the age category the adult belongs to. Describe how the data presented are organized in the same way as data are organized for one-way analysis of variance.

The average amount of cash donated to charities during the holiday season varies greatly from person to person. This USA Snapshot® reports that men and women give different average amounts, and the average amount donated changes as the person's age changes.

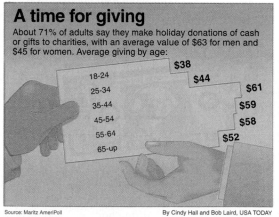

USA SNAPSHOTS®

A look at statistics that shape the nation

A time for giving

About 71% of adults say they make holiday donations of cash or gifts to charities, with an average value of $63 for men and $45 for women. Average giving by age:

Age	Amount
18-24	$38
25-34	$44
35-44	$61
45-54	$59
55-64	$58
65-up	$52

Source: Maritz AmeriPoll   By Cindy Hall and Bob Laird, USA TODAY

Source: Copyright 1998, USA TODAY. Reprinted with permission.

ILLUSTRATION 12.4 ▼

A rifle club performed an experiment on a randomly selected group of first-time shooters. The purpose of the experiment was to determine whether shooting accuracy is affected by the method of sighting used: only the right eye open, only the left eye open, or both eyes open. Fifteen beginning shooters were selected and split into three groups. Each group experienced the same training and practicing procedures with one exception: the method of sighting used. After completing training, each student was given the same number of rounds and asked to shoot at a target. Their scores appear in Table 12.7.

Table 12.7 Sample Results for Illustration 12.4

Method of Sighting		
Right Eye	**Left Eye**	**Both Eyes**
12	10	16
10	17	14
18	16	16
12	13	11
14		20
		21

At the 0.05 level of significance, is there sufficient evidence to reject the claim that these methods of sighting are equally effective?

Solution

In this experiment the factor is "method of sighting" and the levels are the three different methods of sighting (right eye, left eye, and both eyes open). The replicates are the scores received by the students in each group. The null hypothesis to be tested is "the three methods of sighting are equally effective," or "the mean scores attained using each of the three methods are the same."

STEP 1 **The Set-Up:**
    a. Describe the population parameter of interest.
        The "mean" at each level of the test factor is of interest: the mean score using the right eye μ_R, the mean score using left eye μ_L, and the mean score using both eyes μ_B. The factor being tested, "method of sighting," has three levels: right, left, or both.
    b. State the null hypothesis (H_o) and the alternative hypothesis (H_a).

        $H_o: \mu_R = \mu_L = \mu_B$

        H_a: **The means are not all equal** (that is, at least one mean is different).

STEP 2 **The Hypothesis Test Criteria:**
    a. Check the assumptions.
        The shooters were randomly assigned to the method and their scores are independent of each other. The effects due to chance and untested factors are assumed to be normally distributed.
    b. Identify the probability distribution and the test statistic to be used.
        The F-distribution and formula (12.12) will be used with df(numerator) = df(method) = 2 and df(denominator) = df(error) = 12.
    c. Determine the level of significance, α.
        $\alpha = 0.05$.

STEP 3   **The Sample Evidence:**

    a. **Collect the sample information.**

        See Table 12.8 below.

    b. **Calculate the value of the test statistic.**

        Calculate the test statistic F ★: Table 12.8 is used to find column totals.

Table 12.8   Sample Results for Illustration 12.4

Replicates	Factor Levels: Method of Sighting		
	Right Eye	Left Eye	Both Eyes
$k = 1$	12	10	16
$k = 2$	10	17	14
$k = 3$	18	16	16
$k = 4$	12	13	11
$k = 5$	14		20
$k = 6$			21
Totals	$C_R = 66$	$C_L = 56$	$C_B = 98$

First, the summations $\sum x$ and $\sum x^2$ need to be calculated:

$$\sum x = 12 + 10 + 18 + 12 + 14 + 10 + 17 + \cdots + 21 = \mathbf{220}$$
$$(66 + 56 + 98 = \mathbf{220} \; \text{ⓒⓚ})$$

$$\sum x^2 = 12^2 + 10^2 + 18^2 + 12^2 + 14^2 + 10^2 + \ldots + 21^2 = \mathbf{3392}$$

Using formula (12.2), we find

$$\text{SS(total)} = \sum(x^2) - \frac{(\sum x)^2}{n}: \quad \text{SS(total)} = 3392 - \frac{(220)^2}{15} = 3392 - 3226.67 = \mathbf{165.33}$$

Using formula (12.3), we find

$$\text{SS(method)} = \left(\frac{C_1^2}{k_1} + \frac{C_2^2}{k_2} + \frac{C_3^2}{k_3} + \cdots\right) - \frac{(\sum x)^2}{n}:$$

$$\text{SS(method)} = \left(\frac{66^2}{5} + \frac{56^2}{4} + \frac{98^2}{6}\right) - \frac{(220)^2}{15}$$

$$= (871.20 + 784.00 + 1600.67) - 3226.27$$
$$= 3255.87 - 3226.67 = \mathbf{29.20}$$

Using formula (12.4), we find

    First: $\sum(x^2) = 3392$ (found previously)

$$\left(\frac{C_1^2}{k_1} + \frac{C_2^2}{k_2} + \frac{C_3^2}{k_3} + \cdots\right) = 3255.87 \text{ (found previously)}$$

Then using formula 12.4:

$$\text{SS(error)} = \sum(x^2) - \left(\frac{C_1^2}{k_1} + \frac{C_2^2}{k_2} + \frac{C_3^2}{k_3} + \cdots\right): \quad \text{SS(error)} = 3392 - 3255.87 = \mathbf{136.13}$$

Use formula (12.8) to check the sum of squares.

$$\text{SS(method)} + \text{SS(error)} = \text{SS(total)}: \quad 29.20 + 136.13 = 165.33$$

The number of degrees of freedom is found using formulas (12.5), (12.6), and (12.7):

$$\text{df(method)} = c - 1: \qquad \text{df(method)} = 3 - 1 = \mathbf{2}$$

$$\text{df(total)} = n - 1: \qquad \text{df(total)} = 15 - 1 = \mathbf{14}$$

$$\text{df(error)} = n - c: \qquad \text{df(error)} = 15 - 3 = \mathbf{12}$$

Using formulas (12.10) and (12.11), we find

$$\text{MS(method)} = \frac{\text{SS(method)}}{\text{df(method)}}: \quad \text{MS(method)} = \frac{29.20}{2} = \mathbf{14.60}$$

$$\text{MS(error)} = \frac{\text{SS(error)}}{\text{df(error)}}: \quad \text{MS(error)} = \frac{136.13}{12} = \mathbf{11.34}$$

The results of these computations are recorded in the ANOVA table shown in Table 12.9.

Table 12.9   ANOVA Table for Illustration 12.4

Source	df	SS	MS
Method	2	29.20	14.60
Error	12	136.13	11.34
Total	14	165.33	

The calculated value of the test statistic is then found using formula (12.12).

$$F\bigstar = \frac{\text{MS(factor)}}{\text{MS(error)}}: \quad F\bigstar = \frac{\text{MS(method)}}{\text{MS(error)}} = \frac{14.60}{11.34} = \mathbf{1.287}$$

STEP 4 **The Probability Distribution:**

_p_-VALUE PROCEDURE:

a. Calculate the _p_-value.
Use the right-hand tail:
P = $P(F > 1.287$, with $\text{df}_n = 2$ and $\text{df}_d = 12)$ as shown in the figure.

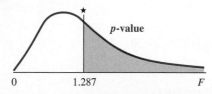

To find the _p_-value, you have two options:
1. Use Table 9(a) (Appendix B) to place bounds on the _p_-value: **P > 0.05**
2. Use a computer or calculator to calculate the _p_-value: **P = 0.312**
For additional instructions, see page 520.

b. Determine whether or not the _p_-value is smaller than α.
The _p_-value **is not smaller** than the level of significance, α (0.05).

OR   **CLASSICAL PROCEDURE:**

a. Determine the critical region and critical value(s).
The critical region is the right-hand tail; the critical value is obtained from Table 9(a):
$F_{(2, 12, 0.05)} = \mathbf{3.89}.$

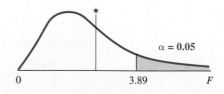

For additional instructions, see page 517.

b. Determine whether or not the calculated test statistic is in the critical region.
$F\bigstar$ **is not in** the critical region, as shown in red in the figure above.

STEP 5   **The Results:**

    a. State the decision about H_o.

      Fail to reject H_o.

    b. State the conclusion about H_a.

      The data show no evidence that would give reason to reject the null hypothesis that the three methods are equally effective.   ▲

The following commands will calculate a one-way analysis of variance table to complete a hypothesis test for several means.

MINITAB

```
Input the data for each level into columns C1, C2, . . .;
then continue with:
```

    Choose: **Stat > ANOVA > Oneway (unstacked)**

    Enter: Responses: **C1 C2 . . .***

```
OR
```

```
Input all of the data in C1 with its corresponding levels of
factors in C2; then continue with:
```

    Choose: **Stat > ANOVA > Oneway**

    Enter: Response: **C1**

                   Factor: **C2***

*Optional for either method:

    Choose: **Graphs. . .**

    Select: **Dotplots of data** and/or **Boxplots of data**   **> OK**

EXCEL

```
Input the data for each level into
columns A, B, . . .; then continue
with:
```

    Choose: **Tools > Data Analysis > ANOVA:Single Factor > OK**

    Enter: Input Range: **(A1:C4 or select cells)**

    Select: Grouped By: **Columns Labels in First Row** (if necessary)

    Enter: Alpha: **α**

    Select: **Output Range:**

    Enter: **(D1 or select cell)**

```
To make the output more readable con-
tinue with: Format > Column > Autofit
Selection.
```

TI-83

```
Input the data for each level into
lists L1, L2, . . .; then continue
with:
```

    Choose: **STAT > TESTS > F:ANOVA(**

    Enter: **L1, L2, . . .)**

```
(Commands for side-by-side dotplots can
be found in Chapter 2, p. 137.)
```

NOTE Side-by-side dotplots are very useful in visualizing the within-sample variation, the between-sample variation, and the relationship between them.

COMPUTER SOLUTION MINITAB Printout for Illustration 12.4:

Information given to computer ⟶

Row	Right eye	Left eye	Both eyes
1	12	10	16
2	10	17	14
3	18	16	16
4	12	13	11
5	14		20
6			21

ANALYSIS OF VARIANCE

The ANOVA table ⟶ compare to Table 12.9

SOURCE	DF	SS	MS	F	P
FACTOR	2	29.2	14.6	1.29	0.312
ERROR	12	136.1	11.3		
TOTAL	14	165.3			

The calculated value of $F, F\star$

Sample statistics for each factor level ⟶

LEVEL	N	MEAN	ST. DEV.
1	5	13.200	3.033
2	4	14.000	3.162
3	6	16.333	3.724

The calculated p-value

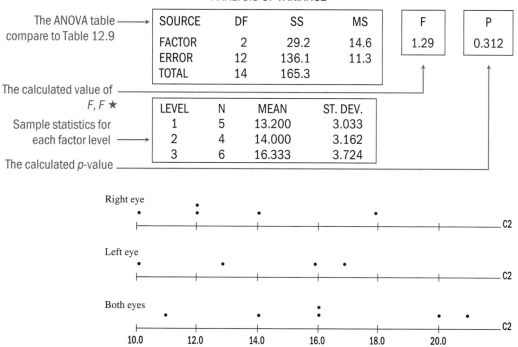

Recall the null hypothesis, that there is no difference between the levels of the factor being tested. A "fail to reject H_o" decision must be interpreted as the conclusion that there is no evidence of a difference due to the levels of the tested factor, whereas the rejection of H_o implies that there is a difference between the levels. That is, at least one level is different from the others. If there is a difference, the next problem is to locate the level or levels that are different. Locating this difference may be the main object of the analysis. In order to find the difference, the only method that is appropriate at this stage is to inspect the data. It may be obvious which level(s) caused the rejection of H_o. In Illustration 12.1 it seems quite obvious that at least one of the levels [level 1 (68°F) or level 3 (76°F), because they have the largest and smallest sample means] is different from the other two. If the higher values are more desirable for finding the "best" level to use, we would choose that corresponding level of the factor.

Thus far we have discussed analysis of variance for data dealing with one factor. It is not unusual for problems to have several factors of interest. The ANOVA techniques presented in this chapter can be developed further and applied to more complex cases.

EXERCISES • • • • • • • • • • • • •

12.7 State the null hypothesis H_o and the alternative hypothesis H_a that would be used to test the following statements.

   a. The mean value of x is the same at all five levels of the experiment.

   b. The scores are the same at all four locations.

   c. The four levels of the test factor do not significantly affect the data.

   d. The three different methods of treatment do affect the variable.

12.8 Find the p-value for each of the following situations.

   a. $F\bigstar = 3.852$, df(Factor) = 3, df(Error) = 12

   b. $F\bigstar = 4.152$, df(Factor) = 5, df(Error) = 18

12.9 Determine the critical region(s) and critical value(s) that would be used to test, using the classical approach, the following null hypotheses.

   a. H_o: $\mu_1 = \mu_2 = \mu_3 = \mu_4$ with $n = 18$, $\alpha = 0.05$

   b. H_o: $\mu_1 = \mu_2 = \mu_3 = \mu_4 = \mu_5$ with $n = 15$, $\alpha = 0.01$

   c. H_o: $\mu_1 = \mu_2 = \mu_3$ with $n = 25$, $\alpha = 0.05$

12.10 Why does df(factor), the number of degrees of freedom associated with the factor, always appear first in the critical value notation $F_{[df(factor),\ df(error),\ \alpha]}$?

12.11 Suppose that an F-test (as described in this chapter using the p-value approach) has a p-value of 0.04.

   a. What is the interpretation of p-value = 0.04?

   b. What is the interpretation of the situation if you had previously decided on a 0.05 level of significance?

   c. What is the interpretation of the situation if you had previously decided on a 0.02 level of significance?

12.12 Suppose that an F-test (as described in this chapter using the classical approach) has a critical value of 2.2, as shown in this figure:

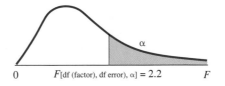

a. What is the interpretation of a calculated value of F larger than 2.2?

b. What is the interpretation of a calculated value of F smaller than 2.2?

c. What is the interpretation if the calculated F were 0.1? 0.01?

12.13 a. State the null hypothesis, in a general form, for the one-way ANOVA.

   b. State the alternative hypothesis, in a general form, for the one-way ANOVA.

   c. What must happen in order to "reject H_o"? (If using p-value approach; classical approach.)

   d. How would a decision of "reject H_o" be interpreted?

   e. What must happen in order to "fail to reject H_o"? (If using p-value approach; classical approach.)

   f. How would a decision of "fail to reject H_o" be interpreted?

12.14 The following table of data is to be used for single-factor ANOVA. Find each of the following:

   a. $x_{3,2}$ **b.** $x_{4,3}$ **c.** C_3 **d.** Σx **e.** $\Sigma(C_i)^2$

(continued)

Replicates	Level of Factor			
	1	2	3	4
1	13	12	16	14
2	17	8	18 .	11
3	9	15	10	19

12.15 The article "An Investigation of High School Preparation As Predictors of the Cultural Literacy of Developmental, Nondevelopmental and ESL College Students" (*RTDE,* Fall 1990) reported on a study that examined the cultural literacy of developmental, nondevelopmental, and ESL (English as a Second Language) college freshmen.

Analysis of Variance by Group for Total Score

Source	df	SS	MS	F	P
Group	2	4062.06	2031.03	14.49	0.0001
Error	117	16394.53	104.12		
Total	119	20456.59			

Analysis of Variance by Group for Foreign Language Preparation

Source	df	SS	MS	F	P
Group	2	0.95	0.475	1.93	0.1493
Error	117	28.75	0.246		
Total	119	29.70			

a. How many student scores were in the samples?

b. The students were divided into how many groups?

c. Given the sum of squares (SS) and degree of freedom (df) values, verify the mean square (MS), the calculated *F*-value, and the *p*-value for each figure.

d. Do the statistics in the first table show that the total scores were different for the groups involved? Explain.

e. Do the statistics in the second table show that the foreign-language preparation scores were different for the groups involved? Explain.

12.16 An article titled "The Effectiveness of Biofeedback and Home Relaxation Training on Reduction of Borderline Hypertension" (*Health Education,* October/November 1988) compared different methods of reducing blood pressure. Biofeedback ($n = 13$ subjects), Biofeedback/Relaxation ($n = 15$), and Relaxation ($n = 14$) were the three methods compared. There were no differences among the three groups on pretest diastolic or systolic blood pressures. There was a significant posttest difference between groups on the systolic measure, $F_{(2, 39)} = 4.14$, **P** < 0.025, and diastolic measure, $F_{(2, 39)} = 5.56$, **P** < 0.008.

a. Verify that df(method) = 2 and df(error) = 39.

b. Use Tables 9a, 9b, and 9c in Appendix B to verify that for systolic, **P** < 0.025 and for diastolic, **P** < 0.008.

12.17 A new operator was recently assigned to a crew of workers who perform a certain job. From the records of the number of units of work completed by each worker each day last month, a sample of size five was randomly selected for each of the two experienced workers and the new worker. At the 0.05 level of significance, does the evidence provide sufficient reason to reject the claim that there is no difference in the amount of work done by the three workers?

(continued)

	Workers		
	New	**A**	**B**
Units of Work	8	11	10
(replicates)	10	12	13
	9	10	9
	11	12	12
	8	13	13

a. Solve using the *p*-value approach.   **b.** Solve using the classical approach.

12.18 An employment agency wants to see which of three types of ads in the help-wanted section of local newspapers is the most effective. Three types of ads (big headline, straight-forward, and bold print) were randomly alternated over a period of weeks, and the number of people responding to the ads was noted each week. Do these data support the null hypothesis that there is no difference in the effectiveness of the ads, as measured by the mean number responding, at the 0.01 level of significance?

	Type of Advertisement		
	Big Headline	**Straightforward**	**Bold Print**
Number of Responses	23	19	28
(replicates)	42	31	33
	36	18	46
	48	24	29
	33	26	34
	26		34

a. Solve using the *p*-value approach.   **b.** Solve using the classical approach.

12.19 Cities across the United States have restaurants that offer themes associated with foreign countries. The United States embargo of Cuba keeps most Americans away from that nation, but domestic Cuban restaurants can be found that serve authentic samples of the island's food and beverages. The following ratings, based upon three categorical judgments of food quality, décor, and service, were assembled from different restaurants located in different cities. The ratings were made on the same scale from 0 to 30 (the higher the better):

Restaurant Rating Category		
Food Quality	**Décor**	**Service**
20	16	18
21	12	16
19	18	18
24	21	21
20	17	19
22	18	21

Source: Fortune, "Feed Your Face Like Fidel," July 6, 1998.

Is there any significant difference in the ratings given to the Cuban restaurants in each category? Construct a one-way ANOVA table and test for the difference at the 0.05 level of significance.
a. Solve using the *p*-value approach.   **b.** Solve using the classical approach.

12.20 An article titled "A Contrast in Images: Nursing and Nonnursing College Students" (*Nursing Education,* March 1994) described a seven-point Likert scale questionnaire to compare nursing students, business majors, engineering majors, human service majors,

(continued)

and social science majors with respect to several attitudes towards nursing. The nonnursing students found nursing a significantly less dangerous profession than did the nursing students. A sample of five students from each of the five majors were asked to respond to the dangers associated with the nursing profession. The scores for each of the 25 students are given below.

Nursing	Business	Engineering	Human Services	Social Science
6	4	5	4	3
5	4	4	4	3
5	3	4	5	2
7	3	6	4	4
4	2	2	6	2

Complete an ANOVA table for the above data. Test the null hypothesis that the mean score is the same for each of the five groups. Use the 0.01 level of significance.
a. Solve using the *p*-value approach.　　**b.** Solve using the classical approach.

12.21 A number of sports enthusiasts have argued that major league baseball players from teams in the Central Division have an unfair advantage over coastal players in the Western and Eastern Divisions. This is because the impact due to the differences in time is likely to be greater when playing on the road (i.e., games away from home). Players from teams on the coasts could gain (going west) or lose (going east) up to three hours, whereas Central Division players would seldom gain or lose more than one hour. The following data show the won/loss percentages within divisions for games played on the road by all three divisions of major league teams at the end of the 1997 season:

Major League Division

Eastern	Central	Western
63.0	46.9	51.9
49.4	44.4	50.6
46.9	44.4	44.4
40.7	39.5	45.7
37.0	32.1	55.5
64.2	51.9	46.9
60.5	43.2	46.9
45.7	38.3	37.0
48.1	40.7	
42.0	42.0	

Source: The World Almanac and Book of Facts 1998, "Sports—Baseball."

Complete an ANOVA table for won/loss percentages by teams representing each division. Test the null hypothesis that when teams play on the road, the mean won/loss percentage is the same for each of the three divisions. Use the 0.05 level of significance.
a. Solve using the *p*-value approach.　　**b.** Solve using the classical approach.

 12.22 A study was conducted to assess the effectiveness of treating vertigo (motion sickness) with the Transdermal Therapeutic System (TTS—patch worn on skin). Two other treatments, both oral (one pill containing a drug and one a placebo) were used. The age and the gender of the patients for each treatment are listed below.

(continued)

TTS		Antivert		Placebo	
47-f	53-m	51-f	43-f	67-f	38-m
41-f	58-f	53-f	56-f	52-m	59-m
63-m	62-f	27-m	48-m	47-m	33-f
59-f	34-f	29-f	52-f	35-f	32-f
62-f	47-f	31-f	19-f	37-f	26-f
24-m	35-f	25-f	31-f	40-f	37-m
43-m	34-f	52-f	48-f	31-f	49-f
20-m	63-m	55-f	53-m	45-f	49-m
55 f	46 f	32 f	63-m	41-f	38-f
		51-f	54-m	49-m	
		21-f			

Is there a significant difference between the mean age of the three test groups? Use $\alpha = 0.05$.

a. Solve using the *p*-value approach.   **b.** Solve using the classical approach.

12.23 Snacks are a concern of people who are trying not to eat too much between meals. "The trick is to give your body what it's asking for, but without sacrificing good nutrition," according to an article written by a nutritionist for *Woman's Day* ("50 Snacks Under 100 Calories," 11-1-98). The author assembled a list of 50 snacks containing 100 calories or less, together with the fat content. The list was then divided into four categories: Crunchy Choices, Salty Sensations, Creamy Concoctions, and Sinless Sweets. The table below summarizes the nutritional data:

Food Snack Category

(1) Crunchy Choices		(2) Salty Sensations		(3) Creamy Concoctions		(4) Sinless Sweets	
Calories	Fat (grams)	Calories	Fat (grams)	Calories	Fat (grams)	Calories	Fat (grams)
89	1.5	100	0.0	99	5.0	90	1.0
99	1.3	32	0.0	97	0.2	99	0.5
91	0.5	60	0.0	65	5.0	94	0.0
76	0.6	65	7.0	79	2.0	91	1.0
90	2.0	100	1.3	99	0.0	100	0.0
90	2.0	70	0.8	90	0.0	54	2.5
97	8.5	5	0.0	90	1.0	87	0.5
52	0.9	14	0.1	100	5.0	88	0.0
93	5.0	52	0.4	50	0.0	96	1.5
80	1.5	59	2.8	94	8.0	50	0.4
82	0.5			100	1.5	70	1.0
90	1.0			70	0.5	90	0.0
						75	0.4
						100	0.3
						70	0.0
						88	3.0

Complete an ANOVA table for (i) calories and (ii) fat content. Test the null hypotheses that the calorie content and the fat content are the same for each of the four categories of snacks. Use the 0.01 level of significance.

a. Solve using the *p*-value approach.   **b.** Solve using the classical approach.

Return to CHAPTER CASE STUDY

How much time did you spend reading the newspaper yesterday? Do you think that a person's age will have any effect on the amount of time spent reading the newspaper yesterday? The USA Snapshot® "Speed readers" that appeared in *USA Today,* July 14, 1998 seems to suggest that older adults spent more time reading the newspaper than did younger adults. (See page 575.)

EXERCISE 12.24

Twenty-five newspaper readers were asked how much time, estimated to the nearest minute, they spent reading yesterday's newspaper. Their times are listed by age category of the reader.

	\multicolumn{6}{c}{Age of Adult}
	18–24	**25–29**	**30–34**	**35–49**	**50–64**	**65–up**
x, time	10	13	14	20	20	8
spent	5	8	7	12	12	30
reading	15	20	20	5	18	35
newspaper	4	10		28	15	20
yesterday				15	30	

Does the sample show that a person's age has an effect on the amount of time spent daily reading the newspaper? Does the sample show that a person's age has an effect on the average amount of time spent daily reading the newspaper?

a. Are these different questions? Explain.

b. Using the ANOVA technique learned in this chapter, do these data show sufficient evidence to claim that a person's age has an effect on the average amount of time spent daily reading the newspaper? Use $\alpha = 0.05$.

IN RETROSPECT

In this chapter we have presented an introduction to the statistical techniques known as analysis of variance. The techniques studied here were restricted to the test of a hypothesis that dealt with questions about means from several populations. We were restricted to normal populations and populations with homogeneous (equal) variances. The test of multiple means is accomplished by partitioning the sum of squares into two segments: (1) the sum of squares due to variation between the levels of the factor being tested and (2) the sum of squares due to the variation between the replicates within each level. The null hypothesis about means is then tested by using the appropriate variance measurements.

Note that we restricted our development to one-factor experiments. This one-factor technique represents only a beginning to the study of analysis of variance techniques.

CHAPTER EXERCISES

12.25 Samples of peanut butter produced by three different manufacturers were tested for salt content with the following results.

Brand 1:	2.5	8.3	3.1	4.7	7.5	6.3
Brand 2:	4.5	3.8	5.6	7.2	3.2	2.7
Brand 3:	5.3	3.5	2.4	6.8	4.2	3.0

Hint: Each level of data is entered into a separate column.

(continued)

Is there a significant difference in the mean amount of salt in these samples? Use $\alpha = 0.05$.

a. State the null and alternative hypotheses.
b. Determine the test criteria: assumptions, level of significance, test statistic.
c. Using the information on the computer printout next page, state the decision and conclusion to the hypothesis test.
d. What does the *p*-value tell you? Explain.

```
Analysis of Variance

Source    DF       SS        MS       F        P
Factor     2      4.68      2.34    0.64     0.541
Error     15     54.88      3.66
Total     17     59.56
                                    Individual 95% CIs For Mean
                                    Based on Pooled StDev
Level     N     Mean     StDev   ----+------+-------+------+--
Brand1     6    5.400    2.359             (--------*--------)
Brand2     6    4.500    1.669       (--------*----------)
Brand3     6    4.200    1.621     (--------*----------)
                                   ----+------+-------+-----+--
Pooled StDev = 1.913                  3.0    4.5     6.0    7.5
```

12.26 A new all-purpose cleaner is being test-marketed by placing sales displays in three different locations within various supermarkets. The number of bottles sold from each location within each of the supermarkets tested is reported below.

	I	40	35	44	38
Locations	**II**	32	38	30	35
	III	45	48	50	52

a. State the null and alternative hypotheses for testing "the location of the sales display had no effect on the number of bottles sold."
b. Using $\alpha = 0.01$, determine the test criteria: assumptions, level of significance, test statistic.
c. Using the information on the computer printout below, state the decision and conclusion to the hypothesis test.
d. What does the *p*-value tell you? Explain.

Hint: Each level of data is entered into a separate column.

```
Analysis of Variance

Source    DF       SS        MS       F       P
Factor     2      460.7     230.3   19.51   0.001
Error      9      106.2      11.8
Total     11      566.9
                              Individual 95% CIs For Mean
                              Based on Pooled StDev
Level      N     Mean    StDev   -----------+------------+-----------+----------
Location   4    39.250   3.775                (----*----)
Location   4    33.750   3.500        (----*----)
Location   4    48.750   2.986                              (----*----)
                                 -----------+------------+-----------+----------
Pooled StDev = 3.436                        35.0         42.0        49.0
```

12.27 An experiment was designed to compare the lengths of time that four different drugs provided pain relief following surgery. The results (in hours) are shown in the following table.

		Drug	
A	**B**	**C**	**D**
8	6	8	4
6	6	10	4
4	4	10	2
2	4	10	
		12	

Is there enough evidence to reject the null hypothesis that there is no significant difference in the length of pain relief for the four drugs at $\alpha = 0.05$?
a. Solve using the p-value approach.   **b.** Solve using the classical approach.

12.28 The distance required to stop a vehicle on wet pavement was measured to compare the stopping power of four major brands of tires. A tire of each brand was tested on the same vehicle on a controlled wet pavement. The resulting distances are shown in the following table. At $\alpha = 0.05$, is there sufficient evidence to conclude that there is a difference in the mean stopping distance?

	Brand A	**Brand B**	**Brand C**	**Brand D**
Distance	37	37	33	41
(replicate)	34	40	34	41
	38	37	38	40
	36	42	35	39
	40	38	42	41
	32		34	43

a. Solve using the p-value approach.   **b.** Solve using the classical approach.

12.29 A certain vending company's soft-drink dispensing machines are supposed to serve six ounces of beverage. Various machines were sampled and the resulting amounts of dispensed drink were recorded, as shown in the following table. Does this sample evidence provide sufficient reason to reject the null hypothesis that all five machines dispense the same average amount of soft drink? Use $\alpha = 0.01$.

Machines	A	B	C	D	E
Amounts of Soft	3.8	6.8	4.4	6.5	6.2
Drink Dispensed	4.2	7.1	4.1	6.4	4.5
	4.1	6.7	3.9	6.2	5.3
	4.4		4.5		5.8

a. Solve using the p-value approach.   **b.** Solve using the classical approach.

12.30 Suburbs, each with its own attributes, are located around every metropolitan area. There is always the "rich" one (the most expensive one), the least expensive one, and so on. Does the suburb affect the transfer value of its homes? x = transfer value, the amount on which county transfer taxes are paid.
a. Do the sample data show sufficient evidence to conclude that the suburbs represented do have a significant effect on the transfer value of their homes? Use $\alpha = 0.01$.
b. Construct a graph that demonstrates the conclusion reached in (a).

(continued)

Suburb A	Suburb B	Suburb C	Suburb D	Suburb E
105	101	95	74	79
114	88	107	135	89
85	105	101	165	140
177	100	92	114	114
104	161	91	80	80
135	113	89	115	86
	94			94
				102

12.31 "All tillage tools are not created equal," according to Larry Reichenberger as he reported in *Successful Farming,* February 1979. The table below presents the yield per plot obtained in an experiment designed to compare six different methods of tilling the ground.

Tillage tool:	Plow	VChisel	CoulChis	StdChis	HvyDisk	LtDisk
Replicates:	118.3	115.8	124.1	109.2	118.1	118.3
	125.6	122.5	118.5	114.0	117.5	113.7
	123.8	118.9	113.3	122.5	121.4	113.7

a. Construct a dotplot showing the six samples separately and side by side.
Using one-way ANOVA, test the claim that "all tillage tools are not created equal." Use $\alpha = 0.05$.
b. State the null and alternative hypotheses and describe the meaning of each.
c. Complete using the *p*-value approach. **d.** Complete using the classical approach.

12.32 Stock funds are a popular form of mutual fund investment, and several types of funds are available to investors. Like other forms of investment, stock funds are thought to vary in the degree of risk associated with them. The table below shows five types of stock funds: Aggressive Growth, Small Company, Growth, Growth and Income, and International. Ten funds were selected for each type, based upon the highest returns they provided to their investors over the past five years. The numbers in the table are the measure of risk (volatility) exhibited by the stock fund.

Stock Fund Type

(1) Aggressive Growth	(2) Small Company	(3) Growth	(4) Growth and Income	(5) International
20.6	28.5	19.7	13.3	19.9
18.0	17.1	17.3	12.0	14.4
18.3	19.8	22.8	13.3	15.5
22.0	18.0	21.9	14.1	12.7
17.6	18.2	15.7	15.0	12.2
18.4	13.0	20.0	13.3	14.3
13.2	13.5	15.7	13.3	14.7
20.6	16.2	20.0	13.0	12.9
18.4	12.9	16.7	13.7	15.2
25.1	13.5	13.0	13.7	14.1

Source: Fortune, "The Best Mutual Funds," August 17, 1998.

Does this sample indicate that risk differs from one category of stock fund to the next? Using one-way ANOVA, test at the 0.01 level of significance to determine if there is a difference in the mean risk levels measured for the five types of stock funds.
a. Solve using the *p*-value approach. **b.** Solve using the classical approach.

12.33 The question arises every year when the playoffs begin in the National Football League, "Which division teams are the toughest, Eastern, Central or Western?" Two ways to measure the strength of the football teams that play are the number of points they score and the number of points their opponents score. The final results for the 16 games played in the 1996–97 season are shown in the table below:

NFL Conference Division

Eastern		Central		Western	
Points	**Opp. Points**	**Points**	**Opp. Points**	**Points**	**Opp. Points**
418	313	344	257	391	275
319	266	325	335	297	300
317	334	372	369	310	376
339	325	345	319	340	293
279	454	371	441	317	376
286	250	456	210	367	218
363	341	298	315	398	257
364	312	283	305	303	409
300	397	221	293	309	461
242	297	302	368	229	339

Source: The World Almanac and Book of Facts 1998, "National Football League."

Complete an ANOVA table for (a) points scored and (b) points scored by opposing teams. In each case, test the null hypothesis that the mean points scored is the same for each of the three divisions. Use the 0.05 level of significance.
a. Solve using the *p*-value approach.   **b.** Solve using the classical approach.

12.34 To compare the effectiveness of three different methods of teaching reading, 26 children of equal reading aptitude were divided into three groups. Each group was instructed for a given period of time using one of the three methods. After completing the instruction period, all students were tested. The test results are shown in the following table. Is the evidence sufficient to reject the hypothesis that all three instruction-methods are equally effective? Use $\alpha = 0.05$.

	Method I	Method II	Method III
Test Scores:	45	45	44
(replicates)	51	44	50
	48	46	45
	50	44	55
	46	41	51
	48	43	51
	45	46	45
	48	49	47
	47	44	

a. Solve using the *p*-value approach.   **b.** Solve using the classical approach.

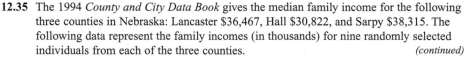

12.35 The 1994 *County and City Data Book* gives the median family income for the following three counties in Nebraska: Lancaster $36,467, Hall $30,822, and Sarpy $38,315. The following data represent the family incomes (in thousands) for nine randomly selected individuals from each of the three counties. *(continued)*

Lancaster	Hall	Sarpy
45.0	32.0	40.0
39.5	30.0	42.0
42.0	37.0	45.0
35.0	35.0	39.5
40.0	28.5	40.0
37.0	37.5	38.0
44.0	31.0	51.0
48.5	37.6	47.5
50.0	25.0	41.0

Complete an ANOVA for the above data, and test the null hypothesis that the mean family income is the same for each of the three counties. Use a 0.05 level of significance.
a. Solve using the *p*-value approach.   **b.** Solve using the classical approach.

 12.36 The following table shows the number of arrests made last year for violations of the narcotic drug laws in 25 communities. The data given are rates of arrest per 10,000 inhabitants. At $\alpha = 0.05$, is there sufficient evidence to reject the hypothesis that the mean rates of arrests are the same in all four sizes of communities?

Cities (over 250,000)	Cities (under 250,000)	Suburban Communities	Rural Communities
45	23	25	8
34	18	17	16
41	27	19	14
42	21	28	17
37	26	31	10
28	34	37	23

a. Solve using the *p*-value approach.   **b.** Solve using the classical approach.

12.37 The May 16, 1994, issue of *Fortune* magazine gave the percent change in home prices during 1993 for the top five markets as follows: Denver 12.7%, Salt Lake City–Ogden 9.7%, Miami–Hialeah 8.3%, Nashville 7.0%, and Portland 7.0%. The following data represent the percent change in home prices for randomly selected homes in St. Louis, Kansas City, and Oklahoma City.

St. Louis	Kansas City	Oklahoma City
3.0	4.5	1.0
2.5	2.5	−2.5
−1.5	7.0	−3.5
4.0	9.0	2.0
−1.0	1.5	4.6
5.5	2.0	0.5

Complete an ANOVA table for the above data. Test the null hypothesis that the mean percent change is equal for the three cities. Use a 0.01 level of significance.
a. Solve using the *p*-value approach.   **b.** Solve using the classical approach.

12.38 For the following data, show that

$$SS(factor) = k_1(\bar{x}_1 - \bar{X})^2 + k_2(\bar{x}_2 - \bar{X})^2 + k_3(\bar{x}_3 - \bar{X})^2$$

(continued)

where $\bar{x}_1, \bar{x}_2, \bar{x}_3$ are the means for the three factor levels and $\bar{X}$ is the overall mean.

Factor Level

1	2	3
6	13	9
8	12	11
10	14	7

12.39 For the following data, find SS(error) and show that

$$SS(error) = [(k_1 - 1)s_1^2 + (k_2 - 1)s_2^2 + (k_3 - 1)s_3^2]$$

where s_i^2 is the variance for the ith factor level.

Factor Level

1	2	3
8	6	10
4	6	12
2	4	14

12.40 A study reported in the *Journal of Research and Development in Education* (Summer 1989) evaluates the effectiveness of social skills training and cross-age tutoring for improving academic skills and social communication behaviors among boys with learning disabilities. Twenty boys were divided into three groups, and their scores on the Test of Written Spelling (TWS) may be summarized as follows:

> Use the information in Exercises 12.38 and 12.39.

Group	n	TWS Mean	St. Dev.
Social skills training and tutoring components	7	21.43	9.48
Social skills training only	7	20.00	8.91
Neither component	6	20.83	9.06

Calculate the entries of the ANOVA table using these results.

12.41 An article in the *Journal of Pharmaceutical Sciences* (Dec. 1987) discusses the change of plasma protein binding of Diazepam at various concentrations of Imipramine. Suppose the results were reported as follows:

Diazepam Alone(1.25 mg/mL)	Diazepam with Imipramine		
	1.25	2.50	5.00
97.99	97.68	96.29	93.92

The values given represent mean plasma protein binding and $n = 8$ for each of the four groups. Find the sum of squares among the four groups.

12.42 Seven golf balls from each of six manufacturers were randomly selected and tested for durability. Each ball was hit 300 times or until failure occurred, whichever came first. Do the following sample data show sufficient reason to reject the null hypothesis that the six different brands tested withstood the durability test equally well? Use $\alpha = 0.05$.

(continued)

A	B	C	D	E	F
300	190	228	276	162	264
300	164	300	296	175	168
300	238	268	62	157	254
260	200	280	300	262	216
300	221	300	230	200	257
261	132	300	175	256	183
300	156	300	211	92	93

a. Solve using the *p*-value approach.   **b.** Solve using the classical approach.

VOCABULARY LIST

Be able to define each term. Pay special attention to the key terms, which are printed in **red.** In addition, describe in your own words, and give an example of, each term. Your examples should not be ones given in class or in the textbook.

The bracketed numbers indicate the chapters in which the term previously appeared, but you should define the terms again to show increased understanding of their meaning. Page numbers indicate the first appearance of the term in Chapter 12.

analysis of variance (ANOVA) (p. 576)
between-sample variation (p. 578)
degrees of freedom [9, 10, 11] (p. 579)
experimental error (p. 585)
levels of the tested factor (p. 577)
mathematical model (p. 584)
mean square, MS(Factor), MS(error)
  (p. 580)
partitioning (p. 578)
randomize [2] (p. 584)

replicate (p. 576)
response variable [1] (p. 582)
sample variance [2, 9, 10] (p. 578)
sum of squares (p. 578)
test statistic, F★ (p. 580)
total sum of squares, SS(total) (p. 578)
variation between levels, MS(factor)
  (p. 578, 582)
variation within a level, MS(error) (p. 579, 582)
within-sample variation (p. 579)

CHAPTER PRACTICE TEST

Part I: Knowing the Definitions

Answer "True" if the statement is always true. If the statement is not always true, replace the words shown in bold with words that make the statement always true.

12.1   To partition the sum of squares for the total is to separate the numerical value of SS(total) into two values such that the **sum** of these two values is equal to SS(total).

12.2   A **sum of squares** is actually a measure of variance.

12.3   **Experimental error** is the name given to the variability that takes place between the levels of the test factor.

12.4   **Experimental error** is the name given to the variability that takes place among the replicates of an experiment as it is repeated under constant conditions.

12.5   **Fail to reject** H_o is the desired decision when the means for the levels of the factor being tested are all different.

12.6   The **mathematical model** for a particular problem is an equational statement showing the anticipated makeup of an individual piece of data.   *(continued)*

12.7 The degrees of freedom for the factor are equal to the **number of factors tested.**

12.8 The measure of a specific level of a factor being tested in an ANOVA is the **variance** of that factor level.

12.9 We **need not** assume that the observations are independent to do analysis of variance.

12.10 The rejection of H_o **indicates** that you have identified the level(s) of the factor that is (are) different from the others.

Part II: Applying the Concepts

12.11 Determine the truth (T/F) for each statement with regard to the one-factor analysis of variance technique.

_____ **a.** The mean squares are measures of variance.

_____ **b.** "There is no difference between the mean values of the random variable at the various levels of the test factor" is a possible interpretation of the null hypothesis.

_____ **c.** "The factor being tested has no effect on the random variable x" is a possible interpretation of the alternative hypothesis.

_____ **d.** "There is no variance among the mean values of x for each of the different factor levels" is a possible interpretation of the null hypothesis.

_____ **e.** The "partitioning" of the variance occurs when the SS(total) is separated into SS(factor) and SS(error).

_____ **f.** We will want to reject the null hypothesis and conclude that the factor has an effect on the variable when the amount of variance assigned to the factor is significantly larger than the variance assigned to error.

_____ **g.** In order to apply the F-test, the sample size from each factor level must be the same.

_____ **h.** In order to apply the F-test, the sample standard deviation from each factor level must be the same.

_____ **i.** If 20 is subtracted from every data value, then the calculated value of the $F \star$ statistic is also reduced by 20.

When the calculated value of F, $F \star$, is greater than the table value for F,

_____ **j.** the decision will be: "Fail to reject H_o."

_____ **k.** the conclusion will be: "The factor being tested does have an effect on the variable."

Independent samples were collected in order to test the effect a factor had on a variable. The data are summarized in the following ANOVA table:

	SS	df
Factor	810	2
Error	720	8
Total	1530	10

(continued)

Is there sufficient evidence to reject the null hypothesis that all levels of the test factor have the same effect on the variable?

_____ **l.** The null hypothesis could be: $\mu_A = \mu_B = \mu_C = \mu_D$

_____ **m.** The calculated value of F is 1.125.

_____ **n.** The critical value of F for $\alpha = 0.05$ is 6.06.

_____ **o.** The null hypothesis can be rejected at $\alpha = 0.05$.

12.12 Determine the values A, B, C, D, and E missing in the table.

	SS	df	MS	F★
Factor	A	4	18	E
Error	B	18	D	
Total	144	C		

Find the values of

a. A **b.** B **c.** C **d.** D **e.** E

Part III: Understanding the Concepts

12.13 In 50 words or less, explain what a single-factor ANOVA experiment is.

12.14 A state environmental agency tested three different scrubbers used to reduce the resulting air pollution in the generation of electricity. The primary concern was the emission of particulate matter. Several trials were run with each scrubber. The amount of particulate emission was recorded for each trial.

		Replicates Amounts of Emission					
Scrubbers	**I**	11	10	12	9	13	12
Tested	**II**	12	10	12	8	9	
	III	9	11	10	7	8	

a. State the mathematical model for this experiment.
b. State the null and alternative hypotheses.
c. Calculate and form the ANOVA table.
d. Complete the testing of H_o using a 0.05 level of significance. State the decision and conclusion clearly.
e. Construct a graph representing the data that is helpful in picturing the results of the hypothesis test.

Chapter

13

Linear Correlation and Regression Analysis

CHAPTER OUTLINE

CHAPTER CASE STUDY

Charting the Pounds

Height and weight charts, similar to the one below, have been around for many years. They are used in research, by insurance companies, by doctors, and by each of us to determine the "ideal" weight for our height.

This is the standard chart used in a major study of women's weight and heart attacks. The study found that women weighing at least 6% below these "ideals" had a 23% less risk of heart attacks.

	Women's Weight (lb.)		
Height	Small Frame	Medium Frame	Large Frame
4'10"	102–111	109–121	118–131
4'11"	103–113	111–123	120–134
5'0"	104–115	113–126	122–137
5'1"	106–118	155–129	125–140
5'2"	108–121	118–132	128–143
5'3"	111–124	121–135	131–147
5'4"	114–127	124–138	134–151
5'5"	117–130	127–141	137–155
5'6"	120–133	130–144	140–159
5'7"	123–136	133–147	143–163
5'8"	126–139	136–150	146–167
5'9"	129–142	139–153	149–170
5'10"	132–145	142–156	152–173
5'11"	135–148	145–159	155–176
6'0"	138–151	148–162	158–179

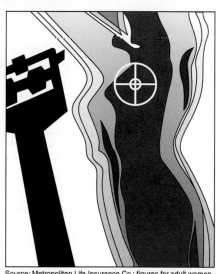

Source: Metropolitan Life Insurance Co.; figures for adult women ages 25–59, in 3 pounds of clothing, wearing shoes with one-inch heels

By Suzy Parker, USA TODAY

Draw a scatter diagram showing the information for each frame size, using height as the input variable. Graph all three sets of data on one graph using a different color for each frame size. How many pounds does each inch of height add to the ideal weight? How does the increase in weight for each inch increase of height relate to the line of best fit? How do your women friends' weights compare to the ideal weight intervals? (See Exercise 13.59, p. 649.)

CHAPTER OBJECTIVES

In Chapter 3 the basic ideas of regression and linear correlation analysis were introduced. (If these concepts are not fresh in your mind, review Chapter 3 before beginning this chapter.) Chapter 3 was only a first look: a presentation of the basic graphic (the **scatter diagram**) and descriptive statistical aspects of linear correlation and regression analysis. In this chapter we take a second, more detailed look at **linear correlation** and **regression analysis**.

Previously we used the linear correlation coefficient to measure the strength of the linear relationship between two variables. Now we will determine whether there is a linear relationship by using a hypothesis test in which the probability of a type I error is fixed by the value assigned to α. In Chapter 3 we introduced a set of formulas for finding the equation of the straight **line of best fit**. Now we wish to ask, "Is the linear equation of any real use?" Previously we used the equation of the line of best fit to make point predictions. Now we will make confidence interval estimations. In short, this second look at linear correlation and regression analysis will be a much more complete presentation than that in Chapter 3.

Recall that **bivariate data** are ordered pairs of numerical values. The two values are each response variables. They are paired with each other as a result of a common bond (see pp. 131, 137).

13.1 Linear Correlation Analysis

In Chapter 3 the linear correlation coefficient was presented as a quantity that measures the strength of a linear relationship (dependency). Now let's take a second look at this concept and see how r, the coefficient of linear correlation, works. Intuitively, we want to think about how to measure the mathematical linear dependency of one variable on another. As x increases, does y tend to increase (decrease)? How strong (consistent) is this tendency? We are going to use two measures of dependence—covariance and the coefficient of linear correlation—to measure the relationship between two variables. We'll begin our discussion by examining a set of bivariate data and identifying some related facts as we prepare to define covariance.

Figure 13.1

Graph of Data for Illustration 13.1

Remember to adjust the scales so the resulting data window is approximately square.

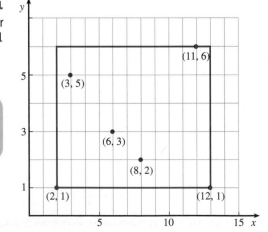

ILLUSTRATION 13.1 ▼

Let's consider the following sample of six pieces of bivariate data: (2, 1), (3, 5), (6, 3), (8, 2), (11, 6), (12, 1). (See Figure 13.1.) The mean of the six x-values (2, 3, 6, 8, 11, 12) is $\bar{x} = 7$. The mean of the six y-values (1, 5, 3, 2, 6, 1) is $\bar{y} = 3$.

The point $(\bar{x}, \bar{y})$, which is (7, 3), is located as shown on the graph of the sample points in Figure 13.2. The point $(\bar{x}, \bar{y})$ is called the **centroid** of the data. If a vertical and a horizontal line are drawn through the centroid, the graph is divided into four sections, as shown in Figure 13.2. Each point (x, y) lies a certain distance from each of these two lines. $(x - \bar{x})$ is the horizontal distance from (x, y) to the vertical line passing through the centroid. $(y - \bar{y})$ is the vertical distance from (x, y) to the horizontal line passing through the centroid. Both the horizontal and vertical distances of each data point from the centroid can be measured, as shown in Figure 13.3. The distances may be positive, negative, or zero, depending on the position of the point (x, y) in reference to $(\bar{x}, \bar{y})$. $[(x - \bar{x})$ and $(y - \bar{y})$ are represented by means of braces, with positive or negative signs, as shown in Figure 13.3.]

Figure 13.2

The Point (7, 3) Is the Centroid

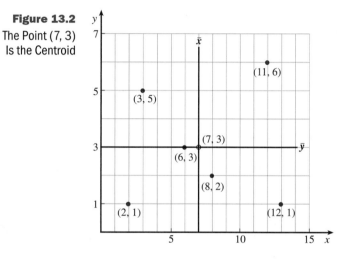

Figure 13.3

Measuring the Distance of Each Data Point from the Centroid

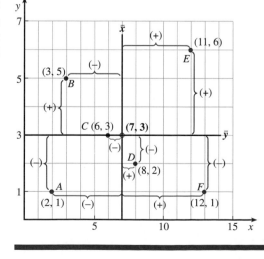

One measure of linear dependency is the covariance. The **covariance of x and y** is defined as the sum of the products of the distances of all values of x and y from the centroid, $\sum[(x - \bar{x})(y - \bar{y})]$, divided by $n - 1$:

$$\text{covar}(x, y) = \sum \frac{(x - \bar{x})(y - \bar{y})}{n - 1} \qquad (13.1)$$

Table 13.1 Calculations for Finding Covar(x, y) for the Data of Illustration 13.1

Points	$x - \bar{x}$	$y - \bar{y}$	$(x - \bar{x})(y - \bar{y})$
$A(2, 1)$	−5	−2	10
$B(3, 5)$	−4	2	−8
$C(6, 3)$	−1	0	0
$D(8, 2)$	1	−1	−1
$E(11, 6)$	4	3	12
$F(12, 1)$	5	−2	−10
Total	0 (ck)	0 (ck)	3

The covariance for the data given in Illustration 13.1 is calculated in Table 13.1. The covariance, written as covar(x, y), of the data is $\dfrac{3}{5} = \mathbf{0.6.}$

NOTE 1 $\sum(x - \bar{x}) = 0$ and $\sum(y - \bar{y}) = 0$. This will always happen. Why? (See p. 77.)

NOTE 2 Even though the variance of a single set of data is always positive, the covariance of bivariate data can be negative.

The covariance is positive if the graph is dominated by points to the upper right and to the lower left of the centroid. The products of $(x - \bar{x})$ and $(y - \bar{y})$ are positive in these two sections. If the majority of the points are in the upper-left and lower-right sections relative to the centroid, the sum of the products is negative. Figure 13.4 shows data that represent a positive dependency (a), a negative dependency (b), and little or no dependency (c). The covariances for these three situations would definitely be positive in part (a), negative in (b), and near zero in (c). (The sign of the covariance is always the same as the sign of the slope of the regression line.)

Figure 13.4
Data and Covariance

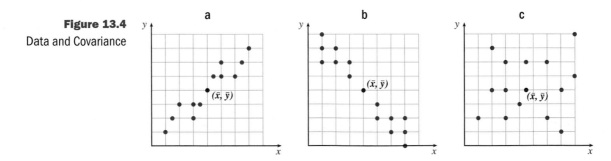

The biggest disadvantage of covariance as a measure of linear dependency is that it does not have a standardized unit of measure. One reason for this is that the spread of the data is a strong factor in the size of the covariance. For example, if we were to multiply each data point in Illustration 13.1 by 10, we would have (20, 10), (30, 50), (60, 30), (80, 20), (110, 60), and (120, 10). The relationship of the points to each other would be changed only in that they would be much more spread out. However, the covariance for this new set of data is 60. Does this mean that the amount of dependency between the x and y variables is stronger than in the original case? No, it does not; the relation-

This calculation is assigned in Exercise 13.9, page 613.

ship is the same, even though each data value has been multiplied by 10. This is the trouble with covariance as a measure. We must find some way to eliminate the effect of the spread of the data when we measure dependency.

If we standardize x and y by dividing the distance of each from the respective mean by the respective standard deviation,

$$x' = \frac{x - \bar{x}}{s_x} \quad \text{and} \quad y' = \frac{y - \bar{y}}{s_y}$$

and then compute the covariance of x' and y', we will have a covariance that is not affected by the spread of the data. This is exactly what is accomplished by the linear correlation coefficient. It divides the covariance of x and y by a measure of the spread of x and by a measure of the spread of y (the standard deviations of x and of y are used as measures of spread). Therefore, by definition, the **coefficient of linear correlation** is

$$r = \text{covar}(x', y') = \frac{\text{covar}(x, y)}{s_x \cdot s_y} \tag{13.2}$$

The coefficient of linear correlation standardizes the measure of dependency and allows us to compare the relative strengths of dependency of different sets of data. [Formula (13.2) for linear correlation is also commonly referred to as **Pearson's product moment, r**].

The value of r, the coefficient of linear correlation, for the data in Illustration 13.1 can be found by calculating the two standard deviations and then dividing:

$$s_x = 4.099 \quad \text{and} \quad s_y = 2.098$$

Refer to Chapter 3, page 148 for an illustration of the use of this formula.

$$r = \frac{\text{covar}(x, y)}{s_x \cdot s_y}: \quad r = \frac{0.6}{(4.099)(2.098)} = \textbf{0.07}$$

Finding the correlation coefficient by using formula (13.2) can be a very tedious arithmetic process. The formula can be written in a more workable form, as it was in Chapter 3:

Computer and/or calculator commands to calculate the correlation coefficient can be found on page 149 in Chapter 3.

$$r = \frac{\text{covar}(x, y)}{s_x \cdot s_y} = \frac{\dfrac{\sum[(x - \bar{x}) \cdot (y - \bar{y})]}{n - 1}}{s_x \cdot s_y} = \frac{\text{SS}(xy)}{\sqrt{\text{SS}(x) \cdot \text{SS}(y)}} \tag{13.3}$$

Formula (13.3) avoids the separate calculations of $\bar{x}$, $\bar{y}$, s_x, and s_y, and, more important, the calculations of the deviations from the means. Therefore, formula (13.3) is much easier to use and more accurate when decimals are involved as it minimizes round-off error.

EXERCISES

13.1 Explain why $\sum(x - \bar{x}) = 0$ and $\sum(y - \bar{y}) = 0$.

13.2 Consider a set of paired bivariate data. Describe the relationship of the ordered pairs that will cause $\sum[(x - \bar{x}) \cdot (y - \bar{y})]$ to be
   a. positive   **b.** negative   **c.** near zero

13.3 **a.** Construct a scatter diagram of the following bivariate data.

Point	A	B	C	D	E	F	G	H	I	J
x	1	1	3	3	5	5	7	7	9	9
y	1	2	2	3	3	4	4	5	5	6

(continued)

b. Calculate the covariance. **c.** Calculate s_x and s_y.
d. Calculate r using formula (13.2). **e.** Calculate r using formula (13.3).

13.4 Consider the accompanying bivariate data.

Point	A	B	C	D	E	F	G	H	I	J
x	0	1	1	2	3	4	5	6	6	7
y	6	6	7	4	5	2	3	0	1	1

a. Draw a scatter diagram for the data. **b.** Calculate the covariance.
c. Calculate s_x and s_y. **d.** Calculate r by formula (13.2).
e. Calculate r by formula (13.3).

13.5 A computer was used to complete the preliminary calculations : form the extensions table, calculate the summations Σx, Σy, Σx^2, Σxy, Σy^2, and find the SS(x), SS(y), SS(xy) for the following set of bivariate data. Verify the results by calculating the values yourself.

x	45	52	49	60	67	61
y	22	26	21	28	33	32

MINITAB output:

Row	x	y	XSQ	XY	YSQ
1	45	22	2025	990	484
2	52	26	2704	1352	676
3	49	21	2401	1029	441
4	60	28	3600	1680	784
5	67	33	4489	2211	1089
6	61	32	3721	1952	1024

Row	C6	C7	C8	C9	C10
1	334	162	18940	9214	4498

```
SS(X)    347.333
SS(Y)    124.000
SS(XY)   196.000
```

13.6 Use a computer to form the extensions table, calculate the summations Σx, Σy, Σx^2, Σxy, Σy^2, and find the SS(x), SS(y), SS(xy) for the following set of bivariate data.

x	11.4	9.4	6.5	7.3	7.9	9.0	9.3	10.6
y	8.1	8.2	5.8	6.4	5.9	6.5	7.1	7.8

13.7 Professional golfers are commonly rated on the basis of how much money they earn while playing on the tour. They are also given a world point ranking that considers their performances for the past two years, with an emphasis on the last year. Money winnings in 1998 through August of that year and the world point rankings for 16 players are shown in the table below:

Player	Money Earnings ($)	Points
David Duval	1,668,678	9.42
Mark O'Meara	1,523,295	9.51
Fred Couples	1,495,698	7.50
Tiger Woods	1,388,542	11.82
Jim Furyk	1,292,346	6.59

(continued)

See page. 149 for information about using MINITAB, EXCEL or TI-83 to find the correlation coefficient.

Player	Money Earnings ($)	Points
Justin Leonard	1,253,129	7.14
Lee Janzen	1,052,622	5.45
Mark Calcavecchia	910,254	5.51
Scott Hoch	878,623	5.76
Phil Mickelson	873,477	7.61
Jesper Parnevik	857,956	5.73
Tom Watson	832,385	5.47
Davis Love III	815,766	10.59
Tom Lehman	812,114	5.91
Vijay Singh	745,661	6.25
Nick Price	484,737	7.94

Source: Golf, "Stats +," September 1998.

a. Calculate the linear correlation coefficient (Pearson's product moment, r) between the money earnings and the world point rankings.

b. What conclusion might you draw from your answer in (a)?

13.8 The performances of personal computers are rated using a number of different dimensions and applications. Three of the most popular are multimedia, word processing, and spreadsheet. *Windows®* magazine's technical staff conducted benchmark tests in the summer of 1998 on a collection of 12 machines representing both desktop and notebook designs. Results of the tests are shown in the table of scores shown below:

(1) Multimedia	(2) Word Processing	(3) Spreadsheet
121	115	112
115	83	116
100	95	92
78	87	86
110	72	51
80	79	74
76	76	68
83	45	49
57	50	42
131	123	120
120	118	115
119	115	115

Source: Windows®, "Winscore 2.0 Results," September 1998.

a. Calculate the linear correlation coefficient (Pearson's product moment, r) between (1) multimedia and word processing, (2) multimedia and spreadsheet, and (3) word processing and spreadsheet.

b. What conclusions might you draw from your answers in (a)?

13.9 **a.** Calculate the covariance of the set of data (20, 10), (30, 50), (60, 30), (80, 20), (110, 60), and (120, 10).

b. Calculate the standard deviation of the six x-values and the standard deviation of the six y-values.

c. Calculate r, the coefficient of linear correlation, for the data in part (a).

d. Compare these results to those found in the text for Illustration 13.1, page 609.

13.10 A formula that is sometimes given for computing the correlation coefficient is

$$r = \frac{n(\sum xy) - (\sum x)(\sum y)}{\sqrt{n(\sum x^2) - (\sum x)^2} \, \sqrt{n(\sum y^2) - (\sum y)^2}}$$

(continued)

Use this expression as well as the formula

$$r = \frac{SS(xy)}{\sqrt{SS(x) \cdot SS(y)}}$$

to compute r for the data in the table.

x	2	4	3	4	0
y	6	7	5	6	3

13.2 Inferences About the Linear Correlation Coefficient

In Section 13.1, we learned that covariance is a measure of linear dependency. Also noted was the fact that its value is affected by the spread of the data; therefore, the covariance is standardized by dividing it by the standard deviations of both x and y. This standardized form is known as r, the coefficient of linear correlation. Standardizing allows for comparisons of different sets of data thereby allowing r to play a role much like z or t does for $\bar{x}$. Therefore the calculated r-value becomes $r \star$, the test statistic for inferences about ρ, $\boldsymbol{\rho}$ the population correlation coefficient.

ASSUMPTIONS FOR INFERENCES ABOUT LINEAR CORRELATION COEFFICIENT:

The set of (x, y) ordered pairs forms a random sample and the y-values at each x have a normal distribution. Inferences use the t-distribution with $n - 2$ degrees of freedom.

CAUTION

The inferences about the linear correlation coefficient are about the pattern of behavior of the two variables involved and the usefulness of one variable in predicting the other. *Significance of the linear correlation coefficient does not mean that you have established a cause-and-effect relationship.* Cause-and-effect is a separate issue.

Confidence Interval Procedure

Be sure to use "sample size" not degrees of freedom.

As with other parameters, a **confidence interval** may be used to estimate the value of ρ, the linear correlation coefficient of the population. Usually this is accomplished by using a table showing confidence belts. Table 10 in Appendix B gives **confidence belts** for 95% confidence intervals. This table is a bit tricky to read and utilizes n, the sample size; so be extra careful when you use it. The next illustration demonstrates the procedure for estimating ρ.

ILLUSTRATION 13.2 ▼

A random sample of 15 ordered pairs of data have a calculated r-value of 0.35. Find the 95% confidence interval for ρ, the population linear correlation coefficient.

Solution

STEP 1 **The Set-Up:**
Describe the population parameter of interest.
The linear correlation coefficient for the population, ρ

STEP 2 **The Confidence Interval Criteria:**
a. Check the assumptions.
The ordered pairs form a random sample and we will assume that the y-values at each x have a mounded distribution.
b. Identify the probability distribution and the formula to be used.
The calculated linear correlation coefficient, r, is used.
c. State the level of confidence, $1 - \alpha$.
$1 - \alpha = 0.95$

STEP 3 **The Sample Evidence:**
Collect the sample information.
$n = 15$ and $r = 0.35$

STEP 4 **The Confidence Interval:**

The confidence interval is read from Table 10 in Appendix B. Find $r = 0.35$ at the bottom of Table 10. (See the arrow under Figure 13.5.) Visualize a vertical line through that point. Find the two points where the belts marked for the correct sample size cross the vertical line. The sample size is 15. These two points are circled in Figure 13.5.

Figure 13.5

Using Table 10 of Appendix B, Confidence Belts for the Correlation Coefficient

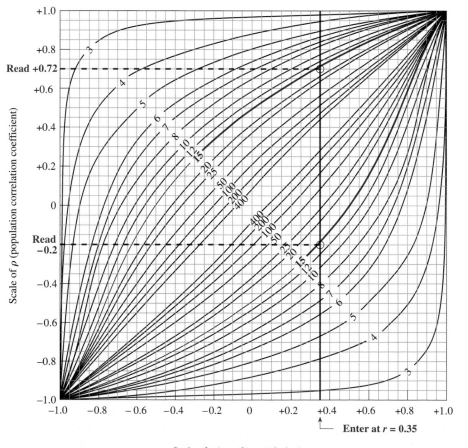

Scale of r (sample correlation)

Exercise 13.11
Find the 95% interval when a sample of $n = 25$ results in $r = 0.35$.

Now look horizontally from the two circled points to the vertical scale on the left and read the confidence interval. The values are **0.72** and **−0.20.**

STEP 5 **The Results:**
State the confidence interval.
Thus, **the 95% confidence interval for ρ, the population coefficient of linear correlation, is −0.20 to 0.72.** ▲

Hypothesis-Testing Procedure

After the linear correlation coefficient, r, has been calculated for the sample data, it seems necessary to ask this question: Does the value of r indicate that there is a linear relationship between the two variables in the population from which the sample was drawn? To answer this question we can perform a hypothesis test. The null hypothesis is "the two variables are linearly unrelated" ($\rho = 0$), where ρ is the linear correlation coefficient for the population. The alternative hypothesis may be either one-tailed or two-tailed. Most frequently it is two-tailed, $\rho \neq 0$. However, when we suspect that there is only a positive or only a negative correlation, we should use a one-tailed test. The alternative hypothesis of a one-tailed test is: $\rho > 0$ or $\rho < 0$.

The area under the curve that represents the p-value or the critical region for the test is on the right when a positive correlation is expected and on the left when a negative correlation is expected. The test statistic used to test the null hypothesis is the calculated value of r from the sample data. Probability bounds for the p-value or critical values for r are found in Table 11 of Appendix B. The number of degrees of freedom for the r-statistic is 2 less than the sample size, df $= n - 2$. Specific details for using Table 11 follow Illustration 13.3.

The rejection of the null hypothesis means that there is evidence of a linear relationship between the two variables in the population. Failure to reject the null hypothesis is interpreted as meaning that a linear relationship between the two variables in the population has not been shown.

Now let's look at an illustration of a hypothesis test.

ILLUSTRATION 13.3 ▼

In a study of 15 randomly selected ordered pairs, $r = 0.548$. Is this linear correlation coefficient significantly different from zero at the 0.02 level of significance?

Solution

STEP 1 **The Set-Up:**
a. Describe the population parameter of interest.
The linear correlation coefficient for the population, ρ (rho)
b. State the null hypothesis (H_o) and the alternative hypothesis (H_a).

H_o: $\rho = 0$

H_a: $\rho \neq 0$

STEP 2 **The Hypothesis Test Criteria:**
a. Check the assumptions.
The ordered pairs form a random sample and we will assume that the y-values at each x have a mounded distribution.
b. Identify the probability distribution and the test statistic to be used.
The test statistic is r ★, formula (13.3), with df $= n - 2 = 15 - 2 = 13$.
c. Determine the level of significance, α.
$\alpha = 0.02$ (given in statement of problem).

STEP 3   **The Sample Evidence:**
   a. $n = 15$ and $r = 0.548$.
   The calculated sample linear correlation coefficient is the test statistic;
   b. The test statistic, $r \star = 0.548$

STEP 4   **The Probability Distribution:**

USING THE *p*-VALUE PROCEDURE:

a. Calculate the *p*-value.
Use both tails since the H_a expresses concern for values related to "different than."
$\mathbf{P} = P(r < -0.548) + P(r > 0.548) = 2 \cdot P(r > 0.548)$, with df $= 13$ as shown in the figure.

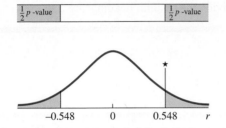

Use Table 11 (Appendix B) to place bounds on the *p*-value:
0.02 < P < 0.05

Specific details follow this illustration.
b. Determine whether or not the *p*-value is smaller than α.
The *p*-value **is not smaller** than the level of significance, α (α = 0.02)

OR

USING THE CLASSICAL PROCEDURE:

a. Determine the critical region and critical value(s).
The critical region is both tails since the H_a expresses concern for values related to "different than." The critical value is found at the intersection of the df $= 13$ row and the two-tailed 0.02 column of Table 11: **0.592**

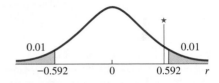

Specific details follow this illustration.
b. Determine whether or not the calculated test statistic is in the critical region.
$r \star$ **is not in** the critical region, as shown in **red** in the figure above.

STEP 5   **The Results:**
   a. State the decision about H_o.
   Fail to reject H_o.
   b. State the conclusion about H_a.
   At the 0.02 level of significance, we have failed to show that *x* and *y* are correlated.

   Using Table 11 in Appendix B to "place bounds" on the p-*value* : By inspecting the df $= 13$ row of Table 11, you can determine an interval within which the *p*-value lies. Locate the $r \star$ along the row labeled df $= 13$. If $r \star$ is not listed, locate the two table values it falls between, and read the bounds for the *p*-value from the top of the table. In this case, $r \star = 0.548$ is between 0.514 and 0.592; therefore, **P** is between 0.02 and 0.05. Table 11 shows only two-tailed values and the alternative hypothesis is two-tailed; therefore the bounds for the *p*-value are read directly from the table.

Exercise 13.12

Place bounds on the *p*-value resulting from a sample with $n = 18$ and $r = 0.444$,

a. If H_a is two-tailed.
b. If H_a is one-tailed.

Portion of Table 11

		Amount of α in two-tails				
df	...	**0.05**	**P**	**0.02**	...	→ **0.02 < P < 0.05**
⋮		⋮	↑	⋮		
13		0.514	**0.548**	0.592		

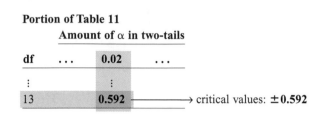

Exercise 13.13
What are the critical values of *r* for $\alpha = 0.05$ and *n* = 20?

a. If H_a is two-tailed.
b. If H_a is one-tailed.

NOTE When H_a is one-tailed, divide the column headings by 2 to place bounds on the *p*-value.

Using Table 11 in Appendix B to find the critical values. The critical value is at the intersection of the df = 13 and the two-tailed $\alpha = 0.02$ column. Table 11 shows only two-tailed values and the alternative hypothesis is two-tailed; therefore the critical values are read directly from the table.

Portion of Table 11
Amount of α in two-tails

df	...	0.02	...
⋮		⋮	
13		0.592	

→ critical values: ± 0.592

NOTE When H_a is one-tailed, divide the column headings by 2. ▲

Case Study 13.1

DNA Quantitation by Image Cytometry of Touch Preparation from Fresh and Frozen Tissue

A scatter plot comparing the DNA indices obtained from fresh and frozen tissue samples is shown in figure. In 54 of 59 cases (91.5% there was excellent agreement between the two methods (Spearman *r* = 0.91; Pearson *r* = 0.94).

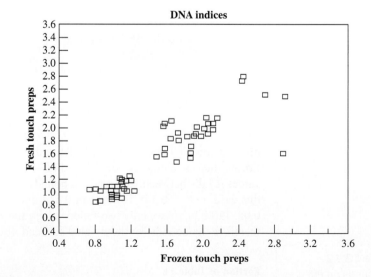

Exercise 13.14
The linear relationship between DNA indices appears strong, and the reported correlation coefficients are 0.91 and 0.94. What is the difference? Are they significant? Explain.

Source: Paula F. Suit, M.D., and Thomas W. Bauer, M.D., Ph.D., *American Journal of Clinical Pathology*, July 1990. Reprinted by permission.

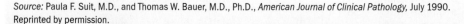

EXERCISES • • • • • • • • • • • • •

13.15 Using graphs to illustrate, explain the meaning of a correlation coefficient whose value is
    a. -1.0
    b. 0.0
    c. $+1.0$
    d. $+0.5$
    e. -0.6

13.16 Use Table 10 of Appendix B to determine a 95% confidence interval for the true population linear correlation coefficient based on the following sample statistics:
    a. $n = 8, r = 0.20$
    b. $n = 100, r = -0.40$
    c. $n = 25, r = +0.65$
    d. $n = 15, r = -0.23$

13.17 The Test-Retest Method is one way of establishing the reliability of a test. The test is administered and then, at a later date, the same test is readministered to the same individuals. The correlation coefficient is computed between the two sets of scores. The following test scores were obtained in a Test-Retest situation.

First Score	75	87	60	75	98	80	68	84	47	72
Second Score	72	90	52	75	94	78	72	80	53	70

Find r and set a 95% confidence interval for ρ.

13.18 Three measures of a company's performance are its net income (earnings), revenues (sales), and the market value of its stock. But are these related? The following table shows these data for 15 of the top 100 fastest growing companies listed by *Fortune* in mid-1998 ($ millions):

Company	(1) Net Income	(2) Revenues	(3) Market Value of Stock
Noble Drilling	175.2	764.8	1,986
Funco	8.1	172.2	94
Marine Drilling	71.7	230.6	645
Vitesse Semiconductor	46.5	151.9	2,360
Central Garden & Pet	33.4	1,181.3	641
Jabil Circuit	74.5	1,265.0	1,178
Cliff's Drilling	54.9	316.7	325
Pairgain Technologies	48.5	291.2	895
RMI Titanium	82.5	355.4	423
Sanmina	59.8	642.6	1,717
UTI Energy	15.6	202.4	165
Veritas DGC	62.4	497.3	611
Fidelity National Financial	73.1	935.2	827
Waste Management	365.9	3,148.5	29,447
TJX	376.4	7,770.6	8,425

Source: Fortune, "America's Fastest Growing Companies," September 28, 1998.

Calculate the correlation coefficient and use it and Table 10 of Appendix B to determine a 95% confidence interval on ρ for each of the following cases:
    a. Net Income and Revenues
    b. Net Income and Market Value of Stock
    c. Revenues and Market Value of Stock

13.19 The *State and Metropolitan Area Data Book—1991* gives the 1988 rate per 100,000 population for physicians and nurses for the West North Central states as follows:

State	Physicians	Nurses
MN	212	787
IA	145	805
MO	190	745
ND	167	935
SD	138	809
NE	168	725
KS	169	676

a. Calculate r.   **b.** Set a 95% confidence interval of ρ.
c. Describe the meaning of answer (b).
d. Explain the meaning of the width of the interval answer in (b).

13.20 The *1994 World Almanac and Book of Facts* gives the Nielsen TV ratings for the favorite syndicated programs for 1992–1993. The following table gives these ratings for women, men, teenagers, and children.

Women	Men	Teenagers	Children
11.8	8.0	3.3	3.4
9.9	6.9	3.0	2.3
7.1	9.2	6.4	5.1
6.5	8.2	5.8	4.8
8.4	3.4	2.7	1.7
6.4	5.0	2.1	1.9
4.9	4.7	4.6	5.8
4.1	4.5	6.8	4.7
6.3	4.5	1.7	1.6
5.1	4.2	2.3	1.8
5.6	4.0	1.8	1.6
4.8	4.7	3.1	2.9
4.4	4.1	3.4	3.8
4.2	4.2	4.0	3.4
4.1	3.2	4.6	7.4
4.4	3.1	5.8	4.3
3.7	4.9	3.4	3.2
4.3	2.7	3.4	2.2
3.3	4.6	2.1	2.2
2.5	4.0	5.2	5.6

Calculate r and use it to determine a 95% confidence interval on ρ for each of the following cases.
a. Women and Men   **b.** Women and Teenagers   **c.** Women and Children
d. Men and Teenagers   **e.** Men and Children   **f.** Teenagers and Children
g. What can be concluded from the above answers? Be specific.

13.21 State the null hypothesis, H_o, and the alternative hypothesis, H_a, that would be used to test the following statements.
a. The linear correlation coefficient is positive.
b. There is no linear correlation.
c. There is evidence of negative correlation.
d. There is positive linear relationship.

13.22 Determine the critical values that would be used in testing each of the following null hypotheses using the classical approach.
  a. H_o: $\rho = 0$ vs. H_a: $\rho \neq 0$, with $n = 18$, $\alpha = 0.05$
  b. H_o: $\rho = 0$ vs. H_a: $\rho > 0$, with $n = 32$, $\alpha = 0.01$
  c. H_o: $\rho = 0$ vs. H_a: $\rho < 0$, with $n = 16$, $\alpha = 0.05$

13.23 A sample of 20 pieces of bivariate data has a linear correlation coefficient of $r = 0.43$. Does this provide sufficient evidence to reject the null hypothesis that $\rho = 0$ in favor of a two-sided alternative? Use $\alpha = 0.10$.

13.24 If a sample of size 18 has a linear correlation coefficient of -0.50, is there significant reason to conclude that the linear correlation coefficient of the population is negative? Use $\alpha = 0.01$.

13.25 In a study involving 24 coastal drainage basins in the Mendocino triple junction region of northern California (*Geological Society of America Bulletin,* Nov. 1989), it is reported that the Pearson correlation coefficient between uplift rate and the length of the drainage basin equals 0.16942. Use Table 11 in Appendix B to determine whether these data provide evidence sufficient to reject H_o: $\rho = 0$ in favor of H_a: $\rho \neq 0$ at $\alpha = 0.05$.

13.26 Is a value of $r = +0.24$ significant in trying to show that ρ is greater than zero for a sample of 62 data at the 0.05 level of significance?

13.27 The population (in millions) and the violent crime rate (per 1000) were recorded for ten metropolitan areas. The data are shown in the following table.

Population	10.0	1.3	2.1	7.0	4.4	0.3	0.3	0.2	0.2	0.4
Crime Rate	12.0	9.5	9.2	8.4	8.2	7.3	7.1	7.0	6.9	6.9

Do these data provide evidence to reject the null hypothesis that $\rho = 0$ in favor of $\rho \neq 0$ at $\alpha = 0.05$?

13.28 Going to work for a particular company is a major decision in anyone's career path. The popularity of any company could be indicated by both the number of job applications it considers each year and the number of people it already employs. The following table was extracted from *Fortune's* 100 best companies to work for, published in early 1998:

Company	U.S. Employees	Applicants
Southwest Airlines	24,757	150,000
Kingston Technology	552	4000
SAS Institute	3154	12,000
W.L. Gore	4118	23,717
Microsoft	14,936	150,000
Merck	31,767	165,000
Hewlett-Packard	66,300	255,000
Goldman Sachs	6546	8000
Corning	8127	60,000
Harley-Davidson	5288	8000

Source: Fortune, "The 100 Best Companies to Work For in America," January 12, 1998.

  a. Do these data provide evidence to reject the null hypothesis that $\rho = 0$ in favor of $\rho > 0$ at $\alpha = 0.01$?
  b. Explain the meaning of the apparent positive correlation.

13.29 Two indicators of the level of economic activity in a given geographical area are its total personal income and the value of the construction contracts. The following table shows the related data for seven states during the third quarter of 1997:   *(continued)*

State	Personal Income ($ millions)	Construction Contracts (Index, 1980 = 100)
Colorado	93,706	274.0
Kansas	56,092	202.4
Missouri	115,588	239.3
Nebraska	35,588	263.9
New Mexico	30,254	162.0
Oklahoma	60,819	125.1
Wyoming	9678	88.8

Source: Regional Economic Digest, "Economic Indicators," First Quarter 1998.

a. Calculate the correlation coefficient between the two variables.
b. Test for a significant correlation at the 0.05 level of significance and draw your conclusion.

13.30 State finances have always fascinated economists, and political leaders seeking office have often used state financial statistics to raise debatable issues at election time. Do states with higher taxes spend more? Do states with higher taxes also acquire more debt? Do states with higher expenditures acquire more debt? The following data (all per capita) from the 50 states in fiscal year 1995 is offered to help study and provide an insight to these questions:

State	(1) Debt	(2) Taxes	(3) Expenditures	State	(1) Debt	(2) Taxes	(3) Expenditures
AL	$ 884	$1194	$2714	MT	2540	1396	$3434
AK	5351	3183	9270	NE	836	1356	2596
AZ	720	1475	2646	NV	1305	1764	2994
AR	798	1365	2663	NH	5306	800	2697
CA	1526	1686	3458	NJ	3066	1713	4104
CO	899	1209	2616	NM	1083	1688	3776
CT	4719	2282	4145	NY	3775	1891	4487
DE	4916	2224	4156	NC	632	1588	2840
FL	1085	1311	2453	ND	1334	1496	3452
GA	781	1317	2660	OH	1103	1362	3138
HI	4377	2422	5067	OK	1140	1347	2742
ID	1120	1490	2889	OR	1745	1365	3512
IL	1855	1402	2789	PA	1184	1513	3263
IN	940	1386	2634	RI	5571	1505	4308
IA	743	1549	3021	SC	1367	1297	3164
KS	447	1468	2774	SD	2282	952	2578
KY	1839	1628	2952	TN	537	1124	2556
LA	1962	1077	3331	TX	530	1084	2384
ME	2451	1461	3368	UT	1056	1371	2963
MD	1872	1599	2989	VT	2851	1370	3442
MA	4566	1910	3998	VA	1317	1327	2575
MI	1313	1856	3631	WA	1624	1877	3904
MN	975	2023	3553	WV	1415	1494	3425
MS	713	1335	2749	WI	1608	1763	3182
MO	1261	1268	2344	WY	1642	1389	4261

Source: Census Bureau, U.S. Department of Commerce, 1996.

Calculate the correlation coefficient and use it and Table 10 of Appendix B to determine a 95% confidence interval on ρ for each of the following cases:
a. Taxes and Debt   **b.** Debt and Expenditures
c. Taxes and Expenditures

13.3 Linear Regression Analysis

Recall that the **line of best fit** results from an analysis of two (or more) related quantitative variables. (We will restrict our work to two variables.) When two variables are studied jointly, we often would like to control one variable by means of controlling the other. Or we might want to predict the value of a variable based on knowledge about another variable. In both cases we want to find the line of best fit, provided one exists, that will best predict the value of the dependent, or output, variable from a value of the independent or input variable . Recall that the variable we know or can control is called the *independent, or input, variable*; the variable resulting from using the equation of the line of best fit is called the *dependent, or predicted , variable.*

Recall that in Chapter 3 the method of least squares was developed. From this concept formulas (3.6) and (3.7) were obtained and are used to calculate b_0 (the **y-intercept**) and b_1 (the **slope of the line of best fit**).

$$b_0 = \frac{1}{n}\left(\sum y - b_1 \cdot \sum x\right) \tag{3.6}$$

$$b_1 = \frac{SS(xy)}{SS(x)} \tag{3.7}$$

Then these two coefficients are used to write the equation for the line of best fit in the form

$$\hat{y} = b_0 + b_1 x$$

When the line of best fit is plotted, it does more than just show us a pictorial representation of the line. It tells us two things: (1) whether or not there really is a linear relationship between the two variables and (2) the quantitative (equation) relationship between the two variables. When there is no relationship between the variables, a horizontal line of best fit will result. A horizontal line has a slope of zero, which implies that the value of the input variable has no effect on the output variable. (This idea will be amplified later in this chapter.)

The result of regression analysis is the mathematical equation that is the equation of the line of best fit. We will, as mentioned before, restrict our work to the simple linear case, that is, one input variable and one output variable where the line of best fit is straight. However, you should be aware that not all relationships are of this nature. If the scatter diagram suggests something other than a straight line, the relationship may be **curvilinear regression.** In cases of this type we might need to introduce terms to higher powers, x^2, x^3, and so on; or other functions, e^x, log x, and so on; or we might need to introduce other input variables. Maybe two or three input variables would improve the usefulness of our regression equation. These possibilities are examples of curvilinear regression and **multiple regression.**

The linear model used to explain the behavior of linear bivariate data in the population is

$$y = \beta_0 + \beta_1 x + \epsilon \tag{13.4}$$

This equation represents the linear relationship between two variables in a population. β_0 is the y-intercept and β_1 is the slope . ϵ (lowercase Greek letter epsilon) is the random **experimental error** in the observed value of y at a given value of x.

The **regression line** from the sample data gives us b_0, which is our estimate of β_0 and b_1, which is our estimate of β_1. The error ϵ is approximated by $e = y - \hat{y}$, the difference between the observed value of y and the predicted value of y, $\hat{y}$, at a given value of x.

$$e = y - \hat{y} \tag{13.5}$$

The random variable e (also known as the residual) is positive when the observed value of y is greater than the predicted value, $\hat{y}$; e is negative when y is less than $\hat{y}$. The sum of the errors (residuals) for all values of y for a given value of x is exactly zero. (This is part of the least-squares criteria .) The mean value of the experimental error is zero; its variance is σ_ϵ^2. Our next goal is to estimate this **variance of the experimental error.**

Before we estimate the variance of ϵ, let's try to understand exactly what the error represents. ϵ is the amount of error in our observed value of y. That is, it is the difference between the observed value of y and the mean value of y at that particular value of x. Since we do not know the mean value of y, we will use the regression equation and estimate it with $\hat{y}$, the predicted value of y at this same value of x. Thus the best estimate that we have for ϵ is $e = y - \hat{y}$, as shown in Figure 13.6.

Figure 13.6
The Error e Is $y - \hat{y}$

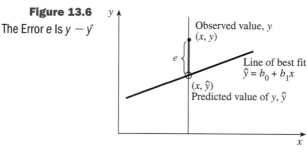

NOTE e is the observed error in measuring y at a specified value of x.

If we were to observe several values of y at a given value of x, we could plot a distribution of y-values about the line of best fit (about $\hat{y}$, in particular). Figure 13.7 shows a sample of bivariate values that share a common x-value. Figure 13.8 shows the theoretical distribution of all possible y-values at a given x-value. A similar distribution occurs at each different value of x. The mean of the observed y's at a given value of x varies, but it can be estimated by $\hat{y}$.

Figure 13.7
Sample of y-Values
at a Given x

Figure 13.8
Theoretical Distribution of
y-Values for a Given x

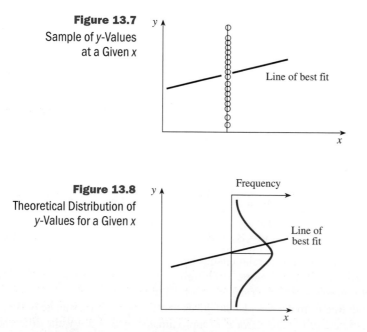

Before we can make any inferences about a regression line, we must assume that the distribution of y's is approximately normal and that the variances of the distributions of y at all values of x are the same. That is, the standard deviation of the distribution of y about $\hat{y}$ is the same for all values of x, as shown in Figure 13.9.

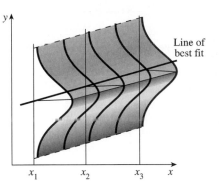

Before looking at the variance of e, let's review the definition of sample variance. The sample variance, s^2, is defined as $\dfrac{\sum(x - \bar{x})^2}{(n - 1)}$, the sum of the squares of each deviation divided by the number of degrees of freedom, $n - 1$, associated with a sample of size n. The variance of y involves an additional complication: there is a different mean for y at each value of x. (Notice the many distributions in Figure 13.9.) However, each of these "means" is actually the predicted value, $\hat{y}$, that corresponds to the x that fixes the distribution. The variance of the error e is estimated by the formula

$$s_e^2 = \frac{\sum(y - \hat{y})^2}{n - 2} \tag{13.6}$$

where $n - 2$ is the number of degrees of freedom.

NOTE The variance of y about the line of best fit is the same as the variance of the error e. Recall that $e = y - \hat{y}$.

Formula (13.6) can be rewritten by substituting $b_0 + b_1 x$ for $\hat{y}$. Since $\hat{y} = b_0 + b_1 x$, then s_e^2 becomes

$$s_e^2 = \frac{\sum(y - b_0 - b_1 x)^2}{n - 2} \tag{13.7}$$

With some algebra, and some patience, this formula can be rewritten once again into a more workable form. The form we will use is

$$s_e^2 = \frac{(\sum y^2) - (b_0)(\sum y) - (b_1)(\sum xy)}{n - 2} \tag{13.8}$$

For ease of discussion let's agree to call the numerator of formulas (13.6), (13.7), and (13.8) the **sum of squares for error** (SSE).

Now let's see how all of this information can be used.

ILLUSTRATION 13.4 ▼

Suppose that you move into a new city and take a job. You will, of course, be concerned about the problems you will face commuting to and from work. For example, you would like to know how long (in minutes) it will take you to drive to work each morning. Let's use "one-way distance to work" as a measure of where you live. You live x miles away from work and want to know how long it will take you to commute each day. Your new employer, foreseeing this question, has already collected a random sample of data to be used in answering your question. Fifteen of your co-workers were asked to give their one-way travel time and distance to work. The resulting data are shown in Table 13.2. (For convenience the data have been arranged so that the x-values are in numerical order.) Find the line of best fit and the variance of y about the line of best fit, s_e^2.

Table 13.2   Data for Illustration 13.4

Co-worker	Miles (x)	Minutes (y)	x^2	xy	y^2
1	3	7	9	21	49
2	5	20	25	100	400
3	7	20	49	140	400
4	8	15	64	120	225
5	10	25	100	250	625
6	11	17	121	187	289
7	12	20	144	240	400
8	12	35	144	420	1225
9	13	26	169	338	676
10	15	25	225	375	625
11	15	35	225	525	1225
12	16	32	256	512	1024
13	18	44	324	792	1936
14	19	37	361	703	1369
15	20	45	400	900	2025
Total	184	403	2616	5623	12,493

Solution

The extensions and summations needed for this problem are shown in Table 13.2. The line of best fit can now be calculated using formulas (2.9), (3.4), (3.6), and (3.7).

Using formula (2.9),

$$\text{SS}(x) = \sum x^2 - \frac{(\sum x)^2}{n}: \quad \text{SS}(x) = 2616 - \frac{(184)^2}{15} = 358.9333$$

Using formula (3.4),

$$\text{SS}(xy) = \sum xy - \frac{\sum x \sum y}{n}: \quad \text{SS}(xy) = 5623 - \frac{(184)(403)}{15} = 679.5333$$

> Use extra decimal place during these calculations.

slope: using formula (3.6),

$$b_1 = \frac{\text{SS}(xy)}{\text{SS}(x)}: \quad b_1 = \frac{679.5333}{358.9333} = 1.893202 = \mathbf{1.89}$$

y-intercept: using formula (3.7),

$$b_0 = \frac{\sum y - (b_1 \cdot \sum x)}{n}: \quad b_0 = \frac{403 - (1.893202)(184)}{15} = 3.643387 = \mathbf{3.64}$$

Therefore, the equation for the line of best fit is

$$\hat{y} = \mathbf{3.64 + 1.89}x$$

The variance of y about the regression line is calculated by using formula (13.8).

$$s_e^2 = \frac{(\sum y^2) - (b_0)(\sum y) - (b_1)(\sum xy)}{n-2}:$$

$$s_e^2 = \frac{(12{,}493) - (3.643387)(403) - (1.893202)(5{,}623)}{15-2} = \frac{379.2402}{13} = \mathbf{29.17}$$

$s_e^2 = 29.17$ is the variance of the 15 e's. In Figure 13.10 the 15 e's are shown as vertical line segments.

NOTE   Extra decimal places are often needed for this type of calculation. Notice that b_1 (1.893202) was multiplied by 5623. If 1.89 had been used instead, that one product would have changed the numerator by approximately 18. That, in turn, would have changed the final answer by almost 1.4, and that is a sizable round-off error.

Figure 13.10

The 15 Random Errors as Line Segments

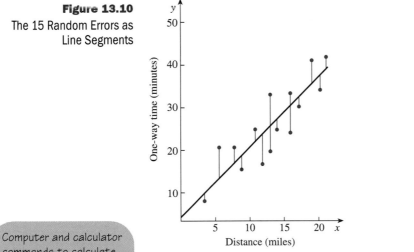

Distance (miles)

> Computer and calculator commands to calculate the regression line for a set of bivariate data can be found on page 160, 161 in Chapter 3.

In the sections that follow, the variance of e will be used in much the same way as the variance of x (as calculated in Chapter 2) was used in Chapters 8, 9, and 10 to complete the statistical inferences studied there.

EXERCISES

13.31   Ten salespeople were surveyed and the average number of client contacts per month, x, and the sales volume, y (in thousands), were recorded for each:

x	12	14	16	20	23	46	50	48	50	55
y	15	25	30	30	30	80	90	95	110	130

Refer to the following computer output and verify that the equation of the line of best fit is $\hat{y} = -13.4 + 2.3x$ and that $s_e = 10.17$ by calculating these values yourself.

```
The regression equation is y = −13.4 + 2.30 x

Predictor   Coef
Constant   −13.414
x            2.3028
s = 10.17
```

13.32 A paper titled "Blue Grama Response to Zinc Source and Rates" by E. M. White (*Journal of Range Management,* January 1991) described five different experiments involving herbage yield and zinc application. Some of the data in experiment number two were as follows:

Grams of Zinc per kg Soil	0.0	0.1	0.2	0.4	0.8
Grams of Herbage	3.2	2.8	2.6	2.0	0.1

The paper quotes the correlation coefficient as equal to -0.99 and the line of best fit as $\hat{y} = -3.825x + 3.29$, where x = grams of zinc and y = grams of herbage.
a. Verify the value for r.   **b.** Verify the equation of the line of best fit.

13.33 There's an old adage that more fat content in foods means more calories. A list of restaurant foods appeared in the December 1997 *Nutrition Action Healthletter* that indicated both the calorie and fat content of foods that people typically consume at restaurants:

Restaurant Foods	Calories	Fat (g)
Kung Pao chicken with rice	1620	76
Chicken burrito with extras	1530	68
Spaghetti with meatballs	1155	39
Porterhouse steak (20 oz.)	1100	82
Tuna salad sandwich w/mayo	835	56
Turkey club sandwich	735	34
Pizza Hut Pan Pizza	700	34
Movie theater popcorn w/butter	630	50
Baked potato with sour cream	620	31
Meat loaf	570	38
Burger King BK Broiler	550	29
Ham & cheese omelette	510	39
Grilled cheese sandwich	510	33
Au Bon Pain Blueberry Muffin	430	18
McDonald's Cheese Danish	410	22
KFC Chicken Breast	400	24
McDonald's Grilled Chicken Salad	350	23

Source: Nutrition Action Healthletter, Vol. 24, No. 10, December 1997.

a. Draw a scatter diagram of the data. Use calories as the dependent variable (y) and fat content as the independent variable (x).
b. Find the equation of the line of best fit and graph it on the scatter diagram.
c. What is the average increase in calories for each additional gram of fat in the food?
d. Find the variance of y about the regression line.

13.34 The computer-science aptitude score, x, and the achievement score, y (measured by a comprehensive final), were measured for 20 students in a beginning computer-science course. The results were as follows. Find the equation of the line of best fit and s_e^2.

x	4	16	20	13	22	21	15	20	19	16	18	17	8	6	5	20	18	11	19	14
y	19	19	24	36	27	26	25	28	17	27	21	24	18	18	14	28	21	22	20	21

13.35 a. Using the ten points shown in the following table, find the equation of the line of best fit, $\hat{y} = b_0 + b_1 x$ and graph it on a scatter diagram.

Point	A	B	C	D	E	F	G	H	I	J
x	1	1	3	3	5	5	7	7	9	9
y	1	2	2	3	3	4	4	5	5	6

b. Find the ordinates $\hat{y}$ for the points on the line of best fit whose abscissas are $x = 1, 3, 5, 7,$ and 9.
(continued)

    c. Find the value of e for each of the points in the given data ($e = y - \hat{y}$).
    d. Find the variance s_e^2 by using formula (13.6).
    e. Find the variance s_e^2 by using formula (13.8). [Answers to (d) and (e) should be the same.]

13.36 The following data show the number of hours studied for an exam, x, and the grade received on the exam, y (y is measured in 10's; that is, $y = 8$ means that the grade, rounded to the nearest 10 points, is 80).

x	2	3	3	4	4	5	5	6	6	6	7	7	7	8	8
y	5	5	7	5	7	7	8	6	9	8	7	9	10	8	9

    a. Draw a scatter diagram of the data.
    b. Find the equation of the line of best fit and graph it on the scatter diagram.
    c. Find the ordinates $\hat{y}$ that correspond to $x = 2, 3, 4, 5, 6, 7$, and 8.
    d. Find the five values of e that are associated with the points where $x = 3$ and $x = 6$.
    e. Find the variance s_e^2 of all the points about the line of best fit.

13.4 Inferences Concerning the Slope of the Regression Line

Now that the equation of the line of best fit has been determined and the linear model has been verified (by inspection of the scatter diagram), we are ready to determine whether we can use the equation to predict y. We will test the null hypothesis "the equation of the line of best fit is of no value in predicting y given x." That is, the null hypothesis to be tested is "β_1 (the slope of the relationship in the population) is zero." If $\beta_1 = 0$, then the linear equation will be of no real use in predicting y.

Before we look at the confidence interval or the hypothesis test, let's discuss the **sampling distribution** of the slope. If random samples of size n are repeatedly taken from a bivariate population, the calculated slopes, the b_1's, would form a sampling distribution that is normally distributed with a mean of β_1, the population value of the slope, and with a variance of $\sigma_{b_1}^2$, where

$$\sigma_{b_1}^2 = \frac{\sigma_\epsilon^2}{\sum(x - \bar{x})^2} \tag{13.9}$$

> "Lack of fit" means a different model is needed.

provided there is no lack of fit. An appropriate estimator for $\sigma_{b_1}^2$ is obtained by replacing σ_ϵ^2 by s_e^2, the estimate of the variance of the error about the regression line:

$$s_{b_1}^2 = \frac{s_e^2}{\sum(x - \bar{x})^2} = \frac{s_e^2}{\text{SS}(x)} \tag{13.10}$$

> Recall we have previously found SS(x), formula (2.9).

This formula may be rewritten in the following, more manageable form:

$$s_{b_1}^2 = \frac{s_e^2}{\sum x^2 - \frac{(\sum x)^2}{n}} \tag{13.11}$$

NOTE The "**standard error** of _____" is the standard deviation of the sampling distribution of _____. Therefore, the standard error of regression (slope) is σ_{b_1} and is estimated by s_{b_1}.

In our illustration of travel times and distances, the variance among the b_1's is estimated by use of formula (13.11):

$$s_{b_1}^2 = \frac{s_e^2}{\sum x^2 - \frac{(\sum x)^2}{n}}: \quad s_{b_1}^2 = \frac{29.1723}{358.9333} = 0.081275 = \mathbf{0.0813}$$

ASSUMPTIONS FOR INFERENCES ABOUT LINEAR REGRESSION

The set of (x, y) ordered pairs forms a random sample and the y-values at each x have a normal distribution. Since the population standard deviation is unknown and replaced with the sample standard deviation, the t-distribution will be used with $n - 2$ degrees of freedom.

Confidence Interval Procedure

The slope β_1 of the regression line of the population can be estimated by means of a confidence interval. The confidence interval is determined by

$$b_1 \pm t_{(n-2,\alpha/2)} \cdot s_{b_1} \tag{13.12}$$

ILLUSTRATION 13.5 ▼

Find the 95% confidence interval for the population's slope, β_1, for Illustration 13.4.

Solution

STEP 1 **The Set-Up:**
Describe the population parameter of interest.
The slope, β_1, for the line of best fit for the population.

STEP 2 **The Confidence Interval Criteria:**
a. **Check the assumptions.**
The ordered pairs form a random sample and we will assume that the y-values (minutes) at each x (miles) have a mounded distribution.
b. **Identify the probability distribution and the formula to be used.**
The Student's t-distribution will be used with formula (13.12).
c. **State the level of confidence, $1 - \alpha$.**
$1 - \alpha = 0.95$

STEP 3 **The Sample Evidence:**
Collect the sample information.
$n = 15$, $b_1 = 1.89$ and $s_{b_1}^2 = 0.0813$

STEP 4 **The Confidence Interval:**
a. **Determine the confidence coefficients.**
From Table 6, we find $t_{(df, \alpha/2)} = t_{(13, 0.025)} = 2.16$.
b. **Find the maximum error of estimate.**
Using formula 13.12:
$$E = t_{(n-2, \alpha/2)} \cdot s_{b_1}: \quad E = 2.16 \cdot \sqrt{0.0813} = 0.6159$$
c. **Find the lower and upper confidence limits.**

$$b_1 - E \quad \text{to} \quad b_1 + E$$
$$1.89 - 0.62 \quad \text{to} \quad 1.89 + 0.62$$
1.27 to 2.51, the 0.95 confidence interval for β_1

> Since confidence interval does not include zero, we may conclude there is a linear relationship between x and y.

STEP 5 **The Results:**
State the confidence interval.

(continued)

Thus, we can say that the slope of the line of best fit of the population from which the sample was drawn is between 1.27 and 2.51 with 95% confidence.   ▲

Hypothesis-Testing Procedure

We are now ready to test the hypothesis $\beta_1 = 0$. That is, we want to determine whether the equation for the line of best fit is of any real value in predicting y. For this hypothesis test, the null hypothesis is always $H_o: \beta_1 = 0$. It will be tested using the Student's t-distribution with df $= n - 2$ degrees of freedom and the test statistic $t \star$ found using formula 13.13.

$$t \star = \frac{b_1 - \beta_1}{s_{b_1}}$$ (13.13)

Case Study 13.2

Reexamining the Use of Seriousness Weights in an Index of Crime

Regression of the Arizona UCR index on the average seriousness index produces the linear relationship depicted in figure. Also shown is the ninety-five percent confidence interval (3.001, 3.262), which is based upon a standard error of .065 on the estimate of the slope. The regression equation for this relationship is

$$S_t = -3953.85 + 3.13A_t.$$

Exercise 13.37

a. The vertical scale on the figure in Case Study 13.2 is drawn at $A_t = 12,600$, and the line of best fit appears to intersect the vertical scale at approximately 35,500. Verify the coordinates of this point of intersection.

b. The article also gives an interval estimate of (3.001, 3.262). Verify this 95% interval using the information given in the article.

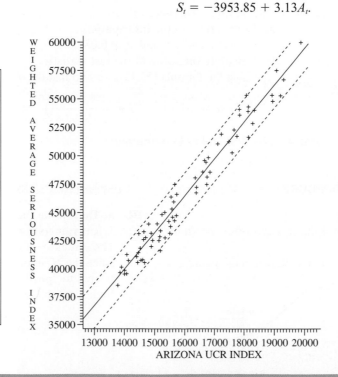

Source: Reprinted with permission from the *Journal of Criminal Justice,* Volume 17, Thomas Epperlein and Barbara C. Nienstedt, "Reexamining the Use of Seriousness Weights in an Index of Crime," 1989, Pergamon Press, Inc.

ILLUSTRATION 13.6 ▼

Is the slope for the line of best fit significant enough to show that one-way distance is useful in predicting one-way travel time in Illustration 13.4? Use $\alpha = 0.05$.

Solution

STEP 1 The Set-Up:
   a. Describe the population parameter of interest.
   The parameter of interest is β_1, the slope of the line of best fit for the population.
   b. State the null hypothesis (H_o) and the alternative hypothesis (H_a).

   H_o: $\beta_1 = 0$ (this implies that x is of no use in predicting y: that is, that $\hat{y} = \bar{y}$ would be as effective.)

The alternative hypothesis can be either one-tailed or two-tailed. If we suspect that the slope is positive, as in Illustration 13.4, a one-tailed test is appropriate.

   H_a: $\beta_1 > 0$ (we would expect travel time y to increase as the distance x increased)

STEP 2 The Hypothesis Test Criteria:
   a. Check the assumptions.
   The ordered pairs form a random sample and we will assume that the y-values (minutes) at each x (miles) have a mounded distribution.
   b. Identify the probability distribution and the test statistic to be used.
   The t-distribution with df $= n - 2 = 13$, and the test statistic is $t \bigstar$, formula (13.13).
   c. Determine the level of significance, α.
   $\alpha = 0.05$.

STEP 3 The Sample Evidence:
   a. Collect the sample information.
   $n = 15$, $b_1 = 1.89$ and $s_{b_1}^2 = 0.0813$
   b. Calculate the value of the test statistic.
   Using the formula (13.13), we find that the observed value of t:

$$t \bigstar = \frac{b_1 - \beta_1}{s_{b_1}}: \quad t \bigstar = \frac{1.89 - 0.0}{\sqrt{0.0813}} = 6.629 = \mathbf{6.63}$$

STEP 4 The Probability Distribution:

USING THE p-VALUE PROCEDURE:

a. Calculate the p-value.
Use the right-hand tail since the H_a expresses concern for values related to "positive."
$\mathbf{P} = P(t > 6.63$, with df $= 13)$ as shown in the figure.

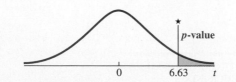

OR

USING THE CLASSICAL PROCEDURE:

a. Determine the critical region and critical value(s).
The critical region is the right-hand tail since the H_a expresses concern for values related to "positive." The critical value is found in Table 6:

$$t(13, 0.05) = \mathbf{1.77.}$$

USING THE *p*-VALUE PROCEDURE:

To find the *p*-value, use one of three methods:
1. Usc Tablc 6 (Appendix B) to place bounds on the
   p-value: **P < 0.005**
2. Use Table 7 (Appendix B) to place bounds on the
   p-value: **P < 0.001**
3. Use a computer or calculator to calculate the *p*-value:
   P = 0.0000076

Specific details are on page 420, 421.

b. Determine whether or not the *p*-value is smaller than α.

The *p*-value **is smaller** than the level of significance, α.

USING THE CLASSICAL PROCEDURE:

OR

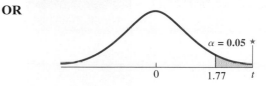

Specific instructions are on page 419.

b. Determine whether or not the calculated test statistic is in the critical region.

t ★ **is in** the critical region, as shown in **red** in the figure above.

STEP 5 **The Results:**
   a. State the decision about H_o.
      Reject H_o.
   b. State the conclusion about H_a.
      At the 0.05 level of significance, we conclude that the slope of the line of best fit in the population is greater than zero. The evidence indicates that there is a linear relationship and that the one-way distance (*x*) is useful in predicting the travel time to work (*y*).

The following commands will provide the equation for the regression line, information for a t-test concerning the slope of the regression line, the standard deviation of error, r and/or r², and a scatter diagram showing the regression line.

MINITAB

```
MINITAB output also includes the predicted y values for
given x values and residuals.

Input the x-variable data into C1 and the corresponding
y-variable data into C2; then continue with:

  Choose:  Stat > Regression > Regression. . .
  Enter:   Response (y): C2
           Predictors (x): C1
  Choose:  Storage
  Select:  Residuals
           Fits    > OK
  Choose:  Graph > Plot
  Enter:   Graph 1: Y: C2 X: C1
  Choose:  Annotation > Title
  Enter:   your title    > OK
  Choose:  Annotation > Line
  Enter:   Points: C1 C3 (whichever column Fits is located)
           Type: Solid
```

EXCEL

Excel output also includes predicted *y* values for given *x* values, residuals, and a $1 - \alpha$ confidence interval for the slope.

Input the *x*-variable data into column A and the corresponding *y*-variable data into column B; then continue with:

Choose: **Tools > Data Analysis > Regression > OK**

Enter: Input *y* Range: **(B1:B10 or select cells)**

Input *x* Range: **(A1:A10 or select cells)**

Select: **Labels** (if necessary)
Confidence Level:

Enter: **95%** (desired level)

Select: **Output Range:**

Enter: **(C1 or select cell)**

Select: **Line Fit Plots**

To make the output more readable continue with: Format > Column > Autofit Selection.

Additional commands to adjust the window can be found on page 139.

TI-83

Input the *x*-variable data into L1 and the corresponding *y*-variable data into L2; then continue with the following, entering the appropriate values and highlighting Calculate:

Choose: **STAT > TESTS > E:LinRegTTest**

To enter Y1, use: VARS > YVARS
> 1:Function. . . > 1:Y1

Enter the following to obtain a scatter diagram with regression line:

Choose: **2nd > STATPLOT > 1: Plot1**

Choose: **ZOOM > 9:ZoomStat > Trace**

Computer Solution: MINITAB Printout for parts of Illustration 13.4

Regression Analysis

The regression equation is
y, minute $= 3.64 + 1.89$ x, miles

Equation of line of best fit,
$\hat{y} = 3.64 + 1.89x$; see p. 626.

Calculated values of b_0
and b_1

Calculated value of s_{b_1},
$s_{b_1} = 0.285$; compare to
$s_{b_1}^2 = 0.0813$ see p. 629
($\sqrt{0.0813} = 0.285$)

Predictor	Coef	StDev	T	P
Constant	3.643	3.765	0.97	0.351
x, miles	1.8932	0.2851	6.64	0.000

$s = 5.401$ R $-$ Sq $= 77.2\%$ R $-$ Sq (adj) $= 75.5\%$

Calculated $t \bigstar$ and p-value
for H_o: $\beta_1 = 0$ as found in
steps 3 and 4 on p. 632–633

Calculated value of s_e,
$s_e = 5.4011$: compare to
$s_b^2 = 29.1723$ as found on p. 627
($\sqrt{29.1723} = 5.4011$)

Given data

Values of $\hat{y}$ for each
given x-value using
$\hat{y} = 3.6434 + 1.8932x$

Obs	x, miles	y, minute	Fit	Residual
1	3.0	7.00	9.32	−2.32
2	5.0	20.00	13.11	6.89
3	7.0	20.00	16.90	3.10
4	8.0	15.00	18.79	−3.79
5	10.0	25.00	22.58	2.42
6	11.0	17.00	24.47	−7.47
7	12.0	20.00	26.36	−6.36
8	12.0	35.00	26.36	8.64
9	13.0	26.00	28.26	−2.26
10	15.0	25.00	32.04	−7.04
11	15.0	35.00	32.04	2.96
12	16.0	32.00	33.93	−1.93
13	18.0	44.00	37.72	6.28
14	19.0	37.00	39.61	−2.61
15	20.0	45.00	41.51	3.49

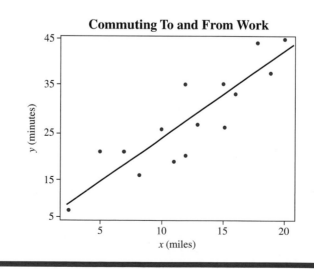

Commuting To and From Work

13.38 State the null hypothesis, H_o, and the alternative hypothesis, H_a, that would be used to test the following statements.
a. The slope for the line of best fit is positive.
b. There is no regression.
c. There is evidence of negative regression.

13.39 Determine the p-value for each of the following situations.
a. $H_a: \beta_1 > 0$, with $n = 18$, $t \star = 2.4$
b. $H_a: \beta_1 \neq 0$, with $n = 15$, $b_1 = 0.16$, $s_{b_1} = 0.08$
c. $H_a: \beta_1 < 0$, with $n = 24$, $b_1 = -1.29$, $s_{b_1} = 0.82$

13.40 Determine the critical value(s) and region(s) that would be used in testing each of the following null hypotheses using the classical approach.
a. $H_o: \beta_1 = 0$ vs. $H_a: \beta_1 \neq 0$, with $n = 18$, $\alpha = 0.05$
b. $H_o: \beta_1 = 0$ vs. $H_a: \beta_1 > 0$, with $n = 28$, $\alpha = 0.01$
c. $H_o: \beta_1 = 0$ vs. $H_a: \beta_1 < 0$, with $n = 16$, $\alpha = 0.05$

13.41 Calculate the estimated standard error of regression, s_{b_1}, for the computer-science aptitude score–achievement score relationship in Exercise 13.34.

13.42 Calculate the estimated standard error of regression, s_{b_1}, for the number of hours studied–exam grade relationship in Exercise 13.36.

13.43 An article titled "Statistical Approach for the Estimation of Strontium Distribution Coefficient" (November 1993 issue of *Environmental Science & Technology*) reports a linear correlation coefficient of 0.55 between the strontium distribution coefficient (mL/g) and the total aluminum (mmol/100 g-soil) for soils collected from the surface throughout Japan. Consider the following data for ten such samples.

Soil Sample	Strontium Distribution Coefficient	Total Aluminum
1	100	200
2	120	225
3	300	325
4	250	310
5	400	350
6	500	400
7	450	375
8	445	385
9	310	350
10	200	290

Let Y represent the strontium distribution coefficient and X represent the total aluminum.
a. Find the equation of the line of best fit.
b. Find a 95% confidence interval for β_1.
c. Explain the meaning of the interval in (b).

13.44 The relationship between the diameter of a spot weld, x, and the shear strength of the weld, y, is very useful. The diameter of the spot weld can be measured after the weld is completed. The shear strength of the weld can be measured only by applying force to the weld until it breaks. Thus it would be very useful to be able to predict the shear strength based only on the diameter. The following data were obtained from several sample welds.

x, Dia. of Weld (.001 in.)	190	215	200	230	209	250	215	265	215	250
y, Shear Strength (lb)	680	1025	800	1100	780	1030	885	1175	975	1300

(continued)

Complete these questions with the aid of a computer.
a. Draw a scatter diagram.
b. Find the equation for the line of best fit.
c. Is the value of b_1 significantly greater than zero at the 0.05 level?
d. Find the 95% confidence interval for β_1.

13.45 A sample of ten students were asked for the distance and the time required to commute to college yesterday. The data collected are shown in the following table.

Distance	1	3	5	5	7	7	8	10	10	12
Time	5	10	15	20	15	25	20	25	35	35

a. Draw a scatter diagram of these data.
b. Find the equation that describes the regression line for these data.
c. Does the value of b_1 show sufficient strength to conclude that β_1 is greater than zero at the $\alpha = 0.05$ level?
d. Find the 98% confidence interval for the estimation of β_1. (Retain these answers for use in Exercise 13.49.)

13.46 A stockbroker once claimed that an investor should always refrain from buying expensive stocks. Why? "Lower-priced stocks will jump faster, percentage-wise, in a bull market than higher-priced stocks because they have more room to climb in the first place," he touted. "It's simple mathematics." But wait a minute. Isn't that a double-edged sword? Won't lower-priced stocks fall faster, percentage-wise, especially when the bear takes over? Let the data speak for themselves. The following table of 30 stocks was assembled from *The Wall Street Journal*'s list of the biggest price percentage gainers and losers following trading on August 17, 1998. Closing prices have been converted from fractions to decimals:

Price Percentage Gainers		Price Percentage Losers	
Closing Price	**Percent Change**	**Closing Price**	**Percent Change**
7.437	17.8	5.187	-40.7
12.312	16.6	18.750	-30.6
4.000	14.3	1.562	-21.9
24.562	14.2	9.437	-17.0
63.125	13.7	5.500	-12.9
32.875	13.4	6.250	-11.5
6.937	13.3	6.000	-11.1
15.500	12.7	5.625	-10.9
5.062	12.5	3.312	-10.2
2.937	11.9	8.125	-9.7
14.125	11.9	11.625	-9.7
31.500	11.8	3.562	-9.5
9.750	11.4	19.125	-9.5
24.750	11.2	9.750	-9.3
17.750	10.9	1.937	-8.8

Source: The Wall Street Journal, Vol. CII, No. 35, August 19. 1998.

a. Draw two scatter diagrams of the data, one for losers and the other for gainers. In each case, use percent change in price as the dependent variable (y) and closing price as the independent variable (x).
b. Find the two equations for the lines of best fit and graph them on their scatter diagrams.
c. Are the values of the coefficients for the slopes of the two regression lines not equal to zero? Use $\alpha = 0.05$ and draw your conclusions in each case.

13.47 The September 1994 issue of *Popular Mechanics* gives specifications and dimensions for various jet boats. The following table summarizes some of this information.

(continued)

Model	Base Price	Engine Horsepower
Baja Blast	8395	120
Bayliner Jazz	8495	90
Boston Whaler Rage 15	11,495	115
Dynasty Jet Storm	8495	90
Four Winds Fling	9568	115
Regal Rush	9995	90
Sea-Doo Speedster	11,499	160
Sea Ray Sea Rayder	8495	90
Seaswirl Squirt	8495	115
Suga Sand Mirage	8395	120

a. Find the equation for the line of best fit. Let X equal the horsepower, and let Y equal the base price.
b. Find the standard deviation along the line of best fit and the standard error of slope.
c. Describe the meaning of the two answers in (b).
d. Is horsepower an effective predictor of the base price? Explain your response using statistical evidence. (Use $\alpha = 0.05$.)

13.48 Politicians have often debated over how to improve the quality of education in the United States. Some argue that teachers' salaries should be increased, whereas others claim that more teachers should be hired to reduce the number of pupils per teacher. The percentage of students graduating (graduation rate) from public high school is frequently used as an overall measure of the quality of education in a given area. The following table summarizes all three variables of interest measured in 1995 for each state and the District of Columbia.

State	Pupils/ teacher	Avg. Pay	Grad. Rate	State	Pupils/ teacher	Avg. Pay	Grad. Rate
AL	16.9	32,549	60.2	MT	16.4	29,950	85.6
AK	17.3	50,647	68.2	NE	14.5	31,768	84.3
AZ	19.6	33,350	63.2	NV	19.1	37,340	65.1
AR	17.1	29,975	73.1	NH	15.7	36,867	74.9
CA	24.0	43,474	64.0	NJ	13.8	49,349	83.5
CO	18.5	36,175	73.1	NM	17.0	29,715	64.0
CT	14.4	50,426	75.0	NY	15.5	49,560	61.8
DE	16.8	41,436	64.7	NC	16.2	31,225	65.5
DC	15.0	45,012	60.1	ND	15.9	27,711	86.8
FL	18.9	33,881	59.1	OH	17.1	38,831	74.6
GA	16.5	36,042	56.6	OK	15.7	29,270	75.3
HI	17.8	35,842	75.0	OR	19.8	40,900	68.9
ID	19.0	31,818	79.5	PA	17.0	47,429	77.3
IL	17.1	42,679	75.5	RI	14.3	43,019	72.6
IN	17.5	38,575	70.1	SC	16.2	32,659	55.1
IA	15.5	33,275	85.1	SD	15.0	26,764	86.6
KS	15.1	35,837	77.4	TN	16.7	33,789	63.8
KY	16.9	33,950	70.3	TX	15.6	32,644	59.7
LA	17.0	28,347	58.7	UT	23.8	31,750	79.1
ME	13.9	33,800	72.3	VT	13.8	37,200	89.4
MD	16.8	41,148	73.9	VA	14.4	35,837	71.9
MA	14.6	43,806	76.0	WA	20.4	37,860	73.4
MI	19.7	44,251	68.9	WV	14.6	33,159	75.4
MN	17.8	37,975	86.8	WI	15.8	38,950	81.7
MS	17.5	27,720	60.1	WY	14.8	31,721	78.2
MO	15.4	34,342	72.7				

Source: National Center for Education Statistics, U.S. Dept. of Education, National Education Association

(continued)

Find s_e: $s_e^2 - 29.17$ (found in Illustration 13.4)

$$s_e = \sqrt{29.17} = \textbf{5.40}$$

Find $y_{x\,=\,7}$: $\hat{y} = 3.64 + 1.89x = 3.64 + 1.89(7) = \textbf{16.87}$

STEP 4 **The Confidence Interval:**

  a. Determine the confidence coefficients.

   Confidence coefficient: $t_{(13,\,0.025)} = 2.16$ (from Table 6 in Appendix B)

  b. Find the maximum error of estimate.

   Using formula 13.15:

$$E = t_{(n-2,\,\alpha/2)} \cdot s_e \cdot \sqrt{\frac{1}{n} + \frac{(x_0 - \bar{x})^2}{SS(x)}}: \quad E = (2.16)(5.40)\sqrt{\frac{1}{15} + \frac{(7 - 12.27)^2}{358.933}}$$

$$E = (2.16)(5.40)\sqrt{0.06667 + 0.07738}$$

$$E = (2.16)(5.40)(0.38) = 4.43$$

  c. Find the lower and upper confidence limits.

$$\hat{y} - E \quad\quad \text{to} \quad\quad \hat{y} + E$$

$$16.87 - 4.43 \quad\quad \text{to} \quad\quad 16.87 + 4.43$$

12.44 to 21.30, 95% confidence interval for $\mu_{y|x\,=\,7}$

This confidence interval is shown in Figure 13.12 by the heavy red vertical line. The confidence interval belt showing the upper and lower boundaries of all intervals at 95% confidence is also shown in red. Notice that the boundary lines for x-values far away from $\bar{x}$ become close to the two lines that represent the equations having slopes equal to the extreme values of the 95% confidence interval for the slope (see Figure 13.12).

Figure 13.12

Confidence Belts for $\mu_{y|x_0}$

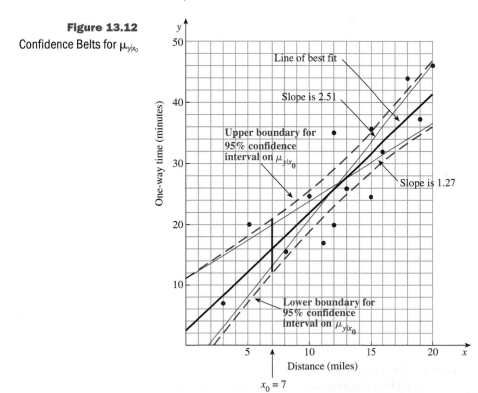

Often when making a prediction, we want to predict the value of an individual y. For example, you live seven miles from your place of business and you are interested in an estimate of how long it will take you to get to work. You are somewhat less interested in the average time for all of those who live seven miles away. The formula for the prediction interval of the value of a single randomly selected y is

$$\hat{y} \pm t_{(n-2, \alpha/2)} \cdot s_e \cdot \sqrt{1 + \frac{1}{n} + \frac{(x_0 - \bar{x})^2}{\text{SS}(x)}} \tag{13.16}$$

ILLUSTRATION 13.8 ▼

What is the 95% prediction interval for the time it will take you to commute to work if you live seven miles away?

Solution

STEP 1 **The Set-Up:**
Describe the population parameter of interest.
$y_{x=7}$, the travel time for one co-worker who travels seven miles.

STEP 2 **The Prediction Interval Criteria:**
    a. Check the assumptions.
        The ordered pairs form a random sample and we will assume that the y-values (minutes) at each x (miles) have a mounded distribution.
    b. Identify the probability distribution and the formula to be used.
        The Student's t-distribution will be used with formula (13.16).
    c. State the level of confidence, $1 - \alpha$.
        $1 - \alpha = 0.95$

STEP 3 **The Sample Evidence:**
Collect the sample information.
See Illustration 13.7: $s_e = 5.40$, $\hat{y}_{x=7} = 16.87$

STEP 4 **The Confidence Interval:**
    a. Determine the confidence coefficients.
        Confidence coefficient: $t_{(13, 0.025)} = 2.16$ (from Table 6 in Appendix B)
    b. Find the maximum error of estimate.

Using formula 13.16:

$$E = t_{(n-2, \alpha/2)} \cdot s_e \cdot \sqrt{1 + \frac{1}{n} + \frac{(x_0 - \bar{x})^2}{\text{SS}(x)}}: \quad E = (2.16)(5.40) \sqrt{1 + \frac{1}{15} + \frac{(7 - 12.27)^2}{358.933}}$$

$$E = (2.16)(5.40) \sqrt{1 + 0.06667 + 0.07738}$$

$$E = (2.16)(5.40) \sqrt{1.14405}$$

$$E = (2.16)(5.40)(1.0696) = 12.48$$

    c. Find lower and upper confidence limits.

$$\hat{y} - E \quad \text{to} \quad \hat{y} + E$$

$$16.87 - 12.48 \quad \text{to} \quad 16.87 + 12.48$$

4.39 to 29.35, 95% prediction interval for $y_{x=7}$

The prediction interval is shown in Figure 13.13 as the blue vertical line segment at $x_0 = 7$. Notice that it is much longer than the confidence interval for $\mu_{y|x=7}$. The dashed blue lines represent the prediction belts, the upper and lower boundaries of the prediction intervals for individual y-values for all given x-values.

Figure 13.13
Prediction Belts for y_{x_0}

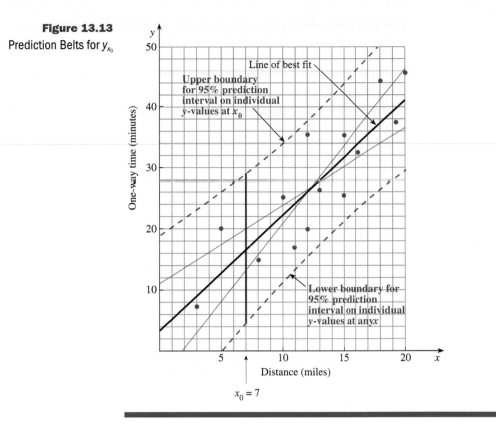

Figure 13.14
Confidence Belts for
the Mean Value of y
and Prediction Belts
for Individual y's.

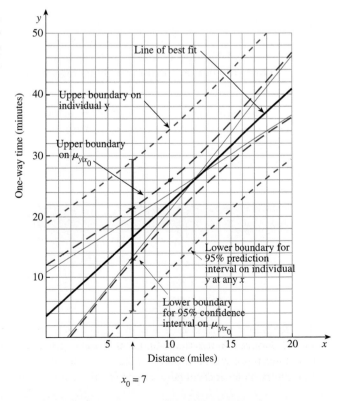

Can you justify the fact that the prediction interval for individual values of y is wider than the confidence interval for the mean values? Think about "individual values" and "mean values" and study Figure 13.14.

There are three basic precautions that you need to be aware of as you work with regression analysis:

1. Remember that the regression equation is meaningful only in the domain of the x variable studied. Estimation outside this domain is extremely dangerous; it requires that we know or assume that the relationship between x and y remains the same outside the domain of the sample data. For example, Joe says that he lives 75 miles from work, and he wants to know how long it will take him to commute. We certainly can use $x = 75$ in all the formulas, but we do not expect the answers to carry the confidence or validity of the values of x between 3 and 20, which were in the sample. The 75 miles may represent a distance to the heart of a nearby major city. Do you think the estimated times, which were based on local distances of 3 to 20 miles, would be good predictors in this situation? Also, at $x = 0$ the equation has no real meaning. However, although projections outside the interval may be somewhat dangerous, they may be the best predictors available.

2. Don't get caught by the common fallacy of applying the regression results inappropriately. For example, this fallacy would include applying the results of Illustration 13.4 to another company. But suppose that the second company had a city location, whereas the first company had a rural location or vice versa. Do you think the results for a rural location would also be valid for a city location? Basically, the results of one sample should not be used to make inferences about a population other than the one from which the sample was drawn.

3. Don't jump to the conclusion that the results of the regression prove that x causes y to change. (This is perhaps the most common fallacy.) Regressions only measure movement between x and y; they never prove causation . A judgment of causation can be made only when it is based on theory or knowledge of the relationship separate from the regression results. The most common difficulty in this regard occurs because of what is called the missing variable, or third-variable, effect. That is, we observe a relationship between x and y because a third variable, one that is not in the regression, affects both x and y.

The following commands will give all of the information that the MINITAB Regression command provided (as shown on p. 635) as well as $1 - \alpha$ confidence interval belts for estimating the mean value of y and prediction belts for estimating individual y values.

MINITAB

```
Input the x-variable data into C1 and the corresponding y-
variable data into C2; then continue with:

  Choose:  Stat > Regression > Fitted Line Plot
  Enter:   Response (y): C2
           Predictors (x): C1
  Select:  Type of Regression Model: Linear
```

(continued)

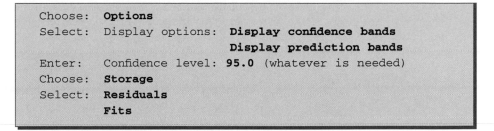

```
Choose:   Options
Select:   Display options:  Display confidence bands
                            Display prediction bands
Enter:    Confidence level: 95.0 (whatever is needed)
Choose:   Storage
Select:   Residuals
          Fits
```

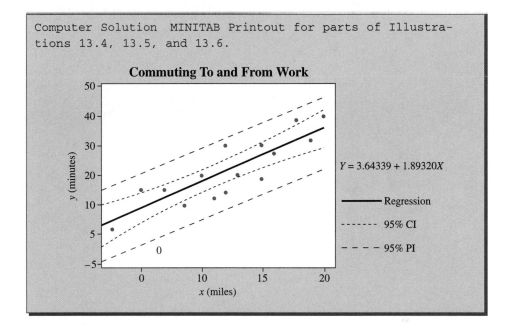

Computer Solution MINITAB Printout for parts of Illustrations 13.4, 13.5, and 13.6.

Commuting To and From Work

$Y = 3.64339 + 1.89320X$

——— Regression

------- 95% CI

- - - - 95% PI

x (miles)

y (minutes)

EXERCISES

13.49 Use the data and the answers found in Exercise 13.45 (p. 637) to make the following estimates.
  a. Give a point estimate for the mean time required to commute four miles.
  b. Give a 90% confidence interval for the mean travel time required to commute four miles.
  c. Give a 90% prediction interval for the travel time required for one person to commute the four miles.
  d. Answer (a), (b), and (c) for $x = 9$.

13.50 A study in *Physical Therapy* (April 1991) reports on seven different methods to determine crutch length plus two new techniques utilizing linear regression. One of the regression techniques uses the patient's reported height. One hundred and seven individuals were in the study. The mean of the self-reported heights was 68.84 in. The regression equation determined was $y = 0.68x + 4.8$, where y = crutch length and x = self-reported height. The MSE (s_e^2) was reported to be 0.50. In addition, the standard deviation of the self-reported heights was 7.35 in. Use this information to determine a 95% confidence interval estimate for the mean crutch length for individuals who say they are 70 in. tall.

13.51 An experiment was conducted to study the effect of a new drug in lowering the heart rate in adults. The data collected are shown in the following table.

x, Drug Dose in mg	0.50	0.75	1.00	1.25	1.50	1.75	2.00	2.25	2.50	2.75
y, Heart Rate Reduction	10	7	5	12	15	14	20	20	18	21

a. Find the 95% confidence interval for the mean heart-rate reduction for a dose of 2.00 mg.

b. Find the 95% prediction interval for the heart-rate reduction expected for an individual receiving a dose of 2.00 mg.

13.52 The relationship between the "strength" and "fineness" of cotton fibers was the subject of a study that produced the following data.

x, Strength	76	69	71	76	83	72	78	74	80	82	90	81	78	80	81	78
y, Fineness	4.4	4.6	4.6	4.1	4.0	4.1	4.9	4.8	4.2	4.4	3.8	4.1	3.8	4.2	3.8	4.2

a. Draw a scatter diagram.

b. Find the 99% confidence interval for the mean measurement of fineness for fibers with a strength of 80.

c. Find the 99% prediction interval for an individual measurement of fineness for fibers with a strength of 75.

13.53 An article titled "Ailing and Well Babies: 'Gap Is Striking'" appeared in the September 8, 1994, issue of the *Omaha World-Herald*. The article gave weighted median household income in 1989 and percent of households with up-to-date immunizations for children age 2 for Douglas County, which was divided into eight sections. The information is as follows:

Section	Median Household Income	% with Up-to-Date Immunization
East/Northeast	17,723	43
West/Northeast	27,005	51
North/Central	33,424	62
Northwest	43,337	66
East/Southeast	19,226	46
West/Southeast	29,775	59
South/Central	40,607	65
Southwest	45,496	62

Let Y represent the percent with up-to-date immunizations and X represent the median household income.

a. Find the equation of the line of best fit.

b. Find a 95% confidence interval on the mean percent with up-to-date immunizations for families with a median household income equal to $40,000.

c. Find a 95% prediction interval for the probability that a family will have up-to-date immunizations if their median household income equals $40,000.

d. Explain the meaning of the intervals found in (b) and (c).

13.54 People not only live longer today but they also are living independently longer; even if an individual becomes temporarily dependent at some age, they still may enjoy years of independent living during their remaining life. The May/June 1989 issue of *Public Health Reports* included an article titled "A Multistate Analysis of Active Life Expectancy." Two of the variables studied were people's age at which they became depen-

(continued)

dent and the number of independent years they had remaining. Suppose the data were as follows:

x, Age When Became Dependent	65	66	67	68	70	72	74	76	78	80	83	85
y, Independent Years Remaining	11.1	10.0	10.4	9.3	8.2	6.8	6.8	4.4	5.4	2.5	2.7	0.9

a. Draw a scatter diagram.

b. Calculate the equation for the line of best fit.

c. Draw the line of best fit on the scatter diagram.

d. For a person who becomes dependent at age 80, how many years of independent living can be expected to remain? Find the answer two different ways; use equation (b) and use the line on the scatter diagram (c).

e. Construct a 99% prediction interval for the number of years of independent living remaining for a person who becomes dependent at age 80.

f. Draw a vertical line segment on the scatter diagram representing the interval found in (e).

13.55 There are at least two ways that special television programs could be rated, and both are of interest to advertisers trying to promote their products—the estimated size of the audience and a rating based on the percentage of TV-owning households that tuned into the program. The table below lists the top 20 most popular televised programs as compiled by Nielsen Media Research that includes broadcasts through January, 1997:

Program	Rating (%)	Audience (1000)
M*A*S*H (last episode)	60.2	50,150
Dallas (Who Shot J.R.?)	53.3	41,470
Roots–Pt. 8	51.1	36,380
Super Bowl XVI	49.1	40,020
Super Bowl XVII	48.6	40,480
XVII Winter Olympics	48.5	45,690
Super Bowl XX	48.3	41,490
Gone with The Wind–Pt 1	47.7	33,960
Gone with The Wind–Pt 2	47.4	33,750
Super Bowl XII	47.2	34,410
Super Bowl XIII	47.1	35,090
Bob Hope Christmas Show	46.6	27,260
Super Bowl XVIII	46.4	38,800
Super Bowl XIX	46.4	39,390
Super Bowl XIV	46.3	35,330
Super Bowl XXX	46.0	44,150
ABC Theater (The Day After)	46.0	38,550
Roots–Pt. 6	45.9	32,680
The Fugitive	45.9	25,700
Super Bowl XXI	45.8	40,030

Source: World Almanac Book of Facts 1998, "All-Time Top Televison Programs."

a. Draw a scatter diagram of these data with audience size as the dependent variable *y*, and rating percentage as the predictor variable, *x*.

b. Calculate the regression equation and draw the regression line on the scatter diagram.

c. If the next Super Bowl obtains a 50.0% rating, what would you estimate the audience size to be? Make your estimate based on the equation, and then draw a line on the scatter diagram to illustrate it.

d. Construct a 95% prediction interval for the estimate you obtained in (c).

13.56 Explain why a 95% confidence interval for the mean value of y at a particular x is much narrower than a 95% prediction interval for an individual y-value at the same value of x.

13.57 When $x_0 = \bar{x}$, is the formula for the standard error of $\hat{y}_{x_0}$ what you might have expected it to be, $s \cdot \dfrac{1}{\sqrt{n}}$?

13.6 Understanding the Relationship Between Correlation and Regression

Now that we have taken a closer look at both correlation and regression analysis, it is necessary to decide when to use them. Do you see any duplication of work?

The primary use of the linear correlation coefficient is in answering the question "Are these two variables linearly related?" There are other words that may be used to ask this basic question. For example, "Is there a linear correlation between the annual consumption of alcoholic beverages and the salary paid to firemen?"

The linear correlation coefficient can be used to indicate the usefulness of x as a predictor of y in the case where the linear model is appropriate. The test concerning the slope of the regression line (H_o: $\beta_1 = 0$) also tests this same basic concept. Either one of the two is sufficient to determine the answer to this query.

The choice of mathematical model can be tested statistically (called a "lack of fit" test); however, these procedures are beyond the scope of this text. We do perform this test informally, or subjectively, when we view the scatter diagram and use the presence of a linear pattern as our reason for using the linear model.

The concepts of linear correlation and regression are quite different, because each measures different characteristics. It is possible to have data that yield a strong linear correlation coefficient and have the wrong model. For example, the straight line can be used to approximate almost any curved line if the interval of domain is restricted sufficiently. In such a case the linear correlation coefficient can become quite high, but the curve will still not be a straight line. Figure 13.15 suggests one such interval where r could be significant, but the scatter diagram does not suggest a straight line.

Regression analysis should be used to answer questions about the relationship between two variables. Such questions as "What is the relationship?" "How are two variables related?" and so on, require this regression analysis.

Figure 13.15

The Value of r Is High but the Relationship Is Not Linear

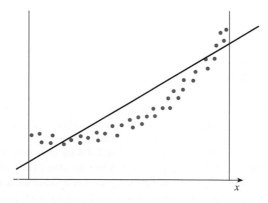

Case Study 13.3

Hotel Chains Cutting Staff to Boost Profits

Exercise 13.58

Refer to "Key to success."

a. Construct a scatter diagram using year as input and number of employees as output variables.

b. Find the equation for the line of best fit.

c. Based on the information you have, predict the number of employees per 100 occupants for the year 2000; year 2025; year 2050; year 2075. Are these predictions meaningful? Explain.

When a variable is observed periodically over time, the resulting bivariate data form a "time series." Although time series and forecasting are statistical topics not included in this textbook, time series data uses the line of best fit for the trend line. Can you "predict" the number of employees per 100 occupants for 1997 in the following example?

If you think you're seeing fewer workers in hotels today than in years ago, it's not your imagination. Many hotels are relying on fewer employees now than ever before.

A 10% reduction in a hotel's staff results in an annual cost savings of about 6%, says Bjorn Hanson, chairman of Coopers & Lybrand's lodging group. These savings, along with record occupancy levels and higher room rates, have contributed to 1997's estimated record profit of $14.5 billion.

Source: USA TODAY, 12-29-97

Key to success

Hotel-worker productivity is improving. Employees per 100 occupied rooms:

Year	70	75	80	Value
1986				80.9
1987				81.4
1988				82.1
1989				81.0
1990				81.3
1991				80.4
1992				77.8
1993				76.5
1994				75.3
1995				74.6
1996				73.5

Source: Coopers & Lybrand   USA TODAY

Return to CHAPTER CASE STUDY

The Chapter Case Study (p. 607) contains the Metropolitan Life Insurance Company weight chart for women using three different "frame" categories and using height as the input variable, *x*.

EXERCISE 13.59

a. Construct a scatter diagram showing the intervals for small-frame women.

b. Construct a scatter diagram showing the intervals for medium-frame women.

c. Construct a scatter diagram showing the intervals for large-frame women.

d. Does it appear that the variables "height" and "weight" have a linear relationship? Why do the scatter diagrams seem to suggest such a strong relationship?

e. One inch of height adds how many pounds to the ideal weight? How is the number of pounds per inch related to the line of best fit?

IN RETROSPECT

In this chapter we have made a more thorough inspection of the linear relationship between two variables. Although the curvilinear and multiple regression situations were only mentioned in passing, the basic techniques and concepts have been explored. We would only have to modify our mathematical model and our formulas if we wanted to deal with these other relationships.

Although it was not directly emphasized, we have applied many of the topics of earlier chapters in this chapter. The ideas of confidence interval and hypothesis testing were applied to the regression problem. Reference was made to the sampling distribution of the sample slope b_1. This allowed us to make inferences about β_1, the slope of the population from which the sample was drawn. We estimated the mean value of y at a fixed value of x by pooling the variance for the slope with the variance of the y's. This was allowable since they are independent. Recall that in Chapter 10 we presented formulas for combining the variances of independent samples. The idea here is much the same. Finally, we added a measure of variance for individual values of y and made estimates for these individual values of y at fixed values of x.

Case Study 13.2 presents the results of regression analysis on data collected to compare two crime-reporting indices. (Take another look at Case Study 13.2, p. 631.) The scatter diagram very convincingly shows that the two crime indices being compared are related to each other in a very strong and predictable pattern. Thus, as stated, "the weighted index contributed no further information to national information coming from them is basically the same. Thus, the introduction of the weighted index seems unnecessary since the Uniform Crime Reports index is a recognized standard.

As this chapter ends, you should be aware of the basic concepts of regression analysis and correlation analysis. You should now be able to collect the data for, and do a complete analysis on, any two-variable linear relationship.

CHAPTER EXERCISES

13.60 Answer the following as "sometimes," "always," or "never." Explain each "never" and "sometimes" response.
   a. The correlation coefficient has the same sign as the slope of the least squares line fitted to the same data.
   b. A correlation coefficient of 0.99 indicates a strong causal relationship between the variables under consideration.
   c. An r-value greater than zero indicates that ordered pairs with high x-values will have low y-values.
   d. The two coefficients for the line of best fit have the same sign.
   e. If x and y are independent, then the population correlation coefficient equals zero.

13.61 A study in the *Journal of Range Management* (Sept. 1990) examines the relationships between elements in Russian wild rye. The correlation coefficient between magnesium and calcium was reported to be 0.69 for a sample of size 45. Is there a significant correlation between magnesium and calcium in Russian wild rye (i.e., is $\rho = 0$)?

13.62 A study concerning the plasma concentration of the drug Ranitidine was reported in the *Journal of Pharmaceutical Sciences* (Dec. 1989). The drug was administered (coded I), and the plasma concentration of Ranitidine was followed for twelve hours. The time to the first peak in concentration was called T_{max1}. The same experiment was repeated one week later (coded II). Twelve subjects participated in the study. The correlation coefficient between T_{max1}, I and T_{max1}, II was reported to be 0.818. Use Table 11 in Appendix B to determine bounds on the p-value for the hypothesis test of H_o: $\rho = 0$ versus H_a: $\rho \neq 0$.

13.63 Shopping for a new personal computer may be one of the most difficult tasks facing consumers today. Paying more for a product usually means higher quality. But is that the situation with PCs? The staff of *PC World* published the results of their study of the top 20 power desktops and top 20 budget desktops in late 1998. The street prices of each machine were listed along with their overall rating and are reproduced in the table below:

(continued)

Power Desktops		Budget Desktops	
(1) Street Price	**(2) Overall Rating**	**(3) Street Price**	**(4) Overall Rating**
$2579	87	$1999	82
2599	86	1499	81
2704	85	1199	81
2299	85	1599	80
2250	85	1748	80
2199	83	1599	80
2783	83	1797	80
2704	83	1849	80
2395	83	1599	79
2675	82	1649	79
2499	82	1299	79
2750	80	999	79
2798	80	1699	78
2499	80	1649	78
2823	80	1397	78
2168	79	1925	77
2609	78	1499	77
2297	77	1879	77
2890	77	1749	76
2099	77	1100	76

Source: PC World, "Top 100," November 1998.

a. Calculate the linear correlation coefficient (Pearson's product moment, r) between street price and overall rating for (1) power desktops and (2) budget desktops.

b. What conclusions might you draw from your answers in (a)?

13.64 About 10,750 athletes competed in the 26th Olympiad held at Atlanta, Georgia, for medals in 271 events. Athletes from 197 nations and territories participated. The table below shows the distribution of gold, silver, and bronze medals awarded to athletes representing the 16 nations that won the most:

Nation	**Gold**	**Silver**	**Bronze**
United States	44	32	25
Germany	20	18	27
Russia	26	21	16
China	16	22	12
Australia	9	9	23
France	15	7	15
Italy	13	10	12
South Korea	7	15	5
Cuba	9	8	8
Ukraine	9	2	12
Canada	3	11	8
Hungary	7	4	10
Romania	4	7	9
Netherlands	4	5	10
Poland	7	5	5
Spain	5	6	6

Source: World Almanac and Book of Facts 1998, "Olympics."   *(continued)*

Calculate the correlation coefficient and use it and Table 10 of Appendix B to determine a 95% confidence interval on ρ for each of the following cases:

a. Gold and Silver   **b.** Gold and Bronze   **c.** Silver and Bronze

13.65 The use of electrical stimulation (ES) to increase muscular strength is discussed in the *Journal of Orthopedic and Sports Physical Therapy* (Sept. 1990). Seventeen healthy volunteers were used in the experiment. Muscular strength, Y, was measured as a torque in foot-pounds, and electrical stimulation, X, was measured in mA (micro-amps). The equation for the line of best fit is given as $Y = 1.8X + 28.7$, and the Pearson correlation coefficient as 0.61.

a. Was the correlation coefficient significantly different from zero? Use $\alpha = 0.05$.

b. Predict the torque for a current equal to 50 mA.

13.66 Innovative companies continue their search for new products to market. One measure of a company's attempts in pioneering new products is the number of U.S. patents it receives. But are the patented products successful in generating earnings for the company after they are launched? The following table lists 13 corporations receiving patents in 1996, together with the 12-month total return to shareholders of their stock as of October, 1998:

Company	No. of Patents	12-mo. Total Return %
IBM	1867	22.2
Canon K.K.	1541	−28.5
Motorola	1064	−38.9
Hitachi	963	−50.2
Sony	855	−26.5
Matsushita	841	−24.3
General Electric	819	18.7
Eastman Kodak	768	21.5
Xerox	703	2.2
Texas Instruments	600	−20.4
MMM	537	−18.2
AT&T	510	35.1
Hewlett-Packard	501	−23.2

Sources: Technology Assessment and Forecast Report, U.S. Patent and Trademark Office, 1997 and *Stock Guide,* Standard & Poor's, October 1998.

a. Calculate the correlation coefficient between the two variables.

b. Test for a significant correlation at the 0.05 level of significance and draw your conclusion.

13.67 An article in *Geology* (Sept. 1989) gives the following equation relating pressure, P, and total aluminum content, AL, for 12 Hornblende rims: $P = -3.46(+0.24) + 4.23(+0.13)AL$. The quantities shown in parentheses are standard errors for the y-intercept and slope estimates. Find a 95% confidence interval for the slope, β_1.

13.68 The tobacco settlement negotiated by an eight-man team of attorneys general on behalf of 41 states resulted in $206 billion dollars to be paid by the tobacco industry to recoup Medicaid costs the states incurred while treating sick smokers. Payments are to be made in annual increments over a 25-year span, starting in 1998. The table below shows the population (in millions) and the amounts (in $billions) awarded to 46 states, the District of Columbia, and Puerto Rico:

State	Settlement	Population	State	Settlement	Population
AL	3.17	4.27	NE	1.17	1.65
AK	0.67	0.61	NV	1.19	1.60
AZ	2.89	4.43	NH	1.30	1.16

(continued)

State	Settlement	Population	State	Settlement	Population
AR	1.62	2.51	NJ	7.58	7.99
CA	25.00	31.88	NM	1.17	1.71
CO	2.69	3.82	NY	25.00	18.18
CT	3.64	3.27	NC	4.57	7.32
DE	7.75	0.72	ND	0.72	0.64
DC	1.19	0.54	OH	9.87	11.17
GA	4.81	7.35	OK	2.03	3.30
HI	1.18	1.18	OR	2.25	3.20
ID	0.71	1.19	PA	11.30	12.06
IL	9.12	11.85	PR	2.20	3.78
IN	4.00	5.84	RI	1.41	0.99
IA	1.70	2.85	SC	2.30	3.70
KS	1.63	2.57	SD	0.68	0.73
KY	3.45	3.88	TN	4.78	5.32
LA	4.42	4.35	UT	0.87	2.00
ME	1.51	1.24	VT	0.81	0.59
MD	4.43	5.07	VA	4.01	6.68
MA	7.91	6.09	WA	4.02	5.53
MI	8.53	9.59	WV	1.74	1.83
MO	4.46	5.36	WI	4.06	5.16
MT	0.83	0.88	WY	0.49	0.48

Sources: Washington State Attorney General Office and Bureau of the Census, U.S. Dept. of Commerce.

a. Draw a scatter diagram of these data with tobacco settlement as the dependent variable y, and population as the predictor variable, x.

b. Calculate the regression equation and draw the regression line on the scatter diagram.

c. If your state's population were equal to 11.5 million people, of all 48 observations shown in the table, what would you estimate the tobacco settlement to be? Make your estimate based on the equation, and then draw a line on the scatter diagram to illustrate it.

d. Construct a 95% prediction interval for the estimate you obtained in (c).

13.69 The following data resulted from an experiment performed for the purpose of regression analysis. The input variable, x, was set at five different levels, and observations were made at each level.

x	0.5	1.0	2.0	3.0	4.0
y	3.8	3.2	2.9	2.4	2.3
	3.5	3.4	2.6	2.5	2.2
	3.8	3.3	2.7	2.7	2.3
		3.6	3.2	2.3	

a. Draw a scatter diagram.

b. Draw the regression line by eye.

c. Place a star, ★, at each level approximately where the mean of the observed y-values is located. Does your regression line look like the line of best fit for these five mean values?

d. Calculate the equation of the regression line.

e. Find the standard deviation of y about the regression line.

f. Construct a 95% confidence interval for the true value of β_1.

g. Construct a 95% confidence interval for the mean value of y at $x = 3.0$. At $x = 3.5$.

h. Construct a 95% prediction interval for an individual value of y at $x = 3.0$. At $x = 3.5$.

13.70 The following set of 25 scores was randomly selected from a teacher's class list. Let x be the prefinal average and y the final examination score. (The final examination had a maximum of 75 points.)

Student	x	y	Student	x	y
1	75	64	14	73	62
2	86	65	15	78	66
3	68	57	16	71	62
4	83	59	17	86	71
5	57	63	18	71	55
6	66	61	19	96	72
7	55	48	20	96	75
8	84	67	21	59	49
9	61	59	22	81	71
10	68	56	23	58	58
11	64	52	24	90	67
12	76	63	25	92	75
13	71	66			

a. Draw a scatter diagram for these data.
b. Draw the regression line (by eye) and estimate its equation.
c. Estimate the value of the coefficient of linear correlation.
d. Calculate the equation of the line of best fit.
e. Draw the line of best fit on your graph. How does it compare with your estimate?
f. Calculate the linear correlation coefficient. How does it compare with your estimate?
g. Test the significance of r at $\alpha = 0.10$.
h. Find the 95% confidence interval for the true value of ρ.
i. Find the standard deviation of the y-values about the regression line.
j. Calculate a 95% confidence interval for the true value of the slope β_1.
k. Test the significance of the slope at $\alpha = 0.05$.
l. Estimate the mean final-exam grade that all students with an 85 prefinal average will obtain (95% confidence interval).
m. Using the 95% prediction interval, predict the grade that John Henry will receive on his final, knowing that his prefinal average is 78.

13.71 Twenty-one mature flowers of a particular species were dissected, and the number of stamens and carpels present in each flower were counted. See the following table.

x, Stamens	y, Carpels	x, Stamens	y, Carpels	x, Stamens	y, Carpels
52	20	65	30	45	27
68	31	43	19	72	21
70	28	37	25	59	35
38	20	36	22	60	27
61	19	74	29	73	33
51	29	38	28	76	35
56	30	35	25	68	34

a. Is there sufficient evidence to claim a linear relationship between these two variables at $\alpha = 0.05$?
b. What is the relationship between the number of stamens and carpels in this variety of flower?
c. Is the slope of the regression line significant at $\alpha = 0.05$?
d. Give the 95% prediction interval for the number of carpels that one would expect to find in a mature flower of this variety if the number of stamens were 64.

13.72 It is believed that the amount of nitrogen fertilizer used per acre has a direct effect on the amount of wheat produced. The following data show the amount of nitrogen fertilizer used per test plot and the amount of wheat harvested per test plot.

x, Pounds of Fertilizer	y, 100 Pounds of Wheat	x, Pounds of Fertilizer	y, 100 Pounds of Wheat
30	5	70	19
30	9	70	23
30	14	70	31
40	6	80	24
40	14	80	32
40	18	80	35
50	12	90	27
50	14	90	32
50	23	90	38
60	18	100	34
60	24	100	35
60	28	100	39

a. Is there sufficient reason to conclude that the use of more fertilizer results in a higher yield? Use $\alpha = 0.05$.

b. Estimate, with a 98% confidence interval, the mean yield that could be expected if 50 pounds of fertilizer were used per plot.

c. Estimate, with a 98% confidence interval, the mean yield that could be expected if 75 pounds of fertilizer were used per plot.

13.73 The correlation coefficient, r, is related to the slope for the line of best fit b_1, by the equation

$$r = b_1 \sqrt{\frac{SS(x)}{SS(y)}}$$

Verify this equation using the following data.

x	1	2	3	4	6
y	4	6	7	9	12

13.74 The following equation is known to be true for any set of data:

$$\sum(y - \bar{y})^2 = \sum(y - \hat{y})^2 + \sum(\hat{y} - \bar{y})^2$$

Verify this equation with the following data.

x	0	1	2
y	1	3	2

VOCABULARY LIST

Be able to define each term. Pay special attention to the key terms, which are printed in **red.** In addition, describe in your own words, and give an example of, each term. Your examples should not be one given in class or in the textbook.

The bracketed numbers indicate the chapters in which the term previously appeared, but you should define the terms again to show increased understanding of their meaning. Page numbers indicate the first appearance of the term in Chapter 13.

bivariate data [3] (p. 608)
centroid (p. 609)

coefficient of linear correlation [3] (p. 611)
confidence belts (p. 614)

CHAPTER PRACTICE TEST

Part I: Knowing the Definitions

Answer "True" if the statement is always true. If the statement is not always true, replace the words shown in bold with words that make the statement always true.

13.1 The error **must be** normally distributed if inferences are to be made.

13.2 Both x and y **must be** normally distributed.

13.3 A high correlation between x and y **proves** that x causes y.

13.4 The value of the input variable **must be** randomly selected to achieve valid results.

13.5 The output variable must be **normally distributed** about the regression line for each value of x.

13.6 **Covariance** measures the strength of the linear relationship and is a standardized measure.

13.7 The **sum of squares for error** is the name given to the numerator portion of the formula used to calculate the variance of y about the line of regression.

13.8 **Correlation** analysis attempts to find the equation of the line of best fit for two variables.

13.9 There are **$n - 3$** degrees of freedom involved with the inferences about the regression line.

13.10 $\hat{y}$ serves as the **point estimate** for both $\mu_{y|x_0}$ and y_{x_0}.

Part II: Applying the Concepts

Answer all questions, showing formulas and work.

It is believed that the amount of nitrogen fertilizer used per acre has a direct effect on the amount of wheat produced. The data show the amount of nitrogen fertilizer used per test plot and the amount of wheat harvested per test plot. All test plots were of the same size.

Pounds of Fertilizer, x	100 Pounds of wheat, y
30	9
30	11
30	14
50	12
50	14
50	23
70	19
70	22
70	31
90	29
90	33
90	35

13.11   Draw a scatter diagram of the data (use graph paper and a straight edge). Be sure to label completely.

13.12   Complete an extensions table.

13.13   Calculate SS(x), SS(xy), SS(y).

13.14   Calculate the linear correlation coefficient, r.

13.15   Determine the 95% confidence interval estimate for the population linear correlation coefficient.

13.16   Calculate the equation for the line of best fit.

13.17   Draw the line of best fit on the scatter diagram (in red ink).

13.18   Calculate the standard deviation of the y-values about the line of best fit.

13.19   Does the value of b_1 show strength significant enough to conclude that the slope is greater than zero at the 0.05 level?

13.20   Determine the 0.95 confidence interval for the mean yield when 85 pounds of fertilizer is used per plot.

13.21   Draw a line on the scatter diagram representing the 95% confidence interval found in question 13.10 (in blue ink).

Part III: Understanding the Concepts

13.22   "There is a high correlation between how frequently skiers have their bindings tested and the incidence of lower-leg injuries, according to researchers at the Rochester Institute of Technology. To make sure your bindings release properly when you begin to fall, you should have them serviced by a ski mechanic every 15 to 30 ski days or at least at the start of each ski season." (University of California, Berkeley, "Wellness Letter," Feb. 1991) Explain what two variables are being discussed in this statement and interpret the "high correlation" mentioned.

13.23   Describe why the method used to define the correlation coefficient is referred to as "a product moment."

13.24   If you know the value of r is very close to zero, what value would you anticipate for b_1? Explain why.

13.25   Describe why the method used to find the line of best fit is referred to as "the method of least squares."

13.26   You wish to study the relationship between the amount of sugar contained in a child's breakfast and the child's hyperactivity in school during the four hours after breakfast. You ask 200 mothers of fifth-grade children to keep a careful record of what the child eats and drinks each morning. The parent's report is analyzed and the sugar consumption is determined. During the same time period, data on hyperactivity are collected at school. What statistic will measure the strength and kind of relationship that exists between the amount of sugar and the amount of hyperactivity? Explain why the statistic you selected is appropriate and what value you expect this statistic might have.

13.27   You are interested in studying the relationship between the length of time a person has been supported by welfare and self-esteem. You believe that the longer a person is supported, the lower the self-esteem. What data would you need to collect and what statistics would you calculate if you wish to predict a person's level of self-esteem after having been on welfare for a certain period of time? Explain in detail.

Chapter

14

Elements of Nonparametric Statistics

CHAPTER OUTLINE

14.1 Nonparametric Statistics
Distribution-free, or **nonparametric,** methods provide test statistics for an unspecified distribution.

14.2 Comparing Statistical Tests
When choosing between parametric and nonparametric tests, we are interested primarily in the **control of error,** the relative **power** of the test, and **efficiency.**

14.3 The Sign Test
A simple **count** of plus and minus signs tells us whether or not to reject the null hypothesis.

14.4 The Mann–Whitney *U* Test
The **rank number** for each piece of data is used to compare **two independent samples.**

14.5 The Runs Test
A **sequence of data** that possesses a common property, **a run,** is used to test the question of randomness.

14.6 Rank Correlation
The linear correlation coefficient's **nonparametric alternative,** rank correlation, uses only rankings to determine whether or not to reject the null hypothesis.

CHAPTER CASE STUDY

Men and Women in the Workplace

Men Are from Mars, Women Are from Venus is a well known book by John Gray, Ph.D., which describes many ways that men and women are different: how they perceive things and how they communicate differently, and so on. Men and women are different when they arrive in the workplace; also, they bring different abilities and skills to the job. The USA Snapshot® "Skills in demand" appeared in *USA Today* on July 7, 1998, and shows how top executives view the relationship between the skills their workers have and the skills needed "to maintain a competitive edge."

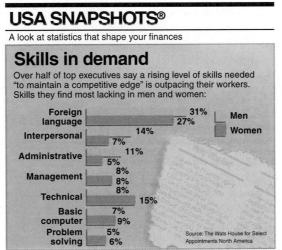

USA SNAPSHOTS®

A look at statistics that shape your finances

Skills in demand

Over half of top executives say a rising level of skills needed "to maintain a competitive edge" is outpacing their workers. Skills they find most lacking in men and women:

Skill	Men	Women
Foreign language	31%	27%
Interpersonal	14%	7%
Administrative	11%	5%
Management	8%	8%
Technical	8%	15%
Basic computer	7%	9%
Problem solving	5%	6%

Source: The Wats House for Select Appointments North America

By Anne R. Carey and Marcy E. Mullins, USA TODAY

At first glance it may appear that men and women are lacking the same skills, but notice that "technical" and "basic computer" are the second and third most lacking skills for women while "interpersonal" and "administrative" are the second and third most lacking skills for men. On closer inspection, maybe there is a difference, just as John Gray suggests in his book; women are better at communication and interpersonal skills, while men are better at mechanical skills and problem solving. Are these rankings for men and women significantly different? Is there a relationship between the men's rankings and the women's rankings? In this chapter, we will learn about statistical methods that can be used to answer these questions. We will return to these questions at the end of the chapter. (See Exercise 14.51, p. 700.)

CHAPTER OBJECTIVES

Many of the statistical tests studied in the previous chapters have required assumptions about the data; most typically, a normal distribution of the data. Sometimes data are not normally distributed; sometimes not even close. In situations like that, nonparametric tests become very important alternatives. Unlike their parametric counterparts, many of the best-known non-parametric tests, also known as distribution-free tests, are founded on a basis of elementary probability theory. The derivation of most of these tests is well within the grasp of the student who is competent in high school algebra and understands binomial probability. Thus the non-mathematical statistics user is much more at ease with nonparametric techniques.

This chapter is intended to give you a feeling for basic concepts involved in nonparametric techniques and to show you that nonparametric methods are extremely versatile and easy to use once a table of critical values is developed for a particular application. The selection of the nonparametric methods presented here is a cross section that includes only a few of the common tests and applications. You will learn about the sign test, the Mann–Whitney U test, the runs test, and Spearman's rank correlation test. These will be used to make inferences corresponding to both one- and two-sample situations.

14.1   Nonparametric Statistics

Most of the statistical procedures we have studied in this book are known as **parametric methods.** For a statistical procedure to be parametric, either we assume that the parent population is at least approximately normally distributed or we rely on the Central Limit Theorem to give us a normal approximation for the statistic. This is particularly true of the statistical methods studied in Chapters 8, 9, and 10.

The **nonparametric methods,** or **distribution-free methods,** as they are also known, do not depend on the distribution of the population being sampled. The nonparametric statistics are usually subject to much less confining restrictions than are their parametric counterparts. Some, for example, require only that the parent population be continuous.

The recent popularity of nonparametric statistics can be attributed to the following characteristics :

1. Nonparametric methods require few assumptions about the parent population.
2. Nonparametric methods are generally easier to apply than their parametric counterparts.
3. Nonparametric methods are relatively easy to understand.
4. Nonparametric methods can be used in situations where the normality assumptions cannot be made.
5. Nonparametric methods are generally only slightly less efficient then their parametric counterparts.

14.2   Comparing Statistical Tests

Only four nonparametric tests are presented in this chapter. They represent a very small sampling of the many different nonparametric tests that exist. Many of the nonparametric tests can be used in place of certain parametric tests. The question is, then, which statistical test do we use, the parametric or the nonparametric? Sometimes there is more than one nonparametric test to choose from.

The decision about which test to use must be based on the answer to the question "Which test will do the job best?" First, let's agree that when comparing two or more tests , they must be equally qualified for use. That is, each test has a set of assumptions that must be satisfied before it can be applied. From this starting point we will attempt to define "best" to mean the test that is best able to control the risks of error and at the same time keep the size of the sample to a number that is reasonable to work with. (Sample size means cost, cost to you or your employer.)

Let's look first at the ability to control the risk of error . The risk associated with a type I error is controlled directly by the level of significance α . Recall that P(type I error) $= \alpha$ and P(type II error) $= \beta$. Therefore, it is β that we must control. Statisticians like to talk about power (as do others), and the **power of a statistical test** is defined to be $1 - \beta$. Thus, the power of a test, $1 - \beta$, is the probability that we reject the null hypothesis when we should have rejected it. If two tests with the same α are equal candidates for use, the one with the greater power is the one you would want to choose.

The other factor is the sample size required to do a job. Suppose that you set the levels of risk you can tolerate, α and β , and then are able to determine the sample size it would take to meet your specified challenge. The test that required the smaller sample size would then seem to have the edge. Statisticians usually use the term efficiency to talk about this concept. **Efficiency** is the ratio of the sample size of the best parametric test to the sample size of the best nonparametric test when compared under a fixed set of risk values. For example, the efficiency rating for the sign test is approximately 0.63. This means that a sample of size 63 with a parametric test will do the same job as a sample of size 100 will do with the sign test.

The power and the efficiency of a test cannot be used alone to determine the choice of test. Sometimes you will be forced to use a certain test because of the data you are given. When there is a decision to be made, the final decision rests in a trade-off of three factors: (1) the power of the test, (2) the efficiency of the test, and (3) the data (and the number of data) available. Table 14.1 shows how the nonparametric tests discussed in this chapter compare with the parametric tests covered in previous chapters.

Table 14.1   Comparison of Parametric and Nonparametric Tests

Test Situation	Parametric Test	Nonparametric Test	Efficiency of Nonparametric Test
One mean	t-test (p. 424)	Sign test (p. 662)	0.63
Two independent means	t-test (p. 493)	U test (p. 674)	0.95
Two dependent means	t-test (p. 481)	Sign test (p. 665)	0.63
Correlation	Pearson's (p. 616)	Spearman test (p. 692)	0.91
Randomness		Runs test (p. 685)	Not meaningful; there is no parametric test for comparison

14.3   The Sign Test

The sign test is a versatile and an exceptionally easy-to-apply nonparametric method that uses only plus and minus signs. Three sign test applications are presented here: (1) a confidence interval for the median of one population, (2) a hypothesis test concerning the value of the median for one population, and (3) a hypothesis test concerning the median difference (paired difference) for two **dependent samples.** These sign tests are carried out using the same basic confidence interval and hypothesis test procedures as described in earlier chapters, and are the nonparametric alternatives to the *t*-tests used for one mean (Section 9.1) and the difference between two dependent means (Section 10.2).

> **ASSUMPTIONS FOR INFERENCES ABOUT THE POPULATION MEDIAN USING THE SIGN TEST**
>
> The *n* random observations forming the sample are selected independently and the population is continuous in the vicinity of the median, *M*.

Single-Sample Confidence Interval Procedure

The sign test can be applied to obtain a confidence interval for the unknown **population median, M.** To accomplish this we will need to arrange the sample data in ascending order (smallest to largest). The data are identified as x_1 (smallest), $x_2, x_3, \ldots, x_n$ (largest). The critical value, k, (known as the "maximum allowable number of signs") is obtained from Table 12 in Appendix B and tells us the number of positions to be dropped from each end of the ordered data. The remaining extreme values become the bounds of the $1 - \alpha$ confidence interval. That is, the lower boundary for the confidence interval is x_{k+1}, the $(k + 1)$th piece of data; the upper boundary is x_{n-k}, the $(n - k)$th piece of data. The following illustration will clarify this procedure.

ILLUSTRATION 14.1 ▼

Suppose that we have 12 pieces of data in ascending order $(x_1, x_2, x_3, \ldots, x_{12})$ and we wish to form a 95% confidence interval for the population median. Table 12 shows a critical value of 2 $(k = 2)$ for $n = 12$ and $\alpha = 0.05$ for a hypothesis test. This means that the last two values on each end $(x_1$ and x_2 on the left; x_{11} and x_{12} on the right) are dropped. The confidence interval is bounded by x_3 and x_{10} inclusively. That is, the 95% confidence interval is x_3 to x_{10}, and is expressed:

$$x_3 \text{ to } x_{10}, \text{ 95\% confidence interval for } M$$

In general, the two pieces of data that bound the confidence interval will occupy positions $k + 1$ and $n - k$, where k is the critical value read from Table 12. Thus

$$x_{k+1} \text{ to } x_{n-k}, \text{ } 1 - \alpha \text{ confidence interval for } M \qquad \blacktriangle$$

Single-Sample Hypothesis Test Procedure

The sign test can be used when the null hypothesis to be tested concerns the value of the population median *M*. The test may be either one- or two-tailed. This test procedure is presented in the following illustration.

ILLUSTRATION 14.2 ▼

A random sample of 75 students was selected, and each student was asked to carefully measure the amount of time required to commute from his or her front door to the college parking lot. The data collected were used to test the hypothesis "the median time required for students to commute is 15 minutes" against the alternative that the median is unequal to 15 minutes. The 75 pieces of data were summarized as follows:

Under 15:	18
15:	12
Over 15:	45

Use the sign test to test the null hypothesis against the alternative hypothesis.

Solution

The data are converted to $(+)$ and $(-)$ signs according to whether the data is more or less than 15. A plus sign will be assigned to each piece of data larger than 15, a minus sign to each piece of data smaller than 15, and a zero to those data equal to 15. The sign test uses only the plus and minus signs; therefore, the zeros are discarded and the usable sample size becomes 63. That is, $n(+) = 45$, $n(-) = 18$, and $n = n(+) + n(-) = 45 + 18 = 63$.

STEP 1   **The Set-Up:**

    a. Describe the population parameter of interest.
      M, population median time to commute.

    b. State the null hypothesis (H_o) and the alternative hypothesis (H_a).

      $H_o: M = 15$

      $H_a: M \neq 15$

STEP 2   **The Hypothesis Test Criteria:**

    a. Check the assumptions.
      The 75 observations were randomly selected and the variable commute time is continuous.

    b. Identify the test statistic to be used.
      The **test statistic** that will be used is the number of the less frequent sign ; the smaller of $n(+)$ and $n(-)$, which is $n(-)$ for our illustration. We will want to reject the null hypothesis whenever the number of the less frequent sign is extremely small. Table 12 in Appendix B gives the maximum allowable number for the less frequent sign, k, that will allow us to reject the null hypothesis. That is, if the number of the less frequent sign is less than or equal to the critical value in the table, we will reject H_o. If the observed value of the less frequent sign is larger than the table value, we will fail to reject H_o. In the table, n is the total number of signs, not including zeros. The test statistic $= x \bigstar = n(-)$.

    c. Determine the level of significance, α.
      $\alpha = 0.05$ for a two-tailed test.

> The "useable" sample size is sometimes smaller than the number of data collected.

STEP 3   **The Sample Evidence:**

    a. The sample information.
      $n = 63$; the observed value of the test statistic is $x = n(-) = $ **18.**

    b. The test statistic.
      $x \bigstar = n(-) = $ **18**

STEP 4   **The Probability Distribution:**

USING THE *p*-VALUE PROCEDURE:

a. Calculate the *p*-value.
Since the concern is for values "not equal to," the *p*-value is the area of both tails. We will find the left tail and double it.

$$\mathbf{P} = 2 \times P(x \leq 18, \text{ for } n = 63)$$

OR   **USING THE CLASSICAL PROCEDURE:**

a. Determine the critical region and critical value(s).
The critical region is split into two equal parts since the H_a expresses concern for

USING THE *p*-VALUE PROCEDURE:

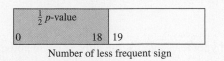

To find **P**, use one of two methods:

1. Use Table 12 (Appendix B) to place bounds on the **P**:
   Table 12 lists only two-tail values (do not double).

   P < 0.01

2. Use a computer or calculator to find **P**: **P = 0.0011**
   Specific instructions follow this illustration.

b. **Determine whether or not the *p*-value is smaller than α.**
   The *p*-value **is smaller** than α.

USING THE CLASSICAL PROCEDURE:

OR values related to "not equal to." Since the table is for two-tailed tests, the critical value is located at the intersection of the α = 0.05 column and the *n* = 63 row of Table 12: **23**

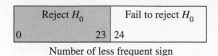

b. **Determine whether or not the calculated test statistic is in the critical region.**
   x★ **is in** the critical region; see figure above.

STEP 5 **The Results:**

   a. **State the decision about *H_o*.**
      Reject *H_o*.
   b. **State the conclusion about *H_a*.**
      The sample shows sufficient evidence at the 0.05 level to conclude that the median commute time is not equal to 15 minutes.

   Calculating the *p*-value when using the sign test:

   Method 1 *Using Table 12 in Appendix B to place bounds on the p-value:* By inspecting the *n* = 63 row of Table 12, you can determine an interval within which the *p*-value lies. Locate the value of *x* along the *n* = 63 row and read the bounds from the top of the table. Table 12 lists only two-tail values (therefore, do not double). **P < 0.01**

   Method 2 If you are doing the hypothesis test with the aid of a computer or calculator, most likely it will calculate the *p*-value for you. Specific instructions are described below.

The following commands will complete a sign test for a single sample hypothesis test of the median. (If original data are not given, just the number of plus and minus signs; input data values above and below the median that will compute into the correct number of each sign.)

MINITAB

```
Input the set of data into C1; then continue with:

   Choose:   Stat > Nonparametrics > 1-Sample Sign
   Enter:    Variables: C1
   Select:   Test median:*
   Enter:    M (hypothesized median value)
             Alternative: less than or not equal or greater
             than
*A confidence interval may also be selected.
```

The following commands (EXCEL and TI-83) will compute the differences between the data values and the hypothesized median. The data will then be sorted so that the number of + and − signs can be easily counted.

EXCEL

Input the data into column A, select cell B1; then continue with:

Choose: **Paste function fₓ > All > SIGN > OK**

Enter: Number: **A1 − hypothesized median value   > OK**

Drag: **Bottom right corner of the B1 cell down to give other differences**

Select the data in columns A and B; then continue with:

Choose: **Data > Sort**
Enter: Sort by: **Column B**
Select: **Ascending**

TI-83

Input the data into L1; then continue with:

Highlight: **L2**
Enter: **L1 − hypothesized median value**
Choose: **STAT > EDIT > 2:SortA(**
Enter: **L2)**

Two-Sample Hypothesis Test Procedure

The sign test may also be applied to a hypothesis test dealing with the median difference between **paired data** that result from **two dependent samples.** A familiar application is the use of before-and-after testing to determine the effectiveness of some activity. In a test of this nature, the signs of the differences are used to carry out the test. Again, zeros are disregarded.

ASSUMPTIONS FOR INFERENCES ABOUT MEDIAN OF PAIRED DIFFERENCES USING SIGN TEST

The paired data are selected independently and the variables are ordinal or numerical.

The following illustration shows this procedure.

ILLUSTRATION 14.3 ▼

A new no-exercise, no-starve weight-reducing plan has been developed and advertised. To test the claim that "you will lose weight within two weeks or . . .," a local statistician obtained the before-and-after weights of 18 people who had used this plan. Table 14.2 lists the people, their weights, a minus (−) for those who lost weight during the two weeks, a 0 for those who remained the same, and a plus (+) for those who actually gained weight.

Table 14.2 Sample Results for Illustration 14.2

| Person | Weight | | Sign of Difference, |
	Before	After	Before to After
Mrs. Smith	146	142	−
Mr. Brown	175	178	+
Mrs. White	150	147	−
Mr. Collins	190	187	−
Mr. Gray	220	212	−
Ms. Collins	157	160	+
Mrs. Allen	136	135	−
Mrs. Noss	146	138	−
Ms. Wagner	128	132	+
Mr. Carroll	187	187	0
Mrs. Black	172	171	−
Mrs. McDonald	138	135	−
Ms. Henry	150	151	+
Ms. Greene	124	126	+
Mr. Tyler	210	208	−
Mrs. Williams	148	148	0
Mrs. Moore	141	138	−
Mrs. Sweeney	164	159	−

The claim being tested is that people are able to lose weight. The null hypothesis that will be tested is that "there is no weight loss (or the median weight loss is zero)," meaning that only a rejection of the null hypothesis will allow us to conclude in favor of the advertised claim. Actually we will be testing to see whether there are significantly more minus signs than plus signs. If the weight-reducing plan is of absolutely no value, we would expect to find an equal number of plus and minus signs. If it works, there should be significantly more minus signs than plus signs. Thus, the test performed here will be a one-tailed test. (We will want to reject the null hypothesis in favor of the advertised claim if there are "many" minus signs.)

Solution

STEP 1 **The Set-Up:**
   a. Describe the population parameter of interest.
   M, median weight loss.
   b. State the null hypothesis (H_o) and the alternative hypothesis (H_a).

   H_o: $M = 0$ (no weight loss)

   H_a: $M < 0$ (weight loss)

STEP 2 **The Hypothesis Test Criteria:**
   b. Check the assumptions.
   The 18 observations were randomly selected and the variables, weight before and weight after, are both continuous.
   b. Identify the test statistic to be used.
   The test statistic that will be used is the number of the less frequent sign: the test statistic $= x\star = n(+)$.
   c. Determine the level of significance, α.
   $\alpha = 0.05$ for a one-tailed test.

STEP 3   **The Sample Evidence:**
$n = 16 [n(+) = 5, n(-) = 11]$; the observed value of the test statistic is
$x \bigstar = n(+) = \mathbf{5.}$

STEP 4   **The Probability Distribution:**

USING THE *p*-VALUE PROCEDURE:

a. Calculate the *p*-value.
Since the concern is for values "less than," the *p*-value is the area to the left.

$$\mathbf{P} = P(x \leq 5, \text{ for } n = 16).$$

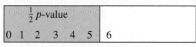

Number of less frequent sign

To find **P**, use one of two methods:
1. Use Table 12 (Appendix B) to estimate **P**: Table 12 lists only two-tail α (this is one-tailed, so divide α by two).

$$\mathbf{P} \approx \mathbf{0.125}$$

2. Use a computer or calculator to find **P**:   **P = 0.1051**
For specific instructions see page 664.

b. Determine whether or not the *p*-value is smaller than α.
The *p*-value **is not smaller** than α.

OR

USING THE CLASSICAL PROCEDURE:

a. Determine the critical region and critical value(s).
The critical region is one-tail since the H_a expresses concern for values related to "less than." Since the table is for two-tailed tests, the critical value is located at the intersection of the α = 0.10 column (i.e., α = 0.05 in each tail) and the *n* = 16 row of Table 12:

$$k = 4$$

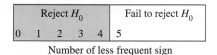

Number of less frequent sign

b. Determine whether or not the calculated test statistic is in the critical region.
$x\bigstar$ **is not in** the critical region; see figure above.

STEP 5   **The Results:**
a. State the decision about H_o.
Fail to reject H_o.
b. State the conclusion about H_a.
The evidence observed is not sufficient to allow us to reject the no-weight-loss null hypothesis at the 0.05 level of significance.

The following commands will complete a sign test for the median difference between paired data from two dependent samples.

```
MINITAB

Input the paired set of data into C1 and C2; then continue with:

  Choose:   Calc > Calculator
  Enter:    Store result in: C3
            Expression: C1 - C2 (whichever order is needed, based on Ha)    > OK
  Choose:   Stat > Nonparametrics > 1-Sample Sign. . .
  Enter:    Variables: C3
  Select:   Test median:*
  Enter:    0 (hypothesized median value)
            Alternative: less than or not equal or greater than
*As before, confidence interval may be selected.
```

The following Excel commands will compute the differences between the paired data values. The data will then be sorted so that the number of + and − signs can be easily counted.

EXCEL

Input the paired data into columns A and B; then continue with:

Choose: **Tools > Data Analysis Plus > Sign Test > OK**

Enter: Block coordinates: **(A1:B20 or select cells) > OK**

Choose: Alternative Hypothesis: **differs** or **right** or **left**

TI-83

Input the paired data into L1 and L2; then continue with:

Highlight: **L3**

Enter: **L1−L2** (whichever order is needed, based on H_a)

Choose: **STAT > EDIT > 2:SortA(**

Enter: **L3**

Normal Approximation

The sign test may be carried out by means of a normal approximation using the standard normal variable z. The normal approximation will be used if Table 12 does not show the particular levels of significance desired or if n is large.

NOTES

1. x may be the number of the less frequent sign or the most frequent sign. You will have to determine this in such a way that the direction is consistent with the interpretation of the situation.

2. x is really a **binomial random variable,** where $p = 0.5$. The sign test statistic satisfies the properties of a binomial experiment (see p. 253). Each sign is the result of an independent trial. There are n trials, and each trial has two possible outcomes ($+$ or $-$). Since the median is used, the probabilities for each outcome are both 0.5. Therefore, the mean, μ_x, is equal to

$$\mu_x = \frac{n}{2} \quad \left[\mu = np = n \cdot \frac{1}{2} = \frac{n}{2} \right]$$

and the standard deviation, σ_x, is equal to

$$\sigma_x = \frac{1}{2}\sqrt{n} \quad \left[\sigma = \sqrt{npq} = \sqrt{n \cdot \frac{1}{2} \cdot \frac{1}{2}} = \frac{1}{2}\sqrt{n} \right]$$

3. x is a discrete variable. But recall that the normal distribution must be used only with continuous variables. However, although the binomial random variable is discrete, it does become approximately normally distributed for large n. Nevertheless, when using the normal distribution for testing, we should make an adjustment in the variable so that the approximation is more accurate. (See Section 6.5, p. 304, on the normal approximation.) This adjustment is illustrated in Figure 14.1 and is called a **continuity correction.** For this discrete variable the area that represents the probability is a rectangular bar. Its width is 1 unit wide, from $\frac{1}{2}$ unit below to $\frac{1}{2}$ unit above the value of interest. Therefore, when z is to be used, we will need to make a $\frac{1}{2}$-unit adjustment before calculating the ob-

served value of z. x' will be the adjusted value for x. If x is larger than $n/2$, $x' = x - \frac{1}{2}$. If x is smaller than $n/2$, $x' = x + \frac{1}{2}$. The test is then completed by the usual procedure, using x'.

Figure 14.1
Continuity Correction

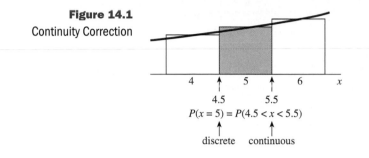

$$P(x = 5) = P(4.5 < x < 5.5)$$

discrete   continuous

If the normal approximation is to be used (including the continuity correction), the position numbers for a $1 - \alpha$ confidence interval for M are found using the formula:

$$\frac{1}{2}(n) \pm (\frac{1}{2} + \frac{1}{2} \cdot z_{(\alpha/2)} \cdot \sqrt{n}) \tag{14.1}$$

The interval is

$$x_L \text{ to } x_U, \quad 1 - \alpha \text{ confidence interval for } M \text{ (median)}$$

where

$$L = \frac{n}{2} - \frac{1}{2} - \frac{z_{(\alpha/2)}}{2} \cdot \sqrt{n} \quad \text{and} \quad U = \frac{n}{2} + \frac{1}{2} + \frac{z_{(\alpha/2)}}{2} \cdot \sqrt{n}$$

NOTE L should be rounded down and U should be rounded up to be sure that the level of confidence is at least $1 - \alpha$.

ILLUSTRATION 14.4 ▼

Estimate the population median with a 95% confidence interval for a given set of 60 pieces of data: $x_1, x_2, x_3, \ldots, x_{59}, x_{60}$.

Solution

When we use formula (14.1), the position numbers L and U are

$$\frac{1}{2}(n) \pm (\frac{1}{2} + \frac{1}{2} \cdot z_{(\alpha/2)} \cdot \sqrt{n}): \quad \frac{1}{2}(60) \pm (\frac{1}{2} + \frac{1}{2} \cdot 1.96 \cdot \sqrt{60}$$

$$30 \pm [0.50 + 7.59]$$

$$30 \pm 8.09$$

Thus,

$$L = 30 - 8.09 = 21.91; \text{ rounded down becomes } \mathbf{21} \quad \text{(21st piece of data)}$$

$$U = 30 + 8.09 = 38.09; \text{ rounded up becomes } \mathbf{39} \quad \text{(39th piece of data)}$$

Therefore

$$\mathbf{x_{21} \text{ to } x_{39},} \quad 95\% \text{ confidence interval for } M \text{ (median)} \qquad \blacktriangle$$

When a hypothesis test is to be completed using the standard normal distribution, z will be calculated by using the formula:

$$z \bigstar = \frac{x' - \dfrac{n}{2}}{\dfrac{1}{2} \cdot \sqrt{n}} \tag{14.2}$$

(See Note 3, (p. 668) with regard to x'.)

ILLUSTRATION 14.5 ▼

Use the sign test to test the hypothesis that the median number of hours, M, worked by students of a certain college is at least 15 hours per week. A survey of 120 students was taken; a plus sign was recorded if the number of hours the student worked last week was equal to or greater than 15, and a minus sign was recorded if the number of hours was less than 15. Totals showed 80 minus signs and 40 plus signs.

Solution

STEP 1 **The Set-Up:**
   a. **Describe the population parameter of interest.**
      M, the median number of hours worked by students.
   b. **State the null hypothesis (H_o) and the alternative hypothesis (H_a).**

   H_o: $M = 15$ ($\geq$) (at least as many plus signs as minus signs)

   H_a: $M < 15$ (fewer plus signs than minus signs)

STEP 2 **The Hypothesis Test Criteria:**
   a. **Check the assumptions.**
      The random sample of 120 adults was independently surveyed and the variable, hours worked, is continuous.
   b. **Identify the probability distribution and the test statistic to be used.**
      The **standard normal z** will be used with **formula (14.2).**
   c. **Determine the level of significance, α.**
      $\alpha = 0.05$.

STEP 3 **The Sample Evidence:**
   a. **Collect the sample information.**
      $n(+) = 40$, $n(-) = 80$; therefore, $n = 120$ and x is the number of plus signs; $x = 40$.
   b. **Calculate the value of the test statistic.**
      Using formula (14.2)

$$z \bigstar = \frac{x' - \dfrac{n}{2}}{\dfrac{1}{2} \cdot \sqrt{n}}: \quad z \bigstar - \frac{40.5 - \dfrac{120}{2}}{\dfrac{1}{2} \cdot \sqrt{120}} = \frac{40.5 - 60}{\dfrac{1}{2} \cdot (10.95)} = \frac{-19.5}{5.475}$$

$$z \bigstar = -3.562 = \mathbf{-3.56}$$

STEP 4   **The Probability Distribution:**

USING THE *p*-VALUE PROCEDURE:

a. Calculate the *p*-value.
Use the left-hand tail since the H_a expresses concern for values related to "fewer than."

$$\mathbf{P} = P(z < -3.56) \text{ as shown in the figure.}$$

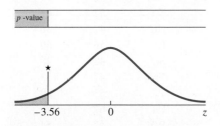

To find the *p*-value, you have three options:
1. Use Table 3 (Appendix B) to calculate *p*-value:
   P = 0.5000 − 0.4998 = 0.0002
2. Use Table 5 (Appendix B) to place bounds on the *p*-value.
   P = 0.0002
3. Use a computer or calculator to calculate *p*-value: **P = 0.0002**
For specific instructions, see page 379.
b. Determine whether or not the *p*-value is smaller than α.
The *p*-value **is smaller** than α.

OR

USING THE CLASSICAL PROCEDURE:

a. Determine the critical region and critical value(s).
The critical region is the left-hand tail since the H_a expresses concern for values related to "fewer than." The critical value is obtained from Table 4A: $-z_{(0.05)} = \mathbf{-1.65.}$

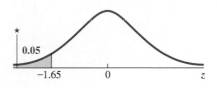

Specific instructions for finding critical values are on pages 394, 395.
b. Determine whether or not the calculated test statistic is in the critical region.
$z\star$ **is in** the critical region, as shown in red in the figure above.

See pages 664 and 665 for computer and calculator commands.

STEP 5   **The Results:**
a. State the decision about H_o.
Reject H_o.
b. State the conclusion about H_a.
At the 0.05 level, there are significantly more minus signs than plus signs, thereby implying that the median is less than the claimed 15 hours.   ▲

EXERCISES • • • • • • • • • • • • • •

14.1   The following daily highs were recorded in the city of Rochester, New York, on 20 randomly selected December days.

47	46	40	40	46	35	34	59	54	33
65	39	48	47	46	46	42	36	45	38

Use the sign test to determine the 95% confidence interval for the median daily high temperature in Rochester during December.

14.2   Find the 75% confidence interval for the median swim time for a swimmer whose recorded times are

24.7	24.7	24.6	25.5	25.7	25.8	26.5	24.5	25.3
26.2	25.5	26.3	24.2	25.3	24.3	24.2	24.2	

14.3   Each year elementary students are given achievement tests in mathematics. The following table lists for 26 states the percentage of eighth-grade students who in 1992 and 1996 scored at or above the proficient level in national tests:

Math Achievement Scores				**Math Achievement Scores**		
State	**1992**	**1996**		**State**	**1992**	**1996**
AL	39	45		MO	62	64
AZ	55	57		NE	70	76
AR	44	52		NM	48	51
CA	50	51		NY	57	61
CO	64	67		NC	47	56
CT	64	70		RI	56	60
DE	52	55		SC	48	48
FL	49	54		TN	47	53
GA	48	51		TX	53	59
IN	60	68		UT	67	70
IA	76	78		WV	47	54
MN	74	75		WI	71	75
MS	33	36		WY	67	68

Source: National Assessment of Educational Progress, National Center of Education Statistics, U.S. Dept. of Education.

   a. Construct a table showing the sign of the difference between the 1992 and 1996 achievement tests.
   b. Find a 95% confidence interval for the median difference in achievement test scores.

14.4   A sample of the daily rental-car rates for a compact car was collected in order to estimate the average daily cost of renting a compact car.

33.93	35.00	36.99	32.99	36.93	29.00	34.95	23.99	43.93	44.95
28.95	22.99	37.93	37.00	35.99	36.99	30.93	28.95	29.99	25.99
39.93	40.50	28.90	23.80	26.93	23.70	26.99	21.94	47.93	40.00
29.94	28.99	23.93	22.70	26.99	25.48	31.93	31.90	31.92	29.99

Find the 99% confidence interval for the median daily rental cost.

14.5   State the null hypothesis, H_o, and the alternative hypothesis, H_a, that would be used to test the following statements.
   a. The median value is at least 32.
   b. People prefer the taste of the bread made with the new recipe.
   c. There is no change in weight from weigh-in until after two weeks of the diet.

14.6   Determine the critical value that would be used to test the null hypothesis for the following situations using the classical approach and the sign test.
   a. $H_o: P(+) = 0.5$ vs. $H_a: P(+) \neq 0.5$, with $n = 18$, $\alpha = 0.05$
   b. $H_o: P(+) = 0.5$ vs. $H_a: P(+) > 0.5$, with $n = 78$, $\alpha = 0.05$
   c. $H_o: P(+) = 0.5$ vs. $H_a: P(+) < 0.5$, with $n = 38$, $\alpha = 0.05$
   d. $H_o: P(+) = 0.5$ vs. $H_a: P(+) \neq 0.5$, with $n = 148$, $\alpha = 0.05$

14.7   Use the sign test to test the hypothesis that the median daily high temperature in the city of Rochester, New York, during the month of December is 48 degrees based on the sample data given in Exercise 14.1.
   a. State the null hypothesis for the test.
   b. State specifically what it is that you are actually testing when using the sign test.
   c. Complete the test for $\alpha = 0.05$, and carefully state your findings.

14.8   *USA Today* reported on May 8, 1991, "Teachers' pay increases 5.4%." The following sample of average teacher salary increases was taken from the annual report of the National Education Association.

1.6	11.4	4.2	8.7	4.5	6.4	5.9	3.7	4.5	4.6	4.4	4.1
7.6	5.5	10.6	2.8	6.8	3.0	3.5	7.9	8.0	4.3	2.4	5.0

Do the sample data suggest that the claim "median pay increase is 5.4%" be rejected at the 0.05 level of significance?

14.9   In order to test the null hypothesis that there in no difference in the ages of husbands and wives, the following data were collected.

Husbands	28	45	40	37	25	42	21	22	54	47	35	62	29	44	45	38	59
Wives	26	36	40	42	28	40	20	24	50	54	37	60	25	40	34	42	49

Does the sign test show that there is a significant difference in the ages of husbands and wives at $\alpha = 0.05$?

14.10   Each year elementary students are given achievement tests in mathematics. The percentage of eighth-grade students who in 1992 and 1996 scored at or above the proficient level in national tests are given in Exercise 14.3. Use the sign test to determine at the 0.025 level of significance if the math achievement test scores improved between 1992 and 1996.

14.11   An article titled "Venocclusive Disease of the Liver: Development of a Model for Predicting Fatal Outcome After Marrow Transplantation" (*Journal of Clinical Oncology*, Sept. 1993) gives the median age of 355 patients who underwent marrow transplantation at the Fred Hutchinson Cancer Research Center as 30 years. A sample of 100 marrow transplantation patients were recently selected for a study, and it was found that 40 of the patients were over 30 and 60 were under 30 years of age. Test the null hypothesis that the median age of the population from which the 100 patients were selected equals 30 years versus the alternative that the median does not equal 30 years. Use $\alpha = 0.05$.

14.12   An article titled "Naturally Occurring Anticoagulants and Bone Marrow Transplantation: Plasma Protein C Predicts the Development of Venocclusive Disease of the Liver" (*Blood*, June 1993) compared baseline values for antithrobin III with antithrobin II values 7 days after a bone marrow transplant for 45 patients. The differences were found to be nonsignificant. Suppose 17 of the differences were positive and 28 were negative. The null hypothesis is that the median difference is zero, and the alternative hypothesis is that the median difference is not zero. Use the 0.05 level of significance. Complete the test and carefully state your conclusion.

14.13   A blind taste test was used to determine people's preference for the taste of the "classic" cola and "new" cola. The results were

> 645 preferred the new
>
> 583 preferred the old
>
> 272 had no preference

Is the preference for the taste of the new cola significantly greater than one-half? Use $\alpha = 0.01$.

14.14   A taste test was conducted with a regular beef pizza. Each of 133 individuals was given two pieces of pizza, one with a whole-wheat crust and the other with a white crust. Each person was then asked whether she or he preferred whole-wheat or white crust. The results were

> 65 preferred whole-wheat to white crust
>
> 53 preferred white to whole-wheat crust
>
> 15 had no preference

(continued)

Is there sufficient evidence to verify the hypothesis that whole-wheat crust is preferred to white crust at the $\alpha = 0.05$ level of significance?

14.15 According to an article in a *Newsweek* special issue (Fall/Winter 1990), 51.1% of 17-year-olds answered the following question correctly:

If $7X + 4 = 5X + 8$, then $X = $ _____   (a) 1   (b) 2   (c) 4   (d) 6

Suppose we wished to test the null hypothesis that one-half of all 16-year-olds could solve the problem above against the alternative hypothesis, "the proportion who can solve differs from one-half." Furthermore, suppose we asked 75 randomly selected 17-year-olds to solve the problem. Let $+$ represent a correct solution and $-$ represent an incorrect solution. Do we have sufficient evidence to show the proportion who can solve is different than one-half? Explain.

a. If we obtain 20 ($+$) signs and 55 ($-$) signs.
b. If we obtain 27 ($+$) signs and 48 ($-$) signs.
c. If we obtain 30 ($+$) signs and 45 ($-$) signs.
d. If we obtain 33 ($+$) signs and 42 ($-$) signs.

14.16 According to an article in *USA Today* (6-7-91), "only 46% of high school seniors can solve problems involving fractions, decimals, and percentages." Suppose we wish to test the null hypothesis "one-half of all seniors can solve problems involving fractions, decimals, and percentages" against an alternative that the proportion who can solve differs from one-half. Let $+$ represent passed and $-$ represent failed the test on fractions, decimals, and percentages. If a random sample of 1500 students is tested, what value of x, the number of the least frequent sign, will be the critical value at the 0.05 level of significance?

14.4   The Mann–Whitney U Test

The Mann–Whitney U test is a nonparametric alternative for the t-test for the difference between two independent means. The usual two-sample situation occurs when the experimenter wants to see if the difference between the two samples is sufficient to reject the null hypothesis that the two sampled populations are identical.

ASSUMPTIONS FOR INFERENCES ABOUT TWO POPULATIONS USING MANN–WHITNEY TEST

The two **independent random samples** are independent within each sample as well as between samples and the random variables are ordinal or numerical.

This test is often applied in situations in which the two samples are drawn from the same population of subjects, but different "treatments" are used on each set. We will demonstrate the procedure in the following illustration.

Hypothesis Test Procedure

ILLUSTRATION 14.6 ▼

In a large lecture class, when the instructor gives a one-hour exam, she gives two "equivalent" examinations. It is reasonable to ask, "Are these two different exams equivalent?" Students in even-numbered seats take exam A, and those in the odd-numbered seats take exam B. To test this "equivalent" hypothesis, the following two random samples were taken.

Table 14.3   Data for Illustration 14.6

A	52	78	56	90	65	86	64	90	49	78
B	72	62	91	88	90	74	98	80	81	71

Assuming that the odd- or even-numbered seats had no effect, does the sample present sufficient evidence to reject the hypothesis "the test forms yielded scores that had identical distributions"? Test using $\alpha = 0.05$.

Solution

STEP 1   **The Set-Up:**

    a. Describe the population parameter of interest.

    The distribution of test scores for each version of the test.

    b. State the null hypothesis (H_o) and the alternative hypothesis (H_a).

    H_o: Form A and Form B have test scores with identical distributions.

    H_a: The two distributions are not the same.

STEP 2   **The Test Criteria:**

    a. Check the assumptions.

    The two samples are independent and the random variable, test score, is numerical.

    b. Identify the probability distribution and the test statistic to be used.

    The Mann–Whitney U statistic will be used and is described below.

    c. Determine the level of significance, α.

    $\alpha = 0.05$

STEP 3   **The Sample Evidence:**

    a. Collect the sample information.

    The sample data is listed above in Table 14.3.

    b. Calculate the value of the test statistic.

The size of the individual samples will be called n_a and n_b; actually, it makes no difference which way these are assigned. In our illustration they both have the value 10.

    The two samples are combined into one sample (all $n_a + n_b$ pieces of data) and ordered from smallest to largest:

49	52	56	62	64	65	71	72	74	78
78	80	81	86	88	90	90	90	91	98

Each piece of data is then assigned a **rank** number. The smallest (49) is assigned rank 1, the next smallest (52) is assigned rank 2, and so on, up to the largest, which is assigned rank $n_a + n_b$ (20). Ties are handled by assigning to each of the tied observations the mean rank of those rank positions that they occupy. For example, in our illustration there are two 78s; they are the 10th and 11th pieces of data. The mean rank for each is then $\dfrac{(10 + 11)}{2} = 10.5$. In the case of the three 90s, the 16th, 17th, and 18th pieces of data, each is assigned 17, since $\dfrac{(16 + 17 + 18)}{3} = 17$. The rankings are shown in Table 14.4.

Equal sample sizes are not necessary when completing a Mann–Whitney U test.

Table 14.4 Ranked Data for Illustration 14.6

Ranked Data	Rank	Source	Ranked Data	Rank	Source
49	1	A	78	10.5	A
52	2	A	80	12	B
56	3	A	81	13	B
62	4	B	86	14	A
64	5	A	88	15	B
65	6	A	90	17	A
71	7	B	90	17	A
72	8	B	90	17	B
74	9	B	91	19	B
78	10.5	A	98	20	B

Figure 14.2 shows the relationship between the two sets of data, first by using the data values and second by comparing the rank numbers for each piece of data.

Figure 14.2

Comparing the Data of Two Samples

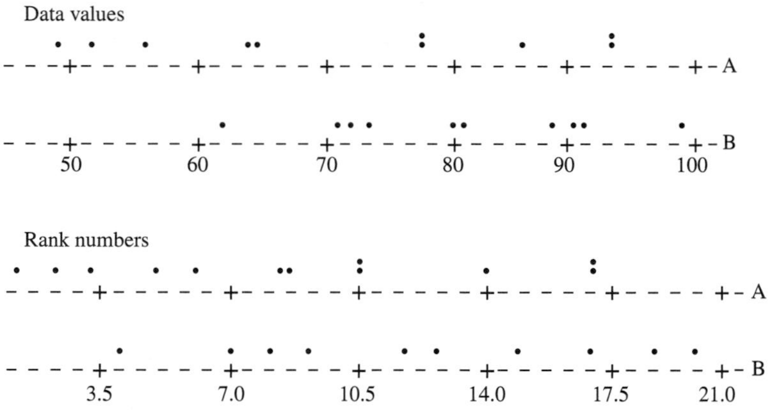

Exercise 14.17
Do you see a different relationship between the two sets of data? Explain.

The calculation of the **test statistic** U is a two-step procedure. We first determine the sum of the ranks for each of the two samples. Then, using the two sums of ranks, we calculate a U score for each sample. The smaller U score is the test statistic.

The sum of ranks for R_a for sample A is computed as

$$R_a = 1 + 2 + 3 + 5 + 6 + 10.5 + 10.5 + 14 + 17 + 17 = \textbf{86}$$

The sum of ranks R_b for sample B is

$$R_b = 4 + 7 + 8 + 9 + 12 + 13 + 15 + 17 + 19 + 20 = \textbf{124}$$

The U score for each sample is obtained by using the following pair of formulas.

$$U_a = n_a \cdot n_b + \frac{(n_b)(n_b + 1)}{2} - R_b \tag{14.3}$$

$$U_b = n_a \cdot n_b + \frac{(n_a)(n_a + 1)}{2} - R_a \tag{14.4}$$

$U \star$, the test statistic, will be the smaller of U_a and U_b.

For our illustration, we obtain

$$U_a = (10)(10) + \frac{(10)(10 + 1)}{2} - 124 = \mathbf{31} \quad \text{and}$$

$$U_b = (10)(10) + \frac{(10)(10 + 1)}{2} - 86 = \mathbf{69}$$

Therefore,

$$U \bigstar = \mathbf{31}$$

Before we carry out the test for this illustration, let's try to understand some of the underlying possibilities. Recall that the null hypothesis is that the distributions are the same and that we will most likely want to conclude from this that the means are approximately equal. Suppose for a moment that they are indeed quite different; say, all of one sample comes before the smallest piece of data in the second sample when they are ranked together. This would certainly mean that we want to reject the null hypothesis. What kind of a value can we expect for *U* in this case? Suppose, in Illustration 14.6, that the ten A values had ranks I through 10 and the ten B values had ranks 11 through 20. Then we would obtain

$$R_a = 55 \qquad R_b = 155$$

$$U_a = (10)(10) + \frac{(10)(10 + 1)}{2} - 155 = 0$$

and

$$U_b = (10)(10) + \frac{(10)(10 + 1)}{2} - 55 = 100$$

Therefore,

$$U \bigstar = \mathbf{0}$$

If this were the case, we certainly would want to reach the decision "reject the null hypothesis."

Suppose, on the other hand, that both samples were perfectly matched, that is, a score in each set identical to one in the other. Now what would happen?

54	54	62	62	71	71	72	72	. . .
A	B	A	B	A	B	A	B	. . .
1.5	1.5	3.5	3.5	5.5	5.5	7.5	7.5	. . .

$$R_a = R_b = 105$$

and

$$U_a = U_b = (10)(10) + \frac{(10)(10 + 1)}{2} - 105 = 50$$

Therefore,

$$U \bigstar = \mathbf{50}$$

If this were the case, we certainly would want to reach the decision "fail to reject the null hypothesis."

NOTE The sum of the two U's ($U_a + U_b$) will always be equal to the product of the two sample sizes (n_a n_b). For this reason we need only to concern ourselves with the smaller U-value.

Now, let's return to the solution of Illustration 14.6.

STEP 4 **The Probability Distribution:**

USING THE p-VALUE PROCEDURE:

a. Calculate the p-value.
Since the concern is for values related to "not the same," the p-value is the probability of both tails. It will be found by finding the probability of the left-tail and doubling.

$$\mathbf{P} = 2 \times P(U \le 31, \text{ for } n_1 = 10 \text{ and } n_2 = 10).$$

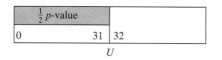

To find **P**, you have two options:
1. Use Table 13 (Appendix B) to place bounds on **P**:
   P > 0.10
2. Use a computer to find **P**:   **P = 0.1612**
Specific instructions follow this illustration.
b. Determine whether or not the p-value is smaller than α.
The p-value **is not smaller** than α.

OR

USING THE CLASSICAL PROCEDURE:

a. Determine the critical region and critical value(s).
The critical region is two-tailed since the H_a expresses concern for values related to "not the same." Use Table 13A for two-tailed $\alpha = 0.05$. The critical value is at the intersection of column $n_1 = 10$ and row $n_2 = 10$: **23**. The critical region is $U \le 23$.

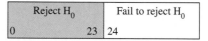

b. Determine whether or not the calculated test statistic is in the critical region.
$U \bigstar$ **is not in** the critical region; see figure above.

STEP 5 **The Results:**
      a. State the decision about H_o.
        Fail to reject H_o.
      b. State the conclusion about H_a.
        We do not have sufficient evidence to reject the "equivalent" hypothesis.

Calculating the p-value when using the Mann–Whitney test:
Method 1   *Using Table 13 in Appendix B to place bounds on the p-value:* By inspecting Tables 13A and 13B at the intersection of column $n_1 = 10$ and row $n_2 = 10$, you can determine that the **p-value is greater than 0.10**; the larger two-tailed value of α is 0.10 in table 13B.
Method 2   If you are doing the hypothesis test with the aid of a computer, most likely it will calculate the p-value for you. Specific instructions are described on pages 681, 682.
▲

Normal Approximation

If the samples are both larger than size 10, we may make the test decision with the aid of the standard normal variable, z . This is possible since the distribution of U is approximately normal with a mean

$$\mu_u = \frac{n_a \cdot n_b}{2}$$

(14.5)

and a standard deviation

$$\sigma_u = \sqrt{\frac{n_a \cdot n_b \cdot (n_a + n_b + 1)}{12}} \tag{14.6}$$

The hypothesis test is then completed using the **test statistic z ★,**

$$z ★ = \frac{U - \mu_U}{\sigma_U} \tag{14.7}$$

The standard normal distribution may be used whenever n_a and n_b are both greater than 10.

ILLUSTRATION 14.7 ▼

A dog-obedience trainer is training 27 dogs to obey a certain command. The trainer is using two different training techniques, the reward-and-encouragement method (I) and the no-reward method (II). The following table shows the number of obedience sessions that were necessary before the dogs would obey the command. Does the trainer have sufficient evidence to claim that the reward method will, on average, require less training time ($\alpha = 0.05$)?

Table 14.5   Data for Illustration 14.7

Method I	29	27	32	25	27	28	23	31	37	28	22	24	28	31	34
Method II	40	44	33	26	31	29	34	31	38	33	42	35			

Solution

STEP 1   The Set-Up:
 a. **Describe the population parameter of interest.**
    The distribution of training times for each technique.
 b. **State the null hypothesis (H_o) and the alternative hypothesis (H_a).**

   H_o: The distributions for the amount of training time required are the same for both methods.

   H_a: The reward method, on the average, requires less time.

STEP 2   The Test Criteria:
 a. **Check the assumptions.**
    The two samples are independent and the random variable, training time, is continuous.
 b. **Identify the probability distribution and the test statistic to be used.**
    The Mann–Whitney U statistic will be used.
 c. **Determine the level of significance, α.**
    $\alpha = 0.05$

STEP 3   The Sample Evidence:
 a. **Collect the sample information.**
    The sample data are listed above in Table 14.5.
 b. **Calculate the value of the test statistic.**
    The two sets of data are ranked jointly and ranks are assigned, as shown in Table 14.6.

Table 14.6   Rankings for Training Methods

Number of Sessions	Group	Rank		Number of Sessions	Group	Rank	
22	I	1		31	II	15 ⎤	14.5
23	I	2		31	II	16 ⎦	14.5
24	I	3		32	I	17	
25	I	4		33	II	18 ⎤	18.5
26	II	5		33	II	19 ⎦	18.5
27	I	6 ⎤	6.5	34	I	20 ⎤	20.5
27	I	7 ⎦	6.5	34	II	21 ⎦	20.5
28	I	8 ⎤	9	35	II	22	
28	I	9	9	37	I	23	
28	I	10 ⎦	9	38	II	24	
29	I	11 ⎤	11.5	40	II	25	
29	II	12 ⎦	11.5	42	II	26	
31	I	13 ⎤	14.5	44	II	27	
31	I	14 ⎦	14.5				

The sums are:

$$R_{\text{I}} = 1 + 2 + 3 + 4 + 6.5 + \cdots + 20.5 + 23 = 151.0$$

$$R_{\text{II}} = 5 + 11.5 + 14.5 + \cdots + 26 + 27 = 227.0$$

The U scores are found using formulas (14.3) and (14.4):

$$U_{\text{I}} = (15)(12) + \frac{(12)(12 + 1)}{2} - 227 = 180 + 78 - 227 = \mathbf{31} \quad \text{and}$$

$$U_{\text{II}} = (15)(12) + \frac{(15)(15 + 1)}{2} - 151 = 180 + 120 - 151 = \mathbf{149}$$

Therefore,

$$U = \mathbf{31}$$

Now we use formulas (14.5), (14.6), and (14.7) to determine the z statistic.

$$\mu_U = \frac{n_a \cdot n_b}{2}: \qquad \mu_u = \frac{12 \cdot 15}{2} = 90$$

$$\sigma_u = \sqrt{\frac{n_a \cdot n_b \cdot (n_a + n_b + 1)}{12}}: \quad \sigma_u = \sqrt{\frac{12 \cdot 15 \cdot (12 + 15 + 1)}{12}}$$

$$= \sqrt{\frac{(180)(28)}{12}} = \sqrt{420} = 20.49$$

$$z \bigstar = \frac{U - \mu_U}{\sigma_U}: \qquad z \bigstar = \frac{31 - 90}{20.49} = \frac{-59}{20.49} = -2.879$$

$$z \bigstar = \mathbf{-2.88}$$

STEP 4   **The Probability Distribution:**

USING THE *p*-VALUE PROCEDURE: **OR**   **USING THE CLASSICAL PROCEDURE:**

a. Calculate the *p*-value.
Use the left-hand tail since the H_a expresses concern for values
related to "less than."

$$\mathbf{P} = \textit{p}\text{-value} = P(z < -2.88) \text{ as shown in the figure.}$$

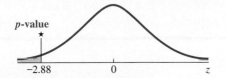

To find the *p*-value, you have three options:
1. Use Table 3 (Appendix B) to calculate *p*-value:
   P = 0.5000 − 0.4980 = 0.0020
2. Use Table 5 (Appendix B) to place bounds on the *p*-value:
   0.0019 < P < 0.0022
3. Use a computer or calculator to calculate *p*-value:
   P = 0.0020
For specific instructions, see page 379.
b. Determine whether or not the *p*-value is smaller than α.
The *p*-value **is smaller** than α.

**a. Determine the critical region and
critical value(s).**
The critical region is the left-hand tail
since the H_a expresses concern for values related to "less than." The critical
value is obtained from Table 4A:
$-z(0.05) = -1.65.$

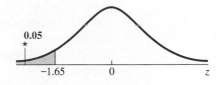

Specific instructions for finding critical values are
on pages 394, 395.
**b. Determine whether or not the cal-
culated test statistic is in the critical region.**
$z \star$ **is in** the critical region, as shown in
red in the figure above.

STEP 5   **The Results:**
a. **State the decision about H_o.**
    Reject H_o.
b. **State the conclusion about H_a.**
    At the 0.05 level of significance, the data show sufficient evidence to con-
    clude that the reward method does, on the average, require less training
    time.   ▲

*The following commands will complete the Mann–Whitney test for the difference between two independent
distributions.*

MINITAB

```
Input the two independent sets of data into C1 and C2; then continue with:

   Choose:   Stat > Nonparametrics > Mann-Whitney
   Enter:    First Sample: C1
             Second Sample: C2
             Alternative: less than or not equal or greater than

With respect to the p-value approach, the p-value is given. With respect to the
classical approach, just the sum of the ranks for one of the samples is given. Use
this to find U for that one sample. The U for the other sample is found by sub-
tracting U from the product of n₁ and n₂.
```

EXCEL

Input all of the data from the two sources into column A and corresponding numbers identifying the source into column B; then continue with:

Choose: **Tools > Data Analysis Plus > Wilcoxon Rank Sum Test* > OK**

Enter: Block Coordinates: **(A1:B20 or select cells) > OK**

Choose: Alternative Hypothesis:
The location of A differs from B
or **The location of A is right of B**
or **The location of A is left of B**

The sum of the ranks is given for both samples and also the *p*-value.

*The Wilcoxon Rank Sum Test is equivalent to the Mann-Whitney test.

TI-83

Input all of the data from the two sources into L1 and corresponding numbers identifying the source into L2; then continue with:

Choose: **STAT > EDIT > 2:SortA(**
Enter: **L1, L2)**

The data will be sorted along with its corresponding source. Ranks can now be assigned.

Case Study 14.1

Health Beliefs and Practices of Runners Versus Nonrunners

Valerie Walsh used the Mann–Whitney U statistic to conclude that runners place a greater value on personal health than do nonrunners.

Returns of mailed questionnaires from 77 runner and 63 nonrunner respondents showed that runners placed a statistically higher value on health and performed greater numbers of health-related behaviors. Major differences were found in nutrition, exercise, and medical awareness and self-care. No major differences were found in addictive substance use, stress management, or safety practices. A number of concerns regarding runners' health practices were identified, including running while ill or in pain, incidence of injuries, negative feelings when unable to run, neglect of a conscious cool-down period, low weight levels, and a tendency to increase workouts following perceived dietary indiscretions. . . .

The purpose of this investigation was to explore the differences between runners and nonrunners in terms of specific health beliefs and behaviors. It was hypothesized that there is a difference between runners and nonrunners in the relative value placed on personal health. . . .

RESULTS

The first hypothesis, that there is a difference between runners and nonrunners in the relative value placed on personal health, was tested using the Mann–Whitney U, with alpha set at .05, and was accepted, $p < .019$. The value of U was found to be 1876.5. Greater value was placed on personal health by the runners than by the nonrunners. . .

Source: Valerie R. Walsh, *Nursing Research,* November/December 1985, Vol. 34, no. 6.

Exercise 14.18
Calculate the value of $z \star$ and the *p*-value that corresponds to 1876.5 in Case Study 14.1 for a two-tailed test using the Mann–Whitney U.

EXERCISES • • • • • • • • • • • • • •

14.19 State the null hypothesis, H_o, and the alternative hypothesis, H_a, that would be used to test the following statements.
   a. There is a difference in the value of the variable between the two groups of subjects.
   b. The average value is not the same for both groups.
   c. The blood pressure for group A is higher than for group B.

14.20 Determine the critical value that would be used to test the following hypotheses for experiments involving two independent samples, using the classical method:
   a. H_o: Average(A) = Average(B)
   H_a: Average(A) > Average(B)
   with $n_A = 18$, $n_B = 15$, and $\alpha = 0.05$.
   b. H_o: The average score is the same for both groups.
   H_a: Group I average scores are less than those for group II,
   with $n_I = 78$, $n_{II} - 45$, and $\alpha - 0.05$.

14.21 An article in the *International Journal of Sports Medicine* (July 1994) discusses the use of the Mann–Whitney *U* test to compare the total cholesterol (mg/dL) of 35 adipose (obese) boys with that of 27 adipose girls. No significant difference was found between the two groups with respect to total cholesterol. A similar study involving 6 adipose boys and 8 adipose girls gave the following total-cholesterol values.

Adipose Boys	175	185	160	200	170	150		
Adipose Girls	160	190	175	190	185	150	140	195

Use the Mann–Whitney *U* test to test the research hypothesis that the total-cholesterol values differ for the two groups, using the 0.05 level of significance.

14.22 The July 4, 1994, issue of *Newsweek* gives quotes by several tobacco executives. Cigarette makers point out that some brands have more nicotine and tar than others do. Consider a study designed to compare the nicotine content of two different brands of cigarettes. The nicotine content was determined for 25 cigarettes of brand A and 25 cigarettes of brand B. The sum of ranks for brand A equals 688, and the sum of ranks for brand B equals 587. Use the Mann–Whitney *U* statistic to test the null hypothesis that the average nicotine content is the same for the two brands versus the alternative that the average nicotine content differs. Use $\alpha = 0.01$.

14.23 Pulse rates were recorded for 16 men and 13 women. The results are shown in the following table.

Males	61	73	58	64	70	64	72	60	65	80	55	72	56	56	74	65
Females	83	58	70	56	76	64	80	68	78	108	76	70	97			

These data were used to test the hypothesis that the distribution of pulse rates differs for men and women. The following MINITAB output printed out the sum of ranks for males (W = 192.0) and the *p*-value of 0.0373. Verify these two values by calculating them yourself.

```
Mann-Whitney Confidence Interval and Test

Males   n = 16  Median = 64.50
Females      n = 13  Median = 76.00

W = 192.0
Test of ETA1 = ETA2  vs  ETA1 not = ETA2  is significant at
0.0373
```

14.24 The following set of data represents the ages of drivers involved in automobile accidents. Do these data present sufficient evidence to conclude that there is a difference in the average age of men and women drivers involved in accidents? Use a two-tailed test at $\alpha = 0.05$.

Men	70	60	77	39	36	28	19	40	23	23	63	31	36	55	24	76
Women	62	46	43	28	21	22	27	42	21	46	33	29	44	29	56	70

a. State the null hypothesis that is being tested.

b. Complete the test using a computer or calculator.

14.25 A study titled "Textbook Pictures and First-Grade Children's Perception of Mathematical Relationships" by Patricia Campbell (*Journal for Research in Mathematics Education,* Nov. 1978) investigated the influence of artistic style and the number of pictures on first-grade children's perception of mathematics textbook pictures. Analysis with the Mann–Whitney U test indicated that students who initially viewed and described sequences of pictures had significantly higher story-response scores than those students who viewed only single pictures. Consider the following data from two such groups. Group 1 is the group who viewed sequences of pictures, and group 2 is the group who viewed only single pictures.

Group 1	30	35	40	42	45	36
Group 2	25	32	27	39	30	

Using the Mann–Whitney U test, determine if the group 1 scores are significantly higher than the group 2 scores. Use $\alpha = 0.05$.

14.26 Commercial airlines are often evaluated on the basis of two major performance categories: on-time arrivals and baggage handling. In 1997 and 1998, Delta Airlines received the following competitive ratings (the lower the better) on each of these dimensions over a 13-month period:

Month	On-Time Arrivals	Baggage Handling
Aug.	7	4
Sept.	8	7
Oct.	8	5
Nov.	9	6
Dec.	6	4
Jan.	4	6
Feb.	6	7
Mar.	4	4
Apr.	7	5
May	4	5
June	2	1
July	2	3
Aug.	2	1

Source: Fortune, "Pulling Delta Out of Its Dive," December 7, 1998.

a. Convert the table to a table of ranks of the on-time arrivals (A) and baggage handling (B) for Delta.

b. Use the Mann–Whitney U test to test the hypothesis that baggage handling obtained higher ratings than on-time arrivals during the period. Use the 0.05 level of significance.

14.5 The Runs Test

The runs test is most frequently used to test the **randomness** of data (or lack of randomness). A **run** is a sequence of data that possesses a common property.

One run ends and another starts when a piece of data does not display the property in question. The **test statistic** in this test is V, the number of runs observed.

ILLUSTRATION 14.8 ▼

To illustrate the idea of runs, let's draw a sample of ten single-digit numbers from the telephone book, using the next-to-last digit from each of the selected telephone numbers.

Sample: 2, 3, 1, 1, 4, 2, 6, 6, 6, 7

Let's consider the property of "odd" (o) or "even" (e). The sample, as it was drawn, becomes e, o, o, o, e, e, e, e, e, o, which displays four runs.

e   **o o o**   *e e e e e*   **o**

Thus, $V\star = 4$.   ▲

In Illustration 14.8, if the sample contained no randomness, there would be only two runs—all the evens, then all the odds, or the other way around. We would also not expect to see them alternate—odd, even, odd, even. The maximum number of possible runs would be $n_1 + n_2$, or less (provided n_1 and n_2 are not equal), where n_1 and n_2 are the number of data that have each of the two properties being identified.

> **ASSUMPTIONS FOR INFERENCES ABOUT RANDOMNESS USING RUNS TEST**
>
> Each sample data can be classified into one of two categories.

The runs test is generally a two-tailed test. We will reject the hypothesis when there are too few runs, since this would indicate that the data has been "separated" according to the two properties. We will also reject when there are too many runs because that will indicate the data alternate between the two properties too often to be random. For example, if the data alternated all the way down the line, we might suspect that the data had been tampered with. There are many aspects to the concept of randomness. The occurrence of odd and even as discussed in Illustration 14.8 is one aspect. Another aspect of randomness that we might wish to check is the ordering of fluctuations of the data above (a) or below (b) the mean or median of the sample.

ILLUSTRATION 14.9 ▼

Consider the following sample and determine whether the data points form a random sequence with regard to being above or below the median value.

| 2 | 5 | 3 | 8 | 4 | 2 | 9 | 3 | 2 | 3 | 7 | 1 | 7 | 3 | 3 |
| 6 | 3 | 4 | 1 | 9 | 5 | 2 | 5 | 5 | 2 | 4 | 3 | 4 | 0 | 4 |

Test the null hypothesis that this sequence is random. Use $\alpha = 0.05$.

Solution

STEP 1   The Set-Up:
   a.   Describe the population parameter of interest.
       Randomness of the values above or below the median.
   b.   State the null hypothesis (H_o) and the alternative hypothesis (H_a).

   H_o: The numbers in the sample form a random sequence with respect to the two properties "above" and "below" the median value.

   H_a: The sequence is not random.

STEP 2 The Test Criteria:
    a. Check the assumptions.
      Each sample data can be classified as "above" or "below" the median.
    b. Identify the probability distribution and the test statistic to be used.
      V, the number of runs present in the sample data.
    c. Determine the level of significance, α.
      $\alpha = 0.05$

STEP 3 The Sample Evidence:
    Calculate the value of the test statistic.

First we must rank the data and find the median. The ranked data are

0	1	1	2	2	2	2	2	3	3	3	3	3	3	3
4	4	4	4	4	5	5	5	5	6	7	7	8	9	9

Since there are 30 pieces of data, the depth of the median is at $d(\tilde{x}) = 15.5$ position.
Thus, $\tilde{x} = \dfrac{3+4}{2} = 3.5$. By comparing each number in the original sample to the value of the median, we obtain the following sequence of a's (above) and b's (below).

$$b\ a\ b\ a\ a\ b\ a\ b\ b\ b\ a\ b\ a\ b\ b\ a\ b\ a\ b\ a\ b\ a\ a\ b\ a\ a\ b\ a\ b\ a\ b\ a\ b\ a$$

We observe $n_a = 15$, $n_b = 15$, and 24 runs. So $V \bigstar = 24$.
    If n_1 and n_2 are both less than or equal to 20 and a two-tailed test at $\alpha = 0.05$ is desired, then Table 14 in Appendix B will be used to complete the hypothesis test.

STEP 4 The Probability Distribution:

USING THE p-VALUE PROCEDURE:

a. Calculate the p-value.
Since the concern is for values related to "not random," the test is two-tailed. The p-value is found by finding the probability of the right-tail and doubling.

$$\mathbf{P} = 2 \times P(V \geq 24,\ \text{for } n_a = 15 \text{ and } n_b = 15).$$

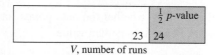

V, number of runs

To find **P**, you have two options:
1. Use Table 14 (Appendix B) to place bounds on the **P**:
   P < 0.05
2. Use a computer to find **P**: **P = 0.003**
Specific instructions follow this illustration.
b. Determine whether or not the p-value is smaller than α.
The p-value **is smaller** than α.

OR

USING THE CLASSICAL PROCEDURE:

a. Determine the critical region and critical value(s).
Since the concern is for values related to "not random," the test is two-tailed. Use Table 14 for two-tailed $\alpha = 0.05$. The critical values are at the intersection of column $n_1 = 15$ and row $n_2 = 15$: **10 and 22.** The critical region is $V \leq 10$ or $V \geq 22$.

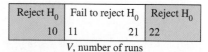

V, number of runs

Specific instructions follow this illustration.
b. Determine whether or not the calculated test statistic is in the critical region.
$V \bigstar$ **is in** the critical region; see figure above.

STEP 5 The Results:
    a. State the decision about H_o.
      Reject H_o.

b. State the conclusion about H_a.

We are able to reject the hypothesis of randomness at the 0.05 level of significance, and conclude that the sequence is not random with regard to above and below the median.

Calculating the *p*-value when using the runs test:

Method 1   *Using Table 14 in Appendix B to place bounds on the* p-*value:* By inspecting Table 14 at the intersection of column $n_1 = 15$ and row $n_2 = 15$, you can determine that the ***p*-value is less than 0.05;** the observed value of $V = 24$ is larger than the larger critical value listed.

Method 2   If you are doing the hypothesis test with the aid of a computer, most likely it will calculate the *p*-value for you. Specific instructions are described below.

The following commands will complete the runs test for testing randomness above and below the median.

MINITAB

Input the set of data into C1; then continue with:

    Choose: **Stat > Nonparametrics > Runs Tests**
    Enter: Variable: **C1**
    Select: **Above and below mean**
    or
    Select: **Above and below:**
    Enter: **Median value**

The following commands (EXCEL and TI-83) will only compute the differences between the data values and the median. Then to complete the runs test you will need to count the number of runs created by the sequence of + and − signs.

EXCEL

Input the data into column A, select cell B1; then continue with:

    Choose: **Edit Formula (=)**
    Enter: **A1 − median   > OK**
    Drag: **Bottom right corner of the
             B1 cell down to give other
             differences**

TI-83

Input the data into L1; then continue with:

    Highlight: **L2**
    Enter: **L1−median*(L1)**
    *[2nd LIST > MATH > 4:median()]

▲

Normal Approximation

To complete the hypothesis test about randomness when n_1 and n_2 are larger than 20 or when α is other than 0.05, we will use z, the standard normal random variable . V is approximately normally distributed with a mean of μ_v and a standard deviation of σ_v. The formulas are as follows:

$$\mu_V = \frac{2n_1 \cdot n_2}{n_1 + n_2} + 1 \tag{14.8}$$

$$\sigma_V = \sqrt{\frac{(2n_1 \cdot n_2) \cdot (2n_1 \cdot n_2 - n_1 - n_2)}{(n_1 + n_2)^2(n_1 + n_2 - 1)}} \tag{14.9}$$

$$z \bigstar = \frac{V - \mu_V}{\sigma_V} \tag{14.10}$$

ILLUSTRATION 14.10 ▼

Test the null hypothesis that the following sequence of sample data is a random sequence with regard to each data being "odd" or "even". Use $\alpha = 0.10$.

1	2	3	0	2	4	3	4	8	1	2	1	2	4	3	9	6	2	4	1
5	6	3	3	2	2	1	2	4	2	3	6	3	5	1	7	3	3	0	1
4	4	1	2	7	2	1	7	5	3										

Solution

STEP 1 **The Set-Up:**
   a. **Describe the population parameter of interest.**
   Randomness of the odd and even numbers.
   b. **State the null hypothesis (H_o) and the alternative hypothesis (H_a).**

   H_o: The sequence of "odd" and "even" numbers is random.

   H_a: The sequence is not random.

STEP 2 **The Test Criteria:**
   a. **Check the assumptions.**
   Each sample value can be classified as "odd" or "even."
   b. **Identify the probability distribution and the test statistic to be used.**
   V, the number of runs present in the sample data.
   c. **Determine the level of significance, α.**
   $\alpha = 0.10$

STEP 3 **The Sample Evidence:**
   Calculate the value of the test statistic.

The sample data when converted to "o" for odd and "e" for even becomes

$$\underline{o}\, e\, \underline{o}\, e\, e\, e\, \underline{o}\, e\, e\, \underline{o}\, e\, \underline{o}\, e\, e\, \underline{o\, o}\, e\, e\, e\, \underline{o\, o}\, e\, \underline{o\, o}\, e$$
$$e\, \underline{o}\, e\, e\, e\, \underline{o}\, e\, \underline{o\, o\, o\, o\, o\, o}\, e\, \underline{o}\, e\, e\, \underline{o}\, e\, \underline{o}\, e\, \underline{o\, o\, o\, o}$$

and reveals: $n_o = 26$, $n_e = 24$, and **29 runs, $V = 29$.**

Now use formulas (14.8), (14.9) and (14.10) to determine the z statistic.

$$\mu_V = \frac{2n_1 \cdot n_2}{n_1 + n_2} + 1: \qquad \mu_V = \frac{2 \cdot 26 \cdot 24}{26 + 24} + 1 = 24.96 + 1 = 25.96$$

$$\sigma_V = \sqrt{\frac{(2n_1 \cdot n_2) \cdot (2n_1 \cdot n_2 - n_1 - n_2)}{(n_1 + n_2)^2(n_1 + n_2 - 1)}} : \sigma_V = \sqrt{\frac{(2 \cdot 26 \cdot 24) \cdot (2 \cdot 26 \cdot 24 - 26 - 24)}{(26 + 24)^2(26 + 24 - 1)}}$$

$$\sigma_V = \sqrt{\frac{(1248)(1198)}{(50)^2 \cdot 49}} = \sqrt{12.20493} = 3.49$$

$$z \bigstar = \frac{V - \mu_V}{\sigma_V}: \quad z \bigstar = \frac{29 - 25.96}{3.49} = \frac{3.04}{3.49} = 0.87$$

$$z \bigstar = \mathbf{0.87}$$

STEP 4   **The Probability Distribution:**

USING THE *p*-VALUE PROCEDURE:

a. Calculate the *p*-value.
A two-tailed test is used.

$$\mathbf{P} = p\text{-value} = 2 \times P(z > 0.87).$$

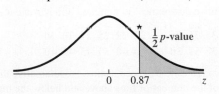

To find the *p*-value, you have three options:
1. Use Table 3 (Appendix B) to calculate *p*-value:
   P = 2(0.5000 − 0.3078) = **0.3844**
2. Use Table 5 (Appendix B) to place bounds on the *p*-value:
   0.3682 < P < 0.3954
3. Use a computer or calculator to calculate *p*-value: **P = 0.3844**
For specific instructions, see page 379.
b. Determine whether or not the *p*-value is smaller than α.
The *p*-value **is smaller** than α.

OR

USING THE CLASSICAL PROCEDURE:

a. Determine the critical region and critical value(s).
A two-tailed test is used. The critical values are obtained from Table 4A:
$-z_{(0.05)} = \mathbf{-1.65}$ and $z_{(0.05)} = \mathbf{1.65}.$

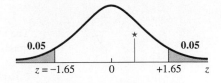

Specific instructions for finding critical values are on pages 394, 395.
b. Determine whether or not the calculated test statistic is in the critical region.
$z \bigstar$ **is not in** the critical region, as shown in red in the figure above.

STEP 5   **The Results:**
   a. State the decision about H_o.
      Fail to reject H_o.
   b. State the conclusion about H_a.
      At the 0.10 level of significance, we are unable to reject the hypothesis of randomness and conclude that this data is a random sequence.   ▲

EXERCISES   •   •   •   •   •   •   •   •   •   •   •   •   •

14.27   State the null hypothesis, H_o, and the alternative hypothesis, H_a, that would be used to test the following statements.
   a. The data did not occur in a random order about the median.
   b. The sequence of odd and even is not random.
   c. The gender of customers entering a grocery store was recorded; the entry is not random in order.

14.28   Determine the critical values that would be used to complete these hypothesis tests for the following runs tests using the classical approach.
   a. H_o: The results collected occurred in random order above and below the median.
      H_a: The results were not random;
      with $n(A) = 14$, $n(B) = 15$, and $\alpha = 0.05$.
   b. H_o: The two properties alternated randomly.
      H_a: The two properties did not occur in random fashion;
      with $n(I) = 78$, $n(II) = 45$, and $\alpha = 0.05$.

14.29 A manufacturing firm hires both men and women. The following shows the sex of the last 20 individuals hired (M = male, F = female).

 M M F M F   F M M M M   M M F M M   F M M M M

At the $\alpha = 0.05$ level of significance, are we correct in concluding that this sequence is not random?

14.30 A student was asked to perform an experiment that involved tossing a coin 25 times. After each toss, the student recorded the results. The following data were reported (H = heads, T = tails).

 H T H T H   T H T H H   T T H H T   T H T H T   H T H T H

Use the runs test at a 5% level of significance to test the student's claim that the results reported are random.

14.31 The following are 24 consecutive downtimes (in minutes) of a particular machine.

Downtime:	20 33 33 35 36 36	22 22 25 27 30 30
	30 31 31 32 32 36	40 40 50 45 45 40

The null hypothesis of randomness is to be tested against the alternative that there is a trend. A MINITAB analysis of the number of runs above and below the median follows.

```
Runs Test
Median of Downtime    K = 32.5000

The observed number of runs = 4
The expected number of runs = 13.0000
12 Observations above K 12 below
          The test is significant at 0.0002
```

a. Confirm the values reported for the median and the number of runs by calculating them yourself.
b. Compute the value of $z \star$ and the p-value.
c. Would you reject the hypothesis of randomness? Explain.
d. Construct a graph that displays the sample data that visually supports answer (c).

14.32 The article "Water Boosts Hemoglobin's Love for Oxygen" (*Science News,* March 30, 1991) discusses the ability of the iron-rich protein pigment hemoglobin to carry oxygen throughout the body. The article states that the protein's conversion to an oxygen-loving state involves between 60 and 80 water molecules. Suppose 20 different determinations of the number of water molecules needed resulted in the following data.

number:	79 75 69 70 70	65 75 75 65 70
	60 62 63 63 67	65 70 60 65 62

a. Determine the median and the number of runs above and below the median.
b. Use the runs test to test these data for randomness about the median.
c. State your conclusion.

14.33 The following data were collected in an attempt to show that the number of minutes the city bus is late is steadily growing larger. The data are in order of occurrence.

 Minutes: 6 1 3 9 10   10 2 5 5 6   12 3 7 8 9   4 5 8 11 14

At $\alpha = 0.05$, do these data show sufficient lack of randomness to support the claim?

14.34 Stock quotations are influenced by numerous variables, including the earnings of the companies, economic conditions, stock market trading volume, press releases, and splits. The table below shows the August 18, 1998, closing prices of 20 stocks selected from the New York Stock Exchange. These were the first 20 stocks whose company names started with the letter B. All prices have been rounded to the nearest dollar. *(continued)*

Company	Closing Price	Company	Closing Price
1	15	11	25
2	34	12	30
3	38	13	38
4	26	14	17
5	27	15	12
6	28	16	6
7	117	17	28
8	8	18	8
9	9	19	25
10	40	20	7

Source: The Wall Street Journal, Vol. CII, No. 35, August 19, 1998.

a. Determine the median closing price and the number of runs above and below the median.
b. Use the runs test to test whether these stock prices are listed in a random sequence about the median.
c. State your conclusion.

14.35 A USA Snapshot® (*USA Today,* 10-14-94) titled "School buildings aging" gives the average age for schools in five cities as follows: Washington, D.C., 75 years; St. Louis 74 years; San Diego 30 years; Baltimore 30 years; and Fresno, California, 30 years. The following ages of school buildings were collected in the sequence given for Spokane, Washington: 5, 13, 25, 45, 15, 17, 22, 35, 16, 23, 36, 22, 35, and 35.
a. Determine the median and the number of runs above and below the median.
b. Use the runs test to test these data for randomness about the median. Use $\alpha = 0.05$.

14.36 An article titled "Clintonomics Hurt Middle Class, Poor" (*USA Today,* 10-17-94) states that the median income for 1993 equals $36,959. A random sample of 250 incomes has a median value different from any of the 250 incomes in the sample. The data contains 105 runs above and below the median. Use the above information to test the null hypothesis that the incomes in the sample form a random sequence with respect to the two properties above and below the median value versus the alternative that the sequence is not random at $\alpha = 0.05$.

14.37 The number of absences recorded at a lecture that met at 8 AM on Mondays and Thursdays last semester were (in order of occurrence)

n(absences):	5	16	6	9	18	11	16	21	14	17	12	14	10
	6	8	12	13	4	5	5	6	1	7	18	26	6

Do these data show a randomness about the median value at $\alpha = 0.05$? Complete this test by using (a) critical values from Table 14 in Appendix B and (b) the standard normal distribution.

14.38 In an attempt to answer the question "Does the husband (h) or wife (w) do the family banking?" the results of a sample of 28 married customers doing the family banking show the following sequence of arrivals at the bank.

w w w w h w h h h h w w w w w h h w w w w h h h h w h h w

Do these data show lack of randomness with regard to whether the husband or wife does the family banking? Use $\alpha = 0.05$.

14.6 Rank Correlation

Charles Spearman developed the rank correlation coefficient in the early 1900s. It is a nonparametric alternative to the linear correlation coefficient [Pearson's product moment, r] that was discussed in Chapters 3 and 13.

The **Spearman rank correlation coefficient,** r_s, is found by using the formula

$$r_s = 1 - \frac{6\sum(d)^2}{n(n^2 - 1)} \tag{14.11}$$

where d is the difference in the **paired rankings** and n is the number of pairs of data. The value of r_s will range from -1 to $+1$ and will be used in much the same manner as Pearson's linear correlation coefficient r was used previously.

The Spearman rank coefficient is defined by using formula (3.1) with data rankings being used in place of quantitative x- and y-values. The original data may be rankings or if the data are quantitative, each variable must be ranked separately, then the rankings are used as pairs. If there are no ties in the rankings, formula (14.11) is equivalent to formula (3.1). Formula (14.11) provides us with an easier procedure to use for calculating the r_s statistic.

> The subscript s is used in honor of Spearman, the originator.

ASSUMPTIONS FOR INFERENCES ABOUT RANK CORRELATION

The n ordered pairs of data form a random sample and the variables are ordinal or numerical.

The null hypothesis that we will be testing is "there is no correlation between the two rankings." The alternative hypothesis may be either two-tailed, "there is correlation," or one-tailed, if we anticipate the existence of either positive or negative correlation. The critical region will be on the side(s) corresponding to the specific alternative that is expected. For example, if we suspect negative correlation, the critical region will be in the left-hand tail.

ILLUSTRATION **14.11** ▼

Let's consider a hypothetical situation in which four judges rank five contestants in a contest. Let's identify the judges as A, B, C, and D and the contestants as a, b, c, d, and e. Table 14.7 lists the awarded rankings.

Table 14.7   Rankings for Five Contestants

Contestant	Judge			
	A	**B**	**C**	**D**
a	1	5	1	5
b	2	4	2	2
c	3	3	3	1
d	4	2	4	4
e	5	1	5	3

When we compare judges A and B, we see that they ranked the contestants in exactly the opposite order; perfect disagreement (see Table 14.8). From our previous work with correlation, we expect the calculated value for r_s to be exactly -1 for these data.

Table 14.8   Rankings of A and B

Contestant	A	B	$d = A - B$	$(d)^2$
a	1	5	−4	16
b	2	4	−2	4
c	3	3	0	0
d	4	2	2	4
e	5	1	4	16
				40

$$r_s = 1 - \frac{6\sum (d)^2}{n(n^2 - 1)}: \quad r_s = 1 - \frac{(6)(40)}{5(5^2 - 1)} = \frac{240}{120} = 1 - 2 = \mathbf{-1}$$

When judges A and C are compared, we see that their rankings of the contestants are identical (see Table 14.9). We would expect to find a calculated correlation coefficient of +1 for these data.

Table 14.9   Rankings of A and C

Contestant	A	C	$d = A - C$	$(d)^2$
a	1	1	0	0
b	2	2	0	0
c	3	3	0	0
d	4	4	0	0
e	5	5	0	0
				0

$$r_s = 1 - \frac{6\sum (d)^2}{n(n^2 - 1)}: \quad r_s = 1 - \frac{(6)(0)}{5(5^2 - 1)} = -\frac{0}{120} = 1 - 0 = \mathbf{1}$$

By comparing the rankings of judge A with those of judge B and then with those of judge C, we have seen the extremes: total agreement and total disagreement. Now let's compare the rankings of judge A with those of judge D (see Table 14.10). There seems to be no real agreement or disagreement here. Let's compute r_s:

Table 14.10   Rankings of A and D

Contestant	A	D	$d = A - D$	$(d)^2$
a	1	5	−4	16
b	2	2	0	0
c	3	1	2	4
d	4	4	0	0
e	5	3	2	4
				24

$$r_s = 1 - \frac{6\sum (d)^2}{n(n^2 - 1)}: \quad r_s = 1 - \frac{(6)(24)}{5(5^2 - 1)} = 1 - \frac{144}{120} = 1 - 1.2 = \mathbf{-0.2}$$

This is fairly close to zero, which is what we should have suspected since there was no real agreement or disagreement.

The test of significance will result in a failure to reject the null hypothesis when r_s is close to zero and will result in a rejection of the null hypothesis in cases where r_s is found to be close to $+1$ or -1. The critical values found in Table 15, Appendix B, are the positive critical values only. Since the null hypothesis is "the population correlation coefficient is zero" (that is, $\rho_s = 0$), we have a symmetric test statistic. Hence we need only add a plus or minus sign to the value found in the table, as appropriate. This will be determined by the specific alternative that we have in mind.

When there are only a few ties, it is common practice to use formula (14.11). Even though the resulting value of r_s is not exactly equal to the value that would occur if formula (3.1) were used, it is generally considered to be an acceptable estimate. Illustration 14.12 below shows the procedure for handling ties and uses formula (14.11) for the calculation of r_s.

When ties occur in either set of the ordered pairs of rankings, assign each tied observation the mean of the ranks that would have been assigned had there been no ties, as was done in the Mann–Whitney U test (see p. 675.).

ILLUSTRATION 14.12 ▼

For comparison, Exercise 14.49 (p. 700) asks you to calculate r_s using formula (3.2).

Students who finish exams more quickly than the rest of the class are often thought to be smarter. The following set of data shows the score and order of finish for 12 students on a recent one-hour exam. At the 0.01 level, do these data support the alternative hypothesis that the first students to complete an exam have higher grades?

Table 14.11 Data for Illustration 14.12

Order of Finish	1	2	3	4	5	6	7	8	9	10	11	12
Exam Score	90	78	76	60	92	86	74	60	78	70	68	64

Solution

STEP 1 **The Set-Up:**
  a. **Describe the population parameter of interest.**
     Rank correlation coefficient between score and order of finish, ρ_s.
  b. **State the null hypothesis (H_o) and the alternative hypothesis (H_a).**

     H_o: Order of finish has no relationship to exam score.

     H_a: First to finish tend to have higher grades.

STEP 2 **The Hypothesis Test Criteria:**
  a. **Check the assumptions.**
     The 12 ordered pairs of data form a random sample; order of finish is an ordinal variable and test score is numerical.
  b. **Identify the test statistic to be used.**
     The test statistic that will be used is the calculated rank correlation coefficient, r_s.
  c. **Determine the level of significance, α.**
     $\alpha = 0.01$ for a one-tailed test.

STEP 3 **The Sample Evidence:**
  a. **The sample information.**

Rank the scores from highest to lowest, assigning the highest score the rank number 1, as shown. (Order of finish is already ranked.)

92	90	86	78	78	76	74	70	68	64	60	60
1	2	3	4	5	6	7	8	9	10	11	12
			4.5	4.5						11.5	11.5

The rankings and preliminary calculations are shown in Table 14.12.

> Notice that the first to finish is ranked 1; the highest test score is ranked 1.

Table 14.12   Rankings for Test Scores and Differences

Order of Finish	Test Score Rank	Difference (d)	$(d)^2$
1	2	-1	1.00
2	4.5	-2.5	6.25
3	6	-3	9.00
4	11.5	-7.5	56.25
5	1	4	16.00
6	3	3	9.00
7	7	0	0.00
8	11.5	-3.5	12.25
9	4.5	4.5	20.25
10	8	2	4.00
11	9	2	4.00
12	10	2	4.00
			142.00

Using formula (14.11), we obtain

$$r_s = 1 - \frac{6\sum(d)^2}{n(n^2-1)}: \quad r_s = 1 - \frac{(6)(142.0)}{12(12^2-1)} = 1 - \frac{852}{1716} = 1 - 0.4965 = \mathbf{0.503}$$

b. The test statistic.
$$r_s \bigstar = 0.503$$

STEP 4   **The Probability Distribution:**

USING THE p-VALUE PROCEDURE:

a. Calculate the p-value.
Since the concern is for values "positive," the p-value is the area to the right.

$$\mathbf{P} = P(r_s \geq 0.503, \text{ for } n = 12).$$

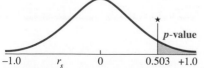

OR

USING THE CLASSICAL PROCEDURE:

a. Determine the critical region and critical value(s).
The critical region is one-tail since the H_a expresses concern for values related to "positively." Since the table is for two-tailed, the critical value is located at the intersection of the $\alpha = 0.02$ column (i.e.: $\alpha = 0.01$ in each tail) and the $n = 12$ row of Table 15: **0.703**

USING THE *p*-VALUE PROCEDURE:

To find **P**, you have two options:
1. Use Table 15 (Appendix B) to place bounds on the **P**:
   Table 15 lists only two-tail α (this test is one-tailed, so divide column heading by 2).   **0.025 < P < 0.05**
2. Use a computer or calculator to find **P**: **P = 0.048**
Specific instructions follow this illustration.

b. Determine whether or not the *p*-value is smaller than α.
The *p*-value **is not smaller** than α.

USING THE CLASSICAL PROCEDURE:

OR

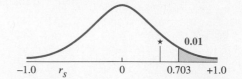

b. Determine whether or not the calculated test statistic is in the critical region.

r_s ★ **is not in** the critical region; see figure above.

STEP 5   **The Results:**
  a. State the decision about H_o.
  Fail to reject H_o.
  b. State the conclusion about H_a.
  There is not sufficient evidence presented by these sample data to enable us to conclude that the first students to finish have the highest grades at the 0.01 level of significance.

  Calculating the *p*-value for the Spearman rank correlation test:
  Method 1   *Using Table 15 in Appendix B to place bounds on the* p-value: By inspecting the $n = 12$ row of Table 15, you can determine an interval within which the *p*-value lies. Locate the value of r_s along the $n = 12$ row and read the bounds from the top of the table. Table 15 lists only two-tail values (therefore, you must divide by 2 for a one-tail test).   **0.025 < P < 0.05**
  Method 2   If you are doing the hypothesis test with the aid of a computer or calculator, most likely it will calculate the *p*-value for you. Specific instructions are described below. MINITAB and EXCEL calculate a two-tailed *p*-value, therefore you must divide by 2 when the test is one-tailed.

The following commands will calculate the Spearman rank correlation coefficient. In MINITAB and the TI-83, commands to rank each set of data (use only if needed) are given first. (Calculated p-value shown in the output only applies to the t-test for Pearson's correlation coefficient.)

MINITAB

```
Input the set of data for the first
variable into C1 and the corresponding
data values for the second variable
into C2; then continue with:

  Choose:  Manip > Rank. . .
  Enter:   Rank data in: C1
           Store ranks in: C3   > OK

Repeat above commands for the data in
C2 and store in C4.

  Choose:  Stat > Basic Statistics >
           Correlation
  Enter:   Variables: C3 C4
(See Method 2 above.)
```

EXCEL

```
Input the set of data for the first
variable into column A and the corre-
sponding data values for the second
variable into column B; then continue
with:

  Choose:  Tools > Data Analysis Plus >
           Spearman Rank Correlation > OK
  Enter:   Block coordinates: (A1:B10
           or select cells)
(See Method 2 above.)
```

```
TI-83
```

Input the set of data for the first variable into L1 and the corresponding data values for the second variable into L3; then continue with:

Highlight: **L2**
Enter: **L1 > ENTER**
Highlight: **L2**
Choose: **STAT > EDIT > 2:SORTA(**
Enter: **L2**

Use the sorted data to assign ranks to the data in L1 (can just replace data values with ranks). Repeat above commands for the data in L3, using L4. Watch for ties.

Continue with the following and highlight Calculate.
Choose: **STAT > TESTS > E:LinRegTTest**

```
LinRegTTest
 Xlist:L1
 Ylist:L3
 Freq:1
 B & ρ:≠0  <0  >0
 RegEQ:Y1
 Calculate
```

Case Study 14.2

Effects of Therapeutic Touch on Tension Headache Pain

Elizabeth Keller and Virginia Bzdek made use of the Spearman rank correlation coefficient to determine the influence of the subjects' education, age, and several other attributes on the results of therapeutic touch. They reported several observed values and the associated p-values.

Exercise 14.39

Explain the meaning of ". . . was inversely correlated with years of education, $r = -0.53$, $p = 0.002$," a quote from Case Study 14.2.

Therapeutic touch (TT) is a modern derivative of the laying on of hands that involves touching with the intent to help or heal. This study investigated the effects of TT on tension headache pain in comparison with a placebo simulation of TT. Sixty volunteer subjects with tension headaches were randomly divided into treatment and placebo groups. The McGill–Melzack Pain Questionnaire was used to measure headache pain levels before each intervention, immediately afterward, and 4 hours later. . . .

FINDINGS

Spearman correlation coefficients were calculated to determine the influence of the subject's education, age, sex, practice of meditation, religion, and level of initial skepticism toward TT. The differences in the MMPQ pain scores were not significantly correlated with any belief system or demographic variable in either group. The one exception was the posttest difference on the PPI scale in the placebo group, which was inversely correlated with years of education, $r = -.53$, $\rho = .002$.

Source: Elizabeth Keller and Virginia M. Bzdek, *Nursing Research,* March/April 1986, Vol. 35, no. 2.

```
EXERCISES  •  •  •  •  •  •  •  •  •  •  •  •  •  •  •
```

14.40 State the null hypothesis, H_o, and the alternative hypothesis, H_a, that would be used to test the following statements.
  a. There is no relationship between the two rankings.
  b. The two variables are unrelated.
  c. There is a positive correlation between the two variables.
  d. Age has a decreasing effect on monetary value.

14.41 Determine the test criteria that would be used to test the null hypothesis for the following Spearman rank correlation experiments.
  a. H_o: No relationship between the two variables.
  H_a: There is a relationship, with $n = 14$ and $\alpha = 0.05$.
  b. H_o: No correlation.
  H_a: Positively correlated, with $n = 27$ and $\alpha = 0.05$.
  c. H_o: Variable A has no effect on variable B.
  H_a: Variable B decreases as A increases, with $n = 18$ and $\alpha = 0.01$.

14.42 The following data represent the ages of 12 subjects and the mineral concentration (in parts per million) in their tissue samples.

Age, x	82	83	64	53	47	50	70	62	34	27	75	28
Mineral Concentration, y	170	40	64	5	15	5	48	34	3	7	50	10

Refer to the following MINITAB output and verify that the Spearman rank correlation coefficient equals 0.753 by calculating it yourself.

> **Correlations (Pearson)**
>
> Correlation of Rank, x and Rank, r = 0.753, P-value = 0.005

14.43 Many people are concerned about eating foods that contain a high sodium content. They are also advised of the benefits of obtaining sufficient fiber in their diets. Do foods high in fiber tend to have more sodium? The following table was obtained by selecting 11 soups from a list published in *Nutrition Action Healthletter*. The soups were measured on the basis of both sodium content and fiber:

Soup	Sodium	Fiber
A	480	12
B	830	0
C	510	1
D	460	5
E	490	3
F	580	7
G	420	2
H	290	4
I	450	10
J	430	6
K	390	9

Source: Nutrition Action Healthletter, Vol. 24, No. 7, December 1997.

  a. Rank the soups in ascending order based on the basis of their sodium content and on their fiber content, and show your results in a table.
  b. Compute the Spearman rank order correlation coefficient for the two sets of rankings.
  c. Does higher sodium content accompany foods that are higher in fiber? Test the null hypothesis that there is no relationship between the fiber and sodium content of the soups versus the alternative that there is a relationship between them. Use $\alpha = 0.05$.

14.44 The December 30, 1998, issue of *USA Today* gives the Nielsen rankings for America's favorite prime-time television programs for the last week of 1998. The top ten programs are listed below along with the rankings assigned by a panel of educators.

Program	Nielsen Ranking	Panel Ranking
Monday Night Football	1.0	8.0
60 Minutes	2.0	1.5
Touched by an Angel	3.0	3.0
Barbara Walters Presents	4.0	4.5

(continued)

Program	Nielsen Ranking	Panel Ranking
Dateline NBC-Monday	5.0	1.5
Home Improvement	6.0	4.5
CBS Sunday Movie	7.0	9.0
Walker Texas Ranger	8.0	10.0
Everybody Loves Raymond	9.5	6.0
CBS Tuesday Movie	9.5	7.0

Compute the Spearman rank coefficient. Test the null hypothesis that there is no relationship between the Nielsen rankings and the panel rankings versus the alternative that there is a relationship between them. Use $\alpha = 0.05$.

14.45 Consumer product testing groups commonly supply ratings of all sorts of products to consumers in an effort to assist them in their purchase decisions. Different manufacturers' products are usually tested for their performance and then given an overall rating. *PC World* ranked the top ten 17-inch computer monitors in late 1998 and also supplied the street price. The ranks of each are shown in the table below with the highest priced monitor given a rank of 1 and the lowest a 10:

The Top 10 Computer Monitors

Overall Rating	Street Price
1	3
2	4
3	6.5
4	8.5
5	5
6	2
7	8.5
8	6.5
9	10
10	1

Source: PC World, "Top 10 Monitors," December 1998.

a. Compute the Spearman rank correlation coefficient for the overall rating and the street price of the 17-inch monitors.

b. Does a higher price yield a higher rating? Test the null hypothesis that there is no relationship between the overall ratings of the monitors and their street prices versus the alternative that there is a relationship between them. Use $\alpha = 0.05$.

14.46 When it comes to getting workers to produce, money is less than everything; feeling appreciated is more important. Do the rankings assigned by workers and the boss show a significant difference in what each thinks is important? Test using $\alpha = 0.05$.

Component of Job Satisfaction	Worker Ranking	Boss Ranking
Full appreciation of work done	1	8
Feeling of being in on things	2	10
Sympathetic help on personal problems	3	9
Job security	4	2
Good wages	5	1
Interesting work	6	5
Promotion and growth in the organization	7	3
Personal loyalty to employees	8	6
Good working conditions	9	4
Tactful disciping	10	7

Source: Philadelphia Inquirer, 29 December 1976.

14.47 An article titled "The Graduate Record Examination as an Admission Requirement for the Graduate Nursing Program" (*Journal of Professional Nursing,* October 1994) reported a significant correlation between undergraduate GPA and GPA at graduation from a graduate nursing program. The following data were collected on ten nursing students who graduated from a graduate nursing program.

Undergraduate GPA	3.5	3.1	2.7	3.7	2.5	3.3	3.0	2.9	3.8	3.2
GPA at Graduation	3.4	3.2	3.0	3.6	3.1	3.4	3.0	3.4	3.7	3.8

Compute the Spearman rank correlation coefficient and test the null hypothesis of no relationship versus a positive relationship. Use a level of significance equal to 0.05.

14.48 An article in *Self* magazine (February 1991) discusses the relationship between the pace of life and the coronary heart disease death rate. New York City, for example, was ranked as only the third-fastest-paced city but number one for deadly heart attacks. Suppose the data from another such study involving eight cities were as follows:

City	Rank for Pace-of-Life	Rank for Heart-Disease Death Rate
Salt Lake City	4	7
Buffalo	2	2
Columbus	5	8
Worcester	6	4
Boston	1	6
Providence	8	3
New York	3	1
Paterson	7	5

Find the Spearman rank correlation coefficient.

14.49 Using formula (3.2), calculate the Spearman rank correlation coefficient for the data in Illustration 14.12, page 694. Recall that formula (3.2) is equivalent to the definition formula (3.1) and that rank numbers must be used with this formula in order for the resulting statistic to be the Spearman r_s.

14.50 Refer to the bivariate data shown in the following table.

x	-2	-1	1	2
y	4	1	1	4

a. Construct a scatter diagram.
b. Calculate the Spearman rank correlation coefficient, r_s [formula (14.11)].
c. Calculate Pearson's correlation coefficient, r [formula (3.2)].
d. Compare the two results from (b) and (c). Do the two measures of correlation measure the same thing?

Return to C H A P T E R   C A S E   S T U D Y

The Chapter Case Study, "Men and Women in the Workplace," (p. 659), displays two distributions (one for men and a second for women) listing the percentage of top executives who identified each specific skill as one whose level is rising and outpacing their men and women workers.

EXERCISE 14.51

a. Construct a table listing the seven skills in the first column and the rankings for the men and women in two additional columns.

(continued)

b. Are the rankings of the lacking skills significantly different for men and women? Use the sign test for paired samples and a level of significance equal to 0.05.
c. Is there a relationship between the men's rankings and the women's rankings? Test using Spearman's rank correlation and a 0.05 level of significance.
d. Explain why you are qualified to use the nonparametric tests in (b) and (c).
e. Compare the two results from (b) and (c). Comment on how they are related and how they are different.

IN RETROSPECT

In this chapter you have become acquainted with some of the basic concepts of nonparametric statistics. While learning about the use of nonparametric methods and specific nonparametric tests of significance, you should have also come to realize and understand some of the basic assumptions that are needed when the parametric techniques of the earlier chapters are encountered. You now have seen a variety of tests, many of which somewhat duplicate the job done by others. What you must keep in mind is that you should use the best test for your particular needs. The power of the test and the cost of sampling, as related to size and availability of the desired response variable, will play important roles in determining the specific test to be used.

The article that forms the Chapter Case Study illustrates only one of many situations in which a nonparametric test can be used. Spearman's rank correlation test may be used to statistically compare the two rankings. They appear to be quite different, but are they significantly different? The answer to this is left for you to determine in Exercise 14.51.

CHAPTER EXERCISES

14.52 Is the absentee rate in the 8 AM statistics class the same as in the 11 AM statistics class? The following sample of the daily number of absences was taken from the attendance records of the two classes.

Class	\multicolumn{12}{c}{Day}
	1	2	3	4	5	6	7	8	9	10	11	12
8 AM	0	1	3	1	0	2	4	1	3	5	3	2
11 AM	1	0	1	0	1	2	3	0	1	3	2	1

Is there sufficient reason to conclude that there are more absences in the 8 AM class? Use $\alpha = 0.05$.

14.53 Track coaches, runners, and fans talk a lot about the "speed of the track." The surface of the track is believed to have a direct effect on the amount of time that it takes a runner to cover the required distance. To test this effect, ten runners were asked to run a 220-yard sprint on each of two tracks. Track A is a cinder track, and track B is made of a new synthetic material. The running times are given in the following table. Test the claim that the surface on track B is conducive to faster running times.

Track	\multicolumn{10}{c}{Runner}
	1	2	3	4	5	6	7	8	9	10
A	27.7	26.8	27.0	25.5	26.6	27.4	27.2	27.4	25.8	25.1
B	27.0	26.7	25.3	26.0	26.1	25.3	26.7	27.1	24.8	27.1

(continued)

a. State the null and alternative hypotheses being tested. Complete the test using $\alpha = 0.05$.

b. State your conclusions.

14.54 A candy company has developed two new chocolate-covered candy bars. Six randomly selected people all preferred candy bar I. Is this statistical evidence, at $\alpha = 0.05$, that the general public will prefer candy bar I?

14.55 While trying to decide on the best time to harvest his crop, a commercial apple farmer recorded the day on which the first apple on the top half and the first apple on the bottom half of 20 randomly selected trees were ripe. The variable x was assigned a value of 1 on the first day that the first ripe apple appeared on 1 of the 20 trees. The days were then numbered sequentially. The observed data are shown in the following table. Do these data provide convincing evidence that the apples on the top of the trees start to ripen before the apples on the bottom half? Use $\alpha = 0.05$.

										Tree										
Position	1	2	3	4	5	6	7	8	9	10	11	12	13	14	15	16	17	18	19	20
Top	5	6	1	4	5	3	6	7	8	5	8	6	4	7	8	10	3	2	9	7
Bottom	6	5	5	7	3	6	6	8	9	4	10	7	5	11	6	11	5	6	9	8

14.56 A sample of 32 students received the following grades on an exam.

41	42	48	46	50	54	51	42	51	50	45	42	32	45	43	56
55	47	45	51	60	44	57	57	47	28	41	42	54	48	47	32

a. Does this sample show that the median score for the exam differs from 50? Use $\alpha = 0.05$.

b. Does this sample show that the median score for the exam is less than 50? Use $\alpha = 0.05$.

14.57 An article in the journal *Sedimentary Geology* (Vol. 57, 1988) compares a measure called the roughness coefficient for translucent and opaque quartz sand grains. If you measured the roughness coefficient for 20 sand grains of each type (translucent and opaque), for what values of the Mann–Whitney U statistic would you reject the null hypothesis in a two-tailed test with alpha equal to 0.05?

14.58 Twenty students were randomly divided into two equal groups. Group 1 was taught an anatomy course using a standard lecture approach. Group 2 was taught using a computer-assisted approach. The test scores on a comprehensive final exam were as follows:

Group 1	75	83	60	89	77	92	88	90	55	70
Group 2	77	92	90	85	72	59	65	92	90	79

Test the claim that a computer-assisted approach produces higher achievement (as measured by final exam scores) in anatomy courses than does a lecture approach. Use $\alpha = 0.05$.

14.59 The use of nuclear magnetic resonance (NMR) spectroscopy for detection of malignancy is discussed in the journal *Clinical Chemistry* (Vol. 34, No. 3, 1988). The line width at the half height of peaks in the NMR spectra is measured. The spectra is produced from assaying plasma from an individual. Suppose the following line widths were obtained from a normal group and a group known to have malignancies. Would you reject a two-tailed research hypothesis at the 0.05 level of significance?

Normal Group	35.1	32.9	30.6	30.5	30.9
Malignancy Group	28.5	29.5	30.7	27.5	28.0

14.60 A firm is currently testing two different procedures for adjusting the cutting machines used in the production of greeting cards. The results of two samples show the following recorded adjustment times.

| Method 1 | 17 | 15 | 14 | 18 | 16 | 15 | 17 | 18 | 15 | 14 | 14 | 16 | 15 | | | |
| Method 2 | 14 | 14 | 13 | 13 | 15 | 12 | 16 | 14 | 16 | 13 | 14 | 13 | 12 | 15 | 17 | 13 |

Is there sufficient reason to conclude that method 2 requires less time (on the average) than method 1 at the 0.05 level of significance?

14.61 Two statistics that baseball enthusiasts use to compare the overall strengths of one team against another are team batting (the higher the batting average, the better) and team pitching (the lower the earned run average, the better). The results for the National and American Leagues in 1997 were as follows:

NL Team	Batting Avg.	ERA		AL Team	Batting Avg.	ERA
Colorado	0.288	5.25		Boston	0.291	3.84
San Diego	0.271	4.99		New York	0.287	3.91
Atlanta	0.270	3.18		Cleveland	0.286	4.73
Los Angeles	0.268	3.63		Seattle	0.280	4.79
Chicago	0.263	4.44		Texas	0.274	4.69
New York	0.262	3.95		Chicago	0.273	4.74
Pittsburgh	0.262	4.28		Anaheim	0.272	4.52
Florida	0.259	3.83		Minnesota	0.270	5.02
Houston	0.259	3.67		Baltimore	0.268	3.91
Montreal	0.258	4.14		Kansas City	0.264	4.71
San Francisco	0.258	4.42		Milwaukee	0.260	4.24
Philadelphia	0.255	4.87		Oakland	0.260	5.49
St. Louis	0.255	3.90		Detroit	0.258	4.56
Cincinnati	0.253	4.42		Toronto	0.244	3.92

Source: The World Almanac and Book of Facts 1998.

a. Convert the table to ranks of the (1) batting averages and (2) earned run averages for the National League and the American League, showing the league (A or N) represented by a team's rank.

b. Use the Mann–Whitney U test to test the hypothesis that (1) the batting average of the American League is higher and (2) the earned run average of the National League is lower. Use the 0.05 level of significance.

14.62 Two table-tennis-ball manufacturers have agreed that the quality of their products can be measured by the height to which the balls rebound. A test is arranged, the balls are dropped from a constant height, and the rebound heights are measured. The results (in inches) are shown in the following table. Manufacturer A claims, "The results show my product to be superior." Manufacturer B replies, "I know of no statistical test that supports this claim." Can you find a test that supports A's claim?

| A | 14.0 | 12.5 | 11.5 | 12.2 | 12.4 | 12.3 | 11.8 | 11.9 | 13.7 | 13.2 |
| B | 12.0 | 12.5 | 11.6 | 13.3 | 13.0 | 13.0 | 12.1 | 12.8 | 12.2 | 12.6 |

a. Does the appropriate parametric test show that A's product is superior? [What parametric test (or tests) is appropriate, and what exactly does it show?]

b. Does the appropriate nonparametric test show that A's product is superior?

14.63 Consider the following sequence of defective parts (d) and nondefective parts (n) produced by a machine.

| n | n | n | d | n | n | n | n | n | d | n | n |
| n | n | n | n | n | d | n | d | n | n | n | n |

Can we reject the hypothesis of randomness at $\alpha = 0.05$?

14.64 A patient was given two different types of vitamin pills, one containing iron and one iron-free. The patient was instructed to take the pills on alternate days. To free himself from remembering which pill he needed to take, he mixed all of the pills together in a large bottle. Each morning he took the first pill that came out of the bottle. To see whether this was a random process, for 25 days he recorded an "I" each morning that he took a vitamin with iron and an "N" for no iron.

Day	1	2	3	4	5	6	7	8	9	10	11	12	13	14	15	16	17	18	19	20	21	22	23	24	25
Type	I	I	N	I	I	N	N	I	N	N	N	N	N	I	I	I	N	I	I	I	I	N	I	I	N

Is there sufficient reason to reject the null hypothesis that the vitamins were taken in random order at the 0.05 level of significance?

14.65 What makes a company more attractive to work for than another? One possibility is the growth in new jobs. In 1998, the editors of *Fortune* developed a list of the top 100 companies to work for in America. Included in the list was the percentage change in full-time positions of each company during the past two years. The top 20 are shown in the table below:

Company	Job Growth	Company	Job Growth
1	26	11	23
2	54	12	13
3	34	13	17
4	10	14	23
5	31	15	9
6	48	16	3
7	26	17	15
8	22	18	11
9	24	19	1
10	10	20	122

Source: Fortune, "The 100 Best Companies to Work For in America," January 12, 1998.

a. Determine the median job growth percentage and the number of runs above and below the median.

b. Use the runs test to test whether the growth rates are listed in a random sequence about the median.

c. Do companies ranked higher also have higher job growth rates? State your conclusion.

14.66 Every year *Sports Illustrated* presents its college football preview and ranks the top 25 teams before the season starts based primarily on scouting reports. As the season progresses, other college football polls provide a weekly ranking of the teams, evaluations that are largely influenced by how well the teams are playing and who they play against. The table below shows the ranks of the teams by *SI* prior to the first snap of the 1998 season and the ranks bestowed by *USA Today*/ESPN and the AP Top 25 after 11 weeks of play had transpired.

	Ranks of the Top 25 College Football Teams		
Team	**(1) Sports Illustrated (Preseason)**	**(2) USA Today/ESPN (After 11 Weeks)**	**(3) AP Top 25 (After 11 Weeks)**
Ohio State	1	7	7
Florida State	2	5	5
Kansas State	3	1.5	2
Florida	4	4	4
Nebraska	5	11	11
UCLA	6	3	3
Arizona State	7	30.5	31
Michigan	8	16	15

(continued)

Team	(1) *Sports Illustrated* (Preseason)	(2) *USA Today*/ESPN (After 11 Weeks)	(3) AP Top 25 (After 11 Weeks)
Tennessee	9	1.5	1
LSU	10	26	31
Syracuse	11	25.5	31
Washington	12	30.5	31
Michigan State	13	26	31
Penn State	14	17	19
North Carolina	15	30.5	31
Missouri	16	15	13
Colorado State	17	30.5	31
West Virginia	18	30.5	31
Miami	19	24	24
Georgia	20	18	17
Texas A&M	21	8	6
Wake Forest	22	30.5	31
Auburn	23	30.5	31
Kentucky	24	30.5	31
Notre Dame	25	13	12

Source: Sports Illustrated, "Scouting Reports," August 31, 1998.

Note: The ranks given to teams that were ranked in the preseason poll but no longer ranked after 11 weeks were obtained by $[26 + 27 + 28 + \cdots (25 + n)] \div n$, where n is the number of teams no longer ranked in the top 25.

a. Compute the Spearman rank correlation coefficient for the *SI* preseason poll and the *USA Today*/ESPN; the *SI* preseason poll and the AP Top 25; and the *USA Today*/ESPN poll and the AP Top 25.

b. Test the null hypothesis that there is no relationship between the polls versus the alternative that there is a relationship between them for each of the three possible paired comparisons. Use $\alpha = 0.05$.

 14.67 Can today's high temperature be effectively predicted using yesterday's high? Pairs of yesterday's and today's high temperatures were randomly selected. The results are shown in the following table. Do the data present sufficient evidence to justify the statement "Today's high temperature tends to correlate with yesterday's high temperature"? Use $\alpha = 0.05$.

Reading	1	2	3	4	5	6	7	8	9	10	11	12	13	14	15	16	17	18
Yesterday's	40	58	46	33	40	51	55	81	85	83	89	64	73	63	46	58	28	69
Today's	40	56	34	59	46	51	74	77	83	84	85	68	65	60	54	62	34	66

14.68 U.S. commercial radio stations are classified by the primary format of their broadcasts. As people change their listening preferences, the stations are likely to react to the change by adjusting their formats. The following table shows the percentages of radio stations in 1991 and 1997, broken down by their primary format:

Primary Format	1991	1997
Country	25.61	24.26
Adult Contemporary (AC)	21.76	14.75
News, Talk, Business, Sports	5.49	12.73
Religion (Teaching and Music)	8.33	10.22
Rock	5.51	9.13
Oldies	7.34	7.30
Spanish and Ethnic	3.86	5.32
Adult Standards	4.25	5.20

(Continued)

Primary Format	1991	1997
Urban, Black, Urban AC	3.24	3.47
Top-40	7.04	3.40
Easy Listening	2.19	0.84
Variety	0.84	0.49
Jazz	0.55	0.48
Classical, Fine Arts	0.53	0.45
Pre-Teen	0.04	0.35
All Other	3.42	1.61

Source: M Street Corporation, Nashville, TN, 1997.

   a. Construct a table that shows the ranks of the relative frequency of stations within each format between 1991 and 1997.
   b. Use the Spearman rank order correlation coefficient to test at the 0.01 level of significance the hypothesis that there is no correlation between the ranks of the formats offered by radio stations in 1991 and 1997.

14.69 In a study to see whether spouses are consistent in their preferences for television programs, a market research firm asked several married couples to rank a list of 12 programs (1 represents the highest score; 12 represents the lowest). The average ranks for the programs, rounded to the nearest integer, were as follows:

	Program											
Rank	**1**	**2**	**3**	**4**	**5**	**6**	**7**	**8**	**9**	**10**	**11**	**12**
Husbands	12	2	6	10	3	11	7	1	9	5	8	4
Wives	5	4	1	9	3	12	2	8	6	10	7	11

Is there significant evidence of negative correlation at the 0.01 level of significance?

14.70 Nonparametric tests are also called distribution-free tests. However, the normal distributions are used in the inference-making procedures.
   a. To what does the distribution-free term apply? (The population? The sample? The sampling distribution?) Explain.
   b. What is it that has the normal distribution? Explain.

VOCABULARY LIST

Be able to define each term. Pay special attention to the key terms, which are printed in **red.** In addition, describe in your own words, and give an example of, each term. Your examples should not be ones given in class or in the textbook.

   The bracketed numbers indicate the chapters in which the term previously appeared, but you should define the terms again to show increased understanding of their meaning. Page numbers indicate the first appearance of the term in Chapter 14.

binomial random variable [5, 9] (p. 668)
continuity correction [6] (p. 668)
correlation [3, 13] (p. 691)
dependent sample [10] (p. 662, 665)
distribution-free test (p. 660)
efficiency (p. 661)
independent sample [10] (p. 674)
Mann–Whitney *U* test (p. 674)
median, *M* [2] (p. 662)
nonparametric test (p. 660)
normal approximation [6] (p. 668, 678, 687)

paired data [3, 10, 13] (p. 665)
parametric test (p. 660)
power (p. 661)
randomness [2, 7] (p. 684)
rank (p. 675)
run (p. 684)
runs test (p. 684)
sign test (p. 661)
**Spearman rank correlation
   coefficient** (p. 692)
test statistic [8, 9, . . .] (p. 663, 676, 679, 684)

CHAPTER PRACTICE TEST

Part I: Knowing the Definitions

Answer "True" if the statement is always true. If the statement is not always true, replace the words shown in bold with words that make the statement always true.

14.1 One of the advantages that the nonparametric tests have is the necessity for **less restrictive** assumptions.

14.2 The sign test is a possible replacement for the **F test.**

14.3 The **sign test** can be used to test the randomness of a set of data.

14.4 If a tie occurs in a set of ranked data, the data that form the tie are **removed from the set.**

14.5 Two dependent **means** can be compared nonparametrically by using the sign test.

14.6 The sign test is a possible alternative to the Student's t test for **one mean value.**

14.7 The **run test** is a nonparametric alternative to the difference between two independent means.

14.8 The **confidence level** of a statistical hypothesis test is measured by $1 - \beta$.

14.9 Spearman's rank correlation coefficient is an alternative to using the **linear correlation coefficient.**

14.10 The **efficiency** of a nonparametric test is the probability that a false null hypothesis is rejected.

Part II: Applying the Concepts

14.11 The weights of nine people before they stopped smoking and five weeks after they stopped smoking are as follows:

Person	1	2	3	4	5	6	7	8	9
Before	148	176	153	116	128	129	120	132	154
After	155	178	151	120	130	136	126	128	158

Find the 95% confidence interval estimate for the average weight change.

14.12 The following data show the weight gains for 20 laboratory mice, half of which were fed one diet and half a different diet. Test to determine if the difference in weight gain is significant at $\alpha = 0.05$.

Diet A	41	40	36	43	36	43	39	36	24	41
Diet B	35	34	27	39	31	41	37	34	42	38

14.13 A large textbook publishing company hired nine new sales representatives three years ago. At the time of hire, the nine were rated according to their potential. Now three years later the company president wants to know how well their potential rates correlate with their sales totals for the three years.

Sales Rep.	a	b	c	d	e	f	g	h	i
Potential	2	5	6	1	4	3	9	8	7
Sales Tot.	450	410	350	345	330	400	250	310	270

Is there significant correlation at the 0.05 level?

14.14 The new school principal thought there might be a pattern to the order in which discipline problems arrived at his office. He had his secretary record the grade level of the students as they arrived. *(continued)*

9	10	11	9	12	11	9	10	10	11	10	11	10	10	11
12	12	9	9	11	12	10	9	12	10	11	12	11	10	10

At the 0.05 level, is there significant evidence of randomness?

Part III: Understanding the Concepts

14.15 What advantages do nonparametric statistics have over parametric methods?

14.16 Explain how the sign test is based on the binomial distribution and is often approximated by the normal distribution.

14.17 Why does the sign test use a null hypothesis about the median instead of the mean like a t-test uses?

14.18 Explain why a nonparametric test is not as sensitive to an extreme datum as a parametric test might be.

14.19 A restaurant has collected data on which of two seating arrangements its customers prefer. In a sign test to determine if one seating arrangement is significantly preferred, the null hypothesis would be

    a. $M = 0$ **b.** $M = 0.5$ **c.** $p = 0$ **d.** $p = 0.5$

Explain your choice.

Working with Your Own Data

Many variables in everyday life can be treated as bivariate. Often two such variables have a mathematical relationship that can be approximated by means of a straight line. The following demonstrates such a situation.

A The Age and Value of Peggy's Car

Peggy would like to sell her 1988 Corvette, and she needs to know what price to ask for it in order to write a newspaper advertisement. The car is in average condition and Peggy expects to get an average price for it ("average for a Corvette!"). She must answer the question "What is an average asking price for a 1988 Corvette?"

Inspection of many classified sections of newspapers turned up only three advertisements for 1988 Corvettes. The prices listed varied a great deal. Peggy finally decided that, in order to determine an accurate selling price, she would define two variables and collect several pairs of values including model years 1977 to the present.

POPULATION: Used Chevrolet Corvettes advertised for sale by individual owners, dealers not included. Model years 1977 to present.

INDEPENDENT VARIABLE, x: The age of the car as measured in years and defined by

$$x = (\text{present calendar year}) - (\text{year of manufacture}) + 1$$

EXAMPLE: During 1999 Peggy's 1988 Corvette is considered to be 12 years old

$$x = (1999\text{-}1988) + 1 = 11 + 1 = 12$$

DEPENDENT VARIABLE, y: The advertised asking price.

The table below lists the data collected in December 1998.

Year	1979	1992	1992	1979	1985	1986	1990	1992	1982	1992
Price	11000	17900	19900	11500	16900	15800	27500	16500	11800	18900

Year	1980	1987	1987	1996	1985	1985	1988	1988	1993	1994
Price	9500	15000	13000	28900	10900	12900	10000	13500	23800	23750

Year	1994	1979	1977	1989	1995	1985	1994	1989	1997	1978
Price	22000	11000	9900	14000	21500	13000	22500	14000	34700	8500

Year	1994	1979	1996	1989	1983	1984	1997	1988	1978	1989
Price	23000	7500	30000	12000	14900	10000	34700	14900	8900	14500

1. Construct and label a scatter diagram of Peggy's data.
2. Determine the equation for the line of best fit.
3. Draw the line of best fit on the scatter diagram.
4. Test the equation of the line of best fit to see whether the linear model is appropriate for the data. Use $\alpha = 0.05$.
5. Construct a 95% confidence interval for the mean advertised price for 1990 Corvettes.
6. Draw a line segment on the scatter diagram that represents the interval estimate found for question 5.
7. What does the value of the slope, b_1, represent? Explain.
8. What does the value of the y-intercept, b_0, represent? Explain.

B   Your Own Investigation

Identify a situation of interest to you that can be investigated statistically using bivariate data. (Consult your instructor for specific guidance.)

1. Define the population, the independent variable, the dependent variable, and the purpose for studying these two variables as a regression analysis.
2. Collect 15 to 20 ordered pairs of data.
3. Construct and label a scatter diagram of your data.
4. Determine the equation for the line of best fit.
5. Draw the line of best fit on the scatter diagram.
6. Test the equation of the line of best fit to see whether the linear model is appropriate for the data. Use $\alpha = 0.05$.
7. Construct a 95% confidence interval for the mean value of the dependent variable at the following value of x: Let x be equal to one-third the sum of the lowest value of x in your sample and twice the largest value. That is,

$$x = \frac{L + 2H}{3}$$

8. Draw a line segment on the scatter diagram that represent the interval estimate found for question 7.
9. What does the value of the slope, b_1, represent? Explain.
10. What does the value of the y-intercept, b_0, represent? Explain.

Basic Principles of Counting

In order to find the probability of many events, it is necessary to determine the number of possible outcomes for the experiment involved. This requires us to enumerate (obtain a "count" of) the possibilities. This "count" can be obtained by using one of two methods: (1) list all the possibilities and then proceed to count them (1, 2, 3, . . .); or (2) since it is often not necessary to delineate (obtain a representation of) all possibilities, the count can be determined by calculating its numerical value. In this section, we are going to learn three commonly used methods for obtaining the count by calculation: the fundamental technique and two specific techniques.

ILLUSTRATION A.1 ▼

An automobile dealer offers one of its small sporty models with two transmission options (standard or automatic) and in one of three colors (black, red, or white). How many different choices of transmission and color combinations are there for the customer?

Solution

The number of choices available can easily be found by listing and counting them. There are six.

Standard, black	Automatic, black
Standard, red	Automatic, red
Standard, white	Automatic, white

The possible choices can also be demonstrated by use of a tree diagram.

Trans. Option	Color Option	Possible Choices
standard	black	standard, black
	red	standard, red
	white	standard, white
automatic	black	automatic, black
	red	automatic, red
	white	automatic, white

NOTE More information and additional illustrations of tree diagrams can be found in Chapter 4 and in the *Statistical Tutor.*

Each of the two transmission choices can be paired with any one of three colors; thus there are 2×3 or six different possible choices. This suggests the following rule:

FUNDAMENTAL COUNTING RULE

If an experiment is composed of two trials, where one of the trials (single action or choice) has m possible outcomes (results) and the other trial has n possible outcomes, then when the two trials are performed together, there are

$$m \times n \qquad\qquad\qquad (A.1)$$

possible outcomes for the experiment.

In Illustration A.1, $m = 2$ (the number of transmission choices) and $n = 3$ (the number of color choices). Using the Fundamental Counting Rule (formula A.1), the number of possible choices available to a customer is

$$m \times n = 2 \times 3 = 6 \qquad \blacktriangle$$

This fundamental counting rule may be extended to include experiments that have more than two trials.

GENERAL COUNTING RULE

If an experiment is composed of k trials performed in a definite order, where the first trial has n_1 possible outcomes, the second trial has n_2 possible outcomes, the third trial has n_3 outcomes, and so on, then the number of possible outcomes for the experiment is

$$n_1 \times n_2 \times n_3 \times \cdots \times n_k. \qquad\qquad (A.2)$$

ILLUSTRATION A.2 ▼

In many states, automobile license plates use three letters followed by three numerals to make up the "license plate number." (There are other combinations of letters and numerals used; however, let's focus only on this six-character "number" for now.) If we assume that any one of the 26 letters may be used for each of the first three characters and that any one of the 10 numerals 0 through 9 can be used for each of the last three characters, how many different license plate numbers are possible?

Solution

There are 26 possible choices for the first letter ($n_1 = 26$), 26 possible choices for the second letter ($n_2 = 26$), and 26 possible choices for the third letter ($n_3 = 26$). In similar fashion, there are 10 choices for the numeral to be used for each of the fourth ($n_4 = 10$), fifth ($n_5 = 10$), and sixth ($n_6 = 10$) characters. Therefore, using the General Counting Rule [formula (A.2)] we find there are

$$26 \times 26 \times 26 \times 10 \times 10 \times 10 = \mathbf{17{,}576{,}000}$$

different "license plate numbers" using this six-character scheme.   ▲

ILLUSTRATION A.3 ▼

How many different "license plate numbers" are possible if the non-zero numerals are used for the three leading characters, letters are used for the three trailing characters, and the letters are not allowed to repeat?

Solution

There are 9 possible choices for each of the first three characters (since only 1 through 9 may be used). Thus, $n_1 = 9$, $n_2 = 9$, and $n_3 = 9$. The fourth character may be chosen from any one of the 26 letters ($n_4 = 26$). However, the fifth chaacter must be chosen from any one of the 25 letters not previously used ($n_5 = 25$), and the sixth character must be chosen from the 24 letters not previously used ($n_6 = 24$). Applying the General Countig Rule (A.2), we find there are

$$9 \times 9 \times 9 \times 26 \times 25 \times 24 = \mathbf{11{,}372{,}400}$$

different "license plate numbers" using this second six-character scheme.   ▲

We are now ready to investigate two additional concepts commonly encountered when enumerating possibilities: *permutations* and *combinations*. A permutation is a collection of distinct objects arranged in a specific order, while a combination is a collection of distinct objects without any specific order.

ILLUSTRATION A.4 ▼

There are four flags of different colors (one each of red, white, blue, and green) in a box, and you are asked to select any three. If you select {red, white, green} you have the same combination of colors as {green, red, white}. This question does not require or distinguish between different "orders" or arrangements; thus, each set of flags is one combination.

ILLUSTRATION A.5 ▼

There are four flags of different colors (one each of red, white, blue, and green) in a box, and you are asked to select any three of them and make a "signal" by hanging the three different flags, one above the other, on a flagpole. Since red over green over white is different from green over white over red, order is important and each possible signal is one permutation.

Permutations

ILLUSTRATION A.6 ▼

Select four different letters from the English alphabet and arrange them in any specific order. As a result of following these instructions, Barbara created the "four-letter word" BSJT. Rob created the word EOST. Steve selected KOCM. How many different "four-letter words" can be created?

Each of these "words" is a *permutation* of four letters selected from the set of 26 different letters forming the alphabet.

> **PERMUTATION**
>
> *An ordered arrangement of a set of distinct objects.* That is, there is a first object, a second object, a third object, and so on; and each object is distinctly different from the others.

The number of permutations that can be formed is calculated using an adaptation of the General Counting Rule.

PERMUTATION FORMULA

The number of permutations that can be formed using r different objects selected from a set of n distinct objects (symbolized by $_nP_r$ and read "the number of permutations of n objects selected r at a time") is

$$_nP_r = n \times (n - 1) \times (n - 2) \times \cdots \times (n - r + 1) \qquad \text{(A.3)}$$

or, in factorial notation,

$$_nP_r = \frac{n!}{(n - r)!} \qquad \text{(A.4)}$$

NOTE   More information and additional illustrations of factorial notation can be found in Chapter 5 and in the *Statistical Tutor*. Remember: $0! = 1$.

Let's continue with the solution of Illustration A.6. Since the 4 letters were selected from the 26 letters of the alphabet, the value of $r = 4$ (the number of selections) and $n = 26$ (the number of objects available for selection). Using formula (A.3),

$$_nP_r = \boldsymbol{n} \times (n - 1) \times \cdots \times (n - r + 1):$$

$$_{26}P_4 = 26 \times 25 \times \cdots \times (26 - 4 + 1)$$

$$= 26 \times 25 \times 24 \times 23 = \boldsymbol{358{,}800}$$

Or using formula (A.4),

$$_nP_r = \frac{n!}{(n - r)!}:$$

$$_{26}P_4 = \frac{26!}{(26 - 4)!} = \frac{26!}{22!}$$

$$= \frac{26 \times 25 \times 24 \times 23 \times 22 \times 21 \times \cdots \times 1}{22 \times 21 \times \cdots \times 1}$$

$$= \frac{26 \times 25 \times 24 \times 23 \times (22 \times 21 \times \cdots \times 1)}{(22 \times 21 \times \cdots \times 1)}$$

$$= 26 \times 25 \times 24 \times 23 = \boldsymbol{358{,}800}$$

ILLUSTRATION A.7 ▼

A group of eight finalists in a ceramic art competition are to be awarded five prizes—first, second, and so on. How many different ways are there to award these five prizes?

Solution

Since the prizes are ordered, this is a permutation of $n = 8$ different people, taken 5 at a time (only five prizes, and each prize is distinctly different from the others). Using formula (A.3), we find there are

$$_nP_r = n \times (n - 1) \times \cdots \times (n - r + 1):$$

$$_8P_5 = 8 \times 7 \times \cdots \times (8 - 5 + 1)$$

$$= 8 \times 7 \times 6 \times 5 \times 4 = \boldsymbol{6{,}720}$$

different possible ways of awarding these five prizes.

Combinations

ILLUSTRATION A.8 ▼

Select a set of four different letters from the English alphabet. As a result of following this instruction, Kevin selected A, E, R, and T. Karen selected D, E, N, and Q. Sue selected R, E, A, and T. Notice that Kevin and Sue selected the same set of letters, even though they selected them in different orders. These three people have selected two different sets of four letters. How many different sets of four letters can be selected?

Solution

Each of these "sets" of four letters represents a *combination* of $r = 4$ objects having been selected from a set of $n = 26$ distinct objects.

COMBINATION

A set of distinct objects without regard to an arrangement or an order. That is, the membership of the set is all that matters.

The number of combinations that can be selected is related to the number of permutations. In Illustration A.6, we found that there were 358,800 permutations of four letters possible. Many permutations were "words" formed from the same set of four letters. For example, the set of four letters A, B, C, and D can be used to form many permutations ("words"):

ABCD	ABDC	ACBD	ACDB	ADBC	ADCB
BACD	BADC	BCAD	BCDA	BDAC	BDCA
CBAD	CBDA	CABD	CADB	CDBA	CDAB
DBCA	DBAC	DCBA	DCAB	DABC	DACB

There are 4! ($4 \times 3 \times 2 \times 1$) or 24 different permutations for this set of four letters. Every other set of four letters can also be used to form 24 permutations. Therefore, if we divide the number of permutations possible (358,800) by the number of permutations each set has (24), the quotient will be the number of different sets (combinations) possible. That is, there are 14,950 (358,800/24) combinations of four letters possible. This concept is generalized in the following formula:

COMBINATION FORMULA

The number of combinations of r objects that can be selected from a set of n distinct objects (symbolized by $_nC_r$ and read "the number of combinations of n things taken r at a time") is

$$_nC_r = \frac{n(n-1)(n-2)\cdots(n-r+1)}{r!} \tag{A.5}$$

or, in factorial notation,

$$_nC_r = \frac{n!}{(n-r)! \times r!} \tag{A.6}$$

Let's continue with the solution of Illustration A.8 using these new formulas. First using formula (A.5),

$$_nC_r = \frac{n(n-1)(n-2)\cdots(n-r+1)}{r!}:$$

$$_{26}C_4 = \frac{26(25)(24)\cdots(26-4+1)}{4!} = \frac{26 \times 25 \times 24 \times 23}{4 \times 3 \times 2 \times 1}$$

$$= \frac{358,800}{24} = \mathbf{14,950}$$

Using formula (A.6),

$$_nC_r = \frac{n!}{(n-r)! \times r!}:$$

$$_{26}C_4 = \frac{26!}{(26-4)! \times 4!} = \frac{26!}{22! \times 4!}$$

$$= \frac{26 \times 25 \times 24 \times 23 \times 22 \times 21 \times \cdots \times 2 \times 1}{(22 \times 21 \times \cdots \times 2 \times 1)(4 \times 3 \times 2 \times 1)}$$

$$= \frac{26 \times 25 \times 24 \times 23 \times (22 \times 21 \times \cdots \times 2 \times 1)}{(22 \times 21 \times \cdots \times 2 \times 1)(4 \times 3 \times 2 \times 1)}$$

$$= \frac{358,800}{24} = \mathbf{14,950}$$

ILLUSTRATION A.9 ▼

A department has 30 members and a committee of 5 people is needed to carry out a task. How many different possible committees are there?

Solution

As stated, there is no specific assignment or order to the members of the committee; therefore, each possible committee is a combination and $n = 30$ (the number of people eligible to be selected), and $r = 5$ (the number to be selected).

$$_nC_r = \frac{n!}{(n-r)! \times r!}:$$

$$_{30}C_5 = \frac{30!}{(30-5)! \times 5!} = \frac{30!}{25! \times 5!}$$

$$= \frac{30 \times 29 \times 28 \times 27 \times 26 \times 25 \times 24 \times \cdots \times 2 \times 1}{(25 \times 24 \times \cdots \times 2 \times 1)(5 \times 4 \times 3 \times 2 \times 1)}$$

$$= \frac{30 \times 29 \times 28 \times 27 \times 26 \times (25 \times 24 \times \cdots \times 2 \times 1)}{(25 \times 24 \times \cdots \times 2 \times 1)(5 \times 4 \times 3 \times 2 \times 1)}$$

$$= \frac{17,100,720}{120} = \mathbf{142,506}$$

It is possible to select 142,506 different committees of 5 people from this department of 30 people.   ▲

NOTE   The number of combinations $_nC_r$ and the binomial coefficient $\binom{n}{r}$ or $\binom{n}{x}$ are numerically equivalent.

The three "counting" formulas described above in this section [formulas (A.2), (A.4), and (A.6)] can be and often are used together to solve problems.

ILLUSTRATION A.10 ▼

A department has 30 members and a committee is needed to carry out a task. The committee is to be composed of a chairperson and four members. How many different possible committees are there?

Solution

This problem is solved by treating it in two parts: consider the chairperson position and the committee members are two separate parts to be combined using the Fundamental Counting Rule ($m \times n$). Let m be the number of possible choices for the chairperson. Since any one of the 30 department members could serve as the chair, $m = 30$. Let n be the number of four-person committees that can be selected from the remaining 29 department members. Since these four have no specific assignment, the number of possibilities, n, is the number of combinations of 29 things taken 4 at a time.

$$n(\text{committees}) = m \times n$$

$$= 30 \times {}_{29}C_4$$

$$= 30 \times \frac{29 \times 28 \times 27 \times 26 \times (25 \times 24 \times \cdots \times 2 \times 1)}{(25 \times 24 \times \cdots \times 2 \times 1) \times 4 \times 3 \times 2 \times 1}$$

$$= 30 \times 23{,}751 = \mathbf{712{,}530}$$

712,530 different committees of size 5 with an assigned chair are possible.

ILLUSTRATION A.11 ▼

A department has 30 members and a committee is needed to carry out a task. The committee is to be composed of two co-chairpersons and three members. How many different possible committees are there?

Solution

This problem is solved by treating it in two parts: consider the selecting of two co-chairpersons and then the remaining committee members as two separate parts to be combined using the Fundamental Counting Rule ($m \times n$). Let m be the number of possible choices for the co-chairpersons. This is like a committee of two, since there is no further distinction between them; therefore $m = {}_{30}C_2$, since any two of the 30 department members could serve as the co-chairs. Let n be the number of three-person committees that can be selected from the remaining 28 department members. Since these three have no specific assignment, the number of possibilities, n, is the number of combinations of 28 things taken 3 at a time.

$$n(\text{committees}) = m \times n$$

$$= {}_{30}C_2 \times {}_{28}C_3$$

$$= \frac{30!}{28! \times 2!} \times \frac{28!}{25! \times 3!}$$

$$= \frac{(30 \times 29 \times 28 \times 27 \times 26 \times 25 \times \cdots \times 1) \times (28 \times 27 \times \cdots \times 1)}{(28 \times 27 \times \cdots \times 1) \times (2 \times 1) \times (25 \times \cdots \times 1) \times (3 \times 2 \times 1)}$$

$$= 15 \times 29 \times 28 \times 9 \times 13$$

$$= \mathbf{1{,}425{,}060}$$

1,425,060 different committees of size 5 with assigned co-chairpersons are possible.   ▲

A.1 A long weekend of three days is being planned. The three days are to be spent taking scenic drives through the countryside and ending each day at a motel where reservations have been previously made for Friday and Saturday nights (will be home Sunday night). There are three scenic routes that may be traveled on Friday, two choices for Saturday, and three scenic route choices for Sunday's return trip. How many different trips are possible if
  a. all the scenic options are considered?
  b. one of the Friday routes has been previously driven, and is not a choice?
  c. on Sunday, it is decided to drive straight home and not take a scenic route?

A.2 **a.** Show that formulas (A.3) and (A.4) are equivalent.
  b. Show that formulas (A.5) and (A.6) are equivalent.

A.3 Explain why each of the following pairs of "counts" are equal:
  a. $_nP_n$ and $_nP_{n-1}$ **b.** $_nP_1$ and $_nC_1$
  c. $_nC_r$ and $_nC_{n-r}$ **d.** $_nC_r$ and the binomial coefficient $\binom{n}{r}$.

A.4 A department of 30 people is to select a committee of 5 persons. How many different committees are possible if the committee is composed of
  a. a chairperson, a secretary, and three others?
  b. two co-chairs and three others?
  c. two co-chairs, a secretary, and two others?

A.5 License plates are to be "numbered" using a combination of letters and numerals. How many different "numbers" are possible if each of the following sets of restrictions is used?
  a. Six characters using any combination or arrangement of the 26 letters and 10 single-digit numerals.
  b. Six characters using any combination or arrangement of letters and single-digit numerals, except that "zero" and "one" are not to be used because they are hard to distinguish from "o" and "i."
  c. Six characters using letters for the two leading characters and the 10 single-digit numerals for the four trailing characters.
  d. Six characters using the 10 single-digit numerals for the four leading characters and letters for the two trailing characters.
  e. Six characters using the 10 single-digit numerals for the four leading characters and letters for the two trailing characters, except that "zero" cannot be the leading character.

A.6 Mathew has six shirts, four pairs of pants, and five pairs of socks clean and ready to wear. How many different "outfits" can he assemble if
  a. he wears one item from each category?
  b. he wears one specific shirt and one item from the other two categories?
  c. he only wears two of the shirts with one specific pair of pants and no socks, but the rest are worn in any complete combination?

A.7 Five cards are to be randomly selected from a standard bridge deck of 52 cards.
  a. How many different "hands" of five cards are possible?
  b. How many different "hands" of five cards are possible if the first drawn is an ace?
  c. How many different "hands" of five cards are possible if the first card drawn is an ace and the remaining four are not aces?
  d. How many different "hands" of five cards are possible if the first card drawn is a club and the remaining four are not clubs?
  e. How many different "hands" of five cards are possible if the first card drawn is an ace and the remaining four are not clubs?

A.8 Millions of people play the large lotteries; some play regularly while others play only when the prize is very large. Powerball is one of the largest and is played in many states. A ticket costs $1 and the players choose five numbers from 1 to 49 and a Powerball from 1 to 42.
  a. Use combinations to calculate the exact odds that someone will match all six numbers.
  b. Does your answer (a) verify the 1 in 80.1 million odds advertised by the game?

Appendix

B *Tables*

Table 1 Random Numbers

10 09 73 25 33	76 52 01 35 86	34 67 35 48 76	80 95 90 91 17	39 29 27 49 45
37 54 20 48 05	64 89 47 42 96	24 80 52 40 37	20 63 61 04 02	00 82 29 16 65
08 42 26 89 53	19 64 50 93 03	23 20 90 25 60	15 95 33 43 64	35 08 03 36 06
99 01 90 25 29	09 37 67 07 15	38 31 13 11 65	88 67 67 43 97	04 43 62 76 59
12 80 79 99 70	80 15 73 61 47	64 03 23 66 53	98 95 11 68 77	12 17 17 68 33
66 06 57 47 17	34 07 27 68 50	36 69 73 61 70	65 81 33 98 85	11 19 92 91 70
31 06 01 08 05	45 57 18 24 06	35 30 34 26 14	86 79 90 74 39	23 40 30 97 32
85 26 97 76 02	02 05 16 56 92	68 66 57 48 18	73 05 38 52 47	18 62 38 85 79
63 57 33 21 35	05 32 54 70 48	90 55 35 75 48	28 46 82 87 09	83 49 12 56 24
73 79 64 57 53	03 52 96 47 78	35 80 83 42 82	60 93 52 03 44	35 27 38 84 35
98 52 01 77 67	14 90 56 86 07	22 10 94 05 58	60 97 09 34 33	50 50 07 39 98
11 80 50 54 31	39 80 82 77 32	50 72 56 82 48	29 40 52 42 01	52 77 56 78 51
83 45 29 96 34	06 28 89 80 83	13 74 67 00 78	18 47 54 06 10	68 71 17 78 17
88 68 54 02 00	86 50 75 84 01	36 76 66 79 51	90 36 47 64 93	29 60 91 10 62
99 59 46 73 48	87 51 76 49 69	91 82 60 89 28	93 78 56 13 68	23 47 83 41 13
65 48 11 76 74	17 46 85 09 50	58 04 77 69 74	73 03 95 71 86	40 21 81 65 44
80 12 43 56 35	17 72 70 80 15	45 31 82 23 74	21 11 57 82 53	14 38 55 37 63
74 35 09 98 17	77 40 27 72 14	43 23 60 02 10	45 52 16 42 37	96 28 60 26 55
69 91 62 68 03	66 25 22 91 48	36 93 68 72 03	76 62 11 39 90	94 40 05 64 18
09 89 32 05 05	14 22 56 85 14	46 42 75 67 88	96 29 77 88 22	54 38 21 45 98
91 49 91 45 23	68 47 92 76 86	46 16 28 35 54	94 75 08 99 23	37 08 92 00 48
80 33 69 45 98	26 94 03 68 58	70 29 73 41 35	54 14 03 33 40	42 05 08 23 41
44 10 48 19 49	85 15 74 79 54	32 97 92 65 75	57 60 04 08 81	22 22 20 64 13
12 55 07 37 42	11 10 00 20 40	12 86 07 46 97	96 64 48 94 39	28 70 72 58 15
63 60 64 93 29	16 50 53 44 84	40 21 95 25 63	43 65 17 70 82	07 20 73 17 90
61 19 69 04 46	26 45 74 77 74	51 92 43 37 29	65 39 45 95 93	42 58 26 05 27
15 47 44 52 66	95 27 07 99 53	59 36 78 38 48	82 39 61 01 18	33 21 15 94 66
94 55 72 85 73	67 89 75 43 87	54 62 24 44 31	91 19 04 25 92	92 92 74 59 73
42 48 11 62 13	97 34 40 87 21	16 86 84 87 67	03 07 11 20 59	25 70 14 66 70
23 52 37 83 17	73 20 88 98 37	68 93 59 14 16	26 25 22 96 63	05 52 28 25 62
04 49 35 24 94	75 24 63 38 24	45 86 25 10 25	61 96 27 93 35	65 33 71 24 72
00 54 99 76 54	64 05 18 81 59	96 11 96 38 96	54 69 28 23 91	23 28 72 95 29
35 96 31 53 07	26 89 80 93 54	33 35 13 54 62	77 97 45 00 24	90 10 33 93 33
59 80 80 83 91	45 42 72 68 42	83 60 94 97 00	13 02 12 48 92	78 56 52 01 06
46 05 88 52 36	01 39 09 22 86	77 28 14 40 77	93 91 08 36 47	70 61 74 29 41
32 17 90 05 97	87 37 92 52 41	05 56 70 70 07	86 74 31 71 57	85 39 41 18 38
69 23 46 14 06	20 11 74 52 04	15 95 66 00 00	18 74 39 24 23	97 11 89 63 38
19 56 54 14 30	01 75 87 53 79	40 41 92 15 85	66 67 43 68 06	84 96 28 52 07
45 15 51 49 38	19 47 60 72 46	43 66 79 45 43	59 04 79 00 33	20 82 66 95 41
94 86 43 19 94	36 16 81 08 51	34 88 88 15 53	01 54 03 54 56	05 01 45 11 76
98 08 62 48 26	45 24 02 84 04	44 99 90 88 96	39 09 47 34 07	35 44 13 18 80
33 18 51 62 32	41 94 15 09 49	89 43 54 85 81	88 69 54 19 94	37 54 87 30 43
80 95 10 04 06	96 38 27 07 74	20 15 12 33 87	25 01 62 52 98	94 62 46 11 71
79 75 24 91 40	71 96 12 82 96	69 86 10 25 91	74 85 22 05 39	00 38 75 95 79
18 63 33 25 37	98 14 50 65 71	31 01 02 46 74	05 45 56 14 27	77 93 89 19 36

For specific details about using this table, see page 23 or the *Statistical Tutor*.

Table 1 (Continued)

74 02 94 39 02	77 55 73 22 70	97 79 01 71 19	52 52 75 80 21	80 81 45 17 48
54 17 84 56 11	80 99 33 71 43	05 33 51 29 69	56 12 71 92 55	36 04 09 03 24
11 66 44 98 83	52 07 98 48 27	59 38 17 15 39	09 97 33 34 40	88 46 12 33 56
48 32 47 79 28	31 24 96 47 10	02 29 53 68 70	32 30 75 75 46	15 02 00 99 94
69 07 49 41 38	87 63 79 19 76	35 58 40 44 01	10 51 82 16 15	01 84 87 69 38
09 18 82 00 97	32 82 53 95 27	04 22 08 63 04	83 38 98 73 74	64 27 85 80 44
90 04 58 54 97	51 98 15 06 54	94 93 88 19 97	91 87 07 61 50	68 47 66 46 59
73 18 95 02 07	47 67 72 62 69	62 29 06 44 64	27 12 46 70 18	41 36 18 27 60
75 76 87 64 90	20 97 18 17 49	90 42 91 22 72	95 37 50 58 71	93 82 34 31 78
54 01 64 40 56	66 28 13 10 03	00 68 22 73 98	20 71 45 32 95	07 70 61 78 13
08 35 86 99 10	78 54 24 27 85	13 66 15 88 73	04 61 89 75 53	31 22 30 84 20
28 30 60 32 64	81 33 31 05 91	40 51 00 78 93	32 60 46 04 75	94 11 90 18 40
53 84 08 62 33	81 59 41 36 28	51 21 59 02 90	28 46 66 87 95	77 76 22 07 91
91 75 75 37 41	61 61 36 22 69	50 26 39 02 12	55 78 17 65 14	83 48 34 70 55
89 41 59 26 94	00 39 75 83 91	12 60 71 76 46	48 94 97 23 06	94 54 13 74 08
77 51 30 38 20	86 83 42 99 01	68 41 48 27 74	51 90 81 39 80	72 89 35 55 07
19 50 23 71 74	69 97 92 02 88	55 21 02 97 73	74 28 77 52 51	65 34 46 74 15
21 81 85 93 13	93 27 88 17 57	05 68 67 31 56	07 08 28 50 46	31 85 33 84 52
51 47 46 64 99	68 10 72 36 21	94 04 99 13 45	42 83 60 91 91	08 00 74 54 49
99 55 96 83 31	62 53 52 41 70	69 77 71 28 30	74 81 97 81 42	43 86 07 28 34
33 71 34 80 07	93 58 47 28 69	51 92 66 47 21	58 30 32 98 22	93 17 49 39 72
85 27 48 68 93	11 30 32 92 70	28 83 43 41 37	73 51 59 04 00	71 14 84 36 43
84 13 38 96 40	44 03 55 21 66	73 85 27 00 91	61 22 26 05 61	62 32 71 84 23
56 73 21 62 34	17 39 59 61 31	10 12 39 16 22	85 49 65 75 60	81 60 41 88 80
65 13 85 68 06	87 60 88 52 61	34 31 36 58 61	45 87 52 10 69	85 64 44 72 77
38 00 10 21 76	81 71 91 17 11	71 60 29 29 37	74 21 96 40 49	65 58 44 96 98
37 40 29 63 97	01 30 47 75 86	56 27 11 00 86	47 32 46 26 05	40 03 03 74 38
97 12 54 03 48	87 08 33 14 17	21 81 53 92 50	75 23 76 20 47	15 50 12 95 78
21 82 64 11 34	47 14 33 40 72	64 63 88 59 02	49 13 90 64 41	03 85 65 45 52
73 13 54 27 42	95 71 90 90 35	85 79 47 42 96	08 78 98 81 56	64 69 11 92 02
07 63 87 79 29	03 06 11 80 72	96 20 74 41 56	23 82 19 95 38	04 71 36 69 94
60 52 88 34 41	07 95 41 98 14	59 17 52 06 95	05 53 35 21 39	61 21 20 64 55
83 59 63 56 55	06 95 89 29 83	05 12 80 97 19	77 43 35 37 83	92 30 15 04 98
10 85 06 27 46	99 59 91 05 07	13 49 90 63 19	53 07 57 18 39	06 41 01 93 62
39 82 09 89 52	43 62 26 31 47	64 42 18 08 14	43 80 00 93 51	31 02 47 31 67
59 58 00 64 78	75 56 97 88 00	88 83 55 44 86	23 76 80 61 56	04 11 10 84 08
38 50 80 73 41	23 79 34 87 63	90 82 29 70 22	17 71 90 42 07	95 95 44 99 53
30 69 27 06 68	94 68 81 61 27	56 19 68 00 91	82 06 76 34 00	05 46 26 92 00
65 44 39 56 59	18 28 82 74 37	49 63 22 40 41	08 33 76 56 76	96 29 99 08 36
27 26 75 02 64	13 19 27 22 94	07 47 74 46 06	17 98 54 89 11	97 34 13 03 58
91 30 70 69 91	19 07 22 42 10	36 69 95 37 28	28 82 53 57 93	28 97 66 62 52
68 43 49 46 88	84 47 31 36 22	62 12 69 84 08	12 84 38 25 90	09 81 59 31 46
48 90 81 58 77	54 74 52 45 91	35 70 00 47 54	83 82 45 26 92	54 13 05 51 60
06 91 34 51 97	42 67 27 86 01	11 88 30 95 28	63 01 19 89 01	14 97 44 03 44
10 45 51 60 19	14 21 03 37 12	91 34 23 78 21	88 32 58 08 51	43 66 77 08 83
12 88 39 73 43	65 02 76 11 84	04 28 50 13 92	17 97 41 50 77	90 71 22 67 69
21 77 83 09 76	38 80 73 69 61	31 64 94 20 96	63 28 10 20 23	08 81 64 74 49
19 52 35 95 15	65 12 25 96 59	86 28 36 82 58	69 57 21 37 98	16 43 59 15 29
67 24 55 26 70	35 58 31 65 63	79 24 68 66 86	76 46 33 42 22	26 65 59 08 02
60 58 44 73 77	07 50 03 79 92	45 13 42 65 29	26 76 08 36 37	41 32 64 43 44
53 85 34 13 77	36 06 69 48 50	58 83 87 38 59	49 36 47 33 31	96 24 04 36 42
24 63 73 97 36	74 38 48 93 42	52 62 30 79 92	12 36 91 86 01	03 74 28 38 73
83 08 01 24 51	38 99 22 28 15	07 75 95 17 77	97 37 72 75 85	51 97 23 78 67
16 44 42 43 34	36 15 19 90 73	27 49 37 09 39	85 13 03 25 52	54 84 65 47 59
60 79 01 81 57	57 17 86 57 62	11 16 17 85 76	45 81 95 29 79	65 13 00 48 60

From tables of the RAND Corporation. Reprinted from Wilfred J. Dixon and Frank J. Massey, Jr., *Introduction to Statistical Analysis.* 3rd ed. (New York: McGraw-Hill, 1969), pp. 446–447. Reprinted by permission of the RAND Corporation.

Table 2 Binomial Probabilities $\left[\binom{n}{x} \cdot p^x \cdot q^{n-x}\right]$

								p							
n	*x*	0.01	0.05	0.10	0.20	0.30	0.40	0.50	0.60	0.70	0.80	0.90	0.95	0.99	*x*
2	0	.980	.902	.810	.640	.490	.360	.250	.160	.090	.040	.010	.002	0+	0
	1	.020	.095	.180	.320	.420	.480	.500	.480	.420	.320	.180	.095	.020	1
	2	0+	.002	.010	.040	.090	.160	.250	.360	.490	.640	.810	.902	.980	2
3	0	.970	.857	.729	.512	.343	.216	.125	.064	.027	.008	.001	0+	0+	0
	1	.029	.135	.243	.384	.441	.432	.375	.288	.189	.096	.027	.007	0+	1
	2	0+	.007	.027	.096	.189	.288	.375	.432	.441	.384	.243	.135	.029	2
	3	0+	0+	.001	.008	.027	.064	.125	.216	.343	.512	.729	.857	.970	3
4	0	.961	.815	.656	.410	.240	.130	.062	.026	.008	.002	0+	0+	0+	0
	1	.039	.171	.292	.410	.412	.346	.250	.154	.076	.026	.004	0+	0+	1
	2	.001	.014	.049	.154	.265	.346	.375	.346	.265	.154	.049	.014	.001	2
	3	0+	0+	.004	.026	.076	.154	.250	.346	.412	.410	.292	.171	.039	3
	4	0+	0+	0+	.002	.008	.026	.062	.130	.240	.410	.656	.815	.961	4
5	0	.951	.774	.590	.328	.168	.078	.031	.010	.002	0+	0+	0+	0+	0
	1	.048	.204	.328	.410	.360	.259	.156	.077	.028	.006	0+	0+	0+	1
	2	.001	.021	.073	.205	.309	.346	.312	.230	.132	.051	.008	.001	0+	2
	3	0+	.001	.008	.051	.132	.230	.312	.346	.309	.205	.073	.021	.001	3
	4	0+	0+	0+	.006	.028	.077	.156	.259	.360	.410	.328	.204	.048	4
	5	0+	0+	0+	0+	.002	.010	.031	.078	.168	.328	.590	.774	.951	5
6	0	.941	.735	.531	.262	.118	.047	.016	.004	.001	0+	0+	0+	0+	0
	1	.057	.232	.354	.393	.303	.187	.094	.037	.010	.002	0+	0+	0+	1
	2	.001	.031	.098	.246	.324	.311	.234	.138	.060	.015	.001	0+	0+	2
	3	0+	.002	.015	.082	.185	.276	.312	.276	.185	.082	.015	.002	0+	3
	4	0+	0+	.001	.015	.060	.138	.234	.311	.324	.246	.098	.031	.001	4
	5	0+	0+	0+	.002	.010	.037	.094	.187	.303	.393	.354	.232	.057	5
	6	0+	0+	0+	0+	.001	.004	.016	.047	.118	.262	.531	.735	.941	6
7	0	.932	.698	.478	.210	.082	.028	.008	.002	0+	0+	0+	0+	0+	0
	1	.066	.257	.372	.367	.247	.131	.055	.017	.004	0+	0+	0+	0+	1
	2	.002	.041	.124	.275	.318	.261	.164	.077	.025	.004	0+	0+	0+	2
	3	0+	.004	.023	.115	.227	.290	.273	.194	.097	.029	.003	0+	0+	3
	4	0+	0+	.003	.029	.097	.194	.273	.290	.227	.115	.023	.004	0+	4
	5	0+	0+	0+	.004	.025	.077	.164	.261	.318	.275	.124	.041	.002	5
	6	0+	0+	0+	0+	.004	.017	.055	.131	.247	.367	.372	.257	.066	6
	7	0+	0+	0+	0+	0+	.002	.008	.028	.082	.210	.478	.698	.932	7
8	0	.923	.663	.430	.168	.058	.017	.004	.001	0+	0+	0+	0+	0+	0
	1	.075	.279	.383	.336	.198	.090	.031	.008	.001	0+	0+	0+	0+	1
	2	.003	.051	.149	.294	.296	.209	.109	.041	.010	.001	0+	0+	0+	2
	3	0+	.005	.033	.147	.254	.279	.219	.124	.047	.009	0+	0+	0+	3
	4	0+	0+	.005	.046	.136	.232	.273	.232	.136	.046	.005	0+	0+	4
	5	0+	0+	0+	.009	.047	.124	.219	.279	.254	.147	.033	.005	0+	5
	6	0+	0+	0+	.001	.010	.041	.109	.209	.296	.294	.149	.051	.003	6
	7	0+	0+	0+	0+	.001	.008	.031	.090	.198	.336	.383	.279	.075	7
	8	0+	0+	0+	0+	0+	.001	.004	.017	.058	.168	.430	.663	.923	8

For specific details about using this table, see page 257.

Table 2 (Continued)

								p								
n	x	0.01	0.05	0.10	0.20	0.30	0.40	0.50	0.60	0.70	0.80	0.90	0.95	0.99	x
9	0	.914	.630	.387	.134	.040	.010	.002	0+	0+	0+	0+	0+	0+	0
	1	.083	.299	.387	.302	.156	.060	.018	.004	0+	0+	0+	0+	0+	1
	2	.003	.063	.172	.302	.267	.161	.070	.021	.004	.0+	0+	0+	0+	2
	3	0+	.008	.045	.176	.267	.251	.164	.074	.021	.003	0+	0+	0+	3
	4	0+	.001	.007	.066	.172	.251	.246	.167	.074	.017	.001	0+	0+	4
	5	0+	0+	.001	.017	.074	.167	.246	.251	.172	.066	.007	.001	0+	5
	6	0+	0+	0+	.003	.021	.074	.164	.251	.267	.176	.045	.008	0+	6
	7	0+	0+	0+	0+	.004	.021	.070	.161	.267	.302	.172	.063	.003	7
	8	0+	0+	0+	0+	0+	.004	.018	.060	.156	.302	.387	.299	.083	8
	9	0+	0+	0+	0+	0+	0+	.002	.010	.040	.134	.387	.630	.914	9
10	0	.904	.599	.349	.107	.028	.006	.001	.0+	0+	0+	0+	0+	0+	0
	1	.091	.315	.387	.268	.121	.040	.010	.002	0+	0+	0+	0+	0+	1
	2	.004	.075	.194	.302	.233	.121	.044	.011	.001	0+	0+	0+	0+	2
	3	0+	.010	.057	.201	.267	.215	.117	.042	.009	.001	0+	0+	0+	3
	4	0+	.001	.011	.088	.200	.251	.205	.111	.037	.006	0+	0+	0+	4
	5	0+	0+	.001	.026	.103	.201	.246	.201	.103	.026	.001	0+	0+	5
	6	0+	0+	0+	.006	.037	.111	.205	.251	.200	.088	.011	.001	0+	6
	7	0+	0+	0+	.001	.009	.042	.117	.215	.267	.201	.057	.010	0+	7
	8	0+	0+	0+	0+	.001	.011	.044	.121	.233	.302	.194	.075	.004	8
	9	0+	0+	0+	0+	0+	.002	.010	.040	.121	.268	.387	.315	.091	9
	10	0+	0+	0+	0+	0+	0+	.001	.006	.028	.107	.349	.599	.904	10
11	0	.895	.569	.314	.086	.020	.004	0+	0+	0+	0+	0+	0+	0+	0
	1	.099	.329	.384	.236	.093	.027	.005	.001	0+	0+	0+	0+	0+	1
	2	.005	.087	.213	.295	.200	.089	.027	.005	.001	0+	0+	0+	0+	1
	3	0+	.014	.071	.221	.257	.177	.081	.023	.004	0+	0+	0+	0+	3
	4	0+	.001	.016	.111	.220	.236	.161	.070	.017	.002	0+	0+	0+	4
	5	0+	0+	.002	.039	.132	.221	.226	.147	.057	.010	0+	0+	0+	5
	6	0+	0+	0+	.010	.057	.147	.226	.221	.132	.039	.002	0+	0+	6
	7	0+	0+	0+	.002	.017	.070	.161	.236	.220	.111	.016	.001	0+	7
	8	0+	0+	0+	0+	.004	.023	.081	.177	.257	.221	.071	.014	0+	8
	9	0+	0+	0+	0+	.001	.005	.027	.089	.200	.295	.213	.087	.005	9
	10	0+	0+	0+	0+	0+	.001	.005	.027	.093	.236	.384	.329	.099	10
	11	0+	0+	0+	0+	0+	0+	0+	.004	.020	.086	.314	.569	.895	11
12	0	.886	.540	.282	.069	.014	.002	0+	0+	0+	0+	0+	0+	0+	0
	1	.107	.341	.377	.206	.071	.017	.003	0+	0+	0+	0+	0+	0+	1
	2	.006	.099	.230	.283	.168	.064	.016	.002	0+	0+	0+	0+	0+	2
	3	0+	.017	.085	.236	.240	.142	.054	.012	.001	0+	0+	0+	0+	3
	4	0+	.002	.021	.133	.231	.213	.121	.042	.008	.001	0+	0+	0+	4
	5	0+	0+	.004	.053	.158	.227	.193	.101	.029	.003	0+	0+	0+	5
	6	0+	0+	0+	.016	.079	.177	.226	.177	.079	.016	0+	0+	0+	6
	7	0+	0+	0+	.003	.029	.101	.193	.227	.158	.053	.004	0+	0+	7
	8	0+	0+	0+	.001	.008	.042	.121	.213	.231	.133	.021	.002	0+	8
	9	0+	0+	0+	0+	.001	.012	.054	.142	.240	.236	.085	.017	0+	9
	10	0+	0+	0+	0+	0+	.002	.016	.064	.168	.283	.230	.099	.006	10
	11	0+	0+	0+	0+	0+	0+	.003	.017	.071	.206	.377	.341	.107	11
	12	0+	0+	0+	0+	0+	0+	0+	.002	.014	.069	.282	.540	.886	12

Table 2   (Continued)

n	x	0.01	0.05	0.10	0.20	0.30	0.40	0.50	0.60	0.70	0.80	0.90	0.95	0.99	x
13	0	.878	.513	.254	.055	.010	.001	0+	0+	0+	0+	0+	0+	0+	0
	1	.115	.351	.367	.179	.054	.011	.002	0+	0+	0+	0+	0+	0+	1
	2	.007	.111	.245	.268	.139	.045	.010	.001	0+	0+	0+	0+	0+	2
	3	0+	.021	.100	.246	.218	.111	.035	.006	.001	0+	0+	0+	0+	3
	4	0+	.003	.028	.154	.234	.184	.087	.024	.003	0+	0+	0+	0+	4
	5	0+	0+	.006	.069	.180	.221	.157	.066	.014	.001	0+	0+	0+	5
	6	0+	0+	.001	.023	.103	.197	.209	.131	.044	.006	0+	0+	0+	6
	7	0+	0+	0+	.006	.044	.131	.209	.197	.103	.023	.001	0+	0+	7
	8	0+	0+	0+	.001	.014	.066	.157	.221	.180	.069	.006	0+	0+	8
	9	0+	0+	0+	0+	.003	.024	.087	.184	.234	.154	.028	.003	0+	9
	10	0+	0+	0+	0+	.001	.006	.035	.111	.218	.246	.100	.021	0+	10
	11	0+	0+	0+	0+	0+	.001	.010	.045	.139	.268	.245	.111	.007	11
	12	0+	0+	0+	0+	0+	0+	.002	.011	.054	.179	.367	.351	.115	12
	13	0+	0+	0+	0+	0+	0+	0+	.001	.010	.055	.254	.513	.878	13
14	0	.869	.488	.229	.044	.007	.001	0+	0+	0+	0+	0+	0+	0+	0
	1	.123	.359	.356	.154	.041	.007	.001	0+	0+	0+	0+	0+	0+	1
	2	.008	.123	.257	.250	.113	.032	.006	.001	0+	0+	0+	0+	0+	2
	3	0+	.026	.114	.250	.194	.085	.022	.003	0+	0+	0+	0+	0+	3
	4	0+	.004	.035	.172	.229	.155	.061	.014	.001	0+	0+	0+	0+	4
	5	0+	0+	.008	.086	.196	.207	.122	.041	.007	0+	0+	0+	0+	5
	6	0+	0+	.001	.032	.126	.207	.183	.092	.023	.002	0+	0+	0+	6
	7	0+	0+	0+	.009	.062	.157	.209	.157	.062	.009	0+	0+	0+	7
	8	0+	0+	0+	.002	.023	.092	.183	.207	.126	.032	.001	0+	0+	8
	9	0+	0+	0+	0+	.007	.041	.122	.207	.196	.086	.008	0+	0+	9
	10	0+	0+	0+	0+	.001	.014	.061	.155	.229	.172	.035	.004	0+	10
	11	0+	0+	0+	0+	0+	.003	.022	.085	.194	.250	.114	.026	0+	11
	12	0+	0+	0+	0+	0+	.001	.006	.032	.113	.250	.257	.123	.008	12
	13	0+	0+	0+	0+	0+	0+	.001	.007	.041	.154	.356	.359	.123	13
	14	0+	0+	0+	0+	0+	0+	0+	.001	.007	.044	.229	.488	.869	14
15	0	.860	.463	.206	.035	.005	0+	0+	0+	0+	0+	0+	0+	0+	0
	1	.130	.366	.343	.132	.031	.005	0+	0+	0+	0+	0+	0+	0+	1
	2	.009	.135	.267	.231	.092	.022	.003	0+	0+	0+	0+	0+	0+	2
	3	0+	.031	.129	.250	.170	.063	.014	.002	0+	0+	0+	0+	0+	3
	4	0+	.005	.043	.188	.219	.127	.042	.007	.001	0+	0+	0+	0+	4
	5	0+	.001	.010	.103	.206	.186	.092	.024	.003	0+	0+	0+	0+	5
	6	0+	0+	.002	.043	.147	.207	.153	.061	.012	.001	0+	0+	0+	6
	7	0+	0+	0+	.014	.081	.177	.196	.118	.035	.003	0+	0+	0+	7
	8	0+	0+	0+	.003	.035	.118	.196	.177	.081	.014	0+	0+	0+	8
	9	0+	0+	0+	.001	.012	.061	.153	.207	.147	.043	.002	0+	0+	9
	10	0+	0+	0+	0+	.003	.024	.092	.186	.206	.103	.010	.001	0+	10
	11	0+	0+	0+	0+	.001	.007	.042	.127	.219	.188	.043	.005	0+	11
	12	0+	0+	0+	0+	0+	.002	.014	.063	.170	.250	.129	.031	0+	12
	13	0+	0+	0+	0+	0+	0+	.003	.022	.092	.231	.267	.135	.009	13
	14	0+	0+	0+	0+	0+	0+	0+	.005	.031	.132	.343	.366	.130	14
	15	0+	0+	0+	0+	0+	0+	0+	0+	.005	.035	.206	.463	.860	15

Table 3   Areas of the Standard Normal Distribution

The entries in this table are the probabilities that a random variable with a standard normal distribution assumes a value between 0 and z; the probability is represented by the shaded area under the curve in the accompanying figure. Areas for negative values of z are obtained by symmetry.

	Second Decimal Place in z									
z	0.00	0.01	0.02	0.03	0.04	0.05	0.06	0.07	0.08	0.09
0.0	0.0000	0.0040	0.0080	0.0120	0.0160	0.0199	0.0239	0.0279	0.0319	0.0359
0.1	0.0398	0.0438	0.0478	0.0517	0.0557	0.0596	0.0636	0.0675	0.0714	0.0753
0.2	0.0793	0.0832	0.0871	0.0910	0.0948	0.0987	0.1026	0.1064	0.1103	0.1141
0.3	0.1179	0.1217	0.1255	0.1293	0.1331	0.1368	0.1406	0.1443	0.1480	0.1517
0.4	0.1554	0.1591	0.1628	0.1664	0.1700	0.1736	0.1772	0.1808	0.1844	0.1879
0.5	0.1915	0.1950	0.1985	0.2019	0.2054	0.2088	0.2123	0.2157	0.2190	0.2224
0.6	0.2257	0.2291	0.2324	0.2357	0.2389	0.2422	0.2454	0.2486	0.2517	0.2549
0.7	0.2580	0.2611	0.2642	0.2673	0.2704	0.2734	0.2764	0.2794	0.2823	0.2852
0.8	0.2881	0.2910	0.2939	0.2967	0.2995	0.3023	0.3051	0.3078	0.3106	0.3133
0.9	0.3159	0.3186	0.3212	0.3238	0.3264	0.3289	0.3315	0.3340	0.3365	0.3389
1.0	0.3413	0.3438	0.3461	0.3485	0.3508	0.3531	0.3554	0.3577	0.3599	0.3621
1.1	0.3643	0.3665	0.3686	0.3708	0.3729	0.3749	0.3770	0.3790	0.3810	0.3830
1.2	0.3849	0.3869	0.3888	0.3907	0.3925	0.3944	0.3962	0.3980	0.3997	0.4015
1.3	0.4032	0.4049	0.4066	0.4082	0.4099	0.4115	0.4131	0.4147	0.4162	0.4177
1.4	0.4192	0.4207	0.4222	0.4236	0.4251	0.4265	0.4279	0.4292	0.4306	0.4319
1.5	0.4332	0.4345	0.4357	0.4370	0.4382	0.4394	0.4406	0.4418	0.4429	0.4441
1.6	0.4452	0.4463	0.4474	0.4484	0.4495	0.4505	0.4515	0.4525	0.4535	0.4545
1.7	0.4554	0.4564	0.4573	0.4582	0.4591	0.4599	0.4608	0.4616	0.4625	0.4633
1.8	0.4641	0.4649	0.4656	0.4664	0.4671	0.4678	0.4686	0.4693	0.4699	0.4706
1.9	0.4713	0.4719	0.4726	0.4732	0.4738	0.4744	0.4750	0.4756	0.4761	0.4767
2.0	0.4772	0.4778	0.4783	0.4788	0.4793	0.4798	0.4803	0.4808	0.4812	0.4817
2.1	0.4821	0.4826	0.4830	0.4834	0.4838	0.4842	0.4846	0.4850	0.4854	0.4857
2.2	0.4861	0.4864	0.4868	0.4871	0.4875	0.4878	0.4881	0.4884	0.4887	0.4890
2.3	0.4893	0.4896	0.4898	0.4901	0.4904	0.4906	0.4909	0.4911	0.4913	0.4916
2.4	0.4918	0.4920	0.4922	0.4925	0.4927	0.4929	0.4931	0.4932	0.4934	0.4936
2.5	0.4938	0.4940	0.4941	0.4943	0.4945	0.4946	0.4948	0.4949	0.4951	0.4952
2.6	0.4953	0.4955	0.4956	0.4957	0.4959	0.4960	0.4961	0.4962	0.4963	0.4964
2.7	0.4965	0.4966	0.4967	0.4968	0.4969	0.4970	0.4971	0.4972	0.4973	0.4974
2.8	0.4974	0.4975	0.4976	0.4977	0.4977	0.4978	0.4979	0.4979	0.4980	0.4981
2.9	0.4981	0.4982	0.4982	0.4983	0.4984	0.4984	0.4985	0.4985	0.4986	0.4986
3.0	0.4987	0.4987	0.4987	0.4988	0.4988	0.4989	0.4989	0.4989	0.4990	0.4990
3.1	0.4990	0.4991	0.4991	0.4991	0.4992	0.4992	0.4992	0.4992	0.4993	0.4993
3.2	0.4993	0.4993	0.4994	0.4994	0.4994	0.4994	0.4994	0.4995	0.4995	0.4995
3.3	0.4995	0.4995	0.4995	0.4996	0.4996	0.4996	0.4996	0.4996	0.4996	0.4997
3.4	0.4997	0.4997	0.4997	0.4997	0.4997	0.4997	0.4997	0.4997	0.4997	0.4998
3.5	0.4998	0.4998	0.4998	0.4998	0.4998	0.4998	0.4998	0.4998	0.4998	0.4998
3.6	0.4998	0.4998	0.4999	0.4999	0.4999	0.4999	0.4999	0.4999	0.4999	0.4999
3.7	0.4999									
4.0	0.49997									
4.5	0.499997									
5.0	0.4999997									

For specific details about using this table to find: probabilities, see page 280; confidence coefficients, page 357; p-values, pages 379, 382; critical values, page 299.

Table 4   Critical Values of Standard Normal Distribution

A   O̅ne-Tailed Situations

The entries in this table are the critical values for z for which the area under the curve representing α is in the right-hand tail. Critical values for the left-hand tail are found by symmetry.

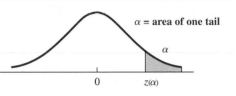

α = area of one tail

Amount of α in one tail

α	0.25	0.10	0.05	0.025	0.02	0.01	0.005
$z(\alpha)$	0.67	1.28	1.65	1.96	2.05	2.33	2.58

One-tailed example:
$\alpha = 0.05$
$z(\alpha) = z(0.05) = 1.65$

B   T̅wo-Tailed Situations

The entries in this table are the critical values for z for which the area under the curve representing α is split equally between the two tails.

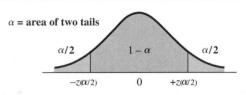

α = area of two tails

Amount of α in two tails

α	0.25	0.20	0.10	0.05	0.02	0.01
$z(\alpha/2)$	1.15	1.28	1.65	1.96	2.33	2.58
$1 - \alpha$	0.75	0.80	0.90	0.95	0.98	0.99

Area in the "center"

Two-tailed example:
$\alpha = 0.05$ or $1 - \alpha = 0.95$
$\alpha/2 = 0.025$
$z(\alpha/2) = z(0.025) = 1.96$

For specific details about using this table to find: confidence coefficients, see page 357; critical values, pages 395, 398.

Table 5 *p*-Values for Standard Normal Distribution

The entries in this table are the *p*-values related to the right-hand tail for the calculated $z\star$ for the standard normal distribution.

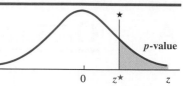

$z \star$	*p*-value	$z \star$	*p*-value	$z \star$	*p*-value	$z \star$	*p*-value
0.00	0.5000	1.00	0.1587	2.00	0.0228	3.00	0.0013
0.05	0.4801	1.05	0.1469	2.05	0.0202	3.05	0.0011
0.10	0.4602	1.10	0.1357	2.10	0.0179	3.10	0.0010
0.15	0.4404	1.15	0.1251	2.15	0.0158	3.15	0.0008
0.20	0.4207	1.20	0.1151	2.20	0.0139	3.20	0.0007
0.25	0.4013	1.25	0.1056	2.25	0.0122	3.25	0.0006
0.30	0.3821	1.30	0.0968	2.30	0.0107	3.30	0.0005
0.35	0.3632	1.35	0.0885	2.35	0.0094	3.35	0.0004
0.40	0.3446	1.40	0.0808	2.40	0.0082	3.40	0.0003
0.45	0.3264	1.45	0.0735	2.45	0.0071	3.45	0.0003
0.50	0.3085	1.50	0.0668	2.50	0.0062	3.50	0.0002
0.55	0.2912	1.55	0.0606	2.55	0.0054	3.55	0.0002
0.60	0.2743	1.60	0.0548	2.60	0.0047	3.60	0.0002
0.65	0.2578	1.65	0.0495	2.65	0.0040	3.65	0.0001
0.70	0.2420	1.70	0.0446	2.70	0.0035	3.70	0.0001
0.75	0.2266	1.75	0.0401	2.75	0.0030	3.75	0.0001
0.80	0.2119	1.80	0.0359	2.80	0.0026	3.80	0.0001
0.85	0.1977	1.85	0.0322	2.85	0.0022	3.85	0.0001
0.90	0.1841	1.90	0.0287	2.90	0.0019	3.90	0+
0.95	0.1711	1.95	0.0256	2.95	0.0016	3.95	0+

For specific details about using this table to find *p*-values, see pages 379, 382.

Table 6   Critical Values of Student's *t*-Distribution

The entries in this table, $t(\text{df}, \alpha)$, are the critical values for Student's *t*-distribution for which the area under the curve in the right-hand tail is α. Critical values for the left-hand tail are found by symmetry.

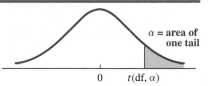

α = area of one tail

$0 \qquad t(\text{df}, \alpha)$

	Amount of α in One Tail					
	0.25	0.10	0.05	0.025	0.01	0.005
	Amount of α in Two Tails					
df	0.50	0.20	0.10	0.05	0.02	0.01
3	0.765	1.64	2.35	3.18	4.54	5.84
4	0.741	1.53	2.13	2.78	3.75	4.60
5	0.729	1.48	2.02	2.57	3.37	4.03
6	0.718	1.44	1.94	2.45	3.14	3.71
7	0.711	1.42	1.89	2.36	3.00	3.50
8	0.706	1.40	1.86	2.31	2.90	3.36
9	0.703	1.38	1.83	2.26	2.82	3.25
10	0.700	1.37	1.81	2.23	2.76	3.17
11	0.697	1.36	1.80	2.20	2.72	3.11
12	0.696	1.36	1.78	2.18	2.68	3.05
13	0.694	1.35	1.77	2.16	2.65	3.01
14	0.692	1.35	1.76	2.14	2.62	2.98
15	0.691	1.34	1.75	2.13	2.60	2.95
16	0.690	1.34	1.75	2.12	2.58	2.92
17	0.689	1.33	1.74	2.11	2.57	2.90
18	0.688	1.33	1.73	2.10	2.55	2.88
19	0.688	1.33	1.73	2.09	2.54	2.86
20	0.687	1.33	1.72	2.09	2.53	2.85
21	0.686	1.32	1.72	2.08	2.52	2.83
22	0.686	1.32	1.72	2.07	2.51	2.82
23	0.685	1.32	1.71	2.07	2.50	2.81
24	0.685	1.32	1.71	2.06	2.49	2.80
25	0.684	1.32	1.71	2.06	2.49	2.79
26	0.684	1.32	1.71	2.06	2.48	2.78
27	0.684	1.31	1.70	2.05	2.47	2.77
28	0.683	1.31	1.70	2.05	2.47	2.76
29	0.683	1.31	1.70	2.05	2.46	2.76
30	0.683	1.31	1.70	2.04	2.46	2.75
35	0.682	1.31	1.69	2.03	2.44	2.73
40	0.681	1.30	1.68	2.02	2.42	2.70
50	0.679	1.30	1.68	2.01	2.40	2.68
70	0.678	1.29	1.67	1.99	2.38	2.65
100	0.677	1.29	1.66	1.98	2.36	2.63
df > 100	0.675	1.28	1.65	1.96	2.33	2.58

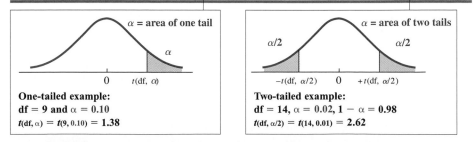

α = area of one tail

α

$0 \qquad t(\text{df}, \alpha)$

One-tailed example:
df = 9 and α = 0.10
$t(\text{df}, \alpha) = t(9, 0.10) = \mathbf{1.38}$

α = area of two tails

$\alpha/2 \qquad\qquad \alpha/2$

$-t(\text{df}, \alpha/2) \quad 0 \quad +t(\text{df}, \alpha/2)$

Two-tailed example:
df = 14, α = 0.02, 1 − α = 0.98
$t(\text{df}, \alpha/2) = t(14, 0.01) = \mathbf{2.62}$

For specific details about using this table to find: confidence coefficients, see page 422; *p*-values, pages 425, 428; critical values, page 418.

Table 7 Probability-Values for Student's *t*-distribution

The entries in this table are the *p*-values related to the right-hand tail for the calculated *t*★ value for the *t*-distribution of df degrees of freedom.

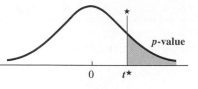

Degrees of Freedom

t★	3	4	5	6	7	8	10	12	15	18	21	25	29	35	df ≥ 45
0.0	0.500	0.500	0.500	0.500	0.500	0.500	0.500	0.500	0.500	0.500	0.500	0.500	0.500	0.500	0.500
0.1	0.463	0.463	0.462	0.462	0.462	0.461	0.461	0.461	0.461	0.461	0.461	0.461	0.461	0.460	0.460
0.2	0.427	0.426	0.425	0.424	0.424	0.423	0.423	0.422	0.422	0.422	0.422	0.422	0.421	0.421	0.421
0.3	0.392	0.390	0.388	0.387	0.386	0.386	0.385	0.385	0.384	0.384	0.384	0.383	0.383	0.383	0.383
0.4	0.358	0.355	0.353	0.352	0.351	0.350	0.349	0.348	0.347	0.347	0.347	0.346	0.346	0.346	0.346
0.5	0.326	0.322	0.319	0.317	0.316	0.315	0.314	0.313	0.312	0.312	0.311	0.311	0.310	0.310	0.310
0.6	0.295	0.290	0.287	0.285	0.284	0.283	0.281	0.280	0.279	0.278	0.277	0.277	0.277	0.276	0.276
0.7	0.267	0.261	0.258	0.255	0.253	0.252	0.250	0.249	0.247	0.246	0.246	0.245	0.245	0.244	0.244
0.8	0.241	0.234	0.230	0.227	0.225	0.223	0.221	0.220	0.218	0.217	0.216	0.216	0.215	0.215	0.214
0.9	0.217	0.210	0.205	0.201	0.199	0.197	0.195	0.193	0.191	0.190	0.189	0.188	0.188	0.187	0.186
1.0	0.196	0.187	0.182	0.178	0.175	0.173	0.170	0.169	0.167	0.165	0.164	0.163	0.163	0.162	0.161
1.1	0.176	0.167	0.161	0.157	0.154	0.152	0.149	0.146	0.144	0.143	0.142	0.141	0.140	0.139	0.139
1.2	0.158	0.148	0.142	0.138	0.135	0.132	0.129	0.127	0.124	0.123	0.122	0.121	0.120	0.119	0.118
1.3	0.142	0.132	0.125	0.121	0.117	0.115	0.111	0.109	0.107	0.105	0.104	0.103	0.102	0.101	0.100
1.4	0.128	0.117	0.110	0.106	0.102	0.100	0.096	0.093	0.091	0.089	0.088	0.087	0.086	0.085	0.084
1.5	0.115	0.104	0.097	0.092	0.089	0.086	0.082	0.080	0.077	0.075	0.074	0.073	0.072	0.071	0.070
1.6	0.104	0.092	0.085	0.080	0.077	0.074	0.070	0.068	0.065	0.064	0.062	0.061	0.060	0.059	0.058
1.7	0.094	0.082	0.075	0.070	0.066	0.064	0.060	0.057	0.055	0.053	0.052	0.051	0.050	0.049	0.048
1.8	0.085	0.073	0.066	0.061	0.057	0.055	0.051	0.049	0.046	0.044	0.043	0.042	0.041	0.040	0.039
1.9	0.077	0.065	0.058	0.053	0.050	0.047	0.043	0.041	0.038	0.037	0.036	0.035	0.034	0.033	0.032
2.0	0.070	0.058	0.051	0.046	0.043	0.040	0.037	0.034	0.032	0.030	0.029	0.028	0.027	0.027	0.026
2.1	0.063	0.052	0.045	0.040	0.037	0.034	0.031	0.029	0.027	0.025	0.024	0.023	0.022	0.022	0.021
2.2	0.058	0.046	0.040	0.035	0.032	0.029	0.026	0.024	0.022	0.021	0.020	0.019	0.018	0.017	0.016
2.3	0.052	0.041	0.035	0.031	0.027	0.025	0.022	0.020	0.018	0.017	0.016	0.015	0.014	0.014	0.013
2.4	0.048	0.037	0.031	0.027	0.024	0.022	0.019	0.017	0.015	0.014	0.013	0.012	0.012	0.011	0.010
2.5	0.044	0.033	0.027	0.023	0.020	0.018	0.016	0.014	0.012	0.011	0.010	0.010	0.009	0.009	0.008
2.6	0.040	0.030	0.024	0.020	0.018	0.016	0.013	0.012	0.010	0.009	0.008	0.008	0.007	0.007	0.006
2.7	0.037	0.027	0.021	0.018	0.015	0.014	0.011	0.010	0.008	0.007	0.007	0.006	0.006	0.005	0.005
2.8	0.034	0.024	0.019	0.016	0.013	0.012	0.009	0.008	0.007	0.006	0.005	0.005	0.005	0.004	0.004
2.9	0.031	0.022	0.017	0.014	0.011	0.010	0.008	0.007	0.005	0.005	0.004	0.004	0.004	0.003	0.003
3.0	0.029	0.020	0.015	0.012	0.010	0.009	0.007	0.006	0.004	0.004	0.003	0.003	0.003	0.002	0.002
3.1	0.027	0.018	0.013	0.011	0.009	0.007	0.006	0.005	0.004	0.003	0.003	0.002	0.002	0.002	0.002
3.2	0.025	0.016	0.012	0.009	0.008	0.006	0.005	0.004	0.003	0.002	0.002	0.002	0.002	0.001	0.001
3.3	0.023	0.015	0.011	0.008	0.007	0.005	0.004	0.003	0.002	0.002	0.002	0.001	0.001	0.001	0.001
3.4	0.021	0.014	0.010	0.007	0.006	0.005	0.003	0.003	0.002	0.002	0.001	0.001	0.001	0.001	0.001
3.5	0.020	0.012	0.009	0.006	0.005	0.004	0.003	0.002	0.002	0.001	0.001	0.001	0.001	0.001	0.001
3.6	0.018	0.011	0.008	0.006	0.004	0.004	0.002	0.002	0.001	0.001	0.001	0.001	0.001	0+	0+
3.7	0.017	0.010	0.007	0.005	0.004	0.003	0.002	0.002	0.001	0.001	0.001	0.001	0+	0+	0+
3.8	0.016	0.010	0.006	0.004	0.003	0.003	0.002	0.001	0.001	0.001	0.001	0+	0+	0+	0+
3.9	0.015	0.009	0.006	0.004	0.003	0.002	0.001	0.001	0.001	0.001	0+	0+	0+	0+	0+
4.0	0.014	0.008	0.005	0.004	0.003	0.002	0.001	0.001	0.001	0+	0+	0+	0+	0+	0+

For specific details about using this table to find *p*-values, see pages 426, 428.

Table 8   Critical Values of χ^2 ("Chi-Square") Distribution

The entries in this table, χ^2 (df, α), are the critical values for the χ^2 distribution for which the area under the curve to the right is α.

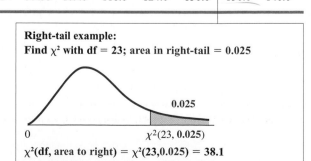

area to right

χ^2(df, area to right)

					Area to the Right							
0.995	0.99	0.975	0.95	0.90	0.75	0.50	0.25	0.10	0.05	0.025	0.01	0.005
	Area in Left-hand Tail					Median			Area in Right-hand Tail			
df	0.005	0.01	0.025	0.05	0.10	0.25	0.50	0.25	0.10	0.05	0.025	0.01	0.005

df	0.005	0.01	0.025	0.05	0.10	0.25	0.50	0.25	0.10	0.05	0.025	0.01	0.005
1	0.0000393	0.000157	0.000982	0.00393	0.0158	0.101	0.455	1.32	2.71	3.84	5.02	6.63	7.88
2	0.0100	0.0201	0.0506	0.103	0.211	0.575	1.39	2.77	4.61	5.99	7.38	9.21	10.6
3	0.0717	0.115	0.216	0.352	0.584	1.21	2.37	4.11	6.25	7.82	9.35	11.3	12.8
4	0.207	0.297	0.484	0.711	1.06	1.92	3.36	5.39	7.78	9.49	11.1	13.3	14.9
5	0.412	0.554	0.831	1.15	1.61	2.67	4.35	6.63	9.24	11.1	12.8	15.1	16.8
6	0.676	0.872	1.24	1.64	2.20	3.45	5.35	7.84	10.6	12.6	14.5	16.8	18.6
7	0.990	1.24	1.69	2.17	2.83	4.25	6.35	9.04	12.0	14.1	16.0	18.5	20.3
8	1.34	1.65	2.18	2.73	3.49	5.07	7.34	10.2	13.4	15.5	17.5	20.1	22.0
9	1.73	2.09	2.70	3.33	4.17	5.90	8.34	11.4	14.7	16.9	19.0	21.7	23.6
10	2.16	2.56	3.25	3.94	4.87	6.74	9.34	12.5	16.0	18.3	20.5	23.2	25.2
11	2.60	3.05	3.82	4.57	5.58	7.58	10.34	13.7	17.3	19.7	21.9	24.7	26.8
12	3.07	3.57	4.40	5.23	6.30	8.44	11.34	14.8	18.5	21.0	23.3	26.2	28.3
13	3.57	4.11	5.01	5.89	7.04	9.30	12.34	16.0	19.8	22.4	24.7	27.7	29.8
14	4.07	4.66	5.63	6.57	7.79	10.2	13.34	17.1	21.1	23.7	26.1	29.1	31.3
15	4.60	5.23	6.26	7.26	8.55	11.0	14.34	18.2	22.3	25.0	27.5	30.6	32.8
16	5.14	5.81	6.91	7.96	9.31	11.9	15.34	19.4	23.5	26.3	28.8	32.0	34.3
17	5.70	6.41	7.56	8.67	10.1	12.8	16.34	20.5	24.8	27.6	30.2	33.4	35.7
18	6.26	7.01	8.23	9.39	10.9	13.7	17.34	21.6	26.0	28.9	31.5	34.8	37.2
19	6.84	7.63	8.91	10.1	11.7	14.6	18.34	22.7	27.2	30.1	32.9	36.2	38.6
20	7.43	8.26	9.59	10.9	12.4	15.5	19.34	23.8	28.4	31.4	34.2	37.6	40.0
21	8.03	8.90	10.3	11.6	13.2	16.3	20.34	24.9	29.6	32.7	35.5	38.9	41.4
22	8.64	9.54	11.0	12.3	14.0	17.2	21.34	26.0	30.8	33.9	36.8	40.3	42.8
23	9.26	10.2	11.7	13.1	14.8	18.1	22.34	27.1	32.0	35.2	38.1	41.6	44.2
24	9.89	10.9	12.4	13.8	15.7	19.0	23.34	28.2	33.2	36.4	39.4	43.0	45.6
25	10.5	11.5	13.1	14.6	16.5	19.9	24.34	29.3	34.4	37.7	40.6	44.3	46.9
26	11.2	12.2	13.8	15.4	17.3	20.8	25.34	30.4	35.6	38.9	41.9	45.6	48.3
27	11.8	12.9	14.6	16.2	18.1	21.7	26.34	31.5	36.7	40.1	43.2	47.0	49.6
28	12.5	13.6	15.3	16.9	18.9	22.7	27.34	32.6	37.9	41.3	44.5	48.3	51.0
29	13.1	14.3	16.0	17.7	19.8	23.6	28.34	33.7	39.1	42.6	45.7	49.6	52.3
30	13.8	15.0	16.8	18.5	20.6	24.5	29.34	34.8	40.3	43.8	47.0	50.9	53.7
40	20.7	22.2	24.4	26.5	29.1	33.7	39.34	45.6	51.8	55.8	59.3	63.7	66.8
50	28.0	29.7	32.4	34.8	37.7	42.9	49.33	56.3	63.2	67.5	71.4	76.2	79.5
60	35.5	37.5	40.5	43.2	46.5	52.3	59.33	67.0	74.4	79.1	83.3	88.4	92.0
70	43.3	45.4	48.8	51.7	55.3	61.7	69.33	77.6	85.5	90.5	95.0	100.0	104.0
80	51.2	53.5	57.2	60.4	64.3	71.1	79.33	88.1	96.6	102.0	107.0	112.0	116.0
90	59.2	61.8	65.6	69.1	73.3	80.6	89.33	98.6	108.0	113.0	118.0	124.0	128.0
100	67.3	70.1	74.2	77.9	82.4	90.1	99.33	109.0	118.0	124.0	130.0	136.0	140.0

Left-tail example:
Find χ^2 with df = 28; area in left-tail = 0.10.

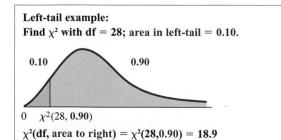

0.10   0.90

$0\quad \chi^2(28, 0.90)$
χ^2(df, area to right) = $\chi^2(28, 0.90)$ = **18.9**

Right-tail example:
Find χ^2 with df = 23; area in right-tail = 0.025

0.025

$0\qquad\qquad \chi^2(23, 0.025)$
χ^2(df, area to right) = $\chi^2(23, 0.025)$ = **38.1**

For specific details about using this table to find: p-values, see page 456; critical values, page 451.

Table 9a Critical Values of the F Distribution ($\alpha = 0.05$)

The entries in this table are critical values of F for which the area under the curve to the right is equal to 0.05.

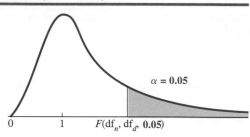

$\alpha = 0.05$

$F(\mathrm{df}_n, \mathrm{df}_d, 0.05)$

Degrees of Freedom for Numerator

	1	2	3	4	5	6	7	8	9	10
1	161.	200.	216.	225.	230.	234.	237.	239.	241.	242.
2	18.5	19.0	19.2	19.2	19.3	19.3	19.4	19.4	19.4	19.4
3	10.1	9.55	9.28	9.12	9.01	8.94	8.89	8.85	8.81	8.79
4	7.71	6.94	6.59	6.39	6.26	6.16	6.09	6.04	6.00	5.96
5	6.61	5.79	5.41	5.19	5.05	4.95	4.88	4.82	4.77	4.74
6	5.99	5.14	4.76	4.53	4.39	4.28	4.21	4.15	4.10	4.06
7	5.59	4.74	4.35	4.12	3.97	3.87	3.79	3.73	3.68	3.64
8	5.32	4.46	4.07	3.84	3.69	3.58	3.50	3.44	3.39	3.35
9	5.12	4.26	3.86	3.63	3.48	3.37	3.29	3.23	3.18	3.14
10	4.96	4.10	3.71	3.48	3.33	3.22	3.14	3.07	3.02	2.98
11	4.84	3.98	3.59	3.36	3.20	3.09	3.01	2.95	2.90	2.85
12	4.75	3.89	3.49	3.26	3.11	3.00	2.91	2.85	2.80	2.75
13	4.67	3.81	3.41	3.18	3.03	2.92	2.83	2.77	2.71	2.67
14	4.60	3.74	3.34	3.11	2.96	2.85	2.76	2.70	2.65	2.60
15	4.54	3.68	3.29	3.06	2.90	2.79	2.71	2.64	2.59	2.54
16	4.49	3.63	3.24	3.01	2.85	2.74	2.66	2.59	2.54	2.49
17	4.45	3.59	3.20	2.96	2.81	2.70	2.61	2.55	2.49	2.45
18	4.41	3.55	3.16	2.93	2.77	2.66	2.58	2.51	2.46	2.41
19	4.38	3.52	3.13	2.90	2.74	2.63	2.54	2.48	2.42	2.38
20	4.35	3.49	3.10	2.87	2.71	2.60	2.51	2.45	2.39	2.35
21	4.32	3.47	3.07	2.84	2.68	2.57	2.49	2.42	2.37	2.32
22	4.30	3.44	3.05	2.82	2.66	2.55	2.46	2.40	2.34	2.30
23	4.28	3.42	3.03	2.80	2.64	2.53	2.44	2.37	2.32	2.27
24	4.26	3.40	3.01	2.78	2.62	2.51	2.42	2.36	2.30	2.25
25	4.24	3.39	2.99	2.76	2.60	2.49	2.40	2.34	2.28	2.24
30	4.17	3.32	2.92	2.69	2.53	2.42	2.33	2.27	2.21	2.16
40	4.08	3.23	2.84	2.61	2.45	2.34	2.25	2.18	2.12	2.08
60	4.00	3.15	2.76	2.53	2.37	2.25	2.17	2.10	2.04	1.99
120	3.92	3.07	2.68	2.45	2.29	2.18	2.09	2.02	1.96	1.91
∞	3.84	3.00	2.60	2.37	2.21	2.10	2.01	1.94	1.88	1.83

Degrees of Freedom for Denominator (vertical label at left of table)

For specific details about using this table to find: p-values, see page 520; critical values, page 516.

Table 9a   (Continued)

<table>
<tr><th colspan="10">Degrees of Freedom for Numerator</th></tr>
<tr><th></th><th>12</th><th>15</th><th>20</th><th>24</th><th>30</th><th>40</th><th>60</th><th>120</th><th>∞</th></tr>
<tr><td>1</td><td>244.</td><td>246.</td><td>248.</td><td>249.</td><td>250.</td><td>251.</td><td>252.</td><td>253.</td><td>254.</td></tr>
<tr><td>2</td><td>19.4</td><td>19.4</td><td>19.4</td><td>19.5</td><td>19.5</td><td>19.5</td><td>19.5</td><td>19.5</td><td>19.5</td></tr>
<tr><td>3</td><td>8.74</td><td>8.70</td><td>8.66</td><td>8.64</td><td>8.62</td><td>8.59</td><td>8.57</td><td>8.55</td><td>8.53</td></tr>
<tr><td>4</td><td>5.91</td><td>5.86</td><td>5.80</td><td>5.77</td><td>5.75</td><td>5.72</td><td>5.69</td><td>5.66</td><td>5.63</td></tr>
<tr><td>5</td><td>4.68</td><td>4.62</td><td>4.56</td><td>4.53</td><td>4.50</td><td>4.46</td><td>4.43</td><td>4.40</td><td>4.37</td></tr>
<tr><td>6</td><td>4.00</td><td>3.94</td><td>3.87</td><td>3.84</td><td>3.81</td><td>3.77</td><td>3.74</td><td>3.70</td><td>3.67</td></tr>
<tr><td>7</td><td>3.57</td><td>3.51</td><td>3.44</td><td>3.41</td><td>3.38</td><td>3.34</td><td>3.30</td><td>3.27</td><td>3.23</td></tr>
<tr><td>8</td><td>3.28</td><td>3.22</td><td>3.15</td><td>3.12</td><td>3.08</td><td>3.04</td><td>3.01</td><td>2.97</td><td>2.93</td></tr>
<tr><td>9</td><td>3.07</td><td>3.01</td><td>2.94</td><td>2.90</td><td>2.86</td><td>2.83</td><td>2.79</td><td>2.75</td><td>2.71</td></tr>
<tr><td>10</td><td>2.91</td><td>2.85</td><td>2.77</td><td>2.74</td><td>2.70</td><td>2.66</td><td>2.62</td><td>2.58</td><td>2.54</td></tr>
<tr><td>11</td><td>2.79</td><td>2.72</td><td>2.65</td><td>2.61</td><td>2.57</td><td>2.53</td><td>2.49</td><td>2.45</td><td>2.40</td></tr>
<tr><td>12</td><td>2.69</td><td>2.62</td><td>2.54</td><td>2.51</td><td>2.47</td><td>2.43</td><td>2.38</td><td>2.34</td><td>2.30</td></tr>
<tr><td>13</td><td>2.60</td><td>2.53</td><td>2.46</td><td>2.42</td><td>2.38</td><td>2.34</td><td>2.30</td><td>2.25</td><td>2.21</td></tr>
<tr><td>14</td><td>2.53</td><td>2.46</td><td>2.39</td><td>2.35</td><td>2.31</td><td>2.27</td><td>2.22</td><td>2.18</td><td>2.13</td></tr>
<tr><td>15</td><td>2.48</td><td>2.40</td><td>2.33</td><td>2.29</td><td>2.25</td><td>2.20</td><td>2.16</td><td>2.11</td><td>2.07</td></tr>
<tr><td>16</td><td>2.42</td><td>2.35</td><td>2.28</td><td>2.24</td><td>2.19</td><td>2.15</td><td>2.11</td><td>2.06</td><td>2.01</td></tr>
<tr><td>17</td><td>2.38</td><td>2.31</td><td>2.23</td><td>2.19</td><td>2.15</td><td>2.10</td><td>2.06</td><td>2.01</td><td>1.96</td></tr>
<tr><td>18</td><td>2.34</td><td>2.27</td><td>2.19</td><td>2.15</td><td>2.11</td><td>2.06</td><td>2.02</td><td>1.97</td><td>1.92</td></tr>
<tr><td>19</td><td>2.31</td><td>2.23</td><td>2.16</td><td>2.11</td><td>2.07</td><td>2.03</td><td>1.98</td><td>1.93</td><td>1.88</td></tr>
<tr><td>20</td><td>2.28</td><td>2.20</td><td>2.12</td><td>2.08</td><td>2.04</td><td>1.99</td><td>1.95</td><td>1.90</td><td>1.84</td></tr>
<tr><td>21</td><td>2.25</td><td>2.18</td><td>2.10</td><td>2.05</td><td>2.01</td><td>1.96</td><td>1.92</td><td>1.87</td><td>1.81</td></tr>
<tr><td>22</td><td>2.23</td><td>2.15</td><td>2.07</td><td>2.03</td><td>1.98</td><td>1.94</td><td>1.89</td><td>1.84</td><td>1.78</td></tr>
<tr><td>23</td><td>2.20</td><td>2.13</td><td>2.05</td><td>2.01</td><td>1.96</td><td>1.91</td><td>1.86</td><td>1.81</td><td>1.76</td></tr>
<tr><td>24</td><td>2.18</td><td>2.11</td><td>2.03</td><td>1.98</td><td>1.94</td><td>1.89</td><td>1.84</td><td>1.79</td><td>1.73</td></tr>
<tr><td>25</td><td>2.16</td><td>2.09</td><td>2.01</td><td>1.96</td><td>1.92</td><td>1.87</td><td>1.82</td><td>1.77</td><td>1.71</td></tr>
<tr><td>30</td><td>2.09</td><td>2.01</td><td>1.93</td><td>1.89</td><td>1.84</td><td>1.79</td><td>1.74</td><td>1.68</td><td>1.62</td></tr>
<tr><td>40</td><td>2.00</td><td>1.92</td><td>1.84</td><td>1.79</td><td>1.74</td><td>1.69</td><td>1.64</td><td>1.58</td><td>1.51</td></tr>
<tr><td>60</td><td>1.92</td><td>1.84</td><td>1.75</td><td>1.70</td><td>1.65</td><td>1.59</td><td>1.53</td><td>1.47</td><td>1.39</td></tr>
<tr><td>120</td><td>1.83</td><td>1.75</td><td>1.66</td><td>1.61</td><td>1.55</td><td>1.50</td><td>1.43</td><td>1.35</td><td>1.25</td></tr>
<tr><td>∞</td><td>1.75</td><td>1.67</td><td>1.57</td><td>1.52</td><td>1.46</td><td>1.39</td><td>1.32</td><td>1.22</td><td>1.00</td></tr>
</table>

Degrees of Freedom for Denominator (left vertical label)

From E. S. Pearson and H. O Hartley, *Biometrika Tables for Statisticians,* vol. 1 (1958), pp. 159–163. Reprinted by permission of the Biometrika Trustees.

Table 9b Critical Values of the F Distribution ($\alpha = 0.025$)

The entries in this table are critical values of F for which the area under the curve to the right is equal to 0.025.

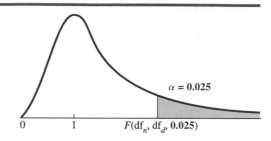

$\alpha = 0.025$

$F(\text{df}_n, \text{df}_d, 0.025)$

Degrees of Freedom for Numerator

		1	2	3	4	5	6	7	8	9	10
Degrees of Freedom for Denominator	1	648.	800.	864.	900.	922.	937.	948.	957.	963.	969.
	2	38.5	39.0	39.2	39.2	39.3	39.3	39.4	39.4	39.4	39.4
	3	17.4	16.0	15.4	15.1	14.9	14.7	14.6	14.5	14.5	14.4
	4	12.2	10.6	9.98	9.60	9.36	9.20	9.07	8.98	8.90	8.84
	5	10.0	8.43	7.76	7.39	7.15	6.98	6.85	6.76	6.68	6.62
	6	8.81	7.26	6.60	6.23	5.99	5.82	5.70	5.60	5.52	5.46
	7	8.07	6.54	5.89	5.52	5.29	5.12	4.99	4.90	4.82	4.76
	8	7.57	6.06	5.42	5.05	4.82	4.65	4.53	4.43	4.36	4.30
	9	7.21	5.71	5.08	4.72	4.48	4.32	4.20	4.10	4.03	3.96
	10	6.94	5.46	4.83	4.47	4.24	4.07	3.95	3.85	3.78	3.72
	11	6.72	5.26	4.63	4.28	4.04	3.88	3.76	3.66	3.59	3.53
	12	6.55	5.10	4.47	4.12	3.89	3.73	3.61	3.51	3.44	3.37
	13	6.41	4.97	4.35	4.00	3.77	3.60	3.48	3.39	3.31	3.25
	14	6.30	4.86	4.24	3.89	3.66	3.50	3.38	3.28	3.21	3.15
	15	6.20	4.77	4.15	3.80	3.58	3.41	3.29	3.20	3.12	3.06
	16	6.12	4.69	4.08	3.73	3.50	3.34	3.22	3.12	3.05	2.99
	17	6.04	4.62	4.01	3.66	3.44	3.28	3.16	3.06	2.98	2.92
	18	5.98	4.56	3.95	3.61	3.38	3.22	3.10	3.01	2.93	2.87
	19	5.92	4.51	3.90	3.56	3.33	3.17	3.05	2.96	2.88	2.82
	20	5.87	4.46	3.86	3.51	3.29	3.13	3.01	2.91	2.84	2.77
	21	5.83	4.42	3.82	3.48	3.25	3.09	2.97	2.87	2.80	2.73
	22	5.79	4.38	3.78	3.44	3.22	3.05	2.93	2.84	2.76	2.70
	23	5.75	4.35	3.75	3.41	3.18	3.02	2.90	2.81	2.73	2.67
	24	5.72	4.32	3.72	3.38	3.15	2.99	2.87	2.78	2.70	2.64
	25	5.69	4.29	3.69	3.35	3.13	2.97	2.85	2.75	2.68	2.61
	30	5.57	4.18	3.59	3.25	3.03	2.87	2.75	2.65	2.57	2.51
	40	5.42	4.05	3.46	3.13	2.90	2.74	2.62	2.53	2.45	2.39
	60	5.29	3.93	3.34	3.01	2.79	2.63	2.51	2.41	2.33	2.27
	120	5.15	3.80	3.23	2.89	2.67	2.52	2.39	2.30	2.22	2.16
	∞	5.02	3.69	3.12	2.79	2.57	2.41	2.29	2.19	2.11	2.05

For specific details about using this table to find: p-values, see page 520; critical values, page 516.

Table 9b   (Continued)

	Degrees of Freedom for Numerator								
	12	15	20	24	30	40	60	120	∞
1	977.	985.	993.	997.	1001.	1006.	1010.	1014.	1018.
2	39.4	39.4	39.4	39.5	39.5	39.5	39.5	39.5	39.5
3	14.3	14.3	14.2	14.1	14.1	14.0	14.0	13.9	13.9
4	8.75	8.66	8.56	8.51	8.46	8.41	8.36	8.31	8.26
5	6.52	6.43	6.33	6.28	6.23	6.18	6.12	6.07	6.02
6	5.37	5.27	5.17	5.12	5.07	5.01	4.96	4.90	4.85
7	4.67	4.57	4.47	4.42	4.36	4.31	4.25	4.20	4.14
8	4.20	4.10	4.00	3.95	3.89	3.84	3.78	3.73	3.67
9	3.87	3.77	3.67	3.61	3.56	3.51	3.45	3.39	3.33
10	3.62	3.52	3.42	3.37	3.31	3.26	3.20	3.14	3.08
11	3.43	3.33	3.23	3.17	3.12	3.06	3.00	2.94	2.88
12	3.28	3.18	3.07	3.02	2.96	2.91	2.85	2.79	2.72
13	3.15	3.05	2.95	2.89	2.84	2.78	2.72	2.66	2.60
14	3.05	2.95	2.84	2.79	2.73	2.67	2.61	2.55	2.49
15	2.96	2.86	2.76	2.70	2.64	2.59	2.52	2.46	2.40
16	2.89	2.79	2.68	2.63	2.57	2.51	2.45	2.38	2.32
17	2.82	2.72	2.62	2.56	2.50	2.44	2.38	2.32	2.25
18	2.77	2.67	2.56	2.50	2.44	2.38	2.32	2.26	2.19
19	2.72	2.62	2.51	2.45	2.39	2.33	2.27	2.20	2.13
20	2.68	2.57	2.46	2.41	2.35	2.29	2.22	2.16	2.09
21	2.64	2.53	2.42	2.37	2.31	2.25	2.18	2.11	2.04
22	2.60	2.50	2.39	2.33	2.27	2.21	2.14	2.08	2.00
23	2.57	2.47	2.36	2.30	2.24	2.18	2.11	2.04	1.97
24	2.54	2.44	2.33	2.27	2.21	2.15	2.08	2.01	1.94
25	2.51	2.41	2.30	2.24	2.18	2.12	2.05	1.98	1.91
30	2.41	2.31	2.20	2.14	2.07	2.01	1.94	1.87	1.79
40	2.29	2.18	2.07	2.01	1.94	1.88	1.80	1.72	1.64
60	2.17	2.06	1.94	1.88	1.82	1.74	1.67	1.58	1.48
120	2.05	1.95	1.82	1.76	1.69	1.61	1.53	1.43	1.31
∞	1.94	1.83	1.71	1.64	1.57	1.48	1.39	1.27	1.00

Degrees of Freedom for Denominator (row axis label)

From E. S. Pearson and H. O. Hartley, *Biometrika Tables for Statisticians,* vol. I (1958), pp. 159–163. Reprinted by permission of the Biometrika Trustees.

Table 9c   Critical Values of the F Distribution ($\alpha = 0.01$)

The entries in the table are critical values of F for which the area under the curve to the right is equal to 0.01.

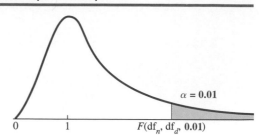

$\alpha = 0.01$

$F(\text{df}_n, \text{df}_d, 0.01)$

Degrees of Freedom for Numerator

		1	2	3	4	5	6	7	8	9	10
Degrees of Freedom for Denominator	1	4052.	5000.	5403.	5625.	5764.	5859.	5928.	5982.	6024.	6056.
	2	98.5	99.0	99.2	99.2	99.3	99.3	99.4	99.4	99.4	99.4
	3	34.1	30.8	29.5	28.7	28.2	27.9	27.7	27.5	27.3	27.2
	4	21.2	18.0	16.7	16.0	15.5	15.2	15.0	14.8	14.7	14.5
	5	16.3	13.3	12.1	11.4	11.0	10.7	10.5	10.3	10.2	10.1
	6	13.7	10.9	9.78	9.15	8.75	8.47	8.26	8.10	7.98	7.87
	7	12.2	9.55	8.45	7.85	7.46	7.19	6.99	6.84	6.72	6.62
	8	11.3	8.65	7.59	7.01	6.63	6.37	6.18	6.03	5.91	5.81
	9	10.6	8.02	6.99	6.42	6.06	5.80	5.61	5.47	5.35	5.26
	10	10.0	7.56	6.55	5.99	5.64	5.39	5.20	5.06	4.94	4.85
	11	9.65	7.21	6.22	5.67	5.32	5.07	4.89	4.74	4.63	4.54
	12	9.33	6.93	5.95	5.41	5.06	4.82	4.64	4.50	4.39	4.30
	13	9.07	6.70	5.74	5.21	4.86	4.62	4.44	4.30	4.19	4.10
	14	8.86	6.51	5.56	5.04	4.70	4.46	4.28	4.14	4.03	3.94
	15	8.68	6.36	5.42	4.89	4.56	4.32	4.14	4.00	3.89	3.80
	16	8.53	6.23	5.29	4.77	4.44	4.20	4.03	3.89	3.78	3.69
	17	8.40	6.11	5.19	4.67	4.34	4.10	3.93	3.79	3.68	3.59
	18	8.29	6.01	5.09	4.58	4.25	4.01	3.84	3.71	3.60	3.51
	19	8.19	5.93	5.01	4.50	4.17	3.94	3.77	3.63	3.52	3.43
	20	8.10	5.85	4.94	4.43	4.10	3.87	3.70	3.56	3.46	3.37
	21	8.02	5.78	4.87	4.37	4.04	3.81	3.64	3.51	3.40	3.31
	22	7.95	5.72	4.82	4.31	3.99	3.76	3.59	3.45	3.35	3.26
	23	7.88	5.66	4.76	4.26	3.94	3.71	3.54	3.41	3.30	3.21
	24	7.82	5.61	4.72	4.22	3.90	3.67	3.50	3.36	3.26	3.17
	25	7.77	5.57	4.68	4.18	3.86	3.63	3.46	3.32	3.22	3.13
	30	7.56	5.39	4.51	4.02	3.70	3.47	3.30	3.17	3.07	2.98
	40	7.31	5.18	4.31	3.83	3.51	3.29	3.12	2.99	2.89	2.80
	60	7.08	4.98	4.13	3.65	3.34	3.12	2.95	2.82	2.72	2.63
	120	6.85	4.79	3.95	3.48	3.17	2.96	2.79	2.66	2.56	2.47
	∞	6.63	4.61	3.78	3.32	3.02	2.80	2.64	2.51	2.41	2.32

For specific details about using this table to find: *p*-values, see page 520; critical values, page 516.

Table 9c   (Continued)

Degrees of Freedom for Numerator

	12	15	20	24	30	40	60	120	∞
1	6106.	6157.	6209.	6235.	6261.	6287.	6313.	6339.	6366.
2	99.4	99.4	99.4	99.5	99.5	99.5	99.5	99.5	99.5
3	27.1	26.9	26.7	26.6	26.5	26.4	26.3	26.2	26.1
4	14.4	14.2	14.0	13.9	13.8	13.7	13.7	13.6	13.5
5	9.89	9.72	9.55	9.47	9.38	9.29	9.20	9.11	9.02
6	7.72	7.56	7.40	7.31	7.23	7.14	7.06	6.97	6.88
7	6.47	6.31	6.16	6.07	5.99	5.91	5.82	5.74	5.65
8	5.67	5.52	5.36	5.28	5.20	5.12	5.03	4.95	4.86
9	5.11	4.96	4.81	4.73	4.65	4.57	4.48	4.40	4.31
10	4.71	4.56	4.41	4.33	4.25	4.17	4.08	4.00	3.91
11	4.40	4.25	4.10	4.02	3.94	3.86	3.78	3.69	3.60
12	4.16	4.01	3.86	3.78	3.70	3.62	3.54	3.45	3.36
13	3.96	3.82	3.66	3.59	3.51	3.43	3.34	3.25	3.17
14	3.80	3.66	3.51	3.43	3.35	3.27	3.18	3.09	3.00
15	3.67	3.52	3.37	3.29	3.21	3.13	3.05	2.96	2.87
16	3.55	3.41	3.26	3.18	3.10	3.02	2.93	2.84	2.75
17	3.46	3.31	3.16	3.08	3.00	2.92	2.83	2.75	2.65
18	3.37	3.23	3.08	3.00	2.92	2.84	2.75	2.66	2.57
19	3.30	3.15	3.00	2.92	2.84	2.76	2.67	2.58	2.49
20	3.23	3.09	2.94	2.86	2.78	2.69	2.61	2.52	2.42
21	3.17	3.03	2.88	2.80	2.72	2.64	2.55	2.46	2.36
22	3.12	2.98	2.83	2.75	2.67	2.58	2.50	2.40	2.31
23	3.07	2.93	2.78	2.70	2.62	2.54	2.45	2.35	2.26
24	3.03	2.89	2.74	2.66	2.58	2.49	2.40	2.31	2.21
25	2.99	2.85	2.70	2.62	2.53	2.45	2.36	2.27	2.17
30	2.84	2.70	2.55	2.47	2.39	2.30	2.21	2.11	2.01
40	2.66	2.52	2.37	2.29	2.20	2.11	2.02	1.92	1.80
60	2.50	2.35	2.20	2.12	2.03	1.94	1.84	1.73	1.60
120	2.34	2.19	2.03	1.95	1.86	1.76	1.66	1.53	1.38
∞	2.18	2.04	1.88	1.79	1.70	1.59	1.47	1.32	1.00

Degrees of Freedom for Denominator (left vertical axis label)

From E. S. Pearson and H. O. Hartley, *Biometrika Tables for Statisticians,* vol. I (1958), pp. 159–163. Reprinted by permission of the Biometrika Trustees.

Table 10 Confidence Belts for the Correlation Coefficient $(1 - \alpha) = 0.95$

The numbers on the curves are sample sizes.

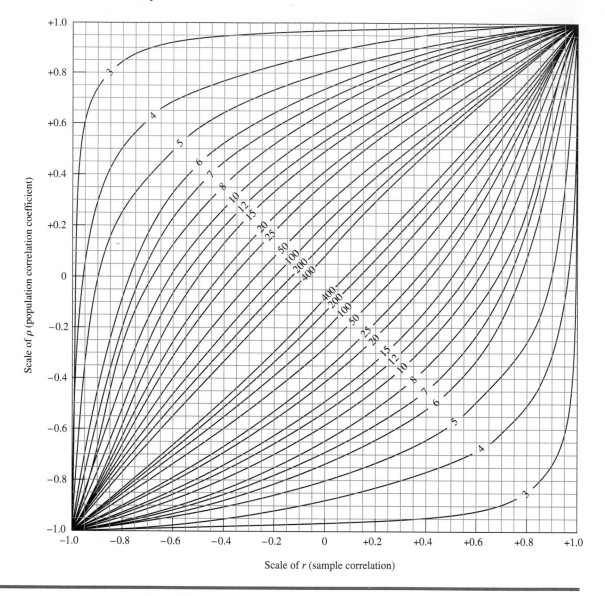

Scale of r (sample correlation)

For specific details about using this table to find confidence intervals, see page 615.

Table 11   Critical Values of *r* When ρ = 0

The entries in this table are the critical values of *r* for a two-tailed test at α. For simple correlation, df = *n* − 2, where *n* is the number of pairs of data in the sample. For a one-tailed test, the value of α shown at the top of the table is double the value of α being used in the hypothesis test.

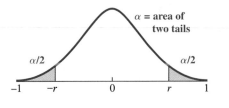

α = area of two tails

df	0.10	0.05	0.02	0.01
1	0.988	0.997	1.000	1.000
2	0.900	0.950	0.980	0.990
3	0.805	0.878	0.934	0.959
4	0.729	0.811	0.882	0.917
5	0.669	0.754	0.833	0.874
6	0.621	0.707	0.789	0.834
7	0.582	0.666	0.750	0.798
8	0.549	0.632	0.716	0.765
9	0.521	0.602	0.685	0.735
10	0.497	0.576	0.658	0.708
11	0.476	0.553	0.634	0.684
12	0.458	0.532	0.612	0.661
13	0.441	0.514	0.592	0.641
14	0.426	0.497	0.574	0.623
15	0.412	0.482	0.558	0.606
16	0.400	0.468	0.542	0.590
17	0.389	0.456	0.528	0.575
18	0.378	0.444	0.516	0.561
19	0.369	0.433	0.503	0.549
20	0.360	0.423	0.492	0.537
25	0.323	0.381	0.445	0.487
30	0.296	0.349	0.409	0.449
35	0.275	0.325	0.381	0.418
40	0.257	0.304	0.358	0.393
45	0.243	0.288	0.338	0.372
50	0.231	0.273	0.322	0.354
60	0.211	0.250	0.295	0.325
70	0.195	0.232	0.274	0.302
80	0.183	0.217	0.256	0.283
90	0.173	0.205	0.242	0.267
100	0.164	0.195	0.230	0.254

From E. S. Pearson and H. O. Hartley, *Biometrika Tables for Statisticians,* vol. 1 (1962), p. 138. Reprinted by permission of the Biometrika Trustees.

For specific details about using this table to find: *p*-values, see page 617; critical values, page 618.

Table 12 Critical Values of the Sign Test

The entries in this table are the critical values for the number of the least frequent sign for a two-tailed test at α for the binomial $p = 0.5$. For a one-tailed test, the value of α shown at the top of the table is double the value of α being used in the hypothesis test.

		α					α		
n	0.01	0.05	0.10	0.25	n	0.01	0.05	0.10	0.25
1					51	15	18	19	20
2					52	16	18	19	21
3				0	53	16	18	20	21
4				0	54	17	19	20	22
5			0	0	55	17	19	20	22
6		0	0	1	56	17	20	21	23
7		0	0	1	57	18	20	21	23
8	0	0	1	1	58	18	21	22	24
9	0	1	1	2	59	19	21	22	24
10	0	1	1	2	60	19	21	23	25
11	0	1	2	3	61	20	22	23	25
12	1	2	2	3	62	20	22	24	25
13	1	2	3	3	63	20	23	24	26
14	1	2	3	4	64	21	23	24	26
15	2	3	3	4	65	21	24	25	27
16	2	3	4	5	66	22	24	25	27
17	2	4	4	5	67	22	25	26	28
18	3	4	5	6	68	22	25	26	28
19	3	4	5	6	69	23	25	27	29
20	3	5	5	6	70	23	26	27	29
21	4	5	6	7	71	24	26	28	30
22	4	5	6	7	72	24	27	28	30
23	4	6	7	8	73	25	27	28	31
24	5	6	7	8	74	25	28	29	31
25	5	7	7	9	75	25	28	29	32
26	6	7	8	9	76	26	28	30	32
27	6	7	8	10	77	26	29	30	32
28	6	8	9	10	78	27	29	31	33
29	7	8	9	10	79	27	30	31	33
30	7	9	10	11	80	28	30	32	34
31	7	9	10	11	81	28	31	32	34
32	8	9	10	12	82	28	31	33	35
33	8	10	11	12	83	29	32	33	35
34	9	10	11	13	84	29	32	33	36
35	9	11	12	13	85	30	32	34	36
36	9	11	12	14	86	30	33	34	37
37	10	12	13	14	87	31	33	35	37
38	10	12	13	14	88	31	34	35	38
39	11	12	13	15	89	31	34	36	38
40	11	13	14	15	90	32	35	36	39
41	11	13	14	16	91	32	35	37	39
42	12	14	15	16	92	33	36	37	39
43	12	14	15	17	93	33	36	38	40
44	13	15	16	17	94	34	37	38	40
45	13	15	16	18	95	34	37	38	41
46	13	15	16	18	96	34	37	39	41
47	14	16	17	19	97	35	38	39	42
48	14	16	17	19	98	35	38	40	42
49	15	17	18	19	99	36	39	40	43
50	15	17	18	20	100	36	39	41	43

From Wilfred J. Dixon and Frank J. Massey, Jr., *Introduction to Statistical Analysis,* 3d ed. (New York: McGraw-Hill, 1969), p. 509. Reprinted by permission.
For specific details about using this table to find: *p*-values, see page 664; critical values, page 662.

Table 13   Critical Values of *U* in the Mann-Whitney Test

A. The entries are the critical values of *U* for a one-tailed test at 0.025 or for a two-tailed test at 0.05.

n_2 \ n_1	1	2	3	4	5	6	7	8	9	10	11	12	13	14	15	16	17	18	19	20
1																				
2								0	0	0	0	1	1	1	1	1	2	2	2	2
3					0	1	1	2	2	3	3	4	4	5	5	6	6	7	7	8
4				0	1	2	3	4	4	5	6	7	8	9	10	11	11	12	13	13
5			0	1	2	3	5	6	7	8	9	11	12	13	14	15	17	18	19	20
6			1	2	3	5	6	8	10	11	13	14	16	17	19	21	22	24	25	27
7			1	3	5	6	8	10	12	14	16	18	20	22	24	26	28	30	32	34
8		0	2	4	6	8	10	13	15	17	19	22	24	26	29	31	34	36	38	41
9		0	2	4	7	10	12	15	17	20	23	26	28	31	34	37	39	42	45	48
10		0	3	5	8	11	14	17	20	23	26	29	33	36	39	42	45	48	52	55
11		0	3	6	9	13	16	19	23	26	30	33	37	40	44	47	51	55	58	62
12		1	4	7	11	14	18	22	26	29	33	37	41	45	49	53	57	61	65	69
13		1	4	8	12	16	20	24	28	33	37	41	45	50	54	59	63	67	72	76
14		1	5	9	13	17	22	26	31	36	40	45	50	55	59	64	67	74	78	83
15		1	5	10	14	19	24	29	34	39	44	49	54	59	64	70	75	80	85	90
16		1	6	11	15	21	26	31	37	42	47	53	59	64	70	75	81	86	92	98
17		2	6	11	17	22	28	34	39	45	51	57	63	67	75	81	87	93	99	105
18		2	7	12	18	24	30	36	42	48	55	61	67	74	80	86	93	99	106	112
19		2	7	13	19	25	32	38	45	52	58	65	72	78	85	92	99	106	113	119
20		2	8	13	20	27	34	41	48	55	62	69	76	83	90	98	105	112	119	127

B. The entries are the critical values of *U* for a one-tailed test at 0.05 or for a two-tailed test at 0.10.

n_2 \ n_1	1	2	3	4	5	6	7	8	9	10	11	12	13	14	15	16	17	18	19	20
1																			0	0
2				0	0	0	1	1	1	1	2	2	2	3	3	3	4	4	4	
3			0	0	1	2	2	3	3	4	5	5	6	7	7	8	9	9	10	11
4			0	1	2	3	4	5	6	7	8	9	10	11	12	14	15	16	17	18
5		0	1	2	4	5	6	8	9	11	12	13	15	16	18	19	20	22	23	25
6		0	2	3	5	7	8	10	12	14	16	17	19	21	23	25	26	28	30	32
7		0	2	4	6	8	11	13	15	17	19	21	24	26	28	30	33	35	37	39
8		1	3	5	8	10	13	15	18	20	23	26	28	31	33	36	39	41	44	47
9		1	3	6	9	12	15	18	21	24	27	30	33	36	39	42	45	48	51	54
10		1	4	7	11	14	17	20	24	27	31	34	37	41	44	48	51	55	58	62
11		1	5	8	12	16	19	23	27	31	34	38	42	46	50	54	57	61	65	69
12		2	5	9	13	17	21	26	30	34	38	42	47	51	55	60	64	68	72	77
13		2	6	10	15	19	24	28	33	37	42	47	51	56	61	65	70	75	80	84
14		2	7	11	16	21	26	31	36	41	46	51	56	61	66	71	77	82	87	92
15		3	7	12	18	23	28	33	39	44	50	55	61	66	72	77	83	88	94	100
16		3	8	14	19	25	30	36	42	48	54	60	65	71	77	83	89	95	101	107
17		3	9	15	20	26	33	39	45	51	57	64	70	77	83	89	96	102	109	115
18		4	9	16	22	28	35	41	48	55	61	68	75	82	88	95	102	109	116	123
19	0	4	10	17	23	30	37	44	51	58	65	72	80	87	94	101	109	116	123	130
20	0	4	11	18	25	32	39	47	54	62	69	77	84	92	100	107	115	123	130	138

Reproduced from the *Bulletin of the Institute of Educational Research at Indiana University,* vol. 1, no. 2; with the permission of the author and the publisher.
For specific details about using this table to find: *p*-values, see page 678; critical values, page 678.

Table 14   Critical Values for Total Number of Runs (V)

The entries in this table are the critical values for a two-tailed test using $\alpha = 0.05$. For a one-tailed test, $\alpha = 0.025$ and use only one of the critical values: the smaller critical value for a left-hand critical region, the larger for a right-hand critical region.

The larger of n_1 and n_2

smaller of n_1 and n_2	5	6	7	8	9	10	11	12	13	14	15	16	17	18	19	20
2								2	2	2	2	2	2	2	2	2
								6	6	6	6	6	6	6	6	6
3		2	2	2	2	2	2	2	2	2	3	3	3	3	3	3
		8	8	8	8	8	8	8	8	8	8	8	8	8	8	8
4	2	2	2	3	3	3	3	3	3	3	3	4	4	4	4	4
	9	9	10	10	10	10	10	10	10	10	10	10	10	10	10	10
5	2	3	3	3	3	3	4	4	4	4	4	4	4	5	5	5
	10	10	11	11	12	12	12	12	12	12	12	12	12	12	12	12
6		3	3	3	4	4	4	4	5	5	5	5	5	5	6	6
		11	12	12	13	13	13	13	14	14	14	14	14	14	14	14
7			3	4	4	5	5	5	5	5	6	6	6	6	6	6
			13	13	14	14	14	14	15	15	15	16	16	16	16	16
8				4	5	5	5	6	6	6	6	6	7	7	7	7
				14	14	15	15	16	16	16	16	17	17	17	17	17
9					5	5	6	6	6	7	7	7	7	8	8	8
					15	16	16	16	17	17	18	18	18	18	18	18
10						6	6	7	7	7	7	8	8	8	8	9
						16	17	17	18	18	18	19	19	19	20	20
11							7	7	7	8	8	8	9	9	9	9
							17	18	19	19	19	20	20	20	21	21
12								7	8	8	8	9	9	9	10	10
								19	19	20	20	21	21	21	22	22
13									8	9	9	9	10	10	10	10
									20	20	21	21	22	22	23	23
14										9	9	10	10	10	11	11
										21	22	22	23	23	23	24
15											10	10	11	11	11	12
											22	23	23	24	24	25
16												11	11	11	12	12
												23	24	25	25	25
17													11	12	12	13
													25	25	26	26
18														12	13	13
														26	26	27
19															13	13
															27	27
20																14
																28

From C. Eisenhart and F. Swed, "Tables for testing randomness of grouping in a sequence of alternatives," *Annals of Statistics,* vol. 14 (1943): 66–87. Reprinted by permission.

For specific details about using this table to find: *p*-values, see page 687; critical values, page 686.

Table 15   Critical Values of Spearman's Rank Correlation Coefficient

The entries in this table are the critical values of r_s for a two-tailed test at α. For a one-tailed test, the value of α shown at the top of the table is double the value of α being used in the hypothesis test.

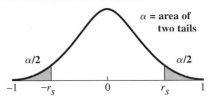

α = area of two tails

n	$\alpha = 0.10$	$\alpha = 0.05$	$\alpha = 0.02$	$\alpha = 0.01$
5	0.900	—	—	—
6	0.829	0.886	0.943	—
7	0.714	0.786	0.893	—
8	0.643	0.738	0.833	0.881
9	0.600	0.700	0.783	0.833
10	0.564	0.648	0.745	0.794
11	0.536	0.618	0.736	0.818
12	0.497	0.591	0.703	0.780
13	0.475	0.566	0.673	0.745
14	0.457	0.545	0.646	0.716
15	0.441	0.525	0.623	0.689
16	0.425	0.507	0.601	0.666
17	0.412	0.490	0.582	0.645
18	0.399	0.476	0.564	0.625
19	0.388	0.462	0.549	0.608
20	0.377	0.450	0.534	0.591
21	0.368	0.438	0.521	0.576
22	0.359	0.428	0.508	0.562
23	0.351	0.418	0.496	0.549
24	0.343	0.409	0.485	0.537
25	0.336	0.400	0.475	0.526
26	0.329	0.392	0.465	0.515
27	0.323	0.385	0.456	0.505
28	0.317	0.377	0.448	0.496
29	0.311	0.370	0.440	0.487
30	0.305	0.364	0.432	0.478

From E. G. Olds, "Distribution of sums of squares of rank differences for small numbers of individuals," *Annals of Statistics,* vol. 9 (1938), pp. 138–148, and amended, vol. 20 (1949), pp. 117–118. Reprinted by permission.
For specific details about using this table to find: *p*-values, see page 696; critical values, page 695.

Answers to Selected Exercises

Chapter 1

1.1 **a.** King-Size Company customers
  b. 10,000
  c. 99% of 10,000 considered airline seating cramped.
  d. Multiple answers allowed.

1.2 **a.** Employed American adults
  b. U.S. Labor Department
  c. Number of cashiers surveyed in 1996; the median hourly pay rate

[handwritten: 1.5]
1.3 **a.** American travelers, trips more than 100 miles
  b. 38-year-old male, income $50,000 or more, travels by car between 100 and 300 miles, usually to visit friends or relatives and stays at their homes.

[handwritten: 1.3]
1.4 **a.** American adults
  b. On September 1, 1998
  c. Within ±4%
  d. Actual percentage could range from 50% to 58%.

1.7 **a.** descriptive
  b. inferential

1.9 **a.** USA adults
  b. 1000
  c. Feels safe; what changed in their lifestyles
  d. 55% of those surveyed carry less cash.
  e. 550
  f. More than one answer is possible.

1.11 **a.** 45%
  b. The percentages are from different groups; all adults married/separated/divorced and those still married.

1.13 population: all U.S. adults;
  sample: the 1200 adults;
  variable: allergy status;
  one data: yes, dust;
  data: set of yes/no responses, set of allergies;
  experiment: method to select the adults;
  parameter: percent of all U.S. adults with an allergy, 36%;
  statistic: 33.2% based on sample

1.14 Parameter has one specific value. Statistics vary by sample.

1.15 ZIP code, gender, highest level of education

1.16 annual income, age, distance to store

1.17 marital status, ZIP code

1.18 highest level of education, rating for first impression of store

1.19 **a.** scores are counted
  b. time is measured

1.20 **a.** Status, Importance, Number of dinners, Length of dinner
  b. nominal, ordinal, discrete, continuous

1.21 **a.** Severity of side effects
  b. attribute (ordinal)

1.23 **a.** All individuals who have hypertension and use prescription drugs to control it
  b. 5000 in study
  c. Proportion of population for which drug is effective
  d. Proportion of sample for which drug is effective, 80%
  e. No

1.25 **a.** All assembled parts from the assembly line
  b. infinite
  c. parts checked
  d. A, B—attribute, C—continuous

1.27 **a.** All people suffering from migraine headaches
  b. 2633 people given the drug
  c. Amount of drug dosage, side effects encountered
  d. Dosage: quantitative, degree of relief and type of side effects: qualitative.

1.29 **a.** numerical   **b.** attribute   **c.** numerical
  d. attribute   **e.** numerical   **f.** numerical

1.31 **a.** Population contains all objects of interest; sample contains only those actually studied
  b. convenience, availability, practicality

1.33 Football players' weights cover a wider range of values

1.35 Price/standard unit makes price the only variable

1.37 Volunteer; yes

1.38 Observational

1.39 **a.** All adults who deal with stockbrokers
b. Experiment

1.40 Senate

1.41 House of Representatives

1.42 Each precinct is a cluster and not all precincts are sampled.

1.43 **a.** Observational or volunteer
b. Yes

1.45 **a.** list from which sample is actually drawn
b. computer list of full-time students
c. Random Number Table; student numbered 1288

1.47 Judgment sampling

1.49 Statistical methods presume the use of random samples.

1.51 probability samples

1.53 convenience sampling

1.55 Proportional sample would work best.

1.57 Not all adults are registered voters.

1.58 A—probability, B—statistics

1.59 **a.** statistics   **b.** probability
c. statistics   **d.** probability

1.61 Performs many of the computations and tests quickly and easily

1.63 Computers cannot determine whether or not a study has been conducted properly.

1.64 **a.** American adults
b. statistics
c. age
d. nominal
e. survey
f. Men (20–34) spend an average of 81 minutes per day behind the wheel; average value determined for each age and gender; different jobs require varying amounts of driving
g. Answers will vary; fans of pro sports
h. Frequency of reading fine print; Ordinal

1.65 **a.** color of hair, major, gender
b. number of courses taken, height, distance from hometown to college

1.67 **a.** piece of data
b. What is average number of times per week people in the sample went shopping?
c. What is average number of times per week people (all people) go shopping?

1.69 **a.** All Alzheimer patients in the U.S.
b. Cost in medical expenses and lost productivity per patient per year
c. Total cost per year for all Alzheimer patients in the U.S.
d. Total cost per year for the Alzheimer patients used as a sample

1.71 **a.** All people in U.S. who died in 1997
b. death from heart disease, state of residence, age, obesity, inactivity
c. mortality rate per 100,000 people in the US
d. death from heart disease—attribute
state of residence—attribute
age at death—numerical
obesity—attribute
inactivity—attribute

1.73 **a.** Attribute data
b. Judgment sample, most likely

Chapter 2

2.1

Methods of Payment

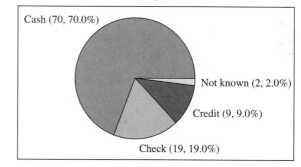

Cash (70, 70.0%)
Not known (2, 2.0%)
Credit (9, 9.0%)
Check (19, 19.0%)

2.2

Method of Payment

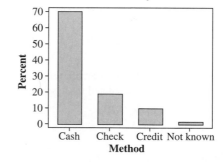

2.3 Circle graph makes it easier to visually compare the relative sizes of the parts to each other and the size of each part to the whole

2.4

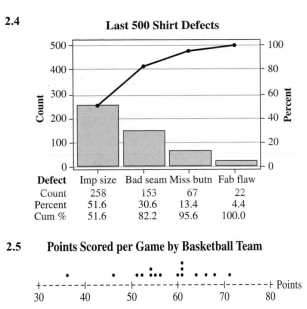

Last 500 Shirt Defects

Defect	Imp size	Bad seam	Miss butn	Fab flaw
Count	258	153	67	22
Percent	51.6	30.6	13.4	4.4
Cum %	51.6	82.2	95.6	100.0

2.5   **Points Scored per Game by Basketball Team**

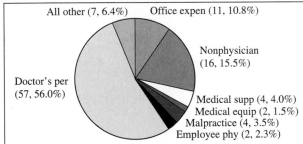

2.6   96: leaf 6 on stem 9
66: leaf 6 on stem 6

2.7   Points per game

```
3 | 6
4 | 6
5 | 6 4 5 4 2 1
6 | 1 1 8 0 6 1 4
7 | 1
```

2.8   Each leaf value is in the ones position.

2.9   a.   **Where Doctors' Fees Go**

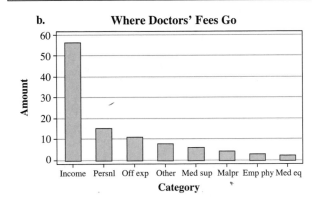

b.   **Where Doctors' Fees Go**

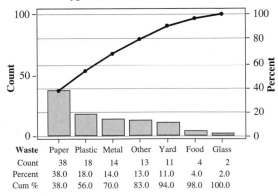

c. Circle graph makes it easy to compare the size of each part to the whole; bar graph makes it easy to compare the sizes of the parts to each other

2.11   a.   **Monster Cookies**

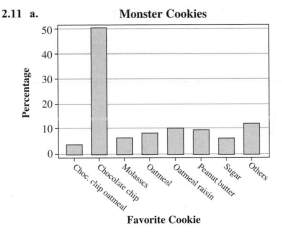

b.   **Monster Cookies**

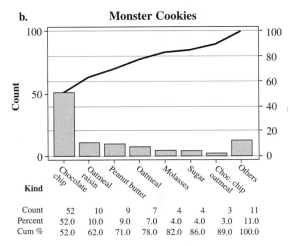

Kind	Chocolate chip	Oatmeal raisin	Peanut butter	Oatmeal	Molasses	Sugar	Choc. chip oatmeal	Others
Count	52	10	9	7	4	4	3	11
Percent	52.0	10.0	9.0	7.0	4.0	4.0	3.0	11.0
Cum %	52.0	62.0	71.0	78.0	82.0	86.0	89.0	100.0

c. stock 4 most popular: Chocolate chip, Oatmeal raisin, Peanut butter, Oatmeal. The Pareto lists them first.

d. Chocolate chip—156, Oatmeal raisin—30, Peanut butter—27, Oatmeal—21, Sugar—12, Molasses—12, Chocolate chip oatmeal—9, Others—33

2.13   a.   **Types of Waste in US Landfills**

Waste	Paper	Plastic	Metal	Other	Yard	Food	Glass
Count	38	18	14	13	11	4	2
Percent	38.0	18.0	14.0	13.0	11.0	4.0	2.0
Cum %	38.0	56.0	70.0	83.0	94.0	98.0	100.0

b. "Other" in middle is inconsistent with Pareto format. We need more complete information on unclassified data.

2.15 a. Heights of Basketball Players

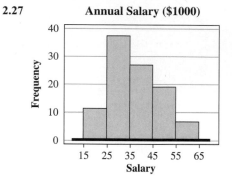

b. shortest—72, tallest—85
c. 81 inches, six players

2.17 a. 15
b. 11.2, 11.2, 11.3, 11.4, 11.7
c. 15.6
d. 13.7; 3

2.19 a. 50 Common Stocks

1	1 7 9 7 6 3 3 7 2 7 1 2 4
2	6 8 6 6 2 4 2 1 8 5 7
3	0 5 1 1 7 6 5 8 2
4	1 7 5 8 3
5	8 2 2 4 2 7 2
6	5
7	2 5 5
8	
9	0

b. Lowest price—$11, highest—$90.
c. More stocks priced between $10 and $20 than in any other interval.

2.21 a. Place value of leaves is hundredths
b. 16
c. 5.97, 6.01, 6.04, 6.08
d. Cumulative frequencies starting at the top and the bottom

2.23

x	f
0	2
1	5
2	3
3	0
4	2
	12

2.24 a. f is frequency; values of 70 or more but less than 80 occurred 8 times.
b. 19
c. Number of data, or sample size

2.25 a. $35 \leq x < 45$
b.

Class Boundaries	Frequency (millions)
$18 \leq x < 25$	2.3
$25 \leq x < 35$	5.4
$35 \leq x < 45$	7.0
$45 \leq x < 55$	6.7
$55 \leq x < 65$	4.7
$65 \leq x$	7.4
	33.5

2.26 a. $65 \leq x < 75$
b. Values greater than or equal to 65 and less than 75
c. Difference between upper and lower class boundaries

2.27   **Annual Salary ($1000)**

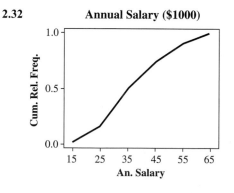

2.28 Shapes are the same; Figure 2.10 uses class marks, frequency; Figure 2.11 uses class boundaries, percentages

2.29 Symmetric: breaking strength of string
Uniform: rolling a die many times
Skewed right: salaries
Skewed left: hour exam scores
Bimodal: weights of both males and females
J-shaped: amount of TV watched per day

2.30

Class Boundaries	Cumulative Frequency
$15 \leq x < 25$	12
$25 \leq x < 35$	49
$35 \leq x < 45$	75
$45 \leq x < 55$	94
$55 \leq x \leq 65$	100

2.31

Class Boundaries	Cum. Rel. Frequency
$15 \leq x < 25$	0.12
$25 \leq x < 35$	0.49
$35 \leq x < 45$	0.75
$45 \leq x < 55$	0.94
$55 \leq x \leq 65$	1.00

2.32   **Annual Salary ($1000)**

2.33 a. "size": Number of Children Living at Home
"frequency": Number of Mexican-American Women

b.

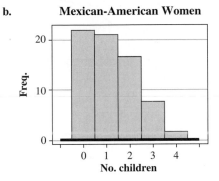

Mexican-American Women

2.35 a.

Age	Frequency	b. Age	Rel. Freq.
17	1	17	0.02
18	3	18	0.06
19	16	19	0.32
20	10	20	0.20
21	12	21	0.24
22	5	22	0.10
23	1	23	0.02
24	2	24	0.04
	50		1.00

c.

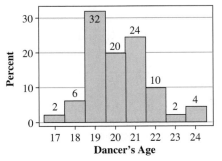

Ages of Dancers at Audition

d.

Age	Cum. Rel. Freq.
17	0.02
18	0.08
19	0.40
20	0.60
21	0.84
22	0.94
23	0.96
24	1.00

e.

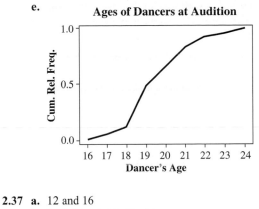

Ages of Dancers at Audition

2.37 a. 12 and 16
b. 2, 6, 10, 14, 18, 22, 26
c. 4.0
d. 0.08, 0.16, 0.16, 0.40, 0.12, 0.06, 0.02
e.

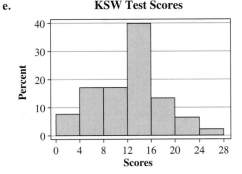

KSW Test Scores

2.39 a.

Class Limits	Frequency
12–18	1
18–24	14
24–30	22
30–36	8
36–42	5
42–48	3
48–54	2

b. 6
c. 27, 24, 30
d.

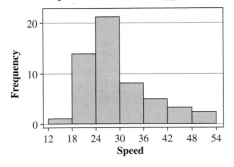

Speed of 55 Cars on City Street

2.41 a.

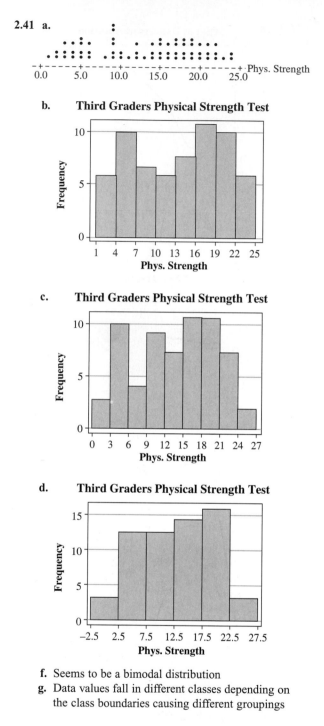

b. **Third Graders Physical Strength Test**

c. **Third Graders Physical Strength Test**

d. **Third Graders Physical Strength Test**

f. Seems to be a bimodal distribution
g. Data values fall in different classes depending on the class boundaries causing different groupings

2.43 a.

Number	Frequency	Rel. Freq.	Cum. Rel. Freq.
0	23	0.329	0.329
1	22	0.314	0.643
2	17	0.243	0.886
3	7	0.100	0.986
4	1	0.014	1.000
	70	1.000	

b.

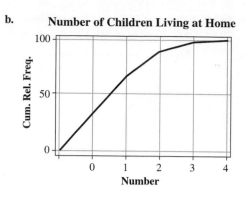

2.45 $\bar{x} = 18/8 = 2.25$

2.46 $d(\tilde{x}) = $ 3rd; $\tilde{x} = 73$

2.47 $d(\tilde{x}) = $ 4.5th; $\tilde{x} = 4.55$

2.48 2

2.49 $5.48

2.50 $\bar{x} = 49/6 = 8.2$
$d(\tilde{x}) = $ 3.5th; $\tilde{x} = 8.5$
mode = 9
midrange = 8.0

2.51 a. $\bar{x} = 354,000/10 = 35,400.00$
$d(\tilde{x}) = $ 5.5th; $\tilde{x} = 33,375$
mode = 31,500
midrange = 39,750
Values all agree with those in the article.
b. The large value of 54,000 is pulling the mean and midrange toward the larger data value.

2.53 a. $\bar{x} = 36/6 = 6.0$
b. $d(\tilde{x}) = $ 3.5th; $\tilde{x} = 6.5$
c. 7
d. 5.5

2.55 a. $\bar{x} = 36709/14 = 2622$
b. $d(\tilde{x}) = $ 7.5th; $\tilde{x} = $ 2494
c. $2998.50
d. no mode

2.57 a. midrange = 54.75
b. Highest and lowest values were the only data given.
c. The distribution of state average greens' fees must be skewed to the right.

2.59 a. See Exer. 2.41(a)
b. mode = 9
c. See Exer. 2.41(b)
d. Bimodal. Modal classes are 4–7, and 16–19.
e. Dotplot shows mode to be 9, which is in the 7–10 class; two modal classes to be 4–7 and 16–19. The mode is not in either modal class.
f. No, only one numerical value per class.
g. The mode is the single data value that occurs most often, while a modal class results from data tending to bunch up.

2.61 a. & b.

	Home	Away	Diff.
Mean	9.77	9.80	−0.03
Med.	9.65	9.78	−0.06
Max.	13.65	11.06	4.89
Min.	7.64	8.67	−1.74
Midr.	10.65	9.87	1.58

    c. At Coors Field the maximum number of runs scored (13.65); away, only 8.76 runs, which ranked second from the bottom.

2.63 a. $\sum x$ needs to be 500.
    b. Need two numbers smaller than 70 and one larger
    c. Need multiple 87's
    d. Need any two numbers that total 140
    e. Need two numbers smaller than 70, total is 330
    f. Need two numbers of 87 and a third so total of all five is 500
    g. Total 500 and total of L and H to be 140; impossible
    h. Two 87's; impossible

2.64 5.14

2.65 a. 12 units above the mean
    b. 20 units below the mean

2.66 $\sum(x - \bar{x}) = \sum x - n\bar{x} = \sum x - n \cdot (\sum x/n) = \sum x - \sum x = 0$

2.67 $n = 5, \sum x = 25, \sum(x - \bar{x})^2 = 46; 11.5$

2.68 $n = 5, \sum x = 25, \sum x^2 = 171; 11.5$

2.69 a. 7
    b. $n = 5, \sum x = 30, \sum(x - \bar{x})^2 = 34; 8.5$
    c. 2.9

2.71 a. $n = 9, \sum x = 72, \sum(x - \bar{x})^2 = 73.60; 8.2$
    b. $n = 9, \sum x = 72, \sum x^2 = 592; 8.2$
    c. 2.9

2.73 $n = 14, \sum x = 36,709, \sum x^2 = 103,772,173$
    a. 578,350.2
    b. 760.5

2.75 a.

$n = 20, \sum x = 601, \sum x^2 = 18,209$
    b. 30.05
    c. 9
    d. 7.8
    e. 2.8
    g. Except for $x = 30$, the distribution looks rectangular.

2.77 a. range $= 83$
    $n = 27, \sum x = 5,255, \sum x^2 = 1,031,585; s = 18.4$

    b. Range $= 83$

16	0
17	5
18	000024555
19	0001225
20	000
21	166
22	0
23	3
24	3

    $s = 18.4$

    c. mounded, slightly skewed

2.79

	$\sum x$	$\sum|x - \bar{x}|$	$\sum(x - \bar{x})^2$	Range
Set 1:	250	14	54	9
Set 2:	250	46	668	35

2.81 a. All data must be the same value.
    b. possible answer: 99, 99.5, 100.5, 100 [100]; $s = 0.57$
    c. possible answer: 107, 95, 94, 108, [100]; $s = 6.53$
    d. possible answer: 75, 78, 123, 124, [100]; $s = 23.53$

2.83 a.

x	f	xf	x^2f
0	1	0	0
1	3	3	3
2	8	16	32
3	5	15	45
4	3	12	48
$\sum$	20	46	128

    b. $\sum f = 20; \sum xf = 46; \sum x^2f = 128$
    c. 4 is one of the possible data values
    8 is the number of times an "x" value occurred
    $\sum f$: sum of the frequencies $=$ sample size
    $\sum xf$: the sum of the data

2.84 a. No meaning unless each value occurred only once.
    b. Each data value is multiplied by how many times it occurred. Summing these products will give the same sum if all data values were listed individually.

2.85 $n = 20, \sum xf = 46; 2.3$

2.86 $n = 20, \sum xf = 46, \sum x^2f = 128; 1.2$

2.87 1.1

2.88 $n = 40, \sum xf = 516, \sum x^2f = 7504; \bar{x} = 12.9, s^2 = 21.7, s = 4.7$

2.89 $n = 73, \sum xf = 134, \sum x^2f = 378; \bar{x} = 1.8, s^2 = 1.8, s = 1.4$

2.91 $n - 123$, $\sum xf = 3{,}387{,}312$, $\sum x^2f = 93{,}827{,}350{,}800$
  a. $3,387,312
  b. $27,539
  c. Academic
  d. $2111

2.93 $n = 40$, $\sum xf = 447.0$, $\sum x^2f = 5472.00$; 11.2, 12.2, 3.5

2.95 $n = 125$, $\sum xf = 1100$, $\sum x^2f = 10{,}376$; 8.8, 2.37

2.97 $n = 100$, $\sum xf = 4210$, $\sum x^2f = 198{,}100$; 42.1, 14.5

2.99 a. Using class marks 9.5, 11.5, . . . ; $n = 50$, $\sum xf = 867.0$, $\sum x^2f = 15{,}448.50$
  b. Grouped: 17.340, 8.464, 2.909
     Ungrouped: 17.260, 8.523, 2.919
  c. 0.5%, −0.7%, −0.3%

2.101 a. $n = 997$, $\sum xf = 63{,}665$, $\sum x^2f = 4{,}302{,}725$; 63.9, 65, 65, 55
  b. 238.25, 15.4

2.103 44th position from L; 7th position from H

2.104 $nk/100 = 10.0$, $d(P_{20}) = 10.5$th, $P_{20} = 64$
  $nk/100 = 17.5$, $d(P_{35}) = 18$th, $P_{35} = 70$

2.105 $nk/100 = 10.0$, $d(P_{80}) = 10.5$th from H, $P_{80} = 88.5$
  $nk/100 = 2.5$, $d(P_{95}) = 3$rd from H, $P_{95} = 95$

2.106 symmetric about the mean

2.107

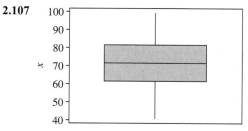

2.108 1.67, −0.75

2.109 a. $nk/100 = 5.0$; $d(P_{25}) = 5.5$th; $Q_1 = 3.8$
     $nk/100 = 15.0$; $d(P_{75}) = 15.5$th; $Q_3 = 5.6$
  b. 4.7
  c. $nk/100 = 3.0$; $d(P_{15}) = 3.5$th; $P_{15} = 3.5$
     $nk/100 = 6.6$; $d(P_{33}) = 7$th; $P_{33} = 4.0$
     $nk/100 = 18.0$; $d(P_{90}) = 18.5$th; $P_{90} = 6.9$

2.111 a. $d(\tilde{x}) = 18$th; $\tilde{x} = 33.0$
  b. 34.8
  c. $nk/100 = 8.75$; $d(P_{25}) = 9$th; $Q_1 = 31.3$
     $nk/100 = 26.25$; $d(P_{75}) = 27$th; $Q_3 = 36.0$
     midquartile = 33.65
  d. 30.1, 31.3, 33.0, 36.0, 39.5

e.

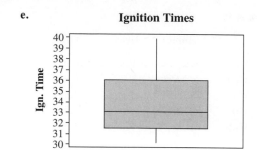

Ignition Times

2.113 a. 1.0
  b. 0.0
  c. 2.25
  d. −1.25

2.115 680

2.117 a. 1.5 standard deviations above the mean
  b. 2.1 standard deviations below the mean
  c. the number of standard deviations from the mean

2.119 1.625, 1.2; A has the higher relative position.

2.121 From 175 through 225 words, inclusive.

2.122 at least 93.75%

2.123 a. 50%
  b. ≈68%
  c. ≈84%

2.124 94

2.125 a. at least 75%
  b. at least 89%

2.127 99.7% lies within 3 standard deviations of the mean.

2.129 a. Range equal to 6 times the standard deviation.
  b. Divide the range by 6.

2.131 a. at most 11%
  b. at most 6.25%

2.133 a. 75%
  b. approximately 95%

2.135 $n = 50$; $\sum x = 3{,}238{,}171$; $\sum x^2 = 240{,}009{,}107{,}515$
  a. $64,763.42, $24,864.56
  b. $39,898.86 and $89,627.98
  c. 33/50 = 66%
  d. $15,034.30 and $114,492.54
  e. 49/50 = 98%
  f. −$9,830.26 and $139,357.10
  g. 50/50 = 100%
  h. 98% and 100% are at least 75% and 89% respectively.
  i. Percentages are close enough to the empirical rule to suggest an approximately normal distribution.

2.139 Yes

2.140 a. Cutting off the bottom

b.

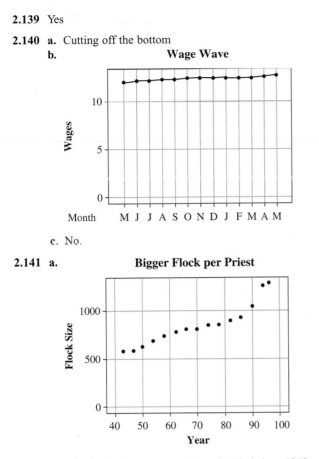

Wage Wave

c. No.

2.141 a.

Bigger Flock per Priest

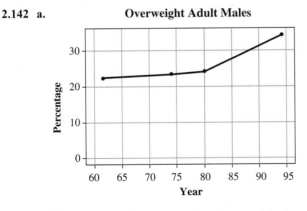

b. The flock size appears to have doubled since 1942, not tripled as suggested by the Case Study graph.

2.142 a.

Overweight Adult Males

b. The percentage has increased greatly; graph in the Case Study seems to overstate the increase.

2.143 The class width is not uniform.

2.146 $n = 50$, $\sum x = 787.00$, $\sum x^2 = 21{,}027$

a. One-Way Commute

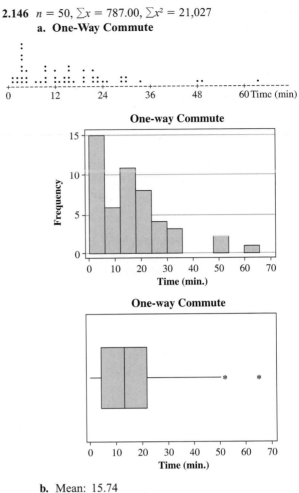

One-way Commute

One-way Commute

b. Mean: 15.74
Med.: $d(\tilde{x}) = 25.5$th; $\tilde{x} = 13.50$
Mode: 3
Midr.: 32.5
Midq.: 5, 22; 13.5

c. $s^2 = 176.319$, $s = 13.28$, range $= 63$

2.147 a. 0.15, 0.34, 0.18, 0.24, 0.09

b. At-Risk Desktop Computers Using CalcuPro

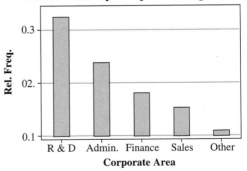

c. The bars in a histogram touch, emphasizing the sequence of the data.

2.149 **a.** Mean increased
   b. Med. is unchanged
   c. Mode is unchanged.
   d. Midr. increased
   e. Range increased
   f. Var. increased
   g. St. dev. increased

2.151 $n = 8, \Sigma x = 36.5, \Sigma x^2 = 179.11$
   a. 4.56
   b. 1.34
   c. very close to 4%

2.153 $n = 118, \Sigma x = 2364$
   a. 20.0   **b.** 17   **c.** 16   **d.** 15, 21   **e.** 14, 43

2.155 $n = 25, \Sigma x = 1997, \Sigma x^2 = 163,205$; 79.9, 12.4

2.157 **a.** Population: all airline passengers; variable:
      passenger luggage status, lost or not.
      Other variables: airline, date of flight.
   b. Statistics
   c. Statistic
   d. No

2.159 $n = 3570; \Sigma xf = 55,155; \Sigma x^2f = 890,655$
   a.

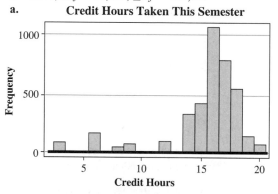

Credit Hours Taken This Semester

   b. $\bar{x} = 15.4; d(\tilde{x}) = 1785.5$th, $\tilde{x} = 16$; mode = 16;
      midrange = 11.5; midquartile = 16
   c. $Q_1 = 15; Q_3 = 17$
   d. $P_{15} = 14; P_{12} = 14$
   e. range = 17; $s^2 = 10.7967; s = 3.3$

2.161 **a.**   **Frequency of Misplaced Remote Control/Week**

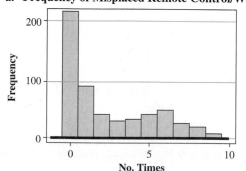

b. $n = \Sigma f = 500, \Sigma xf = 994, \Sigma x^2f = 5200$
   mean = 1.988, med. = 1, mode = 0, midr. = 4.5
c. 6.46, 2.5
d. $Q_1 = 0, Q_3 = 4, P_{90} = 6$
e. 2
f. 0, 0, 1, 4, 9

Mystery of the Remote

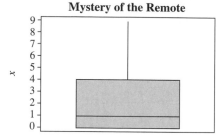

2.163 $n = 220; \Sigma xf = 219,100; \Sigma x^2f = 224,470,000$
   b. 995.9
   c. 169.2

2.165 **b.** 4.6, 13.75, 22.65, 30.10, 41.1
   c. 21.925, 16.35
   d. 1.89, -1.52, 0.40
   e. No

2.167 $n = 1000; \Sigma xf = 142,150; \Sigma x^2f = 50,742,500$;
   \$142; \$175

2.169 **a.** P_{98}
   b. P_{16}

2.171 58, 0, 9.8, 6.9, 181

2.173 $n = 8, \Sigma x = 31,825, \Sigma x^2 = 126,894,839$
   a. 3978.1
   b. 203.9
   c. 3570.3 to 4385.9

2.175 **a.** Calories data: $n = 50, \Sigma x = 5805, \Sigma x^2 = 781,425$
      $\bar{x} = 116.1, s = 46.83$
      Sodium data: $n = 50, \Sigma x = 28,990$,
      $\Sigma x^2 = 19,880,100$
      $\bar{x} = 579.8, s = 250.4$
   b. Calories: between 22.44 and 209.76
      Sodium: between 79.04 and 1,080.56 mg
      Yes; 94%, 98%
   c. 329.42 and 830.18; 64%, 98% ($\pm 2s$);
      approximately normal

2.177 **a.** $n = 40; \Sigma x = 66,693; \Sigma x^2 = 119,230,725$; 1667.3,
      453.81
   b. 768.0, 1392.0, 1699.0, 1942.5, 2527.0
   c. 759.7 to 2574.9; Yes, 100%
   d. 1213.5 to 2121.1; Yes, 65%
   e. All distributions satisfy Chebyshev's theorem;
      however, these data do not have an approximately
      normal distribution.

2.179 Many possible answers; only one is shown.
   a. 70, 77.5, 77.5, 77.5, 85 yields $s = 5.30$
   b. 70, 76, 85, 89, 95 yields $s = 10.02$
   c. 70, 85, 90, 99, 110 yields $s = 15.02$
   d. The data had to become more dispersed.

2.181 a.

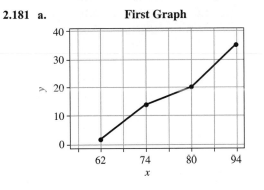

First Graph

b.

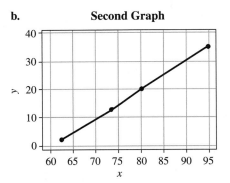

Second Graph

   c. (a) Suggests an accelerated rate from 1980 to 1995, (b) suggests rate of increase has been constant from 1962 to 1995.
   d. By adjusting the horizontal intervals, it is possible to cause the line to show different slopes in those different intervals.

2.183 As sample size increases, the distribution looks more like a normal distribution.

Chapter 3

3.1

	On Airplane	Hotel Room	All Other	Marginal total
Business	35.5%	9.5%	5.0%	50%
Leisure	25.0%	16.5%	8.5%	50%
Marginal total	60.5%	26.0%	13.5%	100%

3.2

	Plane	Room	Other	Marginal total
Business	71.0%	19.0%	10.0%	100%
Leisure	50.0%	33.0%	17.0%	100%
Margin total	60.5%	26.0%	13.5%	100%

Business and leisure are separate distributions.

3.3

	Plane	Room	Other	Marginal total
Business	58.7%	36.5%	37.0%	50%
Leisure	41.3%	63.5%	63.0%	50%
Margin total	100.0%	100.0%	100.0%	100%

Each of the categories is a separate distribution.

3.4

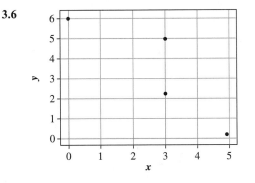

East $\bar{x} = 5.72$; $d(\tilde{x}) = $ 3rd, $\tilde{x} = 5.5$
West $\bar{x} = 6.06$; $d(\tilde{x}) = $ 3rd, $\tilde{x} = 6.0$

3.5 Height; weight is often predicted

3.6

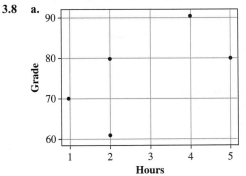

3.7 **a.** Amount of paid vacation, length of employment
   b. One increases, the other increases

3.8 a.

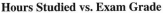

Hours Studied vs. Exam Grade

   b. As hours studied increased it seems that exam grades increased

3.9 **a.** Employed college graduates aged 30–55 who had been out of college 10 or more years; Type of worker; Reason for going to school

b.

Reason for Going to School

Worker	Professional	Personal	Both	Marginal total
Tech	28%	31%	41%	100%
Other	47%	20%	33%	100%

3.11 a.

		Speed Limits (mph)				Row total
	75	70	65	60	55	
Cars	10	16	22	0	2	50
Trucks	9	11	20	3	7	50
Col. total	19	27	42	3	9	100

b.

Interstate Highway Speed Limits (mph)

Vehicle Type	75	70	65	60	55	Row Totals
Cars	10%	16%	22%	0%	2%	50%
Trucks	9%	11%	20%	3%	7%	50%
Column Totals	19%	27%	42%	3%	9%	100%

c.

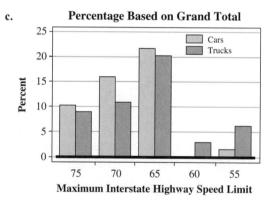

Percentage Based on Grand Total

3.13 a. 3000
  b. Two variables; both are qualitative
  c. 950
  d. 50%
  e. 25%

3.15 a.

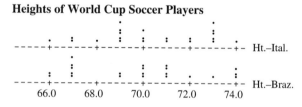

b. Rates for minimum deposits of:

	$500	$1000	$2500
High	5.26	5.45	5.50
Q_3	5.19	5.32	5.24
$\tilde{x}$	5.09	5.15	5.10
Q_1	4.955	5.09	5.00
Low	4.00	5.00	4.75

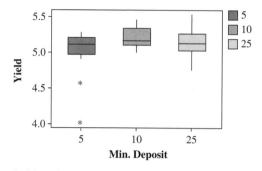

Effect of Minimum Amount

c. $500 set has two extremely low values.

3.17 a.

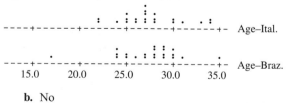

Heights of World Cup Soccer Players

Weights of World Cup Soccer Players

Ages of World Cup Soccer Players

b. No
c. Not paired between teams

3.19

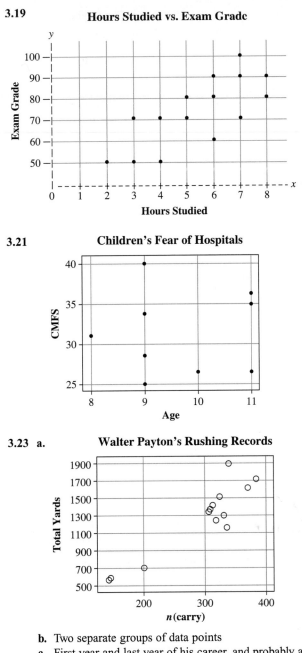

Hours Studied vs. Exam Grade

3.21 Children's Fear of Hospitals

3.23 a. Walter Payton's Rushing Records

b. Two separate groups of data points
c. First year and last year of his career, and probably a year when a serious injury restricted his playing time

3.25 a. $n = 5$, $\sum x = 14$, $\sum y = 380$, $\sum x^2 = 50$, $\sum xy = 1110$, $\sum y^2 = 29,400$; SS$(x) = 10.8$, SS$(y) = 520$, SS$(xy) = 46$
b. 0.61

3.26 a. Second Year Basketball Data

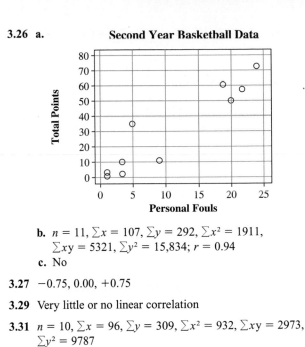

b. $n = 11$, $\sum x = 107$, $\sum y = 292$, $\sum x^2 = 1911$, $\sum xy = 5321$, $\sum y^2 = 15,834$; $r = 0.94$
c. No

3.27 -0.75, 0.00, $+0.75$

3.29 Very little or no linear correlation

3.31 $n = 10$, $\sum x = 96$, $\sum y = 309$, $\sum x^2 = 932$, $\sum xy = 2973$, $\sum y^2 = 9787$
a. 10.4   b. 238.9   c. 6.6   d. 0.13

3.33 a. about 0.7
b. $n = 15$, $\sum x = 81$, $\sum y = 110$, $\sum x^2 = 487$, $\sum xy = 625$, $\sum y^2 = 842$; 0.74

3.35 a. Number of TV Commercials vs. Sales Volume

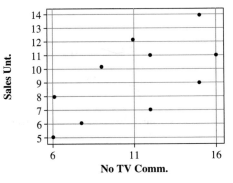

b. 1/2 to 2/3
c. $n = 10$, $\sum x = 110$, $\sum y = 93$, $\sum x^2 = 1332$, $\sum xy = 1085$, $\sum y^2 = 937$; 0.66

3.37 a. Cable Television Video Networks

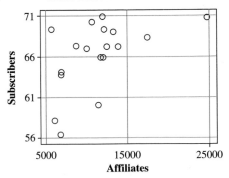

b. $n = 18$, $\sum x = 200,464$, $\sum y = 1193.0$,
$\sum x^2 = 2,689,703,492$, $\sum xy = 13,479,155$,
$\sum y^2 = 79,392$; $r = 0.5025$; positive correlation

3.39 $b_0 = (\sum y - (b_1 \cdot \sum x))/n = \sum y/n - (b_1 \cdot \sum x)/n = \bar{y} - b_1\bar{x}$

3.40 a. $\sum x^2 = 13,717$, $SS(x) = 1396.9$

$\sum y^2 = 15,298$, $SS(y) = 858.0$

$\sum xy = 14,257$, $SS(xy) = 919.0$
b. $\sum$'s are sums of data values, $SS(\ )$'s are parts of complex formulas.

3.41 a. $\hat{y} = 28.1$; $\hat{y} = 47.9$
b. Yes

3.42 a. $n = 5$, $\sum x = 14$, $\sum y = 380$, $\sum x^2 = 50$, $\sum xy = 1110$;
$\hat{y} = 64.1 + 4.26x$
b. Use (1,68.4) and (3,76.9)

Hours Studied vs. Exam Grade

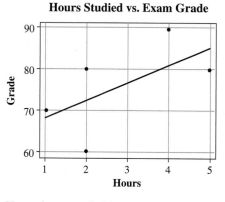

c. Yes, as hours studied increased, exam grades increased.

3.43 a. As height increases by one inch, weight increases by 4.71 pounds.
b. The scale for the y-axis starts at $y = 95$ and the scale for the x-axis starts at $x = 60$.

3.44 a. $\hat{y} = 14.9 + 0.66(40) = 41.3 = 41$
b. No
c. 41 is the average

3.45 a.

35 Years of Car Buying

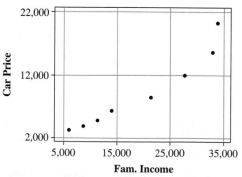

35 Years of Car Buying

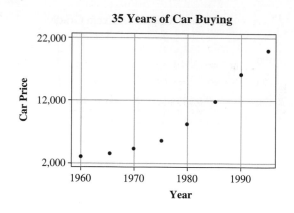

35 Years of Car Buying

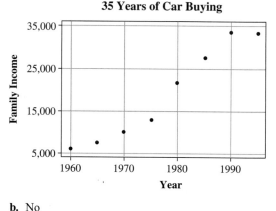

b. No
c. Yes

3.46 Vertical scale is at $x = 58$ and is not the y-axis.

3.47 5.83, -260.61

3.49 a. When no long-distance calls are made there is still the monthly phone charge of \$23.65.
b. \$1.28 is the rate at which the total phone bill will increase for each additional long distance call.

3.51 a. $\hat{y} = 185.7 - 21.52(3) = 121.14$ or \$12,114.
b. $\hat{y} = 185.7 - 21.52(6) = 56.58$ or \$5,658.
c. $21.52(\$100) = \$2152$

3.53 2.75

3.55 a.

Cash Registers

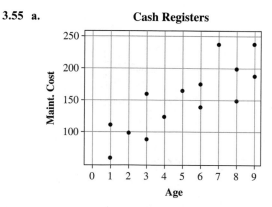

b. $n = 14$, $\sum x = 72$, $\sum y = 2163$, $\sum x^2 = 476$,
$\sum xy = 12,677$; $\hat{y} = 78.95 + 14.69x$

c. $196.47

d. The expected average cost of maintenance for all 8-year-old cash registers

3.57 a.

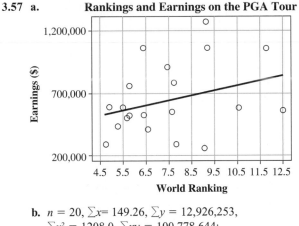

Rankings and Earnings on the PGA Tour

b. $n = 20$, $\sum x = 149.26$, $\sum y = 12,926,253$,
$\sum x^2 = 1208.0$, $\sum xy = 100,778,644$;
$\hat{y} = \$304,230 + 45,837x$

c. See (a)

d. $625,089

3.59 a. Each person will have a different answer.

b.

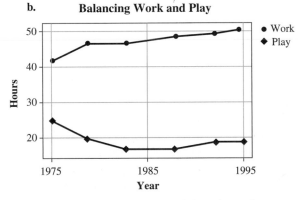

Balancing Work and Play

c. Yes, as work hours increase, leisure hours decrease slightly.

d.

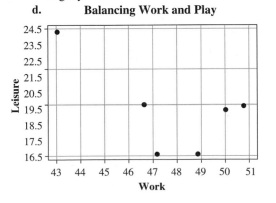

Balancing Work and Play

The scatter diagram seems to suggest that an increase in number of work hours is related to fewer hours of leisure.

3.61 a. Population: Workers aged 25 to 64
Variables: Gender, amount of savings set aside for emergencies
Type: Nominal, ordinal

b.

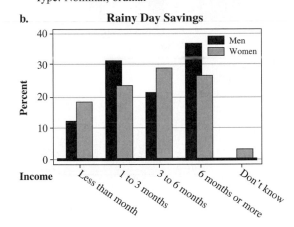

Rainy Day Savings

c. No

3.63 a.

	<6mo	6 mo–1 yr	>1yr	Total
Under 28	413	192	295	900
28–40	574	208	218	1000
Over 40	653	288	259	1200
Total	1640	688	772	3100

b.

	<6mo	6 mo–1 yr	>1yr	Total
Under 28	13.3%	6.2%	9.5%	29.0%
28–40	18.5%	6.7%	7.0%	32.2%
Over 40	21.1%	9.3%	8.4%	38.8%
Total	52.9%	22.2%	24.9%	100%

c.

	<6mo	6 mo–1 yr	>1yr	Total
Under 28	45.9%	21.3%	32.8%	100%
28–40	57.4%	20.8%	21.8%	100%
Over 40	54.4%	24.0%	21.6%	100%
Total	52.9%	22.2%	24.9%	100%

d.

	<6mo	6 mo–1 yr	>1yr	Total
Under 28	25.2%	27.9%	38.2%	29.0%
28–40	35.0%	30.2%	28.2%	32.3%
Over 40	39.8%	41.9%	33.6%	38.7%
Total	100%	100%	100%	100%

e.
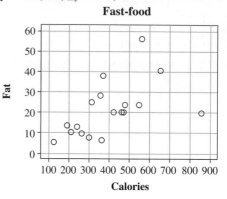

3.65 a. correlation
b. regression
c. correlation
d. regression
e. correlation

3.67

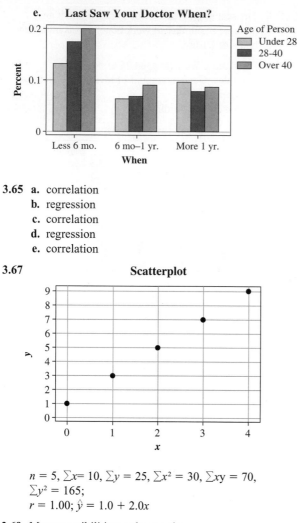

$n = 5$, $\sum x = 10$, $\sum y = 25$, $\sum x^2 = 30$, $\sum xy = 70$,
$\sum y^2 = 165$;
$r = 1.00$; $\hat{y} = 1.0 + 2.0x$

3.69 Many possibilities; only one shown.
a. $(1,5),(2,5),(0,4),(0,6)$ $[(5,5)]$
$n = 5$, $\sum x = 8$, $\sum y = 25$, $\sum x^2 = 30$, $\sum xy = 40$,
$\sum y^2 = 127$
b. $(1,1),(2,2),(3,3),(4,4)$, $[(5,5)]$
$n = 5$, $\sum x = 15$, $\sum y = 15$, $\sum x^2 = 55$, $\sum xy = 55$,
$\sum y^2 = 55$
c. $(6,4),(7,3),(4,6),(3,7)$, $[(5,5)]$
$n = 5$, $\sum x = 25$, $\sum y = 25$, $\sum x^2 = 135$, $\sum xy = 115$,
$\sum y^2 = 135$
d. $(1,5),(5,3),(7,3),(9,5)$ $[(5,5)]$
$n = 5$, $\sum x = 27$, $\sum y = 21$, $\sum x^2 = 181$, $\sum xy = 111$,
$\sum y^2 = 93$; -0.185
e. $(1,2),(3,3),(1,4),(3,5)$, $[(5,5)]$
$n = 5$, $\sum x = 13$, $\sum y = 19$, $\sum x^2 = 45$, $\sum xy = 55$,
$\sum y^2 = 79$; 0.642

3.71 $n = 18$, $\sum x = 7055$, $\sum y = 367$, $\sum x^2 = 3,287,783$,
$\sum xy = 164,625$, $\sum y^2 = 10,309$
a.

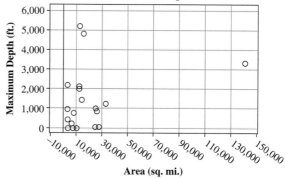

b. 0.541
c. $\hat{y} = 4.80 + 0.04x$
d. If the calories increase so does the fat content.

3.73 a. Scatter Diagram of Areas of Lakes and Their Maximum Depths

b. 0.3709

3.75 a. Numerator $= \sum(x - \bar{x})(y - \bar{y})$
$= \sum[xy - \bar{x}y - x\bar{y} + \bar{x}\bar{y}]$
$= \sum xy - \bar{x} \cdot \sum y - \bar{y} \cdot \sum x + n\bar{x}\bar{y}$
$= \sum xy - [(\sum x/n) \cdot \sum y] - [(\sum y/n) \cdot \sum x]$
$\qquad + [n \cdot (\sum x/n)(\sum y/n)]$
$= \sum xy - [(\sum x \cdot \sum y/n) - (\sum x \cdot \sum y/n)$
$\qquad + (\sum x \cdot \sum y/n)]$
$= \sum xy - [(\sum x \cdot \sum y)/n]$
$= SS(xy)$

Denominator of formula (3-1):
Denominator
$= (n - 1)s_x s_y$
$= (n - 1) \cdot \sqrt{(SS(x)/(n - 1)}$
$\qquad \cdot \sqrt{(SS(y)/(n - 1)}$
$= \sqrt{(SS(x) \cdot SS(y)}$

Chapter 4

4.1 Results will vary but have denominators of 10.

4.3 Results will vary but have denominators of 25.

4.5 a. Results will vary but have denominators of 50.
    b. Results will vary but have denominator of 100.

4.7 0.225

4.8 Divide the experimental count by the sample size.

4.9 Expect a 1 to occur approximately 1/6th of the time.

4.11 a. 0.5397
    b. 0.0849

4.13 Results will vary but have denominators of 50.

4.15 Results will vary.

4.17 {0, 1, 2, 3, 4, 5, 6, 7, 8, 9}

4.18

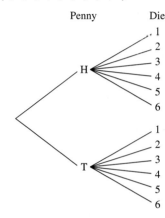

4.19 a.

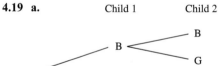

b.

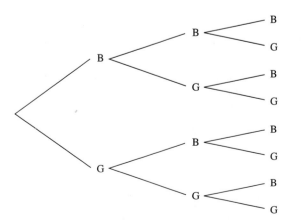

4.20 a. {$1, $5, $10, $20}
    b.

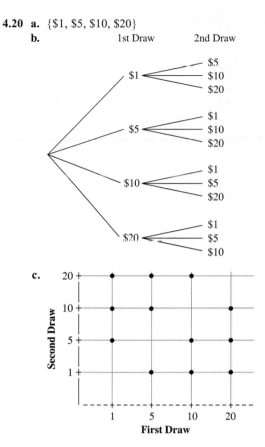

c.

4.21 {JH, JC, JD, JS, QH, QC, QD, QS, KH, KC, KD, KS}

4.23 {*yyy, yyn, yny, ynn, nyy, nyn, nny, nnn*}

4.25 a. {HH, HT, TH, TT} and equally likely
    b. {HH, HT, TH, TT} and not equally likely

4.27 {H1, H2, H3, H4, H5, H6, T1, T2, T3, T4, T5, T6}

4.29 Results will vary.

4.31 $P(5) = 4/36$; $P(6) = 5/36$; $P(7) = 6/36$; $P(8) = 5/36$;
    $P(9) = 4/36$; $P(10) = 3/36$; $P(11) = 2/36$; $P(12) = 1/36$

4.32 4/5

4.33 a. 2/9
    b. 7:2

4.34 a. 0.0004267
    b. 12:7
    c. 37:1 (rounded)

4.35 40/52

4.37 $P(R) = 1/4$, $P(Y) = 1/4$, $P(G) = 2/4 = 1/2$

4.39 a. 1/6
    b. 3/6
    c. 4/6
    d. 3/6

4.41 $P(A) = 1/7$, $P(B) = 2/7$, $P(C) = 4/7$.

4.43 All are inappropriate.

4.45 a. 0.55
b. 0.40

4.47 a. 1 to 232, 232 to 1, 0.00429
b. 1 to 3699, 3699 to 1, 0.00027
c. 1 to 3999, 3999 to 1, 0.00025
d. 1 to 15, 15 to 1, 0.0625
e. 1 to 64, 64 to 1, 0.0154
f. 1 to 129, 129 to 1, 0.0077

4.49 8/9

4.51 a. Yes
b. No
c. No
d. Yes
e. No
f. Yes
g. No

4.52 a. A & C and A & E are mutually exclusive
b. 12/36, 11/36, 10/36

4.53 a. Not mutually exclusive
b. Not mutually exclusive
c. Not mutually exclusive
d. Mutually exclusive

4.55 There is no intersection.

4.57 a. 0.7
b. 0.6
c. 0.7
d. 0.0

4.59 No

4.61 a. 0.4510
b. 0.5000
c. 0.1765
d. 0.7745; 0.2255

4.63 0.04

4.65 a. 0.45
b. 0.40
c. 0.55
d. No, $P(S) \neq P(S|F) \neq P(S|M)$

4.66 0.15

4.67 0.28

4.68 a. 0.40   **b.** Yes

4.71 a. independent
b. not independent
c. independent
d. independent
e. not independent (If you do not own a car, how can your car have a flat tire?)
f. not independent

4.73 a. 0.12
b. 0.4

c. 0.3

4.75 a. 0.5
b. 0.667
c. No

4.77 a. $P(A) = 12/52$, $P(A|B) = 6/26$, independent
b. $P(A) = 12/52$, $P(A|C) = 3/13$, independent
c. $P(B) = 26/52$, $P(B|C) = 13/13$, dependent

4.79 a. 0.51
b. 0.15
c. 0.1326

4.81 a. 0.5041
b. 0.0841
c. 0.357911

4.83 a. 0.5375
b. 0.175
c. 0.6125

4.85 a. 0.36
b. 0.16
c. 0.48

4.87 a. 3/5
b. 4/25, 12/25, 9/25

4.89 a. Are the events part-time and graduate within five years independent?
b. No. Being part-time will have an effect.
c. 0.42336

4.92 $P(\text{owner}|\text{married})$, $P(\$200 + k \text{ income}|\text{married owner})$

4.93 0.7

4.95 a. 0.15
b. 0.65
c. 0.7
d. 0.5
e. 0.7
f. No; independent events

4.97 a. 20/56
b. 30/56
c. 6/56

4.99 a. $P(A \text{ wins on 1st turn}) = 1/2$
$P(B \text{ wins on 1st turn}) = 1/4$
$P(C \text{ wins on 1st turn}) = 1/8$
b. $P(A \text{ wins on 2nd turn}) = 1/16$
$P(A \text{ wins on 1st try or 2nd try}) = 9/16$
$P(B \text{ wins on 1st try or 2nd try}) = 9/32$
$P(C \text{ wins on 1st try or 2nd try}) = 9/64$

4.101 8/30

4.103 a. 0.9
b. $P(D|C) = 1/4 = 0.25$; $P(\overline{D}|C) = 1 - 1/4 = 3/4$;
$P(C \text{ and } \overline{D}) = 0.075$
c. $P(C \text{ and } D) = 0.025$

4.105 a. 0.625
b. 0.25

c. $P(\text{satisfied} \mid \text{skilled woman}) = 0.25$;
$P(\text{satisfied} \mid \text{unskilled woman}) = 0.667$; not independent

4.107 a. $P(M \mid \text{maj.gain}) = 0.5294$
b. $P(W \mid \text{slt.gain}) = 0.3774$

4.109	$P(A_i)$	$P[(B) \mid A_i]$	$P(A_i \text{ and } B)$	$P(A_i \mid B)$
A_1, Republican	0.65	0.54	0.351	0.686
A_2, not Rep.	0.35	0.46	0.161	0.314

4.110 a. Each number occurred nearly 10% of the time.

Relative Frequency on First Number

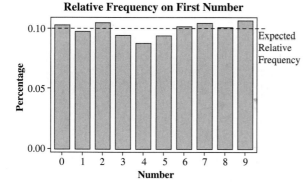

b. Each number occurred nearly 10% of the time in the 2nd, 3rd, and 4th positions also.

Relative Frequency on Second Number

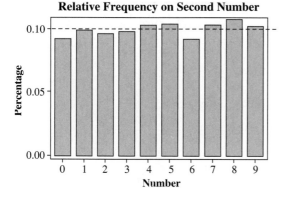

Relative Frequency on Third Number

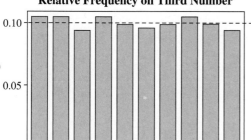

Relative Frequency on Fourth Number

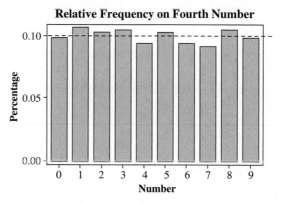

c. The fewest, 1290, is 1290 out of 13,380 or 0.0964; while the most, 1383, is 1383 out of 13,380 or 0.10336. These extremes seem very close to the expected 10%.

4.111 a. 0.0958
b. 0.1172
c. 0.1396
d. 0.4949
e. 0.1606
f. 0.2182
g. 0.0938

4.113 a. 0.30
b. 0.40
c. $P(A \text{ and } B) = 0.233$,
$P(A) \cdot P(B) = 0.12$; not independent
d. mutually exclusive, not complementary, not independent events

4.115 a. False
b. True
c. False
d. False

4.117 a. {GGG, GGR, GRG, GRR, RGG, RGR, RRG, RRR}
b. 3/8
c. 7/8

4.119 a. 0.3168
b. 0.4659
c. No
d. No
e. "Candidate wants job" and "RJB wants candidate" could not both happen.

4.121 0.28

4.123 a. 0.429
b. 0.476
c. 0.905

4.125 0.300

4.127 a. 0.30
b. 0.60
c. 0.10
d. 0.60

e. 0.333
f. 0.25

4.129 a. 0.531
    b. 0.262
    c. 0.047

4.131 0.592

4.133 0.154, 0.769, 0.077

4.135 a. 0.988, 0.012
    b. 0.500, 0.500

4.137 a. 0.60
    b. 0.648
    c. 0.710
    d. (a) 0.70   (b) 0.784   (c) 0.874
    e. (a) 0.90   (b) 0.972   (c) 0.997
    f. More games increase the chance of "best" team winning.
    The greater the difference between the two teams' individual chances, the more likely the "best" team wins.

Chapter 5

5.1 number of siblings, 0, 1, 2, . . . ; length of conversation, 0 to ?

5.2 a. counts
    b. measurements

5.3 discrete, count; continuous, measureable value

5.5 number of children; $x = 0, 1, 2, 3, \ldots, n$; discrete

5.7 distance; $x = 0$ to n, n = radius; continuous

5.9 a. number of home runs
    b. discrete, count

5.11

x	0	1
$P(x)$	1/2	1/2

5.12

x	1	2	3	4	5	6
$P(x)$	1/6	1/6	1/6	1/6	1/6	1/6

5.13 Each unique outcome is assigned a specific numerical value.

5.14 All possible outcomes are accounted for.

5.15 a.

$P(x) = \frac{1}{6}$, for $x = 1, 2, 3, 4, 5, 6$

b. uniform or rectangular

5.16 a.

x	$P(x)$
0	0.05
1	0.12
2	0.15
3	0.25
4	0.21
5	0.10
6	0.05
7	0.04
8	0.02
9	0.01

b. $1/21 \approx 0.05$

5.17

x	0	1	2	3
$P(x)$	0.20	0.30	0.40	0.10

5.19 a.

x	$P(x)$
1	0.12
2	0.18
3	0.28
4	0.42
Σ	1.00

$P(x)$ is a probability function.

b.

$P(x) = (x^2 + 5)/50$, for $x = 1, 2, 3, 4$

5.21 a. Yes.
    b.

5.23 a. No, the information displays all properties of a probability distribution except one; the variable is attribute (not numerical).

b.

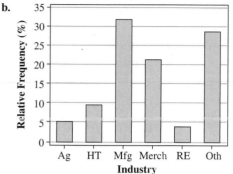

5.25 Answers will vary.

5.27 $\sigma^2 = \sum[(x - \mu)^2 \cdot P(x)]$
$= \sum[(x^2 - 2x\mu + \mu^2) \cdot P(x)]$
$= \sum[x^2 \cdot P(x) - 2x\mu \cdot P(x) + \mu^2 \cdot P(x)]$
$= \sum[x^2 \cdot P(x)] - 2\mu \cdot \sum[x \cdot P(x)] + \mu^2 \cdot [\sum P(x)]$
$= \sum[x^2 \cdot P(x)] - 2\mu \cdot [\mu] + \mu^2 \cdot [1]$
$= \sum[x^2 \cdot P(x)] - 2\mu^2 + \mu^2$
$= \sum[x^2 \cdot P(x)] - \mu^2 \text{ or } \sum[x^2 \cdot P(x)] - \{\sum[x \cdot P(x)]\}^2$

5.28 a.

x	$P(x)$	(b) $xP(x)$	$x^2P(x)$
1	1/6	1/6	1/6
2	2/6	4/6	8/6
3	3/6	9/6	27/6
$\sum$	1.0	(c) 14/6 = 2.33	36/6 = 6.0

5.29 2.33

5.30 0.55556

5.31 0.745

5.32 Nothing of any meaning

5.33 2.0, 1.0

5.35 a.

Random Single Digits; $P(x) = 0.1$, for $x = 0, 1, \ldots, 9$

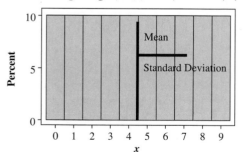

b. 4.5, 2.87
c. See (a)
d. 100%

5.37 a. 2.0, 1.4
  b. The numbers: 1, 2, 3 and 4
  c. 0.9

5.39 a. 2.44, 0.73
  b. Calculated mean and standard deviation is smaller than true values

5.41 Each question is a separate trial.

5.42 Four different ways that one correct and three wrong answers can be obtained

5.43 1/3 is the probability of choosing right answer on each trial, 4 is the number of questions; expected 1/3 correct answers.

5.44 Answers will vary.

5.45 One trial is one coin, repeated $n = 50$ times; independent because P(head) is same for all tosses; Two outcomes on each trial: success = H, failure = T; $p = P(\text{heads}) = \dfrac{1}{2}$ and $q = P(\text{tails}) = \dfrac{1}{2}$; $x = n(\text{heads})$, 0 to 50.

5.46 a. 24
  b. 4

5.47 $P(x) = \dbinom{3}{x}(0.5)^x(0.5)^{3-x}$; 0.125, 0.375, 0.125

5.48 a. 0.0146, 0.00098
  b. $0 \le$ each $P(x) \le 1$; $\sum P(x) = 0.99998 \approx 1$ (round-off error)

5.49 0.007

5.50 a. 0.240
  b. 0.240

5.51 a. 24
  b. 5040
  c. 1
  d. 360
  e. 10
  f. 15
  g. 0.0081
  h. 35
  i. 10
  j. 1
  k. 0.4096
  l. 0.16807

5.53 $n = 100$ trials (shirts), two outcomes (first quality or irregular), $p = P(\text{irregular})$, $x = n(\text{irregular})$; any integer value from 0 to 100.

5.55 a. trials are not independent
  b. $n = 4$, the number of independent trials; two outcomes, ace and not ace; $p = P(\text{ace}) = 4/52$ and $q = P(\text{not ace}) = 48/52$; $x = n(\text{aces})$, any number 0, 1, 2, 3 or 4.

5.57 a.

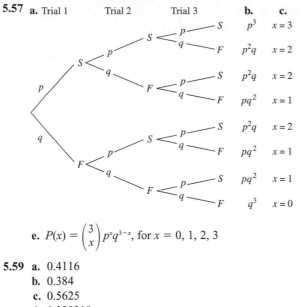

Trial 1   Trial 2   Trial 3   **b.**   **c.**

p^3 $x = 3$
p^2q $x = 2$
p^2q $x = 2$
pq^2 $x = 1$
p^2q $x = 2$
pq^2 $x = 1$
pq^2 $x = 1$
q^3 $x = 0$

e. $P(x) = \binom{3}{x} p^x q^{3-x}$, for $x = 0, 1, 2, 3$

5.59 a. 0.4116
b. 0.384
c. 0.5625
d. 0.329218
e. 0.375
f. 0.0046296

5.61 $n = 5$,
$p = \dfrac{1}{2}$ and $q = \dfrac{1}{2}$,
exponents x and $5 - x$ add up to $n = 5$,
x, any integer value from zero to $n = 5$;
$T(x)$ is a binomial probability distribution
1/32, 5/32, 10/32, 10/32, 5/32, 1/32

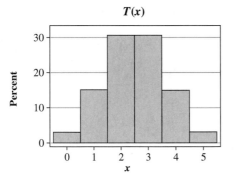

5.63 0.930

5.65 0.668

5.67 a. 0.590
b. 0.918

5.69 0.0011

5.71 a. 0.463
b. 0.537

5.73 0.984

5.75 a. 0.4400
b. 0.0507
c. 0.0048

5.77 a. 0.2757
b. 0.1865
c. 0.5425
d. 0.6435

5.81 a. 0.2122
b. 0.6383

5.83 Number of defective items should be small and easy to count.

5.85 18, 2.7

5.86 a. 0.55, 0.72

b.

x	$P(x)$
0	0.569
1	0.329
2	0.087
3	0.014
4	0.001
5	0+
Σ	1.0

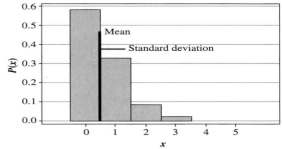

5.87 a. $\Sigma xP(x) = 0.549$, $\Sigma x^2P(x) = 0.819$; $\mu = 0.55$, $\sigma = 0.72$
b. The same

5.89 a. 25.0, 3.5
b. 7.7, 2.7
c. 24.0, 4.7
d. 44.0, 2.3

5.91 400, 0.5

5.93 a. $P(x) = \binom{5}{x} = (0.75)^x(0.25)^{5-x}$ for $x = 0, 1, \ldots, 5$

x	$P(x)$
0	0.00098
1	0.01465
2	0.08789
3	0.26367
4	0.39551
5	0.23730
Σ	1.00000

b.

P(x)

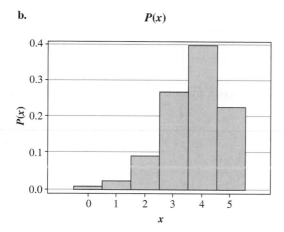

c. 3.75, 0.97

5.95

	Mean	Standard Deviation
Asian	1.50	1.187
Black	6.05	2.141
Hispanic	4.90	1.985

5.97 **a.** Number of evening meals American adults cook at home in an average week.

b.

x	*P(x)*
0	0.08
1	0.05
2	0.10
3	0.13
4	0.15
5	0.21
6	0.09
7	0.19

c. 19%
d. 8%
e. 5 nights
f.

Homecooked Dinner

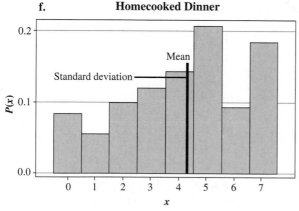

g. Mounded, skewed to left, not normal

h. $\sum[xP(x)] = 4.16$, $\sum[x^2P(x)] = 21.82$; 4.2, 2.1
i. See histogram in (f).
j. $\mu \pm \sigma$: 2.1 to 6.3; 0.58
    $\mu \pm 2\sigma$: 0.00 to 8.4; 1.00
    $\mu \pm 3\sigma$; $P(x) = 1.00$
    Empirical Rule does not apply, Chebyshev's Rule does apply.

5.101 **a.** 0.1
    b. 0.4
    c. 0.6

5.103 **a.** 0.930
    b. 0.264

5.105 0.103

5.107 **a.** 0.070
    b. One trial = one person, $n = 10$, success = "alone," failure = "other than alone," $p = P(\text{alone}) = 0.48$, $q = 0.52$, $x = n(\text{alone}) = 0, 1, \ldots, 10$.

5.109 0.063

5.111 Not binomial, $P(\text{defective})$ changes from selection to selection

5.113 2600

5.115 **a.** $p^3 + 3p^2q$
    b. 0.028
    c. 0.896
    d. when p is greater than 0.5
    e. 0, 0.5, or 1.0

5.117 Tool Shop: mean profit = 7,000.0
    Book Store: mean profit = 6,000.0

5.119 $\mu = \sum[x \cdot P(x)]$
$$= (1) \cdot (1/n) + (2) \cdot (1/n) + \ldots + (n) \cdot (1/n)$$
$$= (1/n) \cdot [1 + 2 + 3 + \ldots + n]$$
$$= (1/n) \cdot [(n)(n+1)/2]$$
$$= (n + 1)/2$$

Chapter 6

6.1 0.4147

6.2 0.0212

6.3 0.9582

6.4 0.4177

6.5 0.0630

6.6 0.8571

6.7 0.2144

6.8 0.84

6.9 -1.15 and $+1.15$

6.11 **a.** 0.4032
    b. 0.3997
    c. 0.4993
    d. 0.4761

6.13 **a.** 0.4394
b. 0.0606
c. 0.9394
d. 0.8788

6.15 **a.** 0.7737
b. 0.8981
c. 0.8983
d. 0.3630

6.17 **a.** 0.5000
b. 0.1469
c. 0.9893
d. 0.9452
e. 0.0548

6.19 **a.** 0.4906
b. 0.9725
c. 0.4483
d. 0.9306

6.21 **a.** 1.14
b. 0.47
c. 1.66
d. 0.86
e. 1.74
f. 2.23

6.23 **a.** 1.65
b. 1.96
c. 2.33

6.25 -1.28 or $+1.28$

6.27 -0.67 and $+0.67$

6.29 1.28, 1.65, 2.33

6.31 2.88

6.32 0.3944

6.33 0.8944

6.34 85.52, or 86

6.35 89.28, or 89

6.37 $29,008

6.38 Answers will vary.

6.39 Answers will vary.

6.40 Answers will vary.

6.41 Answers will vary.

6.42 0.2329 using computer or calculator; 0.2316 using Table 3

6.43 **a.** 0.5000
b. 0.3849
c. 0.6072
d. 0.2946
e. 0.9502
f. 0.0139

6.45 **a.** 0.0038 or 0.38%
b. 0.00003 or 0.003% (practically zero)

6.47 **a.** $0.325 = 32.5\%$
b. $0.7535 = 75.35\%$
c. $0.089 = 8.9\%$
d. $0.058 = 5.8\%$

6.49 **a.** 0.1131
b. 0.0505
c. 4.64 minutes

6.51 **a.** 89.6
b. 79.2
c. 57.3

6.53 20.26

6.55 **a.** 0.0401 or 4%
b. 0.0179 or 1.8%

6.57 **a.** 0.056241
b. 0.505544
c. 0.438215
d. 0.0559, 0.5077, 0.4364
e. Rounding errors

6.59 Answers will vary with a mean and standard deviation close to 100 and 16, respectively, and approximately normally distributed.

6.61

6.62

6.63 1.04

6.64 -0.92

6.65 **a.** $z(0.03)$
b. $z(0.14)$
c. $z(0.75)$
d. $z(0.13)$
e. $z(0.91)$
f. $z(0.82)$

6.67 **a.** 1.96
b. 1.65
c. 2.33

6.69 **a.** 1.28, 1.65, 1.96, 2.05, 2.33, 2.58
b. $-2.58, -2.33, -2.05, -1.96, -1.65, -1.28$

6.71 **a.** area; 0.4602
 b. z-score; 1.28
 c. area; 0.5199
 d. z-score; -1.65

6.73 No; $np = 2$

6.74 **a.** Bin. prob. function: 0.0149, Normal approx.: 0.0202
 b. differ by 0.0053

6.75 **a.** not appropriate, $np = 3$
 b. not appropriate, $np = 0.5$
 c. appropriate, $np = 50$, $nq = 450$
 d. appropriate, $np = 10$, $nq = 40$

6.77 0.1822; 0.177

6.79 0.9429; 0.943

6.81 0.8078

6.83 0.0049

6.85 $P(x \leq 29.5) = 0.6844$

6.87 **a.** 0.9988
 b. 0.9988

6.88 **a.** Yes
 b. 68% within one st. dev. of mean, symmetric, mean is central.
 c. 100
 e. 120 to 130; approx. 23%
 f. 130; "very superior"
 g. above 110; 25%
 h. approximately 2%

6.89 at least 3/4; 0.9544

6.91 **a.** 1.26
 b. 2.16
 c. 1.13

6.93 **a.** 0.0930
 b. 0.9684

6.95 **a.** 1.175 or 1.18
 b. 0.58
 c. -1.04
 d. -2.33

6.97 **a.** 0.0158
 b. 1.446 years

6.99 10.033

6.101 **a.** $n = 25$, $np = 7.5$, $nq = 17.5$
 b. 7.5, 2.29

6.103 **b.** 0.77023
 c. 0.751779

6.105 **a.** $P(0) + P(1) + \ldots + P(75)$
 b. 0.9856
 c. 0.9873

6.107 **a.** 5.534
 c. Answers will vary; most will be 0 or 1.
 d. Yes

6.109 **a.** 0.001
 b. 1.000
 c. 0.995
 d. 0.0668

6.111

Nation	$\mu = np$	$\sigma = \sqrt{npq}$	$P(x \geq 70)$
China	66	7.989	0.3300
Germany	10	3.154	0.0000+
India	130	11.024	0.9999+
Japan	8	2.823	0.0000+
Mexico	64	7.871	0.2420
Russia	34	5.781	0.0000+
S. Africa	98	9.654	0.9984
United States	14	3.729	0.0000+

6.113 **a.** 0.7422
 b. 0.6103
 c. 0.8320
 d. 0.8469

Chapter 7

7.1 one of 25; 3 of the 25

7.3 Classes: $1.8 - 4.6$ using class width of 0.4; Freq.: 3, 5, 6, 6, 5, 4, 1

7.4 $\bar{\bar{x}}$ = observed mean of x-bars; $s_{\bar{x}}$ = observed standard deviation of x-bars

7.5 **a.** Freq.: 1, 1, 4, 2, 1, 3
 b. Airlines have different size fleets.

7.7 **a.**

11	31	51	71	91
13	33	53	73	93
15	35	55	75	95
17	37	57	77	97
19	39	59	79	99

 b.

$\bar{x}$	1	2	3	4	5	6	7	8	9
$P(\bar{x})$	0.04	0.08	0.12	0.16	0.20	0.16	0.12	0.08	0.04

 c.

R	0	2	4	6	8
$P(R)$	0.20	0.32	0.24	0.16	0.08

7.9 Answers will vary.
 c. normal; mounded, symmetric, centered around 4.5
 d. same as part (c)

7.11 Answers will vary.

7.13 Answers will vary.
 d. approximately normal, mounded, symmetric, centered around 100

7.14 6.25, 4.167, 2.50

7.15 **a.** 1.0
   b. as n increases, $\sigma/\sqrt{n}$ gets smaller

7.17 **a.** approximately normal
   b. $31.65
   c. $1.00

7.19 **a.** 43.3 cents
   b. 0.61 cents
   c. approximately normal

7.21 Answers will vary.
   d. $\bar{\bar{x}} \approx 20$, $s_{\bar{x}} \approx 1.84$, approximately normal

7.23 $P(0 < z < 2) = 0.4772$

7.24 0.9699

7.25 38.73

7.27 **a.** approx. normal, $\mu = 69$, $\sigma = 4$
   b. 0.4013
   c. approx. normal
   d. 69; 1.0
   e. 0.1587
   f. 0.0228

7.29 **a.** 0.3830
   b. 0.9938
   c. 0.3085
   d. 0.0031

7.31 **a.** 0.2743
   b. 0.0359
   c. Yes

7.33 **a.** 0.6390
   b. With $n = 250$, CLT holds.
   c. If normal, the median $\approx$ mean.
   d. Probably not; typically skewed
   e. If mean is higher, probability will be less than calculated.

7.35 **a.** 0.6826

7.37 **a & b.**   **Vote For a Woman President?**

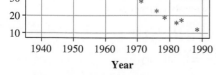

   c. Yes. Yeses steadily rise, no's steadily decrease.
   d. high eighties or in the nineties
   f. Study the variability of a sample statistic.

7.39 **a.** 0.49
   b. 0.049

7.41 **a.** 0.1190
   b. 0.0033

7.43 **a.** 0.1498
   b. 0.0089

7.45 0.0228

7.47 0.0023

7.49 **a.** approximately 1.000
   b. 0.9772

7.53 **a.** 60, 6.48

Chapter 8

8.1 collector fatigue, cost of sampling, time

8.2 $\sigma_{\bar{x}} = \sigma/\sqrt{n} = 18/\sqrt{36} = 18/6 = 3$

8.4 **a.** 33

8.5 **a.** $24 = n$; $4'11'' = \bar{x}$
   b. $16 = \sigma$
   c. $190 = s^2$
   d. $69 = \mu$

8.7 **a.** II has the lower variability.
   b. II has a mean value equal to the parameter.
   c. Neither is a good choice.

8.9 unbiased, $\mu_{\bar{x}} = \mu$; variability decreases as n increases, $\sigma/\sqrt{n}$

8.11 $3,133,948.80

8.14 **a.** 75.92
   b. 0.368
   c. 75.552 to 76.288

8.15 numbers are calculated for samples of snow

8.17 49

8.19 Sampled population is normal or sample is sufficiently large for CLT

8.21 **a.** 25.76 to 31.64
   b. Yes; population is normal

8.23 **a.** 125.58 to 131.42
   b. Yes; CLT

8.25 **a.** Speed
   b. 176.99 to 191.01
   c. 175.67 to 192.33

8.27 **a.** 25.3
   b. 24.29 to 26.31
   c. 23.97 to 26.63

8.29 **a.** 55.20
   b. 8.546
   c. 46.654 to 63.746

8.31 **a.** $5173.97 to $6026.03
    b. Lower level of confidence

8.33 27

8.35 28

8.37 H_o: system is reliable
    H_a: system is not reliable

8.38 H_a: Teaching techniques have a significant effect on student's exam scores.

8.39 A: party will be dud, did not go
    B: party will be a great time, did go
    I: the party will be a great time, did go [party was a dud]
    II: the party will be a dud, did not go [missed great party]

8.40 You missed a great time.

8.41 $\alpha = P$(rejecting a TRUE null hypothesis)

8.42 It is less likely to occur.

8.43 **a.** H_o: Special delivery does not take too much time.
       H_a: Special delivery takes too much time.
    b. H_o: New design is not more comfortable.
       H_a: New design is more comfortable.
    c. H_o: Cigarette smoke has no effect.
       H_a: Cigarette smoke has an effect.
    d. H_o: Hair conditioner is not effective.
       H_a: Hair conditioner is effective.

8.45 **a.** H_o: The victim is alive.
       H_a: The victim is not alive.
    b. A: alive, treated as alive
       I: alive, treated as dead
       II: dead, treated as alive
       B: dead, treated as dead
    c. I: very serious
       II: not as serious

8.47 **a.** I: decide majority do not favor, when majority do favor
       II: decide majority do favor, when they do not
    b. I: decide low salt, when it is not
       II: decide not low salt, when it is
    c. I: decide building must be demolished, when it should not
       II: decide building must not be demolished, when it should
    d. I: decide there is waste, when there is not waste
       II: decide there is no waste, when there is waste.

8.49 **a.** Type I
    b. Type II
    c. Type I
    d. Type II

8.51 **a.** very serious
    b. somewhat serious
    c. not at all serious

8.53 **a.** α
    b. β

8.55 **a.** "See, I told you so!"
    b. "Okay, so this evidence was not significant, I'll try again tomorrow."

8.57 **a.** 0.1151
    b. 0.2119

8.59 H_o: The mean shearing strength is at least 925 lbs.
    H_a: The mean shearing strength is less than 925 lbs.

8.60 H_o: $\mu = 54.4$
    H_a: $\mu \neq 54.4$

8.61 **a.** H_o: $\mu = 1.25$ ($\leq$) vs. H_a: $\mu > 1.25$
    b. H_o: $\mu = 335$ ($\geq$) vs. H_a: $\mu < 335$
    c. H_o: $\mu = 230,000$ vs. H_a: $\mu \neq 230,000$

8.62 A: is at least, decide is at least
    I: is at least, decide is less than
    II: is less than, decide is at least
    B: is less than, decide is less than
    II, buy and use weak rivets

8.63 -1.46

8.64 **a.** 0.0107
    b. 0.0359

8.65 **a.** Fail to reject H_o
    b. Reject H_o

8.66 **a.** 2.0; 0.0228
    b. -1.33; 0.0918
    c. -1.6; 0.1096

8.67 0.2714

8.71 **a.** H_o: $\mu = 26 (\leq)$ vs. H_a: $\mu > 26$
    b. H_o: $\mu = 36.7 (\geq)$ vs. H_a: $\mu < 36.7$
    c. H_o: $\mu = 1600 (\geq)$ vs. H_a: $\mu < 1600$
    d. H_o: $\mu = 210 (\leq)$ vs. H_a: $\mu > 210$
    e. H_o: $\mu = 570$ vs. H_a: $\mu \neq 570$

8.73 **a.** decide mean is less than, when it is at least
    b. decide mean is at least, when it is less than

8.75 **a.** 1.26
    b. 1.35
    c. 2.33
    d. -0.74

8.77 **a.** Reject H_o; Fail to reject H_o
    b. The calculated p-value is smaller than α; the calculated p-value is larger than α.

8.79 **a.** Fail to reject H_o **b.** Reject H_o

8.81 **a.** 0.0694
    b. 0.1977
    c. 0.2420
    d. 0.0174
    e. 0.3524

8.83 **a.** 1.57
    b. -2.13
    c. -2.87 or 2.87

8.85 **a.** H_o: $\mu = 525$ vs. H_a: $\mu < 525$
    b. Fail to reject H_o
    c. $60.0/\sqrt{38} = 9.733$

8.87 **a.** mean test score
    b. H_o: $\mu = 35.70(\geq)$ vs. H_a: $\mu < 35.70$
    c. -5.86; $P(z < -5.86) \approx +0.0000$
    d. Reject H_o

8.89 **a.** mean seconds in error per month
    b. H_o: $\mu = 20(\leq)$ vs. H_a: $\mu > 20$
    c. normality is assumed, $n = 36$
    d. $n = 36$, $\bar{x} = 22.7$
    e. $z^\star = 1.78$; **P** = 0.0375
    f. Reject H_o

8.91 H_o: mean is at least 925
    H_a: mean is less than 925

8.92 H_o: $\mu = 9(\leq)$ vs. H_a: $\mu > 9$

8.93 **a.** H_o: $\mu = 1.25(\geq)$ vs. H_a: $\mu < 1.25$
    b. H_o: $\mu = 335$ vs. H_a: $\mu \neq 335$
    c. H_o: $\mu = 230,000(\leq)$ vs. H_a: $\mu > 230,000$

8.94 A: is at least, decide is at least
    I: is at least, decide is less than
    II: is less than, decide is at least
    B: is less than, decide is less than
    Most serious: II, buy and use weak rivets

8.95 -1.10

8.96 **a.** Reject H_o
    b. Fail to reject H_o

8.97 $z \leq -2.33$

8.98 $z \geq 2.05$

8.99 2.2046 lbs/kg, 119.9

8.101 **a.** H_o: $\mu = 16(\geq)$ vs. H_a: $\mu < 16$
    b. H_o: $\mu = 6$ ft 6 in $(\leq)$ vs. H_a: $\mu > 6$ ft 6 in
    c. H_o: $\mu = 285(\geq)$ vs. H_a: $\mu < 285$
    d. H_o: $\mu = 0.375(\leq)$ vs. H_a: $\mu > 0.375$
    e. H_o: $\mu = 200$ vs. H_a: $\mu \neq 200$

8.103 **a.** Decide mean is greater, when it is not
    b. Decide mean is at most \$85 when it is greater

8.105 **a.** values that will result in reject H_o
    b. boundary value of critical region

8.107 **a.** $z \leq -1.65$, $z \geq 1.65$
    b. $z \geq 2.33$
    c. $z \leq -1.65$
    d. $z \leq -2.58$, $z \geq 2.58$

8.109 21,004.133

8.111 **a.** 3.0 standard errors
    b. yes

8.113 **a.** Reject H_o or Fail to reject H_o
    b. Calculated test statistic falls in the critical region; or calculated test statistic falls in the noncritical region.

8.115 **a.** $\mu = 15.0$; $\mu \neq 15.0$
    b. reject H_o
    c. $0.5/\sqrt{30} = 0.0913$

8.117 **a.** mean gestation period
    b. H_o: $\mu = 44$ vs. Ha: $\mu \neq 44$
    c. normality is assumed, $n = 81$
    d. $n = 81$, $\bar{x} = 42.5$
    e. $z^\star = -2.70$ or $\pm z(0.025) = \pm 1.96$
    f. Reject H_o

8.119 H_o: $\mu = 36.7(\geq)$ vs. H_a: $\mu < 36.7$
    $z^\star = -2.59$
    $z \leq -2.33$
    Reject H_o

8.121 **a.** Mean = 64.17, Median = 50, StDev = 50.29,
        Minimum = 0,
        Maximum = 300, $Q1 = 25$, $Q3 = 100$
    b. 52.96 to 75.37
    c. H_a: $\mu \neq 59.00$; $z^\star = 0.90$; $P = 0.37$ or $z \leq -1.96$, $z \geq 1.96$; fail to reject H_o
    d. normality can be assumed, CLT with $n = 150$

8.123 determines z, the number of standard errors; narrows

8.125 **a.** $\mu = 100$
    b. $\mu \neq 100$
    c. 0.01
    d. 100
    e. 96
    f. 12
    g. 1.70
    h. -2.35
    i. 0.0188
    j. Fail to reject H_o

8.127 **a.** 39.6 to 41.6
    b. H_a: $\mu \neq 40$
        $z^\star = 1.20$
        $P = 0.2302$
        Fail to reject H_o
    c. H_a: $\mu \neq 40$
        $z^\star = 1.20$
        $\pm z(0.025) = \pm 1.96$
        Fail to reject H_o

8.129 **a.** 39.9 to 41.9
    b. $z^\star = 1.80$
        $P = 0.0359$
        Reject H_o
    c. $z^\star = 1.80$
        $z(0.05) = 1.65$
        Reject H_o

8.131 **a.** 9.75 to 9.99
    b. 9.71 to 10.03
    c. widened the interval

8.133 **a.** 69.89 to 75.31
     b. Yes

8.135 3.87 to 4.13

8.137 60

8.139 2.89; 0.0019

8.141 **a.** H_a: $\mu \neq 0.50$
     b. $z^\star = 1.25$
       $P = 0.2112$
     c. $z = \pm 2.33$

8.143 H_a: $\mu > 0.0113$
     $z^\star = 2.47$
     $P = 0.0068$
     $z(0.01) \geq 2.33$
     Reject H_o

8.145 H_a: $\mu > 9$
     $z^\star = 3.14$
     $P = 0.0008$
     $z(0.02) \geq 2.05$
     Reject H_o

8.147 **a.** H_a: $\mu < 129.2$
       $z^\star = -2.02$
       $P = 0.0217$
       Fail to reject H_o
     b. indicates the likelihood of being wrong

8.149 **a.** (2), failure to reject H_o will result in the drug being marketed; burden of proof is on the old ineffective drug.
     b. (1), failure to reject H_o will result in the new drug not being marketed; burden of proof is on the new drug.

8.151 **a.** 125.10 to 132.95

8.153 **a.** H_a: $\mu \neq 18$; Fail to reject H_o

Chapter 9

9.2 identical

9.3 **a.** 2.68
    b. 2.07

9.4 **a.** -1.33
    b. -2.82

9.5 ± 2.18

9.6 0.0241

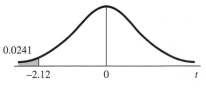

9.7 0.1402

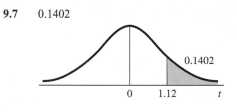

9.8 15.60 to 17.8

9.9 6.073 to 9.427

9.10 1.20

9.11 0.124; Fail to reject H_o

9.12 $t \geq 1.75$; Fail to reject H_o

9.13 -1.30

9.14 0.220; Fail to reject H_o

9.15 $t \leq -2.20$, $t \geq 2.20$; Fail to reject H_o

9.16 0.2202

9.17 $P = 0.091$, Fail to reject H_o

9.18 **a.** $0.026 < P < 0.033$
     b. $P = 0.001$
     c. $0.064 < P < 0.080$

9.19 **a.** 1.71
     b. 1.37
     c. 2.60
     d. 2.08
     e. -1.72
     f. -2.06
     g. -2.47
     h. 2.01

9.21 7

9.23 **a.** -2.49
     b. 1.71
     c. -0.685

9.25 **a.** Symmetric about mean; mean is 0
     b. Standard deviation; t has df

9.27 51.84 to 56.16

9.29 **a.** 2.965 to 4.655
     b. Does not; x most likely is not normal

9.31 $\bar{x} = 86.0$, $s = 11.84$; 79.15 to 92.85

9.35 **a.** H_o: $\mu = 11(\geq)$ vs. H_a: $\mu < 11$
     b. H_o: $\mu = 54(\leq)$ vs. H_a: $\mu < 54$
     c. H_o: $\mu = 75$ vs. H_a: $\mu \neq 75$

9.37 **a.** $t \leq -2.14$, $t \geq 2.14$
     b. $t \geq 2.49$
     c. $t \leq -1.74$
     d. $t \geq 2.42$

9.39 $H_o: \mu = 25 (\geq)$ vs. $H_a: \mu < 25$
$t^\star = -3.25$
$P < 0.005$ or $t \leq -2.46$
Reject H_o

9.41 a. midrange is close to mean, approximately
symmetrical
b. & c.
$H_o: \mu = 38.4\%$ vs. $H_a: \mu \neq 38.4\%$
$t^\star = 1.08$
$0.20 < P < 0.50$ or $t \leq -2.14, t \geq 2.14$
Fail to reject H_o

9.43 $\bar{x} = 37.0, s = 4.817$;
$H_o: \mu = 35$ vs. $H_a: \mu \neq 35$
$t^\star = 1.02$
$0.20 < P < 0.50$ or $t \leq -2.57, t \geq 2.57$
Fail to reject H_o

9.47 $\bar{x} = 18.8142, s = 0.0296$, se mean $= 0.0050$;
$H_o: \mu = 18.810$ vs. $H_a: \mu \neq 18.810$
$t^\star = 0.84$
$P = 0.41$ or $t \leq -2.75, t \geq 2.75$
Fail to reject H_o

9.49 a. Yes
b. mean of p' is p

9.51 a. 0.021
b. 0.189 to 0.271

9.52 2401

9.53 1277

9.54 1.78

9.55 0.0375; Reject H_o

9.56 $z \geq 1.65$; Reject H_o

9.57 $p = 15\%$ is not contained in the interval, $0.052 < p < 0.118$

9.58 a. 0.018
b. 0.9464

9.59 a. n(successes), sample size
b. $45/150 = 0.30$

9.61 a. $p + q = 1$
b. equivalent to (a)
c. 0.4
d. 0.727

9.63 a. 0.21
b. 0.043

9.65 0.206 to 0.528

9.67 0.37 to 0.43

9.69 0.015

9.71 a. 0.162, 0.438
b. 0.562, 0.838
c. 0.239, 0.761

d. 0.418, 0.582
e. 0.474, 0.526
f. symmetric about 0.5

9.73 a. 0.5005
b. 0.003227
c. 0.4942 to 0.5068

9.75 a. 915
b. 229
c. 1825
d. increasing maximum error decreases sample size
e. increasing the level of confidence increases sample size

9.77 a. 0.82
b. 1005
c. 355
d. 273
e. $\dfrac{1}{4}$ the original size

9.79 a. $H_o: p = 0.60 (\leq)$ vs. $H_a: p > 0.60$
b. $H_o: p = 0.50 (\geq)$ vs. $H_a: p < 0.50$
c. $H_o: p = 1/3 (\leq)$ vs. $H_a: p > 1/3$
d. $H_o: p = 0.50 (\geq)$ vs. $H_a: p < 0.50$
e. $H_o: p = 0.50 (\leq)$ vs. $H_a: p > 0.50$
f. $H_o: p = 3/4 (\geq)$ vs. $H_a: p < 3/4$
g. $H_o: p = 0.50$ vs. $H_a: p \neq 0.50$
h. $H_o: p = 0.50$ vs. $H_a: p \neq 0.50$

9.81 a. 0.1388
b. 0.0238
c. 0.1635
d. 0.0559

9.83 a. 0.017
b. 0.085
c. 0.101
d. 0.004

9.85 a. (1)
b. 0.036
c. (4)
d. 0.128

9.87 $H_o: p = 0.90 (\geq)$ vs. $H_a: p < 0.90$
$z^\star = -4.82$
$P = 0.000003$ or $z \leq -1.65$
Reject H_o

9.89 $H_o: p = 0.60 (\geq)$ vs. $H_a: p < 0.60$
$z^\star = -2.04$
$P = 0.0207$ or $z \leq -1.65$
Reject H_o

9.91 $H_o: p = 0.225 (\leq)$ vs. $H_a: p > 0.225$
$z^\star = 2.71$
$P = 0.0034$ or $z \geq 1.65$
Reject H_o

9.93 a. 23.2
b. 23.3

9.94 a. 3.94
　　b. 8.64

9.95 a. 0.8356
　　b. 0.1644

9.96 25.08

9.97 $0.05 < \mathbf{P} < 0.10$; Fail to reject H_o

9.98 $\chi^2 \geq 33.4$; Fail to reject H_o

9.99 60.15

9.100 $0.02 < \mathbf{P} < 0.05$; Reject H_o

9.101 $\chi^2 \leq 24.4$, $\chi^2 \geq 59.3$; Reject H_o

9.102 0.1208

9.103 a. A: 1.72; B: 3.58
　　b. Increased
　　c. Possibly, it doubles the standard deviation

9.105 a. $\chi^2(19,0.05) = 30.1$
　　b. $\chi^2(4,0.01) = 13.3$
　　c. $\chi^2(17,0.975) = 7.56$
　　d. $\chi^2(60,0.95) = 43.2$

9.107 a. 11.1　　**b.** 11.1　　**c.** 9.24

9.109 0.94

9.111 a. H_o: $\sigma = 24$ ($\leq$) vs. H_a: $\sigma > 24$
　　b. H_o: $\sigma = 0.5$ ($\leq$) vs. H_a: $\sigma > 0.5$
　　c. H_o: $\sigma = 10$ vs. H_a: $\sigma \neq 10$
　　d. H_o: $\sigma^2 = 18$ ($\geq$) vs. H_a: $\sigma^2 < 18$
　　e. H_o: $\sigma^2 = 0.025$ vs. H_a: $\sigma^2 \neq 0.025$
　　f. H_o: $\sigma^2 = 34.5$ ($\leq$) vs. H_a: $\sigma^2 > 34.5$

9.113 a. $0.02 < \mathbf{P} < 0.05$
　　b. 0.01
　　c. $0.05 < \mathbf{P} < 0.10$
　　d. $0.025 < \mathbf{P} < 0.05$

9.115 H_o: $\sigma = 8$ vs. H_a: $\sigma \neq 8$
　　$\chi^{2\star} = 29.3$
　　$0.01 < \mathbf{P} < 0.02$ or $\chi^2 \leq 32.4$, $\chi^2 \geq 71.4$
　　Reject H_o

9.117 a. Enables use of the chi-square distribution
　　b. Examine sample
　　c & d. H_o: $\sigma = 0.32$ ($\leq$) vs. H_a: $\sigma > 0.32$
　　　　$\chi^{2\star} = 66.04$
　　　　$\mathbf{P} < 0.005$ or $\chi^2 \geq 45.6$
　　　　Reject H_o

9.119 $s^2 = 17.4595$;
　　H_o: $\sigma = 3.50$ vs. H_a: $\sigma \neq 3.50$
　　$\chi^{2\star} = 19.95$
　　$0.20 < \mathbf{P} < 0.50$ or $\chi^2 \leq 5.63$, $\chi^2 \geq 26.1$
　　Fail to reject H_o

9.123 a. working college students
　　b. 22

c. mean $= \$326.10$, $s = \$164.43$

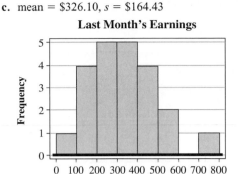

Last Month's Earnings

d. Distribution appears to be approximately normal

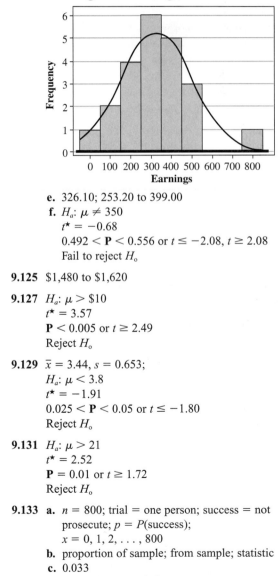

Histogram of Earnings, with Normal Curve

e. 326.10; 253.20 to 399.00
f. H_a: $\mu \neq 350$
　　$t^\star = -0.68$
　　$0.492 < \mathbf{P} < 0.556$ or $t \leq -2.08$, $t \geq 2.08$
　　Fail to reject H_o

9.125 $\$1,480$ to $\$1,620$

9.127 H_a: $\mu > \$10$
　　$t^\star = 3.57$
　　$\mathbf{P} < 0.005$ or $t \geq 2.49$
　　Reject H_o

9.129 $\bar{x} = 3.44$, $s = 0.653$;
　　H_a: $\mu < 3.8$
　　$t^\star = -1.91$
　　$0.025 < \mathbf{P} < 0.05$ or $t \leq -1.80$
　　Reject H_o

9.131 H_a: $\mu > 21$
　　$t^\star = 2.52$
　　$\mathbf{P} = 0.01$ or $t \geq 1.72$
　　Reject H_o

9.133 a. $n = 800$; trial $=$ one person; success $=$ not
　　　　prosecute; $p = P(\text{success})$;
　　　　$x = 0, 1, 2, \ldots, 800$
　　b. proportion of sample; from sample; statistic
　　c. 0.033
　　d. same when rounded

9.135 0.088 to 0.232

9.137

$p =$	0.1	0.2	0.3	0.4	0.5	0.6	0.7	0.8	0.9
$pq =$	0.09	0.16	0.21	0.24	0.25	0.24	0.21	0.16	0.09

9.139 **a.** 0.027
     c. 2213

9.141 0.0261

9.143 13

9.145 H_a: $\sigma \neq 12$
     $\chi^{2\star} = 42.0$
     P = 0.6844 or $\chi^2 \leq 23.7$, $\chi^2 \geq 58.1$
     Fail to reject H_o

9.147 H_a: $\sigma > 0.5$
     $\chi^{2\star} = 87.8$
     P < 0.005 or $\chi^2 \geq 42.6$
     Reject H_o

Chapter 10

10.1 Divide the class into two groups, males and females. Randomly select from each group.

10.2 Randomly select a set of students. Obtaining the two heights from each.

10.3 Twins form pairs

10.5 Dependent, each person provides one data for each sample

10.7 Independent, separate samples

10.11

Pairs	1	2	3	4	5
$d = A - B$	1	1	0	2	-1

10.12 0.6, 1.14

10.13 2.13; number of standard errors

10.14 **a.** 4.24 to 8.36
     b. narrower

10.15 $d =$ After $-$ Before; -0.51 to 6.51

10.16 H_a: $\mu_d > 0$
     $t^\star = 2.45$
     $0.033 < $ **P** $ < 0.037$ or $t \geq 2.13$
     Reject H_o

10.17

Mean	StDev	T	P-value
-2.33	4.41	-1.30	0.13

     Fail to reject H_o

10.18 H_a: $\mu_d \neq 0$
     $t^\star = 1.35$
     $0.234 < $ **P** $ < 0.264$ or $t \leq -4.60$ and $t \geq 4.60$
     Fail to reject H_o

10.19 **a.** 1.0
     b. -1.53 to 3.53

10.21 $d = A - B$; $n = 8$, $\bar{d} = 3.75$, $s_d = 5.726$
     $E = (2.36)(5.726/\sqrt{8}) = 4.78$
     3.75 ± 4.78
     -1.03 to 8.53

10.23 $\bar{d} = 0.8$, $s_d = 1.32$; -0.143 to 1.743

10.25 **a.** H_o: $\mu_d = 10$ ($\leq$) vs. H_a: $\mu_d > 10$; $d = $ post $-$ pre
     b. H_o: $\mu_d = 10$ ($\geq$) vs. H_a: $\mu_d < 10$; $d = a - b$
     c. H_o: $\mu_d = 12$ ($\geq$) vs. H_a: $\mu_d < 12$; $d = b - a$
     d. H_o: $\mu_d = 0$ vs. H_a: $\mu_d \neq 0$; $d = a - b$

10.27 $\bar{d} = 5.5$, $s_d = 11.34$
     H_o: $\mu_d = 0$ ($\leq$) vs. H_a: $\mu_d > 0$
     $t^\star = 3.067$
     P ≈ 0.002 or $t \geq 2.44$
     Reject H_o

10.31 $d = 1.35$, $s_d = 1.631$
     H_o: $\mu_d = 0$ ($\leq$) vs. H_a: $\mu_d > 0$
     $t^\star = 3.70$
     P = 0.001 or $t \geq 2.54$
     Reject H_o

10.33 4.92

10.34 1: df will be between 17 and 40.
     2: df = 17 (smaller df)

10.35 -6.3 to 16.3

10.36 1.24

10.37 A smaller degrees of freedom; requires a higher calculated value to reject H_o.

10.38 **a.** 0.210
     b. ± 2.10

10.39 **a.** Two different sets of schools
     b. Probability of no difference is just about 0.
     c. Using Table 6: $0.678 < t^\star < 1.29$, Table 7: $0.8 < t^* < 0.9$
     These results are fairly typical when null hypothesis is true.
     d. No indication of normality

10.41 0.54 to 0.66

10.43 A: $\bar{x} = 6.0$, $s^2 = 1.333$, B: $\bar{x} = 4.0$, $s^2 = 2.667$; 0.22 to 3.78

10.45 0.787 to 3.733

10.47 -0.75 to 8.75

10.49 **a.** H_o: $\mu_1 - \mu_2 = 0$ vs. H_a: $\mu_1 - \mu_2 \neq 0$
     b. H_o: $\mu_1 - \mu_2 = 0$ ($\leq$) vs. H_a: $\mu_1 - \mu_2 > 0$
     c. H_o: $\mu_1 - \mu_2 = 20$ ($\leq$) vs. H_a: $\mu_1 - \mu_2 > 20$
     d. H_o: $\mu_A - \mu_B = 50$ ($\geq$) vs. H_a: $\mu_A - \mu_B < 50$ or
     H_o: $\mu_B - \mu_A = 50$ ($\leq$) vs. H_a: $\mu_B - \mu_A > 50$

10.51

	Table 6	Table 7
a.	$0.10 < $ **P** $ < 0.25$	0.125
b.	$0.01 < $ **P** $ < 0.025$	0.012
c.	$0.05 < $ **P** $ < 0.10$	0.092
d.	$0.05 < $ **P** $ < 0.10$	0.084

10.53 H_o: $\mu_2 - \mu_1 = 0$ vs. H_a: $\mu_2 - \mu_1 > 0$ $(\mu_1 < \mu_2)$
$t^\star = 2.84$
$0.008 < \mathbf{P} < 0.012$ or $t \geq 1.83$
Reject H_o

10.55 H_o: $\mu_1 - \mu_2 = 0$ vs. H_a: $\mu_1 - \mu_2 > 0$
$t^\star = 4.02$
$\mathbf{P} < 0.005$ or $t \geq 2.44$
Reject H_o

10.57 **b.** $0.554 < \mathbf{P} < 0.624$
c. $0.560 < \mathbf{P} < 0.626$

10.59 $\bar{x}_A = 10.0$, $s_A^2 = 10.44$, $\bar{x}_B = 14.7$, $s_B^2 = 46.01$;
H_o: $\mu_B - \mu_A = 0$ vs. H_a: $\mu_B - \mu_A > 0$
$t^\star = 1.98$
$0.037 < \mathbf{P} < 0.047$ or $t \geq 1.83$
Reject H_o

10.61 **a.** 15.53, 1.98, 1.41
b. 12.53, 1.98, 1.41
c. H_o: $\mu_A - \mu_B = 0$ vs. H_a: $\mu_A - \mu_B \neq 0$
$t^\star = 5.84$
$\mathbf{P} < 0.002$ or $t \leq -2.98$, $t \geq 2.98$
Reject H_o

10.67 75, 250, 0.30, 0.70

10.68 **a.** 36, 45; 4, 5
b. No

10.69 -0.07 to 0.15

10.70 0.076, 0.924

10.71 $p_p' = 0.82$; $z^\star = 2.21$

10.72 0.0901

10.73 H_o: $p_s - p_n = 0$ vs. H_a: $p_s - p_n > 0$
$z^\star = 2.94$
$\mathbf{P} = 0.0016$
Yes, $\mathbf{P} \leq \alpha$

10.75 -0.15 to 0.21

10.77 0.000 to 0.080

10.79 **a.** H_o: $p_m - p_w = 0$ vs. H_a: $p_m - p_w \neq 0$
b. H_o: $p_b - p_g = 0$ vs. H_a: $p_b - p_g > 0$
c. H_o: $p_c - p_{nc} = 0$ vs. H_a: $p_c - p_{nc} > 0$

10.81 H_o: $p_1 - p_2 = 0$ vs. H_a: $p_1 - p_2 \neq 0$
$z^\star = 1.75$
$\mathbf{P} = 0.0802$ or $z \leq -1.96$, $z \geq 1.96$
Fail to reject H_o

10.83 H_o: $p_m - p_c = 0$ vs. H_a: $p_m - p_c \neq 0$
$z^\star = 1.42$
$\mathbf{P} = 0.1556$ or $z \leq -1.96$, $z \geq 1.96$
Fail to reject H_o

10.85 **a.** H_o: $p_m - p_w = 0$ vs. H_a: $p_m - p_w > 0$
$\mathbf{P} = P(z > 2.92) = 0.0018$
b. Reject H_o.

10.87 Divide inequality by σ_p^2

10.88 3.03, 3.78

10.89 $\sigma_m/\sigma_p = 1$ vs. $\sigma_m/\sigma_p > 1$

10.90 1.52

10.91 3.37

10.92 0.495

10.93 Multiply by 2

10.94 **a.** H_o: $\sigma_N^2 = \sigma_A^2$, H_a: $\sigma_N^2 > \sigma_A^2$
b. Sample results are very unlikely if the null hypothsis is true.
c. 0.0058

10.95 **a.** H_o: $\sigma_A^2 = \sigma_B^2$ vs. H_a: $\sigma_A^2 \neq \sigma_B^2$
b. H_o: $\sigma_I = \sigma_{II}$ vs. H_a: $\sigma_I > \sigma_{II}$
c. H_o: $\sigma_A^2/\sigma_B^2 = 1$ vs. H_a: $\sigma_A^2/\sigma_B^2 \neq 1$
d. H_o: $\sigma_C^2/\sigma_D^2 = 1$ vs. H_a: $\sigma_C^2/\sigma_D^2 < 1$ or
H_o: $\sigma_D^2/\sigma_C^2 = 1$ vs. H_a: $\sigma_D^2/\sigma_C^2 > 1$

10.97 **a.** 2.51
b. 2.20
c. 2.91
d. 4.10
e. 2.67
f. 3.77
g. 1.79
h. 2.99

10.99 **a.** 0.10
b. 0.02

10.101 H_o: $\sigma_K^2 = \sigma_M^2$ vs. H_a: $\sigma_K^2 \neq \sigma_M^2$
$F^\star = 1.33$
$\mathbf{P} > 0.10$ or $F \geq 3.73$
Fail to reject H_o

10.103 **a.** H_o: $\sigma_y = \sigma_o$ vs. H_a: $\sigma_y > \sigma_o$
$F^\star = 12.35$
$\mathbf{P} < 0.01$ or $F \geq 2.24$
Reject H_o
b. $\mathbf{P} = 0+$

10.107 **a.** H_o: $p_m - p_w = 0$ vs. H_a: $p_m - p_w \neq 0$
$z^\star = 1.37$
$\mathbf{P} = 0.1706$ or $z \leq -1.96$, $z \geq 1.96$
Fail to reject H_o
b. H_o: $p_u - p_c = 0$ vs. H_a: $p_u - p_c \neq 0$
$z^\star = 1.58$
$\mathbf{P} = 0.1142$ or $z \leq -1.96$, $z \geq 1.96$
Fail to reject H_o
c. The Maritz AmeriPoll found more men than women correctly knew the number of stars on the USA flag. The USA Snapshot seems to imply that it is true for the population, even though it does not specifically say that. In parts (a) and (b), the difference in the sample percentages is deemed not significant.

10.109 $\bar{d} = 3.583$, $s_d = 19.58$; -8.85 to 16.02

10.111 $\bar{d} = 5.2$, $s_d = 7.406$;
H_a: $\mu_d > 0$ (improvement)

$t^\star = 2.22$
$0.022 < P < 0.029$ or $t \geq 1.83$
Reject H_o

10.113 -0.21 to 10.61

10.115 $\bar{x}_A = 57.0$, $s_A^2 = 209.111$, $\bar{x}_B = 62.2$, $s_B^2 = 193.289$;
-9.14 to 19.54

10.117 $\bar{x}_A = 80.021$, $s_A^2 = 0.0005744$, $\bar{x}_B = 79.979$,
$s_B^2 = 0.0009839$; 0.012 to 0.072

10.119 H_a: $\mu_2 - \mu_1 > 0$ (artif. turf yields a lower time)
$t^\star = 0.988$
$0.164 < P < 0.189$ or $t \geq 1.72$
Fail to reject H_o

10.121 $\bar{x}_m = 33.83$, $s_m^2 = 6.38$, $\bar{x}_f = 26.27$, $s_f^2 = 4.02$;
H_a: $\mu_m - \mu_f > 0$
$t^\star = 8.91$
$P < 0.001$ or $t \geq 1.81$
Reject H_o

10.123 -0.065 to 0.155

10.125 **a.** -0.116 to 0.216
b. No; 0 is within interval.

10.127 **a.** No; $z^\star = 1.44$, $P = 0.1498$
b. No; $z^\star = 1.77$, $P = 0.0768$
c. Yes; $z^\star = 2.04$; $P = 0.0414$

10.129 $p'_1 = 0.05$, $p'_r = 0.02$;
H_a: $p_1 - p_r > 0$
$z^\star = 1.90$
$P = 0.0287$
For $\alpha \geq 0.03$, sample results are significant.

10.131 $p'_a = 0.5771$, $p'_1 = 0.5091$;
H_a: $p_a - p_1 \neq 0$
$z^\star = 1.26$
$P = 0.2076$ or $z \leq -2.58$, $z \geq 2.58$
Fail to reject H_o

10.133 H_a: $\sigma_m^2 > \sigma_f^2$
$F^\star = 2.58$
$0.025 < P < 0.05$ or $F \geq 2.53$
Reject H_o

10.135 H_a: $\sigma_n^2 > \sigma_s^2$
$F^\star = 1.28$
$P > 0.05$ or $F \geq 1.79$
Fail to reject H_o

Chapter 11

11.1 **a.** 23.2
b. 23.3
c. 3.94
d. 8.64

11.3 **a.** 50
b. 33 1/3

11.4 $9 + 3 + 3 + 1 = 16$ parts all together

11.5 $556(9/16)$; $315 - 312.75$; $(2.25)^2/312.75$

11.6 **a.** Each adult surveyed
b. Reason we rearrange furniture
c. Bored, moving, new furniture, and so on

11.7 Some use more than one method.

11.9 **a.** 0.0066
b. 0.0503

11.11 **a.** H_o: $P(A) = P(B) = P(C) = P(D) = P(E) = 0.2$
b. χ^2
c. H_a: preferences not all equal
$\chi^{2\star} = 4.40$
$P = 0.355$ or $\chi^2 \geq 7.78$
Fail to reject H_o

11.13 H_a: There is a difference.
$\chi^{2\star} = 0.34$
$P = 0.9523$ or $\chi^2 \geq 7.82$
Fail to reject H_o

11.15 H_a: There is a difference.
$\chi^{2\star} = 193.72$
$P = 0+$ or $\chi^2 \geq 11.3$
Reject H_o

11.17 H_a: Probabilities are not all equal.
$\chi^{2\star} = 6.00$
$P = 0.740$ or $\chi^2 \geq 16.9$
Fail to reject H_o

11.19 **a.** Calculated value of chi-square will be near zero.
b. The calculated chi-square will be 'zero.'
Little brother knew what was expected, so he just reported the expected values.
You want your money back; he didn't roll the die.
"The data and the theory expressed by the null hypothesis are too much alike."
c. Hopefully the experimenter trusts himself to carry out the experiment randomly.

11.21 10

11.22 **a.** 113
b. 122
c. 300
d. 68.23

11.23 **a.**

	West	N.Cent.	South	N.East
Baked	55%	46%	47%	41%
Other	45%	54%	53%	59%

b. The information compares several distributions, a distribution for each region.
c. The distribution of "baked" or "other" are being compared for the 4 different regions.

11.24 a.

Reason	Men	Women
No time	0.36	0.37
In good health	0.39	0.21
Poor health	0.11	0.15
Dislike exercise	0.04	0.08
Too lazy	0.01	0.06
Other	0.09	0.13
Totals	1.00	1.00

b.

Reason	Men	Women	Totals
No time	360	370	730
In good health	390	210	600
Poor health	110	150	260
Dislike exercise	40	80	120
Too lazy	10	60	70
Other	90	130	220
Totals	1000	1000	2000

   c. H_a: Different distributions
$\chi^{2\star} = 116.611$
$P < 0.005$ or $\chi^2 \geq 11.1$
Reject H_o

11.25 a. H_a: Not independent
b. H_a: Distributions are not the same
c. H_a: Proportion of yeses is not the same

11.27 H_a: Not independent
$\chi^{2\star} = 35.749$
$P < 0.005$ or $\chi^2 \geq 13.3$
Reject H_o

11.29 H_a: Not independent
$\chi^{2\star} = 3.390$
$P = 0.335$ or $\chi^2 \geq 6.25$
Fail to reject H_o

11.31 H_a: Not independent
$\chi^{2\star} = 8.548$
$P = 0.074$ or $\chi^2 \geq 9.49$
Fail to reject H_o

11.33 a. & b. H_a: Distributions not same
$\chi^{2\star} = 24.84$
$P < 0.005$ or $\chi^2 \geq 16.8$
Reject H_o
c. Prof. #3

11.35 (1) H_a: Distributions not same
$\chi^{2\star} = 566.915$
$P < 0.005$ or $\chi^2 \geq 6.63$
Reject H_o
(2) H_a: Distributions not same
$\chi^{2\star} = 92.808$
$P < 0.005$ or $\chi^2 \geq 6.63$
Reject H_o

11.37 a. Their favorite way to cool their mouth after eating hot spicy food.
b. Adults professing to love hot, spicy food; method used to cool mouth
c.

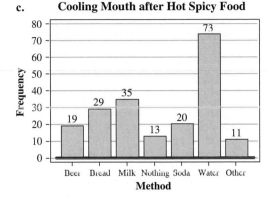

Cooling Mouth after Hot Spicy Food

d. H_a: At least one of the proportions is different
$\chi^{2\star} = 13.537$
$P = 0.0352$ or $\chi^2 \geq 12.6$
Reject H_o

11.39 H_a: Proportions are other than $1:3:4$
$\chi^{2\star} = 10.33$
$P = 0.006$ or $\chi^2 \geq 5.99$
Reject H_o

11.41 H_a: Percentages are different than listed
$\chi^{2\star} = 6.693$
$P = 0.153$ or $\chi^2 \geq 9.49$
Fail to reject H_o

11.43 H_a: Percentages are different than listed
$\chi^{2\star} = 17.92$
$P = 0.003$ or $\chi^2 \geq 15.1$
Reject H_o

11.45 H_a: The proportions are different per age group
$\chi^{2\star} = 3.904$
$P = 0.273$ or $\chi^2 \geq 7.82$
Fail to reject H_o

11.47 H_a: Not independent
$\chi^{2\star} = 23.339$
$P < 0.005$ or $\chi^2 \geq 13.3$
Reject H_o

11.49 H_a: The proportions are different per brand
$\chi^{2\star} = 2.839$
$P = 0.417$ or $\chi^2 \geq 7.82$
Fail to reject H_o

11.51 H_a: Not independent
$\chi^{2\star} = 3.651$
$P = 0.456$ or $\chi^2 \geq 9.49$
Fail to reject H_o

11.55 a. 0.7918
b. 0.6270
c. $(0.7918)^2 = 0.6269$

Chapter 12

12.1   Units produced per hour at each temperature level:

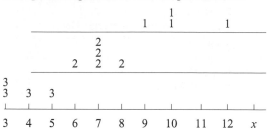

Yes, there appears to be a difference between the three sets.

12.2 H_o: $\mu_1 = \mu_2 = \mu_3$
H_a: Means are not all the same

12.3   0.0052

12.4   **a.** Yes
b. Male and female categories encompass population.

12.5   **a.** 0
b. 2
c. 4
d. 31
e. 393

12.6   "Amount of money donated" is categorized by age.

12.7   **a.** H_o: $\mu_1 = \mu_2 = \mu_3 = \mu_4 = \mu_5$
H_a: Means not all equal
b. H_o: $\mu_1 = \mu_2 = \mu_3 = \mu_4$
H_a: Means not all equal
c. H_o: $\mu_1 = \mu_2 = \mu_3 = \mu_4$
H_a: Means not all equal
d. H_o: $\mu_1 = \mu_2 = \mu_3$
H_a: Means not all equal

12.9   **a.** $F \geq 3.34$
b. $F \geq 5.99$
c. $F \geq 3.44$

12.11   **a.** Depends on whether it is larger or smaller than α
b. Reject H_o
c. Fail to reject H_o

12.13   **a.** Test factor has no effect on mean at tested levels.
b. Test factor does have an effect on mean at tested levels.
c. $\mathbf{P} = P(F > F^\star) \leq \alpha$; $F^\star$ must fall in the critical region.
d. Tested factor has a significant effect on the variable.
e. $\mathbf{P} = P(F > F^\star) > \alpha$; $F^\star$ must fall in the noncritical region.
f. Tested factor does not have a significant effect on the variable.

12.15   **a.** 120
b. 3
d. Yes; p-value is small
e. No; p-value is large

12.17 H_a: Mean values are not all equal

	df	SS	MS	$F^\star$
Work	2	17.73	8.87	4.22
Error	12	25.20	2.10	
Total	14	42.93		

$\mathbf{P} = 0.041$ or $F \geq 3.89$
Reject H_o

12.19 H_a: Means not all equal

	df	SS	MS	$F^\star$
Rating cat.	2	48.11	24.055	4.58
Error	15	78.83	5.255	
Total	17	126.94		

$\mathbf{P} = 0.028$ or $F \geq 3.68$
Reject H_o

12.21 H_a: Means are not all equal

	df	SS	MS	$F^\star$
Division	2	284.46	142.23	2.732
Error	25	1301.93	52.07	
Total	27	1586.39		

$\mathbf{P} = 0.0845$ or $F \geq 3.39$
Fail to reject H_o.

12.23   Calories: H_a: Means are not all equal

	df	SS	MS	$F^\star$
Division	3	6955.56	2318.52	6.088
Error	46	17519.02	380.85	
Total	49	24474.58		

$\mathbf{P} < 0.01$ or $F \geq 4.26$
Reject H_o

Fat content: H_a: Means are not all equal

	df	SS	MS	$F^\star$
Division	3	22.46	7.49	1.745
Error	46	197.26	4.29	
Total	49	219.72		

$\mathbf{P} = 0.171$ or $F \geq 4.26$
Fail to reject H_o

12.24   **a.** They're basically the same question.
b. H_a: Means are not all equal

	df	SS	MS	$F^\star$
Age	5	537.4	107.5	1.76
Error	19	1157.2	60.9	
Total	24	1694.6		

$\mathbf{P} = 0.168$ or $F \geq 2.74$
Fail to reject H_o

12.25 **a.** H_o: Means are same
H_a: Means not the same
b. $F \geq 3.68$
c. Fail to reject H_o; no significant difference
d. This sample data is quite likely to have occurred under the assumed conditions and a true null hypothesis.

12.27 H_a: Means not all the same

	df	SS	MS	$F^\star$
Drug	3	108.33	36.11	12.50
Error	12	34.67	2.89	
Total	15	143.00		

P = 0.001 or $F \geq 3.49$
Reject H_o

12.29 H_a: Means are not all equal

	df	SS	MS	$F^\star$
Machine	4	20.998	5.2495	31.6
Error	13	2.158	0.166	
Total	17	23.156		

P < 0.01 or $F \geq 5.21$
Reject H_o

12.31 **a.**

```
---+------+------+---•---+------+---•---•-+- Plow
---+------+---•--+---•--+------+•------+- VChisel
---+---•--+------+---•--+------+--•---+- CoulChis
•--+------+---•--+------+------+•-----+- StdChis
---+------+------+--••--+----•-+------+- HvyDisk
---+---•--+------+---•--+------+------+- LtDisk
  111.0  114.0  117.0  120.0  123.0  126.0
```

b. H_a: Means not all equal

c. & d.	df	SS	MS	$F^\star$
Tillage	5	114.60	22.921	1.23
Error	12	223.41	18.618	
Total	17	338.01		

P = 0.353 or $F \geq 3.11$
Fail to reject H_o

12.33 **a.** H_a: Means not all equal

	df	SS	MS	$F^\star$
Division	2	413.07	206.535	0.069
Error	27	81335.10	3012.411	
Total	29	81748.17		

P = 0.934 or $F \geq 3.37$
Fail to reject H_o

b. H_a: Means not all equal

	df	SS	MS	$F^\star$
Division	2	487.27	243.635	0.055
Error	27	119698.90	4433.290	
Total	29	120186.17		

P = 0.947 or $F \geq 3.37$
Fail to reject H_o

12.35 H_a: Mean is not the same

	df	SS	MS	$F^\star$
Counties	2	585.9	293.0	13.83
Error	24	508.4	21.2	
Total	26	1094.3		

P < 0.01 or $F \geq 3.40$
Reject H_o

12.37 H_a: Means not all the same

	df	SS	MS	$F^\star$
Cities	2	49.97	24.99	2.91
Error	15	128.59	8.57	
Total	17	178.57		

P = 0.085 or $F \geq 6.36$
Fail to reject H_o

12.39 29.3333

12.41 82.4752

Chapter 13

13.1 In Chapter 2, we learned that the summation of the deviations about the mean was zero.

13.3 **a.**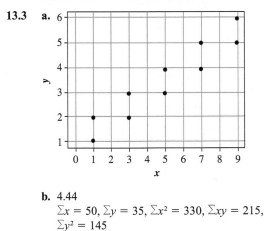

b. 4.44
$\Sigma x = 50$, $\Sigma y = 35$, $\Sigma x^2 = 330$, $\Sigma xy = 215$, $\Sigma y^2 = 145$
c. 2.981, 1.581
d. 0.943
e. 0.943

13.7 **a.** 0.440
b. Slight positive relationship

13.9   **a.** 60
b. 40.99, 20.98
c. 0.0698

13.11   −0.05 to 0.65

13.12   **a.** $0.05 < P < 0.10$
b. $0.025 < P < 0.05$

13.13   **a.** ±0.444
b. −0.378, or 0.378

13.14   significant

13.17 $\Sigma x = 746$, $\Sigma y = 736$, $\Sigma x^2 = 57{,}496$, $\Sigma xy = 56{,}574$, $\Sigma y^2 = 55{,}826$
0.955; 0.78 to 0.98

13.19 $\Sigma x = 1189$, $\Sigma y = 5482$, $\Sigma x^2 = 205{,}787$, $\Sigma xy = 928{,}950$, $\Sigma y^2 = 4{,}333{,}726$
a. −0.177
b. −0.78 to 0.62

13.21   **a.** H_o: ρ = 0, H_a: ρ > 0
b. H_o: ρ = 0, H_a: ρ ≠ 0
c. H_o: ρ = 0, H_a: ρ < 0
d. H_o: ρ = 0, H_a: ρ > 0

13.23 H_o: ρ = 0 vs. H_a: ρ ≠ 0.0
$r^\star = 0.43$
$0.05 < P < 0.10$ or $r \le -0.378$, $r \ge 0.378$
Reject H_o

13.25 H_o: ρ = 0.0 vs. H_a: ρ ≠ 0.0
$r^\star = 0.16942$
$P > 0.10$ or $r \le -0.423$, $r \ge 0.423$
Fail to reject H_o

13.27 $\Sigma x = 26.2$, $\Sigma y = 82.5$, $\Sigma x^2 = 174.88$, $\Sigma xy = 256.41$, $\Sigma y^2 = 704.61$
H_o: ρ = 0.0 vs. H_a: ρ ≠ 0.0
$r^\star = 0.798$
$P < 0.01$ or $r \le -0.632$, $r \ge 0.632$
Reject H_o

13.29   **a.** 0.613
b. H_o: ρ = 0.0 vs. H_a: ρ ≠ 0.0
$r^\star = 0.613$
$P > 0.10$ or $r \le -0.754$, $r \ge 0.754$
Fail to reject H_o

13.33   **a.**

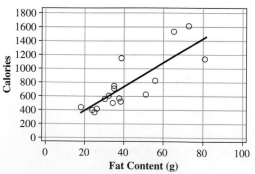

Scatter Diagram

$\Sigma x = 696$, $\Sigma y = 12{,}655$, $\Sigma x^2 = 34{,}362$, $\Sigma xy = 617{,}415$, $\Sigma y^2 = 11{,}801{,}175$
b. $\hat{y} = 16.926x + 51.4$
c. 16.926
d. 46,524.8

13.35 $\Sigma x = 50$, $\Sigma y = 35$, $\Sigma x^2 = 330$, $\Sigma xy = 215$, $\Sigma y^2 = 145$
a. $\hat{y} = 1.0 + 0.5x$
b. 1.5, 2.5, 3.5, 4.5, 5.5
c. −0.5, 0.5, alternately
d. 0.3125
e. 0.3125

13.39   **a.** 0.0145
b. 0.0668
c. 0.0653

13.41   0.1894

13.43   **a.** $\hat{y} = -348 + 2.04x$
b. 1.60 to 2.48

13.45   **a.**

Commuting to College

$\Sigma x = 68$, $\Sigma y = 205$, $\Sigma x^2 = 566$, $\Sigma xy = 1670$, $\Sigma y^2 = 5075$
b. $\hat{y} = 2.38 + 2.664x$
c. H_o: $\beta_1 = 0$ vs. H_a: $\beta_1 > 0$
$t^\star = 6.55$
$P < 0.002$ or $t \ge 1.86$
Reject H_o
d. 1.48 to 3.84

13.47 $\Sigma x = 1105$, $\Sigma y = 93327$, $\Sigma x^2 = 126{,}475$, $\Sigma xy = 10{,}447{,}010$, $\Sigma y^2 = 885{,}420{,}825$
a. $\hat{y} = 5937 + 30.73x$
b. 1135, 17.16
d. H_o: $\beta_1 = 0$ vs. H_a: $\beta_1 > 0$
$t^\star = 1.79$
$0.055 < P < 0.064$ or $t \ge 1.86$
Fail to reject H_o

13.49   **a.** 13.04
b. 9.81 to 16.27
c. 4.69 to 21.39
d. 26.36, 23.41 to 29.31, 18.11 to 34.61

13.51 $\Sigma x = 16.25$, $\Sigma y = 152$, $\Sigma x^2 = 31.5625$, $\Sigma xy = 275$, $\Sigma y^2 = 2504$, $\hat{y} = 6.3758 + 5.4303x$
a. 15.4 to 19.1
b. 11.6 to 22.8

13.53 $\Sigma x = 256,593$, $\Sigma y = 454$, $\Sigma x^2 = 9,013,638,265$,
$\Sigma xy = 15,173,202$, $\Sigma y^2 = 26,316$
  a. $\hat{y} = 31.72 + 0.0007804x$
  b. 59.04% to 66.84%
  c. 53.5% to 72.38%

13.55 a.

Scatter Diagram

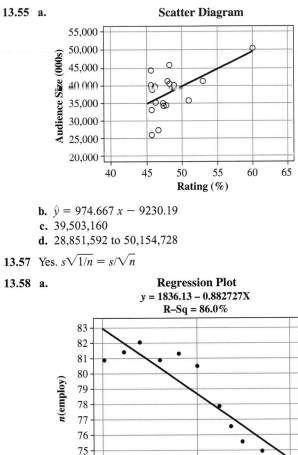

  b. $\hat{y} = 974.667\,x - 9230.19$
  c. 39,503,160
  d. 28,851,592 to 50,154,728

13.57 Yes. $s\sqrt{1/n} = s/\sqrt{n}$

13.58 a.

Regression Plot
$y = 1836.13 - 0.882727X$
R–Sq = 86.0%

  b. $\hat{y} = 1836 - 0.883x$
  c. The pattern demonstrated by the data shows more of
an S pattern than a straight line. The equation found
in (b) should not be used to make predictions.

13.59 a.

Small-Frame Women

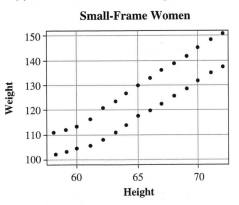

b.

Medium-Frame Women

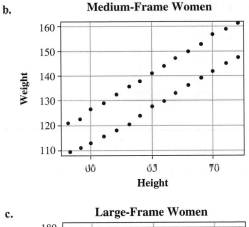

c.

Large-Frame Women

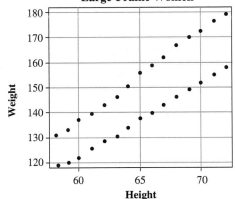

  d. Yes. Height is a dominant factor in determining a
person's weight.
  e. 3, slope

13.61 H_a: $\rho > 0.0$
$r^\star = 0.69$
P < 0.005 or $r \geq 0.29$
Reject H_o

13.63 a. 0.019, 0.096   **b.** Practically no correlation **13.65**
  a. H_a: $\rho \neq 0.0$
    $r^\star = 0.61$
    P < 0.01 or $r \leq -0.482$, $r \geq 0.482$
    Reject H_o
  b. 118.7

13.67 3.94 to 4.52

13.69 a.

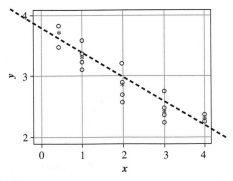

b. See dashed line on (a).

c. See * on (a).

d. $\Sigma x = 37.5$, $\Sigma y = 52.7$, $\Sigma x^2 = 104.75$,
$\Sigma xy = 98.75$, $\Sigma y^2 = 159.49$
$\hat{y} = 3.79 - 0.415x$

e. 0.21045

f. -0.502 to -0.328

g. At $x = 3.0$: 2.42 to 2.68
At $x = 3.5$: 2.18 to 2.50

h. At $x = 3.0$: 2.08 to 3.02
At $x = 3.5$: 1.87 to 2.81

13.71 $\Sigma x = 1177$, $\Sigma y = 567$, $\Sigma x^2 = 70{,}033$, $\Sigma xy = 32{,}548$,
$\Sigma y^2 = 15{,}861$

a. H_a: $\rho \neq 0.0$
$r^\star = 0.513$
$0.01 < P < 0.02$ or $r \leq -0.433$, $r \geq 0.433$
Reject H_o.

b. $\hat{y} = 16.40 + 0.189x$

c. H_a: $\beta_1 > 0$
$t^\star = 0.189/0.07255 = 2.61$
$0.007 < P < 0.009$ or $t \geq 1.73$
Reject H_o

d. 18.53 to 38.47

13.73 $\Sigma x = 16$, $\Sigma y = 38$, $\Sigma x^2 = 66$, $\Sigma xy = 145$, $\Sigma y^2 = 326$
$b_1 = 1.5811$, $r = 0.9973$

Chapter 14

14.1 39 to 47

14.3 **a.** $n(+) = 25$, $n(-) = 0$
b. 3 to 6

14.5 **a.** H_o: Median $= 32$, H_a: Median < 32
b. H_o: $P(\text{prefer}) = 0.50$, H_a: $P(\text{prefer}) < 0.50$
c. H_o: $P(\text{gain}) = 0.5$, H_a: $P(\text{gain}) \neq 0.5$

14.7 **a.** H_o: Median $= 48$
b. $P(\text{above } 48) = 0.5$
c. H_a: Median $\neq 48$
$x = n(+) = 3$
$P \approx 0.01$ or $x \leq 4$
Reject H_o

14.9 H_a: There is a difference in their ages
$x = n(-) = 6$
$P > 0.25$ or $x \leq 3$
Fail to reject H_o.

14.11 H_a: Median $\neq 30$ years
$x = n(+) = 40$
$0.05 < P < 0.10$ or $x \leq 39$
Fail to reject H_o.

14.13 H_a: There is a preference for the new; $p > 0.5$
$z^\star = 1.74$
$P = 0.0409$ or $z \geq 2.33$
Fail to reject H_o.

14.15 H_a: $P(+) \neq 0.5$
If $P \leq \alpha$, reject H_o ; if $P > \alpha$, fail to reject H_o
a. $x = 20$ **b.** $x = 27$
  $P < 0.01$ $0.01 < P < 0.05$
c. $x = 30$ **d.** $x = 33$
  $0.10 < P < 0.25$ $P > 0.25$

14.17 Sample B seems "slid" to the right about 10 points.

14.18 -2.30, 0.0214

14.19 **a.** H_o: is same
  H_a: is different
b. H_o: is same
  H_a: is not same
c. H_o: is same
  H_a: A higher than B

14.21 H_a: not same for both groups
$U^\star = 23$
$P > 0.10$ or $U \leq 8$
Fail to reject H_o.

14.25 H_a: Group 1 scores are higher
$U^\star = 4.5$
$0.025 < P < 0.05$ or $U \leq 6$
Reject H_o

14.27 **a.** H_o: Occur in random order
  H_a: Do not occur in random order
b. H_o: Sequence of odd/even is random
  H_a: Not random
c. H_o: Order of entry was random
  H_a: Was not random

14.29 H_a: Hiring sequence is not random
$V^\star = 9$
$P > 0.05$ or $V \leq 4$, $V \geq 12$
Fail to reject H_o.

14.31 **b.** -3.76, $P < 0.0002$ **c.** Yes, reject
d.

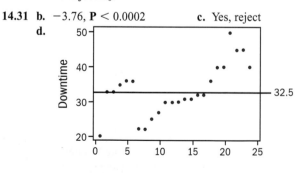

14.33 H_a: Lack of randomness
$V^\star = 8$
$P > 0.05$ or $V \leq 3$, $V \geq 10$
Fail to reject H_o.

14.35 **a.** 22.5; 8
b. H_a: Data did not occur randomly
  $V^\star = 8$
  $P > 0.05$ or $V \leq 3$, $V \geq 13$
  Fail to reject H_o.

14.37 **a.** H_o: Data did not occur randomly
$V^\star = 9$
P > 0.05 or $V \le 8$, $V \ge 20$
Fail to reject H_o

b. H_a: Data did not occur randomly
$z^* = -2.00$
P $= 0.0456$ or $z \le -1.96$, $z \ge 1.96$
Reject H_o.

14.39 As one variable increases, the other decreases; r is negative.

14.41 **a.** $r_s \le -0.545$, $r_s \ge 0.545$
b. $r_s \ge 0.323$ **c.** $r_s \le -0.564$

14.43 **b.** -0.291
c. H_a: $\rho_s > 0$
$r_s{}^\star = -0.291$
P > 0.10 or $r_s \ge 0.618$
Fail to reject H_o.

14.45 **a.** 0.133
b. H_a: $\rho_s > 0$
$r_s{}^\star = 0.133$
P > 0.10 or $r_s \ge 0.564$
Fail to reject H_o.

14.47 H_a: $\rho_s > 0$
$r_s{}^\star = 0.736$
$0.01 < $ **P** < 0.025 or $r_s \ge 0.564$
Reject H_o.

14.49 $\Sigma x = 78$, $\Sigma y = 78$, $\Sigma x^2 = 650$, $\Sigma xy = 578.5$, $\Sigma y^2 = 649$; $r_s = 0.502$

14.51 **a.** $\Sigma d^2 = 41.50$
b. H_a: There is a difference in lacking skills
$x = n(-) = 2$
P > 0.25 or $x \le 0$
Fail to reject H_o.
c. H_a: $\rho_s \ne 0$
$r_s{}^\star = 0.252$
P $= 0.585$ or $r_s \le -0.786$, $r_s \ge 0.786$
Fail to reject H_o.
d. Nothing is known about the population.

14.53 H_a: Average time on B less than A
$x = n(-) = 2$
P > 0.10 or $x \le 1$
Fail to reject H_o.

14.55 H_a: Top ripens first
$x = n(-) = 4$
P ≤ 0.025 or $x \le 5$
Reject H_o.

14.57 Reject for $U \le 127$

14.59 H_a: There is a difference in line width
$U^\star = 2$
P ≈ 0.05 or $U \le 2$
Reject H_o.

14.61 **b.** H_a: Batting averages in AL are higher
$U^* = 54$
P < 0.025 or $U \le 61$
Reject H_o

H_a: Earned run average for NL is lower
$U^\star = 66$
P > 0.05 or $U \le 61$
Fail to reject H_o

14.63 H_a: Lack of randomness.
$V^\star = 9$
P > 0.05 or $V \le 4$, $V \ge 10$
Fail to reject H_o.

14.65 **a.** Median $= 22.5$; Runs above $= 6$, below $= 6$
b. H_a: Lack of randomness. **c.** Not significant
$V^\star - 12$
P > 0.05 or $V \le 6$, $V \ge 16$
Fail to reject H_o.

14.67 H_a: $\rho_s > 0$
$r_s{}^\star = 0.880$
P < 0.01 or $r_s \ge 0.399$
Reject H_o

14.69 H_a: $\rho_s < 0$
$r_s{}^\star = 0.168$
P > 0.05 or $r_s \le -0.703$
Fail to reject H_o.

Appendix A

A.1 **a.** 18 **b.** 12 **c.** 6

A.3 **a.** ${}_nP_n = n!$ and ${}_nP_{n-1} = n \times n-1 \times n-2 \times \cdots \times 2$.
${}_nP_{n-1}$ is the same product as ${}_nP_n$, only the last factor of 1 is missing.
b. Both have value n.
c. ${}_nC_r = \dfrac{n!}{r!(n-r)!}$ and
$${}_nC_{n-r} = \frac{n!}{(n-r)!(n-(n-r))!} = \frac{n!}{(n-r)!r!}$$
d. They are two different names and applications for the same techniques.

A.5 **a.** $2,176,782,336$ assuming zero and 0 can be used as first characters.
b. $1,544,804,416$, assuming "0" can be the first character.
c. $6,760,000$ **d.** $6,760,000$ **e.** $6,084,000$

A.7 **a.** ${}_{32}C_5 = 2,598,960$
b. ${}_4C_1 \times {}_{31}C_4 = 4 \times 249,900 = 999,600$
c. ${}_4C_1 \times {}_{48}C_4 = 4 \times 194,580 = 778,320$
d. ${}_{13}C_1 \times {}_{30}C_4 = 13 \times 82,251 = 1,069,263$
e. ${}_3C_1 \times {}_{39}C_4 + {}_3C_4 \times {}_{38}C_4 = 303,696$

A.8 **a.** ${}_{48}C_5 \times {}_{42}C_1 = 1,906,884 \times 42 = 80,089,128$
b. Yes, one chance in 80.1 million

Answers to Chapter Practice Tests

Part 1: Only the replacement for the word(s) in boldface type is given. (If the statement is true, no answer is shown. If the statement is false, a replacement is given.)

Chapter 1, Page 32

Part I

1.1 descriptive

1.2 inferential

1.4 sample

1.5 population

1.6 attribute or qualitative

1.7 quantitative

1.9 random

Part II

1.11 a. A   **b.** B   **c.** D   **d.** C   **e.** A

1.12 c, g, h, b, e, a, d, f

Part III

1.13 See definitions; examples will vary. Note: *population* is set of ALL possible, while *sample* is the actual set of subjects studied.

1.14 See definitions; examples will vary. Note: *variable* is the idea of interest, while *data* are the actual values obtained.

1.15 See definitions; examples will vary. Note: *data* is the value describing one source, the *statistic* is a value (usually calculated) describing all the data in the sample, the *parameter* is a value describing the entire population (usually unknown).

1.16 Every element of the population has an equal chance of being selected.

Chapter 2, Page 124

Part I

2.1 median

2.2 dispersion

2.3 never

2.5 zero

2.6 higher than

Part II

2.11 a. 30   **b.** 46   **c.** 91   **d.** 15   **e.** 1   **f.** 61
g. 75   **h.** 76   **i.** 91   **j.** 106 or 114

2.12 a. two items purchased
b. Nine people purchased 3 items each.
c. 40   **d.** 120   **e.** 5   **f.** 2   **g.** 3   **h.** 3
i. 3.0   **j.** 1.795   **k.** 1.34

2.13 a. 6.7   **b.** 7   **c.** 8   **d.** 6.5   **e.** 5   **f.** 6
g. 3.0   **h.** 1.7   **i.** 5

2.14 a. -1.5 **b.** 153

Part III

2.15 a. 98   **b.** 50   **c.** 121   **d.** 100

2.16 a. $32,000, $26,500, $20,000, $50,000
b.

Salary ($1000)

c. Mr. VanCott—midrange; business manager—mean; foreman—median; new worker—mode
d. The distribution is J-shaped.

2.17 There is more than one possible answer for these.
  a. 12, 12, 12   **b.** 15, 20, 25
  c. 12, 15, 15, 18   **d.** 12, 15, 16, 25, 25
  e. 12, 12, 15, 16, 17 **f.** 20, 25, 30, 32, 32, 80

2.18 A is right; B is wrong; standard deviation will not change.

2.19 B is correct. For example, if standard deviation is $5, then the variance, (standard deviation)^2, is "25 dollars squared." Who knows what "dollars squared" are?

Chapter 3, Page 171

Part I

3.1 regression

3.2 strength of the

3.3 $+1$ or -1

3.5 positive

3.7 positive

3.8 -1 and $+1$

3.9 output or predicted value

Part II

3.11 **a.** B, D, A, C   **b.** 12   **c.** 10   **d.** 175   **e.** N
  f. (125,13)   **g.** N   **h.** P

3.12 Someone made a mistake in arithmetic, r must be between -1 and $+1$.

3.13 **a.** 12   **b.** 10   **c.** 8   **d.** 0.73   **e.** 0.67
  f. 4.33   **g.** $\hat{y} = 4.33 + 0.67x$

Part III

3.14 Young children have small feet and probably tend to have less mathematics ability, while adults have larger feet and would tend to have more ability.

3.15 Student B is correct. -1.78 can occur only as a result of faulty arithmetic.

3.16 These answers will vary, but should somehow include the basic thought:
  a. strong negative   **b.** strong positive
  c. no correlation   **d.** no correlation
  e. impossible value, bad arithmetic

3.17 There is more than one possible answer for these.
  a. (1,1), (2,1), (3,1)   **b.** (1,1), (3,3), (5,5)
  c. (1,5), (3,3), (5,1)   **d.** (1,1), (5,1), (1,5), (5,5)

Chapter 4, Page 232

Part I

4.1 any number value between 0 and 1, inclusive

4.4 simple

4.5 seldom

4.6 sum to 1.0

4.7 dependent

4.8 complementary

4.9 mutually exclusive or dependent

4.10 multiplication rule

Part II

4.11 **a.** $\dfrac{4}{8}$ **b.** $\dfrac{4}{8}$ **c.** $\dfrac{2}{8}$ **d.** $\dfrac{6}{8}$ **e.** $\dfrac{2}{8}$
  f. $\dfrac{6}{8}$ **g.** 0   **h.** $\dfrac{6}{8}$ **i.** $\dfrac{1}{8}$ **j.** $\dfrac{5}{8}$
  k. $\dfrac{2}{4}$ **l.** 0   **m.** $\dfrac{1}{2}$ **n.** no (e)
  o. yes (g)   **p.** no(i)   **q.** yes (a, k)   **r.** no (b, l)
  s. yes (a, m)

4.12 **a.** 0   **b.** 0.7   **c.** 0   **d.** no (c)

4.13 **a.** 0.14   **b.** 0.76   **c.** 0.2   **d.** no (a)

4.14 **a.** 0.7   **b.** 0.5   **c.** no, $P(E \text{ and } F) = 0.2$
  d. yes, $P(E) = P(E|F)$

4.15 **a.** 0.4   **b.** 0.5   **c.** no, $P(G \text{ and } H) = 0.1$
  d. no, $P(G)$ not equal to $P(G|H)$

4.16 0.51

Part III

4.17 Check the weather reports for a long period of time and determine the relative frequency with which each occurs.

4.18 Student B is right. *Mutually exclusive* means no intersection, while *independence* means one event does not affect the probability of the other.

4.19 These answers will vary, but should somehow include the basic thoughts:
  a. no common occurrence
  b. either event has no effect on the probability of the other
  c. the relative frequency with which the event occurs.
  d. probability that an event will occur even though the conditional event has previously occurred

4.20 **a.** No winners 15% of time; approx. 1 of every 7 drawings
  b. 15% is not that rare
  c. 0.0225 or 2.25%; approx. 1 of every 44 drawings
  d. A fairly unlikely event, but will occur occasionally

Chapter 5, Page 274

Part I

5.1 continuous

5.3 one

5.5 exactly two

5.6   binomial

5.7   one success occurring on 1 trial

5.8   population

5.9   population parameters

Part II

5.11 a. Each $P(x)$ is between zero and 1, and the sum of all $P(x)$ is exactly one.
  b. 0.2   **c.** 0   **d.** 0.8   **e.** 3.2   **f.** 1.25

5.12 a. 0.230   **b.** 0.085   **c.** 1.2   **d.** 1.04

Part III

5.13 n independent repeated trials of two outcomes; the two outcomes are "success" and "failure"; $p = P(\text{success})$ and $q = P(\text{failure})$ and $p + q = 1$; $x = n(\text{success}) = 0, 1, 2, \ldots, n$.

5.14 Student B is correct. The sample mean and standard deviation are statistics found using formulas studied in Chapter 2. The probability distributions studied in Chapter 5 are theoretical populations and their means and standard deviations are parameters.

5.15 Student B is correct. There are no restrictions on the values of the variable x.

Chapter 6, Page 315

Part I

6.1   its mean

6.4   one standard deviation

6.6   right

6.7   zero, 1

6.8   some (many)

6.9   mutually exclusive events

6.10 normal

Part II

6.11 a. 0.4922   **b.** 0.9162   **c.** 0.1020   **d.** 0.9082

6.12 a. 0.63   **b.** -0.95 **c.** 1.75

6.13 a. $z(0.8100)$ **b.** $z(0.2830)$

6.14 0.7910

6.15 28.03

6.16 a. 0.0569   **b.** 0.9890   **c.** 537   **d.** 417   **e.** 605

Part III

6.17 This answer will vary but should somehow include the basic properties: bell-shaped, mean of 0, standard deviation of 1.

6.18 This answer will vary but should somehow include the basic ideas: it is a z-score, α represents the area under the curve and to the right of z.

6.19 All normal distributions have the same shape and probabilities relative to the z-score.

Chapter 7, Page 342

Part I

7.1   is not

7.2   some (many)

7.3   population

7.4   divided by $\sqrt{n}$

7.5   decreases

7.6   approximately normal

7.7   sampling

7.8   means

7.9   random

Part II

7.11 a. 0.4364   **b.** 0.2643

7.12 a. 0.0918   **b.** 0.9525

7.13 0.6247

Part III

7.14 In this case each head produced one piece of data, the estimated length of the line. The CLT assures us that the mean value of a sample is far less variable than individual values of the variable x.

7.15 All samples must be of one fixed size.

7.16 Student A is correct. A population distribution is a distribution formed by all x values that make up the entire population.

7.17 Student A is correct. The standard error is found by dividing the standard deviation by the square root of the *sample size*.

Chapter 8, Page 412

Part I

8.1   alpha

8.2   alpha

8.3   sample distribution of the mean

8.7   type II error

8.8   beta

8.9   correct decision

8.10 critical (rejection) region

Part II

8.11 4.72 to 5.88

8.12 **a.** H_o: $\mu = 245$, H_a: $\mu > 245$

b. H_o: $\mu = 4.5$, H_a: $\mu < 4.5$

c. H_o: $\mu = 35$, H_a: $\mu \neq 35$

8.13 **a.** 0.05, z, $z \leq -1.65$

b. 0.05, z, $z \geq +1.65$

c. 0.05, z, $z \leq -1.96$ or $z \geq +1.96$

8.14 **a.** 1.65 **b.** 2.33 **c.** 1.18 **d.** -1.65

e. -2.05 **f.** -0.67

8.15 **a.** $z^\star = 2.50$ **b.** 0.0062

8.16 H_o: $\mu = 1520$ vs. H_a: $\mu < 1520$, crit. reg. $z \leq -2.33$, $z^\star = -1.61$, fail to reject H_o

Part III

8.17 **a.** No specific effect
b. Reduces it
c. Narrows it
d. No effect
e. Increases it
f. Widens it

8.18 **a.** H_o − (a), H_a − (b) **b.** 3 **c.** 2
d. P(type I error) is alpha, decreases; P(type II error) increases

8.19 The alternative hypothesis expresses the concern; the conclusion answers the concern.

Chapter 9, Page 469

Part I

9.2 Student's t

9.3 chi-square

9.4 to be rejected

9.6 t score

9.7 $n - 1$

9.9 $\sqrt{pq/n}$

9.10 z(normal)

Part II

9.11 **a.** 2.05 **b.** -1.73 **c.** 14.6

9.12 **a.** 28.6 **b.** 1.44 **c.** 27.16 to 30.04

9.13 0.528 to 0.752

9.14 **a.** H_o: $\mu = 225$, H_a: $\mu > 225$
b. H_o: $p = 0.40$, H_a: $p \neq 0.40$
c. H_o: $\sigma = 3.7$, H_a: $\sigma < 3.7$

9.15 **a.** 0.05, z, $z \leq -1.65$
b. 0.05, t, $t \leq -2.08$ or $t \geq +2.08$
c. 0.05, z, $z \geq +1.65$
d. 0.05, χ^2, $\chi^2 \leq 14.6$ or $\chi^2 \geq 43.2$

9.16 H_o: $\mu = 26$ vs. H_a: $\mu < 26$, crit. reg. $t \leq -1.71$, $t^\star = -1.86$, reject H_o

9.17 H_o: $\sigma = 0.1$ vs. H_a: $\sigma > 0.1$, crit. reg. $\chi^2 \geq 21.1$, $\chi^{2\star} = 23.66$, reject H_o

9.18 H_o: $p = 0.50$ vs. H_a: $p > 0.50$, crit. reg. $z \geq 2.05$, $z^\star = 1.29$, fail to reject H_o

Part III

9.19 If the distribution is normal, six standard deviations is approximately equal to the range.

9.20 B

9.21 They are both correct.

9.22 When the sample size, n, is large, the critical value of t is estimated by using the critical value from the standard normal distribution of z.

9.23 Student A

9.24 Student B is right. It is significant at the 0.01 level of significance.

9.25 Student A is correct.

9.26 It depends on what it means to improve the confidence interval. For most purposes, an increased sample size would be the best improvement.

Chapter 10, Page 532

Part I

10.1 two independent means

10.3 F distribution

10.4 Student's t-distribution

10.5 Student's t

10.7 nonsymmetric (or skewed)

10.9 decreases

Part II

10.11 **a.** H_o: $\mu_N - \mu_A = 0$, H_a: $\mu_N - \mu_A \neq 0$
b. H_o: $\sigma_o/\sigma_m = 1.0$, H_a: $\sigma_o/\sigma_m > 1.0$
c. H_o: $p_m - p_f = 0$, H_a: $p_m - p_f \neq 0$
d. H_o: $\mu_d - 0$, H_a: $\mu_d > 0$

10.12 **a.** z, $z \leq -1.96$ or $z \geq 1.96$
b. t, $t \leq -2.05$, $t \geq 2.05$
c. t, df = 7, $t \geq 1.89$
d. t, df = 37, $t \leq -1.69$
e. F, $F \geq 2.11$

10.13 a. 2.05   **b.** 2.13   **c.** 2.51   **d.** 2.18
e. 1.75   **f.** 1.69   **g.** -2.50 **h.** -1.28

10.14 H_o: $\mu_L - \mu_P = 0$ vs. H_a: $\mu_L - \mu_P > 0$, crit. reg.
$t \geq +1.83$, $t^\star = 0.979$, fail to reject H_o

10.15 H_o: $\mu_d = 0$ vs. H_a: $\mu_d > 0$, crit. reg. $t \geq 1.89$, $t^\star = 1.88$, fail to reject H_o

10.16 0.072 to 0.188

Part III

10.17 independent

10.18 One possibility: Test all students before the course starts, then randomly select 20 of those who finish the course and test them afterwards. Use the before scores for these 20 as the before sample.

10.19 For starters, if the two independent samples are of different sizes, the techniques for dependent samples could not be completed. They are testing very different concepts, the "mean of the differences of paired data" and the "difference between two mean values."

10.20 It is only significant if the calculated t-score is in the critical region. The variation among the data and their relative size will play a role.

10.21 The 80 scores actually are two independent samples of size 40. A test to compare the mean scores of the two groups could be completed.

10.22 A fairly large sample of both Catholic and non-Catholic families would need to be taken, and the number of each whose children attended private schools would need to be obtained. The difference between two proportions could then be estimated.

Chapter 11, Page 572

Part I

11.1 one less than

11.3 expected

11.4 contingency table

11.6 test of homogeneity

11.8 approximated by chi-square

Part II

11.11 a. H_o: Digits generated occur with equal probability.
H_a: Digits do not occur with equal probability.
b. H_o: Votes were cast independently of party affiliation.
H_a: Votes were not cast independently of party affiliation.
c. H_o: The crimes distributions are the same for all four cities.
H_a: The crimes distributions are not all the same.

11.12 a. 4.40   **b.** 35.7

11.13 H_o: $P(1) = P(2) = P(3) = \dfrac{1}{3}$

H_a: preferences not all equal, $\chi^{2\star} = 3.78$;
$0.10 < P < 0.25$ or crit. reg. $\chi^2 \geq 5.99$; fail to reject H_o

11.14 a. H_o: The distribution is the same for all types of soil.
H_a: The distributions are not all the same.
b. 25.622   **c.** 13.746
d. $0.005 < P < 0.01$ **e.** $\chi^2 \geq 9.49$
f. Reject H_o. There is sufficient evidence to show that the growth distribution is different for at least one of the three soil types.

Part III

11.15 Similar in that there are n repeated independent trials. Different in that the binomial has two possible outcomes, while the multinomial has several. Each possible outcome has a probability and these probabilities sum to 1 for each different experiment, both for binomial and multinomial.

11.16 The test of homogeneity compares several distributions in a side-by-side comparison, while the test for independence tests the independence of the two factors that create the rows and columns of the contingency table.

11.17 Student A is right in that the calculations are completed in the same manner. Student B is correct in that the test of independence starts with one large sample and homogeneity has several samples.

11.18 a. If a chi-square test is to be used, the results of the four questions would be pooled to estimate the expected probability.
b. Use a chi-square test for homogeneity.

Chapter 12, Page 603

Part I

12.2 mean square

12.3 SS(factor) or MS(factor)

12.5 reject H_o

12.7 the number of factor levels less one

12.8 mean

12.9 need to

12.10 does not indicate

Part II

12.11 a. T   **b.** T   **c.** F   **d.** T   **e.** T   **f.** T   **g.** F
h. F   **i.** F   **j.** F   **k.** T   **l.** F   **m.** F
n. F   **o.** T

12.12 a. 72   **b.** 72   **c.** 22   **d.** 4   **e.** 4.5

Part III

12.13 This answer will vary but should somehow include the basic ideas: It is the comparison of several mean values that result from testing some statistical population by measuring a variable repeatedly at each of the several levels for which the factor is being tested.

12.14 **a.** $x_{r,k} = \mu + F \text{ scrubber} + \epsilon_{k(r)}$
   b. H_o: The mean amount of emissions is the same for all three scrubbers tested.
   H_a: The mean amounts are not all equal.

   c.

Source	df	SS	MS
Scrubber	2	12.80	6.40
Error	13	33.63	2.59
Total	15	46.44	

   d. $F(2, 13, 0.05) = 3.81$, $F^{\star} = 2.47$, fail to reject H_o. The difference in the mean values for the scrubbers is not significant.

   e. I

```
          .  .  .  :  .
--+--+--+--+--+--+--+--
   7     9    11    13
```

   II

```
       .  .  .     :
--+--+--+--+--+--+--+--
   7     9    11    13
```

   III

```
    .  .  .  .  .
--+--+--+--+--+--+--+--
   7     9    11    13
```

Chapter 13, Page 656

Part I

13.2 need not be

13.3 does not prove

13.4 need not be

13.6 the linear correlation coefficient

13.8 regression

13.9 $n - 2$

Part II

13.11

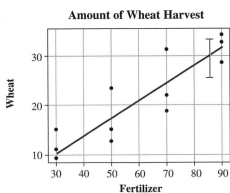

Amount of Wheat Harvest

13.12 $\sum x = 720$, $\sum y = 252$, $\sum x^2 = 49{,}200$,
   $\sum xy = 17{,}240$, $\sum y^2 = 6{,}228$

13.13 $SS(x) = 6000$, $SS(y) = 936$, $SS(xy) = 2120$

13.14 0.895

13.15 0.65 to 0.97

13.16 $\hat{y} = -0.20 + 0.353x$

13.17 See red line in figure in 13.11.

13.18 4.324

13.19 Yes; H_o: $\beta_1 = 0$ vs H_a: $\beta_1 > 0$, $t^{\star} = 6.33$, reject H_o

13.20 25.63 to 33.98

13.21 See blue vertical segment in 13.11.

Part III

13.22 Variable 1: The frequency of skiers having their bindings tested

   Variable 2: The incidence of lower-leg injury
   The statement implies that as the frequency with which the bindings are tested increases, the frequency of lower-leg injury decreases; thus the strong correlation must be negative for these variables.

13.23 A "moment" is the distance from the mean, and the product of both the horizontal moment and the vertical moment is summed in calculating the correlation coefficient.

13.24 A value close to zero, also. The formulas used to calculate both values have the same numerator, namely $SS(xy)$.

13.25 The vertical distance from a potential line of best fit to the data point is measured by $(y - \hat{y})$. The line of best fit is defined to be the line that results in the smallest possible total when the squared values of $(y - \hat{y})$ are totaled. Thus "the method of least squares."

13.26 The strength of the linear relationship could be measured with the correlation coefficient.

13.27 A random sample will be needed from the population of interest. The data collected need to be for the variables length of time on welfare and the measure of current level of self-esteem.

Chapter 14, Page 707

Part I

14.2 t-test

14.3 runs test

14.4 assigned equal ranks

14.7 Mann–Whitney U test

14.8 power

14.10 power

Part II

14.11 -2 to $+7$

14.12 H_o: No difference in weight gain.

H_a: There is a difference in weight gain, crit. val.: 23, $U^\star = 32.5$, fail to reject H_o

14.13 H_o: no correlation

H_a: corelated, crit. val.: ± 0.683, $r_s^\star = -0.70$,

reject H_o. Yes, there is significant correlation.

14.14 $(+) =$ higher grade level than previous problem
$(-) =$ lower grade level than previous problem
H_o: $P(+) = 0.5$
H_a: $P(+) = 0.5$, crit. val.: 7, $x = 11$, fail to reject H_o.
This sample does not show a significant pattern.

Part III

14.15 The nonparametric statistics do not require assumptions about the distribution of the variable.

14.16 The sign test is a binomial experiment of n trials (the n data observations) with two outcomes for each data $[(+)$ or $(-)]$, and $p = (+) = 0.5$. The variable x is the number of the least frequent sign.

14.17 The median is the middle value such that 50% of the distribution is larger in value and 50% is smaller in value.

14.18 The extreme value in a set of data can have a sizeable effect on the mean and standard deviation in the parametric methods. The nonparametric methods typically use rank numbers. The extreme value with ranks is either 1 or n, and neither changes if the value is more extreme.

14.19 d; $p = P(+) = P$(prefer seating arrangement A) $= 0.5$, no preference

Computer and Calculator Instructions

Index

Credits

This page constitutes an extension of the copyright page. We have made every effort to trace the ownership of all copyrighted material and to secure permission from copyright holders. In the event of any question arising as to the use of any material, we will be pleased to make the necessary corrections in future printings. Thanks are due to the following authors, publishers, and agents for permission to use the material indicated.

Chapter 1. **5:** Copyright 1998, USA TODAY. Reprinted with permission. **5:** Copyright 1997, USA TODAY. Reprinted with permission. **6:** Copyright 1994, USA TODAY. Reprinted with permission. **8:** Copyright 1998, USA TODAY. Reprinted with permission. **9:** Copyright 1997, USA TODAY. Reprinted with permission. **10:** Copyright 1998, USA TODAY. Reprinted with permission. **19:** Copyright 1994, USA TODAY. Reprinted with permission.

Chapter 2. **35:** Copyright 1998, USA TODAY. Reprinted with permission. **47:** Copyright 1998, USA TODAY. Reprinted with permission. **53:** Copyright 1997, USA TODAY. Reprinted with permission. **109:** Copyright 1998, USA TODAY. Reprinted with permission. **110:** Copyright 1998, USA TODAY. Reprinted with permission.

Chapter 3. **129:** Copyright 1996, USA TODAY. Reprinted with permission. **138:** Copyright 1997, USA TODAY. Reprinted with permission. **140:** Copyright 1998, USA TODAY. Reprinted with permission. **141:** Copyright 1998, USA TODAY. Reprinted with permission. **167:** Copyright 1998, USA TODAY. Reprinted with permission.

Chapter 5. **235:** Copyright 1996, USA TODAY. Reprinted with permission.

Chapter 8. **349:** Copyright 1998, USA TODAY. Reprinted with permission. **359:** Copyright 1994, USA TODAY. Reprinted with permission.

Chapter 9. **415:** Copyright 1998, USA TODAY. Reprinted with permission.

Chapter 10. **473:** Copyright 1998, USA TODAY. Reprinted with permission.

Chapter 11. **539:** Copyright 1998, USA TODAY. Reprinted with permission.

Chapter 12. **575:** Copyright 1998, USA TODAY. Reprinted with permission.

Chapter 13. **649:** Copyright 1997, USA TODAY. Reprinted with permission.

Chapter 14. **659:** Copyright 1998, USA TODAY. Reprinted with permission.

Photos: **2, 3, 176, 346, 536:** Tony Stone Images.

Glossary of Symbols

$\overline{A}$	Complement of set A
ANOVA	Analysis of variance
α (alpha)	Probability of a type I error
β (beta)	Probability of a type II error
$1 - \beta$	Power of a statistical test
β_0	y-intercept of the true linear relationship
β_1	Slope of the true linear relationship
b_0	y-intercept for the line of best fit for the sample data
b_1	Slope for the line of best fit for the sample data
$_nC_r$	Number of combinations of n things r at a time
C_j	Column total
c	Column number or class width
d	Difference in value between two paired pieces of data or difference in the rankings
$\overline{d}$	Mean value of observed differences d
$d(\ )$	Depth of
df or df( )	Number of degrees of freedom
E	Expected frequency or maximum error of estimate
e	Error (observed)
ϵ (epsilon)	Experimental error
ϵ_{ij}	Amount of experimental error in the value of the jth piece of data in the ith row
F	F distribution statistic
$F(\text{df}_n, \text{df}_d, \alpha)$	Critical value for the F distribution
f	Frequency
H	Value of the largest-valued piece of data in a sample
H_a	Alternative hypothesis
H_o	Null hypothesis
i	Index number when used with $\sum$ notation
i	Position number for a particular data
k	Identifier for the kth percentile
k	Number of cells or variables

L	Value of the smallest-valued piece of data in a sample
m	Number of classes
MS( )	Mean square
μ (mu)	Population mean
μ_d	Mean value of the paired differences
$\mu_{\overline{x}}$	Mean of the distribution of all possible $\overline{x}$'s
$\mu_{y|x_0}$	Mean of all y values at the fixed value of x, x_0
μ_v	Mean number of runs for the sampling distribution of number of runs
M	Population median
n	(Sample size) number of pieces of data in one sample
$n(\ )$	Cardinal number of
$\binom{n}{r}$	Binomial coefficient or number of r successes in n trials
O	Observed frequency
P	Probability value or p-value
$P(A|B)$	Conditional probability, the probability of A given B
$P(a < x < b)$	Probability that x has a value between a and b
P_k	kth percentile
$_nP_r$	Number of permutations of n things r at a time
p or $P(\ )$	Theoretical probability of an event or proportion of time that a particular event occurs
p' or $P'(\ )$	Empirical (experimental) probability or a probability estimate from observed data
p_p	Pooled estimate for the proportion
Q_1	First quartile
Q_3	Third quartile
q	($q = 1 - p$) probability that an event does not occur
q'	($q' = 1 - p'$) observed proportion of time that an event did not occur
R	Range of the data
R_i	Row total

R^2	Coefficient of determination	U	Mann–Whitney U statistic
ρ (rho)	Population linear correlation coefficient	V	Number of runs
r	Linear correlation coefficient for the sample data or row number	χ^2	Chi-square statistic
		$\chi^2_{(df, \alpha)}$	Critical value of chi-square distribution
r_s	Spearman's rank correlation coefficient	x	Value of a single piece of data or class mark
$\sum$ (capital sigma)	Summation notation	$\bar{x}$	Sample mean
SS( )	Sum of squares	$\tilde{x}$	Sample median
s^2	Sample variance	x_{ij}	Value of the jth piece of data in the ith row
σ (lowercase sigma)	Population standard deviation	x_0	A given value of the variable x
$\sigma_{\bar{x}}$	Standard error for means, the standard deviation of the distribution of all possible $\bar{x}$'s	$\hat{y}$	Predicted value of y for a given x
$\sigma_{p'}$	Standard error for proportions	y_{x_0}	Individual value of y at the x value of x_0
$\sigma^2_{\mu r}$	Variance among the means of the r rows (ANOVA)	z	Standard score
σ_v	Standard error for the number of runs for the sampling distribution of number of runs	$z_{(\alpha/2)}$	Critical value of z
		$\bigstar$ (star)	Identifies the calculated value of any test statistic
s	Sample standard deviation	$|\ |$	Absolute value of a number
s_d	Standard deviation of the observed differences d	$=$	Equal to
$s^2_{b_1}$	Square of the standard error for repeated observed values of the slope for the line of best fit	$\neq$	Not equal to
		$<$	Less than
		$\leq$	Less than or equal to
T	Grand total	$>$	Greater than
t	Student's t-distribution statistic	$\geq$	Greater than or equal to
		$\approx$	Approximately equal to
$t_{(df, \alpha/2)}$	Critical value for Student's t-distribution	$\sqrt{}$	Square root